1800   1900

Other Contributors
Pierre de Fermat (1601-1665)   Thomas Simpson (1710-1761)
Michel Rolle (1652-1719)   Pierre-Simon de Laplace (1749-1827)
Brook Taylor (1685-1731)   George Green (1793-1841)
Colin Maclaurin (1698-1746)   George Gabriel Stokes (1819-1903)

Lagrange

Gauss

Cauchy

Riemann

Lebesgue

J. Lagrange (1736-1813)
C. Gauss (1777-1855)
A. Cauchy (1789-1857)
K. Weierstrass (1815-1897)
G. Riemann (1826-1866)
J. Gibbs (1839-1903)
S. Kovalevsky (1850-1891)
H. Lebesgue (1875-1941)

Agnesi

Weierstrass

Kovalevsky

Gibbs

| 1756 | 1799 | 1821 | 1854 | 1873 | 1902 |

Gauss proves Fun-
damental Theorem
of Algebra

Riemann integral

Lebesgue integral

Lagrange
begins
*Mécanique
analytique*

Precise notion of
limit (Cauchy)

e is transcendental
(Hermite)

Calculus
for Biology and Medicine

Calculus
for Biology
and Medicine

Claudia Neuhauser

University of Minnesota

Prentice Hall
Upper Saddle River, New Jersey 07458

Library of Congress Cataloging-in-Publication Data

Neuhauser, Claudia
    Calculus for biology and medicine / Claudia Neuhauser.
       p.   cm.
    ISBN 0-13-085137-X
       1. Biomathematics.   2. Medicine—Mathematics.   3. Calculus.   I. Title.
QH323.5.N46 2000
515—dc21   99-052144
                                                                                                     CIP

Acquisition Editor: George Lobell
Assistant Vice President of Production and Manufacturing: David W. Riccardi
Executive Managing Editor: Kathleen Schiaparelli
Senior Managing Editor: Linda Mihatov Behrens
Production Editor: Brittney Corrigan-McElroy
Manufacturing Buyer: Alan Fischer
Manufacturing Manager: Trudy Pisciotti
Marketing Manager: Melody Marcus
Marketing Assistant: Vince Jansen
Director of Marketing: John Tweeddale
Development Editor: Emily Thompson
Editor in Chief, Development: Carol Trueheart
Associate Editor, Mathematics/Statistics Media: Audra J. Walsh
Editorial Assistant/Supplements Editor: Gale Epps
Art Director: Joseph Sengotta
Assistant to the Art Director: John Christiana
Interior Designer: Maureen Eide
Cover Designer: Maureen Eide
Art Editor: Grace Hazeldine
Art Manager: Gus Vibal
Director of Creative Services: Paul Belfanti
Cover Image: The image on the front and back cover shows a number of inclusions of the bacterium
                Chlamydia trachomatis/Dr. R. Dourmashkin/Photo Researchers, Inc.
Art Studio: Interactive Composition Corporation

© 2000 by Prentice-Hall, Inc.
Upper Saddle River, New Jersey 07458

Printed in the United States of America
10 9 8 7 6 5 4 3 2

ISBN 0-13-085137-X

Prentice-Hall International (UK) Limited, London
Prentice-Hall of Australia Pty. Limited, Sydney
Prentice-Hall Canada Inc., Toronto
Prentice-Hall Hispanoamericana, S.A., Mexico
Prentice-Hall of India Private Limited, New Delhi
Prentice-Hall of Japan, Inc., Tokyo
Pearson Education Asia Pte. Ltd.
Editora Prentice-Hall do Brasil, Ltda., Rio de Janeiro

Contents

3 Differentiation   88

4 Applications of Differentiation   155

10 Multivariable Calculus   506

11 Systems of Differential Equations   560

12 Probability and Statistics   639

Preface

For modeling and analysis of phenomena in the life sciences, calculus is an indispensable tool. This is not obvious in a traditional calculus course, where biology students rarely see how the material is relevant to their training. This text is written exclusively for students in the biological and medical sciences. It makes an effort to demonstrate from the beginning how calculus can help them to understand phenomena in nature. Students find it easier to understand concepts in mathematics if they are related to their field, and this makes mathematics more interesting.

This text differs in a number of ways from traditional calculus texts. First, it is written in a life-science context; concepts are motivated with biological examples, emphasizing that calculus is an important tool in the life sciences. Second, differential equations, one of the most important modeling tools in the life sciences, are introduced very early, immediately after the formal definition of derivatives. Third, biological applications of differentiation and integration are included throughout the text. Fourth, two chapters deal exclusively with differential equations and systems of differential equations; both chapters contain numerous up-to-date applications. Fifth, multivariable calculus is taught in the first year, recognizing that most students in the life sciences will not take the second year of calculus, and that multivariable calculus is needed to analyze systems of differential equations, which they will encounter in their science courses.

This text does not teach modeling; the objective is to teach calculus. Modeling is an art that should be taught in a separate course. However, throughout the text, students encounter mathematical models for biological phenomena; this will facilitate the transition to actual modeling.

Examples Each topic is motivated with biological examples, followed by a thorough discussion outside of the life-science context, to enable students to become familiar with both the meaning and the mechanics of the topic. Examples in the text are completely worked out; steps in calculations are frequently explained in words.

Problems Calculus cannot be learned by watching someone do it. Recognizing this, students are provided with both drill and word problems. Word problems are an integral part of teaching calculus in a life-science context. The word problems are new and up-to-date; they are adapted from either standard biology texts or original research. Because this text is written for college freshmen, the examples were chosen so that no formal training in biology is needed.

Technology This book takes advantage of graphing calculators, which allows students to develop a much better visual understanding of the concepts in calculus. Beyond that, no special software is required.

Chapter Summary

Chapter 1   Basic tools from algebra and trigonometry are summarized in Section 1.1. Section 1.2 contains the basic functions used in text, including exponential and logarithmic functions; their graphical properties and biological relevance are emphasized. Section 1.3 covers log-log and semi-log plots; these are graphical tools that are frequently used in the life sciences.

Chapter 2   Limits and continuity are key concepts for understanding the conceptual parts of calculus. Visual intuition is emphasized before the theory is discussed.

Chapter 3   The geometric definition of a derivative as the slope of a tangent line is given before the formal treatment. After the formal definition of the derivative, differential equations are introduced as models for biological phenomena. Differentiation rules are discussed. These sections give students time to acquaint themselves with the basic rules of differentiation before applications are discussed. Related rates and error propagation, in addition to differential equations, are the main applications.

Chapter 4   This chapter presents biological and more traditional applications of differentiation. Many of the applications are consequences of the mean value theorem. Many of the word problems are adapted from either biology textbooks or original research articles; this puts the traditional applications (such as extrema, monotonicity, and concavity) in a biological context.

Chapter 5   Integration is motivated geometrically. The fundamental theorem of calculus and its consequences are discussed in depth. Both biological and traditional applications of integration are provided before integration techniques are covered.

Chapter 6   This chapter contains integration techniques. However, only the most important techniques are covered. Tables of integrals are used to integrate more complicated integrals. The use of computer software is not covered in the text, though their usefulness in evaluating integrals is acknowledged.

Chapter 7   This chapter provides an introduction to differential equations. The treatment is not complete, but it will equip students with both analytical and graphical skills for analysis. Eigenvalues are introduced early, to facilitate the analytical treatment of systems of differential equations in Chapter 11. Many of the differential equations discussed in the text are important models in biology. Though this text is not a modeling text, students will see how differential equations are used to model biological phenomena, and will be able to interpret differential equations. Chapter 7 contains a large number of up-to-date applications of differential equations in biology.

Chapter 8   This chapter contains additional applications of integration, which help students to understand the importance of integrals. Unless Chapter 12 is included, some or all sections of this chapter can be omitted. (Sections 8.2–4 should be discussed before Chapter 12.)

Chapter 9   Matrix algebra is an indispensible tool for every life scientist. The material in this chapter covers the most basic concepts, tailored to Chapters 10 and 11, where matrix algebra is frequently used. Special emphasis is given to the treatment of eigenvalues and eigenvectors because of their importance in analyzing systems of differential equations.

Chapter 10   This is an introduction to multidimensional calculus. The treatment is brief and tailored to Chapter 11, where systems of differential equations are discussed. The main topics are partial derivatives and linearization of vector-valued functions. The discussion of gradient and diffusion are not required for Chapter 11.

Chapter 11   This material is most relevant for students in the life sciences. Both graphical and analytical tools are developed to enable students to analyze systems of differential equations. The material is divided into linear and nonlinear systems. Understanding the stability of linear systems in terms of vector fields, eigenvectors, and eigenvalues helps students to master the more difficult analysis of nonlinear systems. Theory is explained before applications are given—this allows students to become familiar with the mechanics before delving into applications. An extensive problem set allows students to experience the power of this modeling tool in a biological context.

Chapter 12   This chapter contains basic probabilistic and statistical tools. It cannot replace a full semester course in probability and statistics, but it allows students to see some of the concepts needed in population genetics and experimental design.

How to Use This Book

This book contains more material than can be covered in one year. This was deliberate, and allows for more flexibility in the choice of material. Sections that are noted by asterisks in the table of contents can be omitted; their material is not needed in subsequent sections.

The material can be arranged to suit a one-semester, two-quarter, two-semester, or four-quarter course. Chapters 1–3 must be covered in that order before any of the other sections are covered. In addition to Chapters 1–3, the following arrangements can be chosen:

One semester, emphasis on integration   4.1, 4.2, 4.3, 4.4, 4.5, 4.7, 5.1, 5.2, 5.3 (without 5.3.4), 6.1

One semester, emphasis on differential equations   4.1, 4.2, 4.3, 4.4, 4.5, 4.7, 5.1, 5.2, 7.1 (without 7.1.2, 7.1.3), 7.2

One semester, emphasis on probability   4.1, 4.2, 4.3, 4.4, 4.5, 4.7, 5.1, 5.2, 12.1, 12.2, 12.3, 12.4

Two quarters   4.1, 4.2, 4.3, 4.4, 4.5, 4.7, 5.1, 5.2, 5.3, 6.1, 6.2, 6.3, 6.4, 6.5, 6.7, 7.1, 7.2, 7.3 (one of the subsections), 8.1, 8.2

Two semesters   4.1, 4.2, 4.3, 4.4, 4.5, 4.7, 5.1, 5.2, 5.3, 6.1, 6.2, 6.3, 6.4, 6.5, 6.7, 7.1, 7.2, 7.3 (one of the subsections), 9.1, 9.2 (without 9.2.4), 9.3, 9.4, 10.1, 10.2, 10.3, 10.4, 11.1, 11.2, 11.3, 11.4 (two of the subsections)

Four quarters   All sections that are not labeled optional; optional sections should be chosen as time permits

Acknowledgments

This book would not have been possible without the help of numerous people, and greatly benefited from critical reviews of an earlier draft:

Ramin Vakilian
California State University—Northridge

Steven M. Baer
Arizona State University

Stephen J. Merrill
Marquette University

David Austin
Grand Valley State University

Alexander Levin
The Catholic University of America

Michael Li
Mississippi State University

Thanks for their constructive criticism and enthusiasm to all the students who took this course in 1998–99 at the University of Minnesota, when it was offered for the first time. I thank Mary and Nancy Toscano, who checked the accuracy of all the problems in the book. I owe a special thanks to George Lobell, acquisitions editor at Prentice Hall, who made this book possible in a timely fashion; Brittney Corrigan-McElroy, production editor at Interactive Composition Corporation, who had to work on a rather tight schedule; and Emily Thompson, developmental editor, whose help in the final revision was invaluable. Finally, I wish to thank my husband, Maury Bramson, who patiently put up with my long hours of work on this book.

Claudia Neuhauser
nhauser@math.umn.edu
Minneapolis, Minnesota

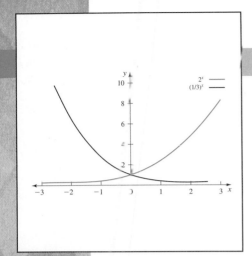

Preview and Review

Isaac Newton (1642–1727) and Gottfried Wilhelm Leibniz (1646–1716) are typically credited with the invention of calculus. They were not the sole inventors but rather the first to systematically develop it.

Calculus is divided into two parts, differential and integral calculus. Historically, the problems that led to the development of differential calculus were concerned with finding tangent lines to curves, and extrema (i.e., maxima and minima) of curves. Integral calculus, on the other hand, has its roots in attempts to determine the areas of regions bounded by curves, or the volumes of solids. It turns out that the two parts of calculus are closely related; in fact, the basic operation of one can be considered the inverse of the other. This result is known as the Fundamental Theorem of Calculus and goes back to Newton and Leibniz. They were the first to understand the meaning of this inverse relationship, and to put this relationship to use in solving difficult problems.

Finding tangents, extrema, and areas are very basic problems, which led to the development of methods that are useful in solving a wide range of scientific problems. For this reason, calculus has been one of the most powerful tools in the mathematical formulation of scientific concepts. For example, many physical laws and many phenomena in biology are formulated in the language of calculus.

In addition to developing the theory of differential and integral calculus, we will consider many examples in which calculus is used to describe or model situations in the biological sciences. The use of mathematics is becoming increasingly more important in biology, for instance, when modeling interactions between species in a community, describing neuron activities, explaining genetic diversity in populations, predicting the impact of global warming on vegetation, and so on.

◆ 1.1 PRELIMINARIES

We provide a brief review of some of the concepts and techniques from precalculus that are frequently used in calculus. (The problems at the end of this section will help you to reacquaint yourself with this material.)

1.1.1 The Real Numbers

▲ **Figure 1.1**

The real number line.

The **real numbers** can most easily be visualized on the **real number line** (see Figure 1.1). Numbers are ordered on the real number line so that if $a < b$, then a is to the left of b. Sets (collections) of real numbers are denoted by capital letters, for instance, A, B, or alike. To describe the set A, we write

$$A = \{x : \text{condition}\}$$

where "condition" tells us which numbers are in the set A. The most important sets in calculus are **intervals**. We use the following notations. If $a < b$, then

$$\text{the } \textbf{open} \text{ interval } (a, b) = \{x : a < x < b\}$$

and

$$\text{the } \textbf{closed} \text{ interval } [a, b] = \{x : a \le x \le b\}$$

We also use **half-open** intervals

$$[a, b) = \{x : a \le x < b\} \qquad \text{and} \qquad (a, b] = \{x : a < x \le b\}$$

Unbounded intervals are sets of the form $\{x : x > a\}$. Here are the possible cases:

$$[a, \infty) = \{x : x \ge a\}$$
$$(-\infty, a] = \{x : x \le a\}$$
$$(a, \infty) = \{x : x > a\}$$
$$(-\infty, a) = \{x : x < a\}$$

The symbols "∞" and "$-\infty$" mean "plus infinity" and "minus infinity;" these are *not* real numbers but merely used for notational convenience. The real number line **R** does not have endpoints and we can write **R** in the following equivalent forms:

$$\mathbf{R} = \{x : -\infty < x < \infty\} = (-\infty, \infty)$$

The location of the number 0 on the real number line is called the **origin**, and we can measure the distance of the number x to the origin. For instance, -5 is 5 units away from the origin. There is a convenient notation for this, namely the absolute value of a real number.

> **Definition**   The **absolute value** of a real number a, denoted by $|a|$, is
>
> $$|a| = \begin{cases} a & \text{if } a \ge 0 \\ -a & \text{if } a < 0 \end{cases}$$

For example, $|-7| = -(-7) = 7$. We can also use absolute values to find the distance between two numbers x_1 and x_2; namely,

$$\text{distance between } x_1 \text{ and } x_2 = |x_1 - x_2|$$

Note that $|x_1 - x_2| = |x_2 - x_1|$. To find the distance between -2 and 4, we compute $|-2 - 4| = |-6| = 6$, or $|4 - (-2)| = |4 + 2| = 6$.

We will frequently need to solve equations containing absolute values, for which the following property is useful. Assume $b \ge 0$; the equation $|a| = b$ is equivalent to $a = b$ if $a \ge 0$, and to $-a = b$ if $a < 0$.

▮ EXAMPLE 1   Solve $|x - 4| = 2$.

Solution   If $x - 4 \geq 0$, then $x - 4 = 2$ and thus $x = 6$. If $x - 4 < 0$, then $-(x - 4) = 2$ and thus $x = 2$. The solutions are therefore $x = 6$ and $x = 2$. This can be graphically illustrated (see Figure 1.2). The points of intersection of $y = |x - 4|$ and $y = 2$ are at $x = 6$ and $x = 2$. Solving $|x - 4| = 2$ can also be interpreted as finding the two numbers that have distance 2 from 4.   ▮

If we need to solve an equation of the form $|a| = |b|$, then we write this as either $a = b$ or $a = -b$, which is illustrated in the next example.

▮ EXAMPLE 2   Solve $|\frac{3}{2}x - 1| = |\frac{1}{2}x + 1|$.

Solution   Either

$$\frac{3}{2}x - 1 = \frac{1}{2}x + 1 \qquad \text{or} \qquad \frac{3}{2}x - 1 = -\left(\frac{1}{2}x + 1\right)$$

$$x = 2 \qquad\qquad\qquad \frac{3}{2}x - 1 = -\frac{1}{2}x - 1$$

$$2x = 0$$

$$x = 0$$

A graphical solution of this example is shown in Figure 1.3.   ▮

Returning to Example 1, where we found the two points whose distance from 4 was equal to 2, we can also try to find those points whose distance from 4 is less than (or greater than) 2. This amounts to solving inequalities with absolute values.

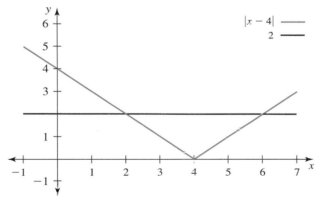

◀ **Figure 1.2**
The graphs of $y = |x - 4|$ and $y = 2$. The points of intersection are at $x = 6$ and $x = 2$.

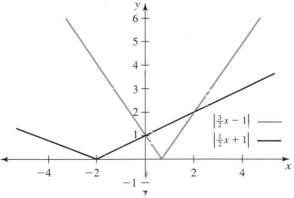

◀ **Figure 1.3**
The graphs of $y = |\frac{3}{2}x - 1|$ and $y = |\frac{1}{2}x + 1|$. The points of intersection are at $x = 0$ and $x = 2$.

Looking back at Figure 1.2, we see that the set of x values whose distance from 4 is less than 2 (i.e., $|x - 4| < 2$) is the interval $(2, 6)$. Similarly, the set of x values whose distance from 4 is greater than 2 (i.e., $|x - 4| > 2$) is the union of the two intervals $(-\infty, 2) \cup (6, \infty)$.

In general, to solve absolute value inequalities, the following two properties are useful. Assume $b > 0$.

---

1. $|a| < b$ is equivalent to $-b < a < b$.

2. $|a| > b$ is equivalent to $a > b$ or $a < -b$.

EXAMPLE 3   **a.** Solve $|2x - 5| < 3$.

b. Solve $|3x - 4| \geq 2$.

Solution

a. We rewrite $|2x - 5| < 3$ as

$$-3 < 2x - 5 < 3$$

Adding 5 to all three parts, we obtain

$$2 < 2x < 8$$

Dividing the result by 2, we find

$$1 < x < 4$$

The solution is therefore the set $\{x : 1 < x < 4\}$. In interval notation, the solution can be written as the open interval $(1, 4)$.

b. To solve $|3x - 4| \geq 2$, we go through the following steps:

$$3x - 4 \geq 2 \qquad \text{or} \qquad 3x - 4 \leq -2$$
$$3x \geq 6 \qquad\qquad\qquad 3x \leq 2$$
$$x \geq 2 \qquad\qquad\qquad x \leq \frac{2}{3}$$

The solution is the set $\{x : x \geq 2 \text{ or } x \leq \frac{2}{3}\}$, or, in interval notation, $\left(-\infty, \frac{2}{3}\right] \cup [2, \infty)$.   ∎

1.1.2 Lines in the Plane

We will frequently encounter situations in which quantities are linearly related, that is, these quantities satisfy a linear equation. For example, when a weight is attached to a helical spring made of some elastic material (and the weight is not too heavy), the length y of the spring will be linearly related to the weight x:

$$y = y_0 + kx \tag{1.1}$$

This is known as Hooke's law. Here, y_0 denotes the length of the spring when no weight is attached to it. The change in length, $y - y_0$, is said to be **proportional** to the attached weight, and we write

$$y - y_0 \propto x$$

or, after introducing the **proportionality factor** k, $y - y_0 = kx$, yielding equation (1.1) after adding y_0 to both sides of the equation.

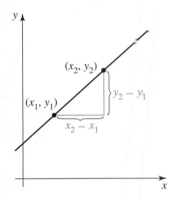

▲ Figure 1.4
The slope of a straight line.

The **standard** form of a linear equation is given by

$$Ax + By + C = 0$$

where A, B, and C are constants, A and B are not both equal to 0, and x and y are the two variables. In algebra you learned that the graph of a linear equation is a straight line.

If you know two points (x_1, y_1) and (x_2, y_2) on a straight line, then the **slope** of the line is

$$m = \frac{y_2 - y_1}{x_2 - x_1}$$

This is illustrated in Figure 1.4. Two points (or one point and the slope) are sufficient to determine the equation of a straight line.

If you are given two points or one point and the slope, the **point-slope** form of a straight line can be used to write down its equation. It is given by

$$y - y_0 = m(x - x_0)$$

where m is the slope and (x_0, y_0) is a point on the line.

Lastly, the most frequently used form of a linear equation is the **slope-intercept** form

$$y = mx + b$$

where m is the slope and b is the y-intercept, which is the point of intersection of the line with the y-axis; it has coordinates $(0, b)$.

We summarize these three forms of linear equations in the following box.

$Ax + By + C = 0$	(Standard Form)
$y - y_0 = m(x - x_0)$	(Point-Slope Form)
$y = mx + b$	(Slope-Intercept Form)

▌**EXAMPLE 4** Determine the equation of the line passing through $(-2, 1)$ and $(3, -\frac{1}{2})$.

Solution The slope of the line is

$$m = \frac{y_2 - y_1}{x_2 - x_1} = \frac{-\frac{1}{2} - 1}{3 - (-2)} = \frac{-\frac{3}{2}}{5} = -\frac{3}{10}$$

Using the point-slope form with $(-2, 1)$, we find

$$y - 1 = -\frac{3}{10}(x - (-2))$$

that is,

$$y = -\frac{3}{10}x + \frac{2}{5}$$

We could have used the other point, $(3, -\frac{1}{2})$, and obtained the same result. ▌

We now recall two special cases that we illustrate in Figure 1.5:

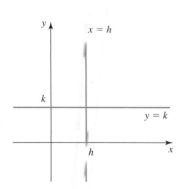

▲ Figure 1.5
The horizontal line $y = k$ and the vertical line $x = h$.

$y = k$	horizontal line (slope 0)
$x = h$	vertical line (no slope)

In the next example, we show how to determine the slope and the y-intercept of a given straight line.

EXAMPLE 5   Determine the slope and the y-intercept of the line $3y - 2x + 9 = 0$.

Solution   We solve for y: $3y = 2x - 9$ and thus $y = \frac{2}{3}x - 3$. We can now read off the slope $m = \frac{2}{3}$ and the y-intercept $b = -3$. ∎

When two lines l_1 and l_2 in the plane have no points in common, they are called **parallel**, denoted by $l_1 \parallel l_2$. The following criterion is useful for deciding whether two lines are parallel. Two lines l_1 and l_2 are parallel ($l_1 \parallel l_2$) if and only if their slopes are identical. For nonvertical lines with slopes m_1 and m_2, respectively, we obtain the following criterion

$$l_1 \parallel l_2 \qquad \text{if and only if} \qquad m_1 = m_2$$

Two lines l_1 and l_2 are called **perpendicular** ($l_1 \perp l_2$) if their intersection forms an angle of $90°$. The following slope criterion is useful when determining equations for perpendicular lines. Lines are perpendicular if and only if their slopes are negative reciprocals. That is, if l_1 and l_2 are nonvertical lines with slopes m_1 and m_2, then

$$l_1 \perp l_2 \qquad \text{if and only if} \qquad m_1 m_2 = -1$$

We will prove this result in Problem 20 at the end of this section.

1.1.3 Equation of the Circle

A **circle** is the set of all points within a given distance, called the **radius**, from a given point, called the **center**. If r is the distance from (x_0, y_0) to (x, y) (see Figure 1.6), then using the Pythagorean theorem,

$$r^2 = (x - x_0)^2 + (y - y_0)^2$$

If $r = 1$ and $(x_0, y_0) = (0, 0)$, the circle is called the **unit circle**.

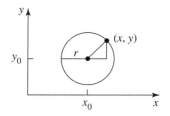

▲ Figure 1.6
Circle with radius r centered at (x_0, y_0).

EXAMPLE 6   Find the equation of the circle with center $(2, 3)$ and passing through $(5, 7)$.

Solution   Using the Pythagorean theorem, we can compute the distance in the plane between $(2, 3)$ and $(5, 7)$, namely

$$\sqrt{(5 - 2)^2 + (7 - 3)^2} = \sqrt{9 + 16} = 5$$

Thus, the circle has radius 5 and center $(2, 3)$, and its equation is given by

$$25 = (x - 2)^2 + (y - 3)^2$$ ∎

1.1.4 Trigonometry

We will need a few results from trigonometry. Recall that angles are measured in either degrees or radians. For reasons which will become clear, the radian measure is preferred in calculus. To convert between degree and radian measure, we use the formula

$$\frac{\theta \text{ measured in degrees}}{360°} = \frac{\theta \text{ measured in radians}}{2\pi}$$

For instance, to convert $23°$ into radian measure, we compute

$$\theta = 23° \frac{2\pi}{360°} = 0.401$$

To convert $\frac{\pi}{6}$ into degrees, we compute

$$\theta = \frac{\pi}{6}\frac{360°}{2\pi} = 30°$$

You should be familiar with sine, cosine, tangent, and secant (the other two, cotangent and cosecant, are rarely used). They are defined on a unit circle (see Figure 1.7). Recall that a positive angle is measured counter clockwise on the unit circle, whereas a negative angle is measured clockwise.

▲ **Figure 1.7**
The trigonometric functions on a unit circle.

$$\sin\theta = \frac{\text{opposite}}{\text{hypotenuse}} = \frac{y}{1} = y \qquad \csc\theta = \frac{1}{\sin\theta} = \frac{1}{y}$$

$$\cos\theta = \frac{\text{adjacent}}{\text{hypotenuse}} = \frac{x}{1} = x \qquad \sec\theta = \frac{1}{\cos\theta} = \frac{1}{x}$$

$$\tan\theta = \frac{\text{opposite}}{\text{adjacent}} = \frac{y}{x} \qquad \cot\theta = \frac{1}{\tan\theta} = \frac{x}{y}$$

The following trigonometric identities are used frequently. Since $\tan\theta = y/x$ with $y = \sin\theta$ and $x = \cos\theta$, it follows that

$$\tan\theta = \frac{\sin\theta}{\cos\theta}$$

Applying the Pythagorean theorem to the triangle in Figure 1.7, we find

$$\sin^2\theta - \cos^2\theta = 1$$

If we divide this identity by $\cos^2\theta$, we find

$$\frac{\sin^2\theta}{\cos^2\theta} + 1 = \frac{1}{\cos^2\theta}$$

Using $\tan\theta = \sin\theta/\cos\theta$ and $\sec\theta = 1/\cos\theta$, this can be written as

$$\tan^2\theta + 1 = \sec^2\theta$$

In the next example, we will solve a trigonometric equation.

∎ EXAMPLE 7   Solve

$$2\sin\theta\cos\theta = \cos\theta \quad \text{on } [0, 2\pi)$$

Solution   Bringing $\cos\theta$ to the left side and factoring $\cos\theta$, we obtain

$$\cos\theta(2\sin\theta - 1) = 0$$

that is,

$$\cos\theta = 0 \quad \text{or} \quad 2\sin\theta - 1 = 0$$

Solving $\cos\theta = 0$, we find

$$\theta = \frac{\pi}{2} \quad \text{or} \quad \theta = \frac{3\pi}{2}$$

Solving $2\sin\theta - 1 = 0$, we get

$$\sin\theta = \frac{1}{2}$$

TABLE 1•1　SOME EXACT TRIGONOMETRIC VALUES

	Angle θ				
	0	$\frac{\pi}{6}$	$\frac{\pi}{4}$	$\frac{\pi}{3}$	$\frac{\pi}{2}$
Function	**(0°)**	**(30°)**	**(45°)**	**(60°)**	**(90°)**
$\sin\theta$	$\frac{1}{2}\sqrt{0}$	$\frac{1}{2}\sqrt{1}$	$\frac{1}{2}\sqrt{2}$	$\frac{1}{2}\sqrt{3}$	$\frac{1}{2}\sqrt{4}$
$\cos\theta$	$\frac{1}{2}\sqrt{4}$	$\frac{1}{2}\sqrt{3}$	$\frac{1}{2}\sqrt{2}$	$\frac{1}{2}\sqrt{1}$	$\frac{1}{2}\sqrt{0}$

which yields

$$\theta = \frac{\pi}{6} \qquad \text{or} \qquad \theta = \frac{5\pi}{6}$$

The solution set is therefore $\left\{ \frac{\pi}{6}, \frac{\pi}{2}, \frac{5\pi}{6}, \frac{3\pi}{2} \right\}$.

The following two identities follow from Figure 1.8 when we compare the two angles θ and $-\theta$. Recall that a positive angle is measured counterclockwise from the positive x-axis, whereas a negative angle is measured clockwise.

$$\sin(-\theta) = -\sin\theta \qquad \text{and} \qquad \cos(-\theta) = \cos\theta$$

You should memorize certain exact trigonometric values; these are collected in Table 1.1. Of course, $\frac{1}{2}\sqrt{0} = 0$, $\frac{1}{2}\sqrt{1} = \frac{1}{2}$, and $\frac{1}{2}\sqrt{4} = 1$, but writing it as in Table 1.1 will make it easier to memorize the values. Using $\tan\theta = \sin\theta / \cos\theta$, you immediately get the values for $\tan\theta$.

1.1.5　Exponentials and Logarithms

Exponentials and logarithms are particularly important in biological contexts.

An exponential is an expression of the form

$$a^r$$

where a is called the **base** and r the **exponent**. We summarize some of its properties.

$$a^r a^s = a^{r+s}$$

$$\frac{a^r}{a^s} = a^{r-s}$$

$$a^{-r} = \frac{1}{a^r}$$

$$(ab)^r = a^r b^r$$

$$\left(\frac{a}{b}\right)^r = \frac{a^r}{b^r}$$

EXAMPLE 8　Evaluate the following exponential expressions.

　a.　$3^2 3^{5/2} = 3^{2+5/2} = 3^{9/2}$

　b.　$\dfrac{2^{-4}2^3}{2^2} = \dfrac{2^{-1}}{2^2} = 2^{-1-2} = 2^{-3} = \dfrac{1}{2^3} = \dfrac{1}{8}$

　c.　$\dfrac{a^k a^{3k}}{a^{5k}} = a^{k+3k-5k} = a^{-k} = \dfrac{1}{a^k}$

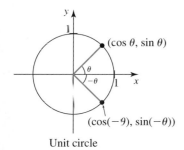

▲ Figure 1.8

Using the unit circle to define trigonometric identities.

Unit circle

$(\cos\theta, \sin\theta)$

$(\cos(-\theta), \sin(-\theta))$

Logarithms allow us to solve equations of the form

$$2^x = 8$$

The solution of this equation is $x = 3$, and we can write this as

$$x = \log_2 8$$

In other words, a logarithm is an exponent. The expression

$$\log_a y$$

is the exponent on the base a that yields the number y. We have the following correspondence between logarithms and exponentials:

$$x = \log_a y \qquad \text{is equivalent to} \qquad y = a^x$$

▌**EXAMPLE 9**   Which real number x satisfies

a. $\log_3 x = -2$ **b.** $\log_{1/2} 8 = x$

Solution

a. We write this in the equivalent form,

$$x = 3^{-2}$$

Hence,

$$x = \frac{1}{3^2} = \frac{1}{9}$$

b. We write this in the equivalent form,

$$\left(\frac{1}{2}\right)^x = 8$$
$$2^{-x} = 2^3$$
$$2^x = 2^{-3}$$

Setting the exponents equal to each other, we find $x = -3$. Note that in order to compare exponents, their bases must be the same.   ▌

Some important properties of the logarithm:

$$\log_a(xy) = \log_a x + \log_a y$$
$$\log_a\left(\frac{x}{y}\right) = \log_a x - \log_a y$$
$$\log_a x^r = r \log_a x$$

The most important logarithm is the **natural logarithm**, which has the number e as its base. The number e is an irrational number whose value is approximately 2.7182818. The natural logarithm is written $\ln x$; that is, $\log_e x = \ln x$.

▌**EXAMPLE 10**   Simplify the following expressions.

a. $\log_3(9x^2) = \log_3 9 + \log_3 x^2 = 2 + 2\log_3 x$

b. $\log_5 \dfrac{x^2 + 3}{5x} = \log_5(x^2 + 3) - \log_5 5 - \log_5 x = \log_5(x^2 + 3) - 1 - \log_5 x$

(Note that $\log_5(x^2 + 3)$ cannot be simplified any further.)

c. $-\ln \frac{1}{2} = \ln \left(\frac{1}{2}\right)^{-1} = \ln 2$

d. $\ln \dfrac{3x^2}{\sqrt{y}} = \ln 3 + \ln x^2 - \ln \sqrt{y} = \ln 3 + 2\ln x - \dfrac{1}{2}\ln y$

(In the last step, we used the fact that $\sqrt{y} = y^{1/2}$.) ▮

In algebra, you learned how to solve equations of the form $e^{2x} = 3$ or $\ln(x+1) = 5$—we will need to do this frequently. The following example illustrates how to solve such equations.

▮ EXAMPLE 11   Solve for x.

    a. $e^{2x} = 3$ **b.** $\ln(x+1) = 5$ **c.** $5^{2x-1} = 2^x$

Solution

    a. To solve $e^{2x} = 3$, we take logarithms to base e on both sides,

$$\ln e^{2x} = \ln 3$$

But $\ln e^{2x} = 2x$; hence,

$$2x = \ln 3 \qquad \text{or} \qquad x = \frac{1}{2}\ln 3$$

    b. To solve $\ln(x+1) = 5$, we write the equation in exponential form,

$$e^{\ln(x+1)} = e^5$$

which simplifies to

$$x + 1 = e^5 \qquad \text{or} \qquad x = e^5 - 1$$

    c. To solve $5^{2x-1} = 2^x$ for x, we need to take logarithms. Since the two bases are different, we take the natural logarithms on both sides. This yields

$$\ln 5^{2x-1} = \ln 2^x$$

or, after simplifying,

$$(2x-1)\ln 5 = x\ln 2$$

Solving for x, we find

$$2x\ln 5 - x\ln 2 = \ln 5$$
$$x(2\ln 5 - \ln 2) = \ln 5$$

hence,

$$x = \frac{\ln 5}{2\ln 5 - \ln 2}$$
▮

◆ 1.1 PROBLEMS

[1.1.1]

1. Find the two numbers that have distance 3 from -1 by **(a)** measuring the distances on the real number line, and **(b)** solving an appropriate equation involving an absolute value.

2. Solve the following equations.

(a) $|2x - 4| = 6$ **(b)** $|x - 3| = 2$
(c) $|2x + 3| = 5$ **(d)** $|7 - 3x| = -2$

3. Solve the following equations.

(a) $|2x + 4| = |5x - 2|$ **(b)** $|5 - 3u| = |3 + 2u|$
(c) $|4 + \frac{t}{2}| = |\frac{3}{2}t - 2|$

4. Solve the following inequalities.

(a) $|5x - 2| \le 4$ **(b)** $|1 - 3x| > 8$
(c) $|7x + 4| \ge 3$

[1.1.2]

In Problems 5–14, determine the equation in standard form for the line that satisfies the stated requirements.

5. The line passing through $(2, 4)$ with slope $-\frac{1}{3}$

6. The line passing through $(1, -2)$ with slope 2

7. The line passing through $(-2, -3)$ and $(1, 4)$

8. The line passing through $(-1, 4)$ and $\left(2, -\frac{1}{2}\right)$

9. The horizontal line through $\left(3, \frac{3}{2}\right)$

10. The vertical line through $\left(-1, \frac{7}{2}\right)$

11. The line with slope 3 and y-intercept $(0, 2)$

12. The line with slope -2 and x-intercept $(1, 0)$

13. The line passing through $(1, 4)$ and perpendicular to

$$2y - 5x + 7 = 0$$

14. The line passing through $(-1, -1)$ and parallel to the line passing through $(0, 1)$ and $(2, 0)$

15. To convert a length measured in feet to a length measured in centimeters, we use the fact that a length measured in feet is proportional to a length measured in centimeters and that 1 ft corresponds to 30.5 cm. If x denotes the length measured in ft and y denotes the length measured in cm, then

$$y = 30.5x$$

(a) Explain this relationship.

(b) Use this relationship to convert the following measurements into cm. (i) 6 ft (ii) 3 ft 2 in (iii) 1 ft 7 in

(c) Use this relationship to convert the following measurements into ft. (i) 173 cm (ii) 75 cm (iii) 48 cm

16. (a) To convert the weight of an object from kilograms (kg) to pounds (lbs), we use the fact that a weight measured in kilograms is proportional to a weight measured in pounds, and that 1 kg corresponds to 2.20 lbs. Find an equation that relates weight measured in kilograms to weight measured in pounds.

(b) Use your answer in (a) to convert the following measurements. (i) 63 lbs (ii) 150 lbs (iii) 2.5 kg (iv) 140 kg

17. Experimental study plots are often squares of length 1 m. If 1 ft corresponds to 0.305 m, express the area of a 1 m by 1 m plot in square feet.

18. Large areas are often measured in hectares (ha) or in acres. If 1 ha = 10,000 m^2 and 1 acre = 4046.86 m^2, how many acres is one hectare?

19. To convert the volume of a liquid measured in ounces to a volume measured in liters, we use the fact that 1 liter equals 33.81 ounces. Denote by x the volume measured in ounces and by y the volume measured in liters. Assume a linear relationship between these two units of measurements.

(a) Find the equation relating x and y.

(b) A typical soda can contains 12 ounces of liquid. How many liters is this?

20. By carrying out the following steps, we will show that if two nonvertical lines l_1 and l_2 with slopes m_1 and m_2 are perpendicular, then $m_1 m_2 = -1$. Assume that $m_1 < 0$ and $m_2 > 0$.

(a) Use a graph to show that if θ_1 and θ_2 are the respective angles of inclination of the lines l_1 and l_2, then $\theta_1 = \theta_2 + \frac{\pi}{2}$. (The angle of inclination of a line is the angle $\theta \in [0, \pi)$ between the line and the positively directed x-axis.)

(b) Show that $m_1 = \tan \theta_1$ and $m_2 = \tan \theta_2$. (Use the fact that $\tan(\pi - x) = -\tan x$.)

(c) Use the fact that $\tan\left(\frac{\pi}{2} - x\right) = \cot x$ and $\cot(-x) = -\cot x$ to show that $m_1 = -\cot \theta_2$. Deduce from this the truth of our claim.

[1.1.3]

21. Find the equation of a circle with center $(-1, 4)$ and radius 3.

22. (a) Find the equation of a circle with center $(2, 5)$ and radius 3.

(b) Where does the circle intersect the y-axis?

(c) Does the circle intersect the x-axis? Explain.

23. Find the center and the radius of the circle given by the equation

$$(x - 2)^2 + y^2 = 16$$

24. Find the center and the radius of the circle given by the equation

$$0 = x^2 + y^2 - 4x + 2y - 11$$

(To do this, you must complete the squares.)

[1.1.4]

25. (a) Convert $75°$ to radian measure.

(b) Convert $\frac{17}{12}\pi$ to degree measure.

26. Evaluate the following expressions without using your calculator.

(a) $\sin\left(-\frac{5\pi}{4}\right)$ **(b)** $\cos\left(\frac{5\pi}{6}\right)$ **(c)** $\tan\left(\frac{\pi}{3}\right)$

27. (a) Find the values of $\alpha \in [0, 2\pi)$ that satisfy

$$\sin \alpha = -\frac{1}{2}\sqrt{3}$$

(b) Find the values of $\alpha \in [0, 2\pi)$ that satisfy

$$\tan \alpha = \sqrt{3}$$

28. Show that the identity

$$1 + \tan^2 \theta = \sec^2 \theta$$

follows from

$$\sin^2 \theta + \cos^2 \theta = 1$$

29. Show that the identity

$$1 + \cot^2 \theta = \csc^2 \theta$$

follows from

$$\sin^2 \theta + \cos^2 \theta = 1$$

30. Solve $2 \cos \theta \sin \theta = \sin \theta$ on $[0, 2\pi)$.

31. Solve $\sec^2 x = \sqrt{3} \tan x + 1$ on $[0, \pi)$.

[1.1.5]

32. Solve for x.

(a) $3^x = 81$ **(b)** $9^{2x+1} = 27$ **(c)** $10^{5x} = 1000$

33. Simplify the following expressions.

(a) $-\ln \frac{1}{5}$ **(b)** $\ln \frac{x^2 - y^2}{\sqrt{x}}$ **(c)** $\log_3 3^{2x+1}$

34. Solve for x.

(a) $e^{3x-1} = 2$ **(b)** $e^{-2x} = 10$ **(c)** $e^{x^2-1} = 10$

35. Solve for x.

(a) $\ln(x - 3) = 5$ **(b)** $\ln(x + 2) + \ln(x - 2) = 1$

(c) $\log_3 x^2 - \log_3 2x = 2$

[1.1.6]

To measure temperature, three scales are commonly used: Fahrenheit, Celsius, and Kelvin. These scales are linearly related. We discuss these scales in the following two problems.

36. (a) The Celsius scale is devised so that $0°C$ is the freezing point of water (at one atmosphere of pressure), and $100°C$ is the boiling point of water (at one atmosphere of pressure). You are probably more familiar with the Fahrenheit scale, and know that water freezes at $32°F$ and boils at $212°F$. Find a linear equation that relates temperature measured in Celsius and temperature measured in Fahrenheit.

(b) The normal body temperature in humans ranges from $97.6°F$ to $99.6°F$. Convert this temperature range into degrees Celsius.

37. (a) The Kelvin (K) scale is an **absolute** scale of temperature. The zero point of the scale (0K) denotes **absolute zero**, the coldest possible temperature; that is, no body can have a temperature below 0K. It has been determined experimentally that 0K corresponds to $-273.15°C$. If 1K denotes the same temperature difference as $1°C$, find an equation that relates the Kelvin and Celsius scales.

(b) Pure nitrogen and pure oxygen can be produced cheaply by first liquefying purified air and then allowing the temperature of the liquid air to rise slowly. Since nitrogen and oxygen have different boiling points, they are distilled at different temperatures. The boiling point of nitrogen is 77.4K; of oxygen, 90.2K. Convert each of the boiling point temperatures into Celsius. If you solved Problem 36(a), also convert the boiling point temperatures into Fahrenheit. Using the techniques described above for distilling nitrogen and oxygen, which gets distilled first?

◆ 1.2 ELEMENTARY FUNCTIONS

1.2.1 What is a Function?

Scientific investigations are often concerned with studying relationships between quantities, such as how seedling density depends on the distance to the adult plant, or how tree ring width depends on the amount of light available to the tree. To describe such relationships mathematically, the concept of a function is useful.

It seems that the word function (or more precisely, its Latin equivalent *functio*, which means "execution") was introduced by Leibniz in 1694 in order to describe curves. Later, Euler (1707–1783) used it to describe any equation involving variables and constants. The modern definition is much broader, and emphasizes the basic idea of expressing relationships between any two sets.

> **Definition**   A **function** f is a rule that assigns each element $x \in A$ exactly one element $y \in B$. The element y is called the **image** (or **value**) of x under f and is denoted by $f(x)$ (read: "f of x"). The set A is called the **domain** of f. The set $f(A) = \{y : y = f(x) \text{ for some } x \in A\}$ is called the **range** of f.

To define a function, we use the notation

$$f : A \to B$$
$$x \to f(x)$$

Frequently, we simply write $y = f(x)$. Of course, $f(x)$ must be specified; for example, $f(x) = x^2$. Note that $f(A) \subset B$, but not every element in B must be contained in $f(A)$. For instance, let

$$f_1 : \mathbf{R} \to \mathbf{R}$$
$$x \to x^2$$

The domain of f is $\mathbf{R}$, but the range of f_1 is only $[0, \infty)$ because the square of a number is nonnegative; that is, $f(\mathbf{R}) = [0, \infty) \neq \mathbf{R}$. Also, the domain of a function need not be the largest possible set on which we can define the function, such as $\mathbf{R}$ in the above example. For instance, we could have defined f_1 on a smaller set, such as [0, 1], calling the new function f_2,

$$f_2 : [0, 1] \to \mathbf{R}$$
$$x \to x^2$$

Though the same rule is used for f_1 and f_2, the two functions are not the same because their respective domains are different.

Two functions f and g are **equal** if and only if

1. f and g are defined on the same domain, and
2. $f(x) = g(x)$ for all x in the domain.

▮EXAMPLE 1   Let

$$f_1 : [0, 1] \to \mathbf{R}$$
$$x \to x^2$$
$$f_2 : [0, 1] \to \mathbf{R}$$
$$x \to \sqrt{x^4}$$

and

$$f_3 : \mathbf{R} \to \mathbf{R}$$
$$x \to x^2$$

Determine which of the functions are equal.

Solution   Because f_1 and f_2 are defined on the same domain, and $f_1(x) = f_2(x) = x^2$ for all $x \in [0, 1]$, it follows that f_1 and f_2 are equal.

Neither f_1 nor f_2 are equal to f_3, because the domain of f_3 is different from the domains of f_1 and f_2. ▮

The choices of domains for the functions that we have considered may look somewhat arbitrary (and they are arbitrary). In applications, however, there is often a natural choice of domain. For instance, if we look at a certain plant response (such as total biomass, or the ratio of above to below biomass) as a function of nitrogen concentration in the soil, then given that nitrogen concentration cannot be negative, the chosen domain for this function could be the set of nonnegative real numbers. To look at yet another example, suppose we define a function that depends on the proportion of a population infected with a certain virus, then a natural choice for the domain of this function would be the interval [0, 1] because a proportion is always a number between 0 and 1.

Sometimes, functions show certain symmetries. For example, looking at Figure 1.9, $f(x) = x$ is symmetric about the origin, that is, $f(x) = -f(-x)$; and $g(x) = x^2$ is symmetric about the y-axis, that is, $g(x) = g(-x)$. In the first case, we say that f is odd; in the second case, that g is even. To check whether a function is even or odd, we use the following criterion.

A function $f : A \to B$ is called

1. **even** if $f(x) = f(-x)$ for all $x \in A$, and
2. **odd** if $f(x) = -f(-x)$ for all $x \in A$.

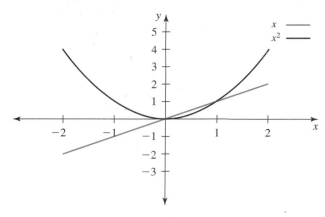

▲ **Figure 1.9**
The graphs of $y = x$ and $y = x^2$.

Using this criterion, we can check that $f(x) = x$, $x \in \mathbf{R}$, is an odd function, namely

$$-f(-x) = -(-x) = x = f(x) \qquad \text{for all } x \in \mathbf{R}$$

Likewise, to check that $g(x) = x^2$, $x \in \mathbf{R}$, is an even function, we compute

$$g(-x) = (-x)^2 = x^2 = g(x) \qquad \text{for all } x \in \mathbf{R}$$

We will now look at the case where one quantity is given as a function of another quantity, which in turn can be written as a function of yet another quantity. To illustrate this, suppose that $y = f(u)$ and $u = g(x)$—we can express f function of x by substituting $g(x)$ for u, that is, $y = f[g(x)]$. Functions tha defined in such a way are called composite functions.

> **Definition**   The **composite function** $f \circ g$ (also called the **composition** of f and g) is defined by
>
> $$(f \circ g)(x) = f[g(x)]$$
>
> for each x in the domain of g for which $g(x)$ is in the domain of f.

▲ **Figure 1.10**
The composition of functions.

The composition of functions is illustrated in Figure 1.10. The phrase "for each x in the domain of g for which $g(x)$ is in the domain of f" is best explained using Figure 1.10. In order to compute $f(u)$, u needs to be in the domain of f. But since $u = g(x)$, we really require that $g(x)$ is in the domain of f for the values of x we use to compute $g(x)$.

EXAMPLE 2   If $f(x) = \sqrt{x}$, $x \geq 0$, and $g(x) = x^2 + 1$, $x \in \mathbf{R}$, find $(f \circ g)(x)$ and $(g \circ f)(x)$.
Solution   To find $(f \circ g)(x)$, we set $f(u) = \sqrt{u}$ and $g(x) = x^2 + 1$. Then

$$y = f(u) = f[g(x)] = f(x^2 + 1) = \sqrt{x^2 + 1} \qquad \text{for } x \in \mathbf{R}$$

To find $(g \circ f)(x)$ we set $g(u) = u^2 + 1$ and $f(x) = \sqrt{x}$. Then

$$y = g(u) = g[f(x)] = g(\sqrt{x}) = (\sqrt{x})^2 + 1 = x + 1 \qquad \text{for } x \geq 0 \qquad ∎$$

In the last example, you should observe that $f \circ g$ is different from $g \circ f$, which implies that the order in which you compose functions is important. The notation

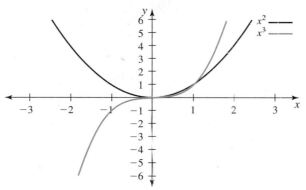

◀ **Figure 1.11**
The graphs of $y = x^n$ for $n = 2$ and 3.

$f \circ g$ means that you apply g first and then f. In addition, you should pay attention to the domains of composite functions. The function $g \circ f$ in Example 2 is only defined for $x \geq 0$ as the domain of f is $\{x : x \geq 0\}$; the range of f is $\{u : u \geq 0\}$ and g is defined for all values in the range of f.

In the following subsections we introduce the basic functions that are used throughout the remainder of this book.

1.2.2 Polynomial Functions

Polynomial functions are the simplest elementary functions.

> **Definition**   A **polynomial** function is a function of the form
> $$f(x) = a_0 + a_1 x + a_2 x^2 + \cdots + a_n x^n$$
> where n is a nonnegative integer and $a_0, a_1, \ldots, a_n$ are (real-valued) constants. If $a_n \neq 0$, then n is called the **degree** of the polynomial function.

We have already encountered two polynomials functions, namely the constant function $f(x) = c$ and the linear function $f(x) = mx + b$. The constant function has degree 0, and the linear function has degree 1. In Figure 1.11, we display $y = x^n$ for $n = 2$ and 3. Looking at Figure 1.11, we see that $y = x^n$ is an even function (i.e., symmetric about the y-axis) when n is even, and an odd function (i.e., symmetric about the origin) when n is odd. This holds in general, and can be shown algebraically using the criterion in Section 1.2.1 (see Problem 24 at the end of this section).

Polynomials arise quite naturally in many situations. We present two examples.

EXAMPLE 3   Suppose that a seed drops from a tree that is 64 ft tall. Ignoring air resistance, one can show that at time t (measured in seconds) the seed is at height $h(t)$ (measured in feet), with

$$h(t) = 64 - 16t^2$$

Show that $h(t)$ is a polynomial and determine its degree. How long will it take the seed to hit the ground?

Solution   The function $h(t)$ is a polynomial of degree 2 with $a_0 = 64$, $a_1 = 0$, and $a_2 = -16$. Its graph is shown in Figure 1.12. The seed will hit the ground when

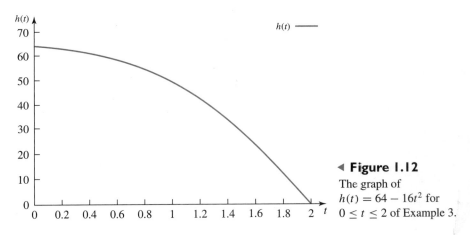

◀ **Figure 1.12**
The graph of
$h(t) = 64 - 16t^2$ for
$0 \le t \le 2$ of Example 3.

$h(t) = 0$. That is, we must solve

$$0 = 64 - 16t^2$$

$$t^2 = \frac{64}{16}$$

$$t = 2$$

We find that it takes the seed 2 seconds to hit the ground (ignoring air resistance). Note that because $h(t) \ge 0$ (where $h(t)$ is the height above the ground, and the height of the ground level is equal to 0), the domain of $h(t)$ should be the interval $[0, 2]$. ▮

EXAMPLE 4   Consider now the speed or reaction rate of the chemical reaction

$$A + B \longrightarrow AB$$

in which the reactants A and B form the product AB. The rate at which this reaction proceeds depends on how often A and B molecules collide. The **law of mass action** states that the rate at which this reaction proceeds is proportional to the product of the respective concentrations of the reactants. Here, "concentration" means the number of molecules per fixed volume. If we denote the reaction rate by R and the concentration of A and B by [A] and [B], respectively, then the law of mass action says that

$$R \propto [A] \cdot [B]$$

Introducing the proportionality factor k, we obtain

$$R = k[A] \cdot [B]$$

Note that $k > 0$ because [A], [B], and R are positive. We assume now that the reaction occurs in a closed vessel; that is, we add specific amounts of A and B to the vessel at the beginning of the reaction and then let the reaction proceed without further additions.

We can express the concentrations of the reactants A and B during the reaction in terms of their initial concentrations a and b, and the concentration of the reaction product [AB]. We set $x = [AB]$, then

$$[A] = a - x \quad \text{for } 0 \le x \le a \qquad \text{and} \qquad [B] = b - x \quad \text{for } 0 \le x \le b$$

The concentration of AB cannot exceed either of the concentrations of A or B. (Suppose five A molecules and seven B molecules are allowed to react; a maximum of five AB molecules can result, at which point all of the A molecules are used up

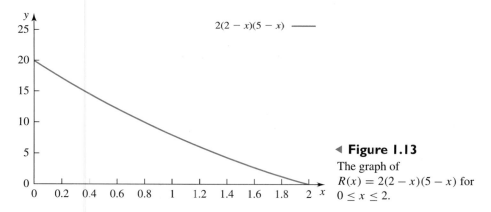

◄ **Figure 1.13**
The graph of
$R(x) = 2(2 - x)(5 - x)$ for
$0 \leq x \leq 2$.

and the reaction ceases. The two B molecules left over have no A molecules to react with.) Therefore, we get

$$R(x) = k(a - x)(b - x) \quad \text{for } 0 \leq x \leq a \text{ and } 0 \leq x \leq b$$

The condition "$0 \leq x \leq a$ and $0 \leq x \leq b$" can be written as $0 \leq x \leq \min(a, b)$, where $\min(a, b)$ denotes the minimum of a and b. To see that $R(x)$ is indeed a polynomial function, we expand the expression for $R(x)$.

$$R(x) = k(ab - ax - bx + x^2)$$
$$= kx^2 - k(a + b)x + kab$$

for $0 \leq x \leq \min(a, b)$. We see that $R(x)$ is a polynomial of degree 2.

A graph of $R(x)$, $0 \leq x \leq a$, is shown in Figure 1.13 for the case $a \leq b$ (we chose $a = 2$ and $b = 5$ in the figure). Notice that when $x = 0$ (that is, when no AB molecules have yet formed), the rate at which the reaction proceeds is at a maximum. As more and more AB molecules form and, consequently, the concentrations of the reactants decline, the reaction rate decreases. This should also be intuitively clear: as fewer and fewer A and B molecules are in the vessel, it becomes less and less likely that they will collide to form the product AB. When $x = a = \min(a, b)$, the reaction rate $R(a) = 0$. This is the point at which all A molecules are exhausted and the reaction necessarily ceases. ∎

1.2.3 Rational Functions

Rational functions are built from polynomial functions.

> **Definition**   A **rational** function is the quotient of two polynomial functions $p(x)$ and $q(x)$:
>
> $$f(x) = \frac{p(x)}{q(x)} \quad \text{for } q(x) \neq 0$$

Since division by 0 is not allowed, we must exclude those values of x for which $q(x) = 0$. Examples of rational functions:

$$y = \frac{1}{x}, \quad x \neq 0$$

$$y = \frac{x^2 + 2x - 1}{x - 3}, \quad x \neq 3$$

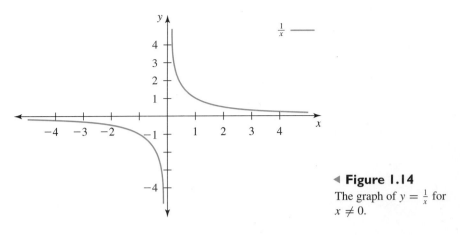

◀ **Figure 1.14**
The graph of $y = \frac{1}{x}$ for $x \neq 0$.

An important example of a rational function is the hyperbola

$$y = \frac{1}{x}, \quad x \neq 0$$

Its graph is shown in Figure 1.14.

Throughout this course we will encounter populations whose sizes change with time. The change in population size is described by the **growth rate**. Roughly speaking, the growth rate tells you the size of a population change during a small time interval. (This is analogous to the velocity of a car: Velocity is also a rate, it tells you how much your position changes in a small time interval. We will give a precise definition of rates in Section 3.1.) The **per capita** growth rate is the growth rate divided by the population size. The per capita growth rate is also called the **specific** growth rate. The following example introduces a function that is frequently used to describe growth rates.

EXAMPLE 5   **Monod Growth Function**

The following function is frequently used to describe the per capita growth rate of organisms when the growth rate depends on the concentration of some nutrient and shows a saturation effect. We denote the concentration of the nutrient by N; the per capita growth rate $r(N)$ is given by the Monod growth function

$$r(N) = a\frac{N}{k + N}, \quad N \geq 0$$

where a and k are positive constants. The graph of $r(N)$ is shown in Figure 1.15. It shows a decelerating rise approaching the saturation level a, which is the maximal specific growth rate. When $N = k$, then $r(N) = a/2$; for this reason, k is called the half-saturation constant. The growth rate increases with nutrient concentration N; however, doubling the nutrient concentration has a much bigger effect on the growth rate for small values of N than when N is already large. The equation describing the growth rate $r(N)$ is often referred to as the Michaelis-Menten equation.

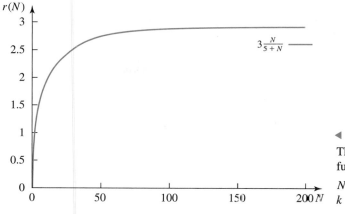

◀ **Figure 1.15**

The graph of the Monod function $r(N) = a\frac{N}{k+N}$ for $N \geq 0$. (Here: $a = 3$ and $k = 5$.)

1.2.4 Power Functions

Definition   A **power** function is of the form

$$f(x) = x^r$$

where r is a real number.

Examples of power functions:

$$y = x^{1/3}, \qquad x \in \mathbf{R}$$
$$y = x^{5/2}, \qquad x \geq 0$$
$$y = x^{1/2}, \qquad x \geq 0$$
$$y = x^{-1/2}, \qquad x > 0$$

Polynomials of the form $y = x^n$, $n = 1, 2, \ldots$ are a special case of power functions. Since power functions may involve even roots, such as in $y = x^{3/2} = \left(\sqrt{x}\right)^3$, we frequently need to restrict the domain of power functions.

In Figure 1.16, we compare some power functions, namely $y = x^{5/2}$, $y = x^{1/2}$, and $y = x^{-1/2}$, which we defined above. You should pay close attention to how the exponent determines the ranking according to size for x between 0 and 1 and for $x > 1$. We find that $x^{5/2} < x^{1/2} < x^{-1/2}$ for $0 < x < 1$, but $x^{5/2} > x^{1/2} > x^{-1/2}$ for $x > 1$.

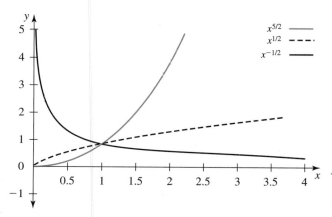

◀ **Figure 1.16**

Some power functions with rational exponents.

EXAMPLE 6 Power functions are frequently found in "scaling relations" between biological variables and organic size. These are relations of the form

$$y \propto x^r$$

where r is a nonzero real number. That is, y is proportional to some power of x. Recall that we can write this as an equation if we introduce the proportionality factor k, namely

$$y = kx^r$$

Finding such relationships is the objective of **allometry**. For instance, in a study of forty-five species of unicellular algae, a relationship between cell volume and cell biomass was sought. It was found (see, for instance, Niklas (1994)) that

$$\text{cell biomass} \propto (\text{cell volume})^{0.794}$$

Most of these scaling relations are to be interpreted in a statistical sense; they are obtained by fitting a curve to data points. The data points are typically scattered about the fitted curve given by the scaling relation (see Figure 1.17). ▍

The following example relates the volume and the surface of a cube. This is not to be understood in a statistical sense, because it is an exact relationship resulting from geometric considerations.

EXAMPLE 7 Suppose that we wish to know the scaling relation between the surface S and the volume V of a cube. For each of these quantities, we know their respective scaling relation with the length L of the cube:

$$S \propto L^2 \quad \text{or} \quad S = k_1 L^2$$
$$V \propto L^3 \quad \text{or} \quad V = k_2 L^3$$

To express S in terms of V, we must first solve L in terms of V and then substitute L in the equation for S. Because $L = (V/k_2)^{1/3}$,

$$S = k_1 \left[\left(\frac{V}{k_2} \right)^{1/3} \right]^2 = \frac{k_1}{k_2^{2/3}} V^{2/3}$$

Introducing the constant of proportionality $k = k_1/k_2^{2/3}$, we find

$$S = kV^{2/3} \quad \text{or simply} \quad S \propto V^{2/3}$$

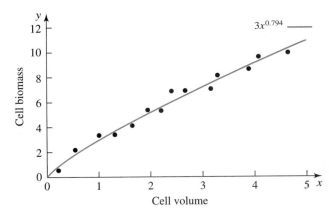

◀ **Figure 1.17**

Some data points and the fitted curve of Example 6. (*Note:* the "data points" aren't real data.)

In words, the surface of a cube scales with the volume in relation to $V^{2/3}$. We can now ask, for instance, by what factor the surface increases when we double the volume. When we double the volume, we find that the resulting surface, denoted by S', is

$$S' = k(2V)^{2/3} = 2^{2/3} \underbrace{kV^{2/3}}_{S}$$

That is, the surface increases by a factor of $2^{2/3} = 1.587$ if we double the volume of the cube. ∎

1.2.5 Exponential Functions

We will first give the precise definition of an exponential function.

> **Definition**   The function f is an **exponential** function with base a if
> $$f(x) = a^x, \quad x \in \mathbf{R}$$
> where a is a positive constant other than 1.

∎EXAMPLE 3   **Exponential Growth**

Most bacteria reproduce by cellular fission, in which the parent cell splits into two daughter cells after duplication of the genetic material. This division may happen as often as every twenty minutes; under ideal conditions, a bacterial colony can double in size in that time.

Let us measure time such that one unit of time corresponds to the doubling time of the colony. If we denote the size of the population at time t by $N(t)$, then the size of the population at time $t + 1$ is twice the population size at time t. We write this as

$$N(t + 1) = 2N(t) \tag{1.2}$$

The function

$$N(t) = 2^t, \quad t \geq 0$$

satisfies equation (1.2):

$$N(t + 1) = 2^{t+1} = 2 \cdot 2^t = 2N(t)$$

The function $N(t) = 2^t$, $t \geq 0$, is an exponential function because the variable t is in the exponent. We chose $t \geq 0$ as the domain for the function $N(t)$ because we think of $t = 0$ as the time when we start measuring the population size. The function $f(t) = 2^t$ can be defined for all $t \in \mathbf{R}$; its graph is shown in Figure 1.18.

When $t = 0$, we find that $N(0) = 1$; that is, there is just one individual in the population at time $t = 0$. If at time $t = 0$ forty individuals were present in the population, we would write $N(0) = 40$, and

$$N(t) = 40 \cdot 2^t, \quad t \geq 0 \tag{1.3}$$

You can check that $N(t)$ in (1.3) also satisfies (1.2).

It is often desirable not to specify the initial number of individuals in the equation describing $N(t)$. This has the advantage that the equation for $N(t)$ describes a more general situation, in the sense that we can use the same equation for different initial population sizes. We often denote the population size at time 0 by N_0 (read: "N sub 0") instead of $N(0)$. The equation for $N(t)$ is then

$$N(t) = N_0 2^t, \quad t \geq 0$$

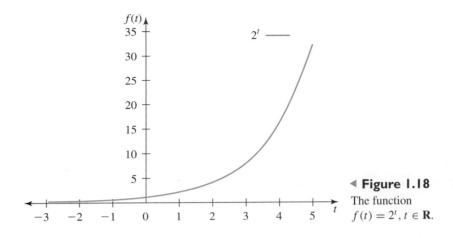

◀ **Figure 1.18**
The function
$f(t) = 2^t, t \in \mathbf{R}.$

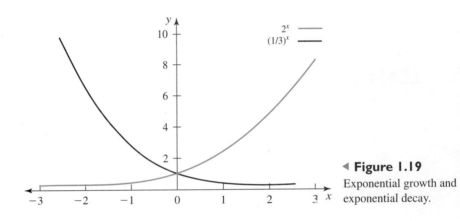

◀ **Figure 1.19**
Exponential growth and
exponential decay.

We can check that $N(0) = N_0 2^0 = N_0$ and that $N(t+1) = N_0 2^{t+1} = 2(N_0 2^t) = 2N(t)$.

The basic shape of the exponential function $f(x) = a^x$ depends on the base a; two examples are shown in Figure 1.19. The graph of $f(x) = 2^x$ shows a rapid increase, whereas the graph of $f(x) = (1/3)^x$ shows a rapid decrease towards 0. We find the rapid increase when $a > 1$, and the rapid decrease when $0 < a < 1$. Therefore, we say that when $a > 1$, we have exponential growth; when $0 < a < 1$, we have exponential decay.

Recall that $a^0 = 1$, $a^{-k} = 1/a^k$, and $a^{1/k} = \sqrt[k]{a}$ where k is a positive integer. Exponential functions have the following properties.

$$a^x a^y = a^{x+y}$$

$$\frac{a^x}{a^y} = a^{x-y}$$

$$(a^x)^y = a^{xy}$$

In many applications, the exponential function is expressed in terms of the base e, where $e = 2.718\ldots$, which we encountered in Section 1.1. The number e is called the **natural exponential base**. The exponential function with base e is alternatively written as $\exp(x)$. That is,

$$\exp(x) = e^x$$

The advantage of this alternative can be seen when we try to write something like $e^{x^2/\sqrt{x^3+1}}$; $\exp\left(x^2/\sqrt{x^3+1}\right)$ is much easier to read. More generally, if $g(x)$ is a function in x, then we can write, equivalently,

$$\exp[g(x)] \qquad \text{or} \qquad e^{g(x)}$$

Base 2 and 10 are also frequently used; in calculus, however, e will turn out to be the most common base.

The next two examples provide an important application of exponential functions.

∎EXAMPLE 9   **Radioactive Decay**

Radioactive isotopes such as Carbon 14 (C^{14}) and Potassium 40 are used to determine the absolute age of fossils or minerals, establishing an absolute chronology of the geological time scale. This technique was discovered in the early years of the twentieth century, and is based on the property of certain atoms to spontaneously transform by giving off protons, neutrons, or electrons. This phenomenon is called radioactive decay; it occurs at a constant rate, and the rate is independent of environmental conditions. The method was used, for instance, to trace the successive emergence of the Hawaiian islands, from the oldest, Kauai, to the youngest, Hawaii (which is about 100,000 years old).

Carbon 14 is formed high in the atmosphere. It is radioactive, and decays into nitrogen (N^{14}). There is an equilibrium between Carbon 12 (C^{12}) and Carbon 14 in the atmosphere; moreover, the ratio of C^{14} to C^{12} in the atmosphere has been relatively constant over a fairly long period. When plants capture CO_2 molecules from the atmosphere and build it into a product (such as cellulose), the initial ratio of C^{14} to C^{12} is the same as in the atmosphere. Once the plants die, their uptake of CO_2 ceases, and the radioactive decay of C^{14} causes the ratio of C^{14} to C^{12} to decline. Since the law of radioactive decay is known, the change in ratio provides an accurate measure of age.

If the amount of C^{14} at time t is denoted by $W(t)$, with $W_0 = W(0)$, then the laws of physics show that

$$W(t) = W_0 e^{-\lambda t}, \quad t \geq 0$$

where $\lambda > 0$ denotes the **decay rate** (λ is the small Greek letter lambda). The function $W(t) = W_0 e^{-\lambda t}$ is an example of an exponential function. Its graph for $\lambda = 1$ is shown in Figure 1.20.

Frequently, the decay rate is expressed in terms of the **half-life** of the material, which is the length of time that it takes for half of the material to decay. If we denote

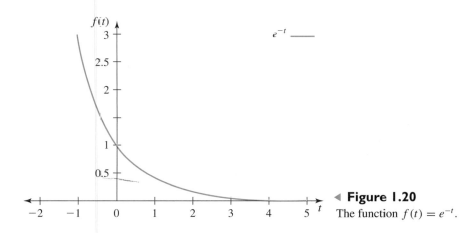

◀ **Figure 1.20**
The function $f(t) = e^{-t}$.

this time by T_h, then

$$W(T_h) = \frac{1}{2}W_0 = W_0 e^{-\lambda T_h}$$

from which we obtain

$$\frac{1}{2} = e^{-\lambda T_h}$$
$$2 = e^{\lambda T_h}$$

Recall from algebra (or Section 1.1) that to solve for the exponent λT_h we must take logarithms on both sides. Since the exponent has base e, we use natural logarithms and find

$$\ln 2 = \lambda T_h$$

Solving for T_h or λ yields

$$T_h = \frac{\ln 2}{\lambda} \qquad \text{or} \qquad \lambda = \frac{\ln 2}{T_h}$$

It is known that the half-life of C^{14} is 5730 years. Hence, expressed in years,

$$\lambda = \frac{\ln 2}{5730} \qquad \qquad ▌$$

An application of the C^{14} method is shown in the next example.

▌**EXAMPLE 10**   Suppose that samples of wood found in an archeological excavation site contain about 23% as much C^{14} (based on their C^{12} content) as living plant material. Determine when the wood was cut.

Solution

$$0.23 = \frac{W(t)}{W(0)} = e^{-\lambda t}$$

$$\lambda t = \ln \frac{1}{0.23}$$

Using $\lambda = \ln 2/5730$ from Example 9, we find

$$t = \frac{5730}{\ln 2} \ln \frac{1}{0.23}$$

Using a calculator to compute this result, we find that the wood was cut about 12,150 years ago.   ▌

1.2.6 Inverse Functions

Before we can introduce logarithmic functions, we need the concept of inverse functions. Roughly speaking, the inverse of a function f reverses the effect of f. That is, if f maps x into $y = f(x)$, the inverse function, denoted by f^{-1} (read: "f inverse"), takes y and maps it back into x. Not every function can have an inverse; because an inverse function is a function itself, we require that every value y in the range of f be mapped into exactly one value x. In other words, we need that whenever $x_1 \neq x_2$, then $f(x_1) \neq f(x_2)$ or, equivalently, that $f(x_1) = f(x_2)$ implies

$x_1 = x_2$. (Recall the definition of a function, where we required that a function assign each element in the domain to *exactly one* element in the range.)

Consider $y = x^3$ and $y = x^2$, for $x \in \mathbf{R}$. The function $y = x^3$ has an inverse function, because $x_1^3 \neq x_2^3$ whenever $x_1 \neq x_2$, whereas $y = x^2$ does not have an inverse function, because $x_1 \neq x_2$ does not imply $x_1^2 \neq x_2^2$ (or, equivalently, $x_1^2 = x_2^2$ does not imply $x_1 = x_2$). The equation $x_1^2 = x_2^2$ only implies $|x_1| = |x_2|$. Since x_1 and x_2 can be positive or negative, we cannot simply drop the absolute value signs. For instance, both -2 and 2 are mapped into 4, and we find that $f(-2) = f(2)$ but $-2 \neq 2$ (note that $|-2| = |2|$). To invert this function, we would have to map 4 into -2 and 2, but then it would no longer be a function by our definition. By restricting the domain of $y = x^2$ to (say) $x \geq 0$, we can define an inverse function of $y = x^2$, $x \geq 0$.

More generally, a function f will have an inverse on its domain A if $x_1 \neq x_2$ implies $f(x_1) \neq f(x_2)$ (or, equivalently, $f(x_1) = f(x_2)$ implies $x_1 = x_2$). Functions that have this property are called **one-to-one**.

If you know what the graph of a function looks like over its domain, then it is easy to determine whether or not the function is one-to-one. Namely, if no horizontal line intersects the graph of f more than once, then f is one-to-one. This is called the horizontal line test. We illustrate how to use this test in Figures 1.21 and 1.22.

We now present the formal definition of an inverse function.

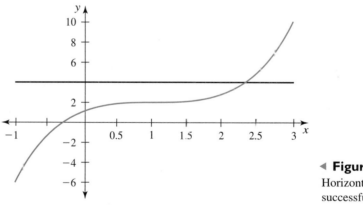

◀ **Figure 1.21**
Horizontal line test successful.

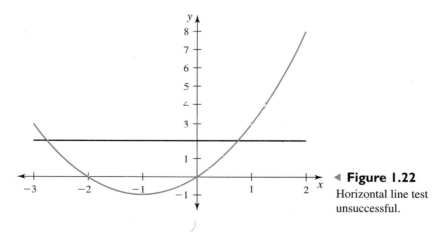

◀ **Figure 1.22**
Horizontal line test unsuccessful.

> **Definition**   Let $f : A \to B$ be a one-to-one function with range $f(A)$. The inverse function f^{-1} has domain $f(A)$ and range A and is defined by
>
> $$f^{-1}(y) = x \text{ if and only if } y = f(x)$$
>
> for all $y \in f(A)$.

▮ EXAMPLE 11   Find the inverse function of $f(x) = x^3 + 1$, $x \geq 0$.

Solution   First note that $f(x)$ is one-to-one. To see this quickly, graph the function and apply the horizontal line test; to demonstrate it, start with $f(x_1) = f(x_2)$ and show that this implies $x_1 = x_2$:

$$f(x_1) = f(x_2)$$
$$x_1^3 + 1 = x_2^3 + 1$$
$$x_1^3 = x_2^3$$

Taking the third root on both sides implies that $x_1 = x_2$, which tells you that $f(x)$ has an inverse. Now we will find f^{-1}.

To find an inverse function there are three steps:

1. Write $y = f(x)$:

$$y = x^3 + 1$$

2. Solve for x:

$$x^3 = y - 1$$
$$x = \sqrt[3]{y - 1}$$

The range of f is $[1, \infty)$, and the range of f becomes the domain of f^{-1}, so we obtain

$$f^{-1}(y) = \sqrt[3]{y - 1}, \quad y \geq 1$$

Typically, we write functions as functions of x. To do this, we simply interchange x and y in $x = f^{-1}(y)$. This is the third step.

3. Interchange x and y:

$$y = f^{-1}(x) = \sqrt[3]{x - 1}, \; x \geq 1$$

Note that switching x and y in step 3 corresponds to reflecting the graph of $y = f(x)$ about the line $y = x$. The graphs of f and f^{-1} are shown in Figure 1.23. Look at the graphs carefully, and see that the two graphs are indeed related to each other by reflecting either one about the line $y = x$. ▮

Here is another property of inverse functions. If we first apply f to x and then f^{-1} to $f(x)$, we obtain the original value x. That is, if $f : A \to B$ has an inverse f^{-1}, then

$$f^{-1}[f(x)] = x \quad \text{for all } x \in A$$
$$f[f^{-1}(y)] = y \quad \text{for all } y \in f(A)$$

A note of warning: the superscript in f^{-1} does *not* mean the reciprocal of f (i.e., $1/f$). This difference is further explained in Problem 62 at the end of this section.

◀ **Figure 1.23**
Inverse functions.

1.2.7 Logarithmic Functions

Recall from algebra (or Section 1.1) that to solve the equation

$$3 = e^x$$

for x, you must take logarithms on both sides:

$$x = \ln 3$$

In other words, the natural logarithm undoes the operation of raising e to the x power—it is inverse of the exponential function. (Consequently, the exponential function is the inverse of the logarithmic function.)

We will now define the inverse of the exponential function $f(x) = a^x$, $x \in \mathbf{R}$. The base a can be any positive number.

Definition   The inverse of $f(x) = a^x$ is called the **logarithm to base a** and is written $f^{-1}(x) = \log_a x$.

The domain of $f(x) = a^x$ is the set of all real numbers, and its range is the set of all positive numbers. Since the range of f is the domain of f^{-1}, we find that the domain of $f^{-1}(x) = \log_a x$ is the set of positive numbers. We can now summarize the relationship between the exponential and the logarithmic functions.

1. $a^{\log_a x} = x$ for $x > 0$

2. $\log_a a^x = x$ for $x \in \mathbf{R}$

It is important to remember that the logarithm is only defined for positive numbers, that is, $y = \log_a x$ is only defined for $x > 0$. The logarithm satisfies the following properties:

$$\log_a (st) = \log_a s + \log_a t$$

$$\log_a \left(\frac{s}{t}\right) = \log_a s - \log_a t$$

$$\log_a s^r = r \log_a s$$

The inverse of the exponential function with the natural base e is denoted by $\ln x$ and is called the natural logarithm of x. The graphs of $y = e^x$ and $y = \ln x$ are shown in Figure 1.24. Note that both e^x and $\ln x$ are increasing functions. However, whereas e^x climbs very quickly for large values of x, $\ln x$ increases very slowly for

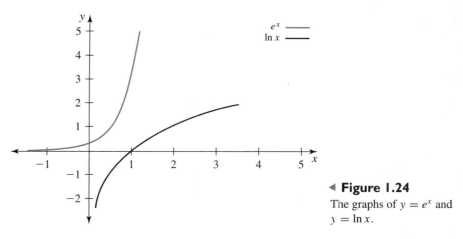

◀ **Figure 1.24**
The graphs of $y = e^x$ and $y = \ln x$.

large values of x. Looking at both graphs, we can see that each can be obtained as the reflection of the other about the line $y = x$.

The logarithm to base 10 is frequently written as $\log x$ (i.e., the notation for base 10 in $\log_{10} x$ is simply dropped).

▌ **EXAMPLE 12**   Simplify the following expressions.

   a. $\log_2(x^2 - 4)$ **b.** $\log_3 9^x$ **c.** $\ln e^{3x^2+1}$

Solution

   a. We must write $x^2 - 4$ as a product before we can simplify the expression.

$$\log_2(x^2 - 4) = \log_2[(x - 2)(x + 2)] = \log_2(x - 2) + \log_2(x + 2)$$

   No further simplification is possible.

   b. Simplifying yields

$$\log_3 9^x = x \log_3 9 = x \log_3 3^2 = 2x$$

   The fact that $\log_3 9 = 2$ can be seen in two ways: we can write $9 = 3^2$ and say that $\log_3$ undoes raising 3 to the second power (as we did in the preceding), or we say that $\log_3 9$ denotes the exponent to which we must raise 3 in order to get 9.

   c. We use the fact that $\ln x$ and e^x are inverse functions, and find that

$$\ln e^{3x^2+1} = 3x^2 + 1$$ ▌

Any exponential function with base a can be written as an exponential function with base e. Likewise, any logarithm to base a can be written in terms of the natural logarithm. The following two identities illustrate this.

$$a^x = \exp[x \ln a]$$
$$\log_a x = \frac{\ln x}{\ln a}$$

The first identity follows from the fact that exp and ln are inverse functions, which implies that $a^x = \exp[\ln a^x]$, and the fact that $\ln a^x = x \ln a$. To understand the second identity, note that

$$y = \log_a x   \quad \text{means} \quad   a^y = x$$

Taking logarithms to base e on both sides of $a^y = x$, we find

$$\ln a^y = \ln x$$

or

$$y \ln a = \ln x$$

hence,

$$y = \frac{\ln x}{\ln a}$$

▮ EXAMPLE 13   Write the following expressions in terms of base e.

a. 2^x **b.** 10^{x^2+1} **c.** $\log_3 x$ **d.** $\log_2(3x - 1)$

Solution

a. $2^x = \exp(\ln 2^x) = \exp(x \ln 2)$

$= e^{x \ln 2}$

b. $10^{x^2+1} = \exp(\ln 10^{x^2+1}) = \exp[(x^2 + 1)\ln 10]$

$= e^{(x^2+1)\ln 10}$

c. $\log_3 x = \dfrac{\ln x}{\ln 3}$

d. $\log_2(3x - 1) = \dfrac{\ln(3x - 1)}{\ln 2}$ ▮

1.2.8 Trigonometric Functions

The trigonometric functions are examples of periodic functions.

> **Definition**   A function $f(x)$ is **periodic** with period a if
> $$f(x + a) = f(x)$$
> for all x in the domain of f.

We begin with the sine and cosine functions. In Section 1.1.4, we recalled the definition of sine and cosine in a unit circle. There, $\sin \theta$ and $\cos \theta$ represented trigonometric functions of angles and θ was either measured in or degrees radians. Now, we define the trigonometric functions as functions of *real* numbers. For instance, we define $f(x) = \sin x$ for $x \in \mathbf{R}$. The value of $\sin x$ is then *by definition* the sine of an angle of x radians. (Similarly for all the other trigonometric functions.)

The graphs of the sine and cosine functions are shown in Figures 1.25 and 1.26, respectively.

The sine function $y = \sin x$ is defined for all $x \in \mathbf{R}$. Its range is $-1 \leq y \leq 1$. Likewise, the cosine function $y = \cos x$ is defined for all $x \in \mathbf{R}$ with range $-1 \leq y \leq 1$. Both functions are periodic with period 2π. (They are also periodic with period $4\pi, 6\pi, \ldots$, but by convention, one uses the smallest possible value to specify the period.) We see from Figures 1.25 and 1.26 that the graph of the cosine function can be obtained by shifting the graph of the sine function $\pi/2$ to the left. (We will discuss such shifts of graphs in more detail in the next section.)

To define the tangent function, $y = \tan x$, recall that

$$\tan x = \frac{\sin x}{\cos x}$$

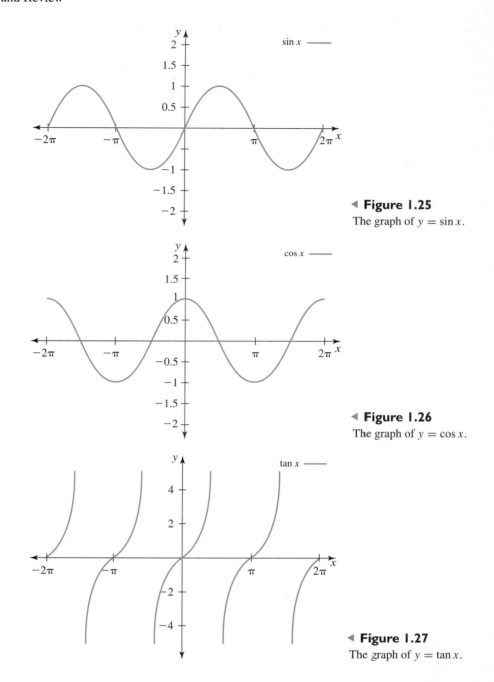

◄ Figure 1.25
The graph of $y = \sin x$.

◄ Figure 1.26
The graph of $y = \cos x$.

◄ Figure 1.27
The graph of $y = \tan x$.

Because $\cos x = 0$ for values of x that are odd integer multiples of $\pi/2$, the domain of $\tan x$ consists of all real numbers with the exception of odd integer multiples of $\pi/2$. The range of $y = \tan x$ is $-\infty < y < \infty$. A graph of $y = \tan x$ is shown in Figure 1.27; there, we see that $\tan x$ is periodic with period π.

The graphs of the remaining three trigonometric functions are shown in Figures 1.28–1.30. The domain of the secant function $y = \sec x$ consists of all real numbers with the exception of odd integer multiples of $\pi/2$; the range is $y \leq -1$ or $y \geq 1$. The domain of the cosecant function $y = \csc x$ consists of all real numbers with the exception of integer multiples of π; the range is $y \leq -1$ or $y \geq 1$. The domain of the cotangent function $y = \cot x$ consists of all real numbers with the exception of integer multiples of π; the range is $-\infty < y < \infty$.

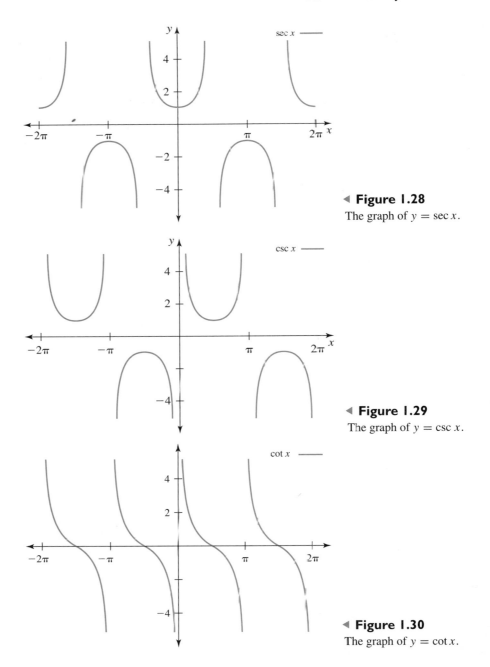

◀ **Figure 1.28**
The graph of $y = \sec x$.

◀ **Figure 1.29**
The graph of $y = \csc x$.

◀ **Figure 1.30**
The graph of $y = \cot x$.

The sine and cosine functions are of particular importance. We therefore describe them in more detail. Consider the function

$$f(x) = a \sin(kx) \qquad \text{for } x \in \mathbf{R}$$

where a is a positive constant and $k \neq 0$. Now, $f(x)$ takes on values between $-a$ and a. We call a the **amplitude**. The function $f(x)$ is periodic. To find the period p of $f(x)$, we set

$$kp = 2\pi \qquad \text{or} \qquad p = \frac{2\pi}{k}$$

Because the cosine function can be obtained from the sine function by a horizontal shift, we can define amplitude and period analogously for the cosine function. That is, $f(x) = a \cos(kx)$ has amplitude a and period $p = 2\pi/k$.

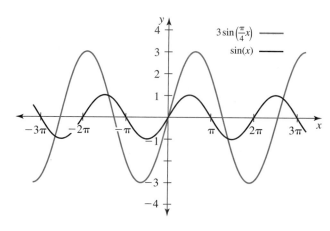

◀ **Figure 1.31**

The graphs of
$f(x) = 3 \sin \left(\frac{\pi}{4} x \right)$ and
$g(x) = \sin x$ in Example 14.

▮ **EXAMPLE 14**   Compare

$$f(x) = 3 \sin \left(\frac{\pi}{4} x \right) \qquad \text{and} \qquad g(x) = \sin x$$

Solution   The amplitude of $f(x)$ is 3, whereas the amplitude of $g(x)$ is 1. The period p of $f(x)$ satisfies $\frac{\pi}{4} p = 2\pi$ or $p = 8$, whereas the period of $g(x)$ is 2π. Graphs of $f(x)$ and $g(x)$ are shown in Figure 1.31.   ▮

Remark.   A number is called **algebraic** if it is the solution of a polynomial equation with rational coefficients. For instance, $\sqrt{2}$ is algebraic as it satisfies the equation $x^2 - 2 = 0$. Numbers that are not algebraic are called **transcendental**. For instance, π and e are transcendental.

A similar distinction is also made for functions. We call a function $y = f(x)$ algebraic if it is the solution of an equation of the form

$$P_n(x) y^n + \cdots + P_1(x) y + P_0(x) = 0$$

in which the coefficients are polynomial functions in x with rational coefficients. For instance, the function $y = 1/(1 + x)$ is algebraic as it satisfies the equation $(x+1) y - 1 = 0$. Here, $P_1(x) = x + 1$ and $P_0(x) = -1$. Other examples of algebraic functions include polynomials and rational functions with rational coefficients.

Functions that are not algebraic are called transcendental. All the trigonometric, exponential, and logarithmic functions that we introduced in this section are examples of transcendental functions.

◆ **1.2 PROBLEMS**

[1.2.1]

In Problems 1–4, state the range for the given functions. Graph each function.

1. $f(x) = x^2, x \in \mathbf{R}$
2. $f(x) = x^2, x \in [0, 1]$
3. $f(x) = x^2, -1 < x \le 0$
4. $f(x) = x^2, -\frac{1}{2} < x < \frac{1}{2}$
5. (a) Show that for $x \ne 1$,

$$\frac{x^2 - 1}{x - 1} = x + 1$$

(b) Are the functions

$$f(x) = \frac{x^2 - 1}{x - 1}, \quad x \ne 1$$

and

$$g(x) = x + 1, \quad x \in \mathbf{R}$$

equal?

6. (a) Show that

$$2|x - 1| = \begin{cases} 2(x - 1) & \text{for } x \ge 1 \\ 2(1 - x) & \text{for } x \le 1 \end{cases}$$

(b) Are the functions

$$f(x) = \begin{cases} 2 - 2x & \text{for } 0 \le x \le 1 \\ 2x - 2 & \text{for } 1 \le x \le 2 \end{cases}$$

and

$$g(x) = 2|x - 1|, \quad x \in [0, 2]$$

equal?

In Problems 7–12, sketch the graph of each function, and decide in each case whether the function is (i) even, (ii) odd, or (iii) does not show any obvious symmetry. Then check your answers using the criteria in Section 1.2.1.

7. $f(x) = 2x$ **8.** $f(x) = 3x^2$

9. $f(x) = |3x|$ **10.** $f(x) = 2x + 1$

11. $f(x) = -|x|$ **12.** $f(x) = 3x^3$

13. Suppose that

$$f(x) = x^2, \quad x \in \mathbf{R}$$

and

$$g(x) = 3 + x, \quad x \in \mathbf{R}$$

(a) Show that

$$(f \circ g)(x) = (3 + x)^2, \quad x \in \mathbf{R}$$

(b) Show that

$$(g \circ f)(x) = 3 + x^2, \quad x \in \mathbf{R}$$

14. Suppose that

$$f(x) = 1 - x^2, \quad x \in \mathbf{R}$$

and

$$g(x) = 2x, \quad x \geq 0$$

(a) Find

$$(f \circ g)(x)$$

together with its domain.

(b) Find

$$(g \circ f)(x)$$

together with its domain.

15. Suppose that

$$f(x) = \frac{1}{x + 1}, \quad x \neq -1$$

and

$$g(x) = 2x^2, \quad x \in \mathbf{R}$$

(a) Find $(f \circ g)(x)$.

(b) Find $(g \circ f)(x)$.

In both (a) and (b), find the domain.

16. Suppose that $f(x) = x^2$, $x \geq 0$, and $g(x) = \sqrt{x}$, $x \geq 0$. Typically, $f \circ g \neq g \circ f$, but here is an example where the order of composition does not matter. Show that $f \circ g = g \circ f$.

[1.2.2]

17. Use a graphing calculator to graph $f(x) = x^2$, $x \geq 0$, and $g(x) = x^4$, $x \geq 0$, together. For which values of x is $f(x) > g(x)$, and for which is $f(x) < g(x)$?

18. Use a graphing calculator to graph $f(x) = x^3$, $x \geq 0$, and $g(x) = x^5$, $x \geq 0$, together. When is $f(x) > g(x)$, and when is $f(x) < g(x)$?

19. Graph $y = x^n$, $x \geq 0$, for $n = 1, 2, 3$, and 4 in one coordinate system. Where do the curves intersect?

20. (a) Graph $f(x) = x$, $x \geq 0$, and $g(x) = x^2$, $x \geq 0$, together, in one coordinate system.

(b) For which values of x is $f(x) \geq g(x)$, and for which values of x is $f(x) \leq g(x)$?

21. (a) Graph $f(x) = x^2$ and $g(x) = x^3$ for $x \geq 0$ together, in one coordinate system.

(b) Show algebraically that

$$x^2 \geq x^3$$

for $0 \leq x \leq 1$.

(c) Show algebraically that

$$x^2 \leq x^3$$

for $x \geq 1$.

22. Show algebraically that if $n \geq m$,

$$x^n \leq x^m \quad \text{for } 0 \leq x \leq 1$$

and

$$x^n \geq x^m \quad \text{for } x \geq 1$$

23. (a) Show that $y = x^2$, $x \in \mathbf{R}$, is an even function.

(b) Show that $y = x^3$, $x \in \mathbf{R}$, is an odd function.

24. Show that

(a) $y = x^n$, $x \in \mathbf{R}$, is an even function when n is an even integer.

(b) $y = x^n$, $x \in \mathbf{R}$, is an odd function when n is an odd integer.

25. In Example 4 of this section, we considered the chemical reaction

$$A + B \rightarrow AB$$

Assume that initially only A and B are in the reaction vessel, and that the initial concentrations are $a = [A] = 3$ and $b = [B] = 4$.

(a) We found that the reaction rate $R(x)$, where x is the concentration of the product AB, is given by

$$R(x) = k(a - x)(b - x)$$

where a is the initial concentration of A, b is the initial concentration of B, and k is the constant of proportionality. Suppose that the reaction rate $R(x)$ is equal to 9 when the concentration of AB is $x = 1$. Use this to find k.

(b) Determine the relevant domain of $R(x)$, and use a graphing calculator to sketch the graph of $R(x)$.

26. An autocatalytic reaction uses its resulting product for the formation of a new product, as in the reaction

$$A + X \rightarrow X$$

If we assume that this reaction occurs in a closed vessel, then the reaction rate is given by

$$R(x) = kx(a - x)$$

for $0 \leq x \leq a$, where $a = $ [A] is the initial concentration of A and x is the concentration of X.

(a) Show that $R(x)$ is a polynomial and determine its degree.

(b) Graph $R(x)$ for $k = 2$ and $a = 6$. Find the value of x at which the reaction rate is maximal.

27. Suppose that a beetle walks up a tree along a straight line at a constant speed of 1 meter per hour. What distance will the beetle have covered after 1 hour, 2 hours, and 3 hours? Write an equation that expresses the distance (in meters) as a function of the time (in hours), and show that this function is a polynomial of degree 1.

28. Suppose that a fungal disease originates in the middle of an orchard, initially affecting only one tree. The disease spreads out radially at a constant speed of 10 feet per day. What area will be affected after 2 days, 4 days, and 8 days? Write an equation that expresses the affected area as a function of time measured in days, and show that this function is a polynomial of degree 2.

[1.2.3]

In Problems 29–32, find the largest possible domain for each function. Determine its range.

29. $f(x) = \dfrac{1}{1 - x}$

30. $f(x) = \dfrac{2x}{(x - 2)(x + 3)}$

31. $f(x) = \dfrac{x - 2}{x^2 - 9}$

32. $f(x) = \dfrac{1}{x^2 + 1}$

33. Compare $y = \frac{1}{x}$ and $y = \frac{1}{x^2}$ for $x > 0$ by graphing the two functions. Where do the curves intersect? Which function is bigger for small values of x (and for large values of x)?

34. Let n and m be two positive integers with $m \leq n$. Compare $y = x^{-n}$ and $y = x^{-m}$ for $x > 0$.

35. Let

$$f(x) = \frac{2x}{3 + x}, \quad x \geq 0$$

(a) Use a graphing calculator to graph $f(x)$.

(b) Explain why $0 \leq f(x) \leq 2$ for all $x \geq 0$.

(c) For which values of x is $f(x) = 1$?

(d) Show that for any value $a \in [0, 2)$, you can find exactly one value $x \geq 0$ so that $f(x) = a$.

In Problems 36–38, we discuss the Monod growth function, which was introduced in Example 5 of this section.

36. Use a graphing calculator to investigate the Monod growth function

$$r(N) = a\frac{N}{k + N}, \quad N \geq 0$$

(a) Graph $r(N)$ for (i) $a = 5$ and $k = 1$, (ii) $a = 5$ and $k = 3$, and (iii) $a = 8$ and $k = 1$. Place all three graphs in one coordinate system.

(b) Based on the graphs in (a), describe in words what happens when you change a.

(c) Based on the graphs in (a), describe in words what happens when you change k.

37. The Monod growth function $r(N)$ describes growth as a function of nutrient concentration N. Assume that

$$r(N) = 5\frac{N}{1 + N}, \quad N \geq 0$$

Find the percentage increase when the nutrient concentration is doubled from $N = 0.1$ to $N = 0.2$. Compare this to your findings when you double the nutrient concentration from $N = 10$ to $N = 20$. This is an example of *diminishing return*.

38. The Monod growth function $r(N)$ describes growth as a function of nutrient concentration N. Assume that

$$r(N) = a\frac{N}{k + N}, \quad N \geq 0$$

where a and k are positive constants.

(a) What happens to $r(N)$ as N increases? Use this to explain why a is called the saturation level.

(b) Show that k is the half-saturation constant; that is, show that if $N = k$, then $r(N) = a/2$.

[1.2.4]

In Problems 39–42, use a graphing calculator to sketch the graphs of the functions.

39. $y = x^{3/2}, x \geq 0$ **40.** $y = x^{1/3}, x \geq 0$

41. $y = x^{-1/4}, x > 0$ **42.** $y = 2x^{-7/8}, x > 0$

43. **(a)** Graph $y = x^{-1/2}, x > 0$, and $y = x^{1/2}, x \geq 0$, together in one coordinate system.

(b) Show algebraically that

$$x^{-1/2} \geq x^{1/2}$$

for $0 < x \leq 1$.

(c) Show algebraically that

$$x^{-1/2} \leq x^{1/2}$$

for $x \geq 1$.

44. **(a)** Graph $y = x^{5/2}, x \geq 0$, and $y = x^{1/2}, x \geq 0$, together in one coordinate system.

(b) Show algebraically that

$$x^{5/2} \leq x^{1/2}$$

for $0 \leq x \leq 1$. (*Hint*: Show that $x^{1/2}/x^{-1/2} = x \leq 1$ for $0 < x \leq 1$.)

(c) Show algebraically that

$$x^{5/2} \geq x^{1/2}$$

for $x \geq 1$.

In Problems 45–47, sketch each scaling relation (Niklas 1994).

45. In a sample based on forty-six species, leaf area was found to be proportional to (stem diameter)$^{1.84}$. Based on your graph, as stem diameter increases, does leaf area increase of decrease?

46. In a sample based on twenty-eight species, the volume fraction of spongy mesophyll was found to be proportional to (leaf thickness)$^{-0.49}$. (The spongy mesophyll is part of the internal tissue of a leaf blade.) Based on your graph, as leaf thickness increases, does the volume fraction of spongy mesophyll increase or decrease?

47. In a sample of sixty species of trees, wood density was found to be proportional to (breaking strength)$^{0.82}$. Based on your graph, when does breaking strength increase: wood density increases or decreases?

48. Suppose that a cube of length L and volume V has mass M, and that $M = 0.35V$. How does the length of the cube depend on its mass?

[1.2.5]

49. Assume that a population size at time t is $N(t)$, and that
$$N(t) = 2^t, \quad t \geq 0$$
(a) Find the population size for $t = 0, 1, 2, 3$, and 4.
(b) Graph $N(t)$ for $t \geq 0$.

50. Assume that a population size at time t is $N(t)$, and that
$$N(t) = 40 \cdot 2^t, \quad t \geq 0$$
(a) Find the population size at time $t = 0$.
(b) Show that
$$N(t) = 40e^{t \ln 2}, \quad t \geq 0$$
(c) How long will it take until the population size reaches 1000? (*Hint*: Find t so that $N(t) = 1000$.)

51. The half-life of C^{14} is 5730 years. If a sample of C^{14} has a mass of 20 gr at time $t = 0$, how much is left after 2000 years?

52. The half-life of C^{14} is 5730 years. If a sample of C^{14} has a mass of 20 grams at time 0, how long will it take until **(a)** 10 grams, and **(b)** 5 grams are left?

53. After 7 days, a particular radioactive substance decays to half of its original amount. Find the decay rate.

54. After 5 days, a particular radioactive substance decays to 37% of its original amount. Find the half-life of this substance.

55. Polonium-210 (Po^{210}) has a half-life of 140 days.
(a) If a sample of Po^{210} has a mass of 300 gr, find a formula for the mass after t days.
(b) How long would it take this sample to decay to 20% of its original amount?
(c) Sketch the graph of the amount of mass left after t days.

56. Suppose that wood found at an archeological excavation site contains about 35% as much C^{14} (in relation to C^{12}) as living plant material. Determine when the wood was cut.

57. Suppose that wood found at an archeological excavation site is 15,000 years old. How much C^{14} (based on C^{12} content) does the wood contain relative to living plant material?

58. The age of rocks of volcanic origin can be estimated using isotopes of Argon 40 (Ar^{40}) and Potassium 40 (K^{40}). K^{40} decays into Ar^{40} over time. If a mineral that contains potassium is buried under the right circumstances, argon forms and is trapped. Because argon is driven off when the mineral is heated to very high temperatures, rocks of volcanic origin do not contain argon when they are formed. The amount of argon found in such rocks can therefore be used to determine the age of the rock. Assume that a sample of volcanic rock contains 0.00047% K^{40}. The sample also contains 0.000079% Ar^{40}. How old is the rock? (The decay rate of K^{40} to Ar^{40} is 5.335×10^{-10}/yr.)

[1.2.6]

59. Which of the following functions is one-to-one? (Use the horizontal line test.)
(a) $f(x) = x^2, x \geq 0$ **(b)** $f(x) = x^2, x \in \mathbf{R}$
(c) $f(x) = \frac{1}{x}, x > 0$ **(d)** $f(x) = e^x, x \in \mathbf{R}$
(e) $f(x) = \frac{1}{x^2}, x \neq 0$

60. **(a)** Show that $f(x) = x^3 - 1, x \in \mathbf{R}$, is one-to-one, and find its inverse together with its domain.
(b) Graph $f(x)$ and $f^{-1}(x)$ in one coordinate system, together with the line $y = x$, and convince yourself that the graph of $f^{-1}(x)$ can be obtained by reflecting the graph of $f(x)$ about the line $y = x$.

61. **(a)** Show that $f(x) = x^2 + 1, x \geq 0$, is one-to-one, and find its inverse together with its domain.
(b) Graph $f(x)$ and $f^{-1}(x)$ in one coordinate system, together with the line $y = x$, and convince yourself that the graph of $f^{-1}(x)$ can be obtained by reflecting the graph of $f(x)$ about the line $y = x$.

62. The reciprocal of a function $f(x)$ can be written as either $1/f(x)$ or $[f(x)]^{-1}$. The point of this problem is to make clear that a reciprocal of a function has nothing to do with the inverse of a function. As an example, let $f(x) = 2x + 1, x \in \mathbf{R}$. Find both $[f(x)]^{-1}$ and $f^{-1}(x)$, and compare the two functions. Graph all three functions together.

[1.2.7]

63. Write the following expressions in terms of base e.
(a) 2^{x^2-1} **(b)** 3^{5x}
(c) $\log(x + 2)$ **(d)** $\log_2(2x^2 - 1)$

64. Simplify the following expressions
(a) $\ln x^2 + \ln x^3$ **(b)** $e^{3 \ln x}$
(c) $\ln(x^2 - 1) - \ln(x + 1)$

65. Let $f(x) = 2^x, x \in \mathbf{R}$.
(a) Show that $f(x)$ is one-to-one.
(b) Find the inverse of $f(x)$ together with its domain.
(c) Graph $f(x) = 2^x, x \in \mathbf{R}$, together with the line $y = x$. Reflect the graph of $f(x)$ about the line $y = x$ to obtain the graph of $f^{-1}(x)$.

66. Show that if $0 < a < 1$, the function $y = a^x$ can be written in the form $y = e^{-\mu x}$, where μ is a positive constant. Write μ in terms of a.

[1.2.8]

In Problems 67–72, use a graphing calculator to compare the following pairs of functions. Describe what you see.

67. $y = \sin x$, and $y = 2 \sin x$

68. $y = \sin x$, and $y = \sin(2x)$

69. $y = \cos x$, and $y = 2 \cos x$

70. $y = \cos x$, and $y = \cos(2x)$

71. $y = \tan x$, and $y = 2\tan x$

72. $y = \tan x$, and $y = \tan(2x)$

[1.2.9]

73. (Adapted from Moss 1980) Hall (1964) investigated the change in population size of the zooplankton species *Daphnia galeata mendota* in Base Line Lake, Michigan. The population size at time t, $N(t)$, was modeled by the equation

$$N(t) = N_0 e^{rt}$$

where N_0 denotes the population size at time 0. The constant r is called the **intrinsic rate of growth.**

(a) Plot $N(t)$ as a function of t if $N_0 = 100$ and $r = 2$. Compare your graph to the graph of $N(t)$ when $N_0 = 100$ and $r = 3$. Which population grows faster?

(b) The constant r is an important quantity since it describes how quickly the population changes. Suppose that you determine the size of the population at the beginning and at the end of a time period of length 1, and you find that in the beginning there were 200 individuals and after one unit of time there were 250 individuals. Determine r. (*Hint*: Consider the ratio $N(t+1)/N(t)$.)

74. Fish are indeterminate growers, that is, they grow throughout their lifetime. The growth can be described by the von Bertalanffy function

$$L(x) = L_\infty(1 - e^{-kx})$$

for $x \geq 0$, where $L(x)$ is the length at age x, and k and L_∞ are positive constants.

(a) Use a graphing calculator to graph $L(x)$ for $L_\infty = 20$, for (i) $k = 1$, and (ii) $k = 0.1$.

(b) For $k = 1$, find x so that the length is 90% of L_∞ (and 99% of L_∞). Can the fish ever attain length L_∞? Interpret the meaning of L_∞.

(c) Compare the graphs obtained in (a). Which growth curve reaches 90% of L_∞ faster? Can you explain what happens to the curve of $L(x)$ when you vary k (for fixed L_∞)?

◆ 1.3 GRAPHING

We introduced the functions most important to our study in the preceding section. You must be able to graph the following functions without a calculator: $y = c, x, x^2, x^3, 1/x, e^x, \ln x, \sin x, \cos x, \sec x,$ and $\tan x$. In this section, you will learn how to obtain new functions from these basic functions, and how to graph them. In addition, we will introduce important transformations that are often used to graphically display data.

1.3.1 Graphing and Basic Transformations of Functions

In this subsection, we will recall some basic transformations: vertical and horizontal translations, reflections about $x = 0$ and $y = 0$, and stretching and compressing.

> **Definition**   The graph of
>
> $$y = f(x) + a$$
>
> is a **vertical translation** of $y = f(x)$. If $a > 0$, the graph of $y = f(x)$ is shifted up a units; if $a < 0$, it is shifted down $|a|$ units.

This is illustrated in Figure 1.32, where we display $y = x^2$, $y = x^2 + 2$, and $y = x^2 - 2$.

> **Definition**   The graph of
>
> $$y = f(x - c)$$
>
> is a **horizontal translation** of $y = f(x)$. If $c > 0$, the graph of $y = f(x)$ is shifted c units to the right; if $c < 0$, it is shifted $|c|$ units to the left.

This is illustrated in Figure 1.33, where we display $y = x^2$, $y = (x - 3)^2$, and $y = (x + 3)^2$. Note that $y = (x + 3)^2$ is shifted to the left, because $y = (x - (-3))^2$ and therefore $c = -3 < 0$.

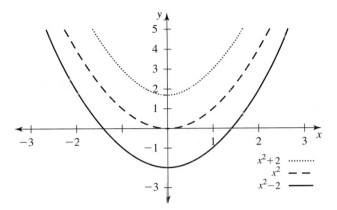

▲ **Figure 1.32**
The graphs of $y = x^2$, $y = x^2 + 2$ and $y = x^2 - 2$.

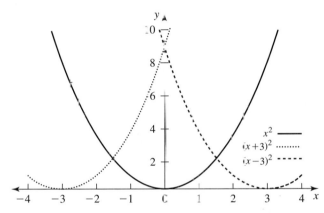

▲ **Figure 1.33**
The graphs of $y = x^2$, $y = (x - 3)^2$ and $y = (x + 3)^2$.

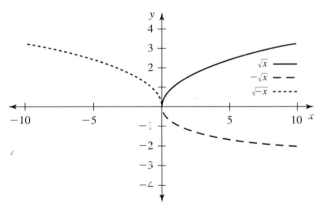

◀ **Figure 1.34**
Reflections about the x-axis
and the y-axis.

Reflections about the x-axis ($y = 0$) and the y-axis ($x = 0$) are illustrated in Figure 1.34. We graph $y = \sqrt{x}$, its reflection about the x-axis, $y = -\sqrt{x}$, and its reflection about the y-axis, $y = \sqrt{-x}$.

Multiplying a function by a factor between 0 and 1 compresses its graph; multiplying a function by a factor greater than 1 stretches its graph. This is illustrated in Figure 1.35, where we graph $y = x^2$ and $y = \frac{1}{2}x^2$, and in Figure 1.36, where we

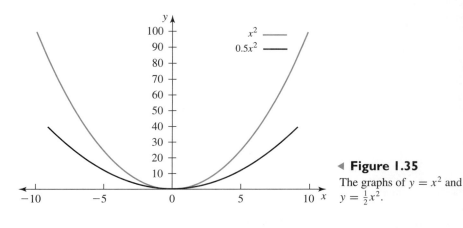

◀ **Figure 1.35**
The graphs of $y = x^2$ and $y = \frac{1}{2}x^2$.

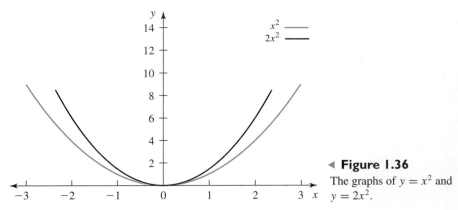

◀ **Figure 1.36**
The graphs of $y = x^2$ and $y = 2x^2$.

graph $y = x^2$ and $y = 2x^2$.

We illustrate these transformations in the following two examples.

▌ **EXAMPLE 1**   Explain how the graph of

$$y = 2\sin\left(x - \frac{\pi}{4}\right) \quad \text{for } x \in \mathbf{R}$$

can be obtained from $y = \sin x$, $x \in \mathbf{R}$.

Solution   We transform $y = \sin x$ using the following two steps, which are illustrated in Figure 1.37. We shift $y = \sin x$ to the right $\frac{\pi}{4}$ units. This yields $y = \sin\left(x - \frac{\pi}{4}\right)$. We then multiply $y = \sin\left(x - \frac{\pi}{4}\right)$ by 2. This corresponds to stretching $y = \sin\left(x - \frac{\pi}{4}\right)$ by the factor 2.   ▌

▌ **EXAMPLE 2**   Explain how the graph of

$$y = -\sqrt{x - 3} - 1, \quad x \geq 3$$

can be obtained from $y = \sqrt{x}$, $x \geq 0$.

Solution   We transform $y = \sqrt{x}$ using the following three steps, which are illustrated in Figure 1.38. We shift $y = \sqrt{x}$ three units to the right, and obtain $y = \sqrt{x - 3}$. We reflect $y = \sqrt{x - 3}$ about the x-axis, which yields $y = -\sqrt{x - 3}$. We shift $y = -\sqrt{x - 3}$ down one unit. This is the graph of $y = -\sqrt{x - 3} - 1$.   ▌

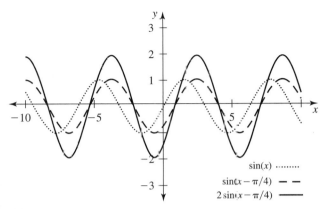

▲ **Figure 1.37**

The graphs of $y = \sin x$, $y = \sin\left(x - \frac{\pi}{4}\right)$ and $y = 2\sin\left(x - \frac{\pi}{4}\right)$.

▲ **Figure 1.38**

The graphs for Example 2.

1.3.2 Transformations into Linear Functions

We begin this subsection with an example from chemistry. The degree of acidity of a solution is measured by the pH value. The acronym pH stands for *pondus hydrogenii*, which is Latin for "weight of hydrogen." It is used to describe the hydrogen-ion activity. For instance, stomach acid has a pH value of 1, pure water has a pH value of 7, and baking soda has a pH value of 9. In dilute solutions, the pH value is approximately equal to the negative logarithm (to base 10) of the concentration of hydrogen atoms. That is, if we denote by $[H^+]$ the concentration of hydrogen ions (measured in moles per liter), then

$$pH = -\log[H^+]$$

A pH value of 5, for instance, means that the concentration of hydrogen ions in solution measured in moles per liter is 10^{-5}.

Since the concentration of hydrogen ions in solution ranges over many orders of magnitude, from about 10^{-14} to 1, it is much more convenient to use a **logarithmic scale**. On the logarithmic scale, the logarithm of the concentration is equally spaced (see Figure 1.39).

In general, when we graph numbers on a logarithmic scale, the number line needs to be scaled appropriately. There are two ways to label the number line: (i) The numbers ..., 0.01, 0.1, 1, 10, 100, 1000, ... appear equally spaced on the number

▲ **Figure 1.39**

The pH scale as a logarithmic scale.

▲ **Figure 1.40**

The logarithmic scale.

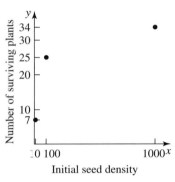

▲ **Figure 1.41**

The data displayed on a linear scale.

▲ **Figure 1.42**

The data displayed on a log linear scale.

line. (ii) The numbers $-2, -1, 0, 1, 2, 3, \ldots$ are equally spaced, but this time the number, n, refers to 10^n. Both ways are shown in Figure 1.40.

When you look at Figure 1.40 and continue number line (i) to the left, the numbers will get closer and closer to 0. In other words, a large negative number on number line (ii) corresponds to a number that is very close to 0. For instance, 10^{-100}, a very small number, is at -100 on number line (ii).

Here is another example that illustrates the usefulness of logarithmic scales. Suppose that you want to study how the survival of plants depends on the initial sowing density. You prepare three plots, the first with 10 seeds per square meter, the second with 100 seeds per square meter, and the third with 1000 seeds per square meter. Suppose that in the first plot 7 plants survive, in the second plot 25 plants survive, and in the third plot 34 plants survive. If you display your results on a graph where both axes are on a linear scale, the graph will not be very useful since the low initial densities all will be bunched up close to the origin. If, instead, you choose a logarithmic scale on the horizontal axis, the graph will be easier to read. (The two graphs are shown in Figures 1.41 and 1.42, respectively.)

Displaying data graphically is often a much more effective way of conveying a relationship between quantities than simply listing data points in a table. Frequently, one wishes to find a functional relationship between experimental quantities. In certain situations, it is possible to transform the functional relationship prior to displaying it, so that a linear relationship results. As straight lines are easy to recognize, this is a particularly useful tool. The most important such transformation is a logarithmic transformation. It can be successfully employed in cases where the functional relationship is in the form of either a power function ($y = cx^r$) or an exponential function ($y = ba^x$). This will be explained in the remainder of this subsection.

Power Functions   We saw that scaling relations in allometry are often expressed in terms of power functions. There are numerous other examples where the relationship between two quantities is best described by a power function. A power function is of the form

$$y = cx^r \tag{1.4}$$

where c and r are constants that depend on the particular example. Taking logarithms to base 10 on both sides of (1.4), we obtain

$$\log y = \log c + r \log x \tag{1.5}$$

If we introduce the new variables

$$Y = \log y$$
$$X = \log x$$

then (1.5) can be written as

$$Y = \log c + rX$$

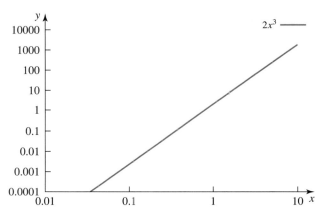

▲ **Figure 1.43**
The graph of $y = 2x^3$ on a linear scale.

▲ **Figure 1.44**
The graph of $y = 2x^3$ on a log-log scale.

which we recognize as the equation of a straight line with slope r. In other words, if we graph a power function in a coordinate system where *both* axes are logarithmically scaled, the graph looks like a straight line. Such graphs or plots are called **log-log plots**, and we say that we graph the function on a log-log scale. This is illustrated in Figures 1.43 and 1.44. Figure 1.43 shows the function $y = 2x^3$ in a coordinate system where both axes are on a linear scale; Figure 1.44 shows the same function on a log-log scale.

The advantage of using logarithms to base 10 is that we can quickly convert the numbers on the axes; for instance, a 2 refers to $10^2 = 100$.

▮ EXAMPLE 3   When growing plants at sufficiently high initial densities, one often observes that the number of plants decreases as the plants grow. This is called self-thinning. When the total dry weight per plant of the aboveground biomass is plotted as a function of the density of survivors on a log-log plot, one frequently finds that the data lie along a straight line with slope $-3/2$. Assume that, for a particular plant, such a relationship holds for plant densities between 10^2 and 10^4 plants per square meter; and that, at a density of 100 plants per square meter, the dry weight per plant is about

10 grams. Find the functional relationship between dry weight and plant density, and graph this function in a log-log plot.

Solution   Since the relationship between density (x) and dry weight (y) follows a straight line with slope $-3/2$ on a log-log plot, we set

$$\log y = C - \frac{3}{2}\log x \quad \text{for } 10^2 \le x \le 10^4$$

where C is a constant. To find C, we use the fact that when $x = 100$, $y = 10$. Therefore,

$$\log 10 = C - \frac{3}{2}\log 100$$

or

$$1 = C - \frac{3}{2}\cdot 2 \quad \text{which implies} \quad C = 4$$

Hence,

$$\log y = 4 - \frac{3}{2}\log x$$

Exponentiating both sides (and remembering that log is the logarithm to base 10), we find that

$$y = 10^4 x^{-3/2} \quad \text{for } 10^2 \le x \le 10^4$$

The graph of this function on a log-log scale is shown in Figure 1.45.   ▮

Exponential Functions   We will now consider exponential functions of the form

$$y = ba^x \tag{1.6}$$

where a and b are positive constants.

If we take logarithms to base 10 on both sides of (1.6), we obtain

$$\log y = \log b + x \log a \tag{1.7}$$

If we introduce

$$Y = \log y$$

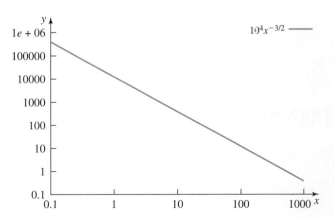

▲ **Figure 1.45**
The graph of $y = 10^4 x^{-3/2}$ on a log-log plot.

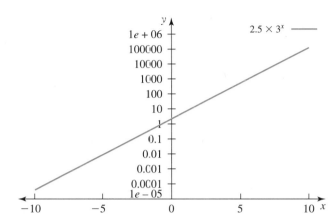

▲ **Figure 1.46**
The graph of $y = 2.5 \times 3^x$ on a semi-log plot.

we can write (1.7) as

$$Y = \log b + x \log a$$

This, again, is in linear form, but note that we only transformed y. If we graph an exponential function in a coordinate system where the vertical axis is on a logarithmic scale and the horizontal axis is on a linear scale, the resulting graph is a straight line that has slope $\log a$. Such plots are called **log-linear** or **semi-log plots**, and we say that we graph the function on a log-linear scale. We illustrate this in Figure 1.46, where $y = (2.5) \times 3^x$. Pay close attention to the scales on the horizontal and the vertical axes.

We can identify the slope of the log-transformed graph of y. Taking logarithms to base 10 on both sides of

$$y = 2.5 \times 3^x$$

we get

$$\log y = \log 2.5 + x \log 3$$

This is the function $y = 2.5 \times 3^x$; graphed on a semi-log plot, it is a straight line with slope $\log 3 = 0.4771$.

EXAMPLE 4   Polonium-210 (Po^{210}) is a radioactive material. To determine the half-life of Po^{210} experimentally, we measure the amount of radioactive material left after time t for various values of t. When we plot the data on a semi-log plot, we find that we can fit a straight line to the curve. The slope of the straight line is -0.0022/days. Find the half-life of Po^{210}.

Solution   Radioactive decay follows the equation

$$W(t) = W(0)e^{-\lambda t} \quad \text{for } t \geq 0$$

where $W(t)$ is the amount of radioactive material left after time t. If we log-transform this equation, we obtain

$$\log W(t) = \log W(0) - \lambda t \log e$$

Note that we use logarithms to base 10. If we plot $W(t)$ as a function of t on a semi-log plot, we obtain a straight line with slope $-\lambda \log e$. Matching this up with the number given in the example, we obtain

$$\lambda \log e = 0.0022/\text{days}$$

Solving for λ yields

$$\lambda = \frac{1}{\log e} 0.0022/\text{days} = 0.0051/\text{days}$$

To find the half-life T_h, we use the formula

$$T_h = \frac{\ln 2}{\lambda} = \frac{\ln 2}{0.0051}\text{days}$$
$$= 136.8\,\text{days}$$ ▌

Note that in the above example, we used logarithms to base 10 to do the log transformation. The radioactive law was given in terms of the natural exponent e. The slope therefore contained the factor $\log e = 0.4343$.

◆ 1.3 PROBLEMS

[1.3.1]

In Problems 1–8, sketch the graph of each function. Do not use a graphing calculator. (Assume the largest possible domain.)

1. $y = x^2 + 1$ **2.** $y = -(x - 2)^2 + 1$

3. $y = x^3 - 2$ **4.** $y = \exp(x - 2)$

5. $y = \ln(x + 1)$ **6.** $y = 3 + 1/x$

7. $y = 2\sin(x + \pi/4)$ **8.** $y = 0.2\cos(-x)$

9. Explain how the following functions can be obtained from $y = x^2$ by basic transformations.

(a) $y = x^2 - 2$ **(b)** $y = (x - 1)^2 + 1$

(c) $y = -2(x + 2)^2$

10. Explain how the following functions can be obtained from $y = 1/x$ by basic transformations.

(a) $y = 1 - \frac{1}{x}$ **(b)** $y = -\frac{1}{x-1}$ **(c)** $y = \frac{x}{x+1}$

[1.3.2]

11. Find the following numbers on a number line that is on a logarithmic scale (base 10): 0.0002, 0.02, 1, 5, 50, 100, 1000, 8000, and 20000.

12. **(a)** Find the following numbers on a number line that is on a logarithmic scale: 10^2, 10^{-3}, 10^{-4}, 10^{-7}, and 10^{-10}.

(b) Can you find 0 on a number line that is on a logarithmic scale?

(c) Can you find negative numbers on a number line that is on a logarithmic scale?

In Problems 13–16, when $\log y$ is graphed as a function of $\log x$, a straight line results. Graph each straight line on a log-log plot, and determine the functional relationship between x and y (before the logarithmic transformation).

13. Slope 0.37, $\log y = 0.89$ when $\log x = 0$

14. Slope -7.8, $\log y = 8.31$ when $\log x = 0$

15. Slope 0.69, $\log y = -2.3$ when $\log x = 0$

16. Slope -0.02, $\log y = -1$ when $\log x = 0$

In Problems 17–20, when $\log y$ is graphed as a function of x, a straight line results. Graph each straight line on a log-linear (semi-log) plot, and determine the functional relationship between x and y (before the logarithmic transformation).

17. Slope -1.73, $\log y = 2.3$ when $x = 0$

18. Slope -0.89, $\log y = -1.7$ when $x = 0$

19. Slope 0.18, $\log y = 0.05$ when $x = 0$

20. Slope 0.05, $\log y = -0.3$ when $x = 0$

In Problems 21–26, use a logarithmic transformation to find a linear relationship between the given quantities, and determine whether a log-log or log-linear plot should be used to graph the resulting linear relationship.

21. $f(x) = 3x^{1.7}$ **22.** $g(s) = 1.8e^{-0.2s}$

23. $N(t) = 130 \times 2^{1.2t}$ **24.** $I(u) = 4.8u^{-0.89}$

25. $R(t) = 3.6t^{1.2}$ **26.** $L(c) = 1.7 \times 10^{2.3c}$

27. Suppose that $N(t)$ denotes a population size at time t, and satisfies the equation

$$N(t) = 2e^{3t} \quad \text{for } t \geq 0$$

(a) If you graph $N(t)$ as a function of t on a semi-log plot, a straight line results. Explain why.

(b) Graph $N(t)$ as a function of t on a semi-log plot, and determine the slope of the resulting straight line.

28. Suppose that you follow a population over time. When you plot your data on a semi-log plot, a straight line with slope 0.03 results. Furthermore, assume that the population size at time 0 was 20. If $N(t)$ denotes the population size at time t, what function best describes the population size at time t?

29. (*Species-area curves*) Many studies have shown that the number of species on an island increases with the area of the island. Frequently, the functional relationship between the number of species (S) and the area (A) is approximated by $S = CA^z$, where z is a constant that depends on the particular species and habitat in the study. (Actual values of z range from about 0.2 to

0.35.) Suppose that the best fit to your data points on a log-log scale is a straight line. Is your model $S = CA^z$ an appropriate description of your data? If yes, how would you find z?

30. (*Michaelis-Menten equation*) Enzymes serve as catalysts in many chemical reactions in living systems. The simplest such reactions transform a single substrate into a product with the help of an enzyme. The Michaelis-Menten equation describes the initial velocity of such enzymatically controlled reactions. The equation gives the relationship between the initial velocity of the reaction (v_0) and the concentration of the substrate (s_0):

$$v_0 = \frac{v_{max} s_0}{s_0 + K_m}$$

where v_{max} is the maximum velocity at which the product may be formed, and K_m is the Michaelis-Menten constant. Note that this equation has the same form as the Monod growth function.

(a) Show that the Michaelis-Menten equation can be written in the form

$$\frac{1}{v_0} = \frac{K_m}{v_{max}} \frac{1}{s_0} + \frac{1}{v_{max}}$$

This is known as the Lineweaver-Burk equation, and shows that there is a linear relationship between $1/v_0$ and $1/s_0$.

(b) Sketch the graph of the Lineweaver-Burk equation. Use a coordinate system where $1/s_0$ is on the horizontal axis and $1/v_0$ on the vertical axis. Show that the resulting graph is a line that intersects the horizontal axis at $-1/K_m$ and the vertical axis at $1/v_{max}$.

(c) To determine K_m and v_{max}, we measure the initial velocity of the reaction, denoted by v_0, as a function of the concentration of the substrate, denoted by s_0, and fit a straight line through the points on a coordinate system, where the horizontal axis is $1/s_0$ and the vertical axis is $1/v_0$. Explain how to determine K_m and v_{max} from the graph.

(Note that this is an example where a non-logarithmic transformation is used to obtain a linear relationship. Since the reciprocals of both quantities of interest are used, the resulting plot is called a double-reciprocal plot.)

31. (*Continuation of Problem 30*) Estimating v_{max} and K_m from the Lineweaver-Burk graph as described in Problem 30 is not always satisfactory. A different transformation typically yields better estimates (Dowd and Riggs 1965). Show that the Michaelis-Menten equation can be written as

$$\frac{v_0}{s_0} = \frac{v_{max}}{K_m} - \frac{1}{K_m} v_0$$

and explain that this transformation results in a straight line when you graph v_0 on the horizontal axis and $\frac{v_0}{s_0}$ on the vertical axis. Explain how you can estimate v_{max} and K_m from this graph.

32. (Adapted from Reiss 1989) In a case study where the maximal rates of oxygen consumption (in ml/sec) for nine species of wild African mammals (Taylor et al. 1980) were plotted against body mass (in kg) on a log-log plot, it was found that the data points fell on a straight line with slope approximately equal to 0.8 and vertical axis intercept of approximately 0.105. Find an equation that relates maximal oxygen consumption and body mass.

33. (Adapted from Benton and Harper 1997) In vertebrates, embryos and juveniles have large heads relative to overall body size. As the animal grows older, proportions change; for instance, the ratio of skull length to body length diminishes. That this is not only the case for living vertebrates but also for fossil vertebrates is shown by the following example.

Ichthyosaurs are a group of marine reptiles that appeared in the early Triassic and died out well before the end of the Cretaceous.[1] They were fish-shaped and comparable in size to dolphins. In a study of twenty fossil skeletons, the following allometric relationship between skull length (S) and backbone length (B) was found.

$$S = 1.162 B^{0.93}$$

(a) Choose suitable transformations of S and B so that the resulting relationship is linear. Plot the transformed relationship, and find the slope and the y-intercept.

(b) Explain why the allometric equation confirms that juveniles had relatively large heads. (*Hint*: Compute the ratio of S to B for a number of different values of B—say, 10 cm, 100 cm, 500 cm—and compare.)

34. Absorption of light in a uniform water column follows an exponential law—that is, the intensity I at depth z is

$$I(z) = I_0 e^{-\epsilon z}$$

where I_0 is the intensity at the surface (i.e., when $z = 0$) and ϵ is the **extinction coefficient**. (In reality, ϵ depends on the wavelength of the light penetrating the surface. We assume here that ϵ is constant.)

(a) Suppose that 10% of the light is absorbed in the uppermost meter. Find ϵ. What are the units of ϵ?

(b) What percentage of the remaining intensity at 1 m is absorbed in the second meter? What percentage of the remaining intensity at 2 m is absorbed in the third meter?

(c) What percentage of the initial intensity remains at 1 m, at 2 m, and at 3 m?

(d) Plot the light intensity as a percentage of the surface intensity, on both a linear plot and a log-linear plot.

(e) Relate the slope of the curve on the log-linear plot to the extinction coefficient ϵ.

(f) The level at which 1% of the surface intensity remains is of biological significance. Approximately, it is the level where algal growth ceases. The zone above this level is called the euphotic zone. Express the depth of the euphotic zone as a function of ϵ.

(g) Compare a very clear lake to a milky glacier stream. Is the extinction coefficient ϵ for the clear lake greater or smaller than the extinction coefficient ϵ for the milky stream?

---

[1]The Triassic is a geologic period that began about 248 million years ago, and ended about 213 million years ago; the Cretaceous began about 144 million years ago, and ended 65 million years ago.

35. (*Logistic transformation*) Suppose that

$$f(x) = \frac{1}{1 + e^{-(b+mx)}} \qquad (1.8)$$

A function of the form (1.8) is called a **logistic** function. The logistic function was introduced by the Dutch mathematical biologist Verhulst around 1840, to describe the growth of populations under limited food resources. Show that

$$\ln \frac{f(x)}{1 - f(x)} = b + mx \qquad (1.9)$$

This transformation is called the logistic transformation. It is a standard transformation for linearizing functions of the form (1.8).

◆ 1.4 KEY TERMS

Chapter 1 Review: Topics   Discuss the following definitions and concepts:

1. Real numbers
2. Intervals: open, closed, half-open
3. Absolute value
4. Proportional
5. Lines: standard form, point-slope form, slope-intercept form
6. Parallel and perpendicular lines
7. Circle: radius, center, equation of circle, unit circle
8. Angle: radians, degrees
9. Trigonometric identities
10. Function: domain, range, image
11. Symmetry of functions: even, odd
12. Composition of functions
13. Polynomial
14. Degree of a polynomial
15. Chemical reaction: law of mass action
16. Rational function
17. Growth rate
18. Specific growth rate and per capita growth rate
19. Monod growth function
20. Power function
21. Allometry and scaling relations
22. Exponential function
23. Exponential growth
24. Natural exponential base
25. Radioactive decay
26. Half-life
27. Inverse Function
28. Logarithmic function
29. Relationship between exponential and logarithmic functions
30. Periodic function
31. Trigonometric functions
32. Amplitude, period
33. Translation: vertical, horizontal
34. Logarithmic scale
35. Logarithmic transformation
36. Log-log plot
37. Semi-log plot

◆ 1.5 REVIEW PROBLEMS

1. Suppose that the number of bacteria in a petri dish is given by

$$B(t) = 10{,}000e^{0.1t}$$

where t is measured in hours.

(a) How many bacteria are present at $t = 0$, 1, 2, 3, and 4?

(b) Find the time t when the number of bacteria reaches 100,000.

2. Suppose that a pathogen is introduced into a population of bacteria at time 0. The number of bacteria then declines as

$$B(t) = 25{,}000e^{-2t}$$

where t is measured in hours.

(a) How many bacteria are left after 3 hours?

(b) How long will it be until only 1% of the initial number of bacteria are left?

3. Consider the chemical reaction

$$A + 2B \rightarrow AB_2$$

Assume that the reaction occurs in a closed vessel, and that the initial concentrations of A and B are $a = [A]$ and $b = [B]$, respectively.

(a) Explain why the reaction rate $R(x)$ is given by

$$R(x) = k(a - x)(b - 2x)^2$$

where $x = [AB_2]$.

(b) Show that $R(x)$ is a polynomial and determine its degree.

(c) Graph $R(x)$ for the relevant values of x when $a = 5, b = 6$, and $k = 0.3$.

4. Euclid, a Greek mathematician who lived around 300 B.C., wrote the *Elements*, which is by far the most important mathematical text of that period. It is a systematic exposition of most of the mathematical knowledge of that time, arranged in thirteen volumes. In Book III, Euclid discusses the construction of a tangent to a circle at a point P on the circle. To phrase the construction in modern terminology, we draw a straight line through the point P that is perpendicular to the line through the center of the circle and the point P on the circle.

(a) Use this geometric construction to find the equation of the line that is tangent to the unit circle at the point $\left(\frac{1}{2}\sqrt{3}, \frac{1}{2}\right)$.

(b) Determine the angle θ between the positive x-axis and the line that goes through $(0, 0)$ and $\left(\frac{1}{2}\sqrt{3}, \frac{1}{2}\right)$. What is the relationship between the angle θ and the slope of the tangent line found in (a)?

5. To compare logarithmic and exponential growth, we consider the following two hypothetical plants that are of the same genus but exhibit rather different growth rates. Both plants produce a single leaf whose length continues to increase as long as the plant is alive. One plant is called *Growthus logarithmiensis*, the other one is called *Growthus exponentialis*. The length L (measured in feet) of the leaf of *G. logarithmiensis* at age t (measured in years), is given by

$$L(t) = \ln(t + 1), \quad t \geq 0$$

The length E (measured in feet) of the leaf of *G. exponentialis* at age t (measured in years), is given by

$$E(t) = e^t - 1, \quad t \geq 0$$

(a) Find the length of each leaf after 1, 10, 100, and 1000 years.

(b) How long would it take for the leaf of *G. exponentialis* to reach a length of 233,810 mi, the average distance from the earth to the moon? (Note that 1 mi = 5280 ft.) How long would the leaf of *G. logarithmiensis* then be?

(c) How many years would it take the leaf of *G. logarithmiensis* to reach a length of 233,810 mi? Compare this to the length of time since life appeared on earth, about 3,500 million years. If *G. logarithmiensis* had appeared 3,500 million years ago, and if there was a plant of this species that had actually survived throughout the entire period, how long would its leaf be today?

(d) Plants only started to conquer land in the late Ordovician; around 450 million years ago.[2] If both *G. exponentialis* and *G. logarithmiensis* had appeared then, and there was a plant of each species that had actually survived throughout the entire period, how long would there respective leaves be today?

6. Charles Darwin asserts in Chapter 3 of *The Origin of Species* (Darwin 1859) that a "struggle for existence inevitably follows from the high rate at which all organic beings tend to increase. ... Although some species may be now increasing, more or less rapidly, in numbers, all cannot do so, for the world would not hold them." To illustrate this point, he continues as follows.

There is no exception to the rule that every organic being naturally increases at so high a rate, that if not destroyed, the earth would soon be covered by the progeny of a single pair. Even slow-breeding man has

doubled in twenty-five years, and at this rate, in a few thousand years, there would literally not be standing room for his progeny.

Starting with a single pair, compute the world's population after 1000 years and after 2000 years under Darwin's assumption that the world's population doubles every twenty-five years, and find the resulting population densities (number of people per square foot). To answer the last part, you need to know that the earth's diameter is about 7900 mi, the surface of a sphere is $4\pi r^2$, where r is the radius of a sphere, and the continents make up about 29% of the earth's surface. (Note that 1 mi = 5280 ft.)

7. Assume that a population grows $q\%$ each year. How many years will it take the population to double in size? Give the functional relationship between the doubling time T and the annual percentage increase q. Produce a table that shows the doubling time T as a function of q for $q = 1, 2, \ldots, 10$, and graph T as a function of q. What happens to T as q gets closer to 0?

8. (*Beverton-Holt recruitment curve*) Many organisms show density-dependent mortality. The following is a simple mathematical model that incorporates this effect. Denote by N_b the density of parents and by N_a the density of *surviving* offspring.

(a) Suppose that without density-dependent mortality the number of surviving offspring per parent is equal to R. Show that if we plot N_b/N_a versus N_b, then this results in a horizontal line with y-intercept $1/R$. That is,

$$\frac{N_a}{N_c} = \frac{1}{R}$$

or

$$N_a = R \cdot N_b$$

The constant R is called the **net reproductive rate**.

(b) To include density-dependent mortality, we assume that N_b/N_a is an increasing function of N_b. The simplest way to do this is to assume that the graph of N_b/N_a versus N_b is a straight line, with y-intercept $1/R$ that goes through the point $(K, 1)$. Show that this implies that

$$N_a = \frac{RN_b}{1 - \frac{(R-1)N_b}{K}}$$

This relationship is called the Beverton-Holt recruitment curve.

(c) Explain in words why the model described by the Beverton-Holt recruitment curve behaves for small initial densities N_b like the model for density-independent mortality described in (a).

(d) Show that if $N_b = K$, then $N_a = K$. Furthermore, show that $N_b < K$ implies $N_b < N_c < K$, and that $N_b > K$ implies $K < N_a < N_b$. Explain in words what this means. (Note that K is called the carrying capacity.)

(e) Plot N_a as a function of N_b for $R = 2$ and $K = 20$. What happens for large values of N_b? Explain in words what this means.

9. (Adapted from Moss 1980) Oglesby (1977) investigated the relationship between annual fish yield (Y) and summer phytoplankton chlorophyll concentration (C). Fish yield was measured in gram dry weight per square meter and per year, and

---

[2]The Ordovician period lasted from about 505 million years ago to about 438 million years ago.

the chlorophyll concentration was measured in micrograms per liter. Data from nineteen lakes, mostly in the northern hemisphere, yielded the following relationship:

$$\log_{10} Y = 1.17 \log_{10} C - 1.92 \qquad (1.10)$$

(a) Plot $\log_{10} Y$ as a function of $\log_{10} C$.

(b) Find the relationship between Y and C, that is, write Y as a function of C. Explain the advantage of the log-log transformation resulting in (1.10); versus writing Y as a function of C. (Hint: Try to plot Y as a function of C and compare with your answer in (a).)

(c) Find the predicted yield (Y_p) as a function of the current yield (Y_c) if the current summer phytoplankton chlorophyll concentration were to double.

(d) By what percentage would the summer phytoplankton chlorophyll concentration need to increase to obtain a 10% increase in fish yield?

10. (Adapted from Moss 1980) To trace the history of a lake, a sample of mud from a core is taken and dated. One such dating method uses radioactive isotopes. The C^{14} method is effective for sediments that are younger than 60,000 years. The $C^{14} : C^{12}$ ratio has been essentially constant in the atmosphere over a long period of time, and living organisms take up carbon in this ratio. Upon death, the uptake of carbon ceases and C^{14} decays, which changes the $C^{14} : C^{12}$ ratio according to

$$\left(\frac{C^{14}}{C^{12}}\right)_t = \left(\frac{C^{14}}{C^{12}}\right)_{initial} e^{-\lambda t}$$

where t is the time since death.

(a) If the $C^{14} : C^{12}$ ratio in the atmosphere is 10^{-12} and the half-life of C^{14} is 5730 years, find an expression for t, the age of the material being dated, as a function of the $C^{14} : C^{12}$ ratio in the material being dated.

(b) Use your answer in (a) to find the age of a mud sample from a core for which the $C^{14} : C^{12}$ ratio is 1.61×10^{-13}.

11. (Adapted from Futuyma 1995 and Dott and Batten 1976) Corals deposit a single layer of lime each day. In addition, seasonal fluctuation in the thickness of the layers allows for grouping these layers into years. In modern corals, one can therefore count 365 layers per year. J. Wells, a paleontologist, counted such growth layers on fossil corals. To his astonishment, he found that Devonian[3] corals that lived about 380 million years ago had about 400 daily layers per year.

(a) Today, the earth rotates about its axis every 24 hours and revolves around the sun every $365\frac{1}{4}$ days. Astronomers have determined that the earth's rotation has slowed down in recent centuries at the rate of about 2 seconds every 100,000 years. That is, 100,000 years ago, a day was 2 seconds shorter than today. Extrapolate this back to the Devonian, and determine the length of a day and the length of a year back when Wells's corals lived. (Hint: The number of hours per year remains constant. Why?)

[3]The Devonian period lasted from about 408 million years ago to about 360 million years ago.

(b) Find a linear equation that relates geologic time (in million of years) to the number of hours per day at a given time.

(c) Algal stromatolites also show daily layers. A sample of some fossil stromatolites showed 400 to 420 daily layers per year. Use your answer in (b) to date the stromatolites.

12. The height in feet (y) of a certain tree as a function of age in years (x) can be approximated by

$$y = 132 e^{-20/x}$$

(a) Use a graphing calculator to plot the graph of this function. Describe in words how the tree grows, paying particular attention to questions like: Does the tree grow equally fast over time? What happens when the tree is young? What happens when the tree is old?

(b) How many years will it take for the tree to reach 100 ft in height?

(c) Can the tree ever reach a height of 200 ft? Is there a final height, that is, a maximum height that the tree will eventually reach?

13. (Model for aging) The probability that an individual lives beyond age t is called the survivorship function and is denoted by $S(t)$. The Weibull model is a popular model in reliability theory and in studies of biological aging. Its survivorship function is described by two parameters, λ and β, and is given by

$$S(t) = \exp[-(\lambda t)^\beta]$$

Mortality data from a Drosophila melanogaster population in Dr. Jim Curtsinger's lab at the University of Minnesota were collected and fitted to this model separately for males and females (Pletcher 1998). The following parameter values were obtained (t was measured in days):

Sex	λ	β
Males	0.019	3.41
Females	0.022	3.24

(a) Use a graphing calculator to sketch the survivorship function for both the female and male populations.

(b) For each population, find the value of t for which the probability of living beyond that age is 1/2.

(c) If you had a male and a female of this species, which would you expect to live longer?

14. Carbon has two stable isotopes, C^{12} and C^{13}. Organic material contains both stable isotopes but the ratio $[C^{13}] : [C^{12}]$ in organic material is smaller than that in inorganic material, reflecting the fact that light carbon (C^{12}) is preferentially taken up by plants during photosynthesis. This is called isotopic fractionation, and is measured as

$$\delta^{13}C = \left[\frac{([C^{13}] : [C^{12}])_{sample}}{([C^{13}] : [C^{12}])_{standard}} - 1\right]$$

The standard is taken from the isotope ratio in the carbon of belemnite shells found in the Cretaceous Pedee formation of South Carolina. Based on the above information, explain why the following quotation from Krauskopf and Bird (1995) makes sense.

the low [negative] values of $\delta^{13}C$ in the hydrocarbons of petroleum are one of the important bits of evidence for ascribing the origin of petroleum to the alteration of organic material rather than to condensation of primeval gases from the Earth's interior.

15. The speed of an enzymatic reaction is frequently described by the Michaelis-Menten equation

$$v = \frac{ax}{k + x}$$

where v is the velocity of the reaction, x is the concentration of the substrate, a is the maximum reaction velocity, and k is the substrate concentration at which the velocity is half of the maximum velocity. This curve describes how the reaction velocity depends on the substrate concentration.

(a) Show that when $x = k$, the velocity of the reaction is half the maximum velocity.

(b) Show that an 81-fold change in substrate concentration is needed to change the velocity from 10% to 90% of the maximum velocity, regardless of the value of k.

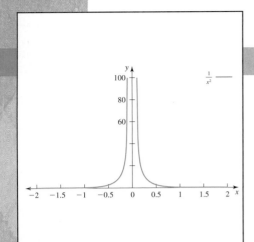

Limits and Continuity

◆ 2.1 LIMITS

Pound cakes have had a long tradition; the recipe is easy to remember and the cake is easy to make. Originally, they were made with a pound each of butter, sugar, flour, and eggs. Since eggs come in discrete units, it seems easier to modify the recipe by weighing a certain number of eggs and then adding the other ingredients at the same weight.

Suppose that you use five eggs whose total weight is 225 grams. You use an old-fashioned balance where you put the eggs on (say) the right scale. You then add (say) sugar to the left scale until the scales are balanced (see Figure 2.1). The angle α measures how unbalanced the scales are. We adopt the convention that α is positive when the balance is tipped to the right, and negative when it is tipped to the left. Then, when α is positive, we need to add more sugar; when α is negative, we need to remove sugar. Suppose now that the angle α (measured in radians) follows the function

$$\alpha = f(x) = 1 - \frac{x}{225}$$

where x (measured in grams) is the amount added to the left scale. The function $f(x)$ is displayed in Figure 2.2.

Our goal is to let x (the amount of sugar) approach 225 (the weight of the eggs), in which case $\alpha = f(x)$ approaches 0. By letting x approach 225, we "take a limit." So, what is a limit and how do we compute it?

◀ **Figure 2.1**
The balance with sugar on the left and eggs on the right: the angle α is positive.

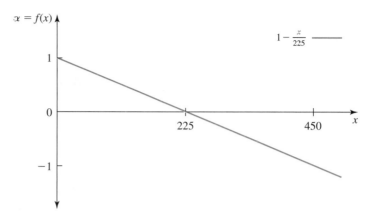

▲ **Figure 2.2**
The angle α as a function of x, the amount added to the left scale, when 225 gr are on the right scale.

2.1.1 An Informal Definition of Limits

> **Definition**   The "**limit of $f(x)$, as x approaches c, is equal to L**" means that $f(x)$ can be made arbitrarily close to L whenever x is sufficiently close to c (but not equal to c). We denote this by
>
> $$\lim_{x \to c} f(x) = L$$
>
> or $f(x) \to L$ as $x \to c$.

If $\lim_{x \to c} f(x) = L$ and L is a finite number, we say that the limit **exists** and that $f(x)$ **converges** to L. If the limit does not exist, we say that $f(x)$ **diverges** as x tends to c.

Note that in the definition we say that we choose x close to c but not equal to c. That is, when finding the limit of $f(x)$ as x approaches c, we do not simply plug c into $f(x)$. In fact, we will see examples where $f(x)$ is not even defined at $x = c$. The value of $f(c)$ is irrelevant when computing the value of $\lim_{x \to c} f(x)$.

Furthermore, when we say "x approaches c," we mean that x approaches c from either side. When x approaches c from only one side, we use the notation

$$\lim_{x \to c+} f(x) \quad \text{when } x \text{ approaches } c \text{ from the right}$$
$$\lim_{x \to c-} f(x) \quad \text{when } x \text{ approaches } c \text{ from the left}$$

and talk about right-handed and left-handed limits. The notation "$x \to c+$" indicates that when x approaches c from the right, values of x are greater than c, and when x approaches c from the left ("$x \to c-$"), values of x are smaller than c.

In the introductory example, we were interested in

$$\lim_{x \to 225} f(x) \qquad \text{where} \qquad f(x) = 1 - \frac{x}{225}$$

Looking at the graph of $f(x)$ (see Figure 2.2), we find that $f(x)$ approaches 0 as x approaches 225; that is, this limit exists and $f(x)$ converges to 0 as x tends to 225. Moreover, we see that x approaches 225 from either side: depending on whether the balance is tipped to the right or the left, we either add or remove sugar.

◀ **Figure 2.3**
As x approaches 2, $f(x) = x^2$ approaches 4.

We will now present a number of examples that will show different cases you may encounter. At this point, we have only an informal definition of limits, therefore the following discussion will be informal, relying on graphs and calculated tables. (A formal discussion will follow in the next subsection.)

Limits That Exist

▐ **EXAMPLE 1**   Find
$$\lim_{x \to 2} x^2$$

Solution   The graph of $y = x^2$ (see Figure 2.3) immediately shows that the limit of x^2 is 4 as x approaches 2 (from either side). We can also see this from the following table where we compute values of x^2 for x close to 2 (but not equal to 2). In the left half of the table, we approach $x = 2$ from the left ($x \to 2-$); in the right half of the table, we approach x from the right ($x \to 2+$).

x	x^2	x	x^2
1.9	3.61	2.1	4.41
1.99	3.9601	2.01	4.0401
1.999	3.996001	2.001	4.004001
1.9999	3.99960001	2.0001	4.00040001

We find that
$$\lim_{x \to 2} x^2 = 4$$ ▐

You might ask, "well, why can't I just plug 2 into $y = x^2$, since this would give me the correct answer, namely 4?" First, the definition of limits explicitly says that we are not allowed to simply plug in the value c that x is supposed to approach. Second, the property that the limit of the function $f(x)$ as x approaches c ($\lim_{x \to c} f(x)$) is equal to $f(c)$ is rather special, and we call such functions continuous at c. (Continuity will be the topic of Section 2.4.) Third, when we take the limit of $f(x)$ as $x \to c$, many times the function $f(x)$ is not defined at $x = c$. We give an example of this important case.

▐ **EXAMPLE 2**   Find
$$\lim_{x \to 3} \frac{x^2 - 9}{x - 3}$$

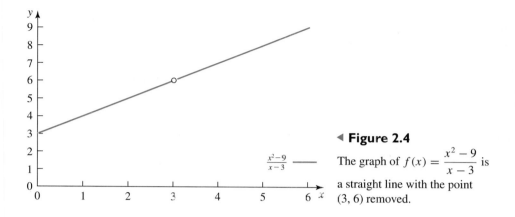

◀ **Figure 2.4**

$\dfrac{x^2-9}{x-3}$ —— The graph of $f(x) = \dfrac{x^2 - 9}{x - 3}$ is a straight line with the point $(3, 6)$ removed.

Solution We define $f(x) = \frac{x^2-9}{x-3}; x \neq 3$. Since the denominator of $f(x)$ is equal to 0 when $x = 3$, we exclude $x = 3$ from its domain. When $x \neq 3$, we can simplify the expression, namely

$$f(x) = \frac{x^2 - 9}{x - 3} = \frac{(x - 3)(x + 3)}{x - 3} = x + 3 \qquad \text{for } x \neq 3$$

We were able to cancel the term $x - 3$, since we assumed $x \neq 3$, and $x - 3 \neq 0$ for $x \neq 3$. (If we allowed $x = 3$, then canceling $x - 3$ would mean dividing by 0.) The graph of $f(x)$ is a straight line with one point deleted at $x = 3$ (see Figure 2.4). Taking the limit, we find

$$\lim_{x \to 3} \frac{x^2 - 9}{x - 3} = \lim_{x \to 3}(x + 3) = 6 \qquad\qquad ∎$$

In the first two examples, we computed limits where x approached a finite value. We will now look at a limit where x tends to infinity. We denote infinity by the symbol ∞ (note that ∞ is not a real number).

∎ **EXAMPLE 3** Find

$$\lim_{x \to \infty} \frac{x}{x + 1}$$

Solution A graph of $f(x) = \frac{x}{x+1}$ reveals that

$$\lim_{x \to \infty} \frac{x}{x - 1} = 1$$

since as x gets larger, the graph approaches the line $y = 1$ (see Figure 2.5). We can also see this by plugging in larger and larger values of x. We call the line $y = 1$ in Example 3 a **horizontal asymptote**. ∎

One-Sided Limits

∎ **EXAMPLE 4** Find

$$\lim_{x \to 0} \frac{|x|}{x}$$

Solution We set $f(x) = \frac{|x|}{x}, x \neq 0$. Since $|x| = x$ for $x \geq 0$ and $|x| = -x$ for $x \leq 0$, we find

$$f(x) = \frac{|x|}{x} = \begin{cases} +1 & \text{for } x > 0 \\ -1 & \text{for } x < 0 \end{cases}$$

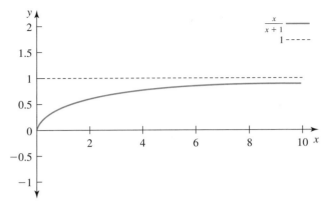

◀ **Figure 2.5**

The graph of $f(x) = \dfrac{x}{x+1}$

in Example 3.

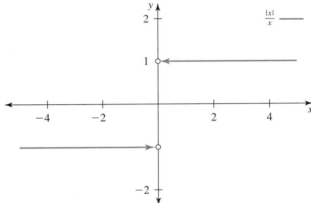

◀ **Figure 2.6**

The graph of $\dfrac{|x|}{x}$ in

Example 4: The function is
not defined at $x = 0$.

A graph of $f(x)$ is shown in Figure 2.6. We see that

$$\lim_{x \to 0+} \frac{|x|}{x} = 1 \qquad \text{and} \qquad \lim_{x \to 0-} \frac{|x|}{x} = -1$$

Since the right-hand limit is different from the left-hand limit, the limit of $f(x)$ as x approaches 0 does not exist (the phrase "x approaches 0" means that x approaches 0 from either side); that is, $f(x)$ diverges as x tends to 0.

Even though the limit does not exist, the one-sided limits exist. We can say $f(x)$ converges to 1 as x tends to 0 from the right (and $f(x)$ converges to -1 as x tends to 0 from the left). ▌

Limits That Do Not Exist

▌**EXAMPLE 5**   Find

$$\lim_{x \to 0} \frac{1}{x^2}$$

Solution   A graph of $f(x) = 1/x^2$, $x \neq 0$, reveals that $f(x)$ increases without bound as $x \to 0$ (see Figure 2.7). We can also see this when we plug in values close to 0. By choosing values sufficiently close to 0, we can get arbitrarily large values of $f(x)$.

x	-0.1	-0.01	-0.001	0.001	0.01	0.1
$f(x)$	100	10,000	10^6	10^6	10,000	100

This indicates that the limit does not exist. ▌

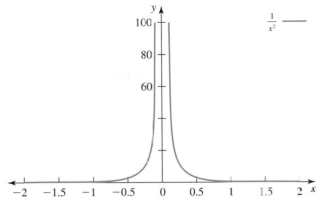

The graph of $f(x) = \dfrac{1}{x^2}$ in

Example 5: The function grows without bound as x tends to 0.

The divergence in Example 5 was such that the function grew without bound. This is an important case, and we summarize it in the following box.

$$\lim_{x \to c} f(x) = +\infty \text{ if } f(x) \text{ grows without bound as } x \to c$$
$$\lim_{x \to c} f(x) = -\infty \text{ if } f(x) \text{ decreases without bound as } x \to c$$

Similar definitions can be given for one-sided limits, which we will need in the next example.

▮ **EXAMPLE 6** Find

$$\lim_{x \to 3} \frac{1}{x - 3}$$

Solution The graph of $f(x) = 1/(x - 3)$, $x \neq 3$, in Figure 2.8 reveals that

$$\lim_{x \to 3+} \frac{1}{x - 3} = +\infty \quad \text{and} \quad \lim_{x \to 3-} \frac{1}{x - 3} = -\infty$$

This also follows when we compute values of $f(x)$ for x close to 3. If x is slightly larger than 3, we see that $f(x)$ is positive and grows without bound as x approaches 3 from the right. Likewise, if x is slightly smaller than 3, $f(x)$ is negative and decreases without bound as x approaches 3 from the left. ▮

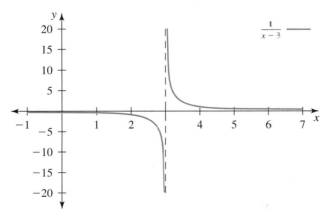

◀ **Figure 2.8**

The graph of $f(x) = \dfrac{1}{x - 3}$

in Example 6 has a vertical asymptote at $x = 3$.

In Example 5, the graph of $f(x)$ gets closer and closer to the vertical line $x = 0$ as $x \to 0$. In Example 6, the graph of $f(x)$ gets closer and closer to the vertical line $x = 3$ as $x \to 3$. Such vertical lines are called **vertical asymptotes**. (We give a precise definition of asymptotes later.)

The next example shows that a function can diverge without having one-sided limits or going to $+\infty$ or $-\infty$.

EXAMPLE 7   Find

$$\lim_{x \to 0} \sin \frac{\pi}{x}$$

Solution   Simply using a calculator and plugging in values to find limits can yield wrong answers if one does not exercise proper caution. If we produced a table of values of $f(x) = \sin \frac{\pi}{x}$ for $x = 0.1, 0.01, 0.001, \ldots$, we would find $\sin \frac{\pi}{0.1} = 0$, $\sin \frac{\pi}{0.01} = 0$, $\sin \frac{\pi}{0.001} = 0$, and so on. (Note that we measure angles in radians.) This might prompt us to conclude that the limit of the function is 0. But let's look at its graph, which is shown in Figure 2.9. The graph does not support our calculator-based conclusion.

What we find instead is that the values of $f(x)$ oscillate infinitely often between -1 and $+1$ as $x \to 0$. We can see why as follows: As $x \to 0+$, the argument in the sine function goes to infinity (likewise, as $x \to 0-$, the argument goes to negative infinity).

$$\lim_{x \to 0+} \frac{\pi}{x} = \infty \qquad \text{and} \qquad \lim_{x \to 0-} \frac{\pi}{x} = -\infty$$

Since the sine function oscillates between -1 and $+1$, we conclude that $\sin \frac{\pi}{x}$ oscillates between -1 and $+1$ as $x \to 0$.

The behavior exhibited in Example 7 is called **divergence by oscillation**.

Pitfalls   The following is an interesting example that shows some limitations of using a calculator when computing limits.

EXAMPLE 8   Find

$$\lim_{x \to 0} \frac{\sqrt{x^2 + 16} - 4}{x^2}$$

Solution   The graph of $f(x) = \frac{\sqrt{x^2+16}-4}{x^2}$, $x \neq 0$, in Figure 2.10 indicates that the limit exists and, based on the graph, we conjecture that it is equal to 0.125. If, instead, we use a calculator to produce a table for values of $f(x)$ close to 0,

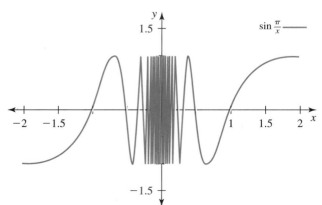

◀ **Figure 2.9**
The graph of $f(x) = \sin \dfrac{\pi}{x}$
in Example 7.

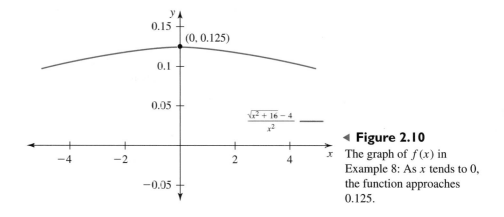

◀ **Figure 2.10**
The graph of $f(x)$ in Example 8: As x tends to 0, the function approaches 0.125.

something strange seems to happen.

x	0.01	0.001	0.0001	0.00001	0.000001	0.0000001
$f(x)$	0.1249998	0.125	0.125	0.125	0.1	0

As we get closer to 0, we first find that $f(x)$ gets closer to 0.125, but when we get very close to 0, $f(x)$ seems to drop to 0. What is going on? First, before you worry too much, $\lim_{x \to 0} f(x) = 0.125$. Later in this section, we will learn how to compute this limit without resorting to the (somewhat dubious) help of the calculator. The reason for the strange behavior of the calculated values lies in the fact that when x is very small, the difference in the numerator is so close to 0 that the calculator can no longer accurately determine the value. The calculator can only accurately compute a certain number of digits, which is good enough for most cases. Here, however, we would need greater accuracy. The same thing happens when you try to graph this function on a graphing calculator. When the x range of the viewing window is too small, the graph is no longer accurate. (Try, for instance, $-0.00001 \le x \le 0.00001$ and $-0.03 \le y \le 0.15$ as the range for the viewing window.)   ▮

2.1.2 A Formal Definition of Limits

The ancient Greeks already used limiting procedures to compute areas, such as the area of a circle. Newton and Leibniz, the inventors of calculus, were very much aware of the importance of taking limits in their development of calculus; however, they did not have a rigorous definition of it. The French mathematician Augustin-Louis Cauchy (1789–1857) was the first to develop a rigorous definition of limits; the definition we will use goes back to the German mathematician Karl Weierstrass (1815–1897).

Before we write the formal definition, let's return to the informal definition. There we stated that $\lim_{x \to c} f(x) = L$ means that the value of $f(x)$ can be made arbitrarily close to L whenever x is sufficiently close to c. But just how close is sufficient? Take Example 1: suppose we wish to actually *show* that

$$\lim_{x \to 2} x^2 = 4$$

What would we have to do? We would need to show that x^2 can be made arbitrarily close to 4 for all values of x sufficiently close to 2 but not equal to 2. (In the following, we will always exclude $x = 2$ from the discussion since the value of x^2 at $x = 2$ is irrelevant for finding the limit.) Suppose we wish to make x^2 close to 4 within 0.01, that is, we want $|x^2 - 4| < 0.01$. Does this hold for all x sufficiently close to 2 but not equal to 2? We begin with

$$|x^2 - 4| < 0.01$$

This is equivalent to

$$-0.01 < x^2 - 4 < 0.01$$
$$3.99 < x^2 < 4.01$$
$$\sqrt{3.99} < |x| < \sqrt{4.01}$$

Now $\sqrt{3.99} = 1.997498\ldots$, and $\sqrt{4.01} = 2.002498\ldots$. We therefore find that values of $x \neq 2$ in the interval $(1.998, 2.002)$ satisfy $|x^2 - 4| < 0.01$. (We chose a somewhat smaller interval than indicated, to get an interval that is symmetric about 2.) That is, for all values of x within 0.002 of 2 but not equal to 2 (i.e., $0 < |x - 2| < 0.002$), x^2 is within the prescribed precision, namely within 0.01 of 4.

You might think about this example in the following way: Suppose that you wish to stake out a square of area 4 ft^2. Each side of your square is 2 ft long. You bring along a stick which you cut to a length of 2 ft. We can then ask how accurately do we need to cut the stick so that the area will be within a prescribed precision. Our prescribed precision was 0.01 and we found that if we cut the stick within 0.002 of 2 ft, we would be able to obtain the prescribed precision.

There is nothing special about 0.01, we could have chosen any other degree of precision and would have found a corresponding interval of x values. We translate this into a formal definition for limits (see Figure 2.11).

Definition   The statement

$$\lim_{x \to c} f(x) = L$$

means that for every $\epsilon > 0$ there corresponds a number $\delta > 0$ so that

$$|f(x) - L| < \epsilon \quad \text{whenever} \quad 0 < |x - c| < \delta$$

Note that, as in the informal definition of limits, we exclude the value $x = c$ from the statement. (This is done in the inequality $0 < |x - c|$.) To apply this definition, we first need to guess the limiting value L. We then choose an $\epsilon > 0$, the prescribed precision, and try to find a $\delta > 0$ so that $f(x)$ is within ϵ of L whenever x is within δ of c but not equal to c. (In our example, $f(x) = x^2$, $L = 4$, $c = 2$, $\epsilon = 0.01$, and $\delta = 0.002$.)

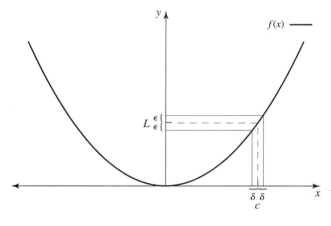

◀ **Figure 2.11**

The $\epsilon - \delta$ definition of limits.

▌ **EXAMPLE 9**   Show that

$$\lim_{x \to 1}(2x - 3) = -1$$

Solution   We let $f(x) = 2x - 3$. Our guess for the limiting value is $L = -1$. Now,

$$\begin{aligned}
|f(x) - L| &= |2x - 3 - (-1)| \\
&= |2x - 2| \\
&= 2|x - 1|
\end{aligned}$$

We choose $\epsilon > 0$ (ϵ is arbitrary, and we do not specify it because our statement needs to hold for all $\epsilon > 0$). Our goal is to find $\delta > 0$ so that $2|x - 1| < \epsilon$ whenever x is within δ of 1 but not equal to 1, that is, $0 < |x - 1| < \delta$. The value of δ will typically depend on our choice of ϵ. Since $|x - 1| < \delta$ implies $2|x - 1| < 2\delta$, this suggests that we should try $2\delta = \epsilon$ or $\delta = \epsilon/2$. If we choose $\delta = \epsilon/2$, then indeed

$$|f(x) - L| = 2|x - 1| < 2\delta = 2\frac{\epsilon}{2} = \epsilon$$

This means that for every $\epsilon > 0$, we can find a number $\delta > 0$ (namely $\delta = \epsilon/2$), so that

$$|f(x) - (-1)| < \epsilon \qquad \text{whenever} \qquad 0 < |x - 1| < \delta$$

But this is exactly the definition of

$$\lim_{x \to 1}(2x - 3) = -1$$

▌

We can also use the definition to show that a limit does not exist.

▌ **EXAMPLE 10**   Show that

$$\lim_{x \to 0}\frac{|x|}{x}$$

does not exist.

Solution   This is trickier (see Figure 2.12). Set $f(x) = |x|/x, x \neq 0$. The approach is as follows: We assume the limit exists and then try to find a contradiction. Suppose there exists an L so that

$$\lim_{x \to 0}\frac{|x|}{x} = L$$

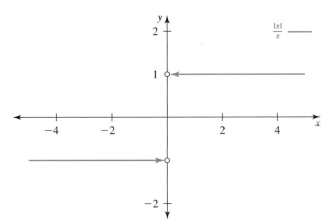

◀ **Figure 2.12**

The graph of $f(x) = \dfrac{|x|}{x}$ in

Example 10: The limit of $\dfrac{|x|}{x}$

as x tends to 0 does not exist.

If we look at Figure 2.12, we see that if, for instance, we choose $L = 1$, then we cannot get close to L when x is less than 0. Similarly, we see that for any value of L, either the distance to $+1$, or the distance to -1 is at least 1. That is, regardless of the value of L, if $\epsilon < 1$, we will not be able to find a value of δ so that if $|x| < \delta$, then $|\frac{|x|}{x} - L| < \epsilon$, since $f(x) \in \{-1, 1\}$. Therefore, $\lim_{x \to 0} \frac{|x|}{x}$ does not exist. ▌

In the previous section, we considered an example where $\lim_{x \to c} f(x) = \infty$. This can be made precise as well.

> **Definition**   The statement
>
> $$\lim_{x \to c} f(x) = \infty$$
>
> means that for every $M > 0$ there exists $\delta > 0$ so that
>
> $$f(x) > M \qquad \text{whenever} \qquad 0 < |x - c| < \delta$$

Similar definitions hold for the case when $f(x)$ decreases without bound as $x \to c$, and for one-sided limits. We will not give definitions for all possible cases but rather illustrate how one would use such a definition.

▌EXAMPLE 11   Show that

$$\lim_{x \to 0} \frac{1}{x^2} = \infty$$

Solution   The graph of $f(x) = 1/x^2$, $x \neq 0$, is shown in Figure 2.13. We fix $M > 0$ (again, M is arbitrary, because our solution must hold for all $M > 0$). We need to find $\delta > 0$ so that $f(x) > M$ whenever $0 < |x| < \delta$ (note that $c = 0$). We start with the inequality $f(x) > M$, and try to determine how to choose δ.

$$\frac{1}{x^2} > M \qquad \text{is the same as} \qquad x^2 < \frac{1}{M}$$

Taking square roots on both sides, we find

$$|x| < \frac{1}{\sqrt{M}}$$

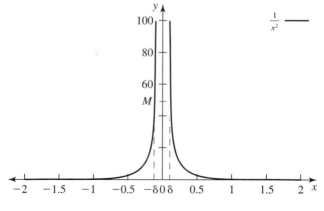

◄ Figure 2.13

The function $f(x) = \dfrac{1}{x^2}$ in Example 11: The limit of $\dfrac{1}{x^2}$ as x tends to 0 does not exist.

This suggests that we should choose $\delta = 1/\sqrt{M}$. Let's try this value for δ; given $M > 0$, we choose $\delta = 1/\sqrt{M}$. If $0 < |x| < \delta$, then this implies that

$$x^2 < \delta^2 \qquad \text{or} \qquad \frac{1}{x^2} > \frac{1}{\delta^2} = M$$

That is, $1/x^2 > M$ whenever $0 < |x| < \delta$. ▮

There is also a formal definition when $x \to \infty$ (and a similar one for $x \to -\infty$).

Definition   The statement

$$\lim_{x \to \infty} f(x) = L$$

means that for every $\epsilon > 0$ there exists an $N > 0$ so that

$$|f(x) - L| < \epsilon \qquad \text{whenever} \qquad x > N$$

▮ EXAMPLE 12   Show that

$$\lim_{x \to \infty} \frac{x}{0.5 + x} = 1$$

Solution   This limit is illustrated in Figure 2.14. You can see that $f(x) = x/(0.5 + x)$, $x \geq 0$, is in the strip of width 2ϵ centered at the limiting value $L = 1$ for all values of x which are greater than N. We still need to determine N. To do this, we try to solve

$$\left| \frac{x}{0.5 + x} - 1 \right| < \epsilon$$

for $\epsilon > 0$. This is equivalent to

$$-\epsilon < \frac{x}{0.5 + x} - 1 < \epsilon$$

or, after adding 1 to all three parts,

$$1 - \epsilon < \frac{x}{0.5 + x} < 1 + \epsilon$$

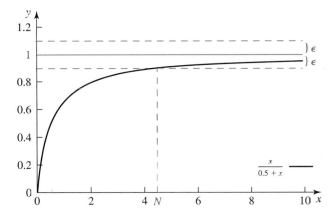

◀ **Figure 2.14**
The function $f(x) = \dfrac{x}{0.5 + x}$ in Example 12: The limit of $f(x)$ as x tends to infinity is 1.

Since $\frac{x}{0.5+x} < 1$ for $x > 0$, the right-hand side of the inequality always holds. We therefore need only consider

$$1 - \epsilon < \frac{x}{0.5 + x}$$

As we are interested in the behavior of $f(x)$ as $x \to \infty$, we need only look at large values of x. Multiplying by $0.5 + x$ (and noticing that we can assume $0.5 + x > 0$, because we let $x \to \infty$), we find

$$(1 - \epsilon)(0.5 + x) < x$$

Solving for x, we find

$$(1 - \epsilon)(0.5) < x - x(1 - \epsilon)$$
$$(1 - \epsilon)(0.5) < \epsilon x$$
$$\frac{1 - \epsilon}{2\epsilon} < x$$

For instance, if $\epsilon = 0.1$, we find

$$x > \frac{0.9}{0.2} = 4.5$$

That is, we would set $N = 4.5$ and conclude that for $x > 4.5$, $|f(x) - 1| < 0.1$.

More generally, we find that for every $\epsilon > 0$, there exists

$$N = \frac{1 - \epsilon}{2\epsilon}$$

so that

$$|f(x) - 1| < \epsilon \qquad \text{whenever} \qquad x > N \qquad\qquad ▮$$

2.1.3 Limit Laws

You saw in the previous subsection that using the formal definition of limits is quite difficult, even if the functions are reasonably straightforward. You can guess from this that it would not be practical to use the formal definition every time we want to compute a limit. Fortunately, there are limit laws, which make computing limits much easier.

LIMIT LAWS

Suppose that a is a constant, and that

$$\lim_{x \to c} f(x) \quad \text{and} \quad \lim_{x \to c} g(x)$$

exist. Then the following hold.

1. $\lim\limits_{x \to c} af(x) = a \lim\limits_{x \to c} f(x)$

2. $\lim\limits_{x \to c}[f(x) + g(x)] = \lim\limits_{x \to c} f(x) + \lim\limits_{x \to c} g(x)$

3. $\lim\limits_{x \to c}[f(x) \cdot g(x)] = \lim\limits_{x \to c} f(x) \cdot \lim\limits_{x \to c} g(x)$

4. $\lim\limits_{x \to c} \dfrac{f(x)}{g(x)} = \dfrac{\lim_{x \to c} f(x)}{\lim_{x \to c} g(x)} \qquad \text{if} \qquad \lim_{x \to c} g(x) \neq 0$

It is not difficult to use the formal definition for limits to show that $\lim_{x \to c} x = c$. (In fact, this is very similar to Example 9.) Starting from this we can show, for instance, that

$$\lim_{x \to c} x^2 = \lim_{x \to c} x \cdot \lim_{x \to c} x = c \cdot c = c^2$$

by using Rule 3. More generally, we find that

> If $f(x)$ is a polynomial, then
>
> $$\lim_{x \to c} f(x) = f(c)$$
>
> If $f(x)$ is a rational function, that is,
>
> $$f(x) = \frac{p(x)}{q(x)}$$
>
> where $p(x)$ and $q(x)$ are polynomials and $q(c) \neq 0$, then
>
> $$\lim_{x \to c} f(x) = \lim_{x \to c} \frac{p(x)}{q(x)} = \frac{p(c)}{q(c)} = f(c)$$

Even though we made a point of demonstrating that we cannot simply substitute the value c into $f(x)$ when we take the limit $x \to c$ of $f(x)$, the preceding results show that in fact there are certain cases where we can do just that.

EXAMPLE 13   Find

$$\lim_{x \to 3}[x^2 - 2x + 1]$$

Solution   Since $f(x) = x^2 - 2x + 1$ is a polynomial, we find

$$\lim_{x \to 3}[x^2 - 2x + 1] = 9 - 6 + 1 = 4$$

EXAMPLE 14   Find

$$\lim_{x \to -1} \frac{2x^3 - x + 5}{x^2 + 3x + 1}$$

Solution   Now,

$$f(x) = \frac{2x^3 - x + 5}{x^2 + 3x + 1}$$

is a rational function that is defined for $x = -1$ (the denominator is not equal to 0 when we substitute $x = -1$). We find

$$\lim_{x \to -1} \frac{2x^3 - x + 5}{x^2 + 3x + 1} = \frac{2(-1)^3 - (-1) + 5}{(-1)^2 + 3(-1) + 1} = \frac{4}{-1} = -4$$

We mention one more limit law, for which we need to assume that when n is even, $f(x) \geq 0$ for all x in an open interval that contains c (except possibly c).

> $$\lim_{x \to c} \sqrt[n]{f(x)} = \sqrt[n]{\lim_{x \to c} f(x)}$$

EXAMPLE 15   We will now recalculate the limit in Example 8:

$$\lim_{x \to 0} \frac{\sqrt{x^2 + 16} - 4}{x^2}$$

Solution   We cannot apply Rule 4, since $f(x) = (\sqrt{x^2 + 16} - 4)/x^2$ is not defined for $x = 0$. (If we plug in 0, we get the expression $0/0$.) We use a trick that will allow us to find the limit; namely, we rationalize the numerator. For $x \neq 0$, we find

$$\frac{\sqrt{x^2 + 16} - 4}{x^2} = \frac{\sqrt{x^2 + 16} - 4}{x^2} \cdot \frac{\sqrt{x^2 + 16} + 4}{\sqrt{x^2 + 16} + 4}$$

$$= \frac{x^2 + 16 - 16}{x^2(\sqrt{x^2 + 16} + 4)} = \frac{x^2}{x^2(\sqrt{x^2 + 16} + 4)}$$

$$= \frac{1}{\sqrt{x^2 + 16} + 4}$$

Note that we are allowed to divide by x^2 in the last step since we assume that $x \neq 0$. We can now apply Rule 4 to $1/(\sqrt{x^2 + 16} + 4)$, and find

$$\lim_{x \to 0} \frac{\sqrt{x^2 + 16} - 4}{x^2} = \lim_{x \to 0} \frac{1}{\sqrt{x^2 + 16} + 4} = \frac{1}{8} = 0.125$$

as we saw in Example 8. In Section 4.5, we will learn another method for finding the limit of expressions of the form 0/0. ▮

◆ 2.1 PROBLEMS

[2.1.1]

In Problems 1–28, evaluate each limit using a table or a graph.

1. $\lim_{x \to 2} (x^2 - 3x + 1)$

2. $\lim_{x \to 2} \dfrac{x^2 + 3}{x + 2}$

3. $\lim_{x \to -1} \dfrac{x}{1 + x^2}$

4. $\lim_{s \to 2} s(s^2 - 4)$

5. $\lim_{x \to \pi} 3 \cos \dfrac{x}{4}$

6. $\lim_{t \to \pi/12} \sin(4t)$

7. $\lim_{x \to \pi/2} 2 \sec \dfrac{x}{3}$

8. $\lim_{x \to \pi/2} \tan \dfrac{x - \pi/2}{2}$

9. $\lim_{x \to -1} e^{-x^2/2}$

10. $\lim_{x \to 0} \dfrac{e^x + 1}{2x + 3}$

11. $\lim_{x \to 0} \ln(x + 1)$

12. $\lim_{t \to e} \ln t^2$

13. $\lim_{x \to 3} \dfrac{x^2 - 16}{x - 4}$

14. $\lim_{x \to 2} \dfrac{x^2 - 4}{x + 2}$

15. $\lim_{x \to \infty} \dfrac{2}{x}$

16. $\lim_{x \to -\infty} \dfrac{3}{x}$

17. $\lim_{x \to \infty} \dfrac{x}{1 + x}$

18. $\lim_{x \to \infty} \dfrac{x^2}{2 - x^2}$

19. $\lim_{x \to -\infty} e^x$

20. $\lim_{x \to \infty} e^{-x}$

21. $\lim_{x \to 4-} \dfrac{2}{x - 4}$

22. $\lim_{x \to 3+} \dfrac{1}{x - 3}$

23. $\lim_{x \to 1-} \dfrac{2}{1 - x}$

24. $\lim_{x \to 2+} \dfrac{3}{2 - x}$

25. $\lim_{x \to 1-} \dfrac{1}{1 - x^2}$

26. $\lim_{x \to 2+} \dfrac{2}{x^2 - 4}$

27. $\lim_{x \to 1} \dfrac{1}{(x - 1)^2}$

28. $\lim_{x \to 0} \dfrac{1 - x^2}{x^2}$

29. Use a table and a graph to check that

$$f(x) = \frac{2}{x^2}$$

has a horizontal asymptote as $x \to \infty$. Give the equation of the horizontal asymptote. What happens when $x \to -\infty$? Does $f(x)$ have a vertical asymptote? If yes, give its equation.

30. Use a table and a graph to check that

$$f(x) = \frac{2x}{x - 1}$$

has a horizontal asymptote as $x \to \infty$. Give the equation of the horizontal asymptote. What happens when $x \to -\infty$? Does $f(x)$ have a vertical asymptote? If yes, give its equation.

31. Use a graphing calculator to investigate

$$\lim_{x \to 1} \sin \frac{1}{x - 1}$$

32. Use a graphing calculator to investigate

$$\lim_{x \to 0} \cos \frac{1}{x}$$

[2.1.2]

33. Find the values of x such that

$$|2x - 1| < 0.01$$

34. Find the values of x such that

$$|x^2 - 9| < 0.01$$

35. Let

$$f(x) = 2x - 1, \ x \in \mathbf{R}$$

(a) Graph $y = f(x)$ for $-3 \leq x \leq 5$.

(b) For which values of x is $y = f(x)$ within 0.1 of 3? (*Hint*: Find values for x so that $|(2x - 1) - 3| < 0.1$.)

(c) Illustrate your result in (b) on the graph that you obtained in (a).

36. Let
$$f(x) = \sqrt{x}, \quad x \geq 0$$

(a) Graph $y = f(x)$ for $0 \leq x \leq 6$.

(b) For which values of x is $y = f(x)$ within 0.2 of 1? (*Hint:* Find values for x so that $|\sqrt{x} - 1| < 0.2$.)

(c) Illustrate your result in (b) on the graph that you obtained in (a).

37. Let
$$f(x) = \frac{1}{x}, \quad x > 0$$

(a) Graph $y = f(x)$ for $0 < x \leq 4$.

(b) For which values of x is $y = f(x)$ greater than 4?

(c) Illustrate your result in (b) on the graph that you obtained in (a).

38. Let
$$f(x) = e^{-x}, \quad x \geq 0$$

(a) Graph $y = f(x)$ for $0 \leq x \leq 6$.

(b) For which values of x is $y = f(x)$ less than 0.1?

(c) Illustrate your result in (b) on the graph that you obtained in (a).

In Problems 39–42, use the formal definition of limits to prove each statement.

39. $\lim_{x \to 2} (2x - 1) = 3$

40. $\lim_{x \to 0} x^2 = 0$

41. $\lim_{x \to 0} \frac{1}{x^4} = \infty$

42. $\lim_{x \to \infty} \frac{1}{x} = 0$

[2.1.3]

In Problems 43–64, use the basic rules to evaluate each limit.

43. $\lim_{x \to -1} (x^3 + 7x - 1)$

44. $\lim_{x \to 3} \left(2x^2 - \frac{1}{x} \right)$

45. $\lim_{x \to -3} \frac{x^3 - 20}{x + 1}$

46. $\lim_{x \to 1} \frac{x^3 - 1}{x + 2}$

47. $\lim_{x \to 3} \frac{3x^2 + 1}{2x - 3}$

48. $\lim_{x \to -2} \frac{1 + x}{1 - x}$

49. $\lim_{x \to 1} \frac{1 - x^2}{1 - x}$

50. $\lim_{u \to 2} \frac{4 - u^2}{2 - u}$

51. $\lim_{x \to 3} \frac{x^2 - 2x - 3}{x - 3}$

52. $\lim_{x \to 1} \frac{(x - 1)^2}{x^2 - 1}$

53. $\lim_{x \to -2} \frac{2x^2 + 3x - 2}{x + 2}$

54. $\lim_{x \to 1/2} \frac{1 - x - 2x^2}{1 - 2x}$

55. $\lim_{x \to \pi/2} \frac{\cos^2 x}{1 - \sin^2 x}$

56. $\lim_{x \to -\pi/2} \frac{1 + \tan^2 x}{\sec^2 x}$

57. $\lim_{x \to 0} \frac{e^{2x} - 1}{e^x - 1}$

58. $\lim_{x \to 0} \frac{e^{-x} - e^x}{e^{-x} + 1}$

59. $\lim_{x \to 1} \sqrt{x^3 + 4x - 1}$

60. $\lim_{x \to -2} \frac{1}{\sqrt{5x^2 - 4}}$

61. $\lim_{x \to 0} \frac{\sqrt{x^2 + 9} - 3}{x^2}$

62. $\lim_{x \to 0} \frac{\sqrt{x^2 + 4} - 2}{x}$

63. $\lim_{x \to 0} \frac{1 - \sqrt{1 - x^2}}{x^2}$

64. $\lim_{x \to 0} \frac{\sqrt{2 - x} - \sqrt{2}}{2x}$

◆ 2.2 THE SANDWICH THEOREM AND SOME TRIGONOMETRIC LIMITS

The following two theorems are both useful and intuitive. We will not prove them.

Comparison Theorem   If $f(x) \leq g(x)$ for all x in an open interval that contains c (except possibly c), then

$$\lim_{x \to c} f(x) \leq \lim_{x \to c} g(x)$$

Sandwich Theorem   If $f(x) \leq g(x) \leq h(x)$ for all x in an open interval that contains c (except possibly c), and

$$\lim_{x \to c} f(x) = \lim_{x \to c} h(x) = L$$

then

$$\lim_{x \to c} g(x) = L$$

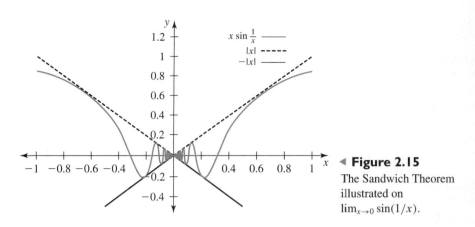

◀ **Figure 2.15**
The Sandwich Theorem
illustrated on
$\lim_{x \to 0} \sin(1/x)$.

It should be clear why the second theorem, illustrated in Figure 2.15, is called the Sandwich Theorem. We "sandwich" the function $g(x)$ between the two functions $f(x)$ and $h(x)$. Since $f(x)$ and $h(x)$ converge to the same value, $g(x)$ also must converge to that value, because it is squeezed in between $f(x)$ and $h(x)$. We demonstrate how to use the Sandwich Theorem in the next example.

▮ **EXAMPLE 1**   Show that

$$\lim_{x \to 0} x \sin \frac{1}{x} = 0$$

Solution   First, note that we cannot use Rule 3, which says that the limit of a product is equal to the product of the limits, because it requires that the limits of both factors exist. We already know that $\lim_{x \to 0} \sin \frac{1}{x}$ does not exist (see Example 7 of the previous section). However, we know that

$$-1 \le \sin \frac{1}{x} \le 1$$

for all $x \ne 0$. Multiplying all three parts by $x > 0$, we find

$$-x \le x \sin \frac{1}{x} \le x$$

Since $\lim_{x \to 0+}(-x) = \lim_{x \to 0+} x = 0$, we can apply the Sandwich Theorem and find

$$\lim_{x \to 0+} x \sin \frac{1}{x} = 0$$

We can repeat the same steps when we multiply by $x < 0$. We now need to reverse the inequality signs. That is, for $x < 0$,

$$-x \ge x \sin \frac{1}{x} \ge x$$

Since $\lim_{x \to 0-}(-x) = \lim_{x \to 0-} x = 0$, we can again apply the Sandwich Theorem and find

$$\lim_{x \to 0-} x \sin \frac{1}{x} = 0$$

Combining the two results, we find

$$\lim_{x \to 0} x \sin \frac{1}{x} = 0$$

This limit is illustrated in Figure 2.15.   ▮

There are two trigonometric limits that are important enough that you should memorize them. Note that the angle x is measured in radians.

$$\lim_{x \to 0} \frac{\sin x}{x} = 1 \quad \text{and} \quad \lim_{x \to 0} \frac{1 - \cos x}{x} = 0$$

We will prove both statements. The proof of the first statement uses a nice geometric argument; the second statement follows from the first.

Proof of $\lim_{x \to 0} \frac{\sin x}{x}$ We assume that $0 < x < \pi/2$. In Figure 2.16, we draw the unit circle together with the triangles OAD and OBC. The angle x is measured in radians. Since $\overline{OB} = 1$, we find

$$\text{arc length of BD} = x$$
$$\overline{OA} = \cos x$$
$$\overline{AD} = \sin x$$
$$\overline{BC} = \tan x$$

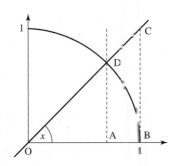

▲ **Figure 2.16**
The unit circle with the triangles OAD and OBC.

Furthermore,

$$\text{area of } \triangle \text{ OAD} \leq \text{area of sector OBD} \leq \text{area of } \triangle \text{ OBC}$$

The area of a sector of central angle x (measured in radians) and radius r is $\frac{1}{2}r^2 x$. Therefore,

$$\frac{1}{2}\overline{OA} \cdot \overline{AD} \leq \frac{1}{2}\overline{OB}^2 \cdot x \leq \frac{1}{2}\overline{OB} \cdot \overline{BC}$$

or

$$\frac{1}{2}\cos x \sin x \leq \frac{1}{2} \cdot 1^2 \cdot x \leq \frac{1}{2} \cdot 1 \cdot \tan x$$

Dividing this by $\frac{1}{2}\sin x$ yields

$$\cos x \leq \frac{x}{\sin x} \leq \frac{1}{\cos x}$$

Here, we used the fact that $\tan x = \frac{\sin x}{\cos x}$. Taking reciprocals and reversing the inequalities yields

$$\frac{1}{\cos x} \geq \frac{\sin x}{x} \geq \cos x$$

We can now take the limit as $x \to 0+$. (Remember, we assumed that $0 < x < \pi/2$, so we can only approach 0 from the right.) Note that

$$\lim_{x \to 0+} \cos x = 1 \quad \text{and} \quad \lim_{x \to 0+} \frac{1}{\cos x} = \frac{1}{\lim_{x \to 0+} \cos x} = 1$$

We are ready to apply the Sandwich Theorem, and find

$$\lim_{x \to 0+} \frac{\sin x}{x} = 1$$

We only showed that $\lim_{x \to 0+} \frac{\sin x}{x} = 1$, but a similar argument can be carried out when $-\frac{\pi}{2} < x < 0$. Combining the two results then shows that

$$\lim_{x \to 0} \frac{\sin x}{x} = 1$$

◆

Proof of $\lim_{x \to 0} \frac{1-\cos x}{x}$ Multiplying both numerator and denominator of $(1-\cos x)/x$ by $1+\cos x$, we can reduce the second statement to the first statement.

$$\lim_{x \to 0} \frac{1-\cos x}{x} = \lim_{x \to 0} \frac{1-\cos x}{x} \frac{1+\cos x}{1+\cos x}$$
$$= \lim_{x \to 0} \frac{1-\cos^2 x}{x(1+\cos x)}$$

Using the identity $\sin^2 x + \cos^2 x = 1$, we can write this as

$$\lim_{x \to 0} \frac{\sin^2 x}{x(1+\cos x)}$$

Rewriting this now as

$$\lim_{x \to 0} \frac{\sin x}{x} \frac{\sin x}{1+\cos x}$$

we can determine the limit. Note that $\lim_{x \to 0} \frac{\sin x}{x}$ exists by the first statement, and $\lim_{x \to 0} \frac{\sin x}{1+\cos x}$ exists because $1+\cos x \neq 0$ for x close to 0. Applying Rule 3 of the limit laws, it follows that the limit of the product is the product of the limits. We therefore find that

$$\lim_{x \to 0} \frac{\sin x}{x} \frac{\sin x}{1+\cos x} = \lim_{x \to 0} \frac{\sin x}{x} \lim_{x \to 0} \frac{\sin x}{1+\cos x} = 1 \cdot 0 = 0 \qquad \blacklozenge$$

EXAMPLE 2   Find the following limits.

a. $\lim_{x \to 0} \dfrac{\sin 3x}{5x}$

b. $\lim_{x \to 0} \dfrac{\sin^2 x}{x^2}$

c. $\lim_{x \to 0} \dfrac{\sec x - 1}{x \sec x}$

Solution

a. We cannot apply the first trigonometric limit directly. The trick is to substitute $z = 3x$ and observe that $z \to 0$ as $x \to 0$. Then

$$\lim_{x \to 0} \frac{\sin 3x}{5x} = \lim_{z \to 0} \frac{\sin z}{5z/3} = \frac{3}{5} \lim_{z \to 0} \frac{\sin z}{z} = \frac{3}{5}$$

b. We find

$$\lim_{x \to 0} \frac{\sin^2 x}{x^2} = \lim_{x \to 0} \left(\frac{\sin x}{x}\right)^2 = \left(\lim_{x \to 0} \frac{\sin x}{x}\right)^2 = 1$$

Here, we used the fact that the limit of a product is the product of the limits, provided that the individual limits exist.

c. We first write $\sec x = 1/\cos x$. Then, we multiply both numerator and denominator by $\cos x$. This yields

$$\lim_{x \to 0} \frac{\sec x - 1}{x \sec x} = \lim_{x \to 0} \frac{\frac{1}{\cos x} - 1}{\frac{x}{\cos x}}$$
$$= \lim_{x \to 0} \frac{\left(\frac{1}{\cos x} - 1\right)\cos x}{\frac{x}{\cos x}\cos x} = \lim_{x \to 0} \frac{1-\cos x}{x} = 0 \qquad \blacksquare$$

◆ 2.2 PROBLEMS

1. Let
$$f(x) = x^2 \cos \frac{1}{x}, \quad x \neq 0$$
(a) Use a graphing calculator to sketch the graph of $y = f(x)$.
(b) Show that for $x \neq 0$,
$$-x^2 \leq x^2 \cos \frac{1}{x} \leq x^2$$
holds.
(c) Use your result in (b) and the Sandwich Theorem to show that
$$\lim_{x \to 0} x^2 \cos \frac{1}{x} = 0$$

2. Let
$$f(x) = x^2 \sin \frac{1}{x}, \quad x \neq 0$$
(a) Use a graphing calculator to sketch the graph of $y = f(x)$.
(b) Show that for $x \neq 0$,
$$-x^2 \leq x^2 \sin \frac{1}{x} \leq x^2$$
holds.
(c) Use your result in (b) and the Sandwich Theorem to show that
$$\lim_{x \to 0} x^2 \sin \frac{1}{x} = 0$$

3. Let
$$f(x) = \frac{\ln x}{x}, \quad x > 0$$
(a) Use a graphing calculator to graph $y = f(x)$.
(b) Use a graphing calculator to investigate the values of x for which
$$\frac{1}{x} \leq \frac{\ln x}{x} \leq \frac{1}{\sqrt{x}}$$
holds.

(c) Use your result in (b) to explain why the following is true:
$$\lim_{x \to \infty} \frac{\ln x}{x} = 0$$
4. Let
$$f(x) = \frac{\sin x}{x}, \quad x > 0$$
(a) Use a graphing calculator to graph $y = f(x)$.
(b) Explain why you cannot use the basic rules for finding limits to compute
$$\lim_{x \to \infty} \frac{\sin x}{x}$$
(c) Show that for $x > 0$
$$-\frac{1}{x} \leq \frac{\sin x}{x} \leq \frac{1}{x}$$
holds, and use the Sandwich Theorem to compute
$$\lim_{x \to \infty} \frac{\sin x}{x}$$

In Problems 5–18, evaluate the trigonometric limits.

5. $\lim\limits_{x \to 0} \dfrac{\sin(2x)}{2x}$

6. $\lim\limits_{x \to 0} \dfrac{\sin(2x)}{3x}$

7. $\lim\limits_{x \to 0} \dfrac{\sin(5x)}{x}$

8. $\lim\limits_{x \to 0} \dfrac{\sin x}{-x}$

9. $\lim\limits_{x \to 0} \dfrac{\sin(\pi x)}{x}$

10. $\lim\limits_{x \to 0} \dfrac{\sin(-\pi x/2)}{2x}$

11. $\lim\limits_{x \to 0} \dfrac{\sin(\pi x)}{\sqrt{x}}$

12. $\lim\limits_{x \to 0} \dfrac{\sin^2 x}{x}$

13. $\lim\limits_{x \to 0} \dfrac{\sin x \cos x}{x(1 - x)}$

14. $\lim\limits_{x \to 0} \dfrac{1 - \cos^2 x}{x^2}$

15. $\lim\limits_{x \to 0} \dfrac{1 - \cos x}{2x}$

16. $\lim\limits_{x \to 0} \dfrac{1 - \cos(2x)}{3x}$

17. $\lim\limits_{x \to 0} \dfrac{\sin x (1 - \cos x)}{x^2}$

18. $\lim\limits_{x \to 0} \dfrac{\csc x - \cot x}{x \csc x}$

◆ 2.3 LIMITS AT INFINITY

The limit laws (as given in Subsection 2.1.3) also hold as x tends to ∞ (or $-\infty$). We begin with the limit we considered in Example 3 of Section 2.1.

❙EXAMPLE 1   Find
$$\lim_{x \to \infty} \frac{x}{x + 1}$$

Solution   We set $f(x) = x$ and $g(x) = x + 1$. Obviously, neither $\lim_{x \to \infty} f(x)$ nor $\lim_{x \to \infty} g(x)$ exists. Thus, the hypothesis of the limit laws does not hold, and we cannot use Rule 4. But we can divide both numerator and denominator by x. We find
$$\lim_{x \to \infty} \frac{x}{x + 1} = \lim_{x \to \infty} \frac{1}{1 + \frac{1}{x}}$$

Since $\lim_{x \to \infty} 1 = 1$ and $\lim_{x \to \infty} (1 + \frac{1}{x}) = 1$, both limits exist. Furthermore, $\lim_{x \to \infty} (1 + \frac{1}{x}) \neq 0$. We can therefore apply Rule 4 after having done the algebraic manipulation, and find

$$\lim_{x\to\infty}\frac{x}{x+1} = \lim_{x\to\infty}\frac{1}{1+\frac{1}{x}} = \frac{\lim_{x\to\infty}1}{\lim_{x\to\infty}(1+\frac{1}{x})} = \frac{1}{1} = 1 \qquad\blacksquare$$

In Example 1, we computed the limit of a rational function as x tended to infinity. Rational functions are ratios of polynomials. To find out how the limit of a rational function behaves as x tends to infinity, we will first compare the relative growth of functions of the form $y = x^n$. We find, for instance, that $x^5 \geq x^2$ for $x \geq 1$, since $x^5/x^2 = x^3 \geq 1$ for $x \geq 1$. More generally, when $n \geq m$, we find

$$x^n \geq x^m \quad \text{for } x \geq 1$$

To see why, note that $x^n/x^m = x^{n-m} \geq 1$, for $x \geq 1$, since $n - m \geq 0$. In fact, one can say quite a bit more: namely, if $n > m$, then x^n dominates x^m for large x, in the sense that

$$\lim_{x\to\infty}\frac{x^n}{x^m} = \infty \quad \text{and} \quad \lim_{x\to\infty}\frac{x^m}{x^n} = 0$$

This follows immediately if one simplifies the fractions. This property is important when we compute limits of rational functions as $x \to \infty$. We compare the following three limits.

a. $\displaystyle\lim_{x\to\infty}\frac{x^2+2x-1}{x^3-3x^2+1}$ **b.** $\displaystyle\lim_{x\to\infty}\frac{2x^3-4x+7}{x^3+7x^2-1}$ **c.** $\displaystyle\lim_{x\to\infty}\frac{x^4+2x-5}{x^2-x+2}$

To determine whether the numerator or the denominator dominates, we look at the **leading term** of the polynomials in the numerator and the denominator. (The leading term is the term with the largest exponent.) The leading term tells us how quickly the function increases for large x.

a. The leading term in the numerator is x^2, and the leading term in the denominator is x^3. As $x \to \infty$, the denominator grows much faster than the numerator. We therefore expect the limit to be equal to 0. We can show this by dividing both numerator and denominator by the highest power, namely x^3. We find

$$\lim_{x\to\infty}\frac{x^2+2x-1}{x^3-3x+1} = \lim_{x\to\infty}\frac{\frac{1}{x}+\frac{2}{x^2}-\frac{1}{x^3}}{1-\frac{3}{x^2}+\frac{1}{x^3}}$$

Since $\lim_{x\to\infty}\left(\frac{1}{x}+\frac{2}{x^2}-\frac{1}{x^3}\right)$ exists (it is equal to 0), and $\lim_{x\to\infty}\left(1-\frac{3}{x^2}+\frac{1}{x^3}\right)$ exists and is not equal to 0 (it is equal to 1), we can apply Rule 4 and find

$$\lim_{x\to\infty}\frac{\frac{1}{x}+\frac{2}{x^2}-\frac{1}{x^3}}{1-\frac{3}{x^2}+\frac{1}{x^3}} = \frac{\lim_{x\to\infty}\left(\frac{1}{x}+\frac{2}{x^2}-\frac{1}{x^3}\right)}{\lim_{x\to\infty}\left(1-\frac{3}{x^2}+\frac{1}{x^3}\right)} = \frac{0}{1} = 0$$

b. The leading term in both the numerator and the denominator is x^3. We divide both numerator and denominator by x^3 and obtain

$$\lim_{x\to\infty}\frac{2x^3-4x+7}{x^3+7x^2-1} = \lim_{x\to\infty}\frac{2-\frac{4}{x^2}+\frac{7}{x^3}}{1+\frac{7}{x}-\frac{1}{x^3}} = \frac{2}{1} = 2$$

In the next-to-last step, we used the facts that the limits in both the numerator and the denominator exist and that the limit in the denominator is not equal to 0. Applying Rule 4 yields the limiting value. Note that the limiting value is equal to the ratio of the coefficients of the leading terms in the numerator and the denominator.

c. The leading term in the numerator is x^4 and the leading term in the denominator is x^2. Since the leading term in the numerator grows much more quickly than the leading term in the denominator, we expect the limit to be ∞. This is indeed the case, and can be seen if we divide the numerator by the denominator. We find

$$\lim_{x \to \infty} \frac{x^4 + 2x - 5}{x^2 - x + 2} = \lim_{x \to \infty} \left(x^2 + x - 1 - \frac{x + 3}{x^2 - x + 2} \right) = \infty$$

Let's summarize our findings. If $f(x)$ is a rational function of the form $f(x) = p(x)/q(x)$, where $p(x)$ is a polynomial of degree $\deg(p)$, and $q(x)$ is a polynomial of $\deg(q)$, we find that

$$\lim_{x \to \infty} f(x) = \lim_{x \to \infty} \frac{p(x)}{q(x)} = \begin{cases} 0 & \text{if } \deg(p) < \deg(q) \\ L \neq 0, \pm\infty & \text{if } \deg(p) = \deg(q) \\ \pm\infty & \text{if } \deg(p) > \deg(q) \end{cases}$$

EXAMPLE 2   Compute

a. $\displaystyle\lim_{x \to -\infty} \frac{1 - x - 2x^2}{3x - 5x^2}$ **b.** $\displaystyle\lim_{x \to \infty} \frac{1 - x^3}{1 + x^5}$ **c.** $\displaystyle\lim_{x \to -\infty} \frac{4 + 3x^2}{1 - 7x}$

Solution

a. Since the degree of the numerator is equal to the degree of the denominator, we find

$$\lim_{x \to -\infty} \frac{1 - x + 2x^2}{3x - 5x^2} = \frac{2}{-5} = -\frac{2}{5}$$

b. Since the degree of the numerator is less than the degree of the denominator, we find

$$\lim_{x \to \infty} \frac{1 - x^3}{1 + x^5} = 0$$

c. Since the degree of the numerator is greater than the degree of the denominator, we find

$$\lim_{x \to -\infty} \frac{4 + 3x^2}{1 - 7x} = -\infty \qquad \text{(does not exist)}$$ ▮

Rational functions are not the only functions that involve limits as $x \to \infty$ (or $x \to -\infty$). Many important applications in biology involve exponential functions. We will use the following result repeatedly; it is one of the most important limits.

$$\lim_{x \to \infty} e^{-x} = 0$$

The graph of $f(x) = e^{-x}$ is given in Figure 2.17. Be certain to familiarize yourself with the basic shape of the function $f(x) = e^{-x}$, and its behavior as $x \to \infty$.

EXAMPLE 3   **Logistic Growth**

The logistic curve describes the density of a population over time, where the rate of growth depends on the population size. We will discuss the exact form of this function in the next chapter. It suffices here to say that the per capita rate of growth decreases with increasing population size. If $N(t)$ denotes the size of the population at time t, then the logistic curve is given by

$$N(t) = \frac{K}{1 + \left(\frac{K}{N(0)} - 1 \right) e^{-rt}} \qquad \text{for } t \geq 0$$

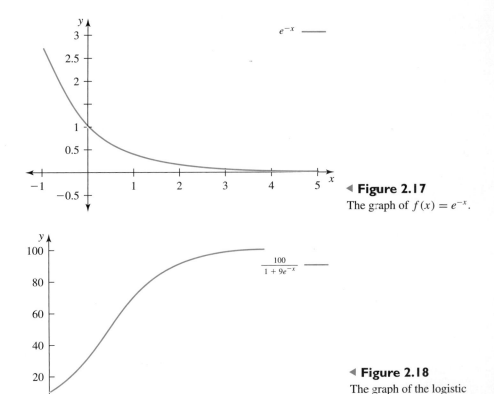

◄ Figure 2.17
The graph of $f(x) = e^{-x}$.

◄ Figure 2.18
The graph of the logistic curve with $K = 100$, $N_0 = 10$ and $r = 1$.

The parameters K and r are positive numbers that describe the population dynamics. The graph of $N(t)$ is shown in Figure 2.18. We will interpret K now; the interpretation of r must wait until the next chapter.

We are interested in the long-term behavior of the population evolving according to the logistic growth curve. That is, we want to investigate what happens to $N(t)$ as $t \to \infty$. We find that

$$\lim_{t \to \infty} \frac{K}{1 + \left(\frac{K}{N(0)} - 1 \right) e^{-rt}} = K$$

since $\lim_{t \to \infty} e^{-rt} = 0$. That is, as $t \to \infty$, the population size approaches K. This value is called the **carrying capacity** of the population. You will encounter logistic growth repeatedly in this book—it is one of the most fundamental equations for describing population growth.

◆ 2.3 PROBLEMS

Evaluate the limits in Problems 1–16.

1. $\displaystyle\lim_{x \to \infty} \frac{2x^2 - 3x + 5}{x^4 - 2x + 1}$

2. $\displaystyle\lim_{x \to \infty} \frac{x^2 + 3}{5x^2 - 2x + 1}$

3. $\displaystyle\lim_{x \to \infty} \frac{x^3 + 3}{x - 2}$

4. $\displaystyle\lim_{x \to -\infty} \frac{2x - 1}{3 - 4x}$

5. $\displaystyle\lim_{x \to \infty} \frac{3x^4 - x^3 + 1}{x^4 + 2x^2}$

6. $\displaystyle\lim_{x \to \infty} \frac{5x^3 - 1}{4x^4 + 1}$

7. $\displaystyle\lim_{x \to \infty} \frac{x^2 - 2}{2x + 1}$

8. $\displaystyle\lim_{x \to -\infty} \frac{3 - x^2}{1 - 2x^2}$

9. $\displaystyle\lim_{x \to \infty} \frac{4}{1 + e^{-2x}}$

10. $\displaystyle\lim_{x \to \infty} \frac{e^{-x}}{1 - e^{-x}}$

11. $\displaystyle\lim_{x \to \infty} \frac{2e^x}{e^x + 3}$

12. $\displaystyle\lim_{x \to \infty} \frac{e^x}{2 - e^x}$

13. $\displaystyle\lim_{x \to -\infty} \exp[x]$

14. $\displaystyle\lim_{x \to \infty} \exp[-\ln x]$

15. $\lim\limits_{x\to\infty} e^{-x}\sin x$

16. $\lim\limits_{x\to\infty} e^{-x}\cos x$

17. In Section 1.2.3, Example 5, we introduced the Monod growth function

$$r(N) = a\frac{N}{k+N}, \quad N \geq 0$$

Find $\lim_{N\to\infty} r(N)$.

18. In Problem 30 of Section 1.3, we discussed the Michaelis-Menten equation, which describes the initial velocity of an enzymatic reaction (v_0) as a function of substrate concentration (s_0). The equation was given by

$$v_0 = \frac{v_{max} s_0}{s_0 + K_m}$$

Find $\lim_{s_0\to\infty} v_0$.

19. Suppose the size of a population at time t is given by

$$N(t) = \frac{500t}{3+t}, \quad t \geq 0$$

(a) Use a graphing calculator to sketch the graph of $N(t)$.

(b) Determine the size of the population as $t \to \infty$. We call this the **limiting population size**.

(c) Show that at time $t = 3$, the size of the population is half its limiting size.

20. (*Logistic Growth*) Suppose that the size of a population at time t is given by

$$N(t) = \frac{100}{1+9e^{-t}}$$

for $t \geq 0$.

(a) Use a graphing calculator to sketch the graph of $N(t)$.

(b) Determine the size of the population as $t \to \infty$, using the basic rules for limits. Compare your answer with the graph that you sketched in (a).

21. **(a)** Use a graphing calculator to sketch the graph of

$$f(x) = e^{ax}\sin x, \quad x \geq 0$$

for $a = -0.1, -0.01, 0, 0.01$ and 0.1.

(b) Which part of the function $f(x)$ produces the oscillations that you see in the graphs sketched in (a)?

(c) Describe in words the effect that the value of a has on the shape of the graph of $f(x)$.

(d) Graph $f(x) = e^{ax}\sin x$, $g(x) = -e^{ax}$, and $h(x) = e^{ax}$ together in one coordinate system for (i) $a = 0.1$, and (ii) $a = -0.1$. (Use separate coordinate systems for (i) and (ii).) Explain what you see in each case. Show that

$$-e^{ax} \leq e^{ax}\sin x \leq e^{ax}$$

Use this to determine the values of a for which

$$\lim_{x\to\infty} f(x)$$

exists, and find the limiting value.

◆ 2.4 CONTINUITY

▲ **Figure 2.19**

A continuous function.

2.4.1 What is Continuity?

We mentioned continuity briefly in Section 2.1, in the context of finding limits. We will now explore the concept of continuous functions in detail. Let us first look at Figures 2.19 and 2.20. It is easy to see the difference between the two graphs. The graph in Figure 2.19 can be drawn without lifting the pencil. In Figure 2.20, a jump occurs at $x = 1$; in order to draw the graph, we must lift the pencil. The function in Figure 2.19 is continuous, the function in Figure 2.20 is discontinuous. Of course, this is not a definition, but it explains intuitively what we mean by a continuous function. To define continuity, we first explain what it means for a function to be continuous at a point $x = c$.

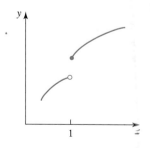

▲ **Figure 2.20**

A discontinuous function: A jump occurs at $x = 1$.

> **Definition**   A function f is said to be **continuous** at $x = c$ if
> $$\lim_{x\to c} f(x) = f(c)$$

In order to check whether a function is continuous at $x = c$, we therefore need to check the following three conditions:

1. $f(x)$ is defined at $x = c$.

2. $\lim\limits_{x\to c} f(x)$ exists.

3. $\lim\limits_{x\to c} f(x)$ is equal to $f(c)$.

If any of these three conditions fails, the function is **discontinuous** at $x = c$.

EXAMPLE 1 Show that $f(x) = 2x - 3$, $x \in \mathbf{R}$, is continuous at $x = 1$.

Solution We must check all three conditions. 1. $f(x)$ is defined at $x = 1$ since $f(1) = 2 \cdot 1 - 3 = -1$. 2. We showed in Example 9 of Section 2.1 that $\lim_{x \to 1} f(x)$ exists. 3. Also in Example 9 of Section 2.1, we showed that $\lim_{x \to 1} f(x) = -1$. This is the same as $f(1)$.

Since all three conditions are satisfied, $f(x) = 2x - 3$ is continuous at $x = 1$. ▮

EXAMPLE 2 Show that

$$f(x) = \begin{cases} \frac{x^2 - x - 6}{x - 3} & \text{if } x \neq 3 \\ 5 & \text{if } x = 3 \end{cases}$$

is continuous at $x = 3$.

Solution Again, we need to check all three conditions. 1. $f(x)$ is defined at $x = 3$, namely, $f(3) = 5$. 2. To check whether

$$\lim_{x \to 3} \frac{x^2 - x - 6}{x - 3}$$

exists, we factor the numerator, $x^2 - x - 6 = (x - 3)(x + 2)$. Hence, for $x \neq 3$,

$$\lim_{x \to 3} \frac{x^2 - x - 6}{x - 3} = \lim_{x \to 3} \frac{(x - 3)(x + 2)}{x - 3} = \lim_{x \to 3}(x + 2)$$

Note that we were able to cancel the term $x - 3$ since $x \neq 3$. Now, $\lim_{x \to 3}(x + 2)$ exists. 3. Finally, $\lim_{x \to 3}(x + 2) = 5$, which is equal to $f(3)$.

Since all three conditions are satisfied, $f(x)$ is continuous at $x = 3$. ▮

We will dwell a bit more on the last example. If we consider

$$g(x) = \frac{x^2 - x - 6}{x - 3}, \quad x \neq 3$$

then $g(x)$ is not continuous at $x = 3$, since it is not even defined at $x = 3$ and therefore condition 1 does not hold. In Example 2, we fixed this problem by simply assigning a value to $f(x)$ at $x = 3$. If we want a continuous function, we need to choose the "correct" value, namely 5. If we had defined the function instead as

$$h(x) = \begin{cases} \frac{x^2 - x - 6}{x - 3} & \text{if } x \neq 3 \\ 7 & \text{if } x = 3 \end{cases}$$

then $h(x)$ would be defined at $x = 3$, namely $h(3) = 7$, but $\lim_{x \to 3} h(x) = 5 \neq 7 = h(3)$ and hence condition 3 would be violated. (Note that condition 2 holds for both $g(x)$ and $h(x)$.) In such cases where we can define a value so that $f(x)$ is continuous at that point, we say that we can *remove the discontinuity*.

The following three examples show that we cannot always remove discontinuities as in Example 2. In the first two examples, the discontinuity is a jump; that is, both the left-hand and the right-hand limit exist at the point where the jump occurs, but the limits differ. In the third example, we encounter a vertical asymptote at the point where the function is discontinuous.

EXAMPLE 3 The bracket function

$$f(x) = [x] = \text{the largest integer less than or equal to } x$$

is graphed in Figure 2.21. The closed circle in the figure indicates that the endpoint is contained in the graph of the function, whereas the open circle indicates that the

$[x]$ ——

◀ **Figure 2.21**
The bracket function $[x]$.

endpoint is not contained in the graph of the function. To explain this function, we compute a few values: $f(2.1) = 2$, $f(2) = 2$, $f(1.9999) = 1$. The function jumps whenever x is an integer. Let k be an integer, then $f(k) = k$ and

$$\lim_{x \to k+} f(x) = k, \quad \lim_{x \to k-} f(x) = k - 1$$

That is, only when x approaches the integer from the right is the limit equal to the value of the function. The function is therefore discontinuous at integer values. If c is not an integer, then $f(x)$ is continuous at $x = c$. ▮

Example 3 motivates the definition of one-sided continuity.

Definition   A function f is said to be continuous from the right at $x = c$ if

$$\lim_{x \to c+} f(x) = f(c)$$

and continuous from the left at $x = c$ if

$$\lim_{x \to c-} f(x) = f(c)$$

The function $f(x) = [x]$, $x \in \mathbf{R}$, of Example 3, is therefore continuous from the right but not from the left. In the next example, the discontinuity is again a jump, but this time we do not even have one-sided continuity.

EXAMPLE 4   Show that

$$f(x) = \frac{|x|}{x}, \quad x \neq 0$$

is discontinuous at $x = 0$, and that the discontinuity at $x = 0$ cannot be removed.

Solution   The graph of $f(x)$ is shown in Figure 2.22. We considered this function in Example 10 of Section 2.1. There, we showed that $\lim_{x \to 0} |x|/x$ does not exist. This is enough to show that $f(x)$ is discontinuous at $x = 0$, as condition 2 does not hold. (Actually, $f(x)$ is not defined at $x = 0$, and so condition 1 is violated as well.) However, we can say a bit more about the function at $x = 0$. We can write

$$f(x) = \begin{cases} 1 & \text{for } x > 0 \\ -1 & \text{for } x < 0 \end{cases}$$

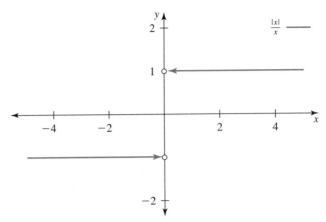

◀ **Figure 2.22**

The function $f(x) = \dfrac{|x|}{x}$ is discontinuous at integers.

since $|x| = x$ for $x > 0$ and $|x| = -x$ for $x < 0$. We therefore get

$$\lim_{x \to 0+} f(x) = 1 \qquad \text{and} \qquad \lim_{x \to 0-} f(x) = -1$$

The one-sided limits exist but they are not equal (which implies that $\lim_{x \to 0} f(x)$ does not exist). When we graph the function, a jump occurs at $x = 0$ (see Figure 2.22). There is no way that we could assign a value to $f(0)$ so that the function would be continuous at $x = 0$. This function does not exhibit even one-sided continuity because $f(x)$ is not defined at $x = 0$.

I EXAMPLE 5   Show that

$$f(x) = \frac{1}{x^2}, \quad x \neq 0$$

is discontinuous at $x = 0$, and that the discontinuity at $x = 0$ cannot be removed.

Solution   The graph of $f(x)$ is displayed in Figure 2.23. To show that $f(x) = \frac{1}{x^2}$ is discontinuous at $x = 0$ is easy, since $f(x)$ is not defined at $x = 0$ and hence condition 1 does not hold. We already know what happens at $x = 0$; we showed in Example 11 of Section 2.1 that

$$\lim_{x \to 0} \frac{1}{x^2} = \infty$$

Because ∞ is not a real number, we cannot assign a value to $f(0)$ so that $f(x)$ would be continuous at $x = 0$.

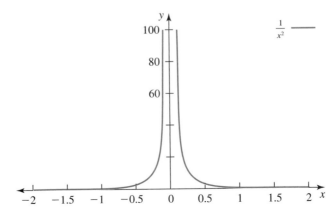

◀ **Figure 2.23**

The function $f(x) = \dfrac{1}{x^2}$ is discontinuous at $x = 0$.

2.4.2 Combinations of Continuous Functions

Using the limit laws, we find that the following hold for combinations of continuous functions.

Suppose that a is a constant and the functions f and g are continuous at $x = c$. Then the following functions are continuous at $x = c$.

1. $a \cdot f$　　**2.** $f + g$　　**3.** $f \cdot g$　　**4.** $\dfrac{f}{g}$　provided $g(c) \neq 0$

Proof　We will prove only the second condition. We must show that conditions 1–3 of the previous subsection hold. 1. Note that $[f + g](x) = f(x) + g(x)$. Therefore, $f + g$ is defined at $x = c$ and $[f + g](c) = f(c) + g(c)$.　2. We assumed that f and g are continuous at $x = c$. This means, in particular, that

$$\lim_{x \to c} f(x) \quad \text{and} \quad \lim_{x \to c} g(x)$$

both exist. That is, the hypothesis in limit laws holds, and we can apply rule 2 for limits and find

$$\lim_{x \to c} [f + g](x) = \lim_{x \to c} [f(x) + g(x)] = \lim_{x \to c} f(x) + \lim_{x \to c} g(x) \tag{2.1}$$

That is, $\lim_{x \to c} [f + g](x)$ exists. and condition 2 holds.　3. Since f and g are continuous at $x = c$,

$$\lim_{x \to c} f(x) = f(c) \quad \text{and} \quad \lim_{x \to c} g(x) = g(c) \tag{2.2}$$

Therefore, combining (2.1) and (2.2),

$$\lim_{x \to c} [f + g](x) = \lim_{x \to c} f(x) + \lim_{x \to c} g(x) = f(c) + g(c)$$

which is equal to $[f + g](c)$ and hence condition 3 holds.

Since we showed that all three conditions hold, it follows that $f + g$ is continuous at $x = c$. The other statements are shown in a similar way, using the basic properties of limits.　　　　　　　　　　　　　　　　◆

We say that a function f is continuous on an interval I if f is continuous for all $x \in I$. Note that if I is a closed interval, then continuity at the left (and respectively, right) endpoint of the interval means continuous from the right (and respectively, left). Many of the elementary functions are indeed continuous wherever they are defined. For polynomials and rational functions, this follows immediately from the fact that certain combinations of continuous functions are continuous. We give a list of the most important cases.

The following functions are continuous wherever they are defined.

1. polynomial functions

2. rational functions

3. power functions

4. trigonometric functions

5. exponential functions of the form a^x, $a > 0$

6. logarithmic functions of the form $\log_a x$, $a > 0$

The phrase "wherever they are defined" is crucial. It helps us to identify points where a function might be discontinuous. For instance, the power function $1/x^2$ is only defined for $x \neq 0$, and the logarithmic function $\log_a x$ is only defined for $x > 0$. We will illustrate these six cases in the following example, paying particular attention to the phrase "wherever they are defined."

▌**EXAMPLE 6**   For which values of $x \in \mathbf{R}$ are the following functions continuous?

    a. $f(x) = 2x^3 - 3x + 1$ **b.** $f(x) = \dfrac{x^2 + x + 1}{x - 2}$

    c. $f(x) = x^{1/4}$ **d.** $f(x) = 3 \sin x$

    e. $f(x) = \tan x$ **f.** $f(x) = 3^x$

    g. $2 \ln(x + 1)$

Solution

    a. $f(x)$ is a polynomial and is defined for all $x \in \mathbf{R}$, it is therefore continuous for all $x \in \mathbf{R}$.

    b. $f(x)$ is a rational function. It is defined for all $x \neq 2$, it is therefore continuous for all $x \neq 2$. Note that since $x^2 + x + 1$ cannot be factored into a product containing the factor $x - 2$, we cannot remove the discontinuity as we were able to do in Example 2.

    c. $f(x) = x^{1/4} = \sqrt[4]{x}$ is a power function that is defined for $x \geq 0$. It is therefore continuous for $x \geq 0$.

    d. $f(x)$ is a trigonometric function. Because $\sin x$ is defined for all $x \in \mathbf{R}$, $3 \sin x$ is continuous for all $x \in \mathbf{R}$.

    e. $f(x)$ is a trigonometric function. The tangent is defined for all $x \neq \frac{\pi}{2} + k\pi$, where k is an integer. It is therefore continuous for all $x \neq \frac{\pi}{2} + k\pi$, where k is an integer.

    f. $f(x)$ is an exponential function. $f(x) = 3^x$ is defined for all $x \in \mathbf{R}$ and is therefore continuous for all $x \in \mathbf{R}$.

    g. $f(x)$ is a logarithmic function. $f(x) = 2 \ln(x + 1)$ is defined as long as $x + 1 > 0$ or $x > -1$. It is therefore continuous for all $x > -1$. ▌

The following result is useful in determining whether a composition of functions is continuous.

> **Theorem**   If $g(x)$ is continuous at $x = c$ with $g(c) = L$, and $f(x)$ is continuous at $x = L$, then
>
> $$\lim_{x \to c}(f \circ g)(x) = \lim_{x \to c} f[g(x)] = f[\lim_{x \to c} g(x)] = f[g(c)] = f(L)$$

To explain this theorem, we recall what it means to compute $(f \circ g)(c) = f[g(c)]$. When we compute $f[g(c)]$, we take the value c, compute $g(c)$, then take the result $g(c)$ and plug it into the function f to obtain $f[g(c)]$. If at each step the functions are continuous, the resulting function will be continuous.

▌**EXAMPLE 7**   Determine where the following functions are continuous.

    a. $h(x) = e^{-x^2}$ **b.** $h(x) = \sin \dfrac{1}{x}$ **c.** $h(x) = \dfrac{1}{1 + 2x^{1/3}}$

Solution

a. Set $g(x) = -x^2$ and $f(x) = e^x$, then $h(x) = (f \circ g)(x)$. Since $g(x)$ is a polynomial, it is continuous for all $x \in \mathbf{R}$, and the range of $g(x)$ is $(-\infty, 0]$. $f(x)$ is continuous for all values in the range of $g(x)$ (in fact, $f(x)$ is continuous for all $x \in \mathbf{R}$). It therefore follows that $h(x)$ is continuous for all $x \in \mathbf{R}$.

b. Set $g(x) = \frac{1}{x}$ and $f(x) = \sin x$. $g(x)$ is continuous for all $x \neq 0$. The range of $g(x)$ is $\mathbf{R} - \{0\}$, that is, the set of real numbers without 0. $f(x)$ is continuous for all x in the range of $g(x)$. Hence, $h(x)$ is continuous for all $x \neq 0$. Recall that we showed in Example 7 of Section 2.1 that

$$\lim_{x \to 0} \sin \frac{1}{x}$$

does not exist. That is, $h(x)$ is discontinuous at $x = 0$.

c. Set $g(x) = x^{1/3}$ and $f(x) = \frac{1}{1+2x}$. Then $h(x) = (f \circ g)(x)$. $g(x)$ is continuous for all $x \in \mathbf{R}$ since $g(x) = x^{1/3} = \sqrt[3]{x}$ and 3 is an odd integer. The range of $g(x)$ is $(-\infty, \infty)$. $f(x)$ is continuous for all $x \in \mathbf{R} - \{-1/2\}$. Since $g\left(-\frac{1}{8}\right) = -\frac{1}{2}$, $h(x)$ is continuous for all $x \in \mathbf{R} - \{-\frac{1}{8}\}$. Another way to see that we need to exclude $-\frac{1}{8}$ from the domain of $h(x)$ is by directly looking at the denominator of $h(x)$. Namely, $1 + 2x^{1/3} = 0$ when $x = -\frac{1}{8}$.   ▮

◆ 2.4 PROBLEMS

[2.4.1]

In Problems 1–4, show that each function is continuous at the given value.

1. $f(x) = x^3 - 2x + 1$, $c = 2$

2. $f(x) = \sqrt{x^2 + 1}$, $c = -1$

3. $f(x) = \sin(2x)$, $c = \dfrac{\pi}{4}$ **4.** $f(x) = e^{-x}$, $c = 1$

5. Show that

$$f(x) = \begin{cases} \frac{x^2 - x - 2}{x-2} & \text{if } x \neq 2 \\ 3 & \text{if } x = 2 \end{cases}$$

is continuous at $x = 2$.

6. Show that

$$f(x) = \begin{cases} \frac{2x^2 + x - 6}{x-2} & \text{if } x \neq -2 \\ -7 & \text{if } x = -2 \end{cases}$$

is continuous at $x = -2$.

7. Let

$$f(x) = \begin{cases} \frac{x^2 - 9}{x-3} & \text{if } x \neq 3 \\ a & \text{if } x = 3 \end{cases}$$

Which value must you assign to a so that $f(x)$ is continuous at $x = 3$?

8. Let

$$f(x) = \begin{cases} \frac{x^2 - x - 2}{x-1} & \text{if } x \neq 1 \\ a & \text{if } x = 1 \end{cases}$$

Which value must you assign to c so that $f(x)$ is continuous at $x = 1$?

9. Show that

$$f(x) = \begin{cases} \frac{1}{x-3} & \text{for } x \neq 3 \\ & \text{for} \\ 0 & \text{for } x = 3 \end{cases}$$

is discontinuous at $x = 3$.

10. Show that

$$f(x) = \begin{cases} \frac{1}{x^2-1} & \text{for } x \neq -1, 1 \\ & \text{for} \\ 0 & \text{for } x = -1 \text{ or } 1 \end{cases}$$

is discontinuous at $x = -1$ and $x = 1$.

11. Show that

$$f(x) = \begin{cases} \frac{x^2 - 3x + 2}{x-2} & \text{if } x \neq 1 \\ & \text{if} \\ 1 & \text{if } x = 1 \end{cases}$$

is discontinuous at $x = 1$.

12. Show that

$$f(x) = \begin{cases} x^2 - 1 & \text{for } x \leq 0 \\ x & \text{for } x > 0 \end{cases}$$

is discontinuous at $x = 0$.

13. Show that the bracket function $f(x) = [x]$ is continuous at $x = 5/2$ but discontinuous at $x = 3$.

14. Show that the bracket function $f(x) = [x]$ is continuous from the right at $x = 2$.

15. (a) Show that

$$f(x) = \sqrt{x - 1}, \quad x \geq 1$$

is continuous from the right at $x = 1$.

(b) Graph $f(x)$.

(c) Does it make sense to look at continuity from the left at $x = 1$?

16. (a) Show that

$$f(x) = \sqrt{x^2 - 4}, \quad |x| \geq 2$$

is continuous from the right at $x = 2$ and continuous from the left at $x = -2$.

(b) Graph $f(x)$.

(c) Does it make sense to look at continuity from the left at $x = 2$ and at continuity from the right at $x = -2$?

[2.4.2]

In Problems 17–26, find the values of $x \in \mathbf{R}$ for which the given functions are continuous.

17. $f(x) = 3x^4 - x^2 + 4$

18. $f(x) = \sqrt{x^2 - 1}$

19. $f(x) = \dfrac{x^2 + 1}{x - 1}$

20. $f(x) = \cos(2x)$

21. $f(x) = e^{-|x|}$

22. $f(x) = \ln(x - 2)$

23. $f(x) = \ln \dfrac{x}{x + 1}$

24. $f(x) = \exp\left[-\sqrt{x - 1}\right]$

25. $f(x) = \tan(2\pi x)$

26. $f(x) = \sin\left(\dfrac{2x}{3 + x}\right)$

27. Let

$$f(x) = \begin{cases} x^2 + 2 & \text{for } x \leq 0 \\ x + c & \text{for } x > 0 \end{cases}$$

(a) Graph $f(x)$ when $c = 1$, and determine whether $f(x)$ is continuous for this choice of c.

(b) How must you choose c so that $f(x)$ is continuous for all $x \in (-\infty, \infty)$?

28. Let

$$f(x) = \begin{cases} \frac{1}{x} & \text{for } x \geq 1 \\ 2x + c & \text{for } x < 1 \end{cases}$$

(a) Graph $f(x)$ when $c = 0$, and determine whether $f(x)$ is continuous for this choice of c.

(b) How must you choose c so that $f(x)$ is continuous for all $x \in (-\infty, \infty)$?

◆ 2.5 THEOREMS ABOUT CONTINUOUS FUNCTIONS

2.5.1 The Extreme Value Theorem

This subsection and the next discuss some important properties of continuous functions on closed intervals.

Suppose that you measure the depth of a creek along a transect between two points A and B (see Figure 2.24). If you look at the resulting profile, you can easily convince yourself that there will be a location of maximum depth and a location of minimum depth. This intuitively represents the content of the extreme value theorem. To formalize this concept we must introduce some terminology.

▲ Figure 2.24

A transect of a creek between the points A and B.

Definition Let f be a function defined on the set D that contains the number c. Then

f has an **absolute maximum** value at c if

$$f(c) \geq f(x) \quad \text{for all } x \in D$$

and

f has an **absolute minimum** value at c if

$$f(c) \leq f(x) \quad \text{for all } x \in D$$

The following result gives conditions under which absolute maxima and absolute minima, collectively called **absolute extrema**, exist.

The Extreme Value Theorem If f is continuous on a closed interval $[a, b]$, $-\infty < a < b < \infty$, then f has an absolute maximum and an absolute minimum.

▲ Figure 2.25

Extreme values at the endpoint.

The proof of this theorem is beyond the scope of this text and will be omitted. However, the result is quite intuitive, and we illustrate it in Figures 2.25 and 2.26.

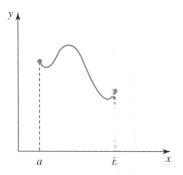

▲ **Figure 2.26**
Extreme values in the interior.

Figure 2.25 shows that the function attains its extreme values at the endpoints of the interval $[a, b]$, whereas in Figure 2.26 the extreme values are attained in the interior of the interval $[a, b]$. The function must be continuous and defined on a closed interval in order to conclude that it has an absolute maximum and an absolute minimum.

Figure 2.25 illustrates the importance of one of the assumptions in the extreme value theorem, namely that the interval is *closed*. If the interval from a to b in Figure 2.25 did not include the endpoints, we would have neither an absolute maximum nor an absolute minimum.

The following two examples illustrate that the theorem cannot be used if either f is discontinuous or the interval is not closed.

▮**EXAMPLE 1**   Let

$$f(x) = \begin{cases} 2x & \text{if } 0 \leq x < 2 \\ 3 - x & \text{if } 2 \leq x \leq 3 \end{cases}$$

Note that $f(x)$ is defined on a *closed* interval, namely $[0, 3]$. However, f is discontinuous at $x = 2$, as can be seen in Figure 2.27. The graph of $f(x)$ shows that there is no value $c \in [0, 3]$ where $f(c)$ attains an absolute maximum. Though

$$\lim_{x \to 2^-} f(x) = 4$$

the function f takes on value 1 at $x = 2$. Hence, the function does not have an absolute maximum. It does, however, have absolute minima, namely at $x = 0$ and $x = 3$, where $f(x)$ takes on the value 0.   ▮

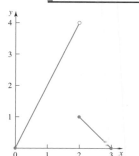

▲ **Figure 2.27**
The graph of $f(x)$ in Example 1.

▮**EXAMPLE 2**   Let

$$f(x) = x \qquad \text{for } 0 < x < 1$$

Note that $f(x)$ is continuous on its domain $(0, 1)$, but it is *not* defined on a closed interval. The function $f(x)$ attains neither an absolute maximum nor an absolute minimum. Though

$$\lim_{x \to 0^+} f(x) = 0 \qquad \text{and} \qquad \lim_{x \to 1^-} f(x) = 1$$

and $0 < f(x) < 1$ for all $x \in (0, 1)$, there is no number c in the open interval $(0, 1)$ where $f(c) = 0$ or $f(c) = 1$.   ▮

2.5.2 The Intermediate Value Theorem

As you hike up a mountain, the temperature decreases with increasing elevation. Suppose the temperature at the bottom of the mountain is 70°F and the temperature at the top of the mountain is 40°F. How do you know that at some time during your hike, you must have crossed a point where the temperature was exactly 50°F? Your answer will probably be something like the following: "To go from 70°F to 40°F, I must have passed through 50°F, since 50°F is between 40°F and 70°F and the temperature changed continuously as I hiked up the mountain." This represents the content of the intermediate value theorem.

The Intermediate Value Theorem   Suppose that f is continuous on the closed interval $[a, b]$. If L is any real number strictly between $f(a)$ and $f(b)$, then there exists at least one number c on the open interval (a, b) such that $f(c) = L$.

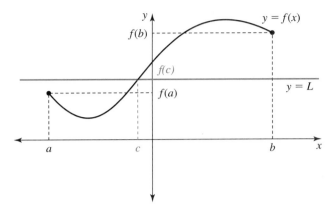

◄ **Figure 2.28**
The Intermediate Value Theorem.

We will not prove this theorem, but Figure 2.28 should convince you that this result is true. Since f is continuous and $f(a) < L < f(b)$, the graph of $f(x)$ must intersect the line $y = L$ at least once in the interval (a, b).

■ **EXAMPLE 3** Let

$$f(x) = 3 + \sin x \quad \text{for } 0 \leq x \leq \frac{3\pi}{2}$$

Show that there exists at least one point c in $(0, 3\pi/2)$ such that $f(c) = 5/2$.

Solution The graph of $f(x)$ is shown in Figure 2.29. First, note that $f(x)$ is defined on a closed interval and is continuous on $[0, 3\pi/2]$. Furthermore, we find that

$$f(0) = 3 + \sin 0 = 3 + 0 = 3$$

$$f\left(\frac{3\pi}{2}\right) = 3 + \sin \frac{3\pi}{2} = 3 + (-1) = 2$$

Given that

$$2 < \frac{5}{2} < 3$$

we conclude from the intermediate value theorem that there exists a number c such that $f(c) = 5/2$. The intermediate value theorem does not tell us where c is, or whether there is more than one such point. ■

When applying the intermediate value theorem, it is important to check that f is continuous. Discontinuous functions can easily miss values; for example,

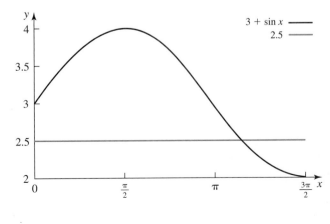

◄ **Figure 2.29**
The Intermediate Value Theorem for $f(x) = 3 + \sin x$, $0 \leq x \leq 3\pi/2$ and $L = 2.5$.

the bracket function in Example 3 of Section 2.4 misses all points which are not integers.

As mentioned in the previous example, the intermediate value theorem only gives us the existence of such a point c, it does not tell us how many such points there are or where they are located.

You might wonder how such a result can be of any use. One important application is that the theorem can be used to find approximate roots of equations. We show how in the following example.

EXAMPLE 4   Find a root of the equation $x^5 - 7x^2 + 3 = 0$.

Solution   Let $f(x) = x^5 - 7x^2 + 3$. Because $f(x)$ is a polynomial, it is continuous for all $x \in \mathbf{R}$. Furthermore,

$$\lim_{x \to -\infty} f(x) = -\infty \quad \text{and} \quad \lim_{x \to \infty} f(x) = \infty$$

That is, if we choose a large enough interval $[a, b]$, then $f(a) < 0$ and $f(b) > 0$ and, therefore, there must be a number $c \in (a, b)$ so that $f(c) = 0$. The existence of such a number c is guaranteed by the intermediate value theorem.

To actually find a number c for which $f(c) = 0$, we use the **bisection method**. We start by finding a and b so that $f(a) < 0$ and $f(b) > 0$. For example,

$$f(-1) = -5 \quad \text{and} \quad f(2) = 7$$

This tells us that there must be a number in $(-1, 2)$ for which $f(c) = 0$. To locate this root with more precision, we take the midpoint of $(-1, 2)$, which is 0.5, and evaluate the function at $x = 0.5$. (The midpoint of the interval (a, b) is $(a + b)/2$.) Now, $f(0.5) = 1.28$ (rounded to two decimals). We thus have

$$f(-1) = -5 \quad f(0.5) = 1.28 \quad f(2) = 7$$

Using the intermediate value theorem again, we can now guarantee a root in $(-1, 0.5)$, since $f(-1) < 0$ and $f(0.5) > 0$. Bisecting the new interval and computing the respective values of $f(x)$, we find

$$f(-1) = -5 \quad f(-0.25) = 2.562 \quad f(0.5) = 1.28$$

Using the intermediate value theorem again, we can guarantee a root in $(-1, -0.25)$, since $f(-1) < 0$ and $f(-0.25) > 0$. Repeating this procedure of bisecting and selecting a new (smaller) interval will eventually produce an interval that is small enough so that we can locate the root to any desired accuracy. The first several steps are summarized in Table 2.1.

TABLE 2•1 BISECTION METHOD

a	$\dfrac{a+b}{2}$	b	$f(a)$	$f\left(\dfrac{a+b}{2}\right)$	$f(b)$
-1	0.5	2	-5	1.28	7
-1	-0.25	0.5	-5	2.562	1.28
-1	-0.625	-0.25	-5	0.170	2.562
-1	-0.8125	-0.625	-5	-1.975	0.170
-0.8125	-0.71875	-0.625	-1.975	-0.808	0.170
-0.71875	-0.671875	-0.625	-0.808	-0.297	0.170
-0.671875	-0.6484375	-0.625	-0.297	-0.0579	0.170
-0.6484375	-0.63671875	-0.625	-0.0579	0.0575	0.170
-0.6484375	-0.642578125	-0.63671875	-0.0579	9.9×10^{-5}	0.0575
-0.6484375		-0.642578125			

After nine steps, we find that there exists a root in

$$(-0.6484375, -0.642578125)$$

The length of this interval is 0.005859375, which gives us a solution that is accurate to two decimal places. If we are satisfied with this precision, we can stop here and choose, for instance, the midpoint of the last interval as an approximate value for a root of the equation $x^5 - 7x^2 + 3 = 0$. The midpoint is

$$\frac{-0.642578125 + (-0.6484375)}{2} = -0.6455078125$$
$$= -0.646$$

after rounding to three decimals.

Note that the length of the interval decreases by a factor of $1/2$ each step. That is, after nine steps, the length of the interval is $(1/2)^9$ of the length of the original interval. In our case, the length of the original interval was 3 and hence the length of the interval after nine steps is

$$3 \cdot \left(\frac{1}{2}\right)^9 = \frac{3}{512} = 0.005859375$$

as we saw.

Figure 2.30 shows a graph of $f(x) = x^5 - 7x^2 + 3$. We see that the graph intersects the x-axis three times. We found the left-most root of the equation $x^5 - 7x^2 + 3$. If we had used another starting interval, say $[1, 2]$, we would have located the right-most root of the equation.

2.5.3 A Final Remark on Continuous Functions

Many functions in biology are in fact discontinuous. For example, if we measure the size of a population over time, then this quantity takes on discrete values only (namely, nonnegative integers) and therefore changes discontinuously. However, if the population size is sufficiently large, an increase or decrease by one changes the population size so slightly that it might be justified to approximate the population size by a continuous function. For example, if we measure the number of bacteria in a petri dish in millions, then the number 2.1 would correspond to 2,100,000 bacteria. An increase by one results in 2,100,001 bacteria, or if we measure the size in millions, in 2.100001, an increase of 10^{-6}.

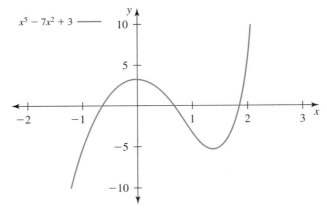

◀ **Figure 2.30**
The graph of
$f(x) = x^5 - 7x^2 + 3$.

We will see many other examples where we will use a continuous function to describe a discontinuous quantity.

◆ 2.5 PROBLEMS

[2.5.1]

In Problems 1–8, each function is continuous and defined on a closed interval. It therefore satisfies the assumptions of the extreme value theorem. With the help of a graphing calculator, graph each function and locate its absolute extrema. (Note that a function may assume an absolute extremum at more than one point.)

1. $f(x) = 2x - 1$, $0 \leq x \leq 1$

2. $f(x) = -x^2 + 1$, $-1 \leq x \leq 1$

3. $f(x) = \sin x$, $0 \leq x \leq 2\pi$

4. $f(x) = \cos x$, $0 \leq x \leq 2\pi$

5. $f(x) = |x|$, $-1 \leq x \leq 1$

6. $f(x) = (x - 1)^2(x + 2)$, $-2 \leq x \leq 2$

7. $f(x) = e^{-x}$, $0 \leq x \leq 2$

8. $f(x) = \ln(x + 1)$, $0 \leq x \leq 2$

9. Let

$$f(x) = x^2 - 1, \quad 0 \leq x \leq 2$$

(a) Graph $y = f(x)$ for $0 \leq x \leq 2$.

(b) Show that

$$f(0) < 0 < f(2)$$

and use the intermediate value theorem to conclude that there exists a number $c \in (0, 2)$ so that $f(c) = 0$.

10. Let

$$f(x) = x^3 - 2x + 3, \quad -3 \leq x \leq -1$$

(a) Graph $y = f(x)$ for $-3 \leq x \leq -1$.

(b) Use the intermediate value theorem to conclude that

$$x^3 - 2x + 3 = 0$$

has a solution in $(-3, -1)$.

11. Let

$$f(x) = \sqrt{x^2 + 2}, \quad 1 \leq x \leq 2$$

(a) Graph $y = f(x)$ for $1 \leq x \leq 2$.

(b) Use the intermediate value theorem to conclude that

$$\sqrt{x^2 + 2} = 2$$

has a solution in $(1, 2)$.

12. Let

$$f(x) = \sin x - x, \quad -1 \leq x \leq 1$$

(a) Graph $y = f(x)$ for $-1 \leq x \leq 1$.

(b) Use the intermediate value theorem to conclude that

$$\sin x = x$$

has a solution in $(-1, 1)$.

13. Use the intermediate value theorem to show that

$$e^{-x} = x$$

has a solution in $(0, 1)$.

14. Use the intermediate value theorem to show that

$$\cos x = x$$

has a solution in $(0, 1)$.

15. Use the bisection method to find a solution of

$$e^{-x} = x$$

that is accurate to two decimal places.

16. Use the bisection method to find a solution of

$$\cos x = x$$

that is accurate to two decimal places.

17. **(a)** Use the bisection method to find a solution of $3x^3 - 4x^2 - x + 2 = 0$.

(b) Graph the function $f(x) = 3x^3 - 4x^2 - x + 2$.

(c) Which root did you locate in (a)? Is it possible in this case to find the other root using the bisection method together with the intermediate value theorem?

18. In Example 4, how many steps are required to guarantee that the approximate root is within 0.0001 of the true value of the root?

19. Suppose that the number of individuals of a population at time t is given by

$$N(t) = \frac{54t}{13 + t}, \quad t \geq 0 \qquad (2.3)$$

(a) Use a calculator to confirm that $N(10)$ is approximately 23.47826. Considering that the number of individuals in a population is an integer, how should you report your answer?

(b) Now, suppose that $N(t)$ is given by the same function (2.3), but that the size of the population is measured in millions. How should you report the population size at time $t = 10$? Make some reasonable assumptions about how accurate a measurement for the size of such a large population might be.

(c) Discuss the use of continuous functions in both (a) and (b).

◆ 2.6 KEY TERMS

Chapter 2 Review: Topics   Discuss the following definitions and concepts:

1. Limit of $f(x)$ as x approaches c
2. One-sided limits
3. Infinite limits
4. Convergence
5. Divergence
6. ϵ-δ definition of limits
7. Limits laws

8. Comparison theorem
9. Sandwich theorem
10. Trigonometric limits
11. Continuity
12. One-sided continuity
13. Continuous functions
14. Extreme value theorem
15. Intermediate value theorem
16. Using the bisection method to find roots

◆ 2.7 REVIEW PROBLEMS

In Problems 1–4, determine where each function is continuous. Investigate the behavior as $x \to \pm\infty$. Use a graphing calculator to sketch the corresponding graphs.

1. $f(x) = e^{-|x|}$

2. $f(x) = \begin{cases} \frac{\sin x}{x} & \text{if } x \neq 0 \\ 1 & \text{if } x = 0 \end{cases}$

3. $f(x) = \dfrac{2}{e^x + e^{-x}}$

4. $f(x) = \dfrac{1}{\sqrt{x^2 - 1}}$

5. Sketch the graph of a function that is discontinuous from the left and continuous from the right at $x = 1$.

6. Sketch the graph of a function that is continuous on $[0, 2]$, and takes on its absolute maximum at $x = 0$ and its absolute minimum at $x = 2$.

7. Sketch the graph of a continuous function on $[0, \infty)$ with $f(0) = 0$ and $\lim_{x\to\infty} f(x) = 1$.

8. Sketch the graph of a continuous function on $(-\infty, \infty)$ with $f(0) = 1$, $f(x) \geq 0$ for all $x \in \mathbf{R}$, and $\lim_{x\to\pm\infty} f(x) = 0$.

9. Show that the bracket function

$$f(x) = [x]$$

is continuous from the right but discontinuous from the left at $x = -2$.

10. Suppose $f(x)$ is continuous on the interval $[1, 3]$. If $f(1) = 0$ and $f(3) = 2$, explain why there must be a number $c \in (1, 3)$ such that $f(c) = 1$.

11. Assume that the size of a population at time t is

$$N(t) = \frac{at}{k + t}, \quad t \geq 0$$

where a and k are positive constants. Suppose that the limiting population size is

$$\lim_{t\to\infty} N(t) = 1.24 \times 10^6$$

and that at time $t = 5$, the population size is half the limiting population size. Use the above information to determine the constants a and k.

12. Suppose that

$$N(t) = 10 + 2e^{-0.3t} \sin t, \quad t \geq 0$$

describes the size of a population (in millions) at time t (measured in weeks).

(a) Use a graphing calculator to sketch the graph of $N(t)$, and describe in words what you see.

(b) Give lower and upper bounds on the size of the population; that is, find N_1 and N_2 so that, for all $t \geq 0$,

$$N_1 \leq N(t) \leq N_2$$

(c) Find $\lim_{t\to\infty} N(t)$. Interpret this expression.

13. Suppose that an organism reacts to a stimulus only when the stimulus exceeds a certain threshold. Assume that the stimulus is a function of time t, and that it is given by

$$s(t) = \sin(\pi t), \quad t \geq 0$$

The organism reacts to the stimulus and shows a certain reaction when $s(t) \geq 1/2$. Define a function $g(t)$ so that $g(t) = 0$ when the organism shows no reaction at time t, and $g(t) = 1$ when the organism shows the reaction.

(a) Plot $s(t)$ and $g(t)$ in the same coordinate system.

(b) Is $s(t)$ continuous? Is $g(t)$ continuous?

14. The following function describes the height of a tree as a function of age:

$$f(x) = 132e^{-20/x}, \quad x \geq 0$$

Find $\lim_{x\to\infty} f(x)$.

15. There are a number of mathematical models that describe predator-prey interactions. Typically, they share the feature that the number of prey eaten per predator increases with the prey density. In the simplest version, the number of encounters with prey per predator is proportional to the product of the total number of prey and the time period over which the predators search for prey. That is, if we let N be the number of prey, P be the number of predators, T be the time period available for searching, and N_e be the number of prey encounters, then

$$\frac{N_e}{P} = aTN \tag{2.4}$$

The quantity N_e/F is the number of prey encountered per predator.

(a) Set $f(N) = aTN$, and sketch the graph of $f(N)$ when $a = 0.1$ and $T = 2$ for $N \geq 0$.

(b) Predators usually spend some time eating the prey that they find. Therefore, not all of the time T can be used for searching. The actual searching time is reduced by the per-prey handling time T_h, and can be written as

$$T - T_h \frac{N_e}{P}$$

Show that if $T - T_h \frac{N_e}{P}$ is substituted for T in (2.4), then

$$\frac{N_e}{P} = \frac{aTN}{1 - aT_hN} \tag{2.5}$$

Define

$$g(N) = \frac{aTN}{1 + aT_hN}$$

and graph $g(N)$ for $N \geq 0$ when $a = 0.1$, $T = 2$, and $T_h = 0.1$.

(c) Show that (2.5) reduces to (2.4) when $T_h = 0$.

(d) Find

$$\lim_{N \to \infty} \frac{N_e}{P}$$

in the cases when $T_h = 0$ and when $T_h > 0$. Explain the difference between the two cases in words.

16. Duarte and Agustí (1998) investigated the CO_2 balance of aquatic ecosystems. They related the community respiration rates (R) to the gross primary production rates (P) of aquatic ecosystems. (Both quantities were measured as g of O_2 m^{-3}day^{-1}.) They make the following statement:

Our results confirm the generality of earlier reports that the relation between community respiration rate and gross production is not linear. Community respiration is scaled as the approximate two-thirds power of gross production ....

(a) Use the above quote to explain why

$$R = aP^b$$

can be used to describe the relationship between the community respiration rates (R) and the gross primary production rates (P). What value would you assign to b based on their quote?

(b) Suppose that you obtained data for gross production and respiration rates of a number of freshwater lakes. How would you display your data graphically to quickly convince an audience that the exponent b in the power equation relating R and P is indeed approximately 2/3? (*Hint*: Use an appropriate log transformation.)

(c) Knowing the ratio R/P for an ecosystem is of importance in assessing its role in the global CO_2 budget. If respiration exceeds production (i.e., $R > P$), then the ecosystem acts as a carbon dioxide source, whereas if production exceeds respiration (i.e., $P > R$), then the ecosystem acts as a carbon dioxide sink. Assume now that the exponent in the power equation relating R and P is 2/3. Show that the ratio R/P as a function of P is continuous for $P > 0$. Furthermore, show that

$$\lim_{P \to 0+} \frac{R}{P} = \infty$$

and

$$\lim_{P \to \infty} \frac{R}{P} = 0$$

Use a graphing calculator to sketch the graph of the ratio R/P as a function of P for $P > 0$. (Experiment with the graphing calculator to see how the value of a affects the slope of the graph.)

(d) Use your results in (c) and the intermediate value theorem to conclude that there exists a value $P^\star$ so that the ratio R/P at $P^\star$ is equal to 1. Based on your graph in (c), is there more than one such value $P^\star$?

(e) Use your results in (d) to identify production rates P where the ratio $R/P > 1$, that is, where respiration exceeds production.

(f) Use your results in (a)–(e) to explain the following quote from Duarte and Agustí:

Unproductive aquatic ecosystems ... tend to be heterotrophic ($R > P$), and act as carbon dioxide sources.

17. Hyperbolic functions are used in the sciences. We take a look at the following three examples: the hyperbolic sine, $\sinh x$; the hyperbolic cosine, $\cosh x$; and the hyperbolic tangent, $\tanh x$. They are defined as

$$\sinh x = \frac{e^x - e^{-x}}{2}, \quad x \in \mathbf{R}$$

$$\cosh x = \frac{e^x + e^{-x}}{2}, \quad x \in \mathbf{R}$$

$$\tanh x = \frac{e^x - e^{-x}}{e^x + e^{-x}}, \quad x \in \mathbf{R}$$

(a) Show that these three hyperbolic functions are continuous for all $x \in \mathbf{R}$. Use a graphing calculator to sketch the graphs of all three functions.

(b) Find

$$\lim_{x \to \infty} \sinh x$$
$$\lim_{x \to -\infty} \sinh x$$
$$\lim_{x \to \infty} \cosh x$$
$$\lim_{x \to -\infty} \cosh x$$
$$\lim_{x \to \infty} \tanh x$$
$$\lim_{x \to -\infty} \tanh x$$

(c) Show that the following two identities,

$$\cosh^2 x - \sinh^2 x = 1$$

and

$$\tanh x = \frac{\sinh x}{\cosh x}$$

are valid.

(d) Show that $\sinh x$ and $\tanh x$ are odd functions, and that $\cosh x$ is even.

(*Note*: It can be shown that if a flexible cable is suspended between two points at equal heights, the shape of the resulting curve is given by the hyperbolic cosine function. This curve is called a catenary.)

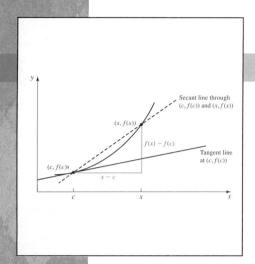

C H A P T E R 3

Differentiation

Differential calculus allows us to solve two of the basic problems that we mentioned in Chapter 1, namely, constructing a tangent line to a curve (see Figure 3.1) and finding maxima and minima of a curve (see Figure 3.2). The solutions to these two problems, by themselves, cannot explain the impact calculus has had on the sciences. In fact, calculus is one of the most important analytical tools for the investigation of dynamical problems in the life sciences. Applications of differential calculus in the life sciences include simple growth models, interactions between organisms,

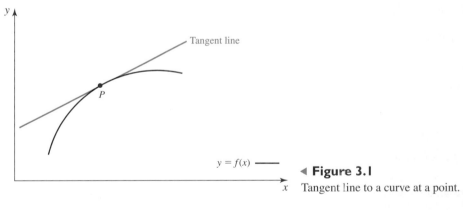

◄ **Figure 3.1**
Tangent line to a curve at a point.

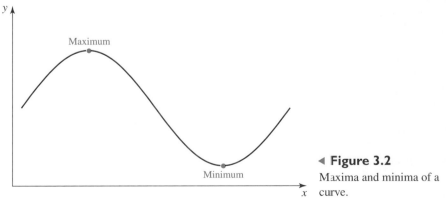

◄ **Figure 3.2**
Maxima and minima of a curve.

invasions of organisms, the working of neurons, enzymatic reactions, harvesting models in fishery, epidemiological modeling, change of gene frequencies under random mating, evolutionary strategies, and many others.

Growth models will be of particular interest to us. Consider a population whose size at time t is given by

$$N(t) = 200\frac{t}{1+t}, \quad t \geq 0$$

We will compute the average growth rate during the time interval $[t, t + h]$. This is given by

$$[\text{average growth rate}] = \frac{[\text{change in population size}]}{[\text{length of time interval}]}$$

The change in population size during the time interval $[t, t + h]$ is

$$\Delta N = N(t + h) - N(t)$$

where the symbol Δ (Δ is the capital Greek letter Delta) indicates that we take differences. The length of the interval $[t, t + h]$ is

$$\Delta t = t + h - t = h$$

The average growth rate is then $\frac{\Delta N}{\Delta t}$. Hence,

$$\frac{\Delta N}{\Delta t} = \frac{N(t + h) - N(t)}{h}$$

We are interested in the geometric interpretation of this quantity. When we draw a straight line through the points $(t, N(t))$ and $(t + h, N(t + h))$, we obtain the **secant line**. The slope of this line is given by the quantity $\Delta N/\Delta t$ (see Figure 3.3).

Instead of the average growth rate $\Delta N/\Delta t$, which tells us something about how the population size changes during a time interval of length Δt, we can ask about the **instantaneous growth rate** that is obtained when the length of the interval $[t, t + h]$ tends to 0. That is, when we compute the instantaneous rate of growth, we take the limit $h \to 0$ in the expression for $\Delta N/\Delta t$. Geometrically, we see that in the limit the secant line converges to the **tangent line** at the point $(t, N(t))$ (see Figure 3.4). The limit of $\Delta N/\Delta t$ as the length of the time interval $[t, t + h]$ goes to 0 (i.e., $\Delta t \to 0$ or $h \to 0$) will therefore be equal to the slope of the tangent line at $(t, N(t))$. We denote the limiting value of $\Delta N/\Delta t$, as $\Delta t \to 0$, by $N'(t)$ (read:

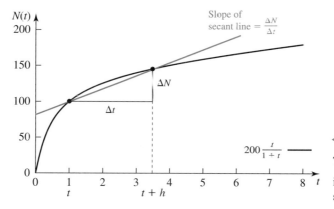

◄ **Figure 3.3**

The average growth rate $\dfrac{\Delta N}{\Delta t}$ is equal to the slope of the secant line.

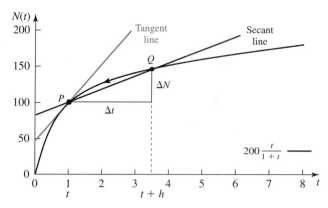

◀ **Figure 3.4**

The instantaneous growth rate is the limit of $\Delta N / \Delta t$ as $\Delta t \to 0$. Geometrically, the point Q moves towards the point P on the graph of $N(t)$, and the secant line through P and Q becomes the tangent line at P. The instantaneous growth rate is then equal to the slope of the tangent line.

N prime of t) and call this quantity the **derivative** of $N(t)$. That is,

$$N'(t) = \lim_{\Delta t \to 0} \frac{\Delta N}{\Delta t} = \lim_{h \to 0} \frac{N(t+h) - N(t)}{h}$$

◆ 3.1 FORMAL DEFINITION OF THE DERIVATIVE

> **Definition**   The derivative of a function f at x, denoted by $f'(x)$, is
> $$f'(x) = \lim_{h \to 0} \frac{f(x+h) - f(x)}{h}$$
> provided that the limit exists.

If the limit exists, then we say f is differentiable at x. In our definition, the phrase "provided the limit exists" is crucial; if we take an arbitrary function f, we cannot expect that the limit exists. In fact, we saw many examples in the previous chapter where limits did not exist. The geometric interpretation will help us to understand when the limit exists, and under which conditions we cannot expect the limit to exist. Notice in the definition that $\lim_{h \to 0}$ is a two-sided limit (that is, we approach 0 from both the negative and the positive side). The quotient

$$\frac{f(x+h) - f(x)}{h}$$

is called the **difference quotient**, and we denote it by $\frac{\Delta f}{\Delta x}$ (see Figure 3.5).

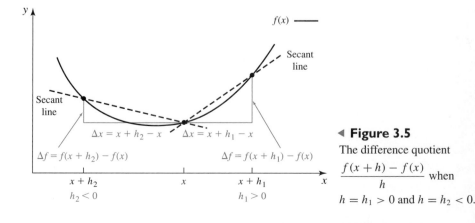

◀ **Figure 3.5**

The difference quotient
$$\frac{f(x+h) - f(x)}{h} \quad \text{when}$$
$h = h_1 > 0$ and $h = h_2 < 0$.

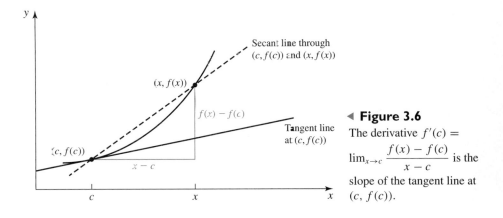

◀ **Figure 3.6**
The derivative $f'(c) =$ $\lim_{x \to c} \dfrac{f(x) - f(c)}{x - c}$ is the slope of the tangent line at $(c, f(c))$.

We say that f is differentiable on $[a, b]$ if it is differentiable at every $x \in (a, b)$. (Since the limit in the definition is two-sided, we cannot compute the derivative at an endpoint of an interval.)

If we want to compute the derivative at $x = c$, we can also write

$$f'(c) = \lim_{x \to c} \frac{f(x) - f(c)}{x - c}$$

which emphasizes that the point $(x, f(x))$ converges to the point $(c, f(c))$ as we take the limit $x \to c$ (see Figure 3.6). This will be important when we discuss the geometric interpretation of the derivative in the next subsection.

There is more than one way to write the derivative of a function $y = f(x)$. The following expressions are equivalent:

$$y' = \frac{dy}{dx} = f'(x) = \frac{df}{dx} = \frac{d}{dx} f(x)$$

The notation $\frac{df}{dx}$ goes back to Leibniz and is called **Leibniz notation** (Leibniz had a real talent for finding good notation). To compute the derivative, we have to compute the limit of the difference quotient

$$\frac{f(x + h) - f(x)}{h}$$

as $h \to 0$. If Δx denotes the difference between $x + h$ and x, and $\Delta y = \Delta f$ denotes the difference between $f(x + h)$ and $f(x)$, then taking the limit $h \to 0$ means that the difference between $x + h$ and x tends to 0, and we can instead write $\Delta x \to 0$. We then get

$$\frac{df}{dx} = \lim_{\Delta x \to 0} \frac{\Delta f}{\Delta x} \qquad \text{or} \qquad \frac{dy}{dx} = \lim_{\Delta x \to 0} \frac{\Delta y}{\Delta x}$$

If we wish to emphasize that we evaluate the derivative of $f(x)$ at $x = c$, we write

$$f'(c) = \frac{df}{dx}\bigg|_{x=c}$$

Newton used different notation to denote the derivative of a function. His notation is still used when we compute the derivative of a function that depends on time, that is, a function of the form $y = f(t)$ where t denotes time. If $y = f(t)$, then Newton wrote $\dot{y}$ (read: y dot) for the derivative of $f(t)$. This notation is frequently used in physics. We will use either Leibniz notation or $f'(x)$.

3.1.1 Geometric Interpretation and Using the Definition

We will look at a simple function, namely $f(x) = x^2$, $x \in \mathbf{R}$ (refer to Figures 3.7 and 3.8 as we go along). We wish to compute the derivative of f at $x = 1$, that is, $f'(1)$. We will first compute the difference quotient at $x = 1$. For $h \neq 0$,

$$\frac{\Delta f}{\Delta x} = \frac{f(1 + h) - f(1)}{h} = \frac{(1 + h)^2 - 1^2}{h}$$
$$= \frac{1 + 2h + h^2 - 1}{h} = \frac{2h + h^2}{h} = \frac{h(2 + h)}{h}$$
$$= 2 + h$$

Looking at Figure 3.7, we see that $\Delta f / \Delta x$ is the slope of the secant line through the points $(1, 1)$ and $(1 + h, (1 + h)^2)$.

To compute the derivative $f'(1)$, we need to take the limit $h \to 0$ (see Figure 3.8):

$$f'(1) = \lim_{h \to 0} \frac{f(1 + h) - f(1)}{h} = \lim_{h \to 0}(2 + h) = 2$$

Taking the limit $h \to 0$ means that the point $(1 + h, (1 + h)^2)$ approaches the point $(1, 1)$. The secant line thus converges to the tangent line at $(1, 1)$. Since $f'(1)$ is the limiting value of the slope of the secant line as the point $(1 + h, (1 + h)^2)$ approaches $(1, 1)$, we find that $f'(1)$ is the slope of the tangent line at the point $(1, 1)$.

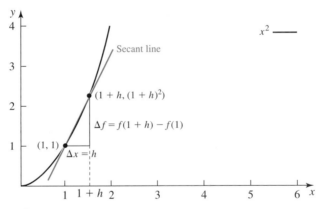

◄ **Figure 3.7**

The slope of the secant line through $(1, 1)$ and $(1 + h, (1 + h)^2)$ is $\Delta f / \Delta x$.

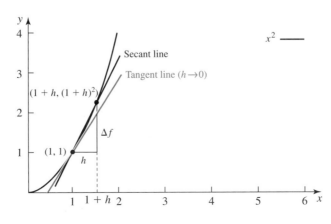

◄ **Figure 3.8**

Taking the limit $h \to 0$, the secant line converges to the tangent line at $(1, 1)$.

$$x^2 \ \underline{\hspace{1.2cm}}$$
$$2x - 1 \ \underline{\hspace{1.2cm}}$$

Tangent line

(1, 1)

◀ **Figure 3.9**
The slope of $f(x) = x^2$ at
$(1,1)$ is $m = 2$. The equation
of the tangent line at $(1, 1)$ is
$y = 2x - 1$.

We can give the equation of the tangent line, since we know the slope and one point on the line. Using the point-slope form

$$y - y_0 = m(x - x_0)$$

where (x_0, y_0) is the point and m is the slope, we find with $(x_0, y_0) = (1, 1)$ and $m = 2$ that the equation of the line is given by

$$y - 1 = 2(x - 1) \qquad \text{or} \qquad y = 2x - 1$$

(see Figure 3.9).

> **Equation of the Tangent Line**   If the derivative of a given function f exists at $x = c$, then $f'(c)$ is the slope of the tangent line at the point $(c, f(c))$. The equation of the tangent line is given by
>
> $$y - f(c) = f'(c)(x - c)$$

Using the geometric interpretation, it is easy to compute the derivatives of linear functions.

▌**EXAMPLE 1**   **The Derivative of a Constant Function**

Let $f(x) = a$ (see Figure 3.10). The graph of $f(x)$ is a horizontal line that intersects the y-axis at $(0, a)$. The slope of a horizontal line is 0; we therefore expect $f'(x) = 0$. We can use the formal definition to arrive at this answer. (Note that $f(x) = a$ and $f(x + h) = a$.)

$$f'(x) = \lim_{h \to 0} \frac{f(x + h) - f(x)}{h} = \lim_{h \to 0} \frac{a - a}{h} = \lim_{h \to 0} \frac{0}{h} = \lim_{h \to 0} 0 = 0$$

Here again, it is important to remember that when we take the limit $h \to 0$, h approaches 0 (from both sides) but is not equal to 0. Since $h \neq 0$, the expression $0/h = 0$. This was used in the second-to-last step.   ▌

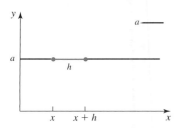

▲ **Figure 3.10**
The slope of a horizontal line is $m = 0$.

▌**EXAMPLE 2**   **The Derivative of a Linear Function**

Let $f(x) = mx + b$ (see Figure 3.11). Then the graph of $f(x)$ is a straight line with slope m and y-intercept b. The derivative of $f(x)$, denoted by $f'(x)$, is the slope of the tangent line at x. Since the graph is a straight line, the tangent line at x coincides with the graph of $f(x)$ and, therefore, the slope of the tangent line at x is equal to

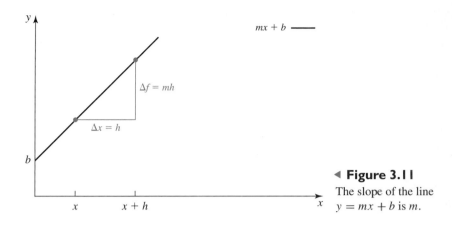

◀ Figure 3.11
The slope of the line
$y = mx + b$ is m.

the slope of the straight line, that is, $f'(x) = m$. We use the formal definition to prove this.

$$f'(x) = \lim_{h \to 0} \frac{f(x+h) - f(x)}{h} = \lim_{h \to 0} \frac{m(x+h) + b - (mx + b)}{h}$$

$$= \lim_{h \to 0} \frac{mx + mh + b - mx - b}{h} = \lim_{h \to 0} \frac{mh}{h} = \lim_{h \to 0} m = m$$

In the second-to-last step we were able to cancel h since $h \neq 0$.

We found that if $f(x) = mx + b$, then $f'(x) = m$. This includes the special case of a constant function where $m = 0$, which we considered in Example 1. ▮

▮ **EXAMPLE 3**   **Using the Definition**

Find the derivative of

$$f(x) = \frac{1}{x} \quad \text{for } x \neq 0$$

Solution   We will use the formal definition of the derivative to compute $f'(x)$ (see Figure 3.12). The main algebraic step is the computation of $f(x+h) - f(x)$; we will do this first. We observe that $f(x+h) = \frac{1}{x+h}$. Then

$$f(x+h) - f(x) = \frac{1}{x+h} - \frac{1}{x}$$

$$= \frac{x - (x+h)}{x(x+h)} = \frac{-h}{x(x+h)}$$

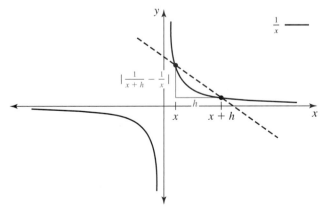

◀ Figure 3.12
The graph of $f(x) = 1/x$ for Example 3.

Using the formal definition of the derivative, we then find

$$\lim_{h \to 0} \frac{f(x+h) - f(x)}{h} = \lim_{h \to 0} \frac{\frac{-h}{x(x+h)}}{h}$$

$$= \lim_{h \to 0} \left(-\frac{h}{x(x+h)} \frac{1}{h} \right)$$

$$= \lim_{h \to 0} \left(-\frac{1}{x(x+h)} \right) = -\frac{1}{x^2}$$

That is, if $f(x) = \frac{1}{x}$, then

$$f'(x) = -\frac{1}{x^2}, x \neq 0$$ ∎

We make two observations: (1) To compute $f'(x)$, we must evaluate the limit

$$\lim_{h \to 0} \frac{f(x+h) - f(x)}{h}$$

Since both $\lim_{h \to 0}[f(x+h) - f(x)]$ and $\lim_{h \to 0} h$ are equal to 0, we cannot simply evaluate the limits in the numerator and the denominator separately, which would result in the undefined expression 0/0. We must first simplify the difference quotient before we can take the limit. (2) We will frequently consider $f'(x)$ itself as a function of x. For instance, in Example 3, $f'(x) = -1/x^2, x \neq 0$, is a function of x.

3.1.2 The Derivative as an Instantaneous Rate of Change – A First Look at Differential Equations

Velocity   Suppose that you ride your bike on a straight road. Your position (in miles) at time t (in hours) is given by

$$s(t) = -t^3 + 6t^2 \quad \text{for } 0 \leq t \leq 6$$

(see Figure 3.13).

You might ask what your average velocity during the time interval (say) [2, 4] is, which is defined as the net change in position during the interval divided by the length of the interval. That is, you find your position at time 2 and at time 4, and take the difference of these two quantities. You then divide this difference by the time that it took you to travel this distance. At time $t = 2$, $s(2) = -8 + 24 = 16$, and at

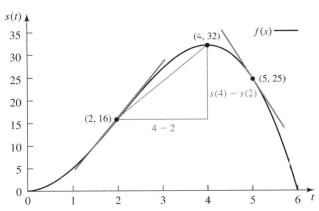

◄ **Figure 3.13**
The average velocity
$$\frac{s(4) - s(2)}{4 - 2}$$ is the slope of
the secant line through (2, 16) and (4, 32). The velocity at time t is slope of the tangent line at t: at $t = 2$, the velocity is positive; at $t = 5$, the velocity is negative.

time $t = 4$, $s(4) = -64 + 96 = 32$. Hence, the average velocity is

$$\frac{s(4) - s(2)}{4 - 0} = \frac{32 - 16}{4 - 2} = 8 \text{ mph}$$

We recognize this ratio as the difference quotient

$$\frac{\Delta s}{\Delta t} = \frac{s(t + h) - s(t)}{h}$$

Hence, we can interpret the difference quotient $\Delta s / \Delta t$ as an **average rate of change**.

When you are on your bike, you might be interested in your velocity at a particular instant. To find your velocity at a particular instant mathematically, you would need to find the ratio between the displacement over a very small time interval and the length of the interval; that is, you would need to find

$$\lim_{\Delta t \to 0} \frac{\Delta s}{\Delta t} = \lim_{h \to 0} \frac{s(t + h) - s(t)}{h}$$

which would give you the velocity at time t. You recognize this as the derivative of $s(t)$ at time t. The quantity

$$\frac{ds}{dt} = \lim_{\Delta t \to 0} \frac{\Delta s}{\Delta t}$$

is the **instantaneous rate of change** or **velocity**. The velocity is thus the slope of the tangent line at a given point of the position function.

Let's look at two points on the graph of $s(t)$, namely, $(2, 16)$ and $(5, 25)$. We find that the slope of the tangent line is positive at $(2, 16)$ and negative at $(5, 25)$. The velocity is therefore positive at time $t = 2$, and negative at time $t = 5$. At $t = 2$, we move away from our starting point, whereas at $t = 5$, we move towards our starting point.

There is a difference between *velocity* and *speed*. If you had a speedometer on your bike, it would tell you the speed and not the velocity. Speed is the absolute value of velocity.

Interpreting the derivative as an instantaneous rate of change will turn out to be extremely important to us. In fact, when you encounter derivatives in your science courses, this will be the interpretation most often used. This interpretation will allow us to describe a quantity in terms of how quickly it changes with respect to another quantity. To illustrate this point, we revisit two previous examples and introduce one new application.

Population Growth   At the beginning of this chapter, we considered the growth of a population at time t. We found that the instantaneous growth of the population was given by dN/dt. That is,

$$\frac{dN}{dt} = [\text{instantaneous population growth rate}]$$

We are frequently interested in the **instantaneous per capita growth rate**. This is the growth rate per individual, and it can be obtained by dividing the instantaneous growth rate of the population by the population size. That is,

$$\frac{1}{N}\frac{dN}{dt} = [\text{instantaneous per capita growth rate}]$$

Note that N is a function of t, that is, $N = N(t)$. We did not explicitly state this dependence in the formula for the per capita growth rate, since it should be clear from the context that N is a function of t. If we want to be explicit, we must write

$$\frac{1}{N(t)}\frac{dN(t)}{dt} \qquad \text{instead of} \qquad \frac{1}{N}\frac{dN}{dt}$$

Equations are typically more readable if we do not write the dependence on t explicitly.

The Rate of a Chemical Reaction   Another illustration of the use of the derivative as an instantaneous rate of change is Example 4 of Subsection 1.2.2, where we discussed the speed of the irreversible chemical reaction

$$A + B \rightarrow AB$$

and found that the reaction rate $R = k[A][B]$. Denoting the concentration of the product AB by x allowed us to write R as a function of x, namely,

$$R(x) = k(a - x)(b - x)$$

where $a = [A]$ was the initial concentration of A, and $b = [B]$ was the initial concentration of B (see Figure 3.14). The reaction rate $R(x)$ tells us how quickly the concentration of x changes with time as the reaction proceeds. That is, we consider x as a function of time t, $x = x(t)$. The reaction rate $R(x)$ is the instantaneous rate of change, that is, $R(x)$ is the change in the concentration x over very small (infinitesimal) time intervals,

$$R(x) = \lim_{\Delta t \to 0} \frac{x(t + \Delta t) - x(t)}{\Delta t}$$

We can identify the limit $\Delta t \to 0$ as the derivative of the function $x(t)$ with respect to t. That is, $R(x) = \frac{dx}{dt}$. We can therefore write

$$\frac{dx}{dt} = k(a - x)(b - x) \tag{3.1}$$

Equation (3.1) is an example of a **differential equation**, that is, an equation that contains the derivative of a function. We will discuss such equations extensively in later chapters.

From this point on, when we say "rate of change" we will always mean "instantaneous rate of change." When we are interested in the average rate of change, we will always state this explicitly.

The rate of change in the chemical reaction is described by a differential equation. It is sometimes possible to solve such differential equations, that is, to explicitly give a function whose derivative satisfies the given equation. We will discuss this in detail later. More often, it is not possible (or not necessary) to explicitly find a solution. Without solving it, one can often obtain useful information about the differential equation itself. We illustrate this in the next application.

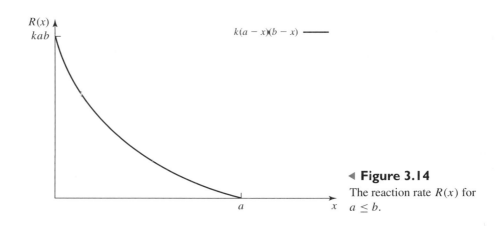

◀ **Figure 3.14**

The reaction rate $R(x)$ for $a \leq b$.

Tilman's Model for Resource Competition   Dr. David Tilman of the University of Minnesota developed a theoretical framework to describe the outcome of competition for limiting resources (Tilman 1982). This applies, for instance, to the grassland habitat at Cedar Creek Natural History Area, Minnesota, where Tilman conducted many experiments to test the predictions of his theory.

For this grassland habitat, nitrogen is a limiting resource; that is, adding nitrogen to the soil will result in an increase in biomass. We will discuss the case where one species competes for a single limiting resource. We assume that the rate of growth depends on the concentration of a single resource. In addition, we assume that plants experience mortality that results in loss of biomass. We thus arrive at the following equation:

$$[\text{rate of biomass change}] = [\text{rate of growth}] - [\text{rate of loss}]$$

We denote the biomass of the plant population at time t by $B(t)$. The **specific rate of change** of biomass is defined as the change of biomass per unit of biomass, that is, $\frac{1}{B}\frac{dB}{dt}$. We denote the resource level by R and the per unit biomass rate of growth (i.e., the specific rate of growth) by $f(R)$. We assume that the per unit biomass rate of loss is constant and denote this quantity by m. We thus obtain

$$\frac{1}{B}\frac{dB}{dt} = f(R) - m \tag{3.2}$$

The function $f(R)$ must be specified. One choice would be the Monod growth function (or Michaelis-Menten equation) that we considered in Example 5 of Subsection 1.2.3. That is,

$$f(R) = a\frac{R}{k + R} \tag{3.3}$$

where a and k are positive constants. We will not solve (3.2); instead, we will use a graphical approach. We graph both $f(R)$ and m together in Figure 3.15. This yields the following observations. When $m < a$, the graphs of the functions $y = f(R)$ and $y = m$ intersect at $R = R^\star$ (read: "R star"). At this resource level, $f(R) = m$, and thus the specific rate of change $\frac{1}{B}\frac{dB}{dt}$ is equal to 0. That is, growth balances loss, and the biomass of the species no longer changes. We say that the biomass is at **equilibrium**. If the resource level R was held at a value less than $R^\star$, $f(R) - m < 0$, the specific rate of growth would be negative, that is, biomass would decrease. If $R > R^\star$, $f(R) - m > 0$, and biomass would increase.

We can compute $R^\star$ in the case when $f(R)$ is given by (3.3), since $R^\star$ satisfies

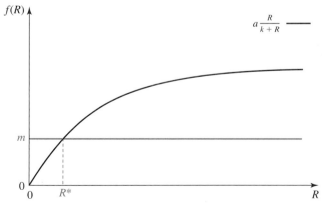

◀ **Figure 3.15**
Growth balances loss when $R = R^\star$.

$f(R) = m$. We find

$$a\frac{R^\star}{k + R^\star} = m \quad \text{or} \quad R^\star = \frac{mk}{a - m}$$

3.1.3 Differentiability and Continuity

Using the geometric interpretation, it is easy to find situations in which $f'(c)$ does not exist.

EXAMPLE 4   **A Function with a "Corner"**

Let

$$f(x) = |x| = \begin{cases} x & \text{for} \quad x \geq 0 \\ -x & \text{for} \quad x < 0 \end{cases}$$

A graph of $f(x)$ is shown in Figure 3.16. It should be clear from the graph that there is no tangent line at $x = 0$. We can define the slope of the secant line when we approach 0 from the right and also when we approach 0 from the left; however, the slopes converge to different values in the limit. The former is $+1$, the latter is -1, as is shown in the following calculation.

$$\lim_{h \to 0+} \frac{f(0 + h) - f(0)}{h} = \frac{h - 0}{h} = 1$$

$$\lim_{h \to 0-} \frac{f(0 + h) - f(0)}{h} = \frac{-h - 0}{h} = -1$$

In the calculation, we used the fact that when $h \to 0+$, $f(0+h) = |0+h| = |h| = h$ since $h > 0$, whereas when $h \to h-$, $f(0+h) = |0+h| = |h| = -h$ since $h < 0$. Since $1 \neq -1$, it follows that

$$\lim_{h \to 0} \frac{f(0 + h) - f(0)}{h}$$

does not exist. Thus, $f'(0)$ does not exist.

At all other points, the derivative exists. We can find the derivative by simply looking at the graph. We see that

$$f'(x) = \begin{cases} +1 & \text{for} \quad x > 0 \\ -1 & \text{for} \quad x < 0 \end{cases}$$

Note that we can consider $f'(x)$ itself as a function of x. ▮

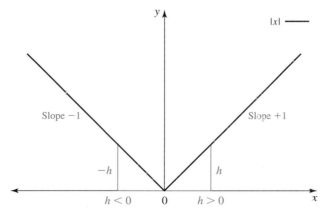

Slope -1 Slope $+1$

$-h$ h

$h < 0$ 0 $h > 0$

$|x|$ ———

◀ Figure 3.16

f is not differentiable at $x = 0$. Since the derivative from the right and from the left are not equal.

We saw in Example 4 that continuity alone is not enough for a function to be differentiable. The function $f(x) = |x|$ is continuous at all values of x but we saw that it was not differentiable at $x = 0$. The example also shows that it is quite easy to draw the graph of a continuous function that is not differentiable at certain points: just put in "corners" at those points where you do not want the function to be differentiable (see Figure 3.17).

However, if a function is differentiable, it is also continuous. This means that continuity is a necessary but not a sufficient condition for differentiability. This result is important enough that we will formulate it as a theorem and prove it.

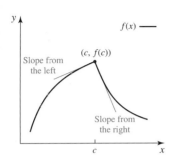

▲ Figure 3.17

$f(x)$ is continuous at $x = c$ but not differentiable at $x = c$: The derivatives from the left and the right are not equal.

Theorem If f is differentiable at $x = c$, it is also continuous at $x = c$.

Proof Since f is differentiable at $x = c$, we know that the limit

$$\lim_{x \to c} \frac{f(x) - f(c)}{x - c} \qquad (3.4)$$

exists and that it is equal to $f'(c)$. To show that f is continuous at $x = c$, we must show that

$$\lim_{x \to c} f(x) = f(c) \qquad \text{or} \qquad \lim_{x \to c}[f(x) - f(c)] = 0$$

First, note that f is defined at $x = c$. (Otherwise, we could not have computed the difference quotient $\frac{f(x) - f(c)}{x - c}$.) Now,

$$\lim_{x \to c}[f(x) - f(c)] = \lim_{x \to c} \frac{f(x) - f(c)}{x - c}(x - c)$$

Given that

$$\lim_{x \to c} \frac{f(x) - f(c)}{x - c}$$

exists and is equal to $f'(c)$ (this is equation (3.4)), and that

$$\lim_{x \to c}(x - c)$$

exists (it is equal to 0), we can apply the product rule for limits and find

$$\lim_{x \to c} \frac{f(x) - f(c)}{x - c}(x - c) = \lim_{x \to c} \frac{f(x) - f(c)}{x - c} \lim_{x \to c}(x - c) = f'(c) \cdot 0 = 0$$

That is,

$$\lim_{x \to c}[f(x) - f(c)] = 0$$

hence, f is continuous at $x = c$. ◆

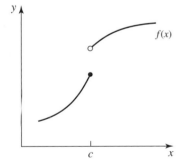

▲ Figure 3.18

The function $y = f(x)$ is not differentiable at $x = c$.

It follows from the theorem that if a function f is not continuous at $x = c$, then f is not differentiable at $x = c$. The function $y = f(x)$ in Figure 3.18 is discontinuous at $x = c$; we cannot draw a tangent line at $x = c$.

Finally, if a function has a vertical tangent at a point, it is not differentiable at that point.

▌EXAMPLE 5 **Vertical Tangent Line**

Show that

$$f(x) = x^{1/3}$$

is not differentiable at $x = 0$.

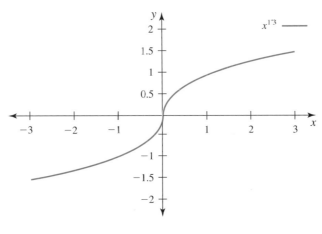

◀ **Figure 3.19**
The function $f(x) = x^{1/3}$
has a vertical tangent line at
$x = 0$. It is therefore not
differentiable at $x = 0$.

Solution   We see from the graph of $f(x)$ in Figure 3.19 that $f(x)$ is continuous
at $x = 0$. Using the formal definition, we find

$$f'(0) = \lim_{h \to 0} \frac{f(h) - f(0)}{h} = \lim_{h \to 0} \frac{h^{1/3} - 0}{h}$$

$$= \lim_{h \to 0} \frac{1}{h^{2/3}} = \infty \qquad \text{does not exist}$$

Since the limit does not exist, $f(x)$ is not differentiable at $x = 0$. We see from the
graph that the tangent line at $x = 0$ is vertical.   ▮

◆ 3.1 PROBLEMS

[3.1.1]

*In Problems 1–8, find the derivative at the indicated point from
the graph of each function*

1. $f(x) = 2$; $x = 1$
2. $f(x) = -3x$; $x = 2$
3. $f(x) = 2x - 5$; $x = -1$
4. $f(x) = -5x + 1$; $x = 0$
5. $f(x) = x^2$; $x = 0$
6. $f(x) = (x - 1)^2$; $x = 1$
7. $f(x) = \cos x$; $x = 0$
8. $f(x) = \sin x$; $x = \frac{\pi}{2}$

*In Problems 9–12, compute $f(c + h) - f(c)$ at the indicated
point.*

9. $f(x) = -2x + 1$; $c = 2$
10. $f(x) = 3x^2$; $c = 1$
11. $f(x) = \sqrt{x}$; $c = 4$
12. $f(x) = \frac{1}{x}$; $c = -2$

13. (a) Use the formal definition of the derivative to find the
derivative of $y = 5x^2$ at $x = -1$.
(b) Show that the point $(-1, 5)$ is on the graph of $y = 5x^2$,
and find the equation of the tangent line at the point $(-1, 5)$.
(c) Graph $y = 5x^2$ and the tangent line at the point $(-1, 5)$ in
the same coordinate system.
14. (a) Use the formal definition to find the derivative of
$y = -2x^2$ at $x = 1$.
(b) Show that the point $(1, -2)$ is on the graph of $y = -2x^2$,
and find the equation of the tangent line at the point $(1, -2)$.
(c) Graph $y = -2x^2$ and the tangent line at the point $(1, -2)$
in the same coordinate system.

15. (a) Use the formal definition to find the derivative of
$y = 1 - x^3$ at $x = 2$.
(b) Show that the point $(2, -7)$ is on the graph of $y = 1 - x^3$,
and find the equation of the normal line at the point $(2, -7)$.
(c) Graph $y = 1 - x^3$ and the tangent line at the point $(2, -7)$
in the same coordinate system.
16. (a) Use the formal definition to find the derivative of $y = \frac{1}{x}$
at $x = 2$.
(b) Show that the point $(2, \frac{1}{2})$ is on the graph of $y = \frac{1}{x}$, and
find the equation of the normal line at the point $(2, \frac{1}{2})$.
(c) Graph $y = \frac{1}{x}$ and the tangent line at the point $(2, \frac{1}{2})$ in the
same coordinate system.
17. Use the formal definition to find the derivative of

$$y = \sqrt{x}$$

for $x > 0$.
18. Use the formal definition to find the derivative of

$$f(x) = \frac{1}{x + 1}$$

for $x \neq -1$.
19. Find the equation of the tangent line to the curve $y = x^2 - 3x + 1$ at the point $(2, -1)$.
20. Find the equation of the normal line to the curve $y = \sqrt{x - 1}$ at the point $(5, 2)$.

21. The following limit represents the derivative of a function f at the point $(a, f(a))$. Find f and a.

$$\lim_{h \to 0} \frac{\frac{1}{(2+h)^2+1} - \frac{1}{5}}{h}$$

22. The following limit represents the derivative of a function f at the point $(a, f(a))$. Find f and a.

$$\lim_{h \to 0} \frac{\sin\left(\frac{\pi}{6} + h\right) - \sin\frac{\pi}{6}}{h}$$

[3.1.2]

23. If $s(t)$ denotes the position of a particle that moves along a straight line, then $\Delta s / \Delta t$ is the average rate of change of $s(t)$, called the average velocity, and $v(t) = ds/dt$ is the instantaneous rate of change of $s(t)$, called the (instantaneous) velocity. The speed of the particle is the absolute value of the velocity, $|v(t)|$.

Suppose that a car moves along a straight road. The location at time t is given by

$$s(t) = \frac{160}{3} t^2$$

where t is measured in hours.

(a) Where is the car at $t = 3/4$, and where is it at $t = 1$?

(b) Find the average velocity of the car between $t = 3/4$ and $t = 1$.

(c) Find the velocity and the speed of the car at $t = 3/4$.

24. Suppose a particle moves along a straight line. The position at time t is given by

$$s(t) = 3t - t^2, \quad t \ge 0$$

(a) Graph $s(t)$ for $t \ge 0$.

(b) Use the graph in (a) to answer the following questions. (i) Where is the particle at time 0? (ii) Is there another time at which the particle visits the location where it was at time 0? (iii) How far to the right on the straight line does the particle travel? (iv) How far to the left on the straight line does the particle travel? (v) Where is the velocity positive, where negative, and where equal to 0?

(c) Find the velocity of the particle.

(d) When is the velocity of the particle equal to 1?

25. In Example 4 of Subsection 3.1.2, we considered Tilman's resource model. Denote by $B(t)$ the biomass at time t, and assume that

$$\frac{1}{B}\frac{dB}{dt} = f(R) - m$$

where R denotes the resource level,

$$f(R) = 200\frac{R}{5 + R}$$

and $m = 40$. Use the graphical approach outlined in Example 4 to find the value $R^\star$ at which $\frac{1}{B}\frac{dB}{dt} = 0$. Then compute $R^\star$ by solving $\frac{1}{B}\frac{dB}{dt} = 0$.

26. Assume that $N(t)$ denotes the size of a population at time t, and that $N(t)$ satisfies the differential equation

$$\frac{dN}{dt} = rN$$

where r is a constant.

(a) Find the per capita growth rate.

(b) Assume that $r > 0$ and that $N(0) = 20$. Is the population size at time 1 greater than 20 or less than 20? Explain your answer.

27. Assume that $N(t)$ denotes the size of a population at time t, and that $N(t)$ satisfies the differential equation

$$\frac{dN}{dt} = 3N \left(1 - \frac{N}{20}\right)$$

Let $f(N) = 3N \left(1 - \frac{N}{20}\right)$ for $N \ge 0$. Graph $f(N)$ as a function of N and identify all equilibria, that is, all points where $\frac{dN}{dt} = 0$.

28. Consider the autocatalytic reaction

$$A + X \longrightarrow X$$

which was introduced in Problem 26 of Section 1.2. Find a differential equation that describes the rate of change of the concentration of the product X.

[3.1.3]

29. Explain the relationship between continuity and differentiability.

In Problems 30–34, graph each function and, based on the graph, guess where the function is not differentiable.

30. $y = |x - 2|$ **31.** $y = \dfrac{1}{1 + x}$

32. $y = |x^2 - 3|$ **33.** $f(x) = \begin{cases} x & \text{for } x \le 0 \\ x + 1 & \text{for } x > 0 \end{cases}$

34. $f(x) = \begin{cases} x^2 + 2 & \text{for } x \le 0 \\ e^{-x} & \text{for } x > 0 \end{cases}$

[3.1.4]

35. Suppose that the rate of change of the size of a population is given by

$$\frac{dN}{dt} = rN \left(1 - \frac{N}{K}\right)$$

where $N = N(t)$ denotes the size of the population at time t, and r and K are positive constants. Find the equilibrium size of the population, that is, the size where the rate of change is equal to 0. Use your answer to explain why K is called the carrying capacity.

36. (*Biotic Diversity*) (Adapted from Valentine 1985.) Walker and Valentine (1984) suggested a model for species diversity that assumes that species extinction rates are independent of diversity, but speciation rates are regulated by competition. They denoted by $N(t)$ the number of species at time t, by b the speciation rate, and by a the extinction rate, and used the following model

$$\frac{dN}{dt} = N\left[b\left(1 - \frac{N}{K}\right) - a\right]$$

where K denotes the number of "niches," or potential places for species in the ecosystem.

(a) Find possible equilibria under the condition $a < b$.

(b) Use your result in (a) to explain the following statement by Valentine (1985):

> In this situation, ecosystems are never "full," with all potential niches occupied by species so long as the extinction rate is above zero.

(c) What happens when $a \geq b$?

◆ 3.2 THE POWER RULE, THE BASIC RULES OF DIFFERENTIATION, AND THE DERIVATIVES OF POLYNOMIALS

In this section, we will begin a systematic treatment of the computation of derivatives. Knowing how to differentiate is fundamental to the understanding of the rest of the course. Although computer software is now available to compute derivatives of many functions (such as $y = cx^n$ or $y = e^{\sin x}$), it is nonetheless important that you master the techniques of differentiation.

3.2.1 The Power Rule

We begin with the simplest of the differentiation rules, namely the derivative of a function of the form $y = x^n$, where n is a positive integer. The power rule in Leibniz notation is as follows.

POWER RULE

Let n be a positive integer; then,

$$\frac{d}{dx}(x^n) = nx^{n-1}$$

We recall the rule for the constant function $f(x) = a$, which we showed in the previous subsection.

If $f(x)$ is the constant function, namely $f(x) = a$, then

$$\frac{d}{dx}f(x) = 0$$

Before we prove the power rule, consider again the example of $f(x) = x^2$ (see Figure 3.20). In Subsection 3.1.1, we computed the derivative of $y = x^2$ at $x = 1$. Instead of $x = 1$, we can compute the difference quotient $\frac{\Delta f}{\Delta x}$ at an arbitrary x,

$$\frac{\Delta f}{\Delta x} = \frac{f(x+h) - f(x)}{h} = \frac{(x+h)^2 - x^2}{h}$$

We use the expansion $(x + h)^2 = x^2 + 2xh + h^2$, and find

$$\frac{\Delta f}{\Delta x} = \frac{x^2 + 2xh + h^2 - x^2}{h} = \frac{2xh + h^2}{h} = 2x + h \qquad (3.5)$$

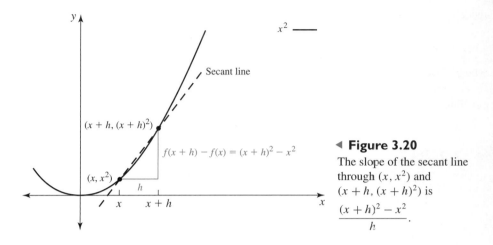

◀ Figure 3.20

The slope of the secant line through (x, x^2) and $(x + h, (x + h)^2)$ is

$$\frac{(x + h)^2 - x^2}{h}.$$

after canceling h in both the numerator and the denominator. To find the derivative, we need to let $h \to 0$.

$$f'(x) = \lim_{\Delta x \to 0} \frac{\Delta f}{\Delta x} = \lim_{h \to 0}(2x + h) = 2x$$

This proves the power rule for $n = 2$. To prove the power rule for other positive integers of n we need the expansion for $(x + h)^n$. This is provided by the binomial theorem.

Binomial Theorem   If n is a positive integer, then

$$(x + y)^n = x^n + nx^{n-1}y + \frac{n(n - 1)}{2 \cdot 1}x^{n-2}y^2$$

$$+ \frac{n(n - 1)(n - 2)}{3 \cdot 2 \cdot 1}x^{n-3}y^3$$

$$+ \cdots + \frac{n(n - 1) \cdots (n - k + 1)}{k(k - 1) \cdots 2 \cdot 1}x^{n-k}y^k$$

$$+ \cdots + nxy^{n-1} + y^n$$

That is, the expansion of $(x + y)^n$ is a sum of terms of the form

$$C_{n,k}x^{n-k}y^k, \quad k = 0, 1, \ldots, n$$

where $C_{n,k}$ is a coefficient that depends on n and k. The exact form of the coefficients $C_{n,k}$ will not be important in the proof of the power rule, except for the two terms $C_{n,0} = 1$ and $C_{n,1} = n$.

Proof of the Power Rule   We use the expansion provided by the binomial theorem to compute the difference in the numerator of the difference quotient.

$$\Delta f = f(x + h) - f(x) = (x + h)^n - x^n$$

$$= \left(C_{n,0}x^n + C_{n,1}x^{n-1}h + C_{n,2}x^{n-2}h^2 + C_{n,3}x^{n-3}h^3\right.$$

$$\left. + \cdots + C_{n,n-1}xh^{n-1} + C_{n,n}h^n\right) - x^n$$

Since $C_{n,0} = 1$, the terms x^n cancel. We can then factor h from the remaining terms, and find

$$f(x+h) - f(x) = h\left[C_{n,1}x^{n-1} + C_{n,2}x^{n-2}h + C_{n,3}x^{n-3}h^2 + \cdots + C_{n,n}h^{n-1}\right]$$

When we divide by h and let $h \to 0$, we see that

$$f'(x) = \lim_{h \to 0} \frac{f(x+h) - f(x)}{h}$$

$$= \lim_{h \to 0} \left[C_{n,1}x^{n-1} + C_{n,2}x^{n-2}h + C_{n,3}x^{n-3}h^2 + \cdots + C_{n,n}h^{n-1}\right]$$

All terms except for the first, have h as a factor and, thus, tend to 0 as $h \to 0$. (The first term does not depend on h.) We find that

$$f'(x) = C_{n,1}x^{a-1}$$

Recall that $C_{n,1} = n$; therefore,

$$f'(x) = nx^{n-1}$$

which proves the power rule.   ◆

▌EXAMPLE 1   We apply the power rule to various examples, and take the opportunity to practice the different notations.

  a. If $f(x) = x^6$, then $f'(x) = 6x^5$.

  b. If $f(x) = x^{300}$, then $f'(x) = 300x^{299}$.

  c. If $g(t) = t^5$, then $\dfrac{d}{dt}g(t) = 5t^4$.

  d. If $z = s^3$, then $\dfrac{dz}{ds} = 3s^2$.

  e. If $x = y^4$, then $\dfrac{dx}{dy} = 4y^3$.   ▐

Example 1 illustrates the importance of knowing the dependence among the variables (see Figure 3.21). If $y = f(x)$, we call x the **independent variable**, and y the **dependent variable** because it does depend on another variable (in this case, x). For instance, in (a), y is a function of x, thus, x is the independent and y is the dependent variable; whereas in (e), x is a function of y, thus, y is now the independent and x the dependent variable.

The Leibniz notation $\frac{dy}{dx}$ emphasizes the variable that we differentiate, with respect to what other variable. That is, when we write $\frac{dy}{dx}$ we consider y as a function of x and differentiate y with respect to x.

Since polynomials and rational functions are built from power functions of the form $y = x^n$, $n = 0, 1, 2, \ldots$, by the basic operations of addition, subtraction, multiplication, and division, we need basic differentiation rules for such operations. We begin with the following rules.

▲ Figure 3.21

If $y = f(x)$, then x is the independent variable and y is the dependent variable.

Theorem   Suppose a is a constant and $f(x)$ and $g(x)$ are differentiable at x. Then the following hold.

  1. $\dfrac{d}{dx}[af(x)] = a\dfrac{d}{dx}f(x)$

  2. $\dfrac{d}{dx}[f(x) + g(x)] = \dfrac{d}{dx}f(x) + \dfrac{d}{dx}g(x)$

Rule 1 says that a constant factor can be pulled out; Rule 2 says that the derivative of a sum of two functions is equal to the sum of the derivatives of the functions. Since $f(x) - g(x) = f(x) + (-1)g(x)$, the derivative of a difference of functions is the difference of the derivatives:

$$\frac{d}{dx}[f(x) - g(x)] = \frac{d}{dx}[f(x) + (-1)g(x)] = \frac{d}{dx}f(x) + \frac{d}{dx}[(-1)g(x)]$$

Using Rule 1 on the right-most term, we find that $\frac{d}{dx}[(-1)g(x)] = (-1)\frac{d}{dx}g(x)$. Therefore,

$$\frac{d}{dx}[f(x) - g(x)] = \frac{d}{dx}f(x) - \frac{d}{dx}g(x)$$

Rules 1 and 2 allow us to differentiate polynomials; as illustrated in the following examples.

▌**EXAMPLE 2**   Differentiate $y = 2x^4 - 3x^3 + x - 7$.

Solution

$$\begin{aligned}\frac{d}{dx}(2x^4 - 3x^3 + x - 7) &= \frac{d}{dx}(2x^4) - \frac{d}{dx}(3x^3) + \frac{d}{dx}x - \frac{d}{dx}7\\ &= 2\frac{d}{dx}x^4 - 3\frac{d}{dx}x^3 + \frac{d}{dx}x - \frac{d}{dx}7\\ &= 2(4x^3) - 3(3x^2) + 1 - 0 = 8x^3 - 9x^2 + 1 \quad ▌\end{aligned}$$

▌**EXAMPLE 3**

a. $\frac{d}{dx}(-5x^7 + 2x^3 - 10) = -35x^6 + 6x^2$

b. $\frac{d}{dt}(t^3 - 8t^2 + 3t) = 3t^2 - 16t + 3$

c. Suppose that n is a positive integer and a is a constant. Then $\frac{d}{ds}(as^n) = ans^{n-1}$.

d. $\frac{d}{dN}(\ln 5 + N \ln 7) = \ln 7$

e. $\frac{d}{dr}\left(r^2 \sin \frac{\pi}{4} - r^3 \cos \frac{\pi}{12} + \sin \frac{\pi}{6}\right) = 2r \sin \frac{\pi}{4} - 3r^2 \cos \frac{\pi}{12}$ ▌

In the previous section we related the derivative to the slope of the tangent line; the following example uses this interpretation.

▌**EXAMPLE 4**   If $f(x) = 2x^3 - 3x + 1$, find the tangent and the normal line at $(-1, 2)$.

Solution   The slope of the tangent line at $(-1, 2)$ is $f'(-1)$. We calculate this as follows:

$$f'(x) = 6x^2 - 3$$

Evaluating $f'(x)$ at $x = -1$, we get

$$f'(-1) = 6(-1)^2 - 3 = 3$$

Therefore the equation of the tangent line at $(-1, 2)$ is

$$y - 2 = 3(x - (-1)) \qquad \text{or} \qquad y = 3x + 5$$

To find the equation of the normal line, recall that the normal line is perpendicular to the tangent line; hence, the slope of the normal line, m, is given by

$$m = -\frac{1}{f'(-1)} = -\frac{1}{3}$$

◀ **Figure 3.22**
The graph of
$f(x) = 2x^3 - 3x + 1$,
together with the tangent and
normal lines at $(-1, 2)$.

The normal line goes through the point $(-1, 2)$ as well. The equation of the normal line is therefore

$$y - 2 = -\frac{1}{3}(x - (-1)) \quad \text{or} \quad y = -\frac{1}{3}x + \frac{5}{3}$$

The graph of $f(x)$, including the tangent and the normal line at $(-1, 2)$, is shown in Figure 3.22. ▍

We point out an important aspect from the last example: When we computed $f'(-1)$, we *first* computed $f'(x)$; the *second* step was to evaluate $f'(x)$ at $x = -1$. It makes no sense to plug -1 into $f(x)$ and then differentiate the result. Since $f(-1) = 2$ is a constant, the derivative would be 0, which is obviously not $f'(-1)$. The notation $f'(-1)$ means that we evaluate the function $f'(x)$ at $x = -1$.

◆ **3.2 PROBLEMS**

Differentiate the functions given in Problems 1–6, with respect to the independent variable.

1. $f(x) = 2x^3 - 3x + 1$ **2.** $g(s) = 5s^7 + 2s^3 - 5s$

3. $h(t) = \frac{1}{2}t^2 - 3t + 2$ **4.** $f(x) = x^2 \sin\frac{\pi}{3} + \tan\frac{\pi}{4}$

5. $f(t) = t^3 e^{-2} + t$ **6.** $f(x) = 20x^3 - 4x^6 + 9x^8$

7. Differentiate

$$f(x) = ax^3$$

with respect to x. Assume that a is a constant.

8. Differentiate

$$h(s) = rs^2 - r$$

with respect to s. Assume that r is a constant.

9. Differentiate

$$f(N) = (b - 1)N^4 - \frac{N^2}{b}$$

with respect to N. Assume that b is a nonzero constant.

10. Differentiate

$$g(t) = a^2 t - at^3$$

with respect to t. Assume that a is a constant.

11. Differentiate

$$V(t) = V_0(1 + \gamma t)$$

with respect to t. Assume that V_0 and γ are positive constants.

12. Differentiate

$$p(T) = \frac{NkT}{V}$$

with respect to T. Assume that N, k, and V are positive constants.

13. Differentiate

$$r(N) = N\left(1 - \frac{N}{K}\right)$$

with respect to N. Assume that K is a positive constant.

14. Differentiate

$$R(T) = \frac{2\pi^5}{15}\frac{k^4}{c^2 h^3}T^4$$

with respect to T. Assume that k, c, and h are positive constants.

In Problems 15–18, find the tangent line, in standard form, to $y = f(x)$ at the indicated point.

15. $y = 3x^2 - 4x + 7$, at $x = 2$

16. $y = 7x^3 + 2x - 1$, at $x = -3$

17. $y = \frac{1}{\sqrt{2}}x^2 - \sqrt{2}$, at $x = 4$

18. $y = 3\pi x^5 - \frac{\pi}{2}x^3$, at $x = -1$

In Problems 19 and 20, find the normal line, in standard form, to $y = f(x)$ at the indicated point.

19. $y = 2x^4 - 3x$, at $x = 2$

20. $y = \sqrt{3}x^4 - 2\sqrt{3}x^2$, at $x = -\sqrt{3}$

21. Find the tangent line to

$$f(x) = ax^2$$

at $x = 1$. Assume that a is a positive constant.

22. Find the normal line to

$$f(x) = ax^3$$

at $x = -1$. Assume that a is a positive constant.

In Problems 23 and 24, find the coordinates of all the points on the graph of $y = f(x)$ that have horizontal tangents.

23. $f(x) = 3x - x^2$ **24.** $f(x) = \frac{1}{2}x^4 - \frac{7}{3}x^3 - 2x^2$

25. Find a point on the curve

$$y = x^3 + 2x + 2$$

whose tangent line is parallel to the line $3x - y = 2$. Is there more than one such point? If so, find all other points with this property.

26. How many tangent lines to the curve

$$y = x^2 + 2x$$

pass through the point $\left(-\frac{1}{2}, -3\right)$?

27. Suppose that $P(x)$ is a polynomial of degree k. Is $P'(x)$ a polynomial as well? If yes, what is its degree?

◆ 3.3 THE PRODUCT AND QUOTIENT RULES, AND THE DERIVATIVES OF RATIONAL AND POWER FUNCTIONS

3.3.1 The Product Rule

The derivative of a sum of differentiable functions is the sum of the derivatives of the functions; the rule for products is not so simple. This can be seen easily from the following example. Consider $y = x^5 = (x^3)(x^2)$. We know that

$$\frac{d}{dx}x^5 = 5x^4$$

$$\frac{d}{dx}x^3 = 3x^2$$

$$\frac{d}{dx}x^2 = 2x$$

This immediately shows that

$$\frac{d}{dx}x^5 \quad \text{is not equal to} \quad \left(\frac{d}{dx}x^3\right)\left(\frac{d}{dx}x^2\right)$$

(Leibniz first thought that the multiplication rule was as simple as that, but he quickly realized his mistake and found the correct formula for differentiating products of functions.)

> **THE PRODUCT RULE**
>
> If $h(x) = f(x)g(x)$, and both $f(x)$ and $g(x)$ are differentiable at x, then
> $$h'(x) = f'(x)g(x) + f(x)g'(x)$$
> If we set $u = f(x)$ and $v = g(x)$, then
> $$(uv)' = u'v + uv'$$

Proof   Since $h(x) = f(x)g(x)$ is a product of two functions, we can visualize $h(x)$ as the area of a rectangle with sides $f(x)$ and $g(x)$. We need $h(x + \Delta x)$ to compute the derivative; this is given by

$$h(x + \Delta x) = f(x + \Delta x)g(x + \Delta x)$$

To compute $h'(x)$, we must compute $h(x + \Delta x) - h(x)$ (see Figure 3.23). We find

$$h(x + \Delta x) - h(x) = \text{area of I} + \text{area of II}$$

$$= [f(x + \Delta x) - f(x)]g(x)$$

$$+ [g(x + \Delta x) - g(x)]f(x + \Delta x)$$

Dividing this result by Δx and taking the limit $\Delta x \to 0$, we obtain

$$h'(x) = \lim_{\Delta x \to 0} \frac{h(x + \Delta x) - h(x)}{\Delta x}$$

$$= \lim_{\Delta x \to 0} \frac{[f(x + \Delta x) - f(x)]g(x) + [g(x + \Delta x) - g(x)]f(x + \Delta x)}{\Delta x}$$

$$= \lim_{\Delta x \to 0} \left[\frac{f(x + \Delta x) - f(x)}{\Delta x} g(x) + \frac{g(x + \Delta x) - g(x)}{\Delta x} f(x + \Delta x) \right]$$

Now, we need the assumption that $f'(x)$ and $g'(x)$ exist. We use the basic rules for limits to write the last expression as

$$\left(\lim_{\Delta x \to 0} \frac{f(x + \Delta x) - f(x)}{\Delta x} \right) g(x) + \left(\lim_{\Delta x \to 0} \frac{g(x + \Delta x) - g(x)}{\Delta x} \right)$$

$$\times \left(\lim_{\Delta x \to 0} f(x + \Delta x) \right)$$

The limits of the difference quotients are the respective derivatives. Using the fact that $f(x)$ is continuous at x (which follows from the fact that $f(x)$ is differentiable at x), we find $\lim_{\Delta x \to 0} f(x + \Delta x) = f(x)$. Therefore,

$$h'(x) = f'(x)g(x) + g'(x)f(x)$$

as claimed.   ◆

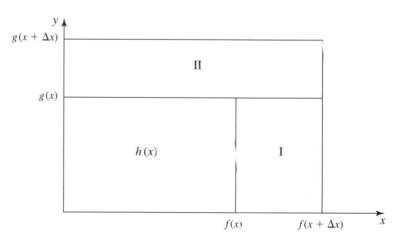

▲ **Figure 3.23**
The product rule.

I EXAMPLE I   Differentiate $f(x) = (3x + 1)(2x^2 - 5)$.

Solution   We write $u = 3x + 1$ and $v = 2x^2 - 5$. The product rule says that $(uv)' = u'v + uv'$. That is, we need the derivatives of both u and v:

$$u = 3x + 1 \qquad v = 2x^2 - 5$$
$$u' = 3 \qquad\quad v' = 4x$$

Then

$$(uv)' = u'v + uv'$$
$$= 3(2x^2 - 5) + (3x + 1)(4x)$$
$$= 6x^2 - 15 + 12x^2 + 4x = 18x^2 + 4x - 15$$

Of course, we could have gotten this result by first multiplying out $(3x + 1)(2x^2 - 5) = 6x^3 - 15x + 2x^2 - 5$, which is simply a polynomial function. We then would have found that

$$\frac{d}{dx}(6x^3 - 15x + 2x^2 - 5) = 18x^2 - 15 + 4x$$

which is the same answer.   **I**

I EXAMPLE 2   Differentiate $f(x) = (3x^3 - 2x)^2$.

Solution   Again, we could simply expand the square and then differentiate the resulting polynomial—but we can also use the product rule. To do so, we write $u = v = 3x^3 - 2x$. Then $f(x) = uv$ and $(uv)' = u'v + uv'$. Since $u = v$, it follows that $u' = v'$ and the formula simplifies to $(uv)' = (u^2)' = u'u + uu' = 2uu'$. Since $u' = 9x^2 - 2$, we find that

$$f'(x) = 2(3x^3 - 2x)(9x^2 - 2)$$ **I**

3.3.2 The Quotient Rule

The quotient rule will allow us to compute the derivative of a quotient of two functions. In particular, this will allow us to compute the derivative of a rational function, because it is the quotient of two polynomial functions.

THE QUOTIENT RULE

If $h(x) = \frac{f(x)}{g(x)}$, $g(x) \neq 0$, and both $f'(x)$ and $g'(x)$ exist, then

$$h'(x) = \frac{f'(x)g(x) - f(x)g'(x)}{[g(x)]^2}$$

In short, with $u = f(x)$ and $v = g(x)$,

$$\left(\frac{u}{v}\right)' = \frac{u'v - uv'}{v^2}$$

We could prove the quotient rule much as we did the product rule, by using the formal definition of derivatives, but that would not be very exciting. Instead, we will give a different proof of the quotient rule in the next subsection.

Carefully note the exact form of the product and the quotient rules. In the product rule, we add $f'g$ and fg', whereas in the quotient rule, we subtract fg' from $f'g$. As mentioned, we can use the quotient rule to find the derivative of rational functions. We illustrate this in the following two examples.

▮ **EXAMPLE 3** Differentiate $y = \frac{x^3-3x+2}{x^2+1}$. (This function is defined for all $x \in \mathbf{R}$, since $x^2 + 1 \neq 0$.)

Solution We set $u = x^3 - 3x + 2$ and $v = x^2 + 1$. Both u and v are polynomials, which we know how to differentiate. We find

$$u = x^3 - 3x + 2 \qquad v = x^2 + 1$$
$$u' = 3x^2 - 3 \qquad v' = 2x$$

Using the quotient rule, we can compute y':

$$y' = \frac{u'v - uv'}{v^2} = \frac{(3x^2 - 3)(x^2 + 1) - (x^3 - 3x + 2)2x}{(x^2 + 1)^2}$$

$$= \frac{3x^4 + 3x^2 - 3x^2 - 3 - 2x^4 + 6x^2 - 4x}{(x^2 + 1)^2}$$

$$= \frac{x^4 + 6x^2 - 4x - 3}{(x^2 + 1)^2}$$ ▮

▮ **EXAMPLE 4** Differentiate the Monod growth function

$$f(R) = a\frac{R}{k + R}, \qquad R \geq 0$$

where a and k are positive constants.

Solution Since a and k are positive constants, $f(R)$ is defined for all $R \geq 0$. We write $u = aR$ and $v = k + R$, and obtain

$$u = aR \qquad v = k - R$$
$$u' = a \qquad v' = 1$$

Hence,

$$\frac{d}{dR}f(R) = \frac{u'v - uv'}{v^2} = \frac{a(k + R) - aR \cdot 1}{(k + R)^2} = \frac{ak}{(k + R)^2}$$

In Figure 3.24, we graph both $f(R)$ and $f'(R)$. We see that the slope of the tangent line at $(R, f(R))$ is positive for all $R \geq 0$. We can also conclude this also from the graph of $f'(R)$ since it is positive for all $R \geq 0$. Furthermore, we see that $f(R)$ becomes less steep as R increases, which is reflected in the fact that $f'(R)$ becomes smaller as R increases. ▮

We can use the quotient rule to extend the power rule to the case where the exponent is a negative integer.

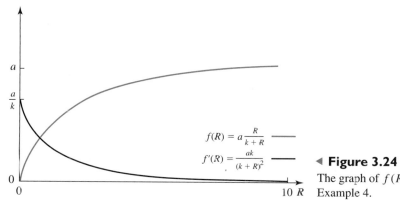

$$f(R) = a\frac{R}{k+R}$$

$$f'(R) = \frac{ak}{(k+R)^2}$$

◀ **Figure 3.24**
The graph of $f(R)$ and $f'(R)$ in Example 4.

POWER RULE (NEGATIVE INTEGER EXPONENTS)

If $f(x) = x^{-n}$, where n is a positive integer, then

$$f'(x) = -nx^{-n-1}$$

Note that the power rule for negative integer exponents works the same way as the power rule for positive integer exponents: we write the old exponent in front and decrease it by 1. We will now prove the power rule for negative integer exponents.

Proof   We write $f(x) = \frac{1}{x^n}$, and set $u = 1$ and $v = x^n$. Then

$$u = 1 \qquad v = x^n$$
$$u' = 0 \qquad v' = nx^{n-1}$$

and, therefore,

$$f'(x) = \frac{u'v - uv'}{v^2} = \frac{0 \cdot x^n - 1 \cdot nx^{n-1}}{(x^n)^2} = -\frac{nx^{n-1}}{x^{2n}} = -nx^{-n-1} \qquad \blacklozenge$$

I EXAMPLE 5   **a.** If $y = \frac{1}{x}$, then

$$y' = \frac{d}{dx}(x^{-1}) = (-1)x^{-1-1} = -\frac{1}{x^2}$$

b. If $g(x) = \frac{3}{x^4}$, then

$$g'(x) = \frac{d}{dx}(3x^{-4}) = 3\frac{d}{dx}x^{-4} = 3(-4)x^{-4-1} = -12x^{-5} = -\frac{12}{x^5} \qquad \blacksquare$$

There is a general form of the power rule, in which the exponent can be any real number. In the next section, we will give the proof for the case when the exponent is rational, and we will prove the general case in Section 3.7.

POWER RULE (GENERAL FORM)

Let $f(x) = x^r$, where r is any real number. Then

$$f'(x) = rx^{r-1}$$

I EXAMPLE 6   **a.** If $y = \sqrt{x}$, then

$$y' = \frac{d}{dx}(x^{1/2}) = \frac{1}{2}x^{\frac{1}{2}-1} = \frac{1}{2}x^{-1/2} = \frac{1}{2\sqrt{x}}$$

b. If $y = \sqrt[5]{x}$, then

$$y' = \frac{d}{dx}(x^{1/5}) = \frac{1}{5}x^{(1/5)-1} = \frac{1}{5}x^{-4/5} = \frac{1}{5x^{4/5}}$$

c. If $g(t) = \frac{1}{\sqrt[3]{t}}$, then

$$g'(t) = \frac{d}{dt}(t^{-1/3}) = \left(-\frac{1}{3}\right)t^{(-1/3)-1} = \left(-\frac{1}{3}\right)t^{-4/3} = -\frac{1}{3t^{4/3}}$$

d. If $h(s) = s^\pi$, then $h'(s) = \pi s^{\pi-1}$.   **I**

The function $f(x) = \sqrt{x}$, $x \geq 0$, appears quite frequently. It is therefore worthwhile to memorize its derivative:

$$\frac{d}{dx}\sqrt{x} = \frac{1}{2\sqrt{x}}$$

▮ EXAMPLE 7   Combining the Rules

Differentiate $f(x) = \sqrt{x}(x^2 - 1)$.

Solution I   We can consider this as a product of two functions. Let

$$u = \sqrt{x} \qquad v = x^2 - 1$$
$$u' = \frac{1}{2\sqrt{x}} \qquad v' = 2x$$

Hence,

$$f'(x) = u'v + uv' = \frac{1}{2\sqrt{x}}(x^2 - 1) + \sqrt{x}(2x)$$

$$= \frac{x^2 - 1 + \sqrt{x}(2x)2\sqrt{x}}{2\sqrt{x}} = \frac{x^2 - 1 + 4x^2}{2\sqrt{x}} = \frac{5x^2 - 1}{2\sqrt{x}}$$

Solution 2   Since $f(x) = \sqrt{x}(x^2 - 1) = x^{5/2} - x^{1/2}$, we can also use the general version of the power rule. We find that

$$f'(x) = \frac{5}{2}x^{(5/2)-1} - \frac{1}{2}x^{(1/2)-1} = \frac{5}{2}x^{3/2} - \frac{1}{2\sqrt{x}} = \frac{5x^{3/2}\sqrt{x} - 1}{2\sqrt{x}} = \frac{5x^2 - 1}{2\sqrt{x}} \qquad ▮$$

▮ EXAMPLE 8   Differentiating a Function, Which Is Not Specified

Suppose that $f(x)$ is differentiable. Find an expression for the derivative of

$$y = \frac{f(x)}{x^2}$$

Solution   We set

$$u = f(x) \qquad v = x^2$$
$$u' = f'(x) \qquad v' = 2x$$

and use the quotient rule. We find that

$$y' = \frac{f'(x)x^2 - f(x)2x}{x^4} = \frac{xf'(x) - 2f(x)}{x^3} \qquad ▮$$

◆ 3.3 PROBLEMS

In Problems 1–8, differentiate with respect to the independent variable.

1. $f(x) = (x^3 + 17)(3x - 4x^2)$

2. $g(s) = (2s^2 - 5s)^2$

3. $h(t) = \dfrac{t^2 - 3t + 1}{t + 1}$

4. $f(s) = \dfrac{2s^3 - 4s^2 + 5s - 7}{(s^2 - 3)^2}$

5. $f(x) = \sqrt{x}(x^4 - 5x^2)$

6. $g(s) = \dfrac{s^{1/7} - s^{2/7}}{s^{3/7} + s^{4/7}}$

7. $f(x) = x^3 - \dfrac{1}{x^3}$

8. $f(x) = (x^3 - 3x^2 + 2)\left(\sqrt{x} + \dfrac{1}{\sqrt{x}} - 1\right)$

In Problems 9–12, find the tangent line, in slope-intercept form, of $y = f(x)$ at the specified point.

9. $f(x) = \dfrac{x^2 + 3}{x^3 + 5}$, at $x = -2$

10. $f(x) = \dfrac{3}{x} - \dfrac{4}{\sqrt{x}} + \dfrac{2}{x^2}$, at $x = 1$

11. $f(x) = \dfrac{2x - 5}{x^3}$, at $x = 2$

12. $f(x) = \sqrt{x}(x^3 - 1)$, at $x = 1$

13. Differentiate

$$f(x) = \frac{ax}{3 + x}$$

with respect to x. Assume that a is a positive constant.

14. Differentiate

$$f(R) = \frac{R^2}{k + R}$$

with respect to R. Assume that k is a positive constant.

15. Differentiate

$$g(t) = (at + 1)^2$$

with respect to t. Assume that a is a positive constant.

16. Differentiate

$$h(t) = \sqrt{at}(1 - a) + a$$

with respect to t. Assume that a is a positive constant.

17. Suppose that $f(2) = -4$, $g(2) = 3$, $f'(2) = 1$, and $g'(2) = -2$. Find

(a) $\left(\frac{1}{f}\right)'(2)$ **(b)** $(fg)'(2)$

18. Suppose that $f(2) = -4$, $g(2) = 3$, $f'(2) = 1$, and $g'(2) = -2$. Find

(a) $(f/2g)'(2)$ **(b)** $(f^2 + g^2)'(2)$

In Problems 19–22, assume that $f(x)$ is differentiable. Find an expression for the derivative of each function.

19. $y = 3x^2 f(x)$

20. $y = \dfrac{x^2 + 4f(x)}{f(x)}$

21. $y = [f(x)]^2 - \dfrac{x}{f(x)}$

22. $y = \dfrac{f(x)}{f(x) + x}$

In Problems 23–26, assume that $f(x)$ and $g(x)$ are differentiable at x. Find an expression for the derivative of each function.

23. $y = [f(x) + 2g(x)]g(x)$

24. $y = \dfrac{f(x)}{[g(x)]^2}$

25. $y = \dfrac{x^2}{f(x) - g(x)}$

26. $y = \sqrt{x}\, f(x)g(x)$

27. Assume that $f(x)$ is a differentiable function. Find the derivative of the reciprocal function $g(x) = 1/f(x)$ at those points x where $f(x) \neq 0$.

28. Find the tangent line to the hyperbola $yx = c$, c a positive constant, at the point (x_1, y_1) with $x_1 > 0$. Show that the tangent line intersects the x-axis at a point that does not depend on c.

29. (Adapted from Roff 1992) The males in the frog species *Eleutherodactylus coqui* (found in Puerto Rico) take care of their brood. While they protect the eggs, they cannot find other mates, and therefore cannot increase their number of offspring. On the other hand, if they do not spend enough time with their brood, then the offspring might not survive. Simple mathematical models are used to give the proportion of hatching offspring per unit time, $w(t)$, as a function of the probability of hatching if time t is spent brooding, $f(t)$, and the cost C associated with the time spent searching for other mates. Given the following relationship

$$w(t) = \frac{f(t)}{C + t}$$

find the derivative of $w(t)$. (We will continue this example in the next chapter.)

◆ 3.4 THE CHAIN RULE AND HIGHER DERIVATIVES

3.4.1 The Chain Rule

In Section 1.2, we defined the composition of functions; in this section, we will discuss how to find the derivative of composite functions.

> **CHAIN RULE**
>
> If g is differentiable at x and f is differentiable at $y = g(x)$, then the composite function $(f \circ g)(x) = f[g(x)]$ is differentiable at x, and the derivative is given by
>
> $$(f \circ g)'(x) = f'[g(x)]g'(x)$$

The derivative of $f \circ g$ can be written in Leibniz notation. If we set $u = g(x)$, then

$$\frac{df[g(x)]}{dx} = \frac{df}{du}\frac{du}{dx}$$

which says that the derivative of a composite function is the product of the derivative of the outer function f evaluated at $u = g(x)$ and the derivative of the inner function g evaluated at x. This form of the chain rule is easy to remember. It suggests that we obtain the solution to the left-hand side by simply canceling the du terms on the right-hand side. (We must be careful, however, since du/dx comes from taking the limit of $\Delta u/\Delta x$ as $\Delta x \to 0$, Δu, and hence du, could be 0 and we cannot divide by 0. Nevertheless, it is a very helpful way to remember the chain rule.) We will give a proof of the chain rule at the end of this subsection.

EXAMPLE 1   **A Polynomial**

Find the derivative of

$$h(x) = (2x + 1)^3$$

Solution   If we set $u = g(x) = 2x + 1$ and $f(u) = u^3$, then $h(x) = (f \circ g)(x)$. We need to find both $f'[g(x)]$ and $g'(x)$ to compute $h'(x)$. Now,

$$g'(x) = 2 \quad \text{and} \quad f'(u) = 3u^2$$

Hence, since $f'[g(x)] = 3(g(x))^2 = 3(2x + 1)^2$,

$$h'(x) = f'[g(x)]g'(x) = 3(2x + 1)^2 \cdot 2$$
$$= 6(2x + 1)^2$$

If we use Leibniz notation, this becomes

$$h'(x) = \frac{df}{du}\frac{du}{dx} = 3u^2 \cdot 2 = 3(2x + 1)^2 \cdot 2$$
$$= 6(2x + 1)^2 \qquad ■$$

EXAMPLE 2   **A Radical**

Find the derivative of $h(x) = \sqrt{x^2 + 1}$.

Solution   If we set $u = g(x) = x^2 + 1$ and $f(u) = \sqrt{u}$, then $h(x) = (f \circ g)(x)$. We find that

$$g'(x) = 2x \quad \text{and} \quad f'(u) = \frac{1}{2\sqrt{u}}$$

We need to evaluate f' at $g(x)$, that is,

$$f'[g(x)] = \frac{1}{2\sqrt{g(x)}} = \frac{1}{2\sqrt{x^2 + 1}}$$

Therefore,

$$h'(x) = f'[g(x)]g'(x) = \frac{1}{2\sqrt{x^2 + 1}}2x = \frac{x}{\sqrt{x^2 + 1}} \qquad ■$$

EXAMPLE 3   **A Radical**

Find the derivative of

$$h(x) = \sqrt[7]{2x^2 + 3x}$$

Solution   We write

$$h(x) = (2x^2 + 3x)^{1/7}$$

The inner function is $u = g(x) = 2x^2 + 3x$ and the outer function is $f(u) = u^{1/7}$. Thus, we find

$$h'(x) = \frac{df}{du}\frac{du}{dx} = \frac{1}{7}u^{1/7-1}(4x + 3)$$

$$= \frac{1}{7}(2x^2 + 3x)^{-6/7}(4x + 3)$$

$$= \frac{4x + 3}{7(2x^2 + 3x)^{6/7}}$$ ▮

▮ **EXAMPLE 4** **A Rational Function**

Find the derivative of $h(x) = \left(\frac{x}{x+1}\right)^2$.

Solution If we set $u = g(x) = \frac{x}{x+1}$ and $f(u) = u^2$, then $h(x) = (f \circ g)(x)$. We use the quotient rule to compute the derivative of $g(x)$:

$$g'(x) = \frac{1 \cdot (x + 1) - x \cdot 1}{(x + 1)^2} = \frac{1}{(x + 1)^2}$$

Since $f'(u) = 2u$, we find

$$h'(x) = f'[g(x)]g'(x) = 2\frac{x}{x + 1}\frac{1}{(x + 1)^2} = \frac{2x}{(x + 1)^3}$$ ▮

The Proof of the Quotient Rule We can use the chain rule to prove the quotient rule. Assume that $g(x) \neq 0$ for all x in the domain of g. If we define $h(x) = \frac{1}{x}$, then

$$(h \circ g)(x) = h[g(x)] = \frac{1}{g(x)}$$

We used the formal definition of the derivative in Example 3 in Section 3.1 to show that $h'(x) = -\frac{1}{x^2}$. This, together with the chain rule, yields

$$(h \circ g)'(x) = -\frac{1}{[g(x)]^2}g'(x) \qquad \text{or} \qquad \left(\frac{1}{g}\right)' = -\frac{g'}{g^2}$$

Since $\frac{f}{g} = f\frac{1}{g}$, we can use the product rule to find the derivative of $\frac{f}{g}$.

$$\left(\frac{f}{g}\right)' = f'\frac{1}{g} + f\left(\frac{1}{g}\right)' = f'\frac{1}{g} + f\left(-\frac{g'}{g^2}\right)$$

$$= \frac{f'g - fg'}{g^2}$$

Note that we did not use the power rule for negative integer exponents (of Subsection 3.3.2) to compute $h'(x)$, but instead used the formal definition of derivatives to compute the derivative of $1/x$. Using the power rule for negative integer exponents would have been circular reasoning; we used the quotient rule to prove the power rule for negative integer exponents, and so we cannot use the power rule for negative integer exponents to prove the quotient rule. ◆

▮ **EXAMPLE 5** **A Function with Parameters**

Find the derivative of

$$h(x) = (ax^2 - 2)^n$$

where $a > 0$ and n is a positive integer.

Solution If we set $u = g(x) = ax^2 - 2$ and $f(u) = u^n$, then $h(x) = (f \circ g)(x)$. Since

$$g'(x) = 2ax \qquad \text{and} \qquad f'(u) = nu^{n-1}$$

we find

$$h'(x) = f'[g(x)]g'(x) = n(ax^2 - 2)^{n-1}2ax$$
$$= 2anx(ax^2 - 2)^{n-1}$$

Looking at $h'(x) = n(ax^2 - 2)^{a-1} \cdot 2ax$, we see that we first differentiated the outer function f, which yielded $n(ax^2 - 2)^{n-1}$ using the power rule, and then multiplied the result by the derivative of the inner function g, which is $2ax$. ▌

▌ **EXAMPLE 6** **Differentiating a Function, Which Is Not Specified**

Suppose $f(x)$ is differentiable. Find

$$\frac{d}{dx} \frac{1}{\sqrt{f(x)}}$$

Solution We set

$$h(x) = \frac{1}{\sqrt{f(x)}} = [f(x)]^{-1/2}$$

Now, $u = f(x)$ is the inner function and $h(u) = u^{-1/2}$ is the outer function, hence

$$\frac{d}{dx}h(x) = \frac{dh}{du}\frac{du}{dx} = -\frac{1}{2}u^{-3/2}f'(x)$$
$$= -\frac{1}{2u^{3/2}}f'(x) = -\frac{f'(x)}{2[f(x)]^{3/2}}$$ ▌

We conclude this subsection with the proof of the chain rule. The first part of the proof follows along the lines of the argument we sketched out in the beginning of this section, but the second part is much more technical and deals with the problem that Δu could be zero.

Proof of the Chain Rule We will use the definition of the derivative to prove the chain rule. Formally,

$$(f \circ g)'(x) = \lim_{x \to c} \frac{f[g(x)] - f[g(c)]}{x - c}$$

We need to show that the right-hand side is equal to $f'[g(c)]g'(c)$. As long as $g(x) \neq g(c)$, we can write

$$\lim_{x \to c} \frac{f[g(x)] - f[g(c)]}{x - c} = \lim_{x \to c} \frac{\frac{f[g(x)]-f[g(c)]}{g(x)-g(c)}[g(x) - g(c)]}{x - c}$$
$$= \lim_{x \to c} \frac{f[g(x)] - f[g(c)]}{g(x) - g(c)} \frac{g(x) - g(c)}{x - c}$$

Since

$$\lim_{x \to c} \frac{f[g(x)] - f[g(c)]}{g(x) - g(c)} = f'[g(c)] \qquad \text{and} \qquad \lim_{x \to c} \frac{g(x) - g(c)}{x - c} = g'(c),$$

these limits exist, and we can use the fact that the limit of a product is the product of the limits. We find

$$\lim_{x \to c} \frac{f[g(x)] - f[g(c)]}{x - c} = \lim_{x \to c} \frac{f[g(x)] - f[g(c)]}{g(x) - g(c)} \lim_{x \to c} \frac{g(x) - g(c)}{x - c}$$
$$= f'[g(c)]g'(c)$$

In the preceding calculation, we needed to assume that $g(x) - g(c) \neq 0$. Of course, when we take the limit $x \to c$, there might be x values with $g(x) = g(c)$, and we must deal with this possibility.

We set $y = g(x)$ and $d = g(c)$. The expression

$$f^{\star}(y) = \frac{f(y) - f(d)}{y - d}$$

is only defined for $y \neq d$. Since

$$\lim_{y \to d} \frac{f(y) - f(d)}{y - d} = f'(d)$$

a natural extension of $f^{\star}[g(x)]$ that makes $f^{\star}[g(x)]$ a continuous function is, therefore,

$$f^{\star}[g(x)] = \begin{cases} \frac{f[g(x)] - f[g(c)]}{g(x) - g(c)} & \text{for } g(x) \neq g(c) \\ f'[g(c)] & \text{for } g(x) = g(c) \end{cases}$$

This means that for all x,

$$f[g(x)] - f[g(c)] = f^{\star}[g(x)][g(x) - g(c)]$$

With this equivalence, we can repeat our calculations and find

$$\lim_{x \to c} \frac{f[g(x)] - f[g(c)]}{x - c} = \lim_{x \to c} \frac{f^{\star}[g(x)][g(x) - g(c)]}{x - c}$$

$$\lim_{x \to c} f^{\star}[g(x)] \lim_{x \to c} \frac{g(x) - g(c)}{x - c} = f'[g(c)] \cdot g'(c)$$

Note that in the last step we used the fact that $f^{\star}[g(x)]$ is continuous at $x = c$. ◆

3.4.2 Implicit Functions and Implicit Differentiation

So far, we have considered only functions of the form $y = f(x)$, which define y *explicitly* as a function of x. It is also possible to define y *implicitly* as a function of x, as in the following equation:

$$y^5 x^2 - yx + 2y^2 = \sqrt{x}$$

Here, y is given of as a function of x, that is, y is the dependent variable. There is no obvious way to solve for y. There is a very useful technique, based on the chain rule, which will allow us to find dy/dx for implicitly defined functions. This technique is called **implicit differentiation**. We explain the procedure in the following example.

▌EXAMPLE 7   Find $\frac{dy}{dx}$ if $x^2 + y^2 = 1$.

Solution   We differentiate both sides of the equation $x^2 + y^2 = 1$ with respect to x. We need to remember that y is a function of x.

$$\frac{d}{dx}(x^2 + y^2) = \frac{d}{dx}(1)$$

Since the derivative of a sum is the sum of the derivatives, we find

$$\frac{d}{dx}(x^2) + \frac{d}{dx}(y^2) = \frac{d}{dx}(1)$$

Starting with the left-hand side and using the power rule, we find $\frac{d}{dx}(x^2) = 2x$. To differentiate y^2 with respect to x, we must apply the chain rule. We find

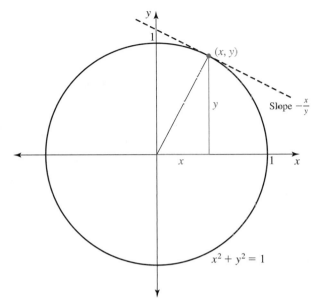

◀ **Figure 3.25**
The slope of the tangent line at
the unit circle $x^2 + y^2 = 1$ at
(x, y) is $m = -\frac{x}{y}$.

$\frac{d}{dx}(y^2) = 2y\frac{dy}{dx}$. On the right-hand side, we obtain $\frac{d}{dx}(1) = 0$. We therefore have

$$2x + 2y\frac{dy}{dx} = 0$$

We can now solve for $\frac{dy}{dx}$ and find

$$\frac{dy}{dx} = -\frac{2x}{2y} = -\frac{x}{y}$$

Since $x^2 + y^2 = 1$ is the equation for the unit circle centered at the origin (see Figure 3.25), we can convince ourselves that this is indeed the correct derivative by solving $x^2 + y^2 = 1$ for y and differentiating y directly with respect to x. When we solve for y, we find

$$y = \begin{cases} \sqrt{1 - x^2} & \text{for } y \geq 0 \\ -\sqrt{1 - x^2} & \text{for } y < 0 \end{cases}$$

Differentiating yields

$$\frac{dy}{dx} = \begin{cases} -\frac{2x}{2\sqrt{1-x^2}} = -\frac{x}{y} & \text{for } y > 0 \\ \frac{2x}{2\sqrt{1-x^2}} = -\frac{x}{-\sqrt{1-x^2}} = -\frac{x}{y} & \text{for } y < 0 \end{cases}$$

This example shows the advantage of using implicit differentiation. Even though we were able to solve for y and thus write y explicitly as a function of x, implicit differentiation was actually quicker. ▌

We summarize the steps to find dy/dx when an equation defines y implicitly as a differentiable function of x.

Step 1. Differentiate both sides of the equation with respect to x, keeping in mind that y is a function of x.

Step 2. Solve the resulting equation for dy/dx.

Note that differentiating terms involving y typically requires the chain rule. Here is one more example.

▮ **EXAMPLE 8**   Find $\frac{dy}{dx}$ when $y^3x^2 - yx + 2y^2 = x$.

Solution   We differentiate both sides of the equation with respect to x:

$$\frac{d}{dx}(y^3x^2) - \frac{d}{dx}(yx) + \frac{d}{dx}(2y^2) = \frac{d}{dx}(x)$$

To differentiate y^3x^2 and yx with respect to x, we use the product rule:

$$\frac{d}{dx}(y^3x^2) = \left(\frac{d}{dx}y^3\right)x^2 + y^3\left(\frac{d}{dx}x^2\right) = \left(\frac{d}{dx}y^3\right)x^2 + (y^3)(2x)$$

$$\frac{d}{dx}(yx) = \left(\frac{d}{dx}y\right)(x) + y\left(\frac{d}{dx}x\right) = \left(\frac{dy}{dx}\right)x + y$$

To differentiate $\frac{d}{dx}y^3$, we use the chain rule and find

$$\frac{d}{dx}y^3 = 3y^2\frac{dy}{dx}$$

Furthermore,

$$\frac{d}{dx}(2y^2) = 4y\frac{dy}{dx}$$

and

$$\frac{d}{dx}(x) = 1$$

Putting the pieces together, we obtain

$$3y^2\frac{dy}{dx}(x^2) + (y^3)(2x) - \left[\left(\frac{dy}{dx}\right)x + y\right] + 4y\frac{dy}{dx} = 1$$

Factoring $\frac{dy}{dx}$ yields

$$\frac{dy}{dx}[3y^2x^2 - x + 4y] + 2xy^3 - y = 1$$

Solving for $\frac{dy}{dx}$ yields

$$\frac{dy}{dx} = \frac{y + 1 - 2xy^3}{3y^2x^2 - x + 4y}$$ ▮

The next example will prepare us for the power rule for rational exponents.

▮ **EXAMPLE 9**   Find $\frac{dy}{dx}$ when $y^2 = x^3$.

Solution   We differentiate both sides with respect to x:

$$\frac{d}{dx}(y^2) = \frac{d}{dx}(x^3)$$

$$2y\frac{dy}{dx} = 3x^2$$

Therefore,

$$\frac{dy}{dx} = \frac{3}{2}\frac{x^2}{y}$$

Since $y = x^{3/2}$, we find

$$\frac{dy}{dx} = \frac{3}{2}\frac{x^2}{y} = \frac{3}{2}\frac{x^2}{x^{3/2}} = \frac{3}{2}x^{1/2}$$

This is the answer we expect from the general version of the power rule:

$$\frac{dy}{dx} = \frac{d}{dx}x^{3/2} = \frac{3}{2}x^{1/2}$$

▮

Power Rule for Rational Exponents   We can generalize this example to functions of the form $y = x^r$, where r is a rational number. This will give a proof of the generalized form of the power rule when the exponent is a rational number, something we promised in the previous section. We write $r = p/q$, where p and q are integers. Then

$$y = x^{p/q} \iff y^q = x^p$$

Differentiating both sides of $y^q = x^p$ with respect to x, we find

$$qy^{q-1}\frac{dy}{dx} = px^{p-1}$$

Hence

$$\frac{dy}{dx} = \frac{p}{q}\frac{x^{p-1}}{y^{q-1}} = \frac{p}{q}\frac{x^{p-1}}{(x^{p/q})^{q-1}}$$

$$= \frac{p}{q}\frac{x^{p-1}}{x^{p-p/q}} = \frac{p}{q}x^{p-1-p+p/q}$$

$$= \frac{p}{q}x^{p/q-1} = rx^{r-1}$$

In other words:

If r is a rational number, then

$$\frac{d}{dx}(x^r) = rx^{r-1}$$

3.4.3 Related Rates

In this subsection, we discuss an important application of implicit differentiation, known as related rates problems. We begin with a motivating example.

We consider the case of a parcel of air rising quickly in the atmosphere. As a consequence, it expands adiabatically, that is, it does not exchange heat with the surrounding air. Laws of physics tell us that the volume (V) and the temperature (T) of the parcel of air are related via

$$TV^{\gamma-1} = C$$

where γ is approximately 1.4 for sufficiently dry air (γ is the small Greek letter gamma) and C is a constant. The temperature is measured in Kelvin[1], a scale that is chosen so that the temperature is always positive (absolute temperature scale). Rising air expands, so the volume of the parcel of air increases with time; that is, $dV/dt > 0$, where t denotes time.

We will now determine how the temperature of the air parcel changes as it rises. Implicit differentiation of $TV^{\gamma-1} = C$ with respect to t yields

$$\frac{dT}{dt}V^{\gamma-1} + T(\gamma-1)V^{\gamma-2}\frac{dV}{dt} = 0$$

---

[1]To compare the Celsius and the Kelvin scale, note that a temperature difference of 1°C is equal to a temperature difference of 1K, and that 0°C = 273.15K and 100°C = 373.15K.

or

$$\frac{dT}{dt} = -\frac{T(\gamma - 1)V^{\gamma - 2}\frac{dV}{dt}}{V^{\gamma - 1}}$$

$$= -T(\gamma - 1)\frac{1}{V}\frac{dV}{dt}$$

If we use $\gamma = 1.4$, then

$$\frac{dT}{dt} = -T(0.4)\frac{1}{V}\frac{dV}{dt}$$

Therefore, when air expands (i.e., $dV/dt > 0$), it follows that $dT/dt < 0$ (since both T and V are positive). That is, the temperature of a parcel of air decreases as the parcel rises. Likewise, the temperature of quickly falling air increases; this can be observed close to high mountains.

In a typical related rates problem, one quantity is expressed in terms of another quantity, and both quantities change with time. We usually know how one of the quantities changes with time and are interested in finding out how the other quantity does. For instance, suppose that y is a function of x, and both y and x depend on time. If we know how x changes with time (i.e., if we know dx/dt), then we might want to know how y changes with time (i.e., dy/dt). We illustrate this in the following example.

▌**EXAMPLE 10**   Find $\frac{dy}{dt}$ when $x^2 + y^3 = 1$ and $\frac{dx}{dt} = 2$ for $x = \sqrt{7/8}$.

Solution   In this example, both x and y are functions of t. Implicit differentiation with respect to t yields

$$\frac{d}{dt}(x^2 + y^3) = \frac{d}{dt}(1)$$

and hence

$$2x\frac{dx}{dt} + 3y^2\frac{dy}{dt} = 0$$

Solving for $\frac{dy}{dt}$ yields

$$\frac{dy}{dt} = -\frac{2}{3}\frac{x}{y^2}\frac{dx}{dt}$$

When $x = \sqrt{7/8}$,

$$y^3 = 1 - x^2 = 1 - \frac{7}{8} = \frac{1}{8}$$

and hence $y = 1/2$. Therefore,

$$\frac{dy}{dt} = -\frac{2}{3}\frac{\sqrt{7/8}}{1/4} \cdot 2 = -\frac{16}{3}\sqrt{\frac{7}{8}} = -\frac{4}{3}\sqrt{14}$$ ▌

We present two more applications of related rates.

▌**EXAMPLE 11**   A spherical balloon is being filled with air. When the radius $r = 6$ in., the radius is increasing at a rate of 2 in./sec. How fast is the volume changing at this time?

Solution   The volume V of a sphere of radius r is given by

$$V = \frac{4}{3}\pi r^3$$

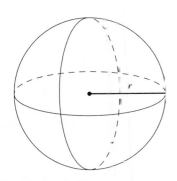

▲ **Figure 3.26**
The volume of a sphere with radius r is $V = \frac{4}{3}\pi r^3$.

(see Figure 3.26). Differentiating both sides with respect to t, we find

$$\frac{dV}{dt} = \frac{4}{3}\pi 3r^2 \frac{dr}{dt}$$

$$= 4\pi r^2 \frac{dr}{dt}$$

When $r = 6$ in. and $dr/dt = 2$ in./sec, then

$$\frac{dV}{dt} = 4\pi 6^2 \text{in.}^2 2\frac{\text{in.}}{\text{sec}} = 288\pi \frac{\text{in.}^3}{\text{sec}}$$

Note the unit of dV/dt is in.^{3}/sec, which is what you should expect because the unit of the volume is in.^{3} and time is measured in seconds. ▮

EXAMPLE 12   (*Adapted from Benton and Harper 1997*)   Ichthyosaurs are a group of marine reptiles that were fish shaped and comparable in size to dolphins. They became extinct during the Cretaceous.[2] Based on a study of twenty fossil skeletons, it was found that the skull length and backbone length of an individual were related through the allometric equation

$$[\text{skull length}] = 1.162[\text{backbone length}]^{0.933}$$

How is the growth rate of the backbone related to the growth rate of the skull?
Solution   Let x denote the age of the ichthyosaur, and set

$$S = S(x) = \text{skull length at age } x$$
$$B = B(x) = \text{backbone length at age } x$$

so that

$$S(x) = (1.162)[B(x)]^{0.933}$$

We are interested in the relationship between dS/dx and dB/dx, the growth rates of the skull and the backbone, respectively. Differentiating the equation for $S(x)$ with respect to x, we find

$$\frac{dS}{dx} = (1.162)(0.933)[B(x)]^{0.933-1}\frac{dB}{dx}$$

Rearranging terms on the right-hand side, we write this as

$$\frac{dS}{dx} = \underbrace{(1.162)[B(x)]^{0.933}}_{S(x)}(0.933)\frac{1}{B(x)}\frac{dB}{dx}$$

hence

$$\frac{1}{S(x)}\frac{dS}{dx} = 0.933\frac{1}{B(x)}\frac{dB}{dx}$$

This equation relates the relative growth rates $\frac{1}{S}\frac{dS}{dx}$ and $\frac{1}{B}\frac{dB}{dx}$. The factor 0.933 is less than 1, which indicates that skulls grow less quickly than backbones. This is familiar to us: Juvenile vertebrates often have larger heads than adults, relative to their body sizes. ▮

---

[2] The Cretaceous period began about 144 million years ago and ended about 65 million years ago.

3.4.4 Higher Derivatives

We have frequently mentioned in previous sections that the derivative of a function f is itself a function. We refer to this derivative as the **first derivative**, denoted f'. Being a function itself, we can therefore define its derivative (where it exists). This derivative is called the **second derivative** and denoted f''. This second derivative is again a function and hence we can define its derivative (where it exists). The result is the **third derivative**, denoted f'''. We can continue in this manner; from the fourth derivative on, we denote the derivatives by $f^{(4)}$, $f^{(5)}$, and so on.

▌ **EXAMPLE 13**   Find the third derivative of $f(x) = x^5$.

Solution   Differentiating $f(x)$, we find the first derivative to be

$$f'(x) = 5x^4$$

Differentiating $f'(x)$, we find the second derivative to be

$$f''(x) = 5(4x^3) = 20x^3$$

Differentiating $f''(x)$, we find the third derivative:

$$f'''(x) = 20(3x^2) = 60x^2$$ ▌

We can write higher order derivatives in Leibniz notation; the nth derivative of $f(x)$ is denoted by

$$\frac{d^n f}{dx^n}$$

▌ **EXAMPLE 14**   Find the second derivative of $f(x) = \sqrt{x}$, $x \geq 0$.

Solution   First, we must find the first derivative,

$$\frac{d}{dx}\sqrt{x} = \frac{d}{dx}x^{1/2} = \frac{1}{2}x^{-1/2} \quad \text{for } x > 0$$

To find the second derivative, we differentiate the first derivative

$$\frac{d^2}{dx^2}\sqrt{x} = \frac{d}{dx}\left(\frac{d}{dx}\sqrt{x}\right) = \frac{d}{dx}\left(\frac{1}{2}x^{-1/2}\right) = \frac{1}{2}\left(-\frac{1}{2}\right)x^{(-1/2)-1}$$

$$= -\frac{1}{4}x^{-3/2}$$ ▌

When functions are implicitly defined, we can use the technique of implicit differentiation to find higher derivatives.

▌ **EXAMPLE 15**   Find $\frac{d^2y}{dx^2}$ when $x^2 + y^2 = 1$.

Solution   We found

$$\frac{dy}{dx} = -\frac{x}{y}$$

in Example 7. Differentiating both sides of this equation with respect to x, we find

$$\frac{d}{dx}\left[\frac{dy}{dx}\right] = \frac{d}{dx}\left[-\frac{x}{y}\right]$$

The left-hand side can be written as

$$\frac{d^2y}{dx^2}$$

On the right-hand side, we use the quotient rule. Hence,

$$\frac{d^2y}{dx^2} = -\frac{1 \cdot y - x\frac{dy}{dx}}{y^2}$$

Substituting $-\frac{x}{y}$ for $\frac{dy}{dx}$, we obtain

$$\frac{d^2y}{dx^2} = -\frac{y - x\left(-\frac{x}{y}\right)}{y^2}$$

$$= -\frac{y - \frac{x^2}{y}}{y^2} = -\frac{y^2 + x^2}{y^3}$$

Since $x^2 + y^2 = 1$, we can simplify this further and obtain

$$\frac{d^2y}{dx^2} = -\frac{1}{y^3}$$

◆ 3.4 PROBLEMS

[3.4.1]

In Problems 1–10, differentiate the functions with respect to the indicated variable.

1. $f(x) = (2x - 1)^3$

2. $f(x) = \sqrt{x^2 + 3}$

3. $f(x) = \dfrac{1}{(x^3 - 2)^4}$

4. $f(s) = \sqrt{s + \sqrt{s}}$

5. $g(t) = \left(\dfrac{t}{t-3}\right)^3$

6. $f(r) = (r^2 - r)^3(r + 3r^3)^{-4}$

7. $h(s) = \dfrac{2(3-s)^2}{s^2 + (7s - 1)^2}$

8. $h(x) = \sqrt[5]{3 - x^4}$

9. $g(s) = (3s^7 - 7s)^\pi$

10. $h(t) = \left(3t + \dfrac{3}{t}\right)^{2/5}$

11. Differentiate

$$f(x) = (ax + 1)^3$$

with respect to x. Assume that a is a positive constant.

12. Differentiate

$$f(z) = \sqrt{ax^2 - 2}$$

with respect to x. Assume that a is a positive constant.

13. Differentiate

$$g(N) = \dfrac{bN}{(k + N)^2}$$

with respect to N. Assume that b and k are positive constants.

14. Differentiate

$$g(T) = a(T_0 - T)^3 - b$$

with respect to T. Assume that a, b, and T_0 are positive constants.

15. Suppose that $f'(x) = 2x + 1$. Find the following.

(a) $\dfrac{d}{dx} f(x^2)$ at $x = -1$.

(b) $\dfrac{d}{dx} f(\sqrt{x})$ at $x = 4$.

16. Suppose that $f'(x) = \frac{1}{x}$. Find the following.

(a) $\dfrac{d}{dx} f(x^2 + 3)$

(b) $\dfrac{d}{dx} f(\sqrt{x} - 1)$

In Problems 17–20, assume that $f(x)$ and $g(x)$ are differentiable.

17. Find $\dfrac{d}{dx}\sqrt{f(x) + g(x)}$

18. Find $\dfrac{d}{dx}\left(\dfrac{f(x)}{g(x)} + 1\right)^2$

19. Find $\dfrac{d}{dx} f\left[\dfrac{1}{g(x)}\right]$

20. Find $\dfrac{d}{dx}\dfrac{[f(x)]^2}{g(2x) + 2x}$

[3.4.2]

In Problems 21–28, find $\dfrac{dy}{dx}$ by implicit differentiation.

21. $x^2 + y^2 = 4$

22. $y = x^2 + yx$

23. $x^{3/4} + y^{3/4} = 1$

24. $xy - y^3 = 1$

25. $\sqrt{xy} = x^2 + 1$

26. $\dfrac{1}{xy} - y^2 = 2$

27. $\dfrac{x}{y} = \dfrac{y}{x}$

28. $\dfrac{x}{xy + 1} = 2xy$

In Problems 29–31, find the lines that are (a) tangential and (b) normal to each curve at the given point.

29. $x^2 + y^2 = 25$, $(4, -3)$ (circle)

30. $\dfrac{x^2}{4} + \dfrac{y^2}{9} = 1$, $\left(1, \frac{3}{2}\sqrt{3}\right)$ (ellipse)

31. $\dfrac{x^2}{25} - \dfrac{y^2}{9} = 1$, $\left(\frac{25}{3}, 4\right)$ (hyperbola)

32. *(Lemniscate)*
(a) The curve with equation $y^2 = x^2 - x^4$ is shaped like the numeral eight. Find $\frac{dy}{dx}$ at $\left(\frac{1}{2}, \frac{1}{4}\sqrt{3}\right)$.
(b) Use a graphing calculator to graph the curve in (a). If the graphing calculator cannot graph implicit functions, graph the upper and the lower half of the curve separately, that is, graph

$$y_1 = \sqrt{x^2 - x^4}$$
$$y_2 = -\sqrt{x^2 - x^4}$$

Choose the viewing rectangle $-2 \le x \le 2$, $-1 \le y \le 1$.

33. *(Astroid)*
(a) Consider the curve with equation $x^{2/3} + y^{2/3} = 4$. Find $\frac{dy}{dx}$ at $(-1, 3\sqrt{3})$.
(b) Use a graphing calculator to graph the curve in (a). If the graphing calculator cannot graph implicit functions, graph the upper and the lower half of the curve separately. To get the left half of the graph, make sure that your calculator evaluates $x^{2/3}$ in the order $(x^2)^{1/3}$. Choose the viewing rectangle $-10 \le x \le 10$, $-10 \le y \le 10$.

34. *(Kampyle of Eudoxus)* **(a)** Consider the curve with equation $y^2 = 10x^4 - x^2$. Find $\frac{dy}{dx}$ at $(1, 3)$.
(b) Use a graphing calculator to graph the curve in (a). If the graphing calculator cannot graph implicit functions, graph the upper and the lower half of the curve separately. Choose the viewing rectangle $-3 \le x \le 3$, $-10 \le y \le 10$.

[3.4.3]
35. Assume that x and y are differentiable functions of t. Find $\frac{dy}{dt}$ when $x^2 + y^2 = 1$, $\frac{dx}{dt} = 2$ for $x = \frac{1}{2}$, and $y > 0$.

36. Assume that x and y are differentiable functions of t. Find $\frac{dy}{dt}$ when $y^2 = x^2 - x^4$, $\frac{dx}{dt} = 1$ for $x = \frac{1}{2}$, and $y > 0$.

37. Assume that x and y are differentiable functions of t. Find $\frac{dy}{dt}$ when $x^2 y = 1$ and $\frac{dx}{dt} = 3$ for $x = 2$.

38. Assume that u and v are differentiable functions of t. Find $\frac{du}{dt}$ when $u^2 + v^3 = 12$, $\frac{dv}{dt} = 2$ for $v = 2$, and $u > 0$.

39. Assume that the side length x and the volume $V = x^3$ of a cube are differentiable functions of t. Express dV/dt in terms of dx/dt.

40. Assume that the radius r and the area $A = \pi r^2$ of a circle are differentiable functions of t. Express dA/dt in terms of dr/dt.

41. Assume that the radius r and the surface area $S = 4\pi r^2$ of a sphere are differentiable functions of t. Express dS/dt in terms of dr/dt.

42. Assume that the radius r and the volume $V = \frac{4}{3}\pi r^3$ of a sphere are differentiable functions of t. Express dV/dt in terms of dr/dt.

43. Suppose that water is stored in a cylindrical tank of radius 5 m. If the height of the water in the tank is h, then the volume of the water is $V = \pi r^2 h = (25\text{m}^2)\pi h = 25\pi h \, \text{m}^2$. If we drain the water at a rate of 250 liters per minute, how quickly does the water level inside the tank drop? (Note that one cubic meter contains 1000 liters.)

44. Suppose that we pump water into an inverted right-circular conical tank at the rate of 5 cubic feet per minute (i.e., the tank stands point down). The tank has a height of 6 ft and the radius on top is 3 ft. How fast is the water level rising when the water is 2 ft deep? (Note that the volume of a right-circular cone of radius r and height h is $V = \frac{1}{3}\pi r^2 h$.)

45. Two people start biking from the same point. One bikes east at 15 mph, the other south at 18 mph. How fast is the distance between the two people changing after 20 minutes, and after 40 minutes?

46. Allometric equations describe the scaling relationship between two measurements, such as skull length versus body length. In vertebrates, one typically finds

$$[\text{skull length}] \propto [\text{body length}]^a$$

for some $a \in (0, 1)$. Express the growth rate of the skull length in terms of the growth rate of the body length.

[3.4.4]
In Problems 47–56, find the first and the second derivative of each function.

47. $f(x) = x^3 - 3x^2 + 1$

48. $f(x) = (x^2 - 3)^2$

49. $g(x) = \dfrac{x-1}{x+1}$

50. $h(s) = \dfrac{1}{s^2 + 2}$

51. $g(t) = \sqrt{3t^3 + 2t}$

52. $f(x) = \dfrac{1}{x^2} + x - 1$

53. $f(s) = \sqrt{s^{3/2} - 1}$

54. $f(x) = \dfrac{x}{x+1}$

55. $g(t) = t^{-5/2} - t^{1/2}$

56. $f(x) = x^3 - \dfrac{1}{x^3}$

57. Find the first ten derivatives of $y = x^5$.

58. Find $f^{(n)}(x)$ and $f^{(n+1)}(x)$ of $f(x) = x^n$.

59. Find a second degree polynomial $p(x) = ax^2 + bx + c$ with $p(0) = 3$, $p'(0) = 2$, and $p''(0) = 6$.

60. The position at time t of a particle that moves along a straight line is given by the function $s(t)$. The first derivative of $s(t)$ is called the velocity, denoted by $v(t)$; that is, the velocity is the rate of change of the position. The rate of change of the velocity is called **acceleration**, denoted by $a(t)$, that is,

$$\frac{d}{dt}v(t) = a(t)$$

Given that $v(t) = s'(t)$ it follows that

$$\frac{d^2}{dt^2}s(t) = a(t)$$

Find the velocity and the acceleration at time $t = 1$ for the following position functions.
(a) $s(t) = t^2 - 3t$ **(b)** $s(t) = \sqrt{t^2 + 1}$ **(c)** $s(t) = t^4 - 2t$

◆ 3.5 DERIVATIVES OF TRIGONOMETRIC FUNCTIONS

3.5.1 Trigonometric Functions

We will need the trigonometric limits from Section 2.2 to compute the derivatives of sine and cosine. Note that all angles are measured in radians.

> **Theorem**   The functions $\sin x$ and $\cos x$ are differentiable for all x, and
>
> $$\frac{d}{dx}\sin x = \cos x \qquad \text{and} \qquad \frac{d}{dx}\cos x = -\sin x$$

A sketch of the derivatives of each of the trigonometric functions based on the geometric interpretation of a derivative as the slope of the tangent line confirms these rules (see Figures 3.27 and 3.28).

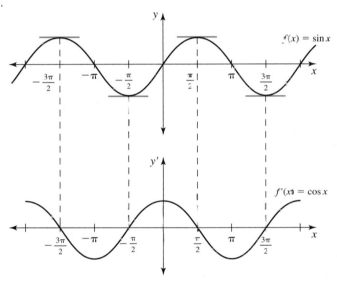

◀ **Figure 3.27**
The function $f(x) = \sin x$ and its derivative $f'(x) = \cos x$. The derivative $f'(x) = 0$ where $f(x)$ has a horizontal tangent line.

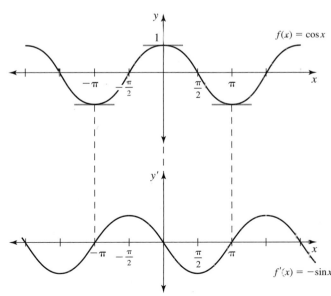

◀ **Figure 3.28**
The function $f(x) = \cos x$ and its derivative $f'(x) = -\sin x$. The derivative $f'(x) = 0$ where $f(x)$ has a horizontal tangent line.

Proof   We prove the first formula; the similar proof of the second formula is discussed in Problem 17. We need the trigonometric identity

$$\sin(\alpha + \beta) = \sin\alpha\cos\beta + \cos\alpha\sin\beta$$

Using the formal definition of derivatives, we find

$$\frac{d}{dx}\sin x = \lim_{h \to 0} \frac{\sin(x + h) - \sin x}{h}$$

$$= \lim_{h \to 0} \frac{\sin x \cos h + \cos x \sin h - \sin x}{h}$$

$$= \lim_{h \to 0} \left[\sin x \frac{\cos h - 1}{h} + \cos x \frac{\sin h}{h}\right]$$

We showed in Section 2.2 that

$$\lim_{h \to 0} \frac{\cos h - 1}{h} = 0 \qquad \text{and} \qquad \lim_{h \to 0} \frac{\sin h}{h} = 1$$

We can therefore apply the basic rules for limits and find

$$\frac{d}{dx}\sin x = \sin x \lim_{h \to 0} \frac{\cos h - 1}{h} + \cos x \lim_{h \to 0} \frac{\sin h}{h}$$

$$= (\sin x)(0) + (\cos x)(1) = \cos x \qquad \blacklozenge$$

▮ **EXAMPLE 1**   Find the derivative of $f(x) = -4\sin x + \cos\frac{\pi}{6}$.

Solution

$$f'(x) = \frac{d}{dx}\left(-4\sin x + \cos\frac{\pi}{6}\right)$$

$$= -4\frac{d}{dx}\sin x + \frac{d}{dx}\cos\frac{\pi}{6}$$

The first term is a trigonometric function and the second term is a constant (namely, $\frac{1}{2}\sqrt{3}$). Differentiating each term, we find

$$= -4(\cos x) + 0$$

$$= -4\cos x \qquad \blacksquare$$

▮ **EXAMPLE 2**   Find the derivative of $y = \cos(x^2 + 1)$.

Solution   We set $u = g(x) = x^2 + 1$ and $f(u) = \cos u$; then $y = f[g(x)]$. Using the chain rule, we find

$$y' = \frac{df}{du}\frac{du}{dx} = \frac{d}{du}(\cos u)\frac{d}{dx}(x^2 + 1) = (-\sin u)(2x)$$

$$= -[\sin(x^2 + 1)]2x = -2x\sin(x^2 + 1) \qquad \blacksquare$$

▮ **EXAMPLE 3**   Find the derivative of $y = x^2\sin(3x) - \cos(5x)$.

Solution   We will use the product rule for the first term; in addition we will need the chain rule for both $\sin(3x)$ and $\cos(5x)$.

$$y' = \frac{d}{dx}[x^2\sin(3x) - \cos(5x)]$$

$$= \frac{d}{dx}[x^2\sin(3x)] - \frac{d}{dx}\cos(5x)$$

$$= \left(\frac{d}{dx}x^2\right)\sin(3x) - x^2\frac{d}{dx}\sin(3x) - \frac{d}{dx}\cos(5x)$$

$$= 2x\sin(3x) + x^2 3\cos(3x) - 5(-\sin(5x))$$

$$= 2x\sin(3x) + 3x^2\cos(3x) + 5\sin(5x) \qquad\blacksquare$$

The derivatives of the other trigonometric functions can be found using the following identities:

$$\tan x = \frac{\sin x}{\cos x} \qquad\qquad \cot x = \frac{\cos x}{\sin x}$$

$$\sec x = \frac{1}{\cos x} \qquad\qquad \csc x = \frac{1}{\sin x}$$

For instance, to find the derivative of the tangent, we use the quotient rule:

$$\frac{d}{dx}\tan x = \frac{d}{dx}\frac{\sin x}{\cos x}$$

$$= \frac{\left(\frac{d}{dx}\sin x\right)\cos x - \sin x\left(\frac{d}{dx}\cos x\right)}{\cos^2 x}$$

$$= \frac{(\cos x)(\cos x) - (\sin x)(-\sin x)}{\cos^2 x}$$

$$= \frac{\cos^2 x + \sin^2 x}{\cos^2 x} = \frac{1}{\cos^2 x} = \sec^2 x$$

In the penultimate step, we used the identity $\cos^2 x + \sin^2 x = 1$.

The other derivatives can be found in a similar fashion, as explained in Problems 18–20.

We summarize the derivatives of all trigonometric functions in the following box.

$$\frac{d}{dx}\sin x = \cos x \qquad\qquad \frac{d}{dx}\cos x = -\sin x$$

$$\frac{d}{dx}\tan x = \sec^2 x \qquad\qquad \frac{d}{dx}\cot x = -\csc^2 x$$

$$\frac{d}{dx}\sec x = \sec x\tan x \qquad\qquad \frac{d}{dx}\csc x = -\csc x\cot x$$

▮ **EXAMPLE 4** Compare the derivatives of $\tan x^2$ and $\tan^2 x$.

Solution

a. If $y = \tan x^2 = \tan(x^2)$, then using the chain rule, we find

$$\frac{dy}{dx} = \frac{d}{dx}\tan(x^2) = (\sec^2(x^2))(2x) = 2x\sec^2(x^2)$$

b. If $y = \tan^2 x = (\tan x)^2$, then using the chain rule, we find

$$\frac{dy}{dx} = \frac{d}{dx}(\tan x)^2 = 2(\tan x)\frac{d}{dx}\tan x = 2\tan x\sec^2 x$$

The two derivatives are clearly different, and you should look again at $\tan x^2$ and $\tan^2 x$ to make sure that you understand which is the inner and which function is the outer function.   **I**

Repeated Application of the Chain Rule

I EXAMPLE 5   Find the derivative of $f(x) = \sec \sqrt{x^2 + 1}$.

Solution   This is a composite function; the inner function is $\sqrt{x^2 + 1}$, and the outer function is $\sec x$. Applying the chain rule once, we find

$$\frac{df}{dx} = \frac{d}{dx} \sec \sqrt{x^2 + 1} = \sec \sqrt{x^2 + 1} \tan \sqrt{x^2 + 1} \frac{d}{dx} \sqrt{x^2 + 1}$$

To evaluate $\frac{d}{dx} \sqrt{x^2 + 1}$, we need to apply the chain rule a second time:

$$\frac{d}{dx} \sqrt{x^2 + 1} = \frac{1}{2\sqrt{x^2 + 1}} 2x = \frac{x}{\sqrt{x^2 + 1}}$$

Combining the two steps, the result is

$$\frac{df}{dx} = \sec \sqrt{x^2 + 1} \tan \sqrt{x^2 + 1} \frac{x}{\sqrt{x^2 + 1}}$$

The function $f(x)$ can be thought of as a composition of three functions. The innermost function is $u = g(x) = x^2 + 1$, the middle is $v = h(u) = \sqrt{u}$, and the outermost function is $f(v) = \sec v$. When we computed the derivative, we applied the chain rule twice in the form

$$\frac{df}{dx} = \frac{df}{dv} \frac{dv}{du} \frac{du}{dx}$$ **I**

◆ 3.5 PROBLEMS

In Problems 1–15 find the derivative with respect to the indicated variable.

1. $f(x) = 3 \sin x + 5 \cos x - 2 \sec x$

2. $f(x) = 3 \sin(x^2)$

3. $f(x) = 3 \sin^2 x$

4. $g(s) = \sqrt{\cos s}$

5. $g(t) = \dfrac{\sin(3t)}{\cos(5t)}$

6. $g(t) = \sec(t^2 - 3)$

7. $f(x) = \cos^2(x^2 - 1)$

8. $f(x) = \sin x \cos x$

9. $f(x) = \dfrac{1}{\sec x + \tan x}$

10. $g(x) = \dfrac{1}{\csc(5x)}$

11. $h(x) = \dfrac{3}{\tan(2x) - x}$

12. $h(x) = \cot(3x) \csc(3x)$

13. $h(s) = \sin^2 s + \cos^2 s$

14. $g(t) = \dfrac{1}{\sqrt{\sin t^2}}$

15. Find the points on the curve $y = \sin\left(\frac{\pi}{3}x\right)$ that have a horizontal tangent.

16. Find the points on the curve $y = \cos^2 x$ that have a horizontal tangent.

17. Use the identity

$$\cos(\alpha + \beta) = \cos \alpha \cos \beta - \sin \alpha \sin \beta$$

and the definition of the derivative to show that

$$\frac{d}{dx} \cos x = -\sin x$$

18. Use the quotient rule to show that

$$\frac{d}{dx} \cot x = -\csc^2 x$$

(*Hint*: Write $\cot x = \frac{\cos x}{\sin x}$.)

19. Use the chain rule to show that

$$\frac{d}{dx} \sec x = \sec x \tan x$$

(*Hint*: Write $\sec x = (\cos x)^{-1}$.)

20. Use the chain rule to show that

$$\frac{d}{dx} \csc x = -\csc x \cot x$$

(*Hint*: Write $\csc x = (\sin x)^{-1}$.)

21. Suppose that the concentration of nitrogen in a lake shows periodic behavior. That is, if we denote by $c(t)$ the concentration of nitrogen at time t, then we assume

$$c(t) = 2 + \sin\left(\frac{\pi}{2}t\right)$$

(a) Find

$$\frac{dc}{dt}$$

(b) Use a graphing calculator to graph both $c(t)$ and $\frac{dc}{dt}$ in the same coordinate system.

(c) By inspecting the graph in (b) answer the following questions:

(i)   When $c(t)$ reaches a maximum, what is the value of dc/dt?

(ii) When dc/dt is positive, is $c(t)$ increasing or decreasing?

(iii) What can you say about $c(t)$ when $dc/dt = 0$?

◆ 3.6 DERIVATIVES OF EXPONENTIAL FUNCTIONS

The basic problem of finding tangent lines can be satisfactorily answered within the framework of the differential calculus. As we have noted, this basic problem was one of the driving forces in the development of the differential calculus.

Pierre de Fermat (1601–1665) devised a method of finding the tangent line at a given point (x, y) on a curve by constructing the **subtangent**. The subtangent is defined as the line segment between the point where the tangent line intersects the x-axis and the point $(x, 0)$ (see Figure 3.29).

Fermat's procedure essentially amounts to finding the slope of the tangent line by considering the secant line through (x, y) and a nearby point. After computing the slope of the secant line, he set the two points he used for computing the slope of the secant line equal to each other, thus obtaining the slope of the tangent line. This sounds very much like the definition of the derivative that we use today and, in fact, it is the same idea. Fermat did not, however, develop and formalize a general framework for the differential calculus; this was done by Leibniz and Newton.

Using the definition of derivatives, we can relate the subtangent to the slope of the tangent at that point. Suppose the tangent line at (x, y) intersects the x-axis at $(p, 0)$, where p is a function of x. We set $c(x) = x - p(x)$. This is the equation of the subtangent. We see from Figure 3.29 that the slope of the tangent line at (x, y) is given by $y/c(x)$. Since the slope of the tangent line at (x, y) is the derivative of the function of the curve evaluated at x (i.e., $f'(x)$), we find

$$\frac{dy}{dx} = \frac{y}{c(x)}$$

A natural problem (which was posed to Descartes by Debaune in 1639) is to find a curve whose subtangent is a given constant. That is, we wish to find the function $y = f(x)$ that satisfies

$$\frac{dy}{dx} = \frac{y}{c}$$

where c is a constant other than 0. (This problem was solved by Leibniz in 1684,

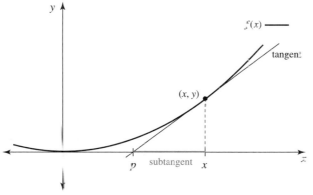

◀ **Figure 3.29**

The subtangent problem.

when he published his differential calculus for the first time.) In words, $y = f(x)$ is a function whose derivative is proportional to the function itself. As we will see in the following, exponential functions are the solutions to this problem.

Recall from Section 1.2 that the function f is an exponential function with base a if

$$f(x) = a^x, \quad x \in \mathbf{R}$$

where a is a positive constant other than 1 (see Figure 3.30). We can use the formal definition of the derivative to compute $f'(x)$, namely,

$$f'(x) = \frac{d}{dx}a^x = \lim_{h \to 0} \frac{f(x+h) - f(x)}{h} = \lim_{h \to 0} \frac{a^{x+h} - a^x}{h}$$

$$= \lim_{h \to 0} \frac{a^x(a^h - 1)}{h} = a^x \lim_{h \to 0} \frac{a^h - 1}{h}$$

In the final step we were able to write the term a^x in front of the limit since a^x does not depend on h. Thus, we are left with investigating

$$\lim_{h \to 0} \frac{a^h - 1}{h}$$

We first note that the limit does not depend on x. If we assume that this limit exists, then it follows that it is equal to $f'(0)$—to see why, use the formal definition of the derivative to compute $f'(0)$. (It can be shown that this limit exists, but this is beyond the scope of this course.)

It follows from the preceding calculation that if $f'(0)$ exists, then $f'(x)$ exists, and

$$f'(x) = a^x f'(0)$$

This shows that the exponential function is a function whose derivative is proportional to the function itself. (The constant of proportionality is $f'(0)$.) That is, exponential functions solve the subtangent problems just mentioned. We single out the case where the value of the base a is such that

$$\lim_{h \to 0} \frac{a^h - 1}{h}$$

is equal to 1. We denote this base by e. The number e is thus defined as

$$\boxed{\lim_{h \to 0} \frac{e^h - 1}{h} = 1} \tag{3.6}$$

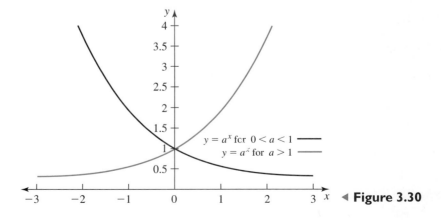

$y = a^x$ for $0 < a < 1$
$y = a^x$ for $a > 1$

◀ **Figure 3.30**

and hence

$$\frac{d}{dx}e^x = e^x \tag{3.7}$$

A graph of $f(x) = e^x$ is shown in Figure 3.31. The domain of this function is **R** and its range is the open interval $(0, \infty)$ (in particular, $e^x > 0$ for all $x \in \mathbf{R}$). Denoting by e the base of the exponential function for which (3.6) and (3.7) hold is no accident. It is indeed the natural exponential base that we introduced in Section 1.2. Though we cannot prove this here, the following table should convince you (recall that $e = 2.71828\ldots$).

h	0.1	0.01	0.001	0.0001
$\frac{e^h-1}{h}$	1.0517	1.0050	1.00050	1.000050

Recall also that there is an alternative notation for e^x, namely $\exp[x]$. Using the identity

$$a^x = \exp[\ln a^x]$$

and the fact that $\ln a^x = x \ln a$, we can find the derivative of a^x with the help of the chain rule, namely,

$$\frac{d}{dx}a^x = \frac{d}{dx}\exp[\ln a^x] = \frac{d}{dx}\exp[x \ln a]$$
$$= \exp[x \ln a]\ln a = (\ln a)a^x$$

That is, we find

$$\frac{d}{dx}a^x = (\ln a)a^x \tag{3.8}$$

which allows us to obtain the following identity:

$$\lim_{h\to 0}\frac{a^h - 1}{h} = \ln a \tag{3.9}$$

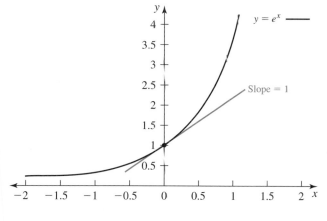

◄ Figure 3.31
The function $f(x) = e^x$. The slope of the tangent line at $x = 1$ is $m = 1$.

EXAMPLE I   Find the derivative of $f(x) = e^{-x^2/2}$.

Solution   We use the chain rule:

$$f'(x) = e^{-x^2/2}\left(-\frac{2x}{2}\right) = -xe^{-x^2/2}$$ **I**

EXAMPLE 2   Find the derivative of $f(x) = 3^{\sqrt{x}}$.

Solution   We can use (3.8) and the chain rule to find

$$\frac{d}{dx}3^{\sqrt{x}} = (\ln 3)3^{\sqrt{x}}\frac{1}{2\sqrt{x}}$$

However, since every exponential function can be written in terms of the base e, and the differentiation rule for e^x is particularly simple $\left(\frac{d}{dx}e^x = e^x\right)$, it is often easier to rewrite the exponential function in terms of e and then to differentiate. That is, we write

$$3^{\sqrt{x}} = \exp[\ln 3^{\sqrt{x}}] = \exp[\sqrt{x}\ln 3]$$

Then, using the chain rule, we find

$$\frac{d}{dx}\exp[\sqrt{x}\ln 3] = \frac{\ln 3}{2\sqrt{x}}\exp[\sqrt{x}\ln 3] = \frac{\ln 3}{2\sqrt{x}}3^{\sqrt{x}}$$ **I**

The exponential function with base e appears in many scientific problems; the following example involves radioactive decay.

EXAMPLE 3   **Radioactive Decay**

Find the derivative of the following radioactive decay function, which describes the amount of material left after t units of time (see Example 9 in Subsection 1.2.5):

$$W(t) = W_0e^{-\lambda t}, \quad t \geq 0$$

and show that it satisfies the differential equation

$$\frac{dW}{dt} = -\lambda W(t)$$

Solution   We use the chain rule to find the derivative of $W(t)$,

$$\frac{d}{dt}W(t) = \underbrace{W_0e^{-\lambda t}}_{W(t)}(-\lambda)$$

that is,

$$\frac{dW}{dt} = -\lambda W(t)$$

In words, the rate of decay is proportional to the amount of material left. This should remind you of the subtangent problem; there, we wanted to find a function whose derivative is proportional to the function itself. This is exactly the situation that we have in this example: The derivative of $W(t)$ is proportional to $W(t)$.   **I**

Since we must frequently differentiate functions of the form $y = e^{g(x)}$, we state this as a separate rule. Using the chain rule, we find

$$\boxed{\frac{d}{dx}e^{g(x)} = g'(x)e^{g(x)}}$$ (3.10)

EXAMPLE 4 Find the derivative of $f(x) = \exp[\sin\sqrt{x}]$.

Solution We set $g(x) = \sin\sqrt{x}$. To differentiate $g(x)$, we must apply the chain rule,

$$\frac{d}{dx}g(x) = (\cos\sqrt{x})\frac{1}{2\sqrt{x}}$$

We can now differentiate $f(x)$ using equation (3.10),

$$\frac{d}{dx}f(x) = (\cos\sqrt{x})\frac{1}{2\sqrt{x}}\exp[\sin\sqrt{x}]$$

Finally, we give an example that shows how (3.9) is used.

EXAMPLE 5 Find

$$\lim_{h\to0}\frac{3^{2h}-1}{h}$$

Solution We make the substitution $l = 2h$, and note that $l \to 0$ as $h \to 0$. Then

$$\lim_{h\to0}\frac{3^{2h}-1}{h} = \lim_{l\to0}\frac{3^l-1}{l/2}$$

$$= 2\lim_{l\to0}\frac{3^l-1}{l} = 2\ln3$$

◆ 3.6 PROBLEMS

Differentiate the functions in Problems 1–12 with respect to the indicated variable.

1. $f(x) = e^{2x}$

2. $f(x) = e^{-4x}$

3. $f(x) = e^{-2x^2+3x-1}$

4. $f(x) = e^{7x^3-\sqrt{x}+3}$

5. $f(x) = e^{\sin x}$

6. $f(x) = \sin(e^x)$

7. $g(s) = \exp[\sec s^2]$

8. $g(s) = \exp[\tan s^3]$

9. $h(t) = 2^{t^2-1}$

10. $h(t) = 5^{\sqrt{t}}$

11. $g(r) = 2^{-3\sin r}$

12. $g(r) = 3^{r^{1/5}}$

Compute the following limits in Problems 13–16.

13. $\lim_{h\to0}\dfrac{e^{2h}-1}{h}$

14. $\lim_{h\to0}\dfrac{e^{5h}-1}{3h}$

15. $\lim_{h\to0}\dfrac{e^h-1}{\sqrt{h}}$

16. $\lim_{h\to0}\dfrac{2^h-1}{h}$

17. Find the length of the subtangent at the point $(1, 2)$ for the curve $y = 2^x$.

18. Find the length of the subtangent at the point $(2, e^4)$ for the curve $y = \exp[x^2]$.

19. Suppose that a bacterial colony grows in such a way that at time t, the population size is

$$N(t) = N(0)2^t$$

where $N(0)$ is the population size at time 0. Find the rate of growth; that is, find dN/dt and express your solution in terms

of $N(t)$. Show that the growth rate of the population is proportional to the population size.

20. (a) Find the derivative of the logistic growth curve (see Example 2 in Section 2.4)

$$N(t) = \frac{K}{1 + \left(\frac{K}{N(0)} - 1\right)e^{-rt}}$$

where r and K are positive constants, and $N(0)$ is the population size at time 0.

(b) Show that $N(t)$ satisfies the equation

$$\frac{dN}{dt} = rN\left(1 - \frac{N}{K}\right)$$

(*Hint*: Use the function $N(t)$ given in (a) for the right-hand side, and simplify until you obtain the derivative of $N(t)$ that you computed in (a).)

(c) Plot the per capita rate of growth, $\frac{1}{N}\frac{dN}{dt}$, as a function of N, and note that it decreases with increasing population size.

21. The following model is used in the fisheries literature to describe the recruitment of fish as a function of the size of the parent stock. If we denote the number of recruits by R and the size of the parent stock by P, then

$$R(P) = \alpha Pe^{-\beta P}, \quad P \geq 0$$

where α and β are positive constants.

(a) Sketch the graph of the function $R(P)$ when $\beta = 1$ and $\alpha = 2$.

(b) Differentiate $R(P)$ with respect to P.

(c) Find all the points on the curve that have a horizontal tangent.

22. The growth of fish can be described by the von Bertalanffy growth function

$$L(x) = L_\infty - (L_\infty - L_0)e^{-kx}$$

where x denotes the age of the fish and k, L_∞, and L_0 are positive constants.

(a) Set $L_0 = 1$ and $L_\infty = 10$. Graph $L(x)$ for $k = 1.0$ and $k = 0.1$.

(b) Interpret L_∞ and L_0.

(c) If you compare the graphs for $k = 0.1$ and $k = 1.0$, in which do fish reach $L = 5$ more quickly?

(d) Show that

$$\frac{d}{dx}L(x) = k(L_\infty - L(x))$$

that is, $dL/dx \propto L_\infty - L$. What does this say about how the rate of growth changes with age?

(e) The constant k is the proportionality constant in (d). What does the value of k tell you about how quickly a fish grows?

23. Suppose $W(t)$ denotes the amount of a radioactive material left after time t. Assume that $W(0) = 10$ and $W(1) = 8$.

(a) Find the differential equation that describes this situation.

(b) How much material is left at time $t = 5$?

(c) What is the half-life of this material?

◆ 3.7 DERIVATIVES OF INVERSE AND LOGARITHMIC FUNCTIONS

Recall that the logarithmic function is the inverse of the exponential function; to find the derivative of the logarithmic function, we must therefore find the derivative of an inverse.

3.7.1 Derivatives of Inverse Functions

We begin with an example (see Figure 3.32). Let $f(x) = x^2$, $x \geq 0$. We computed the inverse function of f in Subsection 1.2.6. First note that $f(x) = x^2$, $x \geq 0$, is one-to-one (use the horizontal line test); hence, we can define its inverse. We repeat the steps to find an inverse function. (Recall that we obtain the inverse function by reflecting $y = f(x)$ about the line $y = x$.)

1. Write $y = f(x)$:

$$y = x^2$$

2. Solve for x:

$$x = \sqrt{y}$$

3. Interchange x and y:

$$y = \sqrt{x}$$

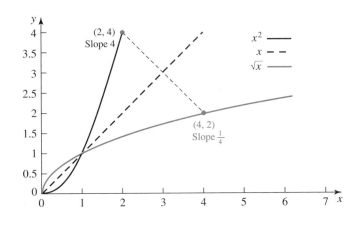

◀ **Figure 3.32**

The function $y = x^2$, $x \geq 0$, and its inverse $y = \sqrt{x^2}$, $x \geq 0$.

Since the range of $f(x)$, which is the interval $[0, \infty)$, becomes the domain for the inverse function, we find

$$f^{-1}(x) = \sqrt{x} \quad \text{for } x \geq 0$$

Let $g(x) = f^{-1}(x)$. Then

$$(f \circ g)(x) = f[g(x)] = (\sqrt{x})^2 = x, \quad x \geq 0$$

Therefore, the derivatives of $(\sqrt{x})^2$ and x must be equal. Applying the chain rule, we find

$$\frac{d}{dx}(\sqrt{x})^2 = 2\sqrt{x}\frac{d}{dx}\sqrt{x}$$

Since $\frac{d}{dx}x = 1$, we obtain

$$2\sqrt{x}\frac{d}{dx}\sqrt{x} = 1$$

or, for $x > 0$,

$$\frac{d}{dx}\sqrt{x} = \frac{1}{2\sqrt{x}}$$

In Figure 3.32, the slope at the point $(2, 4)$ of $y = f(x)$ is $m = 4$, and the slope at the point $(4, 2)$ of $y = f^{-1}(x)$ is $m = 1/4$. We will find this reciprocal relationship of slopes at related points also in the general case.

The steps that led us to the derivative of $\sqrt{x}$ can be used to find a general formula for the derivative of inverse functions. If $g(x)$ is the inverse function of $f(x)$, then $f[g(x)] = x$. Following the same steps, for the general case $f[g(x)] = x$, we find

$$\frac{d}{dx}f[g(x)] = f'([g(x)]g'(x)$$

Since $\frac{d}{dx}x = 1$, we obtain

$$f'[g(x)]g'(x) = 1$$

If $f'[g(x)] \neq 0$, we can divide by $f'[g(x)]$ and find

$$g'(x) = \frac{1}{f'[g(x)]}$$

Since $g(x) = f^{-1}(x)$ and $g'(x) = \frac{d}{dx}g(x) = \frac{d}{dx}f^{-1}(x)$, we obtain

$$\boxed{\frac{d}{dx}f^{-1}(x) = \frac{1}{f'[f^{-1}(x)]}} \tag{3.11}$$

This reciprocal relationship is illustrated in Figure 3.33.

We return to our example, $f(x) = x^2$, $x \geq 0$, where $f^{-1}(x) = \sqrt{x}$, $x \geq 0$. Now, $f'(x) = 2x$. We need to evaluate $f'[f^{-1}(x)]$, that is,

$$f'[f^{-1}(x)] = f'[\sqrt{x}] = 2\sqrt{x}$$

In the above calculation, we needed to assume that $f'[f^{-1}(x)] \neq 0$. We therefore find

$$\frac{d}{dx}\sqrt{x} = \frac{1}{2\sqrt{x}} \quad \text{for } x > 0$$

Looking at the graphs of $y = x^2$ and $y = \sqrt{x}$, $x \geq 0$, it is easy to see why $f^{-1}(x)$ is not differentiable at $x = 0$. (Recall that we obtain the inverse function by reflecting $y = f(x)$ about the line $y = x$.) If we draw the tangent line at $x = 0$ on the curve $y = x^2$, we find that the tangent line is horizontal, that is, its slope is 0. Reflecting

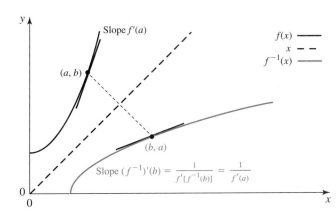

◀ **Figure 3.33**
The graphs of $y = f(x)$ and its inverse $y = f^{-1}(x)$ have reciprocal slopes at the points (a, b) and (b, a).

a horizontal line about $y = x$, results in a vertical line for which the slope is not defined (see Figure 3.32).

The formula for finding derivatives of inverse functions takes on a particularly easy to remember form when we use Leibniz notation. To see this, note that (without interchanging x and y)

$$y = f(x) \qquad \Leftrightarrow \qquad x = f^{-1}(y)$$

and hence

$$\boxed{\frac{dx}{dy} = \frac{1}{\frac{dy}{dx}}}$$

This formula again emphasizes the reciprocal relationship. We illustrate the formula using our example

$$y = x^2 \qquad \Leftrightarrow \qquad x = \sqrt{y}$$

for $x > 0$. Since $\frac{dy}{dx} = 2x$, we find

$$\frac{dx}{dy} = \frac{1}{\frac{dy}{dx}} = \frac{1}{2x} = \frac{1}{2\sqrt{y}}$$

That is,

$$\frac{d}{dy}\sqrt{y} = \frac{1}{2\sqrt{y}}$$

The answer is now in terms of y. This comes from the fact that we did not interchange x and y when we computed the inverse function. If we do so, we again find

$$\frac{d}{dx}\sqrt{x} = \frac{1}{2\sqrt{x}}$$

▮EXAMPLE 1   Let

$$f(x) = \frac{x}{1+x} \quad \text{for } x \geq 0$$

Find $\frac{d}{dx} f^{-1}\left(\frac{1}{3}\right)$.

Solution   To show that $f^{-1}(x)$ exists, we use the horizontal line test and conclude that $f(x)$ is one-to-one on its domain, since each horizontal line intersects the graph of $f(x)$ at most once (see Figure 3.34).

We can actually compute the inverse of $f(x)$. This will give us two different ways to compute the derivative of the inverse of $f(x)$; namely, we can compute the

◀ **Figure 3.34**
The graph of $f(x) = \frac{x}{x-1}$ for $x \geq 0$.

inverse function explicitly and then differentiate the result, or we can use the formula for finding derivatives of inverse functions. We begin with the latter way.

(1) To use the formula (3.11), we need to find $f^{-1}\left(\frac{1}{3}\right)$; this means, we need to find x so that $f(x) = 1/3$. Now,

$$\frac{x}{1+x} = \frac{1}{3} \qquad \text{implies} \qquad 2x = 1 \qquad \text{or} \qquad x = \frac{1}{2}$$

Therefore, $f^{-1}\left(\frac{1}{3}\right) = \frac{1}{2}$. Formula (3.11) becomes

$$\frac{d}{dx} f^{-1}\left(\frac{1}{3}\right) = \frac{1}{f'\left[f^{-1}\left(\frac{1}{3}\right)\right]} = \frac{1}{f'\left(\frac{1}{2}\right)}$$

We use the quotient rule to find the derivative of $f(x)$.

$$f'(x) = \frac{(1)(1+x) - (x)(1)}{(1+x)^2} = \frac{1}{(1+x)^2}$$

At $x = 1/2$, $f'\left(\frac{1}{2}\right) = \frac{1}{(1+1/2)^2} = \frac{4}{9}$. Therefore,

$$\frac{d}{dx} f^{-1}\left(\frac{1}{3}\right) = \frac{1}{\frac{4}{9}} = \frac{9}{4}$$

(2) We can compute the inverse function: Set $y = \frac{x}{1+x}$. Solving for x yields

$$x = \frac{y}{1-y}$$

Interchanging x and y, we find

$$y = \frac{x}{1-x}$$

Since the domain of $f(x)$ is $[0, \infty)$, the range of f is $[0, 1)$. The range of f becomes the domain of the inverse; therefore, the inverse function is

$$f^{-1}(x) = \frac{x}{1-x} \qquad \text{for } 0 \leq x < 1$$

We use the quotient rule to find $\frac{d}{dx} f^{-1}(x)$:

$$\frac{d}{dx} f^{-1}(x) = \frac{(1)(1-x) - (x)(-1)}{(1-x)^2} = \frac{1}{(1-x)^2}$$

and therefore

$$\frac{d}{dx} f^{-1}\left(\frac{1}{3}\right) = \frac{1}{(1-1/3)^2} = \frac{9}{4}$$

which agrees with the answer in (1).

The inverse cannot always be computed explicitly.

EXAMPLE 2   Let

$$f(x) = 2x + e^x \quad \text{for } x > 0$$

Find $\frac{d}{dx} f^{-1}(1)$.

Solution   In this case, it is not possible to solve $y = 2x + e^x$ for x. Therefore, we must use (3.11) if we wish to compute the derivative of the inverse function at a particular point. Equation (3.11) becomes

$$\frac{d}{dx} f^{-1}(1) = \frac{1}{f'[f^{-1}(1)]}$$

when $x = 1$. We need to find $f'(x)$:

$$f'(x) = 2 + e^x$$

Since $f(0) = 1$, it follows that $f^{-1}(1) = 0$ and hence

$$\frac{1}{f'[f^{-1}(1)]} = \frac{1}{f'(0)} = \frac{1}{2+1} = \frac{1}{3} \qquad ∎$$

The next example, in which we again need to use (3.11), involves finding the derivative of the inverse of a trigonometric function.

EXAMPLE 3   Let $f(x) = \tan x$. Find $\frac{d}{dx} f^{-1}(1)$.

Solution   We first need to find $f'(x)$. Recall that

$$f'(x) = \sec^2 x$$

Now, since $f\left(\frac{\pi}{4}\right) = \tan \frac{\pi}{4} = 1$, it follows that $f^{-1}(1) = \frac{\pi}{4}$. We therefore find

$$\frac{d}{dx} f^{-1}(1) = \frac{1}{f'[f^{-1}(1)]} = \frac{1}{f'\left(\frac{\pi}{4}\right)} = \frac{1}{\sec^2\left(\frac{\pi}{4}\right)}$$

$$= \cos^2\left(\frac{\pi}{4}\right) = \left(\frac{1}{2}\sqrt{2}\right)^2 = \frac{1}{2} \qquad ∎$$

If we define $f(x) = \tan x$ on the domain $\left(-\frac{\pi}{2}, \frac{\pi}{2}\right)$, then $f(x)$ is one-to-one, as can be seen from Figure 3.35 (use the horizontal line test). The range of $f(x)$ is $(-\infty, \infty)$. We cannot use algebra to solve $y = \tan x$ for x; instead, the inverse of tan gets its own name. It is called $y = \arctan x$ (or $y = \tan^{-1} x$) and its domain is $(-\infty, \infty)$. In the next example, we will find the derivative of $y = \arctan x$, which will turn out to have a surprisingly simple form.

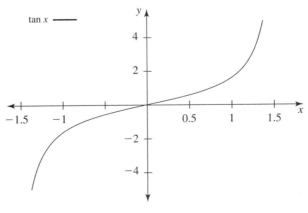

◄ **Figure 3.35**

The function
$f(x) = \tan x$, $(-\frac{\pi}{2}, \frac{\pi}{2})$, is one-to-one on its domain.

EXAMPLE 4   Let $f(x) = \tan x$, $-\frac{\pi}{2} < x < \frac{\pi}{2}$. Find $\frac{d}{dx} f^{-1}(x)$.

Solution   As mentioned above, $f^{-1}(x)$ exists since $f(x)$ is one-to-one on its domain. Recall that

$$\frac{d}{dx} \tan x = \sec^2 x$$

The inverse of tan is denoted by $\tan^{-1} x$ or $\arctan x$. We set $y = \arctan x$ (and hence $x = \tan y$). Then

$$\frac{dy}{dx} = \frac{d}{dx} \arctan x = \frac{1}{\frac{dx}{dy}} = \frac{1}{\frac{d}{dy} \tan y} = \frac{1}{\sec^2 y} = \cos^2 y$$

We can simplify this expression using the following trick, which will allow us to write $\cos^2 y$ in terms of x. Since $x = \tan y$, we find

$$x^2 = \tan^2 y = \frac{\sin^2 y}{\cos^2 y} = \frac{1 - \cos^2 y}{\cos^2 y} = \frac{1}{\cos^2 y} - 1$$

In the second step, we used the identity $\tan y = \frac{\sin y}{\cos y}$, and in the third step, we used the identity $\sin^2 y + \cos^2 y = 1$. Solving the preceding expression for $\cos^2 y$, we get

$$\cos^2 y = \frac{1}{1 + x^2}$$

Therefore,

$$\frac{d}{dx} \arctan x = \cos^2 y = \frac{1}{1 + x^2}$$ ▮

The result in the preceding example is important enough that we summarize it in the following box.

$$\frac{d}{dx} \arctan x = \frac{1}{1 + x^2}$$

The derivative of the inverse sine function, $y = \arcsin x$, is discussed in Problem 12 of this section. The derivatives of the remaining inverse trigonometric functions are listed in the table of derivatives on the inside back cover of this book.

3.7.2 The Derivative of the Logarithmic Function

We introduced the logarithmic function to base a, $\log_a x$, as the inverse function of the exponential function a^x (see Figures 3.36 and 3.37). We can therefore use the formula for derivatives of inverse functions to find the derivative of $y = \log_a x$.

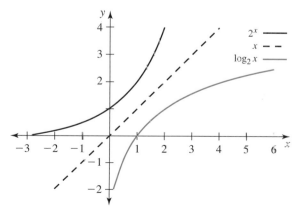

◀ **Figure 3.36**

The function $y = \log_2 x$ as the inverse of $y = 2^x$.

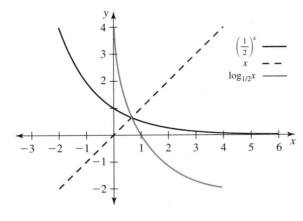

◀ **Figure 3.37**
The function $y = \log_{1/2} x$ as the inverse of $y = \left(\frac{1}{2}\right)^x$.

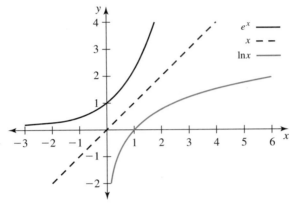

◀ **Figure 3.38**
The natural logarithm $y = \ln x$ as the inverse of the natural exponential function $y = e^x$.

Since

$$\log_a x = \frac{\ln x}{\ln a}$$

it is enough to find the derivative of $\ln x$ (see Figure 3.38). We set $y = e^x$; then $f'(x) = e^x$ and $f^{-1}(x) = \ln x$. Therefore,

$$\frac{d}{dx} \ln x = \frac{d}{dx}[f^{-1}(x)] = \frac{1}{f'[f^{-1}(x)]} = \frac{1}{\exp[\ln x]} = \frac{1}{x}$$

We summarize this in the following box

$$\frac{d}{dx} \ln x = \frac{1}{x}$$

$$\frac{d}{dx} \log_a x = \frac{1}{(\ln a)x}$$

EXAMPLE 5 Find the derivative of $y = \ln(3x)$.

Solution   We use the chain rule with $u = g(x) = 3x$ and $f(u) = \ln u$:

$$\frac{dy}{dx} = \frac{dy}{du}\frac{du}{dx} = \frac{1}{u}3 = \frac{3}{3x} = \frac{1}{x}$$

If you are surprised that the factor 3 disappeared, note that

$$y = \ln(3x) = \ln 3 + \ln x$$

Since $\ln 3$ is a constant, its derivative is 0. Hence,

$$\frac{d}{dx}(\ln 3 + \ln x) = \frac{d}{dx}\ln 3 + \frac{d}{dx}\ln x = 0 + \frac{1}{x} = \frac{1}{x}$$

I EXAMPLE 6   Find the derivative of $y = \ln x^2$, $x \neq 0$.

Solution   We can use the chain rule with $u = g(x) = x^2$ and $f(u) = \ln u$. We find

$$y' = \frac{df}{du}\frac{du}{dx} = \frac{1}{u}2x = \frac{1}{x^2}2x = \frac{2}{x}\quad\text{for } x \neq 0$$

The preceding example is of the form $y = \ln f(x)$. We will frequently encounter such functions; to find their derivatives, we need to use the chain rule, as shown in the following box.

$$\frac{d}{dx}\ln f(x) = \frac{f'(x)}{f(x)}$$

I EXAMPLE 7   Differentiate $y = \ln(\sin x)$.

Solution   This function is also of the form $y = \ln f(x)$, with $f(x) = \sin x$. Since

$$\frac{d}{dx} = \sin x = \cos x$$

we find

$$\frac{dy}{dx} = \frac{\cos x}{\sin x} = \cot x$$

3.7.3 Logarithmic Differentiation

In 1695, Leibniz introduced logarithmic differentiation, following a suggestion from Johann Bernoulli to find derivatives of functions of the form $y = [f(x)]^x$. Bernoulli generalized this method and published his results two years later. The basic idea is simply to take logarithms on both sides and then to differentiate.

I EXAMPLE 8   Find $\frac{dy}{dx}$ when $y = x^x$.

Solution   We take logarithms on both sides of the equation $y = x^x$,

$$\ln y = \ln x^x$$

Applying properties of the logarithm, we can simplify the right-hand side as $\ln x^x = x \ln x$. We can now differentiate both sides with respect to x:

$$\frac{d}{dx}[\ln y] = \frac{d}{dx}[x \ln x]$$

$$\frac{1}{y}\frac{dy}{dx} = 1 \cdot \ln x + x\frac{1}{x}$$

$$\frac{dy}{dx} = y[\ln x + 1]$$

$$\frac{dy}{dx} = (\ln x + 1)x^x$$

If the function $y = x^x$ looks strange, write it as

$$y = x^x = \exp[\ln x^x] = \exp[x \ln x]$$

that is, $y = e^{x \ln x}$. We can differentiate this function without using logarithmic differentiation; namely,

$$\frac{dy}{dx} = e^{x \ln x} \frac{d}{dx}(x \ln x)$$

$$= e^{x \ln x}\left(1 \cdot \ln x + x\frac{1}{x}\right)$$

$$= e^{x \ln x}(\ln x + 1)$$

Either approach will give you the correct answer.

▎**EXAMPLE 9**   Find the derivative of $y = (\sin x)^x$.

Solution   We take logarithms on both sides of the equation and simplify:

$$\ln y = x \ln(\sin x)$$

Differentiating with respect to x yields

$$\frac{d}{dx}[\ln y] = \frac{d}{dx}[x \ln \sin x]$$

$$\frac{1}{y}\frac{dy}{dx} = 1 \cdot \ln \sin x + x\frac{d}{dx}(\ln \sin x)$$

$$= \ln(\sin x) + x\frac{\cos x}{\sin x}$$

$$= \ln(\sin x) + x \cot x$$

Hence, after multiplying by y and substituting $(\sin x)^x$ for y,

$$\frac{dy}{dx} = [\ln(\sin x) + x \cot x](\sin x)^x$$

▎

The following example will probably convince you that logarithmic differentiation can simplify finding the derivatives of complicated expressions.

▎**EXAMPLE 10**   Differentiate

$$y = \frac{e^x x^{3/2}\sqrt{1+x}}{(x^2+3)^4(3x-2)^3}$$

Solution   Without logarithmic differentiation, this would be rather difficult. Taking logarithms on both sides, however, we see that the expression simplifies. It is very important that we apply the properties of the logarithm before differentiating, as this will simplify the expressions that we must differentiate.

$$\ln y = \ln \frac{e^x x^{3/2}\sqrt{1+x}}{(x^2+3)^4(3x-2)^3}$$

$$= \ln e^x + \ln x^{3/2} + \ln \sqrt{1+x} - \ln(x^2+3)^4 - \ln(3x-2)^3$$

$$= x + \frac{3}{2}\ln x + \frac{1}{2}\ln(1+x) - 4\ln(x^2+3) - 3\ln(3x-2)$$

This no longer looks so terrible, and we can differentiate both sides:

$$\frac{d}{dx}[\ln y] = \frac{d}{dx}\left[x + \frac{3}{2}\ln x + \frac{1}{2}\ln(1+x) - 4\ln(x^2+3) - 3\ln(3x-2)\right]$$

$$\frac{1}{y}\frac{dy}{dx} = 1 + \frac{3}{2}\frac{1}{x} + \frac{1}{2}\frac{1}{1+x} - 4\frac{2x}{x^2+3} - 3\frac{3}{3x-2}$$

Finally, solving for dy/dx yields

$$\frac{dy}{dx} = \left(1 + \frac{3}{2x} + \frac{1}{2(1+x)} - \frac{8x}{x^2+3} - \frac{9}{3x-2}\right)\frac{e^x x^{3/2}\sqrt{1+x}}{(x^2+3)^4(3x-2)^3}$$ ▮

We can also use this method to prove the general power rule (as promised in Section 3.3).

POWER RULE (GENERAL FORM)

Let $f(x) = x^r$ where r is any real number. Then

$$\frac{d}{dx}(x^r) = rx^{r-1}$$

Proof   We set $y = x^r$, and use logarithmic differentiation to find

$$\frac{d}{dx}[\ln y] = \frac{d}{dx}[\ln x^r]$$

$$\frac{1}{y}\frac{dy}{dx} = \frac{d}{dx}[r\ln x]$$

$$\frac{1}{y}\frac{dy}{dx} = r\frac{1}{x}$$

Solving for dy/dx yields

$$\frac{dy}{dx} = r\frac{1}{x}y$$

$$= r\frac{1}{x}x^r$$

$$= rx^{r-1}$$ ◆

◆ 3.7 PROBLEMS

[3.7.1]

In Problems 1–4, find the inverse of each function, and differentiate each inverse in two ways: (i) Differentiate the inverse function directly, and (ii) use (3.11) to find the derivative of the inverse.

1. $f(x) = \sqrt{x-1}, x \geq 1$ **2.** $f(x) = x^2 - 2, x \geq 0$

3. $f(x) = 3 - 2x^3, x \geq 0$ **4.** $f(x) = \dfrac{2x^2-1}{x^2-1}, x > 1$

5. Let

$$f(x) = 3x + \ln x, \quad x > 0$$

Find $\frac{d}{dx}f^{-1}(3)$. (Note that $f(1) = 3$.)

6. Let

$$f(x) = x + e^x, \quad x > 0$$

Find $\frac{d}{dx}f^{-1}(1+e^2)$. (Note that $f(1) = 1 + e^2$.)

7. Let

$$f(x) = x - \sin x, \quad x \in \mathbf{R}$$

Find $\frac{d}{dx}f^{-1}(\pi)$. (Note that $f(\pi) = \pi$.)

8. Let

$$f(x) = x^2 + \tan x, \quad x \in \left(-\frac{\pi}{2}, \frac{\pi}{2}\right)$$

Find $\frac{d}{dx}f^{-1}\left(\frac{\pi^2}{16}+1\right)$. (Note that $f\left(\frac{\pi}{4}\right) = \frac{\pi^2}{16}+1$.)

9. Let $f(x) = \tan x$. Find $\frac{d}{dx}f^{-1}(\sqrt{3})$.

10. Let $f(x) = x^5 + x + 1$. Find $\frac{d}{dx}f^{-1}(1)$.

11. Let $f(x) = e^{-x^2} + x$. Find $\frac{d}{dx}f^{-1}(1)$.

12. Denote by $y = \arcsin x$, $-1 \leq x \leq 1$, the inverse of $y = \sin x$, $-\frac{\pi}{2} \leq x \leq \frac{\pi}{2}$. Show that

$$\frac{d}{dx}\arcsin x = \frac{1}{\sqrt{1-x^2}}, \quad -1 < x < 1$$

[3.7.2]

In Problems 13–26, differentiate the functions with respect to the indicated variable. (Note that log denotes the logarithm to base 10.)

13. $f(x) = \ln(x^3)$ **14.** $f(x) = (\ln x)^2$

15. $f(x) = \ln \sqrt{x^2 + 1}$

16. $f(x) = \ln \dfrac{x}{x + 1}$

17. $g(s) = \exp[s^2 + \ln s]$

18. $g(s) = \ln(\sin^2(3s))$

19. $h(t) = \dfrac{\ln t}{1 + t^2}$

20. $h(t) = \sin(\ln(3t))$

21. $h(s) = \ln \ln s$

22. $f(x) = \ln |x^2 - 3|$

23. $f(x) = \log(x^3 - 3x)$

24. $f(x) = \log(\sqrt[3]{\tan x^2})$

25. $f(u) = \log_3(3 + u^4)$

26. $g(s) = \log_5(3^s - 2)$

27. Let $f(x) = \ln x$. We know that $f'(x) = \frac{1}{x}$. We will use this fact and the definition of derivatives to show that

$$\lim_{n \to \infty} \left(1 + \frac{1}{n}\right)^n = e$$

(a) Use the definition of the derivative to show that

$$f'(1) = \lim_{h \to 0} \frac{\ln(1 + h)}{h}$$

(b) Show that (a) implies that

$$\ln[\lim_{h \to 0}(1 + h)^{1/h}] = 1$$

(c) Set $h = \frac{1}{n}$ in (b) and let $n \to \infty$. Show that this implies

$$\lim_{n \to \infty} \left(1 + \frac{1}{n}\right)^n = e$$

28. Assume that $f(x)$ is differentiable with respect to x. Show that

$$\frac{d}{dx} \ln \left[\frac{f(x)}{x}\right] = \frac{f'(x)}{f(x)} - \frac{1}{x}$$

[3.7.3]

In Problems 29–36, use logarithmic differentiation to find the first derivative of the given functions.

29. $f(x) = x^x$

30. $j(x) = (\ln x)^x$

31. $f(x) = x^{\ln x}$

32. $f(x) = x^{1/x}$

33. $y = x^{x^x}$

34. $y = (x^x)^x$

35. $y = x^{\cos x}$

36. $y = (\cos x)^x$

37. Differentiate

$$y = \frac{e^{2x}(9x - 2)^3}{\sqrt[4]{(x^2 + 1)(3x^3 - 7)}}$$

38. Differentiate

$$y = \frac{e^{x-1} \sin^2 x}{(x^2 + 5)^{2x}}$$

◆ 3.8 APPROXIMATION AND LOCAL LINEARITY

The response of a crop yield Y to nitrogen soil level N can often be described by a function of the form

$$Y(N) = Y_{max} \frac{N}{K + N}$$

where Y_{max} is the maximum yield and K is a positive constant. Suppose that the current nitrogen level is N_0. When N_0 changes by a small amount ΔN, the corresponding change in crop yield ΔY is

$$\Delta Y = Y(N_0 + \Delta N) - Y(N_0)$$
$$= \frac{Y(N_0 + \Delta N) - Y(N_0)}{\Delta N} \Delta N$$

When ΔN is sufficiently small, the difference quotient

$$\frac{Y(N_0 + \Delta N) - Y(N_0)}{\Delta N}$$

can be approximated by $Y'(N_0)$ since

$$\lim_{\Delta N \to 0} \frac{Y(N_0 + \Delta N) - Y(N_0)}{\Delta N} = Y'(N_0)$$

that is,

$$\Delta Y \approx Y'(N_0)\Delta N \tag{3.12}$$

where "$\approx$" means "is approximately equal to." A small change in nitrogen level causes a small change in yield. The ratio $\Delta Y / \Delta N$ is approximately equal to $Y'(N)$, which is the slope of the tangent line at the given nitrogen level N. Since $Y(N_0 +$

$$Y(N) = Y_{max} \frac{N}{K + N}$$

◀ **Figure 3.39**

The tangent line approximation: If ΔN is small, the tangent line at $(N_0, Y(N_0))$ closely approximates the function $Y(N)$. The difference between ΔY and $Y'(N_0)\Delta N$ is small.

$\Delta N) = Y(N_0) + \Delta Y$, we can use (3.12) to find the approximate yield at the nitrogen level $N_0 + \Delta N$, namely,

$$Y(N_0 + \Delta N) \approx Y(N_0) + Y'(N_0)\Delta N$$

This is an example of a **linear** or **tangent line approximation**. Graphically, we approximate the function $Y(N)$ for values of N close to N_0 by the tangent line $Y(N_0) + Y'(N_0)\Delta N$ at $N = N_0$ (see Figure 3.39).

The Tangent Line Approximation

Assume that $y = f(x)$ is differentiable at $x = a$; then

$$L(x) = f(a) + f'(a)(x - a)$$

is the linearization of f at $x = a$.

Geometrically, the linearization of f at $x = a$, $L(x) = f(a) + f'(a)(x - a)$, is the equation of the tangent line on the graph of $f(x)$ at the point $(a, f(a))$ (see Figure 3.40).

If $|x - a|$ is sufficiently small, then $f(x)$ can be linearly approximated by $L(x)$, that is,

$$f(x) \approx f(a) + f'(a)(x - a)$$

This is illustrated in Figure 3.41.

EXAMPLE I

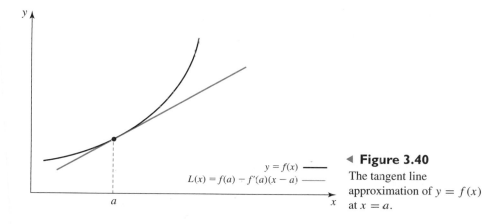

$y = f(x)$ ——
$L(x) = f(a) - f'(a)(x - a)$ ——

◀ **Figure 3.40**

The tangent line approximation of $y = f(x)$ at $x = a$.

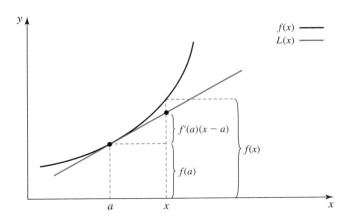

◀ **Figure 3.41**

The linearization of f at $x = a$ can be used to approximate $f(x)$ for x close a.

Find the linear approximation of $f(x) = \sqrt{x}$ at $x = 49$, and use it to find an approximate value of $\sqrt{50}$.

Solution   Since $f(x) = \sqrt{x}$, it follows that $f'(x) = \frac{1}{2\sqrt{x}}$, and the linear approximation at $x = 49$ is

$$
L(x) = f(49) + f'(49)(x - 49)
$$
$$
= \sqrt{49} + \frac{1}{2\sqrt{49}}(x - 49)
$$
$$
= 7 + \frac{x - 49}{14}
$$

(see Figure 3.42). To find the approximate value of $f(50) = \sqrt{50}$, we approximate $f(50)$ by $L(50)$ and find

$$
\sqrt{50} \approx 7 + \frac{50 - 49}{14} = 7.0714
$$

If we compute $\sqrt{50}$ using a calculator, we find that $\sqrt{50} = 7.0711\ldots$, and see that the error using linear approximation is quite small. ▮

▮**EXAMPLE 2**   Find the linear approximation of $f(x) = \sin x$ at $x = 0$.

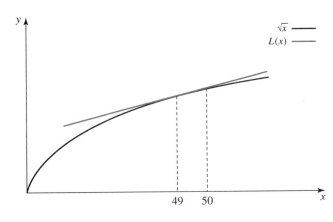

◀ **Figure 3.42**

The linear approximation of $f(x) = \sqrt{x}$ at $x = 49$, is the line $y = L(x)$.

sin x ——— ◀ **Figure 3.43**
x ——— The linear approximation of
y = sin x is the line y = x.

Solution   Since $f'(x) = \cos x$, it follows that

$$L(x) = f(0) + f'(0)(x - 0)$$
$$= \sin 0 + (\cos 0)x = x$$

(see Figure 3.43). That is, for small values of x we can approximate $\sin x$ by x. (Note that x is measured in radians.)

EXAMPLE 3   Let $N(t)$ be the size of a population at time t, and assume that the per capita growth rate is equal to 2%. We can express this as a differential equation, namely,

$$\frac{1}{N}\frac{dN}{dt} = 0.02$$

Suppose we know that $N(10) = 250$. Use a linear approximation to predict the approximate population size at time $t = 10.2$.

Solution   To predict the population size at time 10.2, we use the linearization of $N(t)$ at $t = 10$, namely,

$$L(t) = N(10) + N'(10)(t - 10)$$

To evaluate $L(t)$ at $t = 10.2$, we need to find $N'(10)$.

Using the differential equation, we find

$$N'(t) = (0.02)N(t)$$

When $t = 10$, we obtain

$$N'(10) = (0.02)N(10) = (0.02)(250) = 5$$

hence,

$$L(10.2) = N(10) + N'(10)(10.2 - 10)$$
$$= 250 + (5)(0.2) = 251$$

Thus, we predict that the population size at time 10.2 is approximately 251. Note that this approximation is only good if the time at which we want to predict the population size is very close to the time at which we know the population size.

Error Propagation   Linear approximations are used in problems of **error propagation**. Suppose that you wish to determine the surface area of a spherical cell. Since the surface area S of a sphere with radius r is given by

$$S = 4\pi r^2$$

it suffices to measure the radius r of the cell. If your measurement of the radius is accurate within 2%, how does this affect the accuracy of the surface area?

First we must discuss what it means for a measurement to be accurate within a certain percentage. Suppose that x_0 is the true value of an observation, and x is the measured value. If Δx is the tolerance in measurement, then the true value of x_0 would be in the interval $(x - |\Delta x|, x + |\Delta x|)$. The **relative error** is defined as $\Delta x / x$, and the **percentage error** as $100 \Delta x / x$.

Returning to our example, suppose that you measured the radius r as 3 mm. The surface area would then be $S = 4\pi r^2 = 113.1$ mm^2. We said that the measurement of the radius was accurate within 2%, which means that the true radius may deviate from the measured radius by as much as

$$|\Delta r| = (3 \text{ mm})\frac{2}{100} = 0.06 \text{ mm}$$

Note that Δr may be positive or negative, which is why we use absolute values. The corresponding error when computing the surface area is thus

$$\Delta S = S(r + \Delta r) - S(r)$$

We approximate this by $S'(r)\Delta r$, that is,

$$\Delta S \approx S'(r)\Delta r$$

Since $S'(r) = 8\pi r$, the measured radius $r = 3$ mm, and $\Delta r = \pm 0.06$ mm, it follows that

$$\Delta S \approx 8\pi(3 \text{ mm})(\pm 0.06 \text{ mm}) \approx \pm 4.5 \text{ mm}^2$$

The percentage error is therefore

$$100\frac{\Delta S}{S} = \pm 100\frac{4.5}{113.1} = \pm 4\%$$

which is twice the percentage error of the accuracy in the measurement of the radius. Where the factor of 2 comes from can be seen in the following example.

▮ EXAMPLE 4   Suppose that you wish to determine $f(x)$ from a measurement of x. If $f(x)$ is given by a power function, namely $f(x) = cx^r$, then an error in measurement of x is propagated as follows:

$$\Delta f \approx f'(x)\Delta x = crx^{r-1}\Delta x$$

The percentage error $100\frac{\Delta f}{f}$ is related to the percentage error $100\frac{\Delta x}{x}$ as follows:

$$100\frac{\Delta f}{f} \approx 100\frac{f'(x)\Delta x}{f} = 100\frac{crx^{r-1}\Delta x}{cx^r}$$

$$= r\left(100\frac{\Delta x}{x}\right)$$

In our previous example, $r = 2$, hence the percentage error in the surface area measurement is twice the percentage error in the radius measurement.   ▮

EXAMPLE 5   Suppose that you wish to estimate total leaf area of a tree in a certain plot. Experimental data in your study plot (Niklas 1994) indicate that

$$[\text{leaf area}] \propto [\text{stem diameter}]^{1.84}$$

Instead of trying to measure total leaf area directly, you measure the stem diameter and then use the scaling relationship to estimate total leaf area. How accurately must you measure stem diameter if you want to estimate leaf area within an error of 10%?

Solution   We denote leaf area by A and stem diameter by d. Then

$$A(d) = cd^{1.84}$$

where c is the constant of proportionality. An error in measurement of d is propagated as

$$\Delta A \approx A'(d)\Delta d = c(1.84)d^{0.84}\Delta d$$

The percentage error $100\frac{\Delta A}{A}$ is related to the percentage error $100\frac{\Delta d}{d}$ as

$$100\frac{\Delta A}{A} \approx 100\frac{A'(d)\Delta d}{A}$$

$$= 100\frac{c(1.84)d^{0.84}}{cd^{1.84}}\Delta d$$

$$= (1.84)\left(100\frac{\Delta d}{d}\right)$$

We require $100\frac{\Delta A}{A} = 10$. Hence

$$10 = (1.84)\left(100\frac{\Delta d}{d}\right)$$

or

$$100\frac{\Delta d}{d} = \frac{10}{1.84} = 5.4$$

That is, we must measure stem diameter within an error of 5.4%.

Using the result of Example 4, we could have found this immediately: since

$$A(d) = cd^{1.84}$$

we get $r = 1.84$, where r is the exponent defined in Example 4. Using

$$100\frac{\Delta A}{A} = r\left(100\frac{\Delta d}{d}\right)$$

we obtain

$$100\frac{\Delta d}{d} = \frac{1}{r}\left(100\frac{\Delta A}{A}\right) = \frac{10}{1.84} = 5.4$$

as before.   ∎

3.8 PROBLEMS

In Problems 1–10, use the formula

$$f(x) \approx f(a) + f'(a)(x - a)$$

to approximate the value of the given function, and compare your result with the value from a calculator.

1. $\sqrt{65}$; let $f(x) = \sqrt{x}$, $a = 64$, and $x = 65$
2. $\sqrt{35}$; let $f(x) = \sqrt{x}$, $a = 36$, and $x = 35$
3. $\sqrt[3]{124}$ 4. $(3.9)^3$

5. $(0.99)^{25}$ 6. $\sin(0.01)$

7. $\sin\left(\dfrac{\pi}{2} + 0.02\right)$ 8. $\cos\left(\dfrac{\pi}{4} - 0.01\right)$

9. $\ln(1.01)$ 10. $\dfrac{1}{0.99}$

In Problems 11–18, approximate $f(x)$ at a by the linear approximation

$$L(x) = f(a) + f'(a)(x - a)$$

11. $f(x) = \dfrac{1}{1+x}$ at $a = 0$ **12.** $f(x) = \ln(1+x)$ at $a = 0$

13. $f(x) = e^x$ at $a = 0$ **14.** $f(x) = e^{-x}$ at $a = 0$

15. $f(x) = (1+x)^{-n}$ at $a = 0$. (Assume that n is a positive integer.)

16. $f(x) = (1-x)^{-n}$ at $a = 0$. (Assume that n is a positive integer.)

17. $f(x) = \tan x$ at $a = 0$ **18.** $f(x) = \cos x$ at $a = 0$

19. Suppose that the per capita growth rate of a population is 3%, that is, if $N(t)$ denotes the population size at time t, then

$$\frac{1}{N}\frac{dN}{dt} = 0.03$$

Suppose that the population size at time $t = 4$ is equal to 100. Use a linear approximation to compute the population size at time $t = 4.1$.

20. Suppose that a certain plant is grown along a gradient ranging from nitrogen poor to nitrogen rich soil. Experimental data show that the average mass per plant grown in a soil with total nitrogen content of 1000 mg nitrogen per kg of soil is 2.7 g, and the rate of change of the average mass per plant at this nitrogen level is 1.05×10^{-3} g per mg change in total nitrogen per kg soil. Use a linear approximation to predict the average mass per plant grown in a soil with total nitrogen content of 1100 mg nitrogen per kg of soil.

In Problems 21–26, a measurement error in x affects the accuracy of the value f(x). In each case determine an interval of the form

$$[f(x) - \Delta f, f(x) + \Delta f]$$

that reflects the measurement error Δx.

21. $f(x) = 2x$, $x = 1 \pm 0.1$

22. $f(x) = 1 - 3x$, $x = -2 \pm 0.3$

23. $f(x) = 3x^2$, $x = 2 \pm 0.1$

24. $f(x) = \sqrt{x}$, $x = 10 \pm 0.5$

25. $f(x) = e^x$, $x = 2 \pm 0.2$

26. $f(x) = \sin x$, $x = -1 \pm 0.05$

In Problems 27–30, assume that the measurement of x is accurate within 2%. Determine the error Δf in the calculation of f and the percentage error $100\frac{\Delta f}{f}$ in each case.

27. $f(x) = 4x^3$, $x = 1.5$ **28.** $f(x) = x^{1/4}$, $x = 10$

29. $f(x) = \ln x$, $x = 20$ **30.** $f(x) = \dfrac{1}{1+x}$, $x = 4$

31. The volume V of a spherical cell of radius r is given by

$$V(r) = \frac{4}{3}\pi r^3$$

If you can determine the radius within an accuracy of 3%, how accurate is your calculation of the volume?

32. The speed v of blood flowing along the central axis of an artery of radius R is given by Poiseuille's law

$$v(R) = cR^2$$

where c is a constant. If you can determine the radius of the artery within an accuracy of 5%, how accurate is your calculation of the speed?

33. Suppose that you are studying reproduction in moss. A scaling relation has been found (Niklas 1994) between the number of moss spores (N) and the capsule length (L), namely,

$$N \propto L^{2.11}$$

This relationship is not very accurate but it turns out that it suffices for your purpose. To estimate the number of moss spores you measure capsule length. If you wish to estimate the number of moss spores within an error of 5%, how accurately must you measure capsule length?

34. (*Tilman's resource model*) Suppose that the rate of growth of a plant in a certain habitat depends on a single resource: for instance, nitrogen. Assume that the growth rate $f(R)$ depends on the resource level R as

$$f(R) = a\frac{R}{k+R}$$

where a and k are constants. Express the percentage error of the growth rate, $100\frac{\Delta f}{f}$, as a function of the percentage error of the resource level, $100\frac{\Delta R}{R}$.

35. The reaction rate $R(x)$ of the irreversible reaction

$$A + B \rightarrow AB$$

is a function of the concentration x of the product AB, and is given by

$$R(x) = k(a - x)(b - x)$$

where k is a constant, a is the concentration of A in the beginning of the reaction, and b is the concentration of B in the beginning of the reaction. Express the percentage error of the reaction rate, $100\frac{\Delta R}{R}$, as a function of the percentage error of the concentration x, $100\frac{\Delta x}{x}$.

◆ 3.9 KEY TERMS

Chapter 3 Review: Topics   Discuss the following definitions and concepts:

1. Derivative, formal definition
2. Difference quotient
3. Secant line and tangent line
4. Instantaneous rate of change
5. Average rate of change
6. Differential equation
7. Differentiability and continuity
8. Power rule
9. Basic rules of differentiation
10. Product rule
11. Quotient rule
12. Chain rule

13. Implicit function
14. Implicit differentiation
15. Related rates
16. Higher derivatives
17. Derivatives of trigonometric functions

18. Derivative of the exponential function
19. Derivative of inverse and logarithmic functions
20. Tangent line approximation
21. Error propagation
22. Relative error, percentage error

◆ 3.10 REVIEW PROBLEMS

In Problems 1–8, differentiate with respect to the independent variable.

1. $f(x) = -3x^4 + \dfrac{2}{\sqrt{x}} + 1$

2. $g(x) = \dfrac{1}{\sqrt{x^3 + 4}}$

3. $h(t) = \left(\dfrac{1-t}{1+t}\right)^{1/3}$

4. $f(x) = (x^2 + 1)e^{-x}$

5. $f(x) = e^{2x} \sin\left(\dfrac{\pi}{2}x\right)$

6. $g(s) = \dfrac{\sin(3s + 1)}{\cos(3s)}$

7. $f(x) = 2\dfrac{\ln(x + 1)}{\ln x^2}$

8. $g(x) = e^{-x}\ln(x + 1)$

In Problems 9–12, find the first and second derivatives of the given functions.

9. $f(x) = e^{-x^2/2}$

10. $g(x) = \tan(x^2 + 1)$

11. $h(x) = \dfrac{x}{x + 1}$

12. $f(x) = \dfrac{e^{-x}}{e^{-x} + 1}$

In Problems 13-16, find dy/dx.

13. $x^2 y - y^2 x = \sin x$

14. $e^{x^2 + y^2} = 2x$

15. $\ln(x - y) = 2x$

16. $\tan(x - y) = x^2$

In Problems 17–19, find dy/dx and d^2y/dx^2.

17. $x^2 + y^2 = 16$

18. $x = \tan y$

19. $e^y = \ln x$

20. Assume that x is a function of t. Find $\dfrac{dy}{dt}$ when $y = \cos x$ and $\dfrac{dx}{dt} = \sqrt{3}$ for $x = \dfrac{\pi}{6}$.

21. A flock of birds passes directly overhead, flying horizontally at an altitude of 100 feet and at a speed of 6 feet per second. How quickly is the distance between you and the birds increasing when the distance is 320 feet? (You are on the ground, and are not moving.)

22. Find the derivative of

$$y = \ln|\cos x|$$

23. Suppose that $f(x)$ is differentiable. Find an expression of the derivative for each of the following functions.

(a) $y = e^{f(x)}$

(b) $y = \ln f(x)$

(c) $y = [f(x)]^2$

24. Find the tangent line and the normal line to $y = \ln(x + 1)$ at $x = 1$.

25. Suppose that

$$f(x) = \dfrac{x^2}{1 + x^2}, \quad x \geq 0$$

(a) Use a graphing calculator to graph $f(x)$ for $x \geq 0$. Note that the graph is "S"-shaped.

(b) Find a line through the origin that touches the graph of $f(x)$ at some point $(c, f(c))$ with $c > 0$. This is the tangent line at $(c, f(c))$ that goes through the origin. Graph the tangent line in the same coordinate system that you used in (a).

In Problems 26–29, find an equation for the tangent line to the curve at the specified point.

26. $y = (\sin x)^{\cos x}$ at $x = \dfrac{\pi}{2}$

27. $y = e^{-x^2}\cos x$ at $x = \dfrac{\pi}{3}$

28. $x^2 + y = e^y$ at $x = \sqrt{e - 1}$

29. $x \ln y = y \ln x$ at $x = 1$

30. In Review Problem 17 of Chapter 2, we introduced the hyperbolic functions

$$\sinh x = \dfrac{e^x - e^{-x}}{2}, \quad x \in \mathbf{R}$$

$$\cosh x = \dfrac{e^x + e^{-x}}{2}, \quad x \in \mathbf{R}$$

$$\tanh x = \dfrac{e^x - e^{-x}}{e^x + e^{-x}}, \quad x \in \mathbf{R}$$

(a) Show that

$$\dfrac{d}{dx}\sinh x = \cosh x$$

and

$$\dfrac{d}{dx}\cosh x = \sinh x$$

(b) Use the facts that

$$\tanh x = \dfrac{\sinh x}{\cosh x}$$

and

$$\cosh^2 x - \sinh^2 x = 1$$

together with your results in (a) to show that

$$\dfrac{d}{dx}\tanh x = \dfrac{1}{\cosh^2 x}$$

31. Find a second degree polynomial

$$p(x) = ax^2 + bx + c$$

with $p(-1) = 6$, $p'(1) = 8$, and $p''(0) = 4$.

32. Use the geometric interpretation of the derivative to find the equations of the tangent lines to the curve

$$x^2 + y^2 = 1$$

at the following points.

(a) $(1, 0)$ **(b)** $\left(\frac{1}{2}, \frac{1}{2}\sqrt{3}\right)$

(c) $\left(-\frac{1}{2}\sqrt{2}, -\frac{1}{2}\sqrt{2}\right)$ **(d)** $(0, -1)$

33. Geradedorf and Straightville are connected by a very straight but rather hilly road. Biking from Geradedorf to Straightville, your position at time t (measured in hours) is given by the function

$$s(t) = 3\pi t + 3(1 - \cos(\pi t))$$

for $0 \le t \le 5.5$, where $s(t)$ is measured in miles.

(a) Use a graphing calculator to convince yourself that you didn't backtrack during your trip. How can you check this? Assuming that your trip takes 5.5 hours, find the distance between Geradedorf and Straightville.

(b) Find the velocity $v(t)$ and the acceleration $a(t)$.

(c) Use a graphing calculator to graph $s(t)$, $v(t)$, and $a(t)$. In (a), you used the function $s(t)$ to conclude that you didn't backtrack during your trip. Can you use any of the other two functions to answer the question of backtracking? Explain your answer.

(d) Assuming that you slow down going uphill and speed up going downhill, how many peaks and valleys does this road have?

34. Suppose your position at time t on a straight road is given by

$$s(t) = \cos(\pi t)$$

for $0 \le t \le 2$, where t is measured in hours.

(a) What is your position at the beginning and end of your trip?

(b) Use a graphing calculator to help you to describe your trip in words.

(c) What is the total distance you traveled?

(d) Determine your velocity and your acceleration during the trip. When is your velocity equal to 0? Relate this to your position, and explain what it means.

35. In the following very simple population model, the growth rate at time t depends on the number of individuals at time $t - T$, where T is a positive constant. (That is, the model incorporates a time delay in the birth rate.) This is useful, for instance, if one wishes to take into account the fact that individuals must mature before reproducing.

Denote by $N(t)$ the size of the population at time t, and assume that

$$\frac{dN}{dt} = \frac{\pi}{2T}(K - N(t - T)) \tag{3.13}$$

where K and T are positive constants.

(a) Show that

$$N(t) = K + A\cos\frac{\pi t}{2T}$$

is a solution of (3.13).

(b) Graph $N(t)$ for $K = 100$, $A = 50$, and $T = 1$.

(c) Explain in words how the size of the population changes over time.

36. We denote by $W(t)$ the amount of a radioactive material left at time t if the initial amount present was $W_0 = W(0)$.

(a) Show that

$$W(t) = W_0 e^{-\lambda t}$$

solves the differential equation

$$\frac{dW}{dt} = -\lambda W(t)$$

(b) Show that if you graph $W(t)$ on a semilog paper, then the result is a straight line.

(c) Use your result in (b) to explain why

$$\frac{d \ln W(t)}{dt} = \text{constant}$$

Determine the constant, and relate it to the graph in (b).

(d) Show that

$$\frac{d \ln W(t)}{dt} = \text{constant}$$

implies that

$$\frac{dW}{dt} \propto W(t)$$

37. In Example 12 of Subsection 3.4.3, we introduced an allometric relationship between skull length and backbone length of Ichthyosaurs, a group of extinct marine reptiles. The relationship is

$$S = (1.162)B^{0.933}$$

where S and B denote skull length and backbone length, respectively. Suppose that you only found the skull of an individual, and that based on the skull length you wish to estimate the backbone of this specimen. How accurately must you measure skull length if you want to estimate backbone length within an error of 10%?

CHAPTER 4

Applications of Differentiation

◆ **4.1** LOCAL EXTREMA AND THE MEAN VALUE THEOREM

4.1.1 Local Extrema

In Subsection 2.5.1, we introduced the concept of global (or absolute) extrema. We will now discuss local (or relative) extrema, which are points where a graph is higher or lower than all *nearby* points. This will allow us to identify the peaks and the valleys of the graph of a function (see Figure 4.1). The graph of the function in Figure 4.1 has three peaks, at $x = a$, c, and e, and two valleys, at $x = b$ and d. A peak (or local maximum) has the property that the graph is lower nearby; a valley (or local minimum) has the property that the graph is higher nearby.

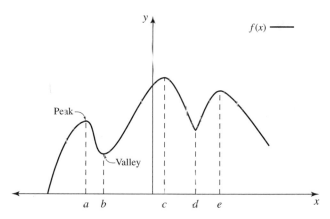

▲ **Figure 4.1**
The function $y = f(x)$ has valleys at $x = b$ and d and peaks at $x = a$, c, and e.

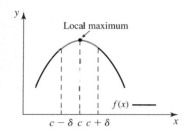

▲ **Figure 4.2**
The function $y = f(x)$ has a local maximum at $x = c$.

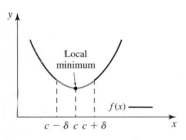

▲ **Figure 4.3**
The function $y = f(x)$ has a local minimum at $x = c$.

The formal definition follows (see Figures 4.2 and 4.3).

A function f defined on a set D has a **local** (or **relative**) **maximum** at a point c if there exists a $\delta > 0$ so that

$$f(c) \geq f(x) \qquad \text{for all } x \in (c - \delta, c + \delta) \cap D$$

A function f defined on a set D has a **local** (or **relative**) **minimum** at a point c if there exists a $\delta > 0$ so that

$$f(c) \leq f(x) \qquad \text{for all } x \in (c - \delta, c + \delta) \cap D$$

Local maxima and local minima are collectively called local (or relative) extrema. If D is an interval and c is in the interior of D (that is, not a boundary point), then the definition simplifies: the function f has a local maximum at c if there exists an open interval I so that $f(c) \geq f(x)$ for all $x \in I$. Likewise, the function f has a local minimum at c if there exists an open interval I so that $f(c) \leq f(x)$ for all $x \in I$. In the definition, we wrote $(c - \delta, c + \delta) \cap D$. If c is an interior point of D, δ can be chosen small enough so that $(c - \delta, c + \delta)$ is contained in D, and we can set $I = (c - \delta, c + \delta)$. Intersecting the interval $(c - \delta, c + \delta)$ with D becomes important when c is a boundary point.

We examine local and global extrema in the following two examples; the discussion is based on looking at the graphs of functions. In the first example, we consider a function that is defined on a closed interval; this allows us to compute the value of the function at both endpoints of its domain. In the second example we consider a function that is defined on a half-open interval; thus, the value of the function can be computed at one endpoint, but not at the other endpoint of its domain.

▌**EXAMPLE 1**   Let

$$f(x) = (x - 1)^2(x + 2) \quad \text{for } -2 \leq x \leq 3$$

a. Use the graph of $f(x)$ to find all local extrema.

b. Find the global extrema.

Solution

a. The graph of $f(x)$ is illustrated in Figure 4.4. The function f is defined on the closed interval $[-2, 3]$. We begin with local extrema that occur at interior

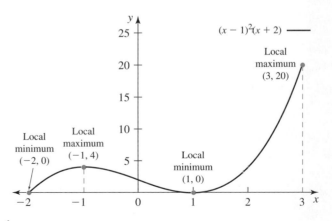

▲ **Figure 4.4**
The graph of $f(x) = (x - 1)^2(x + 2)$ for $-2 \leq x \leq 3$ in Example 1.

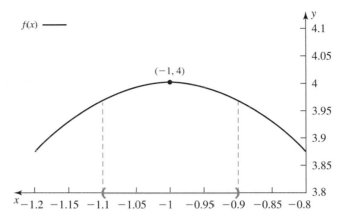

▲ **Figure 4.5**

The graph of $f(x) = (x - 1)^2(x + 2)$ near $x = -1$. The point $(-1, 4)$ is a local maximum: $f(-1) \geq f(x)$ for all x nearby in the domain of f.

points of the domain $D = [-2, 3]$; looking at Figure 4.4, we see that a local maximum occurs at $x = -1$, as there are no greater values of f nearby. That is, we can find a small interval I about $x = -1$ so that $f(-1) \geq f(x)$ for all $x \in I$. For instance, we can choose $\delta = 0.1$ in the above definition and obtain $I = (-1.1, -0.9)$ (see Figure 4.5).

There is a local minimum at $x = 1$, since there are no smaller values of f nearby. This time, we need to find a small interval about $x = 1$ so that $f(1) \leq f(x)$ for all $x \in I$, for example, $I = (0.9, 1.1)$.

A local minimum also occurs at $x = -2$, the endpoint of the domain of f. Note that in the definition of a local minimum, we require an interval I about c so that $f(c) \leq f(x)$ for all $x \in I \cap D$, where D is the domain of the function. If c is an interior point, we can always choose I small enough so that $I \subset D$ and therefore $I \cap D = I$, but this is not possible at an endpoint. To show that $x = -2$ is a local minimum, we must find $\delta > 0$ so that $f(-2) \leq f(x)$ for all $x \in (-2 - \delta, -2 + \delta) \cap D = [-2, -2 + \delta)$. We can again choose $\delta = 0.1$ and see that $f(-2) \leq f(x)$ for all $x \in [-2, -1.9)$ (see Figure 4.6).

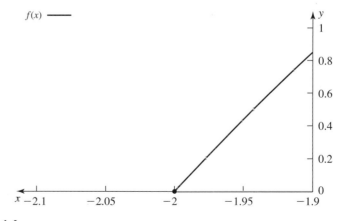

▲ **Figure 4.6**

The graph of $f(x) = (x - 1)^2(x + 2)$ near $x = -2$. The point $(-2, 0)$ is a local minimum: $f(-2) \leq f(x)$ for all x nearby in the domain of f.

Similarly, we see that $x = 3$ is a local maximum, since $f(3) \geq f(x)$ for all $x \in (2.9, 3]$; that is, there is no larger value of f nearby.

b. Recall from Subsection 2.5.1 that global extrema are points at which a function is either largest or smallest. Since f is defined on a closed interval, we know from the extreme value theorem that both a global maximum and a global minimum exist. These global extrema may occur either in the interior or at the endpoints of the domain $D = [-2, 3]$.

To find the global minimum, we compare the local minima. Since $f(-2) = 0$ and $f(1) = 0$, it follows that the global minima occur at $x = -2$ and $x = 1$ (see Figure 4.4). To find the global maximum, we compare the local maxima. Since $f(3) = 20$ and $f(-1) = 4$, it follows that $f(3) > f(-1)$ and, therefore, the global maximum occurs at the endpoint $x = 3$ (see Figure 4.4). ▪

EXAMPLE 2   Let

$$f(x) = |x^2 - 4| \quad \text{for } -2.5 \leq x < 3$$

Find all local and global extrema.

Solution   The graph of $f(x)$, illustrated in Figure 4.7, reveals that local minima occur at $x = -2$ and $x = 2$, and local maxima occur at $x = -2.5$ and $x = 0$. Note that $f(x)$ is not defined at $x = 3$ and, therefore, $x = 3$ cannot be a local maximum. To find the global extrema, however, we need to look at the function values close to the boundary $x = 3$ as well. Candidates for global extrema are all the local extrema, which then must be compared to the value of the function near the boundary $x = 3$. We discuss the global maximum first. Since

$$f(-2.5) = 2.25 \qquad f(0) = 4 \qquad \lim_{x \to 3-} f(x) = 5$$

the function is largest near the point $x = 3$. But since $f(x)$ is not defined at $x = 3$, there exists no global maximum for the function. (This does not contradict the extreme value theorem, since the function is not defined on a closed interval, which is an assumption in the theorem.) To find the global minimum, we need only compare $f(-2)$ and $f(2)$. We find that $f(-2) = 0$ and $f(2) = 0$; therefore, global minima occur at $x = -2$ and $x = 2$. ▪

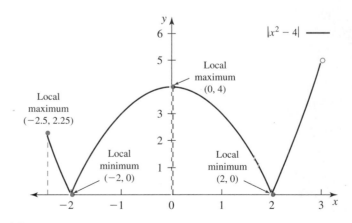

▲ **Figure 4.7**
The graph of $f(x) = |x^2 - 4|$ for $-2.5 \leq x < 3$ in Example 2.

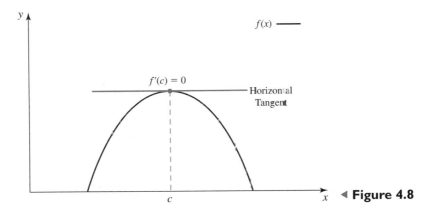

▶ **Figure 4.8**

Looking at Figures 4.4 and 4.7, we see that if the function f is differentiable at an interior point where f has a local extremum, then there is a horizontal tangent line at that point. This is known as Fermat's theorem (see Figure 4.8).

> **Fermat's Theorem**   If f has a local extremum at an interior point c and $f'(c)$ exists, then $f'(c) = 0$.

Proof   We prove Fermat's theorem for the case where the local extremum is a maximum; the proof where the local extremum is a minimum is similar. We need to show that $f'(c) = 0$. To do so, we use the formal definition of the derivative to compute $f'(c)$, namely,

$$f'(c) = \lim_{x \to c} \frac{f(x) - f(c)}{x - c}$$

To compute the limit, we will compute the left-hand limit ($x \to c-$) and the right-hand limit ($x \to c+$). We begin with the following observation (see Figure 4.9).

Suppose that f has a local maximum at an interior point c, then there exists a $\delta > 0$ so that

$$f(x) \le f(c) \quad \text{for all } x \in (c - \delta, c + \delta)$$

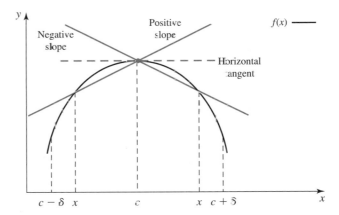

▲ **Figure 4.9**

An illustration of the proof of Fermat's theorem: For $x < c$, the slope of the secant line is positive, for $x > c$, the slope of the secant line is negative. In the limit $x \to c$, the secant lines converge to the horizontal tangent line.

Since $f(x) - f(c) \le 0$ and $x - c < 0$ if $x < c$, we find for the left-hand limit that

$$\lim_{x \to c-} \frac{f(x) - f(c)}{x - c} \ge 0 \tag{4.1}$$

and since $f(x) - f(c) \le 0$ and $x - c > 0$ if $x > c$, we find for the right-hand limit that

$$\lim_{x \to c+} \frac{f(x) - f(c)}{x - c} \le 0 \tag{4.2}$$

Now, since f is differentiable at c,

$$f'(c) = \lim_{x \to c-} \frac{f(x) - f(c)}{x - c} = \lim_{x \to c+} \frac{f(x) - f(c)}{x - c}$$

This, together with equation (4.1), shows that $f'(c) \ge 0$, and together with equation (4.2), that $f'(c) \le 0$. Now, if you have a number that is simultaneously nonnegative and nonpositive, the number must be 0. Therefore, $f'(c) = 0$.   ◆

Caution

1. The condition that $f'(c) = 0$ is a necessary but not sufficient condition for local extrema at interior points where $f'(c)$ exists. In particular, the fact that f is differentiable at c with $f'(c) = 0$ tells us nothing about whether f has a local extremum at $x = c$. For instance, $f(x) = x^3$, $x \in \mathbf{R}$, is differentiable at $x = 0$ and $f'(0) = 0$, but there is no local extremum at $x = 0$. The graph of $y = x^3$ is shown in Figure 4.10. Though there is a horizontal tangent at $x = 0$, there is no local extremum at $x = 0$. Fermat's theorem does tell you, however, that if $x = c$ is an interior point with $f'(c) \ne 0$, then $x = c$ cannot be a local extremum. Interior points with horizontal tangents are *candidates* for local extrema.

2. The function f may not be differentiable at a local extremum. For instance, in Example 2, the function $f(x)$ is not differentiable at $x = -2$ and $x = 2$, but both points turned out to be local extrema. This means that in order to identify candidates for local extrema, it will not be enough to simply look at points with horizontal tangents. You also must look at points, where the function $f(x)$ is not differentiable.

3. Local extrema may occur at endpoints of the domain. Since Fermat's theorem says nothing about what happens at endpoints, you will have to look at endpoints separately.

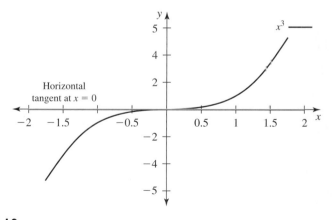

▲ **Figure 4.10**
The graph of $y = x^3$ has a horizontal tangent at $x = 0$ but $(0, 0)$ is not an extremum.

To summarize our discussion, here are some guidelines for finding local extrema:

> **1.** Don't assume that $f'(x) = 0$ gives you a local extremum; these are just candidates.
> **2.** Check points where the derivative is not defined.
> **3.** Check endpoints of the domain.

4.1.2 The Mean Value Theorem

The mean value theorem (MVT) is a very important yet easily understood result in calculus. Its consequences are far reaching, and we will use it in every section in this chapter to derive important results that will help us to analyze functions.

To see what the content of the mean value theorem is, consider

$$f(x) = x^2 \quad \text{for } 0 \leq x \leq 1$$

The secant line connecting the endpoints of the graph of $f(x)$, $(0, 0)$ and $(1, 1)$, has slope

$$m = \frac{f(1) - f(0)}{1 - 0} = \frac{1 - 0}{1 - 0} = 1$$

The graph of $f(x)$ and the secant line are shown in Figure 4.11. Note that $f(x)$ is differentiable in $(0, 1)$; that is, you can draw a tangent line at every point in the open interval $(0, 1)$. If you look at the graph of $f(x)$, you can easily convince yourself that there exists a number $c \in (0, 1)$ such that the slope of the tangent line at $(c, f(c))$ is the same as the slope of the secant line through the points $(0, 0)$ and $(1, 1)$. That is, we claim that there exists a number $c \in (0, 1)$ so that

$$\frac{f(1) - f(0)}{1 - 0} = f'(c)$$

The existence of such a number c is the content of the mean value theorem.

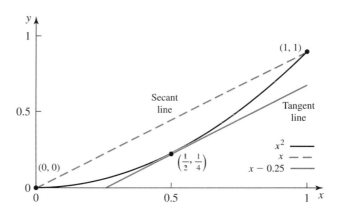

▲ **Figure 4.11**
The graphs of $f(x) = x^2$, the secant line through $(0, 0)$ and $(1, 1)$, and the tangent line parallel to the secant line.

We can compute the value of c in this example. Since $f'(x) = 2x$ and the slope of the secant line is $m = 1$, we must solve

$$1 = 2c \qquad \text{or} \qquad c = \frac{1}{2}$$

Using the point-slope form $(y - y_1 = m(x - x_1))$, we can find the equation of the tangent line at $\left(\frac{1}{2}, f\left(\frac{1}{2}\right)\right) = \left(\frac{1}{2}, \frac{1}{4}\right)$, namely

$$y - \frac{1}{4} = 1\left(x - \frac{1}{2}\right) \qquad \text{or} \qquad y = x - \frac{1}{4}$$

(this tangent line is also shown in Figure 4.11).

The Mean Value Theorem (MVT) If f is continuous on the closed interval $[a, b]$ and differentiable on the open interval (a, b), then there exists at least one number $c \in (a, b)$ such that

$$\frac{f(b) - f(a)}{b - a} = f'(c)$$

The fraction on the left-hand side of the equation in the definition is the slope of the secant line connecting the points $(a, f(a))$ and $(b, f(b))$, and the quantity on the right-hand side is the slope of the tangent line at $(c, f(c))$ (see Figure 4.12).

Geometrically, the MVT is indeed easily understood: It states that there exists a point on the graph between $(a, f(a))$ and $(b, f(b))$, where the tangent line at this point is parallel to the secant line through $(a, f(a))$ and $(b, f(b))$. (We denoted this point by $(c, f(c))$.) The MVT is an "existence" result; it does neither tell us how many such points there are, nor where they are in the interval (a, b).

Going back to our example $f(x) = x^2, 0 \le x \le 1$, we see that $f(x)$ satisfies the assumptions of the MVT, namely, $f(x)$ is continuous on the closed interval $[0, 1]$ and differentiable on the open interval $(0, 1)$. The MVT then guarantees the existence of at least one number $c \in (0, 1)$ such that the slope of the secant line through the points $(0, 0)$ and $(1, 1)$ is equal to the slope of the tangent line at $(c, f(c))$.

At this point, you might be wondering how such a seemingly simple theorem can be so important. For an answer to this question, you must wait until we discuss

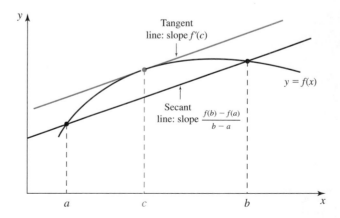

▲ **Figure 4.12**
The mean value theorem guarantees the existence of a number $c \in (a, b)$ such that the tangent line at $(c, f(c))$ has the same slope as the secant line through $(a, f(a))$ and $(b, f(b))$.

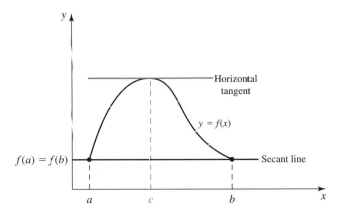

▲ Figure 4.13
An illustration of Rolle's Theorem.

its consequences in the following sections. The rest of this section is devoted to the proof of the MVT. The MVT is typically proved by first showing a special case of the theorem, namely Rolle's theorem.

> **Rolle's Theorem**   If f is continuous on the closed interval $[a, b]$ and differentiable on the open interval (a, b) with $f(a) = f(b)$, then there exists a number $c \in (a, b)$ such that $f'(c) = 0$.

In Figure 4.13, we look at the graph of a function to see why Rolle's theorem is plausible. The function is defined on the closed interval $[a, b]$. Furthermore, it takes on the same values at the two endpoints of $[a, b]$. We see that there is a point with a horizontal tangent.

Before we prove Rolle's theorem, we will see why it is a special case of the MVT. If we compare the assumptions in the two theorems, we see that Rolle's theorem has an additional assumption, $f(a) = f(b)$; that is, the function values must agree at the endpoints of the interval on which f is defined. If we apply the MVT to such a function, it says that there exists a number $c \in (a, b)$ such that

$$f'(c) = \frac{f(b) - f(a)}{b - a}$$

But since $f(b) = f(a)$, it follows that the expression on the right-hand side is equal to 0, that is, $f'(c) = 0$, which is the conclusion of Rolle's theorem.

Proof of Rolle's Theorem   If f is the constant function, then $f'(x) = 0$ for all $x \in (a, b)$ and the result is trivially true. We assume in the following that f is not constant. Since $f(x)$ is continuous on the closed interval $[a, b]$, it follows from the extreme value theorem that the function has a global maximum and a global minimum in the interval $[a, b]$. If f is not constant, then there exists $x_0 \in (a, b)$ so that either $f(x_0) > f(a)$ or $f(x_0) < f(a)$. Since $f(a) = f(b)$, it follows that f attains a global extremum inside the interval (a, b). This global extremum is also a local extremum. Suppose that this local extremum is at $c \in (a, b)$; it follows from Fermat's theorem that $f'(c) = 0$ (see Figure 4.13).   ◆

The mean value theorem follows immediately from Rolle's theorem, and can be thought of as a "tilted" version of Rolle's theorem. (The secant and tangent lines in

Figure 4.12 are no longer necessarily horizontal, as in Rolle's theorem, but "tilted;" they are still parallel, though.)

Proof of the MVT   We define the following function:

$$F(x) = f(x) - \frac{f(b) - f(a)}{b - a}(x - a)$$

The function F is continuous on $[a, b]$ and differentiable on (a, b). Furthermore,

$$F(a) = f(a) - \frac{f(b) - f(a)}{b - a}(a - a) = f(a)$$

$$F(b) = f(b) - \frac{f(b) - f(a)}{b - a}(b - a) = f(a)$$

and, therefore, $F(a) = F(b)$. We can apply Rolle's theorem to the function $F(x)$: There exists $c \in (a, b)$ with $F'(c) = 0$. Since

$$F'(x) = f'(x) - \frac{f(b) - f(a)}{b - a}$$

it follows that

$$0 = F'(c) = f'(c) - \frac{f(b) - f(a)}{b - a}$$

and hence

$$f'(c) = \frac{f(b) - f(a)}{b - a} \qquad \blacklozenge$$

We discuss two consequences of the mean value theorem. The first one is useful in obtaining estimates.

COROLLARY 1

If f is continuous on the closed interval $[a, b]$ and differentiable on the open interval (a, b), such that

$$m \le f'(x) \le M \qquad \text{for all } x \in (a, b)$$

then

$$m(b - a) \le f(b) - f(a) \le M(b - a)$$

If we rewrite the conclusion of the corollary as

$$m \le \frac{f(b) - f(a)}{b - a} \le M$$

we can interpret it geometrically: The slope of the secant line through $(a, f(a))$ and $(b, f(b))$ is somewhere between the smallest and the largest value of the slopes of the tangent lines at points on the graph between a and b (see Figure 4.14).

▌**EXAMPLE 3**   Show that

$$|\sin b - \sin a| \le |b - a|$$

Solution   If $a = b$, then $|\sin a - \sin a| \le |a - a|$ is a true statement. We therefore assume that $a < b$ (the case $a > b$ is similar). Let $f(x) = \sin x$, $a \le x \le b$. Since $f'(x) = \cos x$, it follows that

$$-1 \le f'(x) \le 1$$

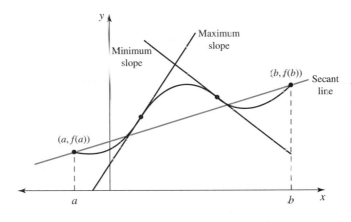

◀ **Figure 4.14**
The slope of the secant line through $(a, f(a))$ and $(b, f(b))$ is between the minimum and the maximum value of the slopes of all tangent lines at the graph of $y = f(x)$ between a and b.

for all $x \in (a, b)$. Applying Corollary 1, with $m = -1$ and $M = 1$, to $f(x) = \sin x$, $a < x < b$, we find

$$-(b - a) \le \sin b - \sin a \le (b - a)$$

which is the same as

$$|\sin b - \sin a| \le |b - a|$$ ▮

The next corollary is quite important, and we will discuss some of its consequences in Section 4.7. We will also use it in Problem 23 to show that the differential equation describing exponential growth has a unique solution.

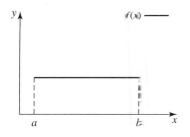

▲ **Figure 4.15**
An illustration of Corollary 2.

COROLLARY 2

If f is continuous on the closed interval $[a, b]$ and differentiable on the open interval (a, b), with $f'(x) = 0$ for all $x \in (a, b)$, then f is constant on $[a, b]$.

Figure 4.15 explains why the result should be true – each point on the graph has a horizontal tangent, so the function must be constant.

▮ **EXAMPLE 4**   Assume that f is continuous on $[-1, 1]$ and differentiable on $(-1, 1)$, with $f(0) = 2$ and $f'(x) = 0$ for all $x \in (-1, 1)$. Find $f(x)$.

Solution   Corollary 2 tells us that $f(x)$ is a constant. Since we know that $f(0) = 2$, this implies that $f(x) = 2$ for all $x \in [-1, 1]$. ▮

Proof of Corollary 2   Let $x_1, x_2 \in (a, b)$, $x_1 \ne x_2$. Then f satisfies the assumptions of the MVT on the closed interval $[x_1, x_2]$. Therefore, there exists $c \in (x_1, x_2)$, such that

$$\frac{f(x_2) - f(x_1)}{x_2 - x_1} = f'(c)$$

Since $f'(c) = 0$, it follows that $f(x_2) = f(x_1)$. Since x_1, x_2 are arbitrary numbers from the interval (a, b), we conclude that f is constant. ◆

◆ **4.1** PROBLEMS

[4.1.1]

In Problems 1–6, use a graphing calculator to determine all local and global extrema of the functions on their respective domains.

1. $f(x) = 2 - x$, $x \in [-1, 3)$

2. $f(x) = 5 + 2x$, $x \in (-2, 1)$

3. $f(x) = x^2 - 2$, $x \in [-1, 1]$

4. $f(x) = x^2 - 4x + 2$, $x \in [0, 3]$

5. $f(x) = -x^2 + 1$, $x \in [-2, 1]$

6. $f(x) = x(x - 1)$, $x \in [0, 1]$

7. Show that $f(x) = x^2$ has a local minimum at $x = 0$, and compute $f'(0)$.

8. Suppose that $f(x) = x^3$. Show that $f'(0) = 0$ but $x = 0$ is not a local extremum.

9. Show that $f(x) = |x|$ has a local minimum at $x = 0$ but $f(x)$ is not differentiable at $x = 0$.

10. Graph

$$f(x) = \big|1 - |x|\big|, \quad -1 \le x \le 2$$

and determine all local and global extrema on $[-1, 2]$.

11. Suppose the size of a population at time t is $N(t)$, and its growth rate is given by the logistic growth function

$$\frac{dN}{dt} = rN\left(1 - \frac{N}{K}\right), \quad t \ge 0$$

where r and K are positive constants.

(a) Graph the growth rate $\frac{dN}{dt}$ as a function of N for $r = 2$ and $K = 100$, and find the population size for which the growth rate is maximal.

(b) Show that $f(N) = rN(1 - N/K)$, $N \ge 0$, is differentiable for $N > 0$, and compute $f'(N)$.

(c) Show that $f'(N) = 0$ for the value of N that you determined in (a) when $r = 2$ and $K = 100$.

12. Suppose that the size of a population at time t is $N(t)$ and its growth rate is given by the logistic growth function

$$\frac{dN}{dt} = rN\left(1 - \frac{N}{K}\right), \quad t \ge 0$$

where r and K are positive constants. The per capita growth rate is defined by

$$g(N) = \frac{1}{N}\frac{dN}{dt}$$

(a) Show that

$$g(N) = r\left(1 - \frac{N}{K}\right)$$

(b) Graph $g(N)$ as a function of N for $N \ge 0$ when $r = 2$ and $K = 100$, and find the population size for which the per capita growth rate is maximal.

[4.1.2]

13. Suppose that $f(x) = x^2 - x - 2$, $x \in [-1, 2]$. Use the MVT to show that there exists a point $c \in [-1, 2]$ with a horizontal tangent. Find c.

14. Suppose that $f(x) = -x^2 + 2$. Explain why there exists a point c in the interval $(-1, 2)$ such that $f'(c) = -1$.

15. Sketch the graph of a function $f(x)$ that is continuous on the closed interval $[0, 1]$ and differentiable on the open interval $(0, 1)$, such that there exist exactly two points on the graph, $(c_1, f(c_1))$ and $(c_2, f(c_2))$, at which the slope of the tangent lines is equal to the slope of the secant line connecting the points $(0, f(0))$ and $(1, f(1))$. Show that the existence of at least one such point is guaranteed by the mean value theorem.

16. Suppose that $f(x) = x^2$, $x \in [a, b]$.

(a) Compute the slope of the secant line through the points $(a, f(a))$ and $(b, f(b))$.

(b) Find the point $c \in (a, b)$ such that the slope of the tangent line to the graph of f at $(c, f(c))$ is equal to the slope of the secant line determined in (a). How do you know that such a point exists? Show that c is the midpoint of the interval (a, b), that is, $c = (a + b)/2$.

17. Assume that f is continuous on $[a, b]$ and differentiable on (a, b). Show that if $f(a) < f(b)$, then f' is positive at some point between a and b.

18. Assume that f is continuous on $[a, b]$ and differentiable on (a, b). Furthermore, assume that $f(a) = f(b) = 0$ but f is not constant on $[a, b]$. Explain why there must be a point $c_1 \in (a, b)$ with $f'(c_1) > 0$, and a point $c_2 \in (a, b)$ with $f'(c_2) < 0$.

19. Suppose that f is differentiable for all $x \in \mathbf{R}$. Furthermore, that f satisfies $f(0) = 0$ and $1 \le f'(x) \le 2$ for all $x > 0$.

(a) Use Corollary 1 of the MVT to show that

$$x \le f(x) \le 2x$$

for all $x \ge 0$.

(b) Use your result in (a) to explain why $f(1)$ cannot be equal to 3.

(c) Find an upper and a lower bound for the value of $f(1)$.

20. Suppose that f is differentiable for all $x \in \mathbf{R}$ with $f(2) = 3$ and $f'(x) = 0$ for all $x \in \mathbf{R}$. Find $f(x)$.

21. Suppose that $f(x) = e^{-|x|}$, $x \in [-2, 2]$.

(a) Show that $f(-2) = f(2)$.

(b) Compute $f'(x)$, where defined.

(c) Show that there is no number $c \in (-2, 2)$ such that $f'(c) = 0$.

(d) Explain why your results in (a) and (c) do not contradict Rolle's theorem.

(e) Use a graphing calculator to sketch the graph of $f(x)$.

22. Use Corollary 2 of the MVT to show that if $f(x)$ is differentiable for all $x \in \mathbf{R}$ and satisfies

$$|f(x) - f(y)| \le |x - y|^2 \tag{4.3}$$

for all $x, y \in \mathbf{R}$, then $f(x)$ is constant. (*Hint:* Show that (4.3) implies that

$$\lim_{x \to y} \frac{f(x) - f(y)}{x - y} = 0 \qquad (4.4)$$

and use the definition of the derivative to interpret the left hand side of (4.4).)

23. We have seen that

$$f(x) = f_0 e^{rx}$$

satisfies the differential equation

$$\frac{df}{dx} = rf(x)$$

with $f(0) = f_0$. This exercise will show that it is in fact the only solution. Suppose that r is a constant and f is a differentiable function, where

$$\frac{df}{dx} = rf(x) \qquad (4.5)$$

for all $x \in \mathbf{R}$ and where $f(0) = f_0$. The following steps will show that $f(x) = f_0 e^{rx}$, $x \in \mathbf{R}$, is the only solution to (4.5).

(a) Define the function

$$F(x) = f(x)e^{-rx}, \quad x \in \mathbf{R}$$

Use the product rule to show that

$$F'(x) = e^{-rx}[f'(x) - rf(x)]$$

(b) Use (a) and (4.5) to show that $F'(x) = 0$ for all $x \in \mathbf{R}$.

(c) Use Corollary 2 to show that $F(x)$ is a constant, and hence $F(x) = F(0) = f_0$.

(d) Show that (c) implies that

$$f_0 = f(x)e^{-rx}$$

and therefore

$$f(x) = f_0 e^{rx}$$

4.2 MONOTONICITY AND CONCAVITY

Fish are indeterminate growers; they increase in body size throughout their life. However, as they become older, they grow proportionately more slowly. Their growth is often described mathematically by the von Bertalanffy equation, which fits a large number of both freshwater and marine fishes. This equation is given by

$$L(x) = L_\infty - (L_\infty - L_0)e^{-Kx}$$

where $L(x)$ denotes the length at age x, L_0 the length at age 0, and L_∞ the asymptotic maximum attainable length. We assume $L_\infty > L_0$. K is related to how quickly the fish grows. Figure 4.16 shows examples for two different values of K; L_∞ and L_0 are the same in both cases. We see from the graphs that the larger K, the more quickly the asymptotic length L_∞ is approached.

The fact that fish increase their body size throughout their life can be expressed mathematically using the first derivative of the function $L(x)$. Looking at the graph, we see that $L(x)$ is an increasing function of x. If we draw the tangent line at any point of the graph, its slope is positive; or, equivalently, $L'(x) > 0$. We can compute $L'(x)$, namely,

$$L'(x) = K(L_\infty - L_0)e^{-Kx}$$

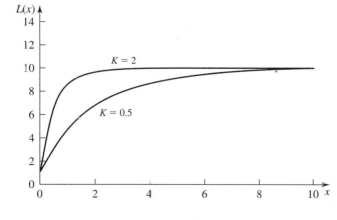

◀ **Figure 4.16**
The function $L(x)$ for $L_0 = 1$ and $L_\infty = 10$ with $K = 0.5$ and $K = 2$.

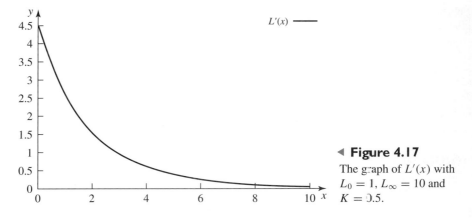

◀ **Figure 4.17**
The graph of $L'(x)$ with $L_0 = 1$, $L_\infty = 10$ and $K = 0.5$.

Since $L_\infty > L_0$, we conclude that $L'(x) > 0$. The graph of $L'(x)$ is shown in Figure 4.17.

The graph of $L'(x)$ shows that $L'(x)$ is a decreasing function of x: though fish increase their body size throughout their life, they do so at a rate that decreases with age. Mathematically, this can be expressed using the second derivative of $L(x)$ – the derivative of the first derivative. The tangent line at any point on the graph of $L'(x)$ has a negative slope, that is, the derivative of $L'(x)$ is negative: $L''(x) < 0$. The fact that the rate of growth decreases with age can also be seen directly from the graph of $L(x)$: it bends downwards. The second derivative thus tells us something about which way the graph of $L(x)$ bends.

This section discusses the important concepts of monotonicity, whether a function is decreasing or increasing, and concavity, whether a function bends upwards or downwards.

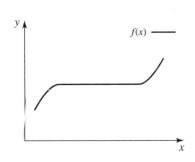

▲ **Figure 4.18**
An increasing function.

▲ **Figure 4.19**
A decreasing function.

4.2.1 Monotonicity

We saw in our motivating example that the first derivative tells us something about whether a function increases or decreases. Not every function is differentiable, however, so we phrase the definitions of increasing and decreasing in terms of the function f alone (see Figures 4.18 and 4.19).

> **Definition**   A function f defined on an interval I is called **(strictly) increasing** on I if
>
> $$f(x_1) < f(x_2) \qquad \text{whenever } x_1 < x_2 \text{ in } I$$
>
> It is called **(strictly) decreasing** on I if
>
> $$f(x_1) > f(x_2) \qquad \text{whenever } x_1 < x_2 \text{ in } I$$

An increasing or decreasing function is called **monotonic**. The word "strictly" in the above definition refers to having a strict inequality ($f(x_1) < f(x_2)$ and $f(x_1) > f(x_2)$). We will frequently drop "strictly." If instead of the strict inequality $f(x_1) < f(x_2)$, we have the inequality $f(x_1) \le f(x_2)$ whenever $x_1 < x_2$ in I, we call f nondecreasing. If $f(x_1) \ge f(x_2)$ whenever $x_1 < x_2$ in I, then f is called nonincreasing (see Figures 4.20 and 4.21).

When the function f is differentiable, then there is a simple test to determine whether a function is increasing or decreasing. This criterion is a consequence of the mean value theorem.

▲ **Figure 4.20**
A nondecreasing function may have regions where the function is constant.

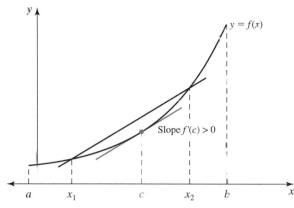

▲ Figure 4.21

A nonincreasing function may have regions where the function is constant.

◀ Figure 4.22

An illustration of the proof of "$f'(x) > 0$ for all $x \in (a, b)$ implies $f(x)$ increasing on $[a, b]$."

Theorem   Suppose f is continuous on $[a, b]$ and differentiable on (a, b).

a. If $f'(x) > 0$ for all $x \in (a, b)$, then f is increasing on $[a, b]$.

b. If $f'(x) < 0$ for all $x \in (a, b)$, then f is decreasing on $[a, b]$.

Proof   (See Figure 4.22.) We choose two numbers x_1 and x_2 in $[a, b]$, $x_1 < x_2$. Then f is continuous on $[x_1, x_2]$ and differentiable on (x_1, x_2). We can therefore apply the mean value theorem to f defined on $[x_1, x_2]$: There exists a number $c \in (x_1, x_2)$ such that

$$\frac{f(x_2) - f(x_1)}{x_2 - x_1} = f'(c)$$

In part (a) of the theorem, we assume $f'(x) > 0$ for all $x \in (a, b)$. Since $c \in (x_1, x_2) \subset (a, b)$, it follows that $f'(c) > 0$. Since, in addition, $x_2 > x_1$, it follows that

$$f(x_2) - f(x_1) = f'(c)(x_2 - x_1) > 0$$

which implies that $f(x_2) > f(x_1)$. Since x_1 and x_2 are arbitrary numbers in $[a, b]$ satisfying $x_1 < x_2$, it follows that f is increasing. The proof of part (b) is similar and relegated to Problem 24.   ◆

▮ EXAMPLE 1   Determine where the function

$$f(x) = x^3 - \frac{3}{2}x^2 - 6x + 3, \quad x \in \mathbf{R}$$

is increasing and where it is decreasing.

Solution   Since $f(x)$ is continuous and differentiable for all $x \in \mathbf{R}$, we can use the first derivative test for monotonic functions. We differentiate $f(x)$ and obtain

$$f'(x) = 3x^2 - 3x - 6 = 3(x - 2)(x + 1), \quad x \in \mathbf{R}$$

The graphs of $f(x)$ and $f'(x)$ are shown in Figure 4.23. The graph of $f'(x)$ is a parabola that intersects the x-axis at $x = 2$ and $x = -1$. The function $f'(x)$ therefore changes sign at $x = -1$ and $x = 2$. We find that

$$f'(x) = \begin{cases} > 0 & \text{if } x < -1 \text{ or } x > 2 \\ < 0 & \text{if } -1 < x < 2 \end{cases}$$

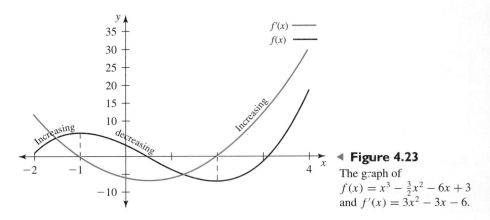

◀ **Figure 4.23**
The graph of
$f(x) = x^3 - \frac{3}{2}x^2 - 6x + 3$
and $f'(x) = 3x^2 - 3x - 6$.

Thus, $f(x)$ is increasing for $x < -1$ or $x > 2$ and decreasing for $-1 < x < 2$. This is also confirmed when looking at the graph of $f(x)$ in Figure 4.23.   ▮

4.2.2 Concavity

We saw in the motivating example at the beginning of this section that the second derivative tells us something about whether a function bends upwards or downwards. We arrived at this conclusion by checking whether the first derivative was increasing or decreasing.

A function is called **concave up** if it bends upwards, and **concave down** if it bends downwards. Before starting a precise definition of concavity for differentiable functions, we give two examples, which are illustrated in Figure 4.24.

First, look at the graph of the differentiable function $y = x^2$: it bends upwards, so we call it concave up. Bending upwards means that the slopes of the tangent lines are increasing as x increases. We can check this quite easily. The slope of the tangent line at x is given by the first derivative, $y' = 2x$. Since $y' = 2x$ is an increasing function, the slopes of the tangent lines are increasing as x increases.

Looking at the graph of the differentiable function $y = -x^2 + 4$, we see that it bends downwards, so we call it concave down. Bending downwards means that the slopes of the tangent lines are decreasing as x increases. We can easily check this; the first derivative of y is $y' = -2x$, which is a decreasing function.

The following definition for differentiable functions is based on the above discussion (see Figures 4.25 and 4.26).

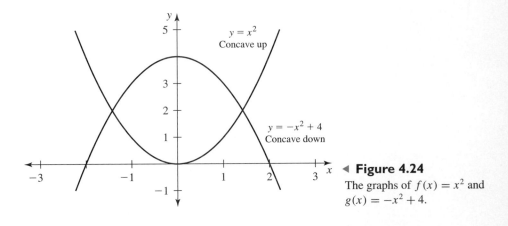

◀ **Figure 4.24**
The graphs of $f(x) = x^2$ and $g(x) = -x^2 + 4$.

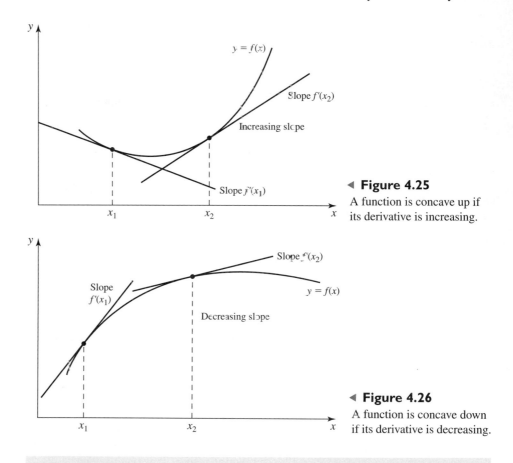

◀ **Figure 4.25**
A function is concave up if its derivative is increasing.

◀ **Figure 4.26**
A function is concave down if its derivative is decreasing.

Definition   A differentiable function $f(x)$ is concave up on an interval I if the first derivative $f'(x)$ is an increasing function on I. It is concave down on an interval I if the first derivative $f'(x)$ is a decreasing function on I.

Note that the definition assumes that $f(x)$ is differentiable. There is a more general definition that does not require differentiability (after all, not all functions are differentiable). The more general definition is more difficult to use. The definition given here suffices for our purposes; it has the added advantage that it provides a criterion we can use to determine whether a twice-differentiable function is concave up or concave down.

Theorem   Suppose that f is twice-differentiable on an open interval I.

a. If $f''(x) > 0$ for all $x \in I$, then f is concave up on I.

b. If $f''(x) < 0$ for all $x \in I$, then f is concave down on I.

Proof   Since f is twice-differentiable, we can apply the first derivative criterion to the function $f'(x)$. The proof of part (a) proceeds, then, as follows. If $f''(x) > 0$ on I, then $f'(x)$ is an increasing function on I. Using the definition of concave up, it follows immediately that f is concave up on I. The proof of part (b) is similar and relegated to Problem 25.

You can use the function $y = x^2$ to remember which functions are concave up. The "u" in concave "up" should remind you of the U-shaped form of the graph of $y = x^2$. You can also use the function $y = x^2$ to remember the second derivative criterion. You already know that the graph of $y = x^2$ is concave up, and you can easily compute the second derivative of $y = x^2$, namely, $y'' = 2 > 0$.   ◆

▌**EXAMPLE 2**   Determine where the function

$$f(x) = x^3 - \frac{3}{2}x^2 - 6x + 3, \quad x \in \mathbf{R}$$

is concave up and where it is concave down.

Solution   This is the same function as in Example 1 (redrawn in Figure 4.27). Since $f(x)$ is a polynomial, it is twice differentiable. In Example 1, we found $f'(x) = 3x^2 - 3x - 6$; differentiating $f'(x)$, we get the second derivative of f,

$$f''(x) = 6x - 3$$

We find

$$f''(x) \begin{cases} > 0 & \text{if } x > \frac{1}{2} \\ < 0 & \text{if } x < \frac{1}{2} \end{cases}$$

Thus, $f(x)$ is concave up for $x > 1/2$ and concave down for $x < 1/2$. A look at Figure 4.27 confirms this result.   ▌

A very common mistake is to associate monotonicity and concavity. One has nothing to do with the other. For instance, an increasing function can bend downwards or upwards (this will be discussed in Problem 21).

There are many biological examples of increasing functions that have a decreasing derivative and are therefore concave down. This is frequently associated with diminishing return. This phenomenon, illustrated in the next example, was also discussed in Section 3.8.

▌**EXAMPLE 3**   The response of crop yield Y to soil nitrogen level N can often be described by a function of the form

$$Y(N) = Y_{\max} \frac{N}{K + N}, \quad N \geq 0$$

where $Y_{\max}$ is the maximum attainable yield and K is a positive constant. The graph of $Y(N)$ is shown in Figure 4.28. It is obvious from the graph that $Y(N)$ is an increasing function of N. The graph bends downwards and hence is concave down. Before continuing, we will check this using the results from this section. Using the

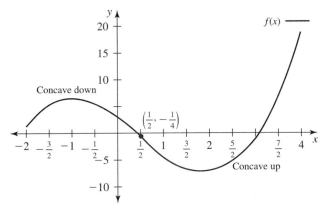

◀ **Figure 4.27**
The graph of
$f(x) = x^3 - \frac{3}{2}x^2 - 6x + 3$.

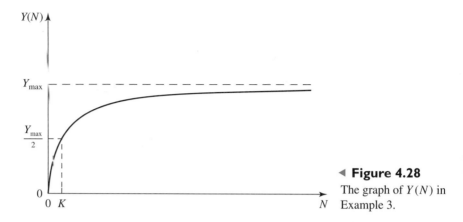

◀ Figure 4.28
The graph of $Y(N)$ in Example 3.

quotient rule, we find

$$Y'(N) = Y_{max} \frac{K + N - N}{(K + N)^2} = Y_{max} \frac{K}{(K + N)^2}$$

and using the chain rule, we find

$$Y''(N) = \frac{d}{dN}\left(Y_{max}K(K + N)^{-2}\right)$$

$$= Y_{max}K(-2)(K + N)^{-3}(1) = -Y_{max}\frac{2K}{(K + N)^3}$$

Since Y_{max} and K are positive constants and $N \geq 0$, it follows immediately that

$$Y'(N) > 0$$

which implies that $Y(N)$ is an increasing function. Furthermore,

$$Y''(N) < 0$$

which implies that $Y(N)$ is concave down. That is, $Y(N)$ is an increasing function but the rate of increase is decreasing. We say that Y is *increasing at a decelerating rate*. What does this mean? As we increase fertilizer levels, the yield will increase but at a less than proportional rate. This is a case of *diminishing return*. To be more concrete, we choose values for Y_{max} and K:

$$Y(N) = 50\frac{N}{5 + N}, \quad N \geq 0$$

The graph of this function is shown in Figure 4.29.

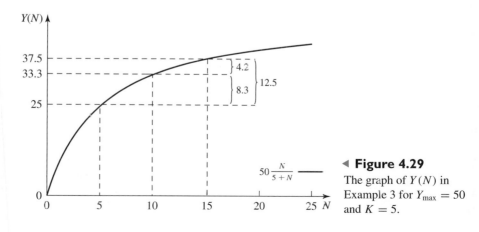

◀ Figure 4.29
The graph of $Y(N)$ in Example 3 for $Y_{max} = 50$ and $K = 5$.

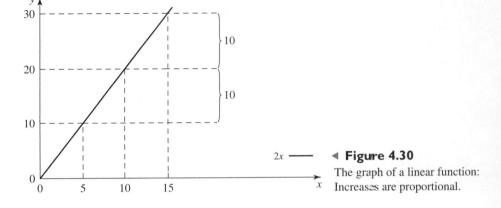

◄ **Figure 4.30**
The graph of a linear function:
Increases are proportional.

Suppose that initially $N = 5$. If we increase N by 5, that is from 5 to 10, then $Y(N)$ changes from $Y(5) = 25$ to $Y(10) = 33.3$, an increase of 8.3. If we increase N by double the amount, namely 10, that is, from 5 to 15, then $Y(15) = 37.5$ and the increase in yield is only 12.5, less than twice 8.3. This can also be understood by comparing successive increments. If we start with $N = 5$ and increase by 5 to $N = 10$, then the change in Y (the Y increment) is 8.3. Increasing N by the same amount starting at 10, the Y increment changes by $Y(15) - Y(10) = 4.2$. Changing N by equal increments has less of an effect for larger values of N, thus we say that the return is diminishing.

You should compare this to a linear function, say $f(x) = 2x$, which is neither concave up nor concave down (see Figure 4.30). If we increase x from 5 to 10, $f(x)$ changes from 10 to 20, that is, $f(x)$ increases by 10. If we increase x from 10 to 15, $f(x)$ changes from 20 to 30, again an increase of 10. That is, for linear functions, the increase is proportional.

◆ **4.2 PROBLEMS**

In Problems 1–20, determine where each function is increasing, decreasing, concave up, and concave down. With the help of a graphing calculator, sketch the graph of each function and label the intervals where it is increasing, decreasing, concave up, and concave down. Make sure that your graphs and your calculations agree.

1. $y = 3x - x^2, x \in \mathbf{R}$

2. $y = x^2 + 5x, x \in \mathbf{R}$

3. $y = x^2 + x - 4, x \in \mathbf{R}$

4. $y = x^3, x \in \mathbf{R}$

5. $y = -\frac{2}{3}x^3 + \frac{7}{2}x^2 - 3x + 4, x \in \mathbf{R}$

6. $y = (x - 2)^3 + 3, x \in \mathbf{R}$ 7. $y = \sqrt{x + 1}, x \geq -1$

8. $y = \sqrt[3]{2x - 3}, x \in \mathbf{R}$ 9. $y = \frac{1}{x}, x \neq 0$

10. $y = \frac{1}{x^2}, x \neq 0$ 11. $(x^2 + 1)^{1/3}, x \in \mathbf{R}$

12. $y = \frac{1}{1 + x}, x \neq -1$ 13. $y = \frac{1}{(1 + x)^2}, x \neq -1$

14. $y = \frac{x}{x + 1}, x \geq 0$ 15. $y = \sin x, 0 \leq x \leq 2\pi$

16. $y = \tan x, -\frac{\pi}{2} < x < \frac{\pi}{2}$ 17. $y = e^x, x \in \mathbf{R}$

18. $y = e^{-x}, x \in \mathbf{R}$ 19. $y = e^{-x^2/2}, x \in \mathbf{R}$

20. $y = \frac{1}{1 + e^{-x}}, x \in \mathbf{R}$

21. Sketch the graph of a function that is

(a) increasing at an accelerating rate; and

(b) increasing at a decelerating rate.

(c) Assume that your functions in (a) and (b) are twice-differentiable. Explain in each case how you could check the respective properties using the first and the second derivatives. Which of the functions is concave up, and which is concave down?

22. Show that if $f(x)$ is the linear function $y = mx + b$, then increases in $f(x)$ are proportional to increases in x. That is, if we increase x by Δx, then $f(x)$ increases by the same amount Δy regardless of the value of x. Compute Δy as a function of Δx. Illustrate this graphically.

23. We frequently must solve equations of the form $f(x) = 0$. When f is a continuous function on $[a, b]$ and $f(a)$ and $f(b)$ have opposite signs, then the intermediate value theorem

guarantees that there exists at least one solution to the equation $f(x) = 0$ in $[a, b]$.

(a) Explain in words why there exists exactly one solution in (a, b) if, in addition f is differentiable in (a, b) and $f'(x)$ is either strictly positive or strictly negative throughout (a, b).

(b) Use the result in (a) to show that

$$x^3 - 4x + 1 = 0$$

has exactly one solution in $[-1, 1]$.

24. Suppose that f is continuous on $[a, b]$ and differentiable on (a, b). Show that if $f'(x) < 0$ for all $x \in (a, b)$, then f is decreasing on $[a, b]$.

25. Suppose that f is twice-differentiable on an open interval I. Show that if $f''(x) < 0$, then f is concave down.

26. Suppose that the height in feet (y) of a tree as a function of tree age in years (x) is given by

$$y = 117e^{-10/x}, \quad x > 0$$

(a) Show that the height of the tree increases with age. What is the maximum attainable height?

(b) Where is the graph of height versus age concave up, and where is it concave down?

(c) Use a graphing calculator to sketch the graph of the height versus age.

(d) Use a graphing calculator to verify that the rate of growth is greatest at the point where the graph in (c) changes concavity.

27. Plants employ two basic reproductive strategies: *polycarpy*, in which reproduction occurs repeatedly during the life time of the organism, and *monocarpy*, in which reproduction occurs only once during the life time of the organism. (Bamboo, for instance, is a monocarpic plant.) The following quote taken from Iwasa et al. (1995)

> The optimal strategy is polycarpy (repeated reproduction) if reproductive success increases with the investment at a decreasing rate, while monocarpy ("big bang" reproduction) or intermittent reproduction if the reproductive success increases at an increasing rate.

(a) Sketch the graph of reproductive success as a function of reproductive investment for the cases of (i) polycarpy and (ii) monocarpy.

(b) Given that the second derivative describes whether a curve bends upwards or downwards, explain the quote in terms of the second derivative of the reproductive success function.

28. Assume that the following formula (Iwasa et al. 1995) expresses the relationship between the number of flowers on a plant, F, and the average number of pollinator visits, $X(F)$:

$$X(F) = cF^{\gamma}$$

where c is a positive constant. Find the range of values for the parameter γ such that the average number of pollinator visits to a plant increases with the number of flowers F but the rate of increase decreases with F. Explain your answer in terms of appropriate derivatives of the function $X(F)$.

29. Assume that the dependence of the average number of pollinator visits to a plant, X, on the number of flowers, F, is given by

$$X(F) = cF^{\gamma}$$

where γ is a positive constant less than 1 and c is a positive constant (Iwasa et al. 1995). How does the average number of pollen grains exported per flower, $E(F)$, change with the number of flowers on the plant, F, if $E(F)$ is proportional to

$$1 - \exp\left[-k\frac{X(F)}{F}\right]$$

where k is a positive constant?

30. (*A simple model of intraspecific competition*) (Adapted from Bellows 1981) Suppose that a study plot contains N annual plants, each of which produces S seeds that are sown within the same plot. The number of surviving plants in the following year is given by

$$A(N) = \frac{NS}{1 + (aN)^b}$$

for some positive constants a and b. This mathematical model incorporates density-dependent mortality; namely, the greater the number of plants in the plot, the lower the number of surviving offspring per plant, which is given by $A(N)/N$ and called the *net reproductive rate*.

(a) Use calculus to show that $A(N)/N$ is a decreasing function of N.

(b) The following quantity, called the *k-value*, can be used to quantify the effects of intraspecific competition (i.e., competition between individuals of the same species):

$$k = \log[\text{initial density}]$$
$$- \log[\text{final density}]$$

where log denotes the logarithm to base 10. Show that for the given model,

$$k = \log[NS] - \log\left[\frac{NS}{1 + (aN)^b}\right]$$
$$= \log[1 + (aN)^b]$$

One typically plots k versus $\log N$; the slope of the resulting curve is then used to quantify the effects of competition.

(i) Show that

$$\frac{d \log N}{dN} = \frac{1}{N \ln 10}$$

where ln denotes the natural logarithm.

(ii) Show that

$$\frac{dk}{d \log N} = (\ln 10)N\frac{dk}{dN}$$
$$= \frac{b}{1 + (aN)^{-b}}$$

(iii) Find

$$\lim_{N \to \infty} \frac{dk}{d \log N}$$

(iv) Show that if

$$\frac{dk}{d \log N} < 1$$

then $A(N)$ is increasing, whereas if

$$\frac{dk}{d \log N} > 1$$

then $A(N)$ is decreasing. (*Hint*: Compute $A'(N)$.) Explain in words what this would mean with respect to varying the initial density of seeds and observing the number of surviving plants in the following year. (*Note*: the first case is called *undercompensation* and the second case is called *overcompensation*.)

(v) The case

$$\frac{dk}{d \log N} = 1$$

is referred to as *exact compensation*. Suppose that you plot k versus $\log N$, and observe that over a certain range of values of N the slope of the resulting curve is equal to 1. Explain what this means.

31. (Adapted from Reiss 1989) Suppose that the rate at which body weight W changes with age x is

$$\frac{dW}{dx} \propto W^a \qquad (4.6)$$

where a is some species-specific positive constant.

(a) The relative growth rate (percentage weight gained per unit of time) is defined as

$$\frac{1}{W}\frac{dW}{dx}$$

What is the relationship between the relative growth rate and body weight? For which values of a is the relative growth rate increasing, and for which values is it decreasing?

(b) As fish grow larger, their weight increases each day but the relative growth rate is decreasing. If the rate of growth is described by (4.6), what values of a can you exclude, based on your results in (a)? Explain in words how the percentage weight increase (relative to the current body weight) differs for juvenile fish and for adult fish.

32. (*Allometric equations*) Allometric equations describe the scaling relationship between two measurements, such as tree height versus tree diameter or skull length versus backbone length. These equations are often of the form

$$Y = bX^a \qquad (4.7)$$

where b is some positive constant and a is a constant that can be positive, negative or zero.

(a) Assume that X and Y are body measurements (and therefore positive) and their relationship is described by an allometric equation of the form (4.7). For what values of a is Y an increasing function of X, but such that the ratio Y/X decreases with increasing X? Is Y as a function of X concave up or concave down in this case?

(b) In vertebrates one typically finds

$$[\text{skull length}] \propto [\text{body length}]^a$$

for some $a \in (0, 1)$. Use your answer in (a) to explain what this means for skull length versus body length in juveniles versus adults; that is, at which developmental stage do vertebrates have larger skulls relative to their body length?

◆ 4.3 EXTREMA, INFLECTION POINTS, AND GRAPHING

4.3.1 Extrema

If f is a continuous function on the closed interval $[a, b]$, then f has a global maximum and a global minimum in $[a, b]$. This is the content of the extreme value theorem, which is an existence result: it only tells us that absolute extrema exist under certain conditions, but it does not tell us how to find them.

Our strategy for finding global extrema in the case where f is a continuous function defined on a closed interval will be, first, to identify all local extrema of the function, and then to select the global extrema from the set of the local extrema. If f is a continuous function defined on an open interval or half-open interval, the existence of global extrema is no longer guaranteed, and we must compare the local extrema with the behavior of the function near the open boundaries of the domain (see Example 2 in Section 4.1).

Local extrema can be found in a very systematic way, using a straightforward recipe to identify candidates for local extrema. We showed in Section 4.1 that if f has a local extremum at an interior point c and $f'(c)$ exists, then $f'(c) = 0$ (Fermat's theorem). That is, those points where f is differentiable and where the first derivative is equal to 0 are certainly candidates for local extrema in the interior of the domain. Of course, these are only candidates, as explained in Section 4.1 (recall that $y = x^3$ has a horizontal tangent at $x = 0$, but $y = x^3$ does not have a local extremum at $x = 0$). In addition to points where the first derivative is equal to 0, we must check all points where the function is not differentiable (for instance, $y = |x|$ has a local minimum at $x = 0$ though it is not differentiable at 0). Points where the first derivative is equal to 0 or does not exist are called **critical points**.

In addition to the critical points, we must always check the endpoints of the interval on which f is defined (provided that there are such endpoints).

There are no other points where local extrema can occur. We are thus equipped with a systematic way of searching for *candidates* for local extrema:

---

1. Find all points c where $f'(c) = 0$.

2. Find all points c where $f'(c)$ does not exist.

3. Find the endpoints of the domain of f.

We illustrate this procedure in the following example. We wish to find all relative and absolute extrema of the function

$$f(x) = |x^2 - 4|, \quad -3 \le x < 2.5$$

Since we know what the graph of the function looks like, we graph it first, which will make it easier to understand the procedure for finding relative extrema (see Figure 4.31). But note that very often we do not know what a graph looks like, and we find relative extrema to gain a better understanding of the graph!

We will first rewrite $f(x)$ as a piecewise-defined function in order to get rid of the absolute value sign

$$f(x) = \begin{cases} x^2 - 4 & \text{for } -3 \le x \le -2 \text{ or } 2 \le x < 2.5 \\ -x^2 + 4 & \text{for } -2 \le x \le 2 \end{cases}$$

This piecewise-defined function is differentiable on the open intervals $(-3, -2)$, $(-2, 2)$ and $(2, 2.5)$. We find

$$f'(x) = \begin{cases} 2x & \text{for } -3 < x < -2 \text{ or } 2 < x < 2.5 \\ -2x & \text{for } -2 < x < 2 \end{cases}$$

Since $f'(x) = 0$ for $x = 0$ and $0 \in (-2, 2)$, $(0, f(0))$ is a critical point and is our first candidate for a local extremum. There are no other points where $f'(x) = 0$.

The second step is to identify interior points where the function is not differentiable. Since the function is differentiable in the open intervals $(-3, -2)$, $(-2, 2)$ and $(2, 2.5)$, we must look at the points where the function is pieced together, namely

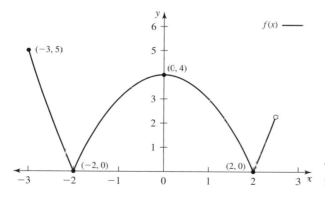

◀ **Figure 4.31**

The graph of $f(x) = |x^2 - 4|$ for $-3 \le x < 2.5$.

at $x = -2$ and at $x = 2$. We find

$$\lim_{x \to -2^-} f'(x) = -4 \qquad \text{and} \qquad \lim_{x \to -2^+} f'(x) = 4$$

and

$$\lim_{x \to 2^-} f'(x) = -4 \qquad \text{and} \qquad \lim_{x \to 2^+} f'(x) = 4$$

This shows that the function is not differentiable at $x = -2$ and $x = 2$. Therefore, there are critical points at $x = -2$ and $x = 2$, and these are also candidates for local extrema.

The third step is to identify endpoints of the domain. Since f is defined on $[-3, 2.5)$, there is an endpoint at $x = -3$. The fourth candidate is thus at $x = -3$. The interval $[-3, 2.5)$ is open at $x = 2.5$ and hence 2.5 is not in the domain of the function. The number 2.5 is therefore not a candidate for an extremum.

Our systematic procedure has provided us with four candidates for local extrema at $x = -3, -2, 0,$ and 2. In each case, we must decide whether they are in fact local extrema and, if so, whether they are local maxima or minima. The following observation, although rather obvious, is the key (see Figures 4.32 and 4.33).

> A continuous function has a local minimum at c if the function is decreasing to the left of c and increasing to the right of c. A function has a local maximum at c if the function is increasing to the left of c and decreasing to the right of c.

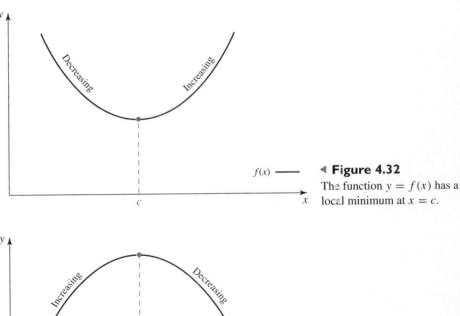

$f(x)$ ——   ◄ **Figure 4.32**
The function $y = f(x)$ has a local minimum at $x = c$.

$f(x)$ ——   ◄ **Figure 4.33**
The function $y = f(x)$ has a local maximum at $x = c$.

If the function is differentiable, as in our example, we can use the first derivative test to identify regions where the function is increasing and where it is decreasing.

Since $f'(x) = 2x$ for $-3 < x < -2$ and $2 < x < 2.5$, we see that $f'(x) > 0$ for $2 < x < 2.5$ and $f'(x) < 0$ for $-3 < x < -2$. Since $f'(x) = -2x$ for $-2 < x < 2$, we find that $f'(x) > 0$ for $x \in (-2, 0)$ and $f'(x) < 0$ for $x \in (0, 2)$. We illustrate this on the following number line. The plus (minus) signs show where $f'(x)$ is positive (negative).

We start with the interior points. At $x = -2$, the function changes from decreasing to increasing, that is, $f(x)$ has a local minimum at $x = -2$. At $x = 0$, the function changes from increasing to decreasing, that is, $f(x)$ has a local maximum at $x = 0$. At $x = 2$, the function changes from decreasing to increasing, that is, $f(x)$ has a local minimum at $x = 2$.

We still need to analyze the endpoint at $x = -3$. We see that the function is decreasing to the right of $x = -3$, that is, $f(x)$ has a local maximum at $x = -3$. You should compare all of our findings with the graph of $f(x)$.

The last step is to select the absolute extrema from the local extrema. But since the domain of f is not a closed interval, we must compare the values of the local extrema against the value at the boundary $x = 2.5$.

$$f(-3) = 5 \qquad f(-2) = 0 \qquad f(0) = 4 \qquad f(2) = 0 \qquad \lim_{x \to 2.5^-} f(x) = 2.25$$

Since 5 is the maximum and 0 is the minimum value, we see that the absolute maximum occurs at $x = -3$ and the absolute minima (there are two) occur at $x = -2$ and $x = 2$.

When a function is twice-differentiable at the point where the first derivative is equal to 0, there is a shortcut for determining whether a local maximum or a local minimum exists. (See Figure 4.34.) We assume that the function $f(x)$ is twice-differentiable. The graph of $f(x)$ in Figure 4.34 has a local maximum at $x = c$, since the function is increasing to the left of $x = c$ and decreasing to the right of $x = c$. If we look at how the slopes of the tangent lines change as we cross $x = c$ from the left, we see that the slopes are decreasing; that is, $f''(c) < 0$. In other words, the function is concave down at $x = c$ (which is immediately apparent when you look at the graph, but remember that typically you don't have the graph in front

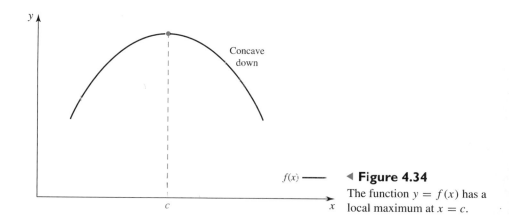

◀ **Figure 4.34**
The function $y = f(x)$ has a local maximum at $x = c$.

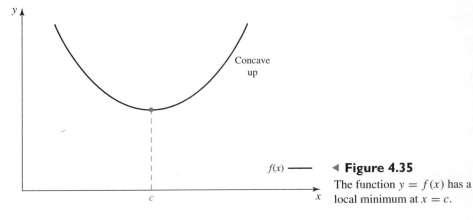

◀ **Figure 4.35**
The function $y = f(x)$ has a local minimum at $x = c$.

of you). There is an analogous result where f has a local minimum at $x = c$ (see Figure 4.35). This yields the following test.

THE SECOND DERIVATIVE TEST FOR LOCAL EXTREMA

If $f'(c) = 0$ and $f''(c) < 0$, then f has a local maximum at $x = c$.
If $f'(c) = 0$ and $f''(c) > 0$, then f has a local minimum at $x = c$.

Note that finding the point c where $f'(c) = 0$ gave us a candidate for local extrema; if the second derivative $f''(c) \neq 0$, it nails it down: Not only does this tell us whether there is a local extremum, it identifies it. This test is easy to apply, as we only have to check the sign of the second derivative at $x = c$, we do not have to check the behavior of the function in a neighborhood of $x = c$. We look at two examples to illustrate this test.

▮ EXAMPLE 1   Find all local and global extrema of

$$f(x) = \frac{3}{2}x^4 - 2x^3 - 6x^2 + 2, \quad x \in \mathbf{R}$$

Solution   (See Figure 4.36 for a graph of $f(x)$.) Since $f(x)$ is twice-differentiable for all $x \in \mathbf{R}$, we begin by finding the first two derivatives of f.

$$f'(x) = 6x^3 - 6x^2 - 12x = 6x(x - 2)(x + 1)$$
$$f''(x) = 18x^2 - 12x - 12$$

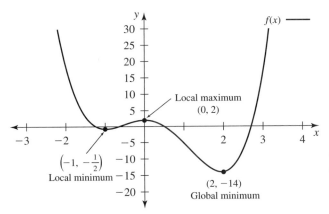

◀ **Figure 4.36**
The graph of
$f(x) = \frac{3}{2}x^4 - 2x^3 - 6x^2 + 2$
in Example 1.

Since $f'(x)$ exists for all $x \in \mathbf{R}$ and there are no endpoints of the domain, the only candidates for local extrema are points where $f'(x) = 0$:

$$6x(x - 2)(x + 1) = 0$$

We find $x = 0$, $x = 2$, and $x = -1$. Since $f''(x)$ exists, we can use the second derivative test to determine whether these candidates are local extrema and of what type. We need to evaluate the second derivative for each candidate.

$$f''(0) = -12 < 0 \Rightarrow \text{local maximum at } x = 0$$
$$f''(2) = 36 > 0 \quad \Rightarrow \text{local minimum at } x = 2$$
$$f''(-1) = 18 > 0 \quad \Rightarrow \text{local minimum at } x = -1$$

To find global extrema, we must check the local extrema and compare their values against each other and against the function values as $x \to \infty$ and $x \to -\infty$.

$$f(0) = 2 \qquad f(2) = -14 \qquad f(-1) = -\frac{1}{2}$$

and

$$\lim_{x \to \infty} f(x) = \infty \qquad \lim_{x \to -\infty} f(x) = \infty$$

Though the local maximum is 2 at $x = 0$, the point $(0, 2)$ is not a global maximum, since the function goes to ∞ as $|x| \to \infty$, so the function certainly exceeds the value 2. In fact, there is no global maximum; there is, however, a global minimum at $x = 2$. ▮

▮ EXAMPLE 2   Find all local and global extrema of

$$f(x) = x(1 - x)^{2/3}, \quad x \in \mathbf{R}$$

Solution   (A graph of $f(x)$ is shown in Figure 4.37.) We differentiate $f(x)$ using the product rule:

$$f'(x) = (1 - x)^{2/3} + x\frac{2}{3}(1 - x)^{-1/3}(-1)$$

$$= (1 - x)^{2/3} - \frac{2x}{3(1 - x)^{1/3}} \quad \text{for } x \neq 1$$

$$f''(x) = \frac{2}{3}(1 - x)^{-1/3}(-1) - \frac{2}{3}\left[(1 - x)^{-1/3} + x\left(-\frac{1}{3}\right)(1 - x)^{-4/3}(-1)\right]$$

$$= -\frac{2}{3(1 - x)^{1/3}} - \frac{2}{3(1 - x)^{1/3}} - \frac{2x}{9(1 - x)^{4/3}}$$

$$= -\frac{4}{3(1 - x)^{1/3}} - \frac{2x}{9(1 - x)^{4/3}} \quad \text{for } x \neq 1$$

We first find points where $f'(x) = 0$. If $x \neq 1$, then $f'(x) = 0$ when

$$(1 - x)^{2/3} = \frac{2x}{3(1 - x)^{1/3}}$$

$$3(1 - x) = 2x$$

$$3 = 5x$$

$$x = \frac{3}{5}$$

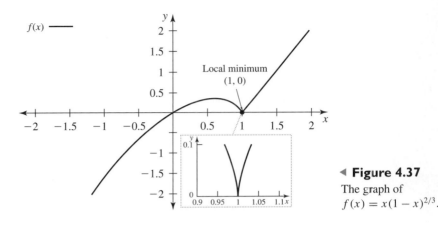

◀ **Figure 4.37**
The graph of
$f(x) = x(1-x)^{2/3}$.

We check the second derivative at $x = \frac{3}{5}$.

$$f''\left(\frac{3}{5}\right) = -\frac{4}{3\left(\frac{2}{5}\right)^{1/3}} - \frac{(2)\left(\frac{3}{5}\right)}{9\left(\frac{2}{5}\right)^{4/3}} < 0$$

that is, $f(x)$ has a local maximum at $x = \frac{3}{5}$.

The first derivative is not defined at $x = 1$, we therefore must investigate the function in the neighborhood of $x = 1$. Note that $f(x)$ is continuous at $x = 1$. We investigate where the function is increasing and where decreasing. The only points at which the derivative of the function can change sign are at $x = \frac{3}{5}$ and $x = 1$. We find that $f'(x)$ is positive to the left of $\frac{3}{5}$, negative between $\frac{3}{5}$ and 1, and positive for $x > 1$. This is summarized in the following number line.

We find that $f(x)$ has a local minimum at $x = 1$. We can say a bit more about what is happening at $x = 1$ (see insert in Figure 4.37).

$$\lim_{x \to 1^-} f'(x) = \lim_{x \to 1^-} \left[(1-x)^{2/3} - \frac{2x}{3(1-x)^{1/3}}\right] = -\infty$$

$$\lim_{x \to 1^+} f'(x) = \lim_{x \to 1^+} \left[(1-x)^{2/3} - \frac{2x}{3(1-x)^{1/3}}\right] = \infty$$

We see that there is a vertical tangent at $x = 1$; such points are called **cusps**.
Since

$$\lim_{x \to \infty} f(x) = \infty \qquad \text{and} \qquad \lim_{x \to -\infty} f(x) = -\infty$$

there are no absolute extrema.

4.3.2 Inflection Points

We begin with a verbal definition of **inflection points**, then give a procedure for locating such inflection points (see Figure 4.38).

> Inflection points are points where the concavity of a function changes – that is, where the function changes from concave up to concave down or from concave down to concave up.

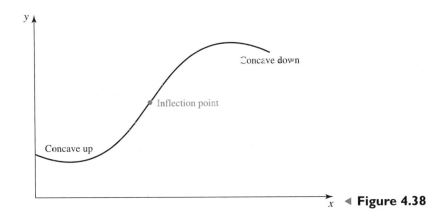

◀ **Figure 4.38**

If the function is twice-differentiable, there is an algebraic condition for finding candidates for inflection points. Recall that if a function f is twice-differentiable, it is concave up if $f'' > 0$ and concave down if $f'' < 0$. At an inflection point, f'' must therefore change sign; that is, the second derivative must be 0 at an inflection point.

> If $f(x)$ is twice-differentiable and has an inflection point at $x = c$, then $f''(c) = 0$.

Note that $f''(c) = 0$ is a necessary but not sufficient condition for an inflection point of a twice-differentiable function. For instance, $f(x) = x^4$ has $f''(0) = 0$ but $f(x)$ does not have an inflection point at $x = 0$. The function $f(x) = x^4$ is concave up and has a local minimum at $x = 0$ (see Figure 4.39). (We encountered a similar situation in Fermat's theorem, where we had a necessary but not sufficient condition for local extrema.) We can therefore use this test only for finding candidates for inflection points. To actually determine whether a candidate is an inflection point, we must check whether the second derivative changes sign.

EXAMPLE 3 Show that the function

$$f(x) = \frac{1}{2}x^3 - \frac{3}{2}x^2 + 2x + 1, \quad x \in \mathbf{R}$$

has an inflection point at $x = 1$.

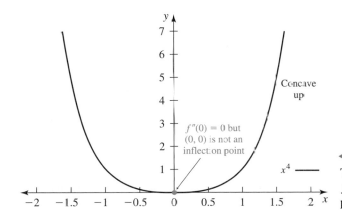

$f''(0) = 0$ but $(0, 0)$ is not an inflection point

x^4 ——

◀ **Figure 4.39**
The function $f(x) = x^4$ has $f''(0) = 0$ but no inflection point at $x = 0$.

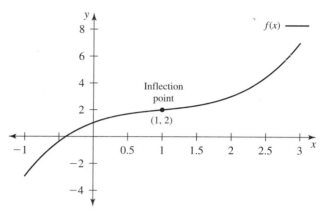

▲ **Figure 4.40**
The graph of $f(x) = \frac{1}{2}x^3 - \frac{3}{2}x^2 + 2x + 1$ has an inflection point at $x = 1$.

Solution   (The graph of $f(x)$ is shown in Figure 4.40.) We compute the first two derivatives:

$$f'(x) = \frac{3}{2}x^2 - 3x + 2$$

$$f''(x) = 3x - 3$$

Now, $f''(x) = 0$ if $x = 1$. Therefore, $x = 1$ is a candidate for an inflection point. Since $f''(x)$ is positive for $x > 1$ and negative for $x < 1$, $f''(x)$ changes sign at $x = 1$. We therefore conclude that $f(x)$ has an inflection point at $x = 1$. ▮

4.3.3 Graphing and Asymptotes

Using the first and the second derivative of a twice-differentiable function, we can obtain a fair amount of information about the function. We can determine intervals where the function is increasing, decreasing, concave up, and concave down. We can identify local and global extrema and find inflection points. To graph the function we also need to know how the function behaves in the neighborhood of points where either the function or its derivative is not defined, and we need to know how the function behaves at the endpoints of its domain (or, if the function is defined for all $x \in \mathbf{R}$, how the function behaves for $x \to \pm\infty$).

We discussed limits in Section 2.1; these operations are now needed to determine the behavior of a function at points where it is not defined, and when $x \to \pm\infty$. We illustrate this in the following example. Consider

$$f(x) = \frac{1}{x}, \quad x \neq 0$$

You are familiar with the graph of $f(x)$ (see Figure 4.41). You can see that the graph of $f(x)$ approaches the line $y = 0$ when $x \to \infty$ and also when $x \to -\infty$. Such a line is called an **asymptote**, and we say that $f(x)$ approaches the line $y = 0$ asymptotically as $x \to \infty$ and also as $x \to -\infty$. Since $y = 0$ is a horizontal line, it is called a horizontal asymptote. We can show this mathematically using limits:

$$\lim_{x \to \infty} f(x) = \lim_{x \to \infty} \frac{1}{x} = 0 \quad \text{and} \quad \lim_{x \to -\infty} f(x) = \lim_{x \to -\infty} \frac{1}{x} = 0$$

The function $f(x) = \frac{1}{x}$ is not defined at $x = 0$. Looking at the graph of the function $f(x) = \frac{1}{x}$, we see that it approaches the line $x = 0$ asymptotically.

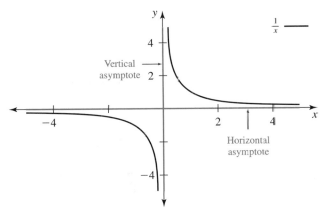

◀ **Figure 4.41**
The graph of $f(x) = \frac{1}{x}$ with horizontal asymptote $y = 0$ and vertical asymptote $x = 0$.

Since $x = 0$ is a vertical line, this is called a vertical asymptote. We can show this mathematically using limits:

$$\lim_{x \to 0^-} f(x) = \lim_{x \to 0^-} \frac{1}{x} = -\infty \qquad \text{and} \qquad \lim_{x \to 0^+} f(x) = \lim_{x \to 0^+} \frac{1}{x} = +\infty$$

> **Definition**   A line $y = b$ is a **horizontal** asymptote if either
>
> $$\lim_{x \to -\infty} f(x) = b \qquad \text{or} \qquad \lim_{x \to \infty} f(x) = b$$
>
> A line $x = c$ is a **vertical** asymptote if
>
> $$\lim_{x \to c^+} f(x) = +\infty \qquad \text{or} \qquad \lim_{x \to c^+} f(x) = -\infty$$
>
> or
>
> $$\lim_{x \to c^-} f(x) = +\infty \qquad \text{or} \qquad \lim_{x \to c^-} f(x) = -\infty$$

In addition to horizontal and vertical asymptotes, there are **oblique** asymptotes (see Figure 4.42). These are straight lines that are neither horizontal nor vertical, such that the graph of the function approaches this line as either $x \to +\infty$ or $x \to -\infty$. Mathematically, we express this in the following way. If

$$\lim_{x \to +\infty} [f(x) - (mx + b)] = 0 \qquad \text{or} \qquad \lim_{x \to -\infty} [f(x) - (mx + b)] = 0$$

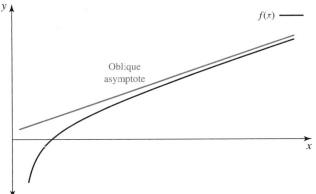

◀ **Figure 4.42**
The function $y = f(x)$ has an oblique asymptote.

then the line $y = mx + b$ is an oblique asymptote. The simplest case of an oblique asymptote is a rational function in which the degree of the numerator is one higher than the degree of the denominator.

EXAMPLE 4   Oblique Asymptote

Let

$$f(x) = \frac{x^2 - 3}{x - 2}, \quad x \neq 2$$

To determine whether $f(x)$ has an oblique asymptote, we use long division. We find

$$\frac{x^2 - 3}{x - 2} = x + 2 + \frac{1}{x - 2}$$

We see that $f(x)$ is the sum of a linear term, namely $x + 2$, and a remainder term, namely $\frac{1}{x-2}$, which goes to 0 as $x \to \pm\infty$. To check that $y = x + 2$ is indeed an oblique asymptote, we carry out the following computation.

$$\lim_{x \to \pm\infty} [f(x) - (x + 2)] = \lim_{x \to \pm\infty} \left[\frac{x^2 - 3}{x - 2} - (x + 2) \right]$$

$$= \lim_{x \to \pm\infty} \left[x + 2 + \frac{1}{x - 2} - (x + 2) \right]$$

$$= \lim_{x \to \pm\infty} \frac{1}{x - 2} = 0$$

Hence, $y = x + 2$ is an oblique asymptote. The graph of $f(x) = \frac{x^2-3}{x-2}$ together with its oblique asymptote is shown in Figure 4.43. (The graph of $f(x)$ also has a vertical asymptote at $x = 2$.)   ▌

We can now combine the results that we have obtained so far to produce a graph of a given function. We illustrate the steps leading to the graph in the following two examples.

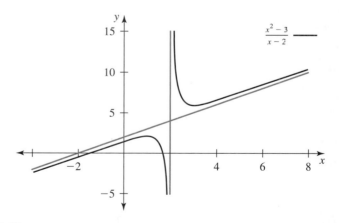

▲ **Figure 4.43**

The graph of $f(x) = \frac{x^2-3}{x-2}$ with oblique asymptote $g(x) = x + 2$ (and a vertical asymptote at $x = 2$).

▌**EXAMPLE 5**   Sketch the graph of the function

$$f(x) = \frac{2}{3}x^3 - 2x + 1, \quad x \in \mathbf{R}$$

Solution

Step 1.   Find $f'(x)$ and $f''(x)$.

$$f'(x) = 2x^2 - 2 = 2(x-1)(x+1), \quad x \in \mathbf{R}$$
$$f''(x) = 4x, \quad x \in \mathbf{R}$$

Step 2.   Find where $f'(x)$ is positive, negative, zero, or undefined. Identify the intervals where the function is increasing and decreasing. Find local extrema.
We begin by setting $f'(x) = 0$, that is,

$$2(x-1)(x+1) = 0$$

We find two solutions, namely $x = 1$ and $x = -1$. We illustrate on the following number line where $f'(x)$ is positive and where negative. Since $f'(x)$ is differentiable for all $x \in \mathbf{R}$, it can only change sign at $x = 1$ and at $x = -1$. This breaks the number line into three intervals, $x < -1, -1 < x < 1$, and $x > 1$. To determine whether $f'(x)$ is positive or negative in each of these intervals, we simply need to evaluate $f'(x)$ at one value in each interval. For instance, $f'(-2) = 2(-3)(-1) = 6 > 0$, that is, $f'(x) > 0$ for $x < -1$. A similar computation can be carried out for the other subintervals.

That is, $f'(x)$ is increasing for $x < -1$ and $x > 1$, and it is decreasing for $-1 < x < 1$. This also shows that the function has a local maximum at $x = -1$, namely $\left(-1, \frac{7}{3}\right)$, and a local minimum at $x = 1$, namely $\left(1, -\frac{1}{3}\right)$.

Step 3.   Find where $f''(x)$ is positive, negative, zero, or undefined. Find inflection points.
We set $f''(x) = 0$, that is,

$$4x = 0$$

We find one solution, namely $x = 0$. Since $f''(x)$ exists for all $x \in \mathbf{R}$, it can only change sign at $x = 0$. Evaluating $f''(x)$ at a value to the left of 0, say at $x = -1$, shows that $f''(x) < 0$ for $x < 0$. Similarly, evaluating $f''(x)$ at a value to the right of 0, say at $x = 1$, shows that $f''(x) > 0$ for $x > 0$. This is illustrated on the following number line.

Since $f''(x)$ changes sign at $x = 0$, we conclude that the function has an inflection point at $x = 0$, namely $(0, 1)$. The function is concave down for $x < 0$ and concave up for $x > 0$.

Step 4.   Determine the behavior at endpoints of the domain.
The domain is $(-\infty, +\infty)$. We therefore need to check the behavior of $f(x)$ as $x \to +\infty$ and as $x \to -\infty$. We find

$$\lim_{x \to -\infty} f(x) = -\infty \quad \text{and} \quad \lim_{x \to +\infty} f(x) = +\infty \qquad ▌$$

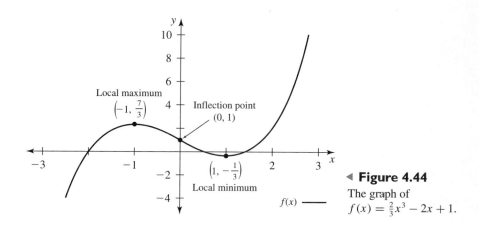

◀ **Figure 4.44**
The graph of
$f(x) = \frac{2}{3}x^3 - 2x + 1.$

Combining the results of these four steps allows us to sketch the graph of the function, as illustrated in Figure 4.44. You should label all extrema and inflection points.

▌EXAMPLE 6   Sketch the graph of the function

$$f(x) = e^{-x^2/2}, \quad x \in \mathbf{R}$$

Solution
Step 1.

$$f'(x) = -xe^{-x^2/2}, \quad x \in \mathbf{R}$$

and

$$f''(x) = (-1)e^{-x^2/2} + (-x)(-x)e^{-x^2/2}$$
$$= e^{-x^2/2}(x^2 - 1), \quad x \in \mathbf{R}$$

Step 2.   Since $f'(x)$ is defined for all $x \in \mathbf{R}$, we only need to identify those points where $f'(x) = 0$.

$$f'(x) = 0 \quad \text{for } x = 0$$

The sign of $f'(x)$ is illustrated on the following number line.

$$\underset{0}{\underline{\quad + + + + + + + + + + + \quad - - - - - - - - - \quad}}$$

We find that $f'(x)$ is increasing for $x < 0$ and decreasing for $x > 0$. Hence, $f(x)$ has a local maximum at $x = 0$, namely $(0, 1)$.

Step 3.

$$f''(x) = 0 \quad \text{for } x = 1 \text{ and } x = -1$$

The sign of $f''(x)$ is illustrated on the following number line.

$$\underset{-1 \qquad\quad 1}{\underline{\quad + + + + + + + \quad - - - - - \quad + + + + + + \quad}}$$

We find that $f(x)$ is concave up for $x < -1$ and $x > 1$, and it is concave down for $-1 < x < 1$. There are two inflection points, one at

◀ **Figure 4.45**
The graph of $f(x) = e^{-x^2/2}$.

$x = -1$, namely $(-1, e^{-1/2})$, and the other at $x = 1$, namely $(1, e^{-1/2})$. There are no other inflection points, since $f''(x)$ is defined for all $x \in \mathbf{R}$.

Step 4.

$$\lim_{x \to -\infty} f(x) = 0 \qquad \text{and} \qquad \lim_{x \to +\infty} f(x) = 0$$

This shows that $y = 0$ is a horizontal asymptote.
The graph of $f(x)$ is shown in Figure 4.45.

◆ 4.3 PROBLEMS

[4.3.1]

Find the local maxima and minima of each of the functions in Problems 1–16. Determine whether each function has absolute maxima and minima and find their coordinates. Find the intervals where each function is increasing and decreasing.

1. $y = 4 - x^2, -2 \leq x \leq 3$

2. $y = \sqrt{x}, 0 \leq x \leq 4$

3. $y = \sin x, 0 \leq x \leq 2\pi$

4. $y = \ln x, x > 0$

5. $y = e^{-x}, x \geq 0$

6. $y = |16 - x^2|, -5 \leq x \leq 8$

7. $y = (x - 1)^3 + 1, x \in \mathbf{R}$

8. $y = \tan x, -\frac{\pi}{2} < x < \frac{\pi}{2}$

9. $y = \cos x, 0 \leq x \leq \pi$

10. $y = \ln(x^2 + 1) - x, x \in \mathbf{R}$

11. $y = e^{-|x|}, x \in \mathbf{R}$

12. $y = e^{-x^2/4}, x \in \mathbf{R}$

13. $y = \frac{1}{3}x^3 + \frac{1}{2}x^2 - 6x + 2, x \in \mathbf{R}$

14. $y = x^2(1 - x), x \in \mathbf{R}$

15. $y = \sin(\pi x^2), -1 < x < 1$

16. $y = \sqrt{1 + x^2}, x \in \mathbf{R}$

17. (This problem illustrates the fact that $f'(c) = 0$ is not a sufficient condition for a local extremum of a differentiable function.) Show that the function $f(x) = x^3$ has a horizontal tangent at $x = 0$; that is, $f'(0) = 0$ but $f'(x)$ does not change sign at $x = 0$ and, hence, $f(x)$ does not have a local extremum at $x = 0$.

18. Suppose that $f(x)$ is twice-differentiable on $\mathbf{R}$ with $f(x) > 0$ for $x \in \mathbf{R}$. Show that if $f(x)$ has a local maximum at $x = c$, then $g(x) = \ln f(x)$ also has a local maximum at $x = c$.

[4.3.2]

In Problems 19–24, determine all inflection points.

19. $f(x) = x^3 - 2, x \in \mathbf{R}$

20. $f(x) = \cos x, 0 \leq x \leq \pi$

21. $f(x) = e^{-x^2}, x \geq 0$

22. $f(x) = xe^{-x}, x \geq 0$

23. $f(x) = \tan x, -\frac{\pi}{2} < x < \frac{\pi}{2}$

24. $f(x) = \ln x + \frac{1}{x}, x > 0$

25. (This problem illustrates that $f''(c) = 0$ is not a sufficient condition for an inflection point of a twice differentiable function.) Show that the function $f(x) = x^4$ has $f''(0) = 0$, but that $f''(x)$ does not change sign at $x = 0$ and, hence, $f(x)$ does not have an inflection point at $x = 0$.

26. (*Logistic equation*) Suppose that the size of a population at time t is denoted by $N(t)$ and satisfies

$$N(t) = \frac{100}{1 + 3e^{-2t}}$$

for $t \geq 0$.

(a) Show that $N(0) = 25$.

(b) Show that $N(t)$ is strictly increasing.

(c) Show that

$$\lim_{t\to\infty} N(t) = 100$$

(d) Show that $N(t)$ has an inflection point when $N(t) = 50$, that is, when the size of the population is at half its limiting value.

(e) Use your results in (a)–(d) to sketch the graph of $N(t)$.

[4.3.3]

Find the local maxima and minima of the functions in Problems 27–34. Determine whether the functions have absolute maxima and minima, and find their coordinates. Find inflection points. Find the intervals where the function is increasing, decreasing, concave up, and concave down. Sketch the graph of each function.

27. $y = \frac{2}{3}x^3 - 2x^2 - 6x + 2$ for $-2 \le x \le 5$

28. $y = x^4 - 2x^2$, $x \in \mathbf{R}$

29. $y = |x^2 - 9|$, $-4 \le x \le 5$

30. $y = \sqrt{|x|}$, $x \in \mathbf{R}$

31. $y = x + \cos x$, $x \in \mathbf{R}$

32. $y = \tan x - x$, $x \in \left(-\frac{\pi}{2}, \frac{\pi}{2}\right)$

33. $y = \dfrac{x^2 - 1}{x^2 + 1}$, $x \in \mathbf{R}$

34. $y = \ln(x^2 + 1)$, $x \in \mathbf{R}$

35. Let

$$f(x) = \frac{x}{x - 1}, \quad x \ne 1$$

(a) Show that

$$\lim_{x\to-\infty} f(x) = 1$$

and

$$\lim_{x\to+\infty} f(x) = 1$$

That is, $y = 1$ is a horizontal asymptote of the curve $y = \frac{x}{x-1}$.

(b) Show that

$$\lim_{x\to1^-} f(x) = -\infty$$

and

$$\lim_{x\to1^+} f(x) = +\infty$$

That is, $x = 1$ is a vertical asymptote of the curve $y = \frac{x}{x-1}$.

(c) Determine where $f(x)$ is increasing and where it is decreasing. Does $f(x)$ have local extrema?

(d) Determine where $f(x)$ is concave up and where it is concave down. Does $f(x)$ have inflection points?

(e) Sketch the graph of $f(x)$ together with its asymptotes.

36. Let

$$f(x) = -\frac{2}{x^2 - 1}, \quad x \ne -1, 1$$

(a) Show that

$$\lim_{x\to+\infty} f(x) = 0$$

and

$$\lim_{x\to-\infty} f(x) = 0$$

That is, $y = 0$ is a horizontal asymptote of $f(x)$.

(b) Show that

$$\lim_{x\to-1^-} f(x) = -\infty$$

and

$$\lim_{x\to-1^+} f(x) = +\infty$$

and that

$$\lim_{x\to1^-} f(x) = +\infty$$

and

$$\lim_{x\to1^+} f(x) = -\infty$$

That is, $x = -1$ and $x = 1$ are vertical asymptotes of $f(x)$.

(c) Determine where $f(x)$ is increasing and where it is decreasing. Does $f(x)$ have local extrema?

(d) Determine where $f(x)$ is concave up and where it is concave down. Does $f(x)$ have inflection points?

(e) Sketch the graph of $f(x)$ together with its asymptotes.

37. Let

$$f(x) = \frac{2x^2 - 5}{x + 2}, \quad x \ne -2$$

(a) Show that $x = -2$ is a vertical asymptote.

(b) Determine where $f(x)$ is increasing and where it is decreasing. Does $f(x)$ have local extrema?

(c) Determine where $f(x)$ is concave up and where it is concave down. Does $f(x)$ have inflection points?

(d) Since the degree of the numerator is one higher than the degree of the denominator, $f(x)$ has an oblique asymptote. Find it.

(e) Sketch the graph of $f(x)$ together with its asymptotes.

38. Let

$$f(x) = \frac{\sin x}{x}, \quad x \ne 0$$

(a) Show that $y = 0$ is a horizontal asymptote.

(b) Since $f(x)$ is not defined at $x = 0$, does this mean that $f(x)$ has a vertical asymptote at $x = 0$? Find $\lim_{x\to0^+} f(x)$ and $\lim_{x\to0^-} f(x)$.

(c) Use a graphing calculator to sketch the graph of $f(x)$.

◆ 4.4 OPTIMIZATION

There are many situations in which one wishes to maximize or minimize certain quantities. For instance, in a chemical reaction, you might wish to know under which conditions the reaction rate is maximized. In an agricultural setting, you might be interested in finding the amount of fertilizer that would maximize yield. Optimization problems arise in the study of the evolution of life histories, involving questions such as when an organism should begin reproduction in order to maximize the number of surviving offspring. In each case, we are interested in finding absolute (or global) extrema.

EXAMPLE 1   **Chemical Reaction**

We consider the chemical reaction

$$A + B \to AB$$

In Example 4 of Subsection 1.2.2, we found that the reaction rate is given by the function

$$R(x) = k(a - x)(b - x), \quad 0 \le x \le \min(a, b)$$

where x is the concentration of the product AB. The constants a and b are the concentrations of the reactants A and B at the beginning of the reaction. To be more concrete, we choose $k = 2$, $a = 2$, and $b = 5$. Then

$$R(x) = 2(2 - x)(5 - x) \quad \text{for } 0 \le x \le 2$$

(see Figure 4.46).

We are interested in finding the value of the concentration x that maximizes the reaction rate; this is the absolute maximum of $R(x)$. Since $R(x)$ is differentiable on $(0, 2)$, we can find all local extrema in $(0, 2)$ by investigating the first derivative. Multiplying out yields

$$R(x) = 20 - 14x + 2x^2 \quad \text{for } 0 \le x \le 2$$

Differentiating with respect to x yields

$$R'(x) = -14 + 4x \quad \text{for } 0 < x < 2$$

To find candidates for local extrema, we set $R'(x) = 0$:

$$-14 + 4x = 0 \quad \text{or} \quad x = \frac{7}{2}$$

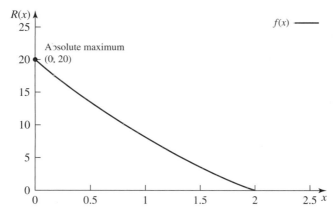

◄ **Figure 4.46**

The chemical reaction rate $R(x)$ in Example 1. The graph of $R(x) = 2(2 - x)(5 - x)$, $0 \le x \le 2$, has an absolute maximum at $(0, 20)$.

Since $\frac{7}{2} \notin (0, 2)$, there are no points in the interval $(0, 2)$ with horizontal tangents. Given that

$$R'(x) = -14 + 4x < 0 \quad \text{for } x \in (0, 2)$$

we conclude that $R(x)$ is decreasing in $(0, 2)$. The absolute maximum is therefore attained at the left endpoint of the interval $[0, 2]$, namely at $x = 0$. Thus, the reaction rate is maximal when the concentration of the product AB is equal to 0. You should compare this result with the graph of $R(x)$ in Figure 4.46. Since the reaction rate is proportional to the product of the concentrations of A and B, and A and B react to form the product AB, their concentrations decrease during the reaction and hence the reaction rate should be highest at the beginning of the reaction, when the concentrations of A and B are highest. ∎

∎EXAMPLE 2   Crop Yield

Let $Y(N)$ be the yield of an agricultural crop as a function of nitrogen level N in the soil. We assume that

$$Y(N) = \frac{N}{1 + N^2} \quad \text{for } N \geq 0$$

(where N is measured in appropriate units.) Find the amount of nitrogen that maximizes yield.

Solution   The function $Y(N)$, shown in Figure 4.47, is differentiable for $N > 0$. We find

$$Y'(N) = \frac{(1 + N^2) - N \cdot 2N}{(1 + N^2)^2} = \frac{1 - N^2}{(1 + N^2)^2}$$

Setting $Y'(N) = 0$, we obtain candidates for local extrema:

$$Y'(N) = 0 \quad \text{if } 1 - N^2 = 0 \text{ or } N = \pm 1$$

Since $N = -1$ is not in the domain of $Y(N)$, we can discard this candidate. The other candidate $N = 1$ is in the domain, and we see that

$$Y'(N) \begin{cases} > 0 & \text{for } 0 < N < 1 \\ < 0 & \text{for } N > 1 \end{cases}$$

Since $Y'(N)$ changes from increasing to decreasing at $N = 1$, $N = 1$ is a local maximum. To find the absolute maximum, we still need to check the endpoints of the domain. We find

$$Y(0) = 0 \quad \text{and} \quad \lim_{N \to \infty} Y(N) = \lim_{N \to \infty} \frac{N}{1 + N^2} = 0$$

Since $Y(1) = \frac{1}{2}$, we conclude that $N = 1$ is the absolute maximum. That is, $N = 1$ is the amount of nitrogen that maximizes yield. ∎

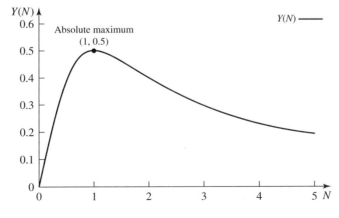

◀ **Figure 4.47**
Cropyield $Y(N)$ in Example 2. The graph of

$$Y(N) = \frac{N}{1 + N^2}, N \geq 0,$$

has an absolute maximum at $\left(1, \frac{1}{2}\right)$.

▌ **EXAMPLE 3**   **Maximizing Area**

A field biologist wants to enclose a rectangular study plot. She has 1600 ft of fencing. Using that material, what are the dimensions of the study plot that will have the largest area?

Solution   Figure 4.48 illustrates the situation. The area A is given by

$$A = xy \qquad (4.8)$$

and the perimeter of the study area is given by

$$1600 = 2x + 2y \qquad (4.9)$$

Solving equation (4.9) for y yields

$$y = 800 - x$$

We can substitute this for y in equation (4.8) and obtain

$$A(x) = x(800 - x) = 800x - x^2 \quad \text{for } 0 \le x \le 800$$

It is important to state the domain of the function. Clearly, the smallest value of x is 0, in which case the enclosed area is also 0 since $A(0) = 0$. The largest possible value for x is 800, this will also produce a rectangle where one side has length 0 and the other side has length 800; the corresponding area is $A(800) = 0$.

We wish to maximize the enclosed area $A(x)$. The function $A(x)$ is differentiable for $x \in (0, 800)$:

$$A'(x) = 800 - 2x \quad \text{for } 0 < x < 800$$
$$A''(x) = -2 \qquad \text{for } 0 < x < 800$$

To find candidates for local extrema, we set $A'(x) = 0$.

$$800 - 2x = 0 \qquad \text{or} \qquad x = 400$$

Since $A''(400) < 0$, $x = 400$ is a local maximum. To find the absolute maximum, we need to check the function $A(x)$ at the endpoints of the interval $[0, 800]$.

$$A(0) = A(800) = 0$$

Since $A(400) = 400^2 = 160,000$, the area is maximized for $x = 400$, which implies that the study plot is a square. This is true in general, for a given perimeter of a rectangle, the square has maximum area (see Problem 2 in this section).   ▌

▲ **Figure 4.48**
The rectangular study plot in Example 3.

▌ **EXAMPLE 4**   **Minimizing Material**

Aluminum soda cans are shaped like a right-circular cylinder, and hold about 12 ounces. Production of aluminum requires a lot of energy, so it is desirable to design soda cans that use the least amount of material. What dimensions would such an optimal soda can have?

Solution   We approximate the soda can by a right-circular cylinder in which the cylinder wall and both ends are made of aluminum. We denote the height of the cylinder by h and the radius by r (see Figure 4.49).

If we measure h and r in centimeters, we must convert the 12 ounces into a volume measured in cubic centimeters (cm^3). For this, we need to know that 1 liter corresponds to a volume of $1000\,cm^3$.

We are now ready to set up the problem. The can with the least amount of material is the cylinder whose surface area is minimal for a given volume. Using formulas from geometry, we find that the surface area A of a right circular cylinder

▲ **Figure 4.49**
A right circular cylinder with height h and radius r.

with top and bottom is given by

$$A = \underbrace{2\pi r h}_{\text{cylinder wall}} + \underbrace{2(\pi r^2)}_{\text{cylinder ends}}$$

The volume V of a right cylinder is given by

$$V = \pi r^2 h$$

Now, 12 ounces are about 0.355 liters, and 1 liter corresponds to 1000 cm³. We therefore need to minimize A when the volume $V = 355$ cm³. Solving $\pi r^2 h = 355$ for h yields

$$h = \frac{355}{\pi r^2}$$

Substituting this for h in the formula for A, we find

$$A(r) = 2\pi r \frac{355}{\pi r^2} + 2\pi r^2 = \frac{710}{r} + 2\pi r^2 \quad \text{for } r > 0$$

The graph of $A(r)$ is shown in Figure 4.50.

To find the absolute minimum, differentiate $A = A(r)$ and set the derivative equal to 0.

$$A'(r) = -\frac{710}{r^2} + 4\pi r = 0$$

Solving for r yields

$$4\pi r = \frac{710}{r^2} \quad \text{or} \quad r^3 = \frac{710}{4\pi} = \frac{355}{2\pi}$$

We thus find that

$$r = \left(\frac{355}{2\pi}\right)^{1/3} = 3.84 \, \text{cm}$$

To check whether this is indeed a minimum, we compute the second derivative of $A(r)$:

$$A''(r) = 2\frac{710}{r^3} + 4\pi > 0 \quad \text{for } r > 0$$

Since $A''(r) > 0$, $r = \left(\frac{355}{2\pi}\right)^{1/3}$ is a local minimum. To determine whether this value of r is the absolute minimum, we need to compare the surface area at the boundaries

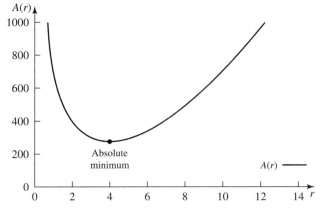

◀ **Figure 4.50**
The surface area A of a right cylinder with given volume as a function of radius r.

of the domain. Given that

$$\lim_{r \to 0+} A(r) = \infty \qquad \text{and} \qquad \lim_{r \to \infty} A(r) = \infty$$

it follows that the absolute minimum is achieved at $r = \left(\frac{355}{2\pi}\right)^{1/3}$.

To find h, we use the fact that $\pi r^2 h = 355$. We obtain

$$h = \frac{355}{\pi r^2} = \frac{355}{\pi \left(\frac{355}{2\pi}\right)^{2/3}} = \frac{\frac{355}{\pi}}{\left(\frac{355}{2\pi}\right)^{2/3}}$$

$$= 2\left(\frac{355}{2\pi}\right)^{1/3} = 2r = 7.67\,\text{cm}$$

We thus find that the can that uses the least amount of material is one where the height of the can is equal to its diameter.

A real soda can has $h = 12.5\,\text{cm}$ and $r = 3.1\,\text{cm}$. In our computation, we assumed that the material used to manufacture the can is of equal thickness. The top of a real soda can, however, is of thicker material, which could explain why soda cans don't have the dimensions we computed in this example.   ■

The previous four examples illustrate the basic types of optimization problems. In the first two examples, the functions that we wished to optimize were given; in the third and fourth examples, we needed to set up equations for the functions that we wished to optimize. Once you obtain a function that you wish to optimize, you must use the results from the previous sections in this chapter to find the absolute extremum. It is important to state the domain of the function, since absolute extrema may be found at the endpoints of the domain. We discuss one more advanced application.

EXAMPLE 5   *(Adapted from Roff 1992)*   Lloyd (1987) proposed the following model to determine optimal clutch size. If clutch size is equal to N and the total amount of resources allocated is R, then the amount of resources allocated to each offspring is $x = R/N$. The survival chances for an individual, denoted by $f(x)$, is related to the investment x per individual. Lloyd proposed an S-shaped or sigmoidal curve; that is, survival chances are very low when the investment is low and the curve shows a saturation effect for large investments. The success of a clutch, or its **fitness**, can be measured as

$$w(x) = [\text{number of offspring}] \times [\text{survival probability of offspring}]$$

$$= Nf(x) = \frac{R}{x} f(x) \quad \text{for } x > 0$$

where R is a positive constant. We wish to find the value of x that maximizes the fitness $w(x)$. If we assume that $f(x)$ is differentiable for $x > 0$, then differentiating $w(x)$ with respect to x yields

$$\frac{dw}{dx} = R\frac{d}{dx}\left(\frac{f(x)}{x}\right)$$

$$= R\frac{f'(x)x - f(x)}{x^2} = \frac{R}{x}\left(f'(x) - \frac{f(x)}{x}\right) \quad \text{for } x > 0$$

Setting $w'(x) = 0$ yields

$$f'(x) = \frac{f(x)}{x} \tag{4.10}$$

We denote the solution of equation (4.10) by $\hat{x}$ (read: "x hat"). If $f(x)$ is twice-differentiable at $\hat{x}$, we can use the second derivative test to test whether $\hat{x}$ is a local maximum or minimum. We find

$$\frac{d^2w}{dx^2} = -\frac{R}{x^2}\left(f'(x) - \frac{f(x)}{x}\right) + \frac{R}{x}\left(f''(x) - \frac{d}{dx}\frac{f(x)}{x}\right)$$

Since $f'(\hat{x}) = f(\hat{x})/\hat{x}$, the first term on the right hand side is equal to 0 when $x = \hat{x}$. Furthermore, $\frac{d}{dx}[f(x)/x] = 0$ when $x = \hat{x}$. Hence,

$$\frac{d^2w}{dx^2}\bigg|_{x=\hat{x}} = \frac{R}{\hat{x}}f''(\hat{x})$$

If $f(x)$ is concave down at $\hat{x}$, then $f''(\hat{x}) < 0$ and $w(x)$ has a local maximum at $\hat{x}$. For concreteness, we choose

$$f(x) = \frac{x^2}{1+x^2} \quad \text{for } x > 0$$

The graph of $f(x)$ is shown in Figure 4.51; the curve is sigmoid. Differentiating $f(x)$ with respect to x yields

$$f'(x) = \frac{2x(1+x^2) - x^2(2x)}{(1+x^2)^2} = \frac{2x}{(1+x^2)^2}$$

Since

$$\frac{dw}{dx} = 0 \quad \text{for } f'(x) = \frac{f(x)}{x}$$

we find

$$\frac{2x}{(1+x^2)^2} = \frac{1}{x}\frac{x^2}{1+x^2}$$

which yields

$$2 = 1 + x^2 \qquad \text{or} \qquad 1 = x^2$$

Since $x \geq 0$, we can discard the solution $x = -1$ and find $\hat{x} = 1$. To see whether $\hat{x} = 1$ is a local maximum, we evaluate $w''(1)$; we need to find $f''(x)$ first.

$$f''(x) = \frac{2(1+x^2)^2 - (2x)2(1+x^2)(2x)}{(1+x^2)^4}$$

$$= \frac{2(1+x^2) - 8x^2}{(1+x^2)^3} = \frac{2 - 6x^2}{(1+x^2)^3}$$

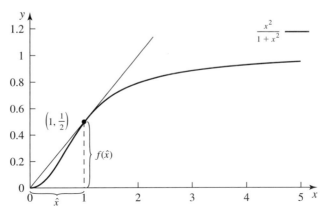

◀ **Figure 4.51**

The graph of $f(x) = \dfrac{x^2}{1+x^2}$

together with the tangent line at $\left(1, \frac{1}{2}\right)$.

Since

$$\left.\frac{d^2w}{dx^2}\right|_{\hat{x}=1} = \left.\frac{R}{x}f''(x)\right|_{\hat{x}=1} = Rf''(1) = R\frac{2-6}{8} < 0$$

we conclude that there is a local maximum at $\hat{x} = 1$. To see whether it is an absolute maximum, we compare $w(1)$ with $w(0)$ and $\lim_{x\to\infty} w(x)$. Since

$$w(x) = \frac{R}{x}f(x) = \frac{R}{x}\frac{x^2}{1+x^2} = R\frac{x}{1+x^2}$$

we find

$$w(0) = 0 \qquad w(1) = \frac{R}{2} \qquad \lim_{x\to\infty} w(x) = 0$$

Hence, $\hat{x} = 1$ is the absolute maximum; for our choice of $f(x) = \frac{x^2}{1+x^2}$, the optimal clutch size N_{opt} satisfies $N_{\text{opt}} = R$. (Other choices of $f(x)$ would give a different result.)

There is a geometric way of finding $\hat{x}$. Namely, since

$$f'(\hat{x}) = \frac{f(\hat{x})}{\hat{x}}$$

the tangent line at $(\hat{x}, f(\hat{x}))$ has slope $\frac{f(\hat{x})}{\hat{x}}$. This line can be obtained by drawing a straight line through the origin that just touches the graph of $y = f(x)$, as illustrated in Figure 4.51. ▌

◆ 4.4 PROBLEMS

1. Find the smallest perimeter possible for a rectangle whose area is 25 in.^{2}.

2. Show that among all rectangles with a given perimeter, the square has the largest area.

3. A rectangle has its base on the x-axis and its upper two vertices on the parabola $y = 3 - x^2$, as shown in Figure 4.52. What is the largest area the rectangle can have?

4. A rectangular study area is to be enclosed by a fence and divided into two equal parts, with a fence running along the division parallel to one of the sides if the total area is 384 ft^2, find the dimensions of the study area that will minimize the total length of the fence. How much fence will be required?

5. A rectangular field is bounded on one side by a river and on the other three sides by a fence. Find the dimensions of the field that will maximize the enclosed area if the fence has total length 320 ft.

6. Find the largest possible area of a right triangle whose hypotenuse is 4 cm long.

7. Suppose that a and b are the side lengths in a right triangle whose hypotenuse is 5 cm long. What is the smallest perimeter possible if the area of the triangle is 2 cm^2?

8. Suppose that a and b are the side lengths in a right triangle whose hypotenuse is 10 cm long. Show that the area of the triangle is largest when $a = b$.

9. A rectangle has its base on the x-axis, its lower left corner at $(0, 0)$, and its upper right corner on the curve $y = 1/x$. What is the smallest perimeter the rectangle can have?

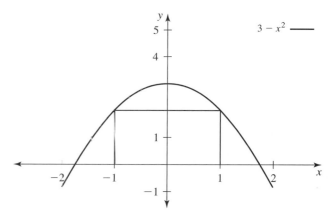

▲ Figure 4.52
The graph of $y = 3 - x^2$ together with the inscribed rectangle in Problem 3.

10. A rectangle has its base on the x-axis and its upper left and right corners on the curve $y = \sqrt{4 - x^2}$, as shown in Figure 4.53. The left and the right corners have equal distance from the vertical axis. What is the largest area the rectangle can have?

11. Denote by (x, y) a point on the straight line $y = 4 - 3x$ (see Figure 4.54)

(a) Show that the distance of (x, y) to the origin is given by

$$f(x) = \sqrt{x^2 + (4 - 3x)^2}$$

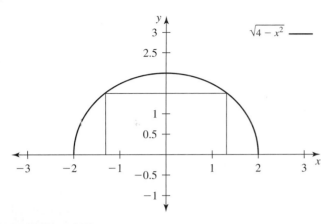

▲ Figure 4.53
The graph of $y = (4 - x^2)^{1/2}$ together with the inscribed rectangle in Problem 10.

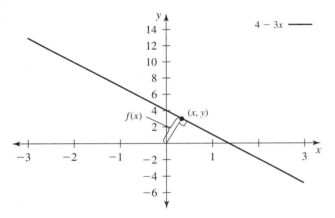

▲ Figure 4.54
The graph of $y = 4 - 3x$ in Problem 11.

(b) Give the coordinates of the point on the line $y = 4 - 3x$ closest to the origin. (*Hint*: Find x so that the distance you computed in (a) is minimized.)

(c) Show that the *square* of the distance between the point (x, y) on the line and the origin is given by

$$g(x) = [f(x)]^2 = x^2 + (4 - 3x)^2$$

and find the minimum of $g(x)$. Show that it agrees with your answer in (b).

12. How close does the curve $y = 1/x$ come to the origin? (*Hint*: Find the point on the curve that minimizes the *square* of the distance between the origin and the point on the curve. If you use the square of the distance instead of the distance you avoid dealing with square roots.)

13. Show that if $f(x)$ is a positive differentiable function that has a local minimum at $x = c$, then $g(x) = [f(x)]^2$ has a local minimum at $x = c$ as well.

14. Show that if $f(x)$ is a differentiable function with $f(x) < 0$ for all $x \in \mathbf{R}$ that has a local maximum at $x = c$, then $g(x) = [f(x)]^2$ has a local minimum at $x = c$.

15. Find the dimensions of a right-circular cylindrical can (bottom and top closed) that has volume 1 liter and that minimizes the amount of material used. (*Note*: One liter corresponds to 1000 cm³.)

16. Find the dimensions of a right-circular cylinder that is open on the top and closed on the bottom, so that the can holds 1 liter and uses the least amount of material.

17. A circular sector with radius r and angle θ has area A. Find r and θ so that the perimeter is smallest when (a) $A = 2$ and (b) $A = 10$. (*Note*: $A = \frac{1}{2}r^2\theta$, and the length of the arc $s = r\theta$, when θ is measured in radians; see Figure 4.55.)

18. A circular sector with radius r and angle θ has area A. Find r and θ so that the perimeter is smallest for a given area A. (*Note*: $A = \frac{1}{2}r^2\theta$, and the length of the arc $s = r\theta$, when θ is measured in radians; see Figure 4.55.)

19. Repeat Example 4 under the assumption that the top of the can is made out of aluminum that is three times as thick as the aluminum used for the wall and the bottom.

20. (*Classical model of viability selection*) We consider a population of diploid organisms (that is, each individual carries two copies of each chromosome). Genes reside on chromosomes, and we call the location of a gene on a chromosome a **locus**. Different versions of the same gene are called **alleles**. We consider the case of one locus with two possible alleles, A_1 and A_2. Since the individuals are diploid, the following types, called **genotypes**, may occur: $A_1 A_1$, $A_1 A_2$, and $A_2 A_2$ (where $A_1 A_2$ and $A_2 A_1$ are considered to be equivalent). If two parents mate and produce an offspring, the offspring receives one gene from each parent. If mating is random, then we can imagine all genes being put in one big gene pool from which we choose two genes at random. If we assume that the frequency of A_1 in the population is p and the frequency of A_2 is $q = 1 - p$, then the combination $A_1 A_1$ is picked with probability p^2, the combination $A_1 A_2$ with probability $2pq$ (the factor 2 appears since A_1 can come from either the father or the mother), and the combination $A_2 A_2$ with probability q^2.

We assume that the survival chances of offspring depend on their genotypes. We define the quantities w_{11}, w_{12}, and w_{22} to describe the differential survival chances of the types $A_1 A_1$, $A_1 A_2$, and $A_2 A_2$. The ratio of $A_1 A_1 : A_1 A_2 : A_2 A_2$ among

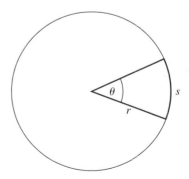

▲ Figure 4.55
The circular sector in Problems 17 and 18.

adults is given by

$$p^2 w_{11} : 2pq w_{12} : q^2 w_{22}$$

The average fitness of this population is defined as

$$\bar{w} = p^2 w_{11} + 2pq w_{12} + q^2 w_{22}$$

We will investigate this function. Since $q = 1 - p$, $\bar{w}$ is a function of p only:

$$\bar{w}(p) = p^2 w_{11} + 2p(1 - p)w_{12} \\ + (1 - p)^2 w_{22}$$

for $0 \le p \le 1$. We consider the following three cases:

(i) Directional selection:

$$w_{11} > w_{12} > w_{22}$$

(ii) Overdominance:

$$w_{12} > w_{11}, w_{22}$$

(iii) Underdominance:

$$w_{12} < w_{11}, w_{22}$$

(a) Show that

$$\bar{w}(p) = p^2(w_{11} - 2w_{12} + w_{22}) \\ - 2p(w_{12} - w_{22}) + w_{22}$$

and graph $\bar{w}(p)$ for each of the three cases, where we choose the parameters as follows:

(i) $w_{11} = 1, w_{12} = 0.7 \ w_{22} = 0.3$

(ii) $w_{11} = 0.7, w_{12} = 1, w_{22} = 0.3$

(iii) $w_{11} = 1, w_{12} = 0.3, w_{22} = 0.7$

(b) Show that

$$\frac{d\bar{w}}{dp} = 2p(w_{11} - 2w_{12} + w_{22}) \\ + 2(w_{12} - w_{22})$$

(c) Find the global maximum of $\bar{w}(p)$ in each of the three cases considered in (a). (Note that the global maximum may occur at the boundary of the domain of $\bar{w}$.)

(d) One can show that under a certain mating scheme the gene frequencies change until $\bar{w}$ reaches its global maximum. Assume that this is the case and state for each of the three cases considered in (a) what the equilibrium frequency will be.

21. (*Continuation of Problem 29 in Subsection 3.3.3*) We discussed the properties of hatching offspring per unit time, $w(t)$, in the species *Eleutherodactylus coqui*. The function $w(t)$ was given by

$$w(t) = \frac{f(t)}{C + t}$$

where $f(t)$ is the proportion of offspring that survive if t is the time spent brooding, and C is the cost associated with the time spent searching for other mates.

We assume now that $f(t)$, $t \ge 0$, is concave down with $f(0) = 0$ and $0 \le f \le 1$. The optimal brooding time is defined as the time that maximizes $w(t)$.

(a) Show that the optimal brooding time can be obtained by finding the point on the curve $f(t)$ where the line through $(-C, 0)$ is tangential to the curve $f(t)$.

(b) Use the procedure in (a) to find the optimal brooding time for $f(t) = \frac{t}{1+t}$ and $C = 2$. Determine the equation through $(-2, 0)$ that is tangential to the curve $f(t) = \frac{t}{1+t}$, and graph both $f(t)$ and the tangent together.

22. (*Optimal age of reproduction, from Roff 1992*) Semelparous organisms breed only once during their lifetime. Examples of this type of reproduction can be found in Pacific salmon and bamboo. The per capita rate of increase, r, can be thought of as a measure of reproductive fitness. The greater r, the more offspring an individual produces. The intrinsic rate of increase is typically a function of age, x. Models for age-structured populations of semelparous organisms predict that the intrinsic rate of increase as a function of x is given by

$$r(x) = \frac{\ln[l(x)m(x)]}{x}$$

where $l(x)$ is the probability of surviving to age x and $m(x)$ is the number of female births at age x. The optimal age of reproduction is the age x that maximizes $r(x)$.

(a) Find the optimal age of reproduction for

$$l(x) = e^{-ax}$$

and

$$m(x) = bx^c$$

where a, b and c are positive constants.

(b) Use a graphing calculator to sketch the graph of $r(x)$ when $a = 0.1, b = 4$, and $c = 0.9$.

23. (*Optimal age at first reproduction, from Lloyd 1987*) Iteroparous organisms breed more than once during their lifetime. We consider a model where the intrinsic rate of increase, r, depends on the age of first reproduction, denoted by x, and satisfies the equation

$$\frac{e^{-x(r(x)+L)}(1 - e^{-kx})^3 c}{1 - e^{-(r(x)+L)}} = 1 \qquad (4.11)$$

where k, L, and c are positive constants describing the life history of the organism. The optimal age of first reproduction is the age x for which $r(x)$ is maximized. Since we cannot separate $r(x)$ in the above equation we must use implicit differentiation to find a candidate for the optimal age of reproduction.

(a) Find an equation for $\frac{dr}{dx}$. (*Hint*: Take logarithms on both sides of (4.11) before differentiating with respect to x.)

(b) Set $\frac{dr}{dx} = 0$ and show that this gives

$$r(x) = \frac{3ke^{-kx}}{1 - e^{-kx}} - L$$

(To find the candidate for the optimal age x you would need to substitute for $r(x)$ in equation (4.11) and solve numerically. You would still need to check that this actually gives you the absolute maximum. This can in fact be done.)

◆ 4.5 L'HOSPITAL'S RULE

Guillaume François l'Hospital was born in France in 1661. He became interested in calculus around 1690, when articles on the new calculus by Leibniz and the Bernoulli brothers began to appear. Johann Bernoulli was in Paris in 1691, and l'Hospital asked Bernoulli to teach him some calculus. Bernoulli left Paris a year later, but continued to provide l'Hospital with new material on calculus. Bernoulli received a monthly salary for his service, and also agreed that he would not give anyone else access to the material. Once l'Hospital thought he understood the material well enough, he decided to write a book on the subject, which was published under his name and met with great success. Bernoulli was not particularly happy about this, as his contributions were hardly acknowledged in the book; l'Hospital perhaps felt that because he had paid for the course material, he had a right to publish it.

Today, l'Hospital is most famous for his treatment of the limits of fractions for which both the numerator and the denominator tend to 0 in the limit. The rule that bears his name was discovered by Johann Bernoulli but published in l'Hospital's book, and it also works when both the numerator and the denominator tend to infinity. We have encountered such examples before; for instance,

$$\lim_{x \to 3} \frac{x^2 - 9}{x - 3} = \lim_{x \to 3}(x + 3) = 6$$

or

$$\lim_{x \to \infty} \frac{kx}{1 + x} = \lim_{x \to \infty} \frac{k}{\frac{1}{x} + 1} = k$$

In both examples we were able to find the limit by algebraic manipulations.

This is not always possible, as in the following example:

$$\lim_{x \to 0} \frac{e^x - 1}{x}$$

Both numerator and denominator tend to 0 as $x \to 0$ (see Figure 4.56). We need to find out how quickly they tend to 0. This is easy for the denominator, which tends to 0 as x (in fact, it is x). For the numerator, we look at the linear approximation of $e^x - 1$ at $x = 0$. Recall from Section 3.8 that the linear approximation of a function $f(x)$ at $x = a$ is defined as

$$L(x) = f(a) + f'(a)(x - a)$$

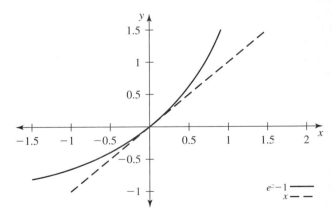

▲ **Figure 4.56**
The graphs of $y = e^x - 1$ and $y = x$.

If $f(x) = e^x - 1$, then $f'(x) = e^x$, $f(0) = 0$, and $f'(0) = 1$. That is, the linear approximation at $a = 0$ is

$$L(x) = 0 + (1)(x - 0) = x$$

We find that the numerator also tends to 0 as x. Since both numerator and denominator tend to 0 as x, and the ratio $\frac{x}{x} = 1$, we expect the limiting value to be 1, which is indeed the case.

To see clearly what we have just done, we look at the general case

$$\lim_{x \to a} \frac{f(x)}{g(x)}$$

and assume that both

$$\lim_{x \to a} f(x) = 0 \qquad \text{and} \qquad \lim_{x \to a} g(x) = 0$$

(see Figure 4.57). Using a linear approximation as before, we find that, for x close to a,

$$\frac{f(x)}{g(x)} \approx \frac{f(a) + f'(a)(x - a)}{g(a) + g'(a)(x - a)}$$

Since $f(a) = g(a) = 0$ and $x \neq a$, the right-hand side is equal to

$$\frac{f'(a)(x - a)}{g'(a)(x - a)} = \frac{f'(a)}{g'(a)}$$

provided that $\frac{f'(a)}{g'(a)}$ is defined. We therefore hope that something like

$$\lim_{x \to a} \frac{f(x)}{g(x)} = \frac{f'(a)}{g'(a)}$$

holds when $\frac{f(a)}{g(a)}$ is of the form $\frac{0}{0}$ and $\frac{f'(a)}{g'(a)}$ is defined. In fact, something like this does hold. It is called l'Hospital's rule and is more general than what we just did. The proof of this result would require a generalized version of the mean value theorem, which is beyond the scope of this book.

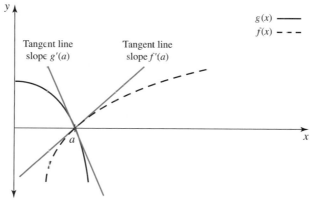

$g(x)$ ———
$f(x)$ - - -

Tangent line
slope $g'(a)$

Tangent line
slope $f'(a)$

◀ **Figure 4.57**

L'HOSPITAL'S RULE

Suppose that f and g are differentiable functions, and that

$$\lim_{x \to a} f(x) = \lim_{x \to a} g(x) = 0$$

or

$$\lim_{x \to a} f(x) = \lim_{x \to a} g(x) = \infty$$

If

$$\lim_{x \to a} \frac{f'(x)}{g'(x)} = L$$

then

$$\lim_{x \to a} \frac{f(x)}{g(x)} = L$$

Using l'Hospital's rule, we can redo the first three examples. In each case, the ratio $\frac{f(x)}{g(x)}$ is an **indeterminate expression**; when a is substituted for x, it is either of the form $\frac{0}{0}$ or $\frac{\infty}{\infty}$. We apply l'Hospital's rule in each case: we differentiate both numerator and denominator and then take the limit of $\frac{f'(x)}{g'(x)}$ as $x \to a$. If this limit exists, it is equal to the limit of $\frac{f(x)}{g(x)}$ as $x \to a$.

$$\lim_{x \to 3} \frac{x^2 - 9}{x - 3} = \lim_{x \to 3} \frac{2x}{1} = 6$$

$$\lim_{x \to \infty} \frac{kx}{1 + x} = \lim_{x \to \infty} \frac{k}{1} = k$$

$$\lim_{x \to 0} \frac{e^x - 1}{x} = \lim_{x \to 0} \frac{e^x}{1} = 1$$

We look at a few more examples; in each case you might to wish to use a graphing calculator to graph the function. The first three examples are straightforward applications of l'Hospital's rule.

▋ **EXAMPLE 1**   **Indeterminate Expression 0/0**

Evaluate

$$\lim_{x \to 2} \frac{x^6 - 64}{x^2 - 4}$$

Solution   This limit is of the form $\frac{0}{0}$, since $2^6 - 64 = 0$ and $2^2 - 4 = 0$. Applying l'Hospital's rule yields

$$\lim_{x \to 2} \frac{x^6 - 64}{x^2 - 4} = \lim_{x \to 2} \frac{6x^5}{2x} = \frac{(6)(2^5)}{(2)(2)} = (6)(2^3) = 48 \qquad ▍$$

▋ **EXAMPLE 2**   **Indeterminate Expression 0/0**

Evaluate

$$\lim_{x \to 0} \frac{1 - \cos^2 x}{\sin x}$$

Solution   This limit is of the form $\frac{0}{0}$, since $1 - \cos^2 0 = 0$ and $\sin 0 = 0$. Applying l'Hospital's rule yields

$$\lim_{x \to 0} \frac{1 - \cos^2 x}{\sin x} = \lim_{x \to 0} \frac{2 \cos x \sin x}{\cos x} = \lim_{x \to 0} 2 \sin x = 0 \qquad ▍$$

EXAMPLE 3   **Indeterminate Expression ∞/∞**

Evaluate

$$\lim_{x \to \infty} \frac{\ln x}{x}$$

Solution   This limit is of the form $\frac{\infty}{\infty}$. We can apply l'Hospital's rule:

$$\lim_{z \to \infty} \frac{\frac{1}{x}}{1} = \lim_{x \to \infty} \frac{1}{x} = 0$$

EXAMPLE 4   **Indeterminate Expression ∞/∞ – One-Sided Limit**

Evaluate

$$\lim_{x \to (\pi/2)^-} \frac{\tan x}{1 + \tan x}$$

Solution   This is of the form $\frac{\infty}{\infty}$. We apply l'Hospital's rule and find

$$\lim_{x \to (\pi/2)^-} \frac{\tan x}{1 + \tan x} = \lim_{x \to (\pi/2)^-} \frac{\sec^2 x}{\sec^2 x} = \lim_{x \to (\pi/2)^-} 1 = 1$$

EXAMPLE 5   **Applying L'Hospital's Rule More than Once**

Evaluate

$$\lim_{x \to \infty} \frac{x^3 - 3x + 1}{3x^3 - 2x^2}$$

Solution   This limit is of the form $\frac{\infty}{\infty}$. Applying l'Hospital's rule yields

$$\lim_{x \to \infty} \frac{x^3 - 3x + 1}{3x^3 - 2x^2} = \lim_{x \to \infty} \frac{3x^2 - 3}{9x^2 - 4x}$$

which is still of the form $\frac{\infty}{\infty}$. Applying l'Hospital's rule again, we find

$$\lim_{x \to \infty} \frac{3x^2 - 3}{9x^2 - 4x} = \lim_{x \to \infty} \frac{6x}{18x - 4}$$

Since this is still of the form $\frac{\infty}{\infty}$, we can apply l'Hospital's rule yet again. We find

$$\lim_{x \to \infty} \frac{6x}{18x - 4} = \lim_{x \to \infty} \frac{6}{18} = \frac{1}{3}$$

In this example, we could have found the answer without using l'Hospital's rule, by dividing both numerator and denominator by x^3. Namely,

$$\lim_{x \to \infty} \frac{x^3 - 3x + 1}{3x^3 - 2x^2} = \lim_{x \to \infty} \frac{x^3 \left(1 - \frac{3}{x^2} + \frac{1}{x^3}\right)}{x^3 \left(3 - \frac{2}{x}\right)}$$

$$= \lim_{x \to \infty} \frac{1 - \frac{3}{x^2} + \frac{1}{x^3}}{3 - \frac{2}{x}} = \frac{1}{3}$$

When you apply l'Hospital's rule, it can certainly happen that the limit is infinite, as in the following example.

▌ **EXAMPLE 6**   **Infinite Limit**

Evaluate

$$\lim_{x \to \infty} \frac{e^x}{x}$$

Solution   This is of the form $\frac{\infty}{\infty}$. We apply l'Hospital's rule.

$$\lim_{x \to \infty} \frac{e^x}{x} = \lim_{x \to \infty} \frac{e^x}{1} = \infty \qquad \text{(limit does not exist)}$$

L'Hospital's rule can sometimes be applied to limits of the form

$$\lim_{x \to a} f(x)g(x)$$

where

$$\lim_{x \to a} f(x) = 0 \qquad \text{and} \qquad \lim_{x \to a} g(x) = \infty$$

since we can write the limit in the form

$$\lim_{x \to a} f(x)g(x) = \lim_{x \to a} \frac{f(x)}{\frac{1}{g(x)}} = \lim_{x \to a} \frac{g(x)}{\frac{1}{f(x)}}$$

which is again of form either $\frac{0}{0}$ or $\frac{\infty}{\infty}$.

▌ **EXAMPLE 7**   **Indeterminate Expression** $0 \cdot \infty$

Evaluate

$$\lim_{x \to 0^+} x \ln x$$

Solution   This is of the form $(0)(-\infty)$. We apply l'Hospital's rule after rewriting it in the form $\frac{\infty}{\infty}$.

$$\lim_{x \to 0^+} x \ln x = \lim_{x \to 0^+} \frac{\ln x}{\frac{1}{x}} = \lim_{x \to 0^+} \frac{\frac{1}{x}}{-\frac{1}{x^2}}$$

$$= \lim_{x \to 0^+} \frac{1}{x} \left(-\frac{x^2}{1} \right) = \lim_{x \to 0^+} (-x) = 0$$

We could have written the limit in the form

$$\lim_{x \to 0^+} \frac{x}{\frac{1}{\ln x}}$$

and then applied l'Hospital's rule. In this case,

$$\lim_{x \to 0^+} \frac{x}{\frac{1}{\ln x}} = \lim_{x \to 0^+} \frac{1}{(-1)(\ln x)\frac{1}{x}} = \lim_{x \to 0^+} \frac{x}{-\ln x} = 0$$

You will probably agree that the first way was easier, because it is more difficult to differentiate $1/\ln x$ than $\ln x$. Before you apply l'Hospital's rule to expressions of the form $0 \cdot \infty$, you should always determine which way will be is easier to evaluate.

▌ **EXAMPLE 8**   **Indeterminate Expression** $0 \cdot \infty$

Evaluate

$$\lim_{x \to 0^+} x \cot x$$

Solution   This is of the form $0 \cdot \infty$. We have two choices: we can evaluate

$$\lim_{x \to 0^+} x \cot x = \lim_{x \to 0^+} \frac{x}{\frac{1}{\cot x}} = \lim_{x \to 0^+} \frac{x}{\tan x}$$

where we use the fact that $\cot x = \frac{1}{\tan x}$, or

$$\lim_{x \to 0^+} x \cot x = \lim_{x \to 0^+} \frac{\cot x}{\frac{1}{x}}$$

It certainly seems easier to apply l'Hospital's rule to the first form.

$$\lim_{x \to 0^+} x \cot x = \lim_{x \to 0^+} \frac{x}{\tan x} = \lim_{x \to 0^+} \frac{1}{\sec^2 x}$$
$$= \lim_{x \to 0^+} \cos^2 x = 1$$

In the penultimate step, we used the fact that $\sec x = \frac{1}{\cos x}$. ❙

Limits of the form $\infty - \infty$ sometimes can be evaluated using l'Hospital's rule, if we can algebraically transform such limits into the form $\frac{0}{0}$ or $\frac{\infty}{\infty}$.

❚ **EXAMPLE 9**   **Indeterminate Expression $\infty - \infty$**

Evaluate

$$\lim_{x \to (\frac{\pi}{2})^-} (\tan x - \sec x)$$

Solution   This limit is of the form $\infty - \infty$. Note that

$$\tan x = \frac{\sin x}{\cos x} \qquad \text{and} \qquad \sec x = \frac{1}{\cos x}$$

Using these two identities, we can write the limit as

$$\lim_{x \to (\frac{\pi}{2})^-} (\tan x - \sec x) = \lim_{x \to (\frac{\pi}{2})^-} \left(\frac{\sin x}{\cos x} - \frac{1}{\cos x} \right)$$
$$= \lim_{x \to (\frac{\pi}{2})^-} \frac{\sin x - 1}{\cos x}$$

This is now of the form $\frac{0}{0}$, and we can apply l'Hospital's rule to find

$$\lim_{x \to (\frac{\pi}{2})^-} \frac{\sin x - 1}{\cos x} = \lim_{x \to (\frac{\pi}{2})^-} \frac{\cos x}{-\sin x} = \frac{0}{-1} = 0 \qquad ❚$$

❚ **EXAMPLE 10**   **Indeterminate Expression $\infty - \infty$**

Evaluate

$$\lim_{x \to \infty} \left(x - \sqrt{x^2 + x} \right)$$

Solution   This is of the form $\infty - \infty$. We need to get a product or a ratio; we can factor x and find

$$\lim_{x \to \infty} \left(x - \sqrt{x^2 + x} \right) = \lim_{x \to \infty} x \left(1 - \sqrt{1 + \frac{1}{x}} \right)$$

This is of the form $\infty \cdot 0$. We can transform it to the form $\frac{0}{0}$ and then apply l'Hospital's rule.

$$\lim_{x \to \infty} x \left(1 - \sqrt{1 + \frac{1}{x}} \right) = \lim_{x \to \infty} \frac{1 - \sqrt{1 + \frac{1}{x}}}{\frac{1}{x}}$$
$$= \lim_{x \to \infty} \frac{-\frac{1}{2}(1 + \frac{1}{x})^{-1/2}(-\frac{1}{x^2})}{-\frac{1}{x^2}}$$
$$= \lim_{x \to \infty} \frac{-1}{2\sqrt{1 + \frac{1}{x}}} = -\frac{1}{2} \qquad ❚$$

Finally, we will consider expressions of the form

$$\lim_{x \to a} [f(x)]^{g(x)}$$

when they are of the types 0^0, ∞^0, or 1^∞. The key to solving such limits is to rewrite them as

$$\lim_{x \to a} [f(x)]^{g(x)} = \lim_{x \to a} \exp \left\{ \ln [f(x)]^{g(x)} \right\}$$

$$= \lim_{x \to a} \exp [g(x) \cdot \ln f(x)]$$

$$= \exp \left[\lim_{x \to a} g(x) \cdot \ln f(x) \right]$$

The last step, where we interchanged lim and exp, uses the fact that the exponential function is continuous. Rewriting the limit in this way transforms

$$0^0 \quad \text{into} \quad \exp[0 \cdot (-\infty)]$$
$$\infty^0 \quad \text{into} \quad \exp[0 \cdot (\infty)]$$
$$1^\infty \quad \text{into} \quad \exp[\infty \cdot \ln 1] = \exp[\infty \cdot 0]$$

Since we know how to deal with limits of the form $0 \cdot \infty$, we are in good shape again. We present a couple of examples.

▍ **EXAMPLE 11**   **Indeterminate Expression 0^0**

Evaluate

$$\lim_{x \to 0^+} x^x$$

Solution   This is of the form 0^0; we rewrite the limit first.

$$\lim_{x \to 0^+} x^x = \lim_{x \to 0^+} \exp[\ln x^x] = \lim_{x \to 0^+} \exp[x \ln x]$$

$$= \exp[\lim_{x \to 0^+} x \ln x]$$

We evaluated this limit in Example 7, and found

$$\lim_{x \to 0^+} x \ln x = 0$$

Hence,

$$\lim_{x \to 0^+} x^x = \exp[\lim_{x \to 0^+} x \ln x] = \exp[0] = 1$$

(see Figure 4.58).   ▍

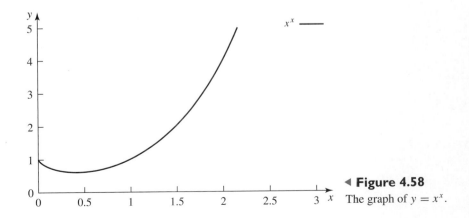

◀ **Figure 4.58**
The graph of $y = x^x$.

❚ EXAMPLE 12   **Indeterminate Expression 1^∞**

Evaluate

$$\lim_{x \to (\pi/4)^-} (\tan x)^{\tan 2x}$$

Solution   Since $\tan \frac{\pi}{4} = 1$ and $\tan\left(2\frac{\pi}{4}\right) = \infty$, this is of the form 1^∞. We rewrite the limit as

$$\lim_{x \to (\pi/4)^-} (\tan x)^{\tan 2x} = \lim_{x \to (\pi/4)^-} \exp[\tan(2x) \cdot \ln \tan x]$$

$$= \exp\left[\lim_{x \to (\pi/4)^-} \tan(2x) \cdot \ln \tan x\right]$$

The limit is now of the form $\infty \cdot 0$ (since $\ln \tan \frac{\pi}{4} = \ln 1 = 0$). We evaluate the limit by writing it in the form $\frac{0}{0}$ and then applying l'Hospital's rule.

$$\lim_{x \to (\pi/4)^-} \tan(2x) \cdot \ln \tan x = \lim_{x \to (\pi/4)^-} \frac{\ln \tan x}{\frac{1}{\tan(2x)}}$$

$$= \lim_{x \to (\pi/4)^-} \frac{\ln \tan x}{\cot(2x)}$$

Since

$$\frac{d}{dx} \ln \tan x = \frac{\sec^2 x}{\tan x} = \frac{\cos x}{\cos^2 x \sin x} = \frac{1}{\sin x \cos x}$$

and

$$\frac{d}{dx} \cot(2x) = -(\csc^2(2x)) \cdot 2 = \frac{-2}{\sin^2(2x)}$$

we find

$$\lim_{x \to (\pi/4)^-} \frac{\ln \tan x}{\cot(2x)} = \lim_{x \to (\pi/4)^-} \frac{\frac{1}{\sin x \cos x}}{\frac{-2}{\sin^2(2x)}} = \lim_{x \to (\pi/4)^-} \frac{\sin^2(2x)}{-2\sin x \cos x}$$

$$= \frac{1}{(-2)\left(\frac{1}{2}\sqrt{2}\right)\left(\frac{1}{2}\sqrt{2}\right)} = -1$$

Therefore

$$\lim_{x \to (\pi/4)^-} (\tan x)^{\tan(2x)} = \exp\left[\lim_{x \to (\pi/4)^-} \tan(2x) \ln \tan x\right]$$

$$= \exp[-1] = e^{-1}$$

The graph of $f(x) = (\tan x)^{\tan(2x)}$ is shown in Figure 4.59.   ❚

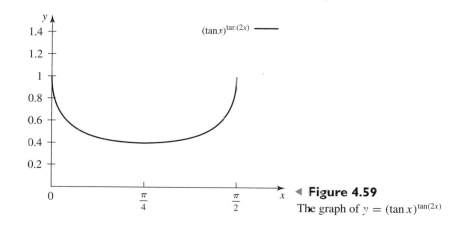

◀ Figure 4.59
The graph of $y = (\tan x)^{\tan(2x)}$

◆ 4.5 PROBLEMS

Use l'Hospital's rule to find the limits in Problems 1–40.

1. $\lim\limits_{x \to 4} \dfrac{x^2 - 16}{x - 4}$

2. $\lim\limits_{x \to 1} \dfrac{x - 1}{x^2 - 1}$

3. $\lim\limits_{x \to -2} \dfrac{2x^2 + x - 6}{x + 2}$

4. $\lim\limits_{x \to -3} \dfrac{x + 3}{x^2 + 2x - 3}$

5. $\lim\limits_{x \to 0} \dfrac{\sqrt{2x + 4} - 2}{x}$

6. $\lim\limits_{x \to 0} \dfrac{3 - \sqrt{2x + 9}}{2x}$

7. $\lim\limits_{x \to 0} \dfrac{\sin x}{x \cos x}$

8. $\lim\limits_{x \to 0} \dfrac{x \sin x}{1 - \cos x}$

9. $\lim\limits_{x \to 0} \dfrac{1 - \cos x}{x \tan x}$

10. $\lim\limits_{x \to \pi/2} \dfrac{\sin \left(\frac{\pi}{2} - x \right)}{\cos x}$

11. $\lim\limits_{x \to 0^+} \dfrac{\sqrt{x}}{\ln(x + 1)}$

12. $\lim\limits_{x \to \infty} \dfrac{\ln x}{\sqrt{x}}$

13. $\lim\limits_{x \to 0} \dfrac{2^x - 1}{3^x - 1}$

14. $\lim\limits_{x \to 0} \dfrac{2^{-x} - 1}{5^x - 1}$

15. $\lim\limits_{x \to 0} \dfrac{e^x - 1 - x}{x^2}$

16. $\lim\limits_{x \to 0} \dfrac{e^x - 1 - x - \frac{x^2}{2}}{x^3}$

17. $\lim\limits_{x \to \infty} \dfrac{(\ln x)^2}{x^2}$

18. $\lim\limits_{x \to \infty} \dfrac{x^4}{e^x}$

19. $\lim\limits_{x \to (\pi/2)^-} \dfrac{\tan x}{\sec^2 x}$

20. $\lim\limits_{x \to 0} \dfrac{e^x - 1}{\sin x}$

21. $\lim\limits_{x \to \infty} x e^{-x}$

22. $\lim\limits_{x \to \infty} x^2 e^{-x}$

23. $\lim\limits_{x \to 0^+} \sqrt{x} \ln x$

24. $\lim\limits_{x \to 0^+} x^2 \ln x$

25. $\lim\limits_{x \to (\pi/2)^-} \left(\dfrac{\pi}{2} - x \right) \sec x$

26. $\lim\limits_{x \to 1^-} (1 - x) \tan \left(\dfrac{\pi}{2} x \right)$

27. $\lim\limits_{x \to \infty} \sqrt{x} \sin \frac{1}{x}$

28. $\lim\limits_{x \to \infty} x^2 \sin \frac{1}{x^2}$

29. $\lim\limits_{x \to 0^+} (\cot x - \csc x)$

30. $\lim\limits_{x \to \infty} (x - \sqrt{x^2 - 1})$

31. $\lim\limits_{x \to 0^+} \left(\dfrac{1}{\sin x} - \dfrac{1}{x} \right)$

32. $\lim\limits_{x \to 0^+} \left(\dfrac{1}{\sin^2 x} - \dfrac{1}{x} \right)$

33. $\lim\limits_{x \to 0^+} x^{2x}$

34. $\lim\limits_{x \to 0^+} x^{\sin x}$

35. $\lim\limits_{x \to \infty} x^{1/x}$

36. $\lim\limits_{x \to \infty} (1 + e^x)^{1/x}$

37. $\lim\limits_{x \to \infty} \left(1 + \dfrac{3}{x} \right)^x$

38. $\lim\limits_{x \to \infty} \left(1 + \dfrac{3}{x^2} \right)^x$

39. $\lim\limits_{x \to \infty} \left(\dfrac{x}{1 + x} \right)^x$

40. $\lim\limits_{x \to 0^+} (\cos(2x))^{3/x}$

Find the limits in Problems 41–50. Be sure to check whether l'Hospital's rule can be applied before you evaluate the limit.

41. $\lim\limits_{x \to 0} x e^x$

42. $\lim\limits_{x \to 0^+} \dfrac{e^x}{x}$

43. $\lim\limits_{x \to (\pi/2)^-} (\tan x + \sec x)$

44. $\lim\limits_{x \to (\pi/2)^-} \dfrac{\tan x}{1 + \sec x}$

45. $\lim\limits_{x \to 1} \dfrac{x^2 + 5}{x + 1}$

46. $\lim\limits_{x \to 0} \dfrac{1 - \cos x}{\sec x}$

47. $\lim\limits_{x \to -\infty} x e^x$

48. $\lim\limits_{x \to 0^+} \left(\dfrac{1}{x} - \dfrac{1}{\sqrt{x}} \right)$

49. $\lim\limits_{x \to 0^+} x^{3x}$

50. $\lim\limits_{x \to \infty} \left(\dfrac{x + 1}{x + 2} \right)^x$

51. Use l'Hospital's rule to find

$$\lim\limits_{x \to 0} \dfrac{a^x - 1}{b^x - 1}$$

where $a, b > 0$.

52. Use l'Hospital's rule to find

$$\lim\limits_{x \to \infty} \left(1 + \dfrac{c}{x} \right)^x$$

where c is a constant.

53. For $p > 0$, determine the values of p for which the following limit is either 1 or ∞, or a constant that is neither 1 nor ∞ (assume that c is a positive constant):

$$\lim\limits_{x \to \infty} \left(1 + \dfrac{c}{x^p} \right)^x$$

54. Show that

$$\lim\limits_{x \to \infty} x^p e^{-x} = 0$$

for any positive number p. Graph $f(x) = x^p e^{-x}$, $x > 0$, for $p = 1/2$, 1, and 2. Since $f(x) = x^p e^{-x} = x^p / e^x$, this shows that the exponential function grows faster than any power of x as $x \to \infty$.

55. Show that

$$\lim\limits_{x \to \infty} \dfrac{\ln x}{x^p} = 0$$

for any number $p > 0$. This shows that the logarithmic function grows more slowly than any positive power of x as $x \to \infty$.

56. When l'Hospital introduced indeterminate limits in his textbook, his *first* example was

$$\lim\limits_{x \to a} \dfrac{\sqrt{2a^3 x - x^4} - a\sqrt[3]{a^2 x}}{a - \sqrt[4]{ax^3}}$$

where a is a positive constant. (This example was communicated to him by Bernoulli.) Show that this limit is equal to $(16/9)a$.

57. The height in feet (y) of a tree as a function of tree age in years (x) is given by

$$y = 121 e^{-17/x} \quad \text{for } x > 0$$

(a) Determine the rate of growth when $x \to 0+$ and the limit as $x \to \infty$.

(b) Find the age at which the growth rate is maximal.

(c) Show that the height of the tree is an increasing function of age. At what age is the height increasing at an accelerating rate, and at what age at a decelerating rate?

(d) Sketch the graph of both the height and the rate of growth as functions of age.

◆ 4.6 NUMERICAL METHODS (OPTIONAL)

Numerical methods are very important in the sciences, as one frequently encounters situations, where exact solutions are impossible. In this section, we discuss numerical methods for solving equations of the form $f(x) = x$ (which are of interest in discrete dynamical systems) and $f(x) = 0$. (The two subsections are independent.)

4.6.1 Discrete Dynamical Systems and Fixed Points

At various junctures, we have discussed differential equations that describe population growth, such as

$$\frac{dN}{dt} = rN$$

This formulation implies that generations overlap, that is, that individuals can reproduce at any time. There are many cases, however, where organisms have breeding seasons, which means that generations do not overlap. In this case, a discrete-time formulation would be more appropriate. The discrete-time models are such that the size of the population at generation k, denoted by N_k, is related to the size of the population at generation $k - 1$ by

$$N_k = f(N_{k-1}) \quad \text{for } k = 1, 2, 3, \dots$$

where the function f describes the population dynamics. This is an example of a **discrete dynamical system**. The simplest example is of the form

$$N_k = bN_{k-1} \tag{4.12}$$

for some $b > 0$: the population size at time k is a constant multiple of the population size at time $k - 1$. This should remind you of exponential growth or decay, which is the continuous-time analogue. We can solve (4.12) iteratively. Since $N_k = bN_{k-1}$, we find

$$N_k = bN_{k-1}$$
$$= b(bN_{k-2}) = b^2 N_{k-2}$$

Using $N_{k-2} = bN_{k-3}$, we continue and find

$$= b^2(bN_{k-3}) = b^3 N_{k-4}$$
$$= \dots = b^k N_0$$

If you find it difficult to see why the exponent is k, plug in values of k until you see the pattern:

$$k = 1 : N_1 = bN_0$$
$$k = 2 : N_2 = bN_1 = b(bN_0) = b^2 N_0$$
$$k = 3 : N_3 = bN_2 = b(b^2 N_0) = b^3 N_0$$

and so on, and we find

$$N_k = b^k N_0 \tag{4.13}$$

This is called **geometric growth** or **decay** according to whether $b > 1$ or $0 < b < 1$. It is the discrete-time analogue of the exponential growth and decay model discussed in Section 3.6.

▎EXAMPLE 1   Assume the discrete time population model described in equation (4.12). Find b when the population increases by 5% in each generation. Suppose that at time 0, $N_0 = 20$. How large is the population at time 10?

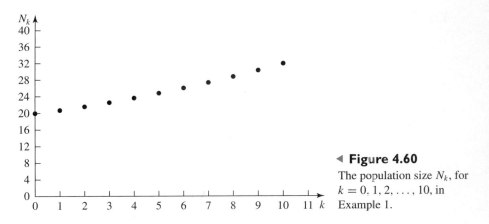

◀ **Figure 4.60**
The population size N_k, for $k = 0, 1, 2, \ldots, 10$, in Example 1.

Solution   Since the population increases by 5% in each generation,

$$N_k = N_{k-1} + 0.05 N_{k-1} = 1.05 N_{k-1} \quad \text{for } k \geq 1$$

we conclude

$$b = 1.05$$

To find the population size at time 10, we can use equation (4.13),

$$N_{10} = b^{10} N_0 = (1.05)^{10}(20) = 32.58$$

As the population size is an integer valued quantity, we round the result to the closest integer and find $N_{10} = 33$. Figure 4.60 shows successive population sizes when $N_0 = 20$. ▮

In the model of geometric growth, if $b > 1$, the population will increase forever. In general, because of limiting resources, it might be more realistic to assume that the population increases from generation to generation as long as the population size is small, but actually decreases once it becomes large enough. A question of interest, then, is whether the population will reach **equilibrium**, that is, a state, where the size of the population remains the same from generation to generation. If the population is described by

$$N_k = f(N_{k-1}) \quad \text{for } k = 1, 2, 3, \ldots$$

then an equilibrium, denoted by N^*, would satisfy the equation

$$N^* = f(N^*) \tag{4.14}$$

The equilibrium N^* is also called a **fixed point** of equation (4.14). (When $f(N^*)$ is evaluated, it returns N^*, that is, N^* is fixed under the function f.) We will illustrate how to find such equilibria in a specific example, namely a recruitment model widely used in the fisheries literature.

Discrete Time Recruitment Models – Ricker Curves   The Ricker curve describes the relationship between the size of the parental stock of some fish and the number of recruits (offspring). If we denote the size of the parental stock by P and the number of recruits by R, then the Ricker curve is given by ($\alpha, \beta > 0$)

$$R(P) = \alpha P e^{-\beta P} \quad \text{for } P \geq 0$$

Note that $R(0) = 0$; without parents there are no offspring. Furthermore, $R(P) > 0$ when $P > 0$. We can phrase this as a discrete-time dynamical system. If the parental stock constitutes the population at time k, then the recruits constitute the population

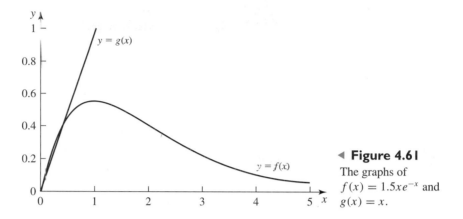

◀ **Figure 4.61**
The graphs of
$f(x) = 1.5xe^{-x}$ and
$g(x) = x$.

at time $k + 1$. That is,

$$N_{k+1} = \alpha N_k e^{-\beta N_k} \quad \text{for } k = 0, 1, 2, \ldots$$

Note that recruits in one generation are the parents in the following generation. Figure 4.61 shows the graph of N_{k+1} as a function of N_k for $\alpha > 1$, together with the line $N_{k+1} = N_k$. (It will soon become clear why we are particularly interested in the case $\alpha > 1$.)

The Ricker curve $f(N) = \alpha N e^{-\beta N}$ has slope $f'(N) = \alpha e^{-\beta N}(1 - \beta N)$. In particular, $f'(0) = \alpha$. We can see from Figure 4.61 that if $\alpha > 1$, the line $N_{k+1} = N_k$ intersects $f(N_k)$ twice, namely at $N_k = 0$ and at $N_k = N^* > 0$, where N^* satisfies

$$N^* = f(N^*)$$

This means that N^* is a nontrivial (nonzero) equilibrium or fixed point. The state $N = 0$ is also an equilibrium or a fixed point. We call 0 a trivial equilibrium; it corresponds to the state where there are no individuals in the population. Note that when $\alpha \leq 1$, the line $N_{k+1} = N_k$ intersects $f(N_k)$ only once at $N_k = 0$. That is, when $\alpha \leq 1$ there is only one equilibrium, whereas when $\alpha > 1$ there are two equilibria.

For $\alpha > 1$, we can compute the nontrivial fixed point N^* as follows:

$$N^* = \alpha N^* e^{-\beta N^*}$$
$$e^{\beta N^*} = \alpha$$
$$\beta N^* = \ln \alpha$$
$$N^* = \frac{1}{\beta} \ln \alpha$$

From the form of N^*, we can also see that we need $\alpha > 1$ to have a nontrivial equilibrium (i.e., $N^* > 0$), since $\ln \alpha > 0$ for $\alpha > 1$.

If we start with a population of size N^*, then the population will remain at N^* for *all* future generations. But what happens when we start with a population size that is different from N^*? We might ask whether the population size will eventually reach N^*. There is a useful graphical method for studying the evolution of such discrete-time models, called **cobwebbing**.

Cobwebbing   The method is illustrated in Figure 4.62. Suppose that we start at N_0. To find N_1, we draw a vertical line through N_0 until the line intersects the curve $f(N_k)$. We then find N_1 on the N_{k+1} axis: we go horizontally from the point of intersection until we hit the N_{k+1} axis. For the next step, we need to locate N_1 on the N_k axis, because the recruits of the first generation are now the parents that will produce the second generation. We use line $N_{k+1} = N_k$. We draw a horizontal line

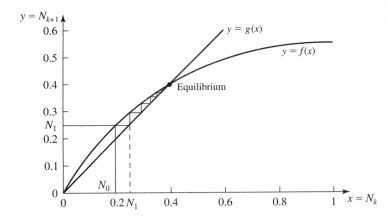

▲ **Figure 4.62**
The graphs of $f(x) = 1.5xe^{-x}$ and $g(x) = x$: The equilibrium is stable.

through N_1 on the N_{k+1} axis, and find the point where the horizontal line intersects the line $N_{k+1} = N_k$. We then draw a vertical line through this point of intersection and locate the point where this vertical line intersects the N_k axis. We can then repeat the procedure starting at the point marked N_1 on the N_k axis. We illustrate successive steps in Figure 4.62, for $\alpha = 1.5$ and $\beta = 1$.

For this choice of parameters, we see from the graph that N_k approaches the fixed point $N^* = \ln(1.5)$. Successive population sizes are shown in Figure 4.63.

It turns out that the derivative of $f(N)$ at N^* determines the behavior of the system near the equilibrium value. We denote the derivative at N^* by λ, that is,

$$\lambda = f'(N^*)$$

The parameter λ is called the **eigenvalue** of the equilibrium N^* of $N_{k+1} = f(N_k)$. (λ is the small Greek letter lambda.)

We say that the equilibrium N^* is locally stable if the system returns to N^* when we start close to N^*. One can show that N^* is locally stable if

$$-1 < \lambda < 1 \qquad (4.15)$$

We write

$$N_k = N^* + v_k$$

where we think of v_k as a small **perturbation** about N^*, that is, $|v_k|$ is a very small number. Then, with $N_{k+1} = N^* + v_{k+1}$,

$$N^* + v_{k+1} = f(N^* + v_k)$$

Since v_k is very small, we use a linear approximation, that is, the linear approximation of the right-hand side is

$$f(N^*) + f'(N^*)v_k$$

Given that $f(N^*) = N^*$, we find

$$v_{k+1} \approx \lambda v_k \quad \text{where } \lambda = f'(N^*)$$

Returning to Example 1, we find that

$$\lim_{k \to \infty} v_{k+1} \approx \lim_{k \to \infty} \lambda^{k+1} v_0 = 0 \quad \text{if } |\lambda| < 1$$

That is, if $|\lambda| < 1$, then a small perturbation about N^* will decay to 0. In this case, we say that N^* is locally stable; otherwise, it is unstable.

We try this out on the Ricker curve. There, we have

$$f(x) = \alpha x e^{-\beta x}$$
$$f'(x) = \alpha e^{-\beta x}(1 - \beta x)$$

Since $N^* = \frac{1}{\beta}\ln\alpha$, we find

$$f'(N^*) = \alpha e^{-\beta(1/\beta)\ln\alpha}\left(1 - \beta\frac{1}{\beta}\ln\alpha\right)$$
$$= \alpha e^{-\ln\alpha}(1 - \ln\alpha)$$
$$= 1 - \ln\alpha$$

The condition $|f'(N^*)| < 1$ thus translates into

$$-1 < 1 - \ln\alpha < 1$$
$$-2 < -\ln\alpha \quad < 0$$
$$0 < \ln\alpha \quad < 2$$
$$1 < \alpha \quad < e^2$$

That is, if $\alpha \in (1, e^2)$, the discrete time recruitment model

$$N_{k+1} = \alpha N_k e^{-\beta N_k}, \quad k = 0, 1, 2, \ldots$$

has a unique nontrivial fixed point $N^* = \frac{1}{\beta}\ln\alpha$. Figure 4.63 illustrates this case for $\alpha = 1.5$.

To see what happens when $\alpha > e^2$, we look at the cobweb for $N_{k+1} = f(N_k)$ when $\alpha = 10$ and $\beta = 1$, and plot the corresponding population sizes as a function of time (see Figure 4.64).

k	0	1	2	3	4	5	6
N_k	2.0	2.71	1.80	2.98	1.51	3.34	1.18

k	7	8	9	10	11	12
N_k	3.63	0.962	3.67	0.934	3.67	0.934

The corresponding cobwebbing graph is shown in Figure 4.65. The population size does not tend to an equilibrium, but eventually oscillates between 0.934 and 3.67. As we increase α further, the behavior will become more and more complex, eventually leading to a situation in which the population size does not settle into a

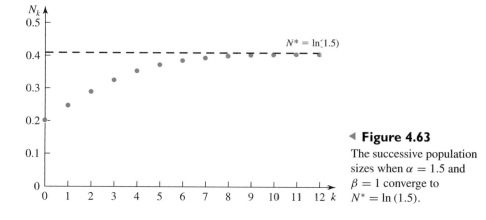

◀ **Figure 4.63**

The successive population sizes when $\alpha = 1.5$ and $\beta = 1$ converge to $N^* = \ln(1.5)$.

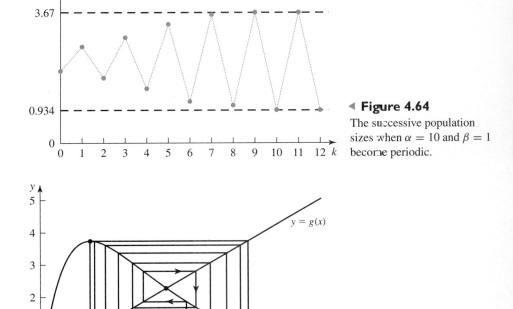

◀ Figure 4.64

The successive population sizes when $\alpha = 10$ and $\beta = 1$ become periodic.

◀ Figure 4.65

The graphs of $f(x) = 10xe^{-x}$ and $g(x) = x$: The system is periodic.

predictable pattern. This is called chaos. (There is a great deal of interest in chaotic behavior of populations models, which is beyond the scope of this text.)

Discrete dynamical systems provide important models in biology; cobwebbing provides a graphical way to investigate such systems, but the behavior of such systems can be very complex.

4.6.2 The Newton-Raphson Method

The Newton-Raphson method allows us to find solutions to equations of the form

$$f(x) = 0$$

The idea behind it can be best explained graphically (see Figure 4.66).

Suppose that we wish to find the roots of $f(x) = 0$. We begin by choosing a value x_0 as our initial guess; then, we replace the graph of $f(x)$ by its tangent line $y = L(x)$ at our initial guess x_0. Since the slope of the tangent line $y = L(x)$ is equal to $f'(x_0)$, and $(x_0, f(x_0))$ is a point on the tangent line, we can use the slope-point form to find the equation of the tangent line. It is given by

$$y - f(x_0) = f'(x_0)(x - x_0)$$

This line will intersect the x-axis at some point $x = x_1$, provided $f'(x_0) \neq 0$. To find x_1, we set $x = x_1$ and $y = 0$ and solve for x_1:

$$0 - f(x_0) = f'(x_0)(x_1 - x_0) \qquad \text{or} \qquad x_1 = x_0 - \frac{f(x_0)}{f'(x_0)}$$

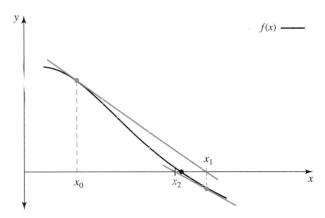

▲ Figure 4.66
The graph of $y = f(x)$ and the first two iterations in the Newton-Raphson method.

We use x_1 as our next guess, and repeat the procedure, which will result in x_2, and so on. The value of x_{n+1} is then given by

$$x_{n+1} = x_n - \frac{f(x_n)}{f'(x_n)} \quad \text{for } n = 0, 1, 2, \dots \tag{4.16}$$

This procedure produces a sequence of numbers $x_0, x_1, x_2, \dots$. If these values get closer and closer to r as $n \to \infty$, where r is the root of $f(x) = 0$, we say that the sequence $x_0, x_1, x_2, \dots$ converges to r and write

$$\lim_{n \to \infty} x_n = r$$

The method will not always converge, but before we discuss situations in which the method fails, we give two examples where the method works.

▮ EXAMPLE 2   Use the Newton-Raphson method to find a numerical approximation to a solution of the equation

$$x^2 - 3 = 0$$

Solution   The equation $x^2 - 3 = 0$ has roots $r = \sqrt{3}$ and $-\sqrt{3}$. Finding a numerical approximation to a root of this equation is therefore the same as finding a numerical approximation to $\sqrt{3}$ or $-\sqrt{3}$. You might ask why we don't just use our calculators and get the numerical value of $\sqrt{3}$, and in fact this is what you would do if you had to solve an equation like $x^2 - 3 = 0$, but we will use this simple example to illustrate the method.

We will find a numerical approximation to $\sqrt{3}$ using the Newton-Raphson method. As our initial value, we choose $x_0 = 2$ since 2 is close to $\sqrt{3}$. The function $f(x)$ is given by

$$f(x) = x^2 - 3$$

and its derivative is given by

$$f'(x) = 2x$$

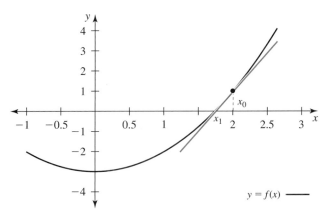

▲ **Figure 4.67**
The graph of $f(x) = x^2 - 3$ and the first step in the Newton-Raphson method.

Using (4.16), we find

$$x_{n+1} = x_n - \frac{x_n^2 - 3}{2x_n}$$

$$= x_n - \frac{x_n}{2} + \frac{3}{2x_n} = \frac{x_n}{2} + \frac{3}{2x_n} \quad \text{for } n = 0, 1, 2, \ldots$$

(See Figure 4.67 for the first step in the approximation.)

The following table shows the results of the procedure. Since we know the result, we can easily compare our numerical approximation to an accurate numerical approximation of $\sqrt{3}$, namely, $\sqrt{3} = 1.732050$ to six decimal places.

n	x_n	$x_{n+1} = \dfrac{x_n}{2} + \dfrac{3}{2x_n}$	$\lvert\sqrt{3} - x_{n+1}\rvert$
0	2	1.75	0.0179
1	1.75	1.7321429	9.2×10^{-5}
2	1.7321429	1.7320508	2.45×10^{-9}

With the starting value $x_0 = 2$, after three steps we obtain the approximation 1.732050, which is correct to six decimal places. As you can see, this method can converge very quickly. In fact, it is used in many calculators to calculate roots. ▮

In the next example, we cannot simply solve for x by algebraic manipulations. To find a solution of the given equation, we must resort to a numerical method.

▮ **EXAMPLE 3**   Solve the equation

$$e^x + 1 + x = 0$$

Solution   It turns out that this can only be done numerically. We will use the Newton-Raphson method to determine the root of $e^x + 1 + x = 0$. Since $e^x + 1 + x$ is positive for $x = 0$ and negative for $x = -2$, we conclude that there is a root of $e^x + 1 + x = 0$ in the interval $(-2, 0)$. We therefore choose $x_0 = -1$. We find

$$f(x) = e^x + x + 1$$
$$f'(x) = e^x + 1$$

▲ **Figure 4.68**
The graph of $f(x) = e^x + x + 1$ and the first step in the approximations of the Newton-Raphson method.

Therefore,

$$x_{n+1} = x_n - \frac{f(x_n)}{f'(x_n)}$$

$$= x_n - \frac{e^{x_n} + x_n + 1}{e^{x_n} + 1}$$

$$= x_n - 1 - \frac{x_n}{e^{x_n} + 1} \quad \text{for } n = 0, 1, 2, \ldots$$

The first step in the approximation is shown in Figure 4.68.
    We summarize the first few steps of the iteration in the following table.

n	x_n	$x_{n+1} = x_n - 1 - \dfrac{x_n}{e^{x_n} + 1}$
0	-1	-1.26894142
1	-1.26894142	-1.27845462
2	-1.27845462	-1.27846454
3	-1.27846454	-1.27846454

Note that x_2 and x_3 agree to four decimal places, and that x_3 and x_4 agree to eight decimal places. We therefore find that $x = -1.278464$ is a root of $e^x + x + 1 = 0$, accurate to six decimal places.

    It is not our goal here to give a complete description of the Newton-Raphson method (when it works, how quickly it converges, etc.). Instead, we give a few examples that illustrate some of the problems one can encounter when using this method.

EXAMPLE 4   This is an example of a situation where the method does not work at all. Let

$$f(x) = \begin{cases} \sqrt{x - 2.5} & \text{for } x \geq 2.5 \\ -\sqrt{2.5 - x} & \text{for } x \leq 2.5 \end{cases}$$

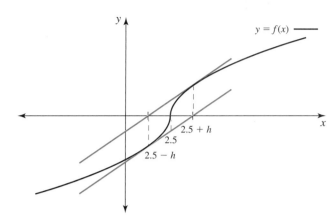

◀ **Figure 4.69**
The graph of $f(x)$ in
Example 4 where the
Newton-Raphson method
does not converge.

We wish to solve the equation $f(x) = 0$. We see immediately that $x = 2.5$ is a
solution. Suppose we wish to use the Newton-Raphson method. Given that

$$f'(x) = \begin{cases} \frac{1}{2\sqrt{x-2.5}} & \text{for } x > 2.5 \\ \frac{1}{2\sqrt{2.5-x}} & \text{for } x < 2.5 \end{cases}$$

we find

$$x_{n+1} = x_n - \frac{f(x_n)}{f'(x_n)}$$

$$= \begin{cases} x_n - \frac{\sqrt{x_n-2.5}}{\frac{1}{2\sqrt{x_n-2.5}}} & \text{for } x_n > 2.5 \\ x_n - \frac{-\sqrt{2.5-x_n}}{\frac{1}{2\sqrt{2.5-x_n}}} & \text{for } x_n < 2.5 \end{cases}$$

$$= \begin{cases} x_n - 2(x_n - 2.5) = -x_n + 5 & \text{for } x_n > 2.5 \\ x_n + 2(2.5 - x_n) = -x_n + 5 & \text{for } x_n < 2.5 \end{cases}$$

$$= -x_n + 5 \quad \text{for } x_n \neq 2.5$$

If $x_0 = 2.5 + h$, then $x_1 = 2.5 - h$, $x_2 = 2.5 + h$, $x_3 = 2.5 - h$, and so on;
that is, successive approximations oscillate between $2.5 + h$ and $2.5 - h$, and never
approach the root $r = 2.5$. A graph of $f(x)$, together with the iterations, is shown
in Figure 4.69. Note that the two tangent lines are parallel and used alternately in
the approximation. ▮

EXAMPLE 5 This example shows that an initial approximation can get worse. Use the Newton-
Raphson method to find the root of

$$x^{1/3} = 0$$

when the starting value is $x_0 = 1$.

Solution We set $f(x) = x^{1/3}$. Then $f'(x) = \frac{1}{3}x^{-2/3}$, and

$$x_{n+1} = x_n - \frac{f(x_n)}{f'(x_n)}$$

$$= x_n - \frac{x_n^{1/3}}{\frac{1}{3}x_n^{-2/3}} = x_n - 3x_n = -2x_n$$

▲ **Figure 4.70**

The graph of $f(x)$ in Example 5 where the Newton-Raphson method does not converge.

The following table shows successive values of x_1.

n	x_n	$x_{n+1} = -2x_n$
0	1	-2
1	-2	4
2	4	-8
3	-8	16

This does not converge to the root $r = 0$. This is graphically illustrated in Figure 4.70.

Graphing calculators have an option in the graphing menu that allows you to find roots of equations $f(x) = 0$ if you graph $f(x)$ and put the cursor close to the root. For this example, the calculator cannot produce an answer. ▌

EXAMPLE 6 This example shows that one does not necessarily converge to the closest root. We wish to find a root of the equation

$$x^4 - x^2 = 0$$

If we set $f(x) = x^4 - x^2$, then $f'(x) = 4x^3 - 2x$ and

$$x_{n+1} = x_n - \frac{f(x_n)}{f'(x_n)}$$

$$= x_n - \frac{x_n^4 - x_n^2}{4x_n^3 - 2x_n} = x_n - \frac{x_n^3 - x_n}{4x_n^2 - 2}$$

We can solve for the roots of $x^4 - x^2 = 0$. Since $x^4 - x^2 = x^2(x^2 - 1)$, there are three roots: $r = 0, -1$, and 1. Suppose that we set $x_0 = -0.7$. The closest root is -1. If we compute x_1, we find

$$x_1 = (-0.7) - \frac{(-0.7)^3 - (-0.7)}{4(-0.7)^2 - 2} = 8.225$$

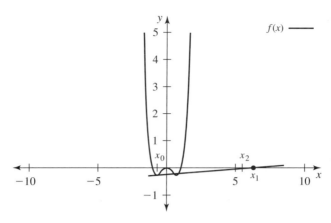

◀ **Figure 4.71**
The graph of $f(x)$ in Example 6 together with the first two approximations.

Successive values are collected in the following list.

$$x_2 = 6.184 \qquad x_7 \ \ = 1.613$$
$$x_3 = 4.659 \qquad x_8 \ \ = 1.306$$
$$x_4 = 3.521 \qquad x_9 \ \ = 1.115$$
$$x_5 = 2.678 \qquad x_{10} = 1.024$$
$$x_6 = 2.059 \qquad x_{11} = 1.001$$

We conclude that the method converges to the root $r = 1$. This is illustrated in Figure 4.71. ▮

◆ 4.6 PROBLEMS

[4.6.1]

1. Assume a discrete-time population whose size at generation k is related to the size of the population at generation $k - 1$ by

$$N_k = (1.03)N_{k-1}, \quad k = 1, 2, \ldots$$

(a) If $N_0 = 10$, how large will the population be at generation 5?
(b) How many generations will it take for the population size to double the size at generation 0?

2. Suppose a discrete-time population evolves according to

$$N_k = (0.9)N_{k-1}, \quad k = 1, 2 \ldots$$

(a) How large will the population be at generation $k = 6$?
(b) After how many generations will the size of the population be half its original size?
(c) What will happen to the population in the long run, that is, as $k \to \infty$?

3. Assume the discrete-time population model

$$N_k = bN_{k-1}, \quad k = 1, 2 \ldots$$

Assume that the population increases by 2% each generation.
(a) Determine b.
(b) Find the size of the population at generation 10 when $N_0 = 20$.
(c) After how many generations will the population size have doubled?

4. Assume the discrete-time population model

$$N_k = bN_{k-1}, \quad k = 1, 2, \ldots$$

Assume that the population decreases by 3% each generation.
(a) Determine b.
(b) Find the size of the population at generation 10 when $N_0 = 50$.
(c) How long will it take until the population is one-half its original size?

5. Assume the discrete-time population model

$$N_k = bN_{k-1}, \quad k = 1, 2, \ldots$$

Assume that the population increases by $x\%$ each generation.
(a) Determine b.
(b) After how many generations will the population size have doubled? Compute the doubling time for $x = 0.1, 0.5, 1, 2, 5,$ and 10.

6. The Ricker curve is given by

$$R(P) = \alpha P e^{-\beta P}$$

for $P \geq 0$, where P denotes the size of the parental stock and $R(P)$ the number of recruits. The parameters α and β are positive constants.

(a) Show that $R(0) = 0$ and $R(P) > 0$ for $P > 0$.

(b) Find
$$\lim_{P \to \infty} R(P)$$

(c) For what size of the parental stock is the number of recruits maximal?

(d) Does $R(P)$ have inflection points? If so, find them.

(e) Sketch the graph of $f(x)$ when $\alpha = 2$ and $\beta = 1/2$.

7. Suppose that the size of a fish population at generation k is given by
$$N_k = 1.5 N_{k-1} e^{-0.001 N_{k-1}}$$

for $k = 1, 2, \ldots$.

(a) Assume that $N_0 = 100$. Find the size of the fish population at generation k for $k = 1, 2, \ldots, 10$.

(b) Assume that $N_0 = 800$. Find the size of the fish population at generation k for $k = 1, 2, \ldots, 10$.

(c) Determine all fixed points. Based on your computations in (a) and (b), can you make a guess as to what will happen to the population in the long run when starting from $N_0 = 100$ and $N_0 = 800$, respectively?

(d) Use the cobwebbing method to illustrate your answer in (a).

(e) Explain why the dynamical system converges to the nontrivial fixed point.

8. Suppose that the size of a fish population at generation k is given by
$$N_k = 10 N_{k-1} e^{-0.01 N_{k-1}}$$

for $k = 1, 2, \ldots$.

(a) Assume that $N_0 = 100$. Find the size of the fish population at generation k for $k = 1, 2, \ldots, 10$.

(b) Show that if $N_0 = 100 \ln 10$, then $N_k = 100 \ln 10$ for all $k \geq 1$; that is, that $N^* = 100 \ln 10$ is a nontrivial fixed point or equilibrium. How would you find N^*? Are there any other equilibria?

(c) Based on your computations in (a), can you make a prediction about the long-term behavior of the fish population when initially $N_0 = 100$? How does this compare to your answer in (b)?

(d) Use the cobwebbing method to illustrate your answer in (c).

9. Use the cobwebbing method to determine the long-term behavior of the following discrete-time dynamical system, which is called the discrete logistic model:
$$N_{k+1} = r N_k \left(1 - \frac{N_k}{100}\right)$$

for $k = 0, 1, 2, \ldots$.

(a) Find all equilibria.

(b) Investigate the system when $N_0 = 10$ and (i) $r = 1/2$, (ii) $r = 2$, and (iii) $r = 3.1$, and describe what you see.

[4.6.2]

10. Use the Newton-Raphson method to find a numerical approximation to the solution of
$$x^2 - 7 = 0$$
that is correct to six decimal places.

11. Use the Newton-Raphson method to find a numerical approximation to the solution of
$$e^{-x} = x$$
that is correct to six decimal places.

12. Use the Newton-Raphson method to find a numerical approximation to the solution of
$$x^2 + \ln x = 0$$
that is correct to six decimal places.

13. The equation
$$x^2 - 5 = 0$$
has two solutions. Use the Newton-Raphson method to approximate the two solutions.

14. Use the Newton-Raphson method to solve the equation
$$\sin x = \frac{1}{2} x$$
in the interval $(0, \pi)$.

15. Let
$$f(x) = \begin{cases} \sqrt{x-1} & \text{for } x \geq 1 \\ -\sqrt{1-x} & \text{for } x \leq 1 \end{cases}$$

(a) Show that if you use the Newton-Raphson method to solve $f(x) = 0$, the following holds: if $x_0 = 1 + h$, then $x_1 = 1 - h$, and if $x_0 = 1 - h$, then $x_1 = 1 + h$.

(b) Does the Newton-Raphson method converge? Use a graph to explain what happens.

16. In Example 5, we discussed the case of finding the root of $x^{1/3} = 0$.

(a) Given x_0, find a formula for $|x_n|$.

(b) Find
$$\lim_{n \to \infty} |x_n|$$

(c) Graph $f(x) = x^{1/3}$ and illustrate what happens when you apply the Newton-Raphson method.

17. In Example 6, we considered the equation
$$x^4 - x^2 = 0$$

(a) What happens if you choose
$$x_0 = -\frac{1}{2}\sqrt{2}$$
in the Newton-Raphson method? Give a graphical illustration.

(b) Repeat the procedure in (a) for $x_0 = -0.71$, and compare with the result we obtained in Example 6 when $x_0 = -0.70$. Give a graphical illustration, and explain it in words. What happens when $x_0 = -0.6$? (This is an example where small changes in the initial value can drastically change the outcome.)

18. Use the Newton-Raphson method to find a numerical approximation to the solution of

$$x^2 - 16 = 0$$

when your initial guess is (a) $x_0 = 3$ and (b) $x_0 = 4$.

19. Suppose that you wish to solve

$$f(x) = 0$$

numerically using the Newton-Raphson method. It just so happens that your initial guess x_0 satisfies $f(x_0) = 0$. What happens to subsequent iterations? Give a graphical illustration of your results. (Assume that $f'(x_0) \neq 0$.)

◆ 4.7 ANTIDERIVATIVES

Throughout this and the previous chapter, we have repeatedly encountered differential equations. Occasionally, we showed that a certain function would solve the given differential equation. In this section, we will discuss a particular type of differential equation and address two important general questions: First, given a differential equation, how can we find solutions? Second, given a solution, how do we know if it is unique?

We will consider differential equations of the form

$$\frac{dy}{dx} = f(x) \quad \text{with } y = y_0 \text{ when } x = x_0 \tag{4.17}$$

that is, where the rate of change of y with respect to x depends *only* on x. Our goal is then to find a function y that satisfies $y' = f(x)$ and whose graph goes through the point (x_0, y_0).

Let's look at an example before we begin a systematic treatment of the solution of differential equations of the form (4.17). Consider a population whose size at time t is denoted by $N(t)$, and assume that the growth rate is given by

$$\frac{dN}{dt} = \frac{1}{2\sqrt{t}} \quad \text{for } t > 0 \tag{4.18}$$

and that the size of the population at time 0 is $N(0) = 20$. Then

$$N(t) = \sqrt{t} + 20 \quad \text{for } t \geq 0 \tag{4.19}$$

is a solution of the differential equation (4.18) that satisfies the condition $N(0) = 20$. This is easy to check: First, note that $N(0) = \sqrt{0} + 20 = 20$. Second, differentiating $N(t)$, we find

$$\frac{dN}{dt} = \frac{d}{dt}(\sqrt{t} + 20) = \frac{1}{2\sqrt{t}} \quad \text{for } t > 0$$

that is, $N(t) = \sqrt{t} + 20$ satisfies (4.18) with $N(0) = 20$.

This example shows that if we have a solution of a differential equation, we can check that the solution indeed satisfies the differential equation by simply differentiating the solution. This suggests that in order to find solutions, we reverse the process of differentiation. We begin with a definition.

> **Definition**   A function F is called an **antiderivative** of f on an interval I if $F'(x) = f(x)$ for all $x \in I$.

We illustrate the process of finding antiderivatives in the following example. Let

$$f(x) = 3x^2 \quad \text{for } x \in \mathbf{R}$$

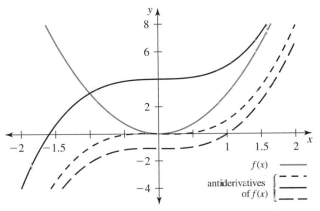

◀ **Figure 4.72**
The function $f(x) = 3x^2$ and some of its antiderivatives.

We wish to find a function whose derivative is $3x^2$. We can easily guess an answer, namely

$$F(x) = x^3 \quad \text{for } x \in \mathbf{R}$$

But this is not the only answer. If we take $F(x) = x^3 + 4$, then $F'(x) = 3x^2$; hence, $x^3 + 4$ is also an antiderivative of $3x^2$. In fact, $F(x) = x^3 + C$, $x \in \mathbf{R}$, where C is any constant, is an antiderivative of $3x^2$. We will soon show that there are no other functions $F(x)$ such that $F'(x) = 3x^2$. The function $f(x)$ and some of its antiderivatives are shown in Figure 4.72. All antiderivatives are related through vertical shifts since they all have the same derivative.

Before we proceed, it is important to note the difficulties that are encountered when finding antiderivatives. Though we will learn rules that will allow us to compute antiderivatives, this process is typically much more difficult than finding derivatives, and sometimes takes ingenuity to come up with the correct answer; in addition, there are cases where it is impossible to find an expression for an antiderivative.

We begin with two corollaries of the mean value theorem that will help us in finding antiderivatives. The first of these is Corollary 2 from Section 4.1, which we recall here.

Corollary 2   If f is continuous on the closed interval $[a, b]$ and differentiable on the open interval (a, b), with $f'(x) = 0$ for all $x \in (a, b)$, then f is constant on $[a, b]$.

Note that Corollary 2 is the converse of the rule that states that $f'(x) = 0$ when $f(x) = c$, where c is a constant. It tells us that *all* antiderivatives of a function that is identically 0 are the constant functions.

The second corollary tells us that functions with identical derivatives differ only by a constant; that is, to find all antiderivatives of a given function, we need only find one.

Corollary 3   If $F(x)$ and $G(x)$ are antiderivatives of the continuous function $f(x)$ on an interval I, then there exists a constant C so that

$$G(x) = F(x) + C \quad \text{for all } x \in I$$

Proof   Since $F(x)$ and $G(x)$ are both antiderivatives of $f(x)$, it follows that $F'(x) = f(x)$ and $G'(x) = f(x)$, that is, $F'(x) = G'(x)$ or $F'(x) - G'(x) = 0$. Since

$$0 = F'(x) - G'(x) = \frac{d}{dx}[F(x) - G(x)]$$

it follows from Corollary 2 applied to the function $F - G$ that $F(x) - G(x) = C$ where C is a constant.   ◆

▮**EXAMPLE 1**   Find general antiderivatives for the given functions. Assume that all functions are defined for $x \in \mathbf{R}$.

    a. $f(x) = 3x^2$

    b. $f(x) = \cos x$

    c. $f(x) = e^x$

Solution

    a. If $F(x) = x^3$, then $F'(x) = 3x^2$, that is, $F(x) = x^3$ is a particular antiderivative of $3x^2$. Using Corollary 3, we find the general antiderivative simply by adding a constant; that is, the general antiderivative of $f(x) = 3x^2$ is the function $G(x) = x^3 + C$, where C is a constant.

    b. If $F(x) = \sin x$, then $F'(x) = \cos x$. Hence, the general antiderivative of $f(x) = \cos x$ is the function $G(x) = \sin x + C$, where C is a constant.

    c. If $F(x) = e^x$, then $F'(x) = e^x$ and the general antiderivative of $f(x) = e^x$ is the function $G(x) = e^x + C$, where C is a constant.   ▮

▮**EXAMPLE 2**   Find general antiderivatives for the given functions. (Assume the largest possible domain.)

    a. $f(x) = 3x^5$

    b. $f(x) = x^2 + 2x - 1$

    c. $f(x) = e^{2x}$

    d. $f(x) = \sec^2(3x)$

Solution

    a. Since $\frac{d}{dx}\left(\frac{1}{2}x^6\right) = 3x^5$, $F(x) = \frac{1}{2}x^6 + C$ is the general antiderivative of $f(x) = 3x^5$.

    b. Since $\frac{d}{dx}\left(\frac{1}{3}x^3 + x^2 - x\right) = x^2 + 2x - 1$, $F(x) = \frac{1}{3}x^3 + x^2 - x + C$ is the general antiderivative of $f(x) = x^2 + 2x - 1$.

    c. Since $\frac{d}{dx}\left(\frac{1}{2}e^{2x}\right) = \frac{1}{2}e^{2x}(2) = e^{2x}$, $F(x) = \frac{1}{2}e^{2x} + C$ is the general antiderivative of $f(x) = e^{2x}$.

    d. Since $\frac{d}{dx}\left(\frac{1}{3}\tan(3x)\right) = \frac{1}{3}(\sec^2(3x))(3) = \sec^2(3x)$, $F(x) = \frac{1}{3}\tan(3x) + C$ is the general antiderivative of $f(x) = \sec^2(3x)$.   ▮

Table 4.1 summarizes some of the rules for finding antiderivatives. We denote functions by $f(x)$ or $g(x)$, and their particular antiderivatives by $F(x)$ or $G(x)$. The general antiderivative is then obtained simply by adding a constant. The quantities a and k denote nonzero constants.

TABLE 4•1 A COLLECTION OF ANTIDERIVATIVES

Function	Particular Antiderivative
$kf(x)$	$kF(x)$
$f(x) + g(x)$	$F(x) - G(x)$
$x^n, \; n \neq -1$	$\dfrac{1}{n+1}x^{n+1}$
$\dfrac{1}{x}$	$\ln|x|$
e^{ax}	$\dfrac{1}{a}e^{ax}$
$\sin(ax)$	$-\dfrac{1}{a}\cos(ax)$
$\cos(ax)$	$\dfrac{1}{a}\sin(ax)$
$\sec^2(ax)$	$\dfrac{1}{a}\tan(ax)$

We can now return to our initial question, namely, how to solve differential equations of the form (4.17)

▌ **EXAMPLE 3**   Find the general solution of

$$\frac{dy}{dx} = \frac{3}{x^2} - 2x^2, \quad x \neq 0$$

Solution   Finding the general solution of this differential equation means finding the antiderivative of the function $f(x) = \frac{3}{x^2} - 2x^2$. Using Table 4.1, we obtain as a particular antiderivative

$$\frac{3}{-1}x^{-1} - \frac{2}{3}x^3 = -\frac{3}{x} - \frac{2}{3}x^3$$

That is, the general solution is

$$y = -\frac{3}{x} - \frac{2}{3}x^3 + C, \quad x \neq 0 \qquad\qquad ▌$$

In Example 3, we found the general solution of the given differential equation. Often, we wish to select a particular solution; for instance, we may know that the solution has to pass through a particular point (x_0, y_0). Such a problem is called an **initial value problem**. The example at the beginning of this section is an initial value problem; we consider it again now.

▌ **EXAMPLE 4**   Solve the initial value problem

$$\frac{dN}{dt} = \frac{1}{2\sqrt{t}} \quad \text{for } t > 0 \text{ with } N(0) = 20$$

Solution   The general antiderivative of $f(t) = \frac{1}{2\sqrt{t}}$ is $F(t) = \sqrt{t} + C$. Since $N(0) = 20$, we find

$$N(0) = \sqrt{0} + C = 20 \qquad \text{or} \qquad C = 20$$

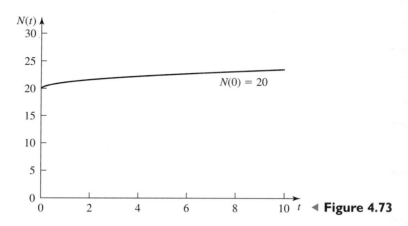

◀ **Figure 4.73**

That is, the function

$$N(t) = \sqrt{t} + 20, \quad t \geq 0$$

solves the initial value problem, and it is the only solution (see Figure 4.73). ▌

EXAMPLE 5   Solve the initial value problem

$$\frac{dy}{dx} = -2x^2 + 3 \quad \text{for } x \in \mathbf{R} \text{ and } y = 10 \text{ when } x = 3$$

Solution   The general antiderivative of $f(x) = -2x^2 + 3$ is $F(x) = -\frac{2}{3}x^3 + 3x + C$. Since

$$F(3) = -\frac{2}{3}3^3 + (3)(3) + C = -9 + C = 10$$

it follows that $C = 19$. That is,

$$y = -\frac{2}{3}x^3 + 3x + 19, \quad x \in \mathbf{R}$$

solves the initial value problem and is the only solution. ▌

EXAMPLE 6   An object that falls freely in a vacuum, close to the surface of the earth, experiences a constant acceleration of

$$g = 9.81 \frac{\text{m}}{\text{sec}^2}$$

If the body is dropped from rest, find its velocity and the distance it travels in t seconds after it is released.

Solution   If the distance function is $s(t)$, then the velocity $v(t)$ is given by

$$v(t) = \frac{d}{dt}s(t)$$

and the acceleration is given by

$$a(t) = \frac{d}{dt}v(t) = \frac{d^2}{dt^2}s(t)$$

We wish to solve the initial value problem

$$\frac{d}{dt}v(t) = 9.81 \frac{\text{m}}{\text{sec}^2} \quad \text{when } v(0) = 0$$

A general solution is

$$v(t) = \left(9.81\frac{m}{sec^2}\right)t + C$$

Since $v(0) = 0$, it follows that $C = 0$. Hence,

$$v(t) = \left(9.81\frac{m}{sec^2}\right)t, \quad t \geq 0$$

To find the distance traveled, note that $s(0) = 0$ and

$$\frac{d}{dt}s(t) = v(t) = \left(9.81\frac{m}{sec^2}\right)t, \quad t \geq 0$$

A general solution is

$$s(t) = \frac{1}{2}\left(9.81\frac{m}{sec^2}\right)t^2 + C$$

Since $s(0) = 0$, it follows that $C = 0$. Hence,

$$s(t) = \frac{1}{2}\left(9.81\frac{m}{sec^2}\right)t^2, \quad t \geq 0$$

Note that if t is measured in seconds, the unit of $v(t)$ is $\frac{m}{sec}$ and the unit of $s(t)$ is m. ▮

◆ 4.7 PROBLEMS

In Problems 1–12 find the general antiderivative for the given function.

1. $f(x) = x^4 - 3x^2 + 1$

2. $f(x) = 2x^2 + x - 5$

3. $f(x) = 4x^3 - 2x + 3$

4. $f(x) = 1 + \dfrac{1}{x}$

5. $f(x) = x^2 - \dfrac{2}{x}$

6. $f(x) = \dfrac{1}{1+x}$

7. $f(x) = e^{-3x}$

8. $f(x) = e^{x/2} + e^{-x/2}$

9. $f(x) = 2e^{2x}$

10. $f(x) = 2\sin(\frac{\pi}{2}x) - 3\cos(\frac{\pi}{2}x)$

11. $f(x) = \sec^2(2x)$

12. $f(x) = \dfrac{\sec x + \cos x}{\cos x}$

In Problems 13–16, find the general solution of the differential equation.

13. $\dfrac{dy}{dx} = \dfrac{2}{x} - x, \; x > 0$

14. $\dfrac{dy}{dt} = t(1 - t), \; t \geq 0$

15. $\dfrac{dy}{ds} = \sin(\pi s), \; 0 \leq s \leq 1$

16. $\dfrac{dy}{dx} = \sec^2\left(\dfrac{x}{2}\right), \; -1 < x < 1$

In Problems 17–22, solve the initial value problem.

17. $\dfrac{dy}{dx} = 2\sqrt{x}$, for $x \geq 0$ with $y = 2$ when $x = 1$.

18. $\dfrac{dN}{dt} = \dfrac{t}{t+1}$, for $t \geq 0$ with $N(0) = 5$.

19. $\dfrac{dW}{dt} = e^{-3t}$, for $t \geq 0$ with $W(0) = 2/3$.

20. $\dfrac{dT}{dt} = \sin(\pi t)$, for $t \geq 0$ with $T(0) = 3$.

21. $\dfrac{dy}{dx} = \dfrac{e^{-x} + e^x}{2}$, for $x \geq 0$ with $y = 0$ when $x = 0$.

22. $\dfrac{dN}{dt} = t^{-1/3}$, for $t > 0$ with $N(0) = 60$.

23. Suppose that the length of a certain organism at age x is given by $L(x)$, which satisfies the differential equation

$$\frac{dL}{dx} = e^{-0.1x}, \quad x \geq 0$$

Find $L(x)$ if the limiting length L_∞ is given by

$$L_\infty = \lim_{x \to \infty} L(x) = 25$$

How big is the organism at age $x = 0$?

24. Fish are indeterminate growers, that is, their length $L(x)$ increases with age x throughout their lifetime. If one plots the growth rate dL/dx versus age x on semilog paper, a straight line with negative slope results. Set up a differential equation that relates growth rate and age. Solve this equation under the assumption that $L(0) = 5$, $L(1) = 10$, and

$$\lim_{x \to \infty} L(x) = 20$$

Graph the solution $L(x)$ as a function of x.

25. An object is dropped from a height of 100 ft. The acceleration is 32 ft/sec². When will the object hit the ground, and what will its speed be at impact?

26. Suppose that the growth rate of a population at time t undergoes seasonal fluctuations in size according to

$$\frac{dN}{dt} = 3\sin(2\pi t)$$

where t is measured in years and $N(t)$ denotes the size of the population at time t. If $N(0) = 10$ (measured in thousands), find an expression for $N(t)$. How are the seasonal fluctuations in the growth rate reflected in the population size?

27. Suppose that the amount of water contained in a plant at time t is denoted by $V(t)$. Due to evaporation, $V(t)$ changes over time. Suppose that the volume change at time t, measured over a 24-hour period, is proportional to $t(24 - t)$ measured in grams per hour. To offset the water loss, you water the plant at a constant rate of 4 grams of water per hour.

(a) Explain why

$$\frac{dV}{dt} = -at(24 - t) + 4$$

$0 \le t \le 24$, for some positive constant a, describes this situation.

(b) Determine the constant a, such that the net water loss over a 24 hour period is equal to 0.

◆ 4.8 KEY TERMS

Chapter 4 Review: Topics   Discuss the following definitions and concepts:

1. Local extrema: local minimum and local maximum
2. Fermat's theorem
3. Mean value theorem
4. Rolle's theorem
5. Increasing and decreasing functions
6. Monotonicity and the first derivative
7. Concavity: concave up and concave down
8. Concavity and the first derivative
9. Concavity and the second derivative
10. Diminishing return
11. Candidates for local extrema
12. Monotonicity and local extrema
13. The second derivative test for local extrema
14. Inflection points
15. Inflection points and the second derivative
16. Asymptotes: horizontal, vertical, and oblique
17. Using calculus to graph functions
18. L'Hospital's rule
19. Dynamical systems: cobwebbing
20. Newton-Raphson method to find roots
21. Antiderivative
22. General antiderivative

◆ 4.9 REVIEW PROBLEMS

1. Suppose that

$$f(x) = xe^{-x}, \quad x \ge 0$$

(a) Show that $f(0) = 0$, $f(x) > 0$ for $x > 0$, and

$$\lim_{x \to \infty} f(x) = 0$$

(b) Find local and absolute extrema.

(c) Find inflection points.

(d) Use the above information to graph $f(x)$.

2. Suppose that

$$f(x) = x \ln x, \quad x > 0$$

(a) Define $f(x)$ at $x = 0$ so that $f(x)$ is continuous for all $x \ge 0$.

(b) Find extrema and inflection points.

(c) Graph $f(x)$.

3. In Review Problem 17 of Chapter 2, we introduced the hyperbolic functions

$$\sinh x = \frac{e^x - e^{-x}}{2}, \quad x \in \mathbf{R}$$

$$\cosh x = \frac{e^x + e^{-x}}{2}, \quad x \in \mathbf{R}$$

$$\tanh x = \frac{e^x - e^{-x}}{e^x + e^{-x}}, \quad x \in \mathbf{R}$$

(a) Show that $f(x) = \tanh x$, $x \in \mathbf{R}$, is a strictly increasing function on $\mathbf{R}$. Evaluate

$$\lim_{x \to -\infty} \tanh x$$

and

$$\lim_{x \to \infty} \tanh x$$

(b) Use your results in (a) to explain why $f(x) = \tanh x$, $x \in \mathbf{R}$, is invertible, and show that its inverse function

$f^{-1}(x) = \tanh^{-1} x$ is given by

$$f^{-1}(x) = \frac{1}{2} \ln \frac{1+x}{1-x}$$

What is the domain of $f^{-1}(x)$?

(c) Show that

$$\frac{d}{dx} f^{-1}(x) = \frac{1}{1-x^2}$$

(d) Use your result in (c) and the facts that

$$\tanh x = \frac{\sinh x}{\cosh x}$$

and

$$\cosh^2 x - \sinh^2 x = 1$$

to show that

$$\frac{d}{dx} \tanh x = \frac{1}{\cosh^2 x}$$

4. Let

$$f(x) = \frac{x}{1 + e^{-x}}, \quad x \in \mathbf{R}$$

(a) Show that $y = 0$ is a horizontal asymptote as $x \to -\infty$.

(b) Show that $y = x$ is an oblique asymptote as $x \to +\infty$.

(c) Show that

$$f'(x) = \frac{1 + e^{-x}(1 + x)}{(1 + e^{-x})^2}$$

(d) Use your result in (c) to show that $f(x)$ has exactly one local extrema at $x = c$, where c satisfies the equation

$$1 + c + e^c = 0$$

(*Hint*: Use your result in (c) to show that $f'(x) = 0$ if and only if $1 + e^{-x}(1 + x) = 0$. Let $g(x) = 1 + e^{-x}(1 + x)$. Show that $g(x)$ is strictly increasing for $x < 0$ and that $g(0) > 0$ and $g(-2) < 0$. This implies that $g(x) = 0$ has exactly one solution in $(-2, 0)$. Since $g(-2) < 0$ and $g(x)$ is strictly increasing for $x < 0$, there are no solutions of $g(x) = 0$ for $x < -2$. Furthermore, $g(x) > 0$ for $x > 0$, hence there are no solutions of $g(x) = 0$ for $x > 0$.)

(e) The equation $1 + c + e^c = 0$ can only be solved numerically for c. With the help of a calculator, find a numerical approximation for c. (*Hint*: From (d) you know that $c \in (-2, 0)$.)

(f) Show that $f(x) < 0$ for $x < 0$. (This implies that for $x < 0$ the graph of $f(x)$ is below the horizontal asymptote $y = 0$.)

(g) Show that $x - f(x) > 0$ for $x > 0$. (This implies that for $x > 0$, the graph of $f(x)$ is below the oblique asymptote $y = x$.)

(h) Use your results in (a)–(g) and the fact that $f(0) = 0$ and $f'(0) = \frac{1}{2}$ to sketch the graph of $f(x)$.

5. (*Recruitment Model*) The **Ricker curve** describes the relationship between the size of the parental stock of some fish and the number of recruits. If we denote the size of the parental stock by P and the number of recruits by R, then the Ricker curve is given by

$$R(P) = \alpha P e^{-\beta P} \quad \text{for } P \geq 0$$

where α and β are positive constants. (Note that $R(0) = 0$, that is, without parents there are no offspring. Furthermore, $R(P) > 0$ when $P > 0$.)

We are interested in the size of the parental stock (P) that maximizes the number of recruits ($R(P)$). Since $R(P)$ is differentiable, we can use the first derivative of $R(P)$ to solve this problem.

(a) Use the product rule to show that for $P > 0$

$$R'(P) = \alpha e^{-\beta P}(1 - \beta P)$$
$$R''(P) = -\alpha \beta e^{-\beta P}(2 - \beta P)$$

(b) Show that $R'(P) = 0$ if $P = 1/\beta$ and that $R''(1/\beta) < 0$. This shows that $R(P)$ has a *local* maximum at $P = \frac{1}{\beta}$. Show that $R(1/\beta) = \frac{\alpha}{\beta} e^{-1} > 0$.

(c) To find the global maximum, you need to check $R(0)$ and $\lim_{P \to \infty} R(P)$. Show that

$$R(0) = 0 \quad \text{and} \quad \lim_{P \to \infty} R(P) = 0$$

and that this implies that there is an absolute maximum at $P = 1/\beta$.

(d) Show that $R(P)$ has an inflection point at $P = 2/\beta$.

(e) Sketch the graph of $R(P)$ for $\alpha = 2$ and $\beta = 1$.

6. The *Gompertz growth curve* is sometimes used to study the growth of populations. Its properties are quite similar to the properties of the logistic growth curve. The Gompertz growth curve is given by

$$N(t) = K \exp[-a e^{-bt}]$$

for $t \geq 0$, where K and b are positive constants.

(a) Show that $N(0) = K e^{-a}$ and hence

$$a = \ln \frac{K}{N_0}$$

if $N_0 = N(0)$.

(b) Show that $y = K$ is a horizontal asymptote and that $N(t) < K$ if $N_0 < K$, $N(t) = K$ if $N_0 = K$, and $N(t) > K$ if $N_0 > K$.

(c) Show that

$$\frac{dN}{dt} = bN(\ln K - \ln N)$$

and

$$\frac{d^2 N}{dt^2} = b\frac{dN}{dt}[\ln K - \ln N - 1]$$

(d) Use your results in (b) and (c) to show that $N(t)$ is strictly increasing if $N_0 < K$ and strictly decreasing if $N_0 > K$.

(e) When does $N(t)$, $t \geq 0$ have an inflection point? Discuss its concavity.

(f) Graph $N(t)$ when $K = 100$ and $b = 1$, if (i) $N_0 = 20$, (ii) $N_0 = 70$, and (iii) $N_0 = 150$, and compare your graphs with your answers in (b)–(e).

7. The Monod growth curve is given by

$$f(x) = \frac{cx}{k + x}$$

for $x \geq 0$, where c and k are positive constants. It can be used to describe the specific growth rate of a species, dependent on a resource level x.

(a) Show that $y = c$ is a horizontal asymptote for $x \to \infty$. The constant c is called the *saturation value*.

(b) Show that $f(x)$, $x \geq 0$, is strictly increasing and concave down. Explain why this implies that the saturation value is equal to the maximal specific growth rate.

(c) Show that if $x = k$, then $f(x)$ is equal to half the saturation value. (For this reason, the constant k is called the *half-saturation constant*.)

(d) Sketch a graph of $f(x)$ for $k = 2$ and $c = 5$, clearly marking the saturation value and the half-saturation constant. Compare this graph to one where $k = 3$ and $c = 5$.

(e) Without graphing the three curves, explain how you can use the saturation value and the half-saturation constant to quickly decide that

$$\frac{10x}{3 + x} > \frac{10x}{5 + x} > \frac{8x}{5 + x}$$

for $x \geq 0$.

8. The logistic growth curve is given by

$$N(t) = \frac{K}{1 + (\frac{K}{N_0} - 1)e^{-rt}}$$

for $t \geq 0$, where K, N_0 and r are positive constants and $N(t)$ denotes the population size at time t.

(a) Show that $N(0) = N_0$, and that $y = K$ is a horizontal asymptote as $t \to \infty$.

(b) Show that $N(t) < K$ if $N_0 < K$, $N(t) = K$ if $N_0 = K$, and $N(t) > K$ if $N_0 > K$.

(c) Show that

$$\frac{dN}{dt} = rN\left(1 - \frac{N}{K}\right)$$

and

$$\frac{d^2N}{dt^2} = r\frac{dN}{dt}\left(1 - \frac{2N}{K}\right)$$

(d) Use your results in (b) and (c) to show that $N(t)$ is strictly increasing if $N_0 < K$ and strictly decreasing if $N_0 > K$.

(e) Show that if $N_0 < K/2$, then $N(t)$, $t \geq 0$, has exactly one inflection point $(t^*, N(t^*))$, with $t^* > 0$ and

$$N(t^*) = \frac{K}{2}$$

(i.e., half the carrying capacity). What happens if $K/2 < N_0 < K$? What if $N_0 > K$? Where is the function $N(t)$, $t \geq 0$, concave up, and where concave down?

(f) Sketch the graphs of $N(t)$ for $t \geq 0$ when

(i) $K = 100$, $N_0 = 10$, $r = 1$

(ii) $K = 100$, $N_0 = 70$, $r = 1$

(iii) $K = 100$, $N_0 = 150$, $r = 1$

together with their respective horizontal asymptotes. Clearly mark the inflection point if it exists.

9. A population is said to be in Hardy-Weinberg equilibrium, with respect to a single gene with two alleles A and a, if the three genotypes AA, Aa, and aa have frequencies $p_{AA} = \theta^2$, $p_{Aa} = 2\theta(1 - \theta)$, and $p_{aa} = (1 - \theta)^2$ for some $\theta \in [0, 1]$. Suppose that we take a random sample of size n from a population. One can show that the probability of observing n_1 individuals of type AA, n_2 individuals of type Aa, and n_3 individuals of type aa is given by

$$\frac{n!}{n_1! \, n_2! \, n_3!} p_{AA}^{n_1} p_{Aa}^{n_2} p_{aa}^{n_3}$$

where $n! = n(n - 1)(n - 2) \cdots 3 \cdot 2 \cdot 1$ (read: "n factorial"). Here $n_1 + n_2 + n_3 = n$. This probability depends on θ. The following method, called the *maximum likelihood method*, can be used to estimate θ. The principle is simple: We find the value of θ that maximizes the probability of the observed data. Since the coefficient

$$\frac{n!}{n_1! n_2! n_3!}$$

does not depend on θ, we need only maximize

$$L(\theta) = p_{AA}^{n_1} p_{Aa}^{n_2} p_{aa}^{n_3}$$

(a) Suppose $n_1 = 8$, $n_2 = 6$ and $n_3 = 3$. Compute $L(\theta)$.

(b) Show that if $L(\theta)$ is maximal for $\theta = \hat{\theta}$ (read: "theta hat"), then $\ln L(\theta)$ is also maximal for $\theta = \hat{\theta}$.

(c) Use your result in (b) to find the value $\hat{\theta}$ that maximizes $L(\theta)$ for the data given in (a). The number $\hat{\theta}$ is the maximum likelihood estimate.

10. Suppose the volume of a cell is increasing at a constant rate of 10^{-12} cm³/sec.

(a) If $V(t)$ denotes the cell volume at time t, set up an initial value problem that describes this situation if the initial volume is 10^{-10} cm³.

(b) Solve the initial value problem given in (a) and determine the volume of the cell after 10 seconds.

11. Suppose the concentration $c(t)$ of a drug in the bloodstream at time t satisfies

$$\frac{dc}{dt} = -0.1e^{-0.3t}$$

for $t \geq 0$.

(a) Solve the differential equation, under the assumption that there will eventually be no trace of the drug in the blood.

(b) How long does it take until the concentration reaches half its initial value?

Integration

◆ 5.1 THE DEFINITE INTEGRAL

Computing the area of a region bounded by curves is an ancient problem that was solved in certain cases by both Greek and Islamic mathematicians. They used a method called exhaustion, which goes back to the Greek mathematician Eudoxus (c. 408–355 BC). The basic idea is to divide an area into very small regions consisting mostly of rectilinear figures of known area (such as triangles), so that the total area of the rectilinear figures is close to the area of the region of interest.

Interest in this problem was renewed in the seventeenth century, when many new curves had been defined, and an attempt was made to determine the areas of regions bounded by such curves. The first curves that were considered were of the form $y = x^n$, where n is a positive integer. Fermat wrote in a letter to Roberval on September 22, 1636, that he had succeeded in computing the area under the curve $y = x^n$. He noted that his method was different from the method employed by the Greeks, notably Archimedes. Whereas Archimedes used triangles to exhaust the area bounded by curved lines, Fermat used rectangles. This sounds like a minor difference, but it enabled Fermat to compute areas bounded by other curves that previously could not have been computed. He found that the area under the parabola $y = x^n$ inscribed in a rectangle of lengths b and b^n is $1/(n+1)$ times the area of the rectangle, that is,

$$\frac{1}{n+1}b \cdot b^n = \frac{1}{n+1}b^{n+1}$$

(see Figure 5.1). (This problem was independently solved around 1640 by Cavalieri, Pascal, Roberval, and Torricelli as well.)

Augustin-Louis Cauchy (1789–1857) was the first to rigorously define areas, based on the limit of the sum of areas of approximating rectangles. The definition we will use goes back to Georg Bernhard Riemann (1826–1866). His definition is more general than Cauchy's, and allows for a larger class of functions to be used as boundary curves for areas.

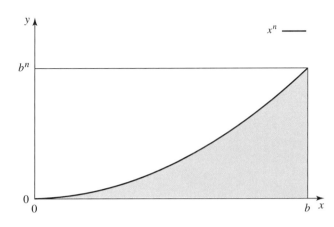

◀ **Figure 5.1**
The area of the shaded region under the parabola $y = x^n$ inscribed in the rectangle is $1/(n + 1)$ times the area of the rectangle.

5.1.1 The Area Problem

We wish to find the surface area of the lake shown in Figure 5.2; to do so, we overlay a grid and count the number of squares that have a nonempty intersection with the lake. This will approximate the area of the lake. The finer the grid, the closer our approximation will be to the true area of the lake.

Dividing a region into smaller regions of known area is the basic principle we will employ in this section to find the area of a region bounded by curves of continuous functions.

▮ **EXAMPLE 1**   We will try to find the area of the region below the parabola $f(x) = x^2$ and above the x-axis between 0 and a with $a > 0$ (see Figure 5.3). To do this, we divide the interval $[0, a]$ into n subintervals of equal length, and approximate the area of interest by a sum of the areas of rectangles whose bases are the subintervals and whose heights are the values of the function at the left endpoints of these subintervals. This is illustrated in Figure 5.4, where $a = 1$ and $n = 5$.

In Figure 5.4, each rectangle has width $1/5 = 0.2$. The height of the first rectangle is $f(0) = 0$, the height of the second rectangle is $f(0.2) = (0.2)^2$, the height of the third rectangle is $f(0.4) = (0.4)^2$, and so on. The area of a rectangle is width times height; adding up the areas of the approximating rectangles in Figure 5.4 yields

$$(0.2)(0)^2 + (0.2)(0.2)^2 + (0.2)(0.4)^2 + (0.2)(0.6)^2 + (0.2)(0.8)^2$$
$$= (0.2)[0^2 + (0.2)^2 + (0.4)^2 + (0.6)^2 + (0.8)^2] = 0.24$$

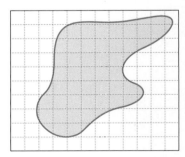

▲ **Figure 5.2**
The outline of a lake with superimposed grid: The finer the grid, the more accurately the area of the lake can be determined.

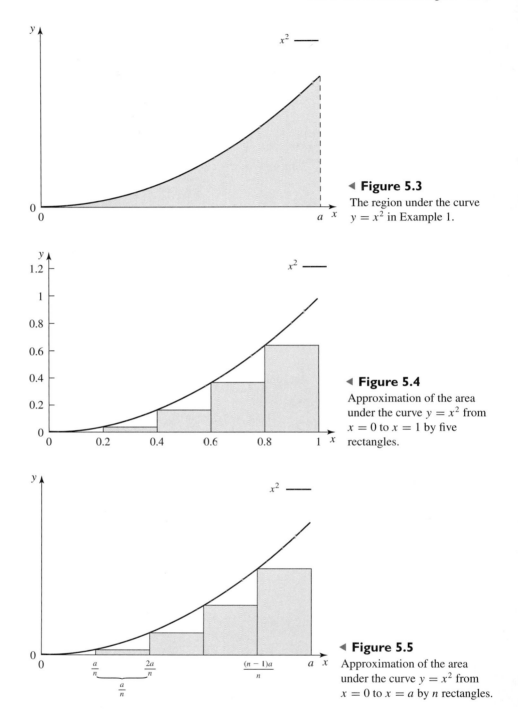

◄ Figure 5.3
The region under the curve $y = x^2$ in Example 1.

◄ Figure 5.4
Approximation of the area under the curve $y = x^2$ from $x = 0$ to $x = 1$ by five rectangles.

◄ Figure 5.5
Approximation of the area under the curve $y = x^2$ from $x = 0$ to $x = a$ by n rectangles.

Thus, an approximation of the area between 0 and 1 is 0.24 when we use five subintervals.

We turn now to the general case (illustrated in Figure 5.5), where the interval is $[0, a]$ and the number of subintervals is n. Since the interval $[0, a]$ has length a and the number of subintervals is n, each subinterval has length a/n. The left endpoints of successive subintervals are therefore $0, a/n, 2a/n, 3a/n, \ldots,$ $(n-1)a/n$. The heights of the successive rectangles are then $f(0) = 0$, $f(a/n) = (a/n)^2$, $f(2a/n) = (2a/n)^2, \ldots, f((n-1)a/n) = ((n-1)a/n)^2$. Denoting the sum of the areas of the n rectangles by S_n, we find that

$$S_n = \frac{a}{n} f(0) + \frac{a}{n} f\left(\frac{a}{n}\right) + \frac{a}{n} f\left(\frac{2a}{n}\right) + \cdots + \frac{a}{n} f\left(\frac{(n-1)a}{n}\right)$$

$$= \frac{a}{n} 0^2 + \frac{a}{n} \frac{a^2}{n^2} + \frac{a}{n} \frac{2^2 a^2}{n^2} + \cdots \frac{a}{n} \frac{(n-1)^2 a^2}{n^2}$$

$$= \frac{a^3}{n^3} [1^2 + 2^2 + \cdots + (n-1)^2]$$

The sum of the squares of the first k integers can be computed:

$$1^2 + 2^2 + 3^2 + \cdots + k^2 = \frac{k(k+1)(2k+1)}{6} \tag{5.1}$$

(For a proof of this formula, see Problem 28.) Using this formula for $k = n - 1$, we find

$$S_n = \frac{a^3}{n^3} \frac{(n-1)(n-1+1)(2(n-1)+1)}{6}$$

$$= \frac{a^3}{n^3} \frac{(n-1)n(2n-1)}{6} = \frac{a^3}{6} \frac{n-1}{n} \frac{n}{n} \frac{2n-1}{n}$$

$$= \frac{a^3}{6} \left(1 - \frac{1}{n}\right) \cdot 1 \cdot \left(2 - \frac{1}{n}\right)$$

The finer the subdivision of $[0, a]$ (that is, the larger n), the more accurate the approximation, as illustrated in Figure 5.6, where we see that the area of the

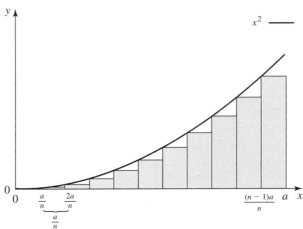

◀ **Figure 5.6**

Increasing the number of approximating rectangles improves the accuracy of the approximation.

region below the parabola and above the x-axis between 0 and a is more accurately approximated when we use a larger number of rectangles. Choosing finer and finer subdivisions means that we let n go to infinity. We find

$$\lim_{n\to\infty} S_n = \lim_{n\to\infty} \frac{a^3}{6}\left(1 - \frac{1}{n}\right)\cdot 1\cdot\left(2 - \frac{1}{n}\right) = \frac{a^3}{6}(1)(1)(2) = \frac{a^3}{3}$$

That is, the area under the parabola $y = x^2$ from 0 to a is equal to $a^3/3$.   ▌

We see from Example 1 that computing areas entails summing a large number of terms. Before we continue our discussion of the computation of areas, we will therefore spend some time on sums of the type we encountered in Example 1.

It will be convenient to have a shorthand notation for sums that involve a large number of terms.

> **Definition (Sigma Notation for Finite Sums)**   Let $a_1, a_2, \ldots, a_n$ be real numbers and n be a positive integer. Then
>
> $$\sum_{k=1}^{n} a_k = a_1 + a_2 + \cdots + a_n$$

The letter Σ is the capital Greek letter sigma and the symbol $\sum_{k=1}^{n}$ means that we sum from $k = 1$ to n, where k is called the index of summation, the number 1 is the lower limit of summation, and the number n is the upper limit of summation.

EXAMPLE 2
  a. Write each sum in expanded form.

  (i) $\displaystyle\sum_{k=1}^{4} k = 1 + 2 + 3 + 4$

  (ii) $\displaystyle\sum_{k=3}^{6} k^2 = 3^2 + 4^2 + 5^2 + 6^2$

  (iii) $\displaystyle\sum_{k=1}^{n} \frac{1}{k} = 1 + \frac{1}{2} + \frac{1}{3} + \cdots + \frac{1}{n}$

  (iv) $\displaystyle\sum_{k=1}^{5} 1 = 1 + 1 + 1 + 1 + 1$

  b. Write each sum in sigma notation.

  (i) $2 + 3 + 4 + 5 = \displaystyle\sum_{k=2}^{5} k$

  (ii) $1^3 + 2^3 + 3^3 + 4^3 = \displaystyle\sum_{k=1}^{4} k^3$

  (iii) $1 + 3 + 5 + 7 + \cdots + (2n + 1) = \displaystyle\sum_{k=0}^{n} (2k + 1)$

  (iv) $x + 2x^2 + 3x^3 + \cdots + nx^n = \displaystyle\sum_{k=1}^{n} kx^k$ ▌

Occasionally, we will need a formula to sum the first n integers. How this is done is shown in the next example.

▮ **EXAMPLE 3**   Show that

$$S_n = \sum_{k=1}^{n} k = 1 + 2 + 3 + \cdots + n = \frac{n(n+1)}{2}$$

Solution   The following trick will enable us to compute the sum S_r. We write the sum in the usual order, and in reverse order.

$$S_n = 1 + 2 + 3 + \cdots + n$$
$$S_n = n + (n-1) + (n-2) + \cdots + 2 + 1$$

Adding vertically, we find

$$2S_n = (1+n) + (2+n-1) + (3+n-2) + \cdots + (n+1)$$
$$= (n+1) + (n+1) + (n+1) + \cdots + (n+1)$$

There are n terms (each $n+1$) on the right-hand side. Hence

$$2S_n = n(n+1) \qquad \text{or} \qquad S_n = \frac{n(n+1)}{2}$$

This method was used by Karl Friedrich Gauss (1777–1855) when he was ten years old: One day, to keep the students busy, his teacher asked them to add all the numbers from 1 to 100. To his teacher's astonishment, Karl Friedrich quickly gave the correct answer, 5050. To find the answer, he did not add the numbers in their numerical order but rather added $1 + 100$, $2 + 99$, $3 + 98, \ldots, 50 + 51$ (just as we did in the above derivation). Each term is equal to 101 and there are fifty such terms, hence the answer is $50 \cdot 101 = 5050$. (Gauss went on to become one of the greatest mathematicians in history, contributing to geometry, number theory, astronomy, and other areas.)   ▮

The following rules are useful when evaluating finite sums.

ALGEBRAIC RULES

1. Constant Value Rule: $\displaystyle\sum_{k=1}^{n} 1 = n$

2. Constant Multiple Rule: $\displaystyle\sum_{k=1}^{n} c \cdot a_k = c \sum_{k=1}^{n} a_k$
if c is a constant that does not depend on k.

3. Sum Rule: $\displaystyle\sum_{k=1}^{n} (a_k + b_k) = \sum_{k=1}^{n} a_k + \sum_{k=1}^{n} b_k$

▮ **EXAMPLE 4**   Use the algebraic rules to simplify the following sums.

a.
$$\sum_{k=2}^{4} (3 + k) = \sum_{k=2}^{4} 3 + \sum_{k=2}^{4} k = (3 + 3 + 3) + (2 + 3 + 4) = 18$$

b.
$$\sum_{k=1}^{n} (k^2 - 2k) = \sum_{k=1}^{n} k^2 - 2 \sum_{k=1}^{n} k$$
$$= \frac{n(n+1)(2n+1)}{6} - 2\frac{n(n+1)}{2} = \frac{n(n+1)(2n-5)}{6}$$

(We used (5.1) to evaluate the first sum and Example 3 to evaluate the second sum.) ∎

5.1.2 Riemann Integrals

We will now develop a more systematic solution to the area problem. Though our approach will be very similar to that in the previous subsection, we will look at a more general situation. We allow the function whose graph makes up the boundary of the region of interest to take on negative values as well. Furthermore, the rectangles that we use for approximating the area may vary in width, and the points that are chosen to compute the heights of the rectangles may be anywhere in their respective subintervals (which form the bases of the rectangles).

Let f be a continuous function on the interval $[a, b]$ (see Figure 5.7). We partition $[a, b]$ into n subintervals by choosing $n - 1$ points $x_1, x_2, \ldots, x_{n-1}$ in (a, b), such that

$$a = x_0 < x_1 < x_2 < \cdots < x_{n-1} < x_n = b$$

The n subintervals

$$[x_0, x_1], [x_1, x_2], \ldots, [x_{n-1}, x_n]$$

form a **partition** of $[a, b]$, which we denote by $P = [x_0, x_1, x_2, \ldots, x_n]$. The length of the kth subinterval $[x_{k-1}, x_k]$ is denoted by Δx_k. The length of the longest subinterval is called the **norm** of P and denoted by $\|P\|$ (read: "norm of P"), that is,

$$\|P\| = \max\{\Delta x_1, \Delta x_2, \ldots, \Delta x_n\}$$

In each subinterval $[x_{k-1}, x_k]$, we choose a point c_k, and construct a rectangle with base Δx_k and height $|f(c_k)|$, as shown in Figure 5.7. If $f(c_k)$ is positive, then $f(c_k)\Delta x_k$ is the area of the rectangle. If $f(c_k)$ is negative, then $f(c_k)\Delta x_k$ is the negative of the rectangle's area. The sum of these products is denoted by S_P. That is,

$$S_P = \sum_{k=1}^{n} f(c_k)\Delta x_k$$

The value of the sum depends on the choice of the partition P (hence the subscript P on S) and the choice of the points $c_k \in [x_{k-1}, x_k]$, and is called a **Riemann sum** for f on $[a, b]$.

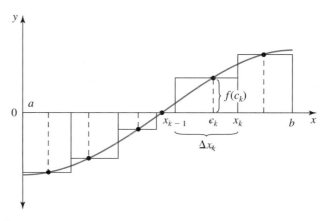

▲ **Figure 5.7**

An illustration of a Riemann sum.

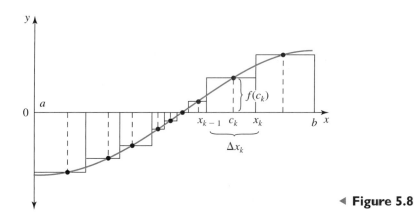

◀ **Figure 5.8**

By choosing a finer partition of $[a, b]$, the rectangles fill out the region between the curve and the x-axis more accurately (see Figure 5.8). A finer partition means that the largest subinterval becomes smaller; that is, the norm of the partition P becomes smaller. The limit of S_P as $\|P\| \to 0$ (if it exists) is called the definite integral of f from a to b.

Definition (Definite Integral)   Let $P = [x_0, x_1, x_2, \ldots, x_n]$ be a partition of $[a, b]$ and set $\Delta x_k = x_k - x_{k-1}$ and $c_k \in [x_{k-1}, x_k]$. The **definite integral** of f from a to b is

$$\int_a^b f(x)\,dx = \lim_{\|P\| \to 0} \sum_{k=1}^n f(c_k)\Delta x_k$$

if the limit exists, in which case f is said to be (Riemann) **integrable** on the interval $[a, b]$.

The symbol $\int$ is an elongated S (as in "sum") and was introduced by Leibniz. It is called the integral sign. In the notation $\int_a^b f(x)\,dx$ (read: "integral from a to b of $f(x)\,dx$"), $f(x)$ is called the **integrand**, the number a is the lower limit and b is the upper limit. Though the symbol dx by itself has no meaning, it should remind you that, as we take the limit, the widths of the subintervals become ever smaller.

The phrase "if the limit exists" means, in particular, that the value of $\lim_{\|P\| \to 0} S_P$ does not depend on how we choose the partitions and the points $c_k \in [x_{k-1}, x_k]$ as we take the limit. An important result tells us that if f is continuous on $[a, b]$, the definite integral of f on $[a, b]$ exists.

Theorem   All continuous functions are Riemann integrable; that is, if $f(x)$ is continuous on $[a, b]$, then

$$\int_a^b f(x)\,dx$$

exists.

The class of functions that are Riemann integrable is quite a bit larger than the set of continuous functions. For instance, functions that are both bounded (for which there exists an $M < \infty$ such that $|f(x)| < M$ for all x over which we wish

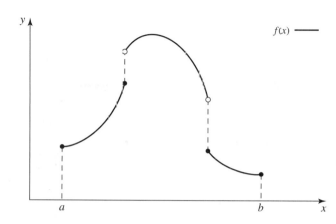

◀ **Figure 5.9**
The function $y = f(x)$ is piecewise continuous and bounded on $[a, b]$.

to integrate) and piecewise continuous (continuous except for a finite number of discontinuities) are integrable (see Figure 5.9). We will primarily be concerned with continuous functions in this text; knowing that continuous functions are Riemann integrable will therefore suffice for the most part.

Note that $\int_a^b f(x)\, dx$ is a number that does not depend on x. We could have written $\int_a^b f(u)\, du$ (or any other letter in place of x) and meant the same thing.

∎ **EXAMPLE 5**   Express the definite integral

$$\int_3^7 (x^2 - 1)\, dx$$

as a limit of Riemann sums.

Solution

$$\int_3^7 (x^2 - 1)\, dx = \lim_{\|P\| \to 0} \sum_{k=1}^n (c_k^2 - 1)\, \Delta x_k$$

where $x_0 = 3 < x_1 < x_2 < \cdots < x_n = 7$ is a partition of $[3, 7]$, $c_k \in [x_{k-1}, x_k]$, $\Delta x_k = x_k - x_{k-1}$, and the limit $\|P\| \to 0$ means that the norm of the partition tends to 0.   ∎

∎ **EXAMPLE 6**   Express the limit

$$\lim_{\|P\| \to 0} \sum_{k=1}^n \sqrt{c_k - 1}\, \Delta x_k$$

as a definite integral, where $P = [x_0, x_1, \ldots, x_n]$ is a partition of $[2, 4]$ into n subintervals, $\Delta x_k = x_k - x_{k-1}$, and $c_k \in [x_{k-1}, x_k]$.

Solution

$$\lim_{\|P\| \to 0} \sum_{k=1}^n \sqrt{c_k - 1}\, \Delta x_k = \int_2^4 \sqrt{x - 1}\, dx$$ ∎

∎ **EXAMPLE 7**   Evaluate

$$\int_0^2 x^2\, dx$$

Solution   We evaluated the Riemann sum and its limit for $y = x^2$ from 0 to a in Example 1, and found

$$\lim_{n \to \infty} S_n = \int_0^a x^2\, dx = \frac{a^3}{3}$$

With $a = 2$, we therefore have

$$\int_0^2 x^2 \, dx = \frac{8}{3}$$

Geometric Interpretation of Definite Integrals   In Example 1, we computed the area of the region below the parabola $y = x^2$ and above the x-axis between 0 and a by approximating the region with n rectangles of equal width, and then taking the limit $n \to \infty$. More generally, we can now define the **area** of a region A above the x-axis, as shown in Figure 5.10, as the limiting value (if it exists) of the Riemann sum of approximating rectangles. (Note that an area is always a positive number.) This definition allows us to interpret the definite integral of a nonnegative function as an area.

If $f(x) \le 0$ on $[a, b]$, then the definite integral $\int_a^b f(x) \, dx$ is less than or equal to 0, and the value is the negative of the area of the region above the graph of f and below the x-axis between a and b (see Figure 5.11). We refer to this as a "signed area." (A signed area may be either positive or negative.)

In general, a definite integral can thus be interpreted as a difference of areas, as illustrated in Figure 5.12. If A_+ denotes the total area of the region above the x-axis and below the graph of f (where $f \ge 0$), and A_- denotes the total area of the region below the x-axis and above the graph of f (where $f \le 0$), then

$$\int_a^b f(x) \, dx = A_+ - A_-$$

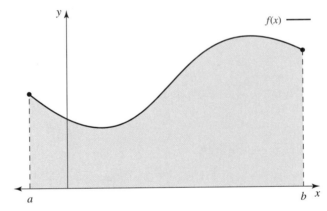

◀ **Figure 5.10**
The area of a region under the curve of a positive function is given by the definite integral $\int_a^b f(x) \, dx$.

◀ **Figure 5.11**
The area of a region under the curve of a negative function is given by $-\int_a^b f(x) \, dx$.

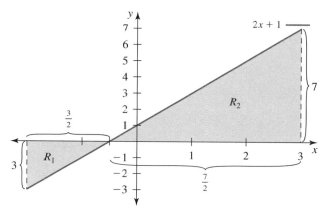

◀ **Figure 5.12**
A_+ is the combined area of
R_1 and R_3, A_- is the area of
R_2. Then
$\int_a^b f(x)\,dx = A_+ - A_-.$

1. If f is integrable on $[a, b]$ and $f(x) \geq 0$ on $[a, b]$, then

$$\int_a^b f(x)\,dx = \begin{bmatrix} \text{the area of the region between the} \\ \text{graph of } f \text{ and the } x\text{-axis from } a \text{ to } b \end{bmatrix}$$

2. If f is integrable on $[a, b]$, then

$$\int_a^b f(x)\,dx = [\text{area above } x\text{-axis}] - [\text{area below } x\text{-axis}]$$

EXAMPLE 8 Find the value of

$$\int_{-2}^3 (2x + 1)\,dx$$

by interpreting it as the signed area of an appropriately chosen region.

Solution We graph $y = 2x + 1$ between -2 and 3 (see Figure 5.13). The line intersects the x-axis at $x = -1/2$. The area of the region to the left of $-1/2$ between the graph of $y = 2x + 1$ and the x-axis is denoted by A_-; the area of the region to the right of $-1/2$ between the graph of $y = 2x + 1$ and the x-axis is denoted by A_+. Both regions are triangles whose areas can be computed using a formula from

◀ **Figure 5.13**
The area of R_1 is A_-; the area
of R_2 is A_+.

geometry.

$$A_- = \frac{1}{2} \cdot \frac{3}{2} \cdot 3 = \frac{9}{4}$$

$$A_+ = \frac{1}{2} \cdot \frac{7}{2} \cdot 7 = \frac{49}{4}$$

Therefore,

$$\int_{-2}^{3} (2x + 1)\,dx = A_+ - A_- = \frac{49}{4} - \frac{9}{4} = \frac{40}{4} = 10$$

▮ **EXAMPLE 9**   Find the value of

$$\int_{0}^{2\pi} \sin x \, dx$$

by interpreting it as the signed area of an appropriately chosen region.

Solution   We graph $y = \sin x$ from 0 to 2π (see Figure 5.14). The function $f(x) = \sin x$ is symmetric about $x = \pi$. It follows from this symmetry that the area of the region below the graph of f and above the x-axis between 0 and π (denoted by A_+) is the same as the area of the region above the graph of f and below the x-axis between π and 2π (denoted by A_-). Therefore, $A_+ = A_-$ and

$$\int_{0}^{2\pi} \sin x \, dx = A_+ - A_- = 0$$ ▮

▮ **EXAMPLE 10**   Find the value of

$$\int_{0}^{2} \sqrt{4 - x^2} \, dx$$

by interpreting it as the signed area of an appropriately chosen region.

Solution   The graph of $y = \sqrt{4 - x^2}, 0 \le x \le 2$, is the quarter-circle with center at $(0, 0)$ and radius 2 in the first quadrant (see Figure 5.15). Since the area of a circle with radius 2 is $\pi(2)^2 = 4\pi$, the area of a quarter-circle is $4\pi/4 = \pi$. Hence,

$$\int_{0}^{2} \sqrt{4 - x^2} \, dx = \pi$$ ▮

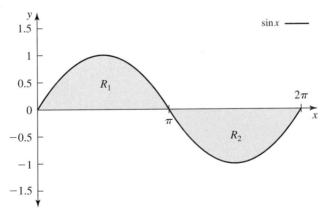

◀ **Figure 5.14**
The graph of
$f(x) = \sin x, 0 \le x \le 2\pi$, in
Example 9. The area of R_1 is
A_+; the area of R_2 is A_-.

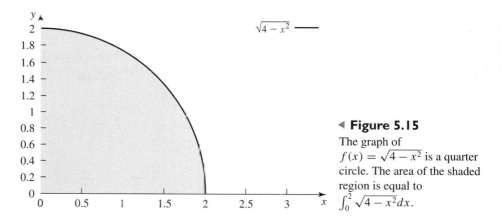

◄ **Figure 5.15**
The graph of
$f(x) = \sqrt{4 - x^2}$ is a quarter
circle. The area of the shaded
region is equal to
$\int_0^2 \sqrt{4 - x^2}\,dx$.

5.1.3 Properties of the Riemann Integral

In this subsection, we will collect important properties that will help us to evaluate definite integrals.

> **Properties**   Assume that f is integrable over $[a, b]$. We set
>
> **1.** $\displaystyle\int_a^a f(x)\,dx = 0$ and
>
> **2.** $\displaystyle\int_b^a f(x)\,dx = -\int_a^b f(x)\,dx$

Both properties make sense geometrically. The first integral says that the signed area between a and a is equal to 0; given that the width of the area is equal to 0, we expect the area to be 0 as well. The second property gives an orientation to the integral; for instance, if $f(x)$ is nonnegative on $[a, b]$, then $\int_a^b f(x)\,dx$ is nonnegative and can be interpreted as the area of the region between the graph of $f(x)$ and the x-axis from a to b. If we reverse the direction of the integration—that is, compute $\int_b^a f(x)\,dx$—we want the integral to be negative.

The next three properties follow immediately from the definition of the definite integral as the limit of a sum of areas of approximating rectangles.

> **Properties**   Assume that f and g are integrable over $[a, b]$.
>
> **3.** If k is a constant, then
>
> $$\int_a^b kf(x)\,dx = k\int_a^b f(x)\,dx$$
>
> **4.** $\displaystyle\int_a^b [f(x) + g(x)]\,dx = \int_a^b f(x)\,dx + \int_a^b g(x)\,dx$
>
> **5.** If f is integrable over an interval containing the three numbers a, b, and c, then
>
> $$\int_a^b f(x)\,dx = \int_a^c f(x)\,dx + \int_c^b f(x)\,dx$$

We prove Property (4) to illustrate how the definition of the definite integral can be used to prove its properties.

Proof of (4)   We choose a partition $P = [x_0, x_1, \ldots, x_n]$ of $[a, b]$ into n subintervals, $\Delta x_k = x_k - x_{k-1}$, and $c_k \in [x_{k-1}, x_k]$, and use the definition of definite integrals.

$$\int_a^b [f(x) + g(x)]\,dx = \lim_{\|P\| \to 0} \sum_{k=1}^n [f(c_k) + g(c_k)]\Delta x_k$$

$$= \lim_{\|P\| \to 0} \sum_{k=1}^n [f(c_k)\Delta x_k + g(c_k)\Delta x_k]$$

Applying the sum rule for finite sums, we find

$$= \lim_{\|P\| \to 0} \left[\sum_{k=1}^n f(c_k)\Delta x_k + \sum_{k=1}^n g(c_k)\Delta x_k \right]$$

Since f and g are integrable, the individual limits exist, and we find

$$= \lim_{\|P\| \to 0} \sum_{k=1}^n f(c_k)\Delta x_k + \lim_{\|P\| \to 0} \sum_{k=1}^n g(c_k)\Delta x_k$$

Using the definition again, we obtain

$$= \int_a^b f(x)\,dx + \int_a^b g(x)\,dx$$

Rule (5) is an addition property, which we illustrate in Figures 5.16 and 5.17. If $a < c < b$, then the rule is clear geometrically (see Figure 5.16), but we do not need this assumption and we will show the result for $a < b < c$ (see Figure 5.17). Note that f need not be positive as in Figures 5.16 and 5.17; property (5) holds for any integrable function.

Proof of (5)   We will prove property (5) for the case $a < b < c$, when f is integrable over $[a, c]$ and $f(x) \geq 0$ for $x \in [a, c]$. From Figure 5.17, we see that

$$\int_a^b f(x)\,dx = \int_a^c f(x)\,dx - \int_b^c f(x)\,dx$$

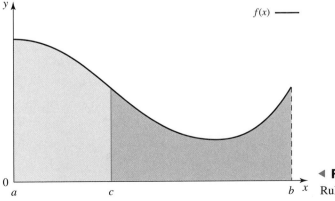

$f(x)$ ———

◀ **Figure 5.16**

Rule (5) when $a < c < b$.

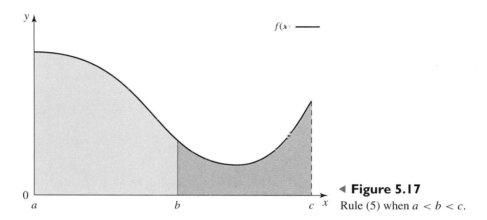

◀ **Figure 5.17**
Rule (5) when $a < b < c$.

But because of (2),

$$\int_b^c f(x)\,dx = -\int_c^b f(x)\,dx$$

and, therefore,

$$\int_a^b f(x)\,dx = \int_a^c f(x)\,dx + \int_c^b f(x)\,dx$$

as stated in (5)   ◆

▮ **EXAMPLE 11**   Given that $\int_0^a x^2 = a^3/3$, evaluate

$$\int_1^4 (3x^2 + 2)\,dx$$

Solution

$$\int_1^4 (3x^2 + 2)\,dx = 3\int_1^4 x^2\,dx + \int_1^4 2\,dx$$

To evaluate $\int_1^4 x^2\,dx$, use the addition property (5), and write

$$\int_1^4 x^2\,dx = \int_1^0 x^2\,dx + \int_0^4 x^2\,dx$$

Since

$$\int_1^0 x^2\,dx = -\int_0^1 x^2\,dx$$

it follows that

$$\int_1^4 x^2\,dx = -\int_0^1 x^2\,dx + \int_0^4 x^2\,dx$$

To evaluate $\int_1^4 2\,dx$, we note that $y = 2$ is a horizontal line that intersects the y-axis at $y = 2$. The region under $y = 2$ from 1 to 4 is therefore a rectangle with base $4 - 1 = 3$ and height 2. Therefore,

$$\int_1^4 (3x^2 + 2)\,dx = 3\left[\int_0^4 x^2\,dx - \int_0^1 x^2\,dx\right] + \int_1^4 2\,dx$$

$$= 3\left(\frac{4^3}{3} - \frac{1^3}{3}\right) + (2)(3)$$

$$= 64 - 1 + 6 = 69$$ ▮

The next three properties are called order properties. They allow us to either compare definite integrals or say something about how big or small a particular definite integral can be. We will first state the properties, and then explain what they mean geometrically.

> **Properties**   Assume that f and g are integrable over $[a, b]$
>
> **6.** If $f(x) \geq 0$ on $[a, b]$, then $\displaystyle\int_a^b f(x)\, dx \geq 0$.
>
> **7.** If $f(x) \leq g(x)$ on $[a, b]$, then $\displaystyle\int_a^b f(x)\, dx \leq \int_a^b g(x)\, dx$.
>
> **8.** If $m \leq f(x) \leq M$ on $[a, b]$, then
>
> $$m(b - a) \leq \int_a^b f(x)\, dx \leq M(b - a)$$

Property (6), illustrated in Figure 5.18, says that if f is nonnegative over the interval $[a, b]$, then the definite integral over this interval is also nonnegative. We can understand this from the geometric interpretation: If $f(x) \geq 0$ for $x \in [a, b]$, then $\int_a^b f(x)\, dx$ is the area between the curve and the x-axis between a and b. But an area must be a nonnegative number.

Property (7) is explained in Figure 5.19 when both f and g are positive functions on $[a, b]$. We use the fact that in this case, the definite integral can be interpreted

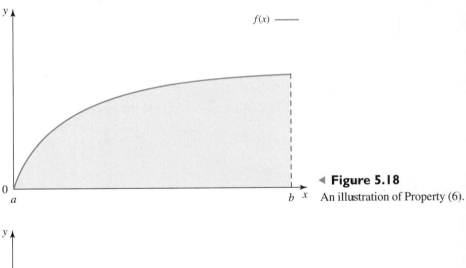

◀ **Figure 5.18**
An illustration of Property (6).

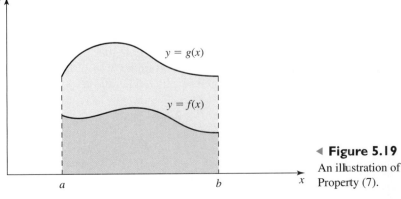

◀ **Figure 5.19**
An illustration of Property (7).

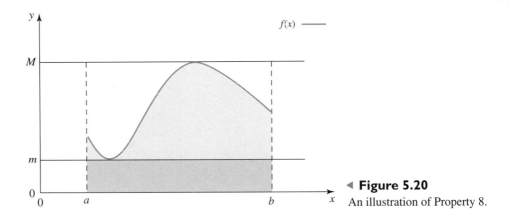

◀ **Figure 5.20**
An illustration of Property 8.

as an area. Looking at Figure 5.19, we see that the function f has a smaller area than g. Property (7) holds without the assumption that both f and g are positive, and one can draw an analogous figure for the general case as well. The definite integral then needs to be interpreted as a signed area.

Property (8) is explained in Figure 5.20 for $f(x) \geq 0$ in $[a, b]$. We see that the rectangle with height m is contained in the area between the graph of f and the x-axis, which in turn is contained in the rectangle with height M. Since $m(b-a)$ is the area of the small rectangle, $\int_a^b f(x)\, dx$ is the area between the graph of f and the x-axis for nonnegative f, and $M(b-a)$ is the area of the big rectangle, the inequalities in (8) follow. Note that the statement does not require that f be nonnegative; you can draw an analogous figure when f is negative on parts or all of $[a, b]$.

The next example illustrates how these order properties are used.

▮EXAMPLE 12 Show that

$$0 \leq \int_0^\pi \sin x\, dx \leq \pi$$

Solution Note that $0 \leq \sin x \leq 1$ for $x \in [0, \pi]$. Using (6), we find

$$\int_0^\pi \sin x\, dx \geq 0$$

Using (8), we find

$$\int_0^\pi \sin x\, dx \leq (1)(\pi) = \pi$$

(see Figure 5.21).

The next example will help us to deepen our understanding of signed areas; it also uses the order properties.

▮EXAMPLE 13 Find the value of $a \geq 0$ that maximizes

$$\int_0^a (1 - x^2)\, dx$$

Solution We graph the integrand $f(x) = 1 - x^2$ for $x \geq 0$ in Figure 5.22. Using the interpretation of the definite integral as the signed area, we see from the graph of $f(x)$ that $a = 1$ maximizes the integral, since the graph of $f(x)$ is positive for $x < 1$ and negative for $x > 1$.

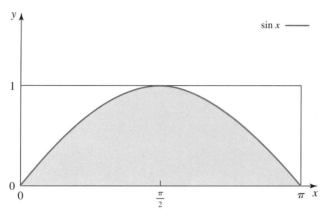

◀ Figure 5.21
An illustration of the integral in Example 12. The shaded area is non-negative and less than the area of the rectangle with base length π and height 1.

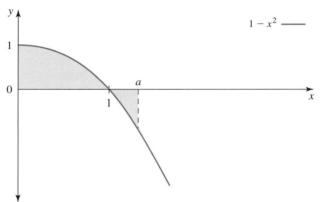

◀ Figure 5.22
An illustration of the integral in Example 13.

We also wish to give a precise argument; our goal is to show that

$$\int_0^1 f(x)\,dx > \int_0^a f(x)\,dx$$

for all $a \geq 0$, provided that $a \neq 1$. This would then imply that $a = 1$ maximizes the integral

$$\int_0^a (1 - x^2)\,dx$$

First, note that $f(x)$ is continuous for $x \geq 0$, and that

$$f(x) \begin{cases} > 0 & \text{for } 0 \leq x < 1 \\ < 0 & \text{for } x > 1 \end{cases}$$

which implies that for $0 \leq a < 1$,

$$\int_a^1 f(x)\,dx > 0$$

and, therefore, for $0 \leq a < 1$,

$$\int_0^1 f(x)\,dx = \int_0^a f(x)\,dx + \underbrace{\int_a^1 f(x)\,dx}_{>0} > \int_0^a f(x)\,dx \qquad (5.2)$$

On the other hand, for $a > 1$,

$$\int_1^a f(x)\,dx < 0$$

and, therefore, for $a > 1$,

$$\int_0^a f(x)\,dx = \int_0^1 f(x)\,dx + \underbrace{\int_1^c f(x)\,dx}_{<0} < \int_0^1 f(x)\,dx \qquad (5.3)$$

Combining (10.13) and (5.3) shows that

$$\int_0^a f(x)\,dx < \int_0^1 f(x)\,dx$$

for all $a \geq 0$ and $a \neq 1$. Hence, $a = 1$ maximizes the integral $\int_0^a f(x)\,dx$. ▌

◆ 5.1 PROBLEMS

[5.1.1]

1. Approximate the area under the parabola $y = x^2$ from 0 to 1, using four equal subintervals with (a) left endpoints, (b) midpoints, and (c) right endpoints.

In Problems 2–11, write each sum in expanded form.

2. $\displaystyle\sum_{k=1}^{4} \sqrt{k}$

3. $\displaystyle\sum_{k=3}^{6} k^2$

4. $\displaystyle\sum_{k=0}^{5} 2^k$

5. $\displaystyle\sum_{k=1}^{3} \frac{k}{k+1}$

6. $\displaystyle\sum_{k=0}^{4} x^k$

7. $\displaystyle\sum_{k=0}^{4} k^x$

8. $\displaystyle\sum_{k=0}^{3} (-1)^k$

9. $\displaystyle\sum_{k=1}^{n} f(c_k)\Delta x_k$

10. $\displaystyle\sum_{k=1}^{n} \left(\frac{k}{n}\right)^2 \frac{1}{n}$

11. $\displaystyle\sum_{k=1}^{n} \cos\left(k\frac{\pi}{n}\right)\frac{\pi}{n}$

In Problems 12–19, write each sum in sigma notation.

12. $1 + 2 + 3 + 4 + 5 + 6$

13. $\frac{1}{\sqrt{1}} + \frac{1}{\sqrt{2}} + \frac{1}{\sqrt{3}} + \frac{1}{\sqrt{4}}$

14. $\ln 2 + \ln 3 + \ln 4 + \ln 5$

15. $\frac{3}{5} + \frac{4}{6} + \frac{5}{7} + \frac{6}{8} + \frac{7}{9}$

16. $2 + 4 + 6 + 8 + \cdots + 2n$

17. $\frac{1}{1} + \frac{1}{2} + \frac{1}{4} + \frac{1}{8} + \frac{1}{16} + \cdots + \frac{1}{2^n}$

18. $1 + q + q^2 + q^3 + q^4 + \cdots + q^{n-1}$

19. $1 - a + a^2 - a^3 + a^4 - a^5 + \cdots - (-1)^n a^n$

In Problems 20–27, use the algebraic rules for sums to evaluate each sum. Recall that

$$\sum_{k=1}^{n} k = \frac{n(n+1)}{2}$$

and

$$\sum_{k=1}^{n} k^2 = \frac{n(n+1)(2n+1)}{6}$$

20. $\displaystyle\sum_{k=1}^{20} (3k+2)$

21. $\displaystyle\sum_{k=1}^{10} (2-k^2)$

22. $\displaystyle\sum_{k=0}^{6} k(k+1)$

23. $\displaystyle\sum_{k=1}^{n} 3k$

24. $\displaystyle\sum_{k=1}^{n} 2k^2$

25. $\displaystyle\sum_{k=1}^{n} (k+2)(k-2)$

26. $\displaystyle\sum_{k=1}^{10} (-1)^k$

27. $\displaystyle\sum_{k=1}^{11} (-1)^k$

28. The following steps will show that

$$\sum_{k=1}^{n} k^2 = \frac{n(n+1)(2n+1)}{6}$$

(a) Show that

$$\sum_{k=1}^{n} [(1+k)^3 - k^3]$$

$$= (2^3 - 1^3) + (3^3 - 2^3) + (4^3 - 3^3)$$
$$+ \cdots + [(1+n)^3 - n^3]$$
$$= (1+n)^3 - 1^3$$

(Sums that "collapse" like this due to cancellation of terms are called telescoping or collapsing sums.)

(b) Use the algebraic rules for sums and Example 3 to show that

$$\sum_{k=1}^{n} [(1+k)^3 - k^3]$$

$$= 3\sum_{k=1}^{n} k^2 + 3\frac{n(n+1)}{2} + n$$

(c) In (a) and (b), we found two expressions for the sum

$$\sum_{k=1}^{n} [(1+k)^3 - k^3]$$

They are therefore equal; that is,

$$(1+n)^3 - 1^3$$

$$= 3\sum_{k=1}^{n} k^2 + 3\frac{n(n+1)}{2} + n$$

Solve this equation for $\sum_{k=1}^{n} k^2$, and show that

$$\sum_{k=1}^{n} k^2 = \frac{n(n+1)(2n+1)}{6}$$

[5.1.2]

29. Approximate

$$\int_{-2}^{2} (1 - x^2) \, dx$$

using four equal subintervals, with (a) left endpoints, (b) midpoints, and (c) right endpoints.

30. **(a)** Assume that $a > 0$. Evaluate $\int_0^a x \, dx$ using the fact that the region bounded by $y = x$ and the x-axis between 0 to a is a triangle (see Figure 5.23).

(b) Assume that $a > 0$. Evaluate $\int_0^a x \, dx$ by approximating the region bounded by $y = x$ and the x-axis from 0 to a with rectangles. Use equal subintervals and take right endpoints. (*Hint:* Use the result in Example 4 to evaluate the sum of the areas of the rectangles.)

31. Assume that $0 < a < b < \infty$. Use a geometric argument to show that

$$\int_a^b x \, dx = \frac{b^2 - a^2}{2}$$

32. Assume that $0 < a < b < \infty$. Use a geometric argument and Example 1 to show that

$$\int_a^b x^2 \, dx = \frac{b^3 - a^3}{3}$$

Express the limits in Problems 33–39 as definite integrals. Note that $P = [x_0, x_1, \ldots, x_n]$ is a partition of the indicated interval, $c_k \in [x_{k-1}, x_k]$, and $\Delta x_k = x_k - x_{k-1}$.

33. $\displaystyle \lim_{\|P\| \to 0} \sum_{k=1}^{n} c_k^2 \Delta x_k$, where P is a partition of $[0, 1]$

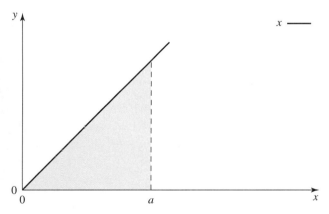

▲ **Figure 5.23**
The region for Problem 30.

34. $\displaystyle \lim_{\|P\| \to 0} \sum_{k=1}^{n} \sqrt{c_k} \Delta x_k$, where P is a partition of $[1, 4]$

35. $\displaystyle \lim_{\|P\| \to 0} \sum_{k=1}^{n} (2c_k - 1) \Delta x_k$, where P is a partition of $[-3, 2]$

36. $\displaystyle \lim_{\|P\| \to 0} \sum_{k=1}^{n} \frac{1}{c_k + 1} \Delta x_k$, where P is a partition of $[1, 2]$

37. $\displaystyle \lim_{\|P\| \to 0} \sum_{k=1}^{n} (c_k - 1)(c_k + 2) \Delta x_k$, where P is a partition of $[-3, 3]$

38. $\displaystyle \lim_{\|P\| \to 0} \sum_{k=1}^{n} (\sin c_k) \Delta x_k$, where P is a partition of $[0, \pi]$

39. $\displaystyle \lim_{\|P\| \to 0} \sum_{k=1}^{n} e^{c_k} \Delta x_k$, where P is a partition of $[-5, 2]$

In Problems 40–45, express the definite integrals as limits of Riemann sums.

40. $\displaystyle \int_{-2}^{-1} \frac{x}{1 - x} \, dx$ **41.** $\displaystyle \int_{2}^{6} \sqrt{x + 1} \, dx$

42. $\displaystyle \int_{1}^{3} e^{-x} \, dx$ **43.** $\displaystyle \int_{1}^{e} \ln x \, dx$

44. $\displaystyle \int_{0}^{\pi} \cos(2x) \, dx$

45. $\displaystyle \int_{0}^{5} g(x) \, dx$, where $g(x)$ is a continuous function on $[0, 5]$

In Problems 46–52, use a graph to interpret the definite integral in terms of areas. Do not compute the integrals.

46. $\displaystyle \int_{-2}^{3} (x - 3) \, dx$ **47.** $\displaystyle \int_{-1}^{2} (x^2 - 1) \, dx$

48. $\displaystyle \int_{-2}^{2} \frac{1}{2} x^3 \, dx$ **49.** $\displaystyle \int_{0}^{5} e^{-x} \, dx$

50. $\displaystyle \int_{-\pi}^{\pi} \cos x \, dx$ **51.** $\displaystyle \int_{1/2}^{4} \ln x \, dx$

52. $\displaystyle \int_{-3}^{2} \left(1 - \frac{1}{2} x\right) dx$

In Problems 53–59, find the value of each integral by interpreting it as the (signed) area under the graph of an appropriately chosen function, and using an area formula from geometry.

53. $\displaystyle \int_{-2}^{3} |x| \, dx$ **54.** $\displaystyle \int_{-3}^{3} \sqrt{9 - x^2} \, dx$

55. $\displaystyle \int_{2}^{5} \left(\frac{1}{2} x - 4\right) dx$ **56.** $\displaystyle \int_{1/2}^{1} \sqrt{1 - x^2} \, dx$

57. $\int_{-2}^{2} (\sqrt{4-x^2} - 2) \, dx$ **58.** $\int_{0}^{1} \sqrt{2-x^2} \, dx$

59. $\int_{-3}^{0} (4 - \sqrt{9-x^2}) \, dx$

[5.1.3]

60. Given that

$$\int_{0}^{a} x^2 \, dx = \frac{1}{3} a^3$$

evaluate the following.

(a) $\int_{0}^{2} \frac{1}{2} x^2 \, dx$ **(b)** $\int_{-3}^{0} 2x^2 \, dx$ **(c)** $\int_{1}^{3} \frac{1}{3} x^2 \, dx$

(d) $\int_{1}^{1} 3x^2 \, dx$ **(e)** $\int_{-2}^{3} \frac{3}{2} x^2 \, dx$ **(f)** $\int_{2}^{4} (x-2)^2 \, dx$

61. Find $\int_{2}^{2} \cos(3z^2) \, dx$.   **62.** Find $\int_{-3}^{-3} e^{-x^2/2} \, dx$.

63. Find $\int_{-2}^{2} (x+1) \, dx$.   **64.** Find $\int_{0}^{-5} (1-x) \, dx$.

65. Find $\int_{-1}^{1} \tan x \, dx$.

66. Explain geometrically why

$$\int_{1}^{2} x^2 \, dx = \int_{0}^{2} x^2 \, dx - \int_{0}^{1} x^2 \, dx \qquad (5.4)$$

and show that (5.4) can be written as

$$\int_{1}^{2} x^2 \, dx = \int_{1}^{0} x^2 \, dx + \int_{0}^{2} x^2 \, dx \qquad (5.5)$$

Relate (5.5) to addition property (5).

In Problems 67–71, verify each inequality without evaluating the integrals.

67. $\int_{0}^{1} x \, dx \geq \int_{0}^{1} x^2 \, dx$ **68.** $\int_{1}^{2} x \, dx \leq \int_{1}^{2} x^2 \, dx$

69. $0 \leq \int_{0}^{4} \sqrt{x} \, dx \leq 8$ **70.** $\frac{1}{2} \leq \int_{0}^{1} \sqrt{1-x^2} \, dx \leq 1$

71. $\frac{\pi}{3} \leq \int_{\pi/6}^{5\pi/6} \sin x \, dx \leq \frac{2\pi}{3}$

72. Find the value of $a \geq 0$ that maximizes $\int_{0}^{a} (4-x^2) \, dx$.

73. Find the value of $a \in [0, 2\pi]$ that maximizes $\int_{0}^{a} \cos x \, dx$.

74. Find $a \in (0, 2\pi]$ such that

$$\int_{0}^{a} \sin x \, dx = 0$$

75. Find $a > 1$ such that

$$\int_{1}^{a} (x-2)^3 \, dx = 0$$

76. Find $a > 0$ such that

$$\int_{-a}^{a} (1 - |x|) \, dx = 0$$

◆ 5.2 THE FUNDAMENTAL THEOREM OF CALCULUS

In Section 5.1, we used the definition of definite integrals to compute $\int_{0}^{a} x^2 \, dx$. This required the summation of a large number of terms, which was facilitated by the explicit summation formula for $\sum_{k=1}^{n} k^2$. Fermat and others were able to carry out similar calculations for the area under curves of the form $y = x^r$, where r was a rational number different from -1. The solution to the case $r = -1$ was found by Belgian mathematician Gregory of St. Vincent (1584–1667) and published in 1647. At that time, it seemed that methods specific to a given function needed to be developed to compute the area under the curve of that function. This would not have been practical.

Fortunately, it turns out that the area problem is related to the tangent problem. This is not at all obvious; among the first to notice this relationship were Isaac Barrow (1630–1677) and James Gregory (1638–1675). Each presented this relationship in geometrical terms, without realizing the importance of their discovery.

Both Newton and Leibniz are to be credited with systematically developing the connection between the tangent and area problems, which ultimately resulted in a method for computing areas and solving problems that can be translated into area problems. The result is known as the fundamental theorem of calculus, which says that the tangent and area problems are inversely related.

The fundamental theorem of calculus has two parts; the first part links antiderivatives and integrals, and the second part provides a method for computing definite integrals.

5.2.1 The Fundamental Theorem of Calculus (Part I)

Let $f(x)$ be a continuous function on $[a, b]$, and let

$$F(x) = \int_a^x f(u)\,du$$

Geometrically, $F(x) = \int_a^x f(u)\,du$ represents the signed area between the graph of $f(u)$ and the horizontal axis between a and x (see Figure 5.24). Note that the independent variable x appears as the upper limit of integration. We can now ask how the signed area $F(x)$ changes as x varies. To answer this question, we compute $\frac{dF}{dx}$ using the definition of the derivative. That is,

$$
\begin{aligned}
\frac{d}{dx}F(x) &= \lim_{h \to 0} \frac{F(x+h) - F(x)}{h} \\
&= \lim_{h \to 0} \frac{1}{h}\left[\int_a^{x+h} f(u)\,du - \int_a^x f(u)\,du\right] \qquad (5.6)\\
&= \lim_{h \to 0} \frac{1}{h}\int_x^{x+h} f(u)\,du
\end{aligned}
$$

(In the last step, we used property (5) of Subsection 5.1.3.) To evaluate

$$\lim_{h \to 0} \frac{1}{h}\int_x^{x+h} f(u)\,du$$

we will resort to the geometric interpretation of definite integrals. The following argument is illustrated in Figure 5.25. Note that

$$\int_x^{x+h} f(u)\,du$$

is the signed area of the region bounded by the graph of $f(u)$ and the horizontal axis between x and $x + h$. If h is small, then this area is closely approximated by the area of the inscribed rectangle with height $|f(x)|$. The signed area of this rectangle is $f(x)h$. Hence,

$$\int_x^{x+h} f(u)\,du \approx f(x)h$$

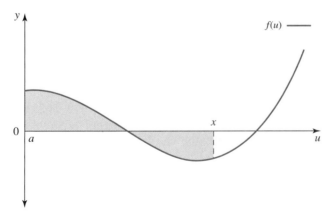

◀ **Figure 5.24**

The shaded signed area is $F(x)$.

◄ **Figure 5.25**

The approximate rectangle in the Fundamental Theorem of Calculus.

If we divide both sides by h and then let h tend to 0, we find

$$\lim_{h \to 0} \frac{1}{h} \int_x^{x+h} f(u)\, du = f(x) \tag{5.7}$$

Combining (5.6) and (5.7), we arrive at the remarkable result

$$\frac{d}{dx} F(x) = f(x)$$

In addition, we see that $F(x)$ is continuous, since it is differentiable.

The preceding argument relies on geometric intuition. To show how one can make this argument mathematically rigorous, we give the complete proof of this result at the end of this subsection. It is summarized in the following theorem.

> **The Fundamental Theorem of Calculus (Part I)**   If f is continuous on $[a, b]$, then the function F defined by
>
> $$F(x) = \int_a^x f(u)\, du, \quad a \le x \le b$$
>
> is continuous on $[a, b]$ and differentiable on (a, b), with
>
> $$\frac{d}{dx} F(x) = f(x)$$

Simply stated, the FTC (part I) says that if we first integrate $f(x)$ and then differentiate the result, we get $f(x)$ again. In this sense, it shows that integration and differentiation are inverse operations.

We begin with an example that can be immediately solved using the fundamental theorem of calculus (FTC).

❙ **EXAMPLE 1**   Compute

$$\frac{d}{dx} \int_0^x (\sin u - e^{-u})\, du$$

for $x > 0$.

Solution   First, note that $f(x) = \sin x - e^{-x}$ is continuous for $x \ge 0$. If we set $F(x) = \int_0^x (\sin u - e^{-u})\, du$ and apply the FTC, then

$$\frac{d}{dx} F(x) = \frac{d}{dx} \int_0^x (\sin u - e^{-u})\, du = \sin x - e^{-x}$$ ❙

In the next example, the upper limit of integration is a function of x.

I EXAMPLE 2   Compute

$$\frac{d}{dx} \int_0^{x^2} (u^3 - 2) \, du, \quad x > 0$$

Solution   We set $F(v) = \int_0^v (u^3 - 2) \, du$, $v > 0$. Then, for $x > 0$,

$$F(x^2) = \int_0^{x^2} (u^3 - 2) \, du$$

We wish to compute $\frac{d}{dx} F(x^2)$. To do so, we need to apply the chain rule. We set $v(x) = x^2$, then

$$\frac{d}{dx} F(x^2) = \frac{dF(v)}{dv} \frac{dv}{dx}$$

To evaluate $\frac{d}{dv} F(v) = \frac{d}{dv} \int_0^v (u^3 - 2) \, du$, we use the FTC:

$$\frac{d}{dv} \int_0^v (u^3 - 2) \, du = v^3 - 2$$

Since $\frac{dv}{dx} = \frac{d}{dx}(x^2) = 2x$, we therefore find that

$$\frac{d}{dx} F(x^2) = (v^3 - 2)2x = [(x^2)^3 - 2]2x$$
$$= (x^6 - 2)2x \qquad \blacksquare$$

Thus far, we have dealt only with the case where the upper limit of integration depends on x. The following example shows what we must do when the lower limit of integration depends on x.

I EXAMPLE 3   Compute

$$\frac{d}{dx} \int_{\sin x}^1 u^2 \, du$$

Solution   We use the fact that

$$\int_{\sin x}^1 u^2 \, du = -\int_1^{\sin x} u^2 \, du$$

The upper limit now depends on x, but we introduced a minus sign. Hence

$$\frac{d}{dx} \int_{\sin x}^1 u^2 \, du = -\frac{d}{dx} \int_1^{\sin x} u^2 \, du$$
$$= -(\sin x)^2 \cos x \qquad \blacksquare$$

The preceding example makes an important point: We need to be careful about whether the upper or the lower limit of integration depends on x. In the following example, we show what we must do when both limits of integration depend on x.

I EXAMPLE 4   For $x \in \mathbf{R}$, compute

$$\frac{d}{dx} \int_{x^2}^{x^3} e^u \, du$$

Solution   Note that $f(u) = e^u$ is continuous for all $u \in \mathbf{R}$. The given integral is therefore defined for all $x \in \mathbf{R}$, and we can split it into two integrals at any $a \in \mathbf{R}$.

We choose $a = 0$, which yields

$$\int_{x^2}^{x^3} e^u\,du = \int_{x^2}^{0} e^u\,du + \int_{0}^{x^3} e^u\,du = -\int_{0}^{x^2} e^u\,du + \int_{0}^{x^3} e^u\,du$$

The right-hand side is now written in a form that we know how to differentiate, and we find that

$$\frac{d}{dx}\int_{x^2}^{x^3} e^u\,du = -\frac{d}{dx}\int_{0}^{x^2} e^u\,du + \frac{d}{dx}\int_{0}^{x^3} e^u\,du$$

$$= -\left[e^{x^2}\frac{d}{dx}x^2\right] + \left[e^{x^3}\frac{d}{dx}x^3\right]$$

$$= -e^{x^2}2x + e^{x^3}3x^2$$

The preceding example illustrates the most general case that we can encounter, namely, when both limits of integration depend on x. We summarize this case in the following box, known as Leibniz's Rule.

LEIBNIZ'S RULE

If $g(x)$ and $h(x)$ are differentiable functions and $f(u)$ is continuous for u between $g(x)$ and $h(x)$, then

$$\frac{d}{dx}\int_{g(x)}^{h(x)} f(u)\,du = f[h(x)]h'(x) - f[g(x)]g'(x)$$

We can check that Examples 1–4 can be solved using this formula; for instance, in Example 3, we have $f(u) = u^2$, $g(x) = \sin x$, and $h(x) = 1$. Then $g'(x) = \cos x$ and $h'(x) = 0$. We therefore find that

$$f[h(x)]h'(x) - f[g(x)]g'(x) = 0 - (\sin x)^2 \cos x$$

which is the answer we found in Example 3.

5.2.2 Antiderivatives and Indefinite Integrals

The first part of the fundamental theorem of calculus tells us that if

$$F(x) = \int_{a}^{x} f(u)\,du$$

then $F'(x) = f(x)$ (provided $f(x)$ is continuous over the range of integration). This says that $F(x)$ is an antiderivative of $f(x)$ (we introduced antiderivatives in Section 4.7). Now, if we let

$$F(x) = \int_{a}^{x} f(u)\,du \qquad \text{and} \qquad G(x) = \int_{b}^{x} f(u)\,du$$

where a and b are two numbers, then both integrals have the same derivative, namely $F'(x) = G'(x) = f(x)$ (provided $f(x)$ is continuous over the range of integration). That is, both $F(x)$ and $G(x)$ are antiderivatives of $f(x)$. We saw in Section 4.7 that antiderivatives of a given function differ only by a constant; we can check this for $F(x)$ and $G(x)$, namely,

$$F(x) = \int_{a}^{x} f(u)\,du = \int_{a}^{b} f(u)\,du + \int_{b}^{x} f(u)\,du = C + G(x)$$

where C is a constant, denoting the number $\int_{a}^{b} f(u)\,du$.

The general antiderivative of a function $f(x)$ is $F(x) + C$, where $F'(x) = f(x)$ and C is a constant. We therefore see that $C + \int_a^x f(u)\,du$ is the general antiderivative of $f(x)$. We will use the notation $\int f(x)\,dx$ to denote both the general antiderivative of $f(x)$ and the function $C + \int_a^x f(u)\,du$, that is,

$$\int f(x)\,dx = C + \int_a^x f(u)\,du \qquad (5.8)$$

We call $\int f(x)\,dx$ an **indefinite integral**. Thus, the first part of the FTC says that indefinite integrals and antiderivatives are the same.

When we write $\int_a^x f(u)\,du$, we use a letter other than x in the integrand because x already appears as the upper limit of integration. However, in the symbolic notation $\int f(x)\,dx$ we write x. This is to be interpreted as in (5.8). The notation $\int f(x)\,dx$ is a convenient shorthand for $C + \int_a^x f(u)\,du$. The choice of value of a for the lower limit of integration on the right side of (5.8) is not important, because different indefinite integrals of the same function $f(x)$ differ only by an additive constant that can be absorbed in the constant C.

Examples 5–7 show how to compute indefinite integrals.

▎ EXAMPLE 5   Compute $\int x^4\,dx$.

Solution   We need to find a function $F(x)$ such that $F'(x) = x^4$. The solution is

$$\int x^4\,dx = \frac{1}{5}x^5 + C$$

where C is a constant. We check that indeed

$$\frac{d}{dx}\left(\frac{1}{5}x^5 + C\right) = x^4 \qquad \blacksquare$$

▎ EXAMPLE 6   Compute $\int (e^x + \sin x)\,dx$.

Solution   We need to find an antiderivative of $f(x) = e^x + \sin x$. Since

$$\frac{d}{dx}(e^x - \cos x) = e^x - (-\sin x) = e^x + \sin x$$

we find that

$$\int (e^x + \sin x)\,dx = e^x - \cos x + C \qquad \blacksquare$$

When we compute the indefinite integral $\int f(x)\,dx$, we want to know the general antiderivative of $f(x)$; this is why we added the constant C in the previous two examples.

▎ EXAMPLE 7   Show that

$$\int \frac{1}{x}\,dx = \ln |x| + C \quad \text{for } x \neq 0$$

Solution   Since the absolute value of x appears on the right-hand side, we split our discussion into two parts, according to whether $x \geq 0$ or $x \leq 0$. Recall that

$$|x| = \begin{cases} x & \text{for } x \geq 0 \\ -x & \text{for } x \leq 0 \end{cases}$$

Since $\ln x$ is not defined at $x = 0$, we consider the two cases $x > 0$ and $x < 0$.
  (i) $x > 0$: Since $\ln |x| = \ln x$ when $x > 0$, we find

$$\frac{d}{dx} \ln x = \frac{1}{x}$$

and hence

$$\int \frac{1}{x} \, dx = \ln x + C \quad \text{for } x > 0$$

(ii) $x < 0$: Since $\ln |x| = \ln(-x)$ when $x < 0$, we find

$$\frac{d}{dx} \ln(-x) = \frac{1}{-x}(-1) = \frac{1}{x}$$

and hence

$$\int \frac{1}{x} \, dx = \ln(-x) - C \quad \text{for } x < 0$$

Combining (i) and (ii), we obtain

$$\int \frac{1}{x} \, dx = \ln |x| + C \quad \text{for } x \neq 0 \qquad\qquad ▌$$

When we introduced logarithmic and exponential functions in Sections 3.6 and 3.7, we had to resort to the calculator to convince ourselves that e^x and $\ln x$ were indeed the functions we knew from precalculus. To give a mathematically rigorous definition of these functions, one typically starts by *defining* $\ln x$ as $\int_1^x \frac{1}{u} \, du$ and then derives the algebraic rules for $\ln x$ from this integral representation. The exponential function e^x is defined as the inverse function of $\ln x$, and the number e is defined so that $\ln e = 1$. This definition is then consistent with the definition in Section 3.6.

We have seen that in order to evaluate indefinite integrals, we must find antiderivatives. We provide a list of indefinite integrals in Table 5.1 (this is a slightly expanded form of the table of antiderivatives from Section 4.7). Examples 8–9 show how to use this list to compute indefinite integrals.

▌EXAMPLE 8   Evaluate

$$\int \frac{1}{\sin^2 x - 1} \, dx$$

TABLE 5•1 A COLLECTION OF INDEFINITE INTEGRALS

$$\int x^n \, dx = \frac{x^{n+1}}{n+1} + C \quad (n \neq -1) \qquad \int \frac{1}{x} \, dx = \ln |x| + C$$

$$\int e^x \, dx = e^x + C \qquad\qquad \int a^x \, dx = \frac{a^x}{\ln a} + C$$

$$\int \cos x \, dx = \sin x - C \qquad\qquad \int \sin x \, dx = -\cos x + C$$

$$\int \sec^2 x \, dx = \tan x + C \qquad\qquad \int \csc^2 x \, dx = -\cot x + C$$

$$\int \sec x \tan x \, dx = \sec x + C \qquad \int \csc x \cot x \, dx = -\csc x + C$$

$$\int \tan x \, dx = \ln |\sec x| + C \qquad \int \cot x \, dx = -\ln |\csc x| + C$$

$$\int \frac{1}{1 + x^2} \, dx = \tan^{-1} x + C \qquad \int \frac{1}{\sqrt{1 - x^2}} \, dx = \sin^{-1} x + C$$

Solution   We first work on the integrand. Using the fact that $\sin^2 x + \cos^2 x = 1$, we find

$$\frac{1}{\sin^2 x - 1} = -\frac{1}{\cos^2 x} = -\sec^2 x$$

Hence,

$$\int \frac{1}{\sin^2 x - 1}\, dx = -\int \sec^2 x\, dx = -\tan x + C \qquad \blacksquare$$

EXAMPLE 9   Evaluate

$$\int \frac{x^2}{x^2 + 1}\, dx$$

Solution   We will first rewrite the integrand. Either by using long division, or by the following algebraic manipulation, we find that

$$\frac{x^2}{x^2 + 1} = \frac{x^2 + 1 - 1}{x^2 + 1} = \frac{x^2 + 1}{x^2 + 1} - \frac{1}{x^2 + 1} = 1 - \frac{1}{x^2 + 1}$$

Hence, using the table of indefinite integrals,

$$\int \frac{x^2}{x^2 + 1}\, dx = \int \left(1 - \frac{1}{x^2 + 1}\right) dx = x - \tan^{-1} x + C \qquad \blacksquare$$

Proof of the Fundamental Theorem of Calculus (Part I)   At the beginning of this subsection, we found that if

$$F(x) = \int_a^x f(u)\, du$$

then

$$\frac{d}{dx} F(x) = \lim_{h \to 0} \frac{1}{h} \int_x^{x+h} f(u)\, du \qquad (5.9)$$

We will now give a mathematically rigorous argument that will show that

$$\frac{d}{dx} F(x) = f(x)$$

We begin with the observation that the continuous function $f(u)$ defined on the closed interval $[x, x + h]$ attains an absolute minimum and an absolute maximum on $[x, x + h]$, according to the extreme value theorem. That is, there exist m and M such that m is the minimum of f on $[x, x + h]$ and M is the maximum of f on $[x, x + h]$, which implies that

$$m \le f(u) \le M \quad \text{for all } u \in [x, x + h] \qquad (5.10)$$

Of course, m and M depend on both x and h. Applying property (8) from Subsection 5.1.3 to (5.10), we find

$$\int_x^{x+h} m\, du \le \int_x^{x+h} f(u)\, du \le \int_x^{x+h} M\, du$$

and hence

$$mh \le \int_x^{x+h} f(u)\, du \le Mh$$

Dividing by h, we obtain

$$m \leq \frac{1}{h} \int_x^{x+h} f(u) \, du \leq M \tag{5.11}$$

We set

$$I = \frac{1}{h} \int_x^{x+h} f(u) \, du$$

then (5.11) reads $m \leq I \leq M$. That is, I is a number between m and M. We compare this with (5.10). which says that $f(u)$ also lies between m, the minimum of f on $[x, x+h]$, and M, the maximum of f on $[x, x+h]$, for all $u \in [x, x+h]$. The intermediate value theorem applied to $f(u)$ tells us that any value between m and M is attained by $f(u)$ for some number in the interval $[x, x+h]$. Specifically, there must exist a number $c_h \in [x, x+h]$ such that $f(c_h) = I$, since I lies between m and M; that is,

$$f(c_h) = \frac{1}{h} \int_x^{x+h} f(u) \, du \tag{5.12}$$

Since $x \leq c_h \leq x + h$, it follows that

$$\lim_{h \to 0} c_h = x$$

Since f is continuous,

$$\lim_{h \to 0} f(c_h) = f(\lim_{h \to 0} c_h) = f(x) \tag{5.13}$$

Combining (5.9), (5.12), and (5.13), we find the result

$$\frac{d}{dx} F(x) = f(x) \qquad \blacklozenge$$

5.2.3 The Fundamental Theorem of Calculus (Part II)

The first part of the fundamental theorem of calculus only allows us to compute integrals of the form $\int_a^x f(u) \, du$ up to an additive constant, for instance,

$$F(x) = \int_1^x u^2 \, du = \frac{1}{3} x^3 + C$$

To evaluate the definite integral $F(2) = \int_1^2 u^2 \, du$, which represents the area under the graph of $f(x) = x^2$ between $x = 1$ and $x = 2$, we would need to know the value of the constant. This value is provided by the second part of the fundamental theorem of calculus, which allows us to evaluate definite integrals. The following calculation shows us how this constant is determined.

We saw in the last subsection that if we set

$$G(x) = \int_a^x f(u) \, du$$

then $G(x)$ is an antiderivative of $f(x)$. Furthermore, if $F(x)$ is another antiderivative of $f(x)$, then $G(x)$ and $F(x)$ differ only by an additive constant. That is,

$$G(x) = F(x) + C$$

where C is a constant. Now,

$$G(a) = \int_a^a f(u) \, du = 0$$

hence,

$$0 = G(a) = F(a) + C$$

which implies that $C = -F(a)$ and, therefore, $G(x) = F(x) - F(a)$, or, using the integral representation of $G(x)$,

$$\int_a^x f(u)\,du = F(x) - F(a)$$

If we set $x = b$, then

$$\int_a^b f(u)\,du = F(b) - F(a)$$

This formula allows us to evaluate definite integrals, and is the content of the second part of the fundamental theorem of calculus.

> **The Fundamental Theorem of Calculus (Part II)**   Assume that f is continuous on $[a, b]$; then
>
> $$\int_a^b f(x)\,dx = F(b) - F(a)$$
>
> where $F(x)$ is an antiderivative of $f(x)$, that is, $F'(x) = f(x)$.

So, how do we use this theorem? To compute the definite integral $\int_a^b f(x)\,dx$, when f is continuous on $[a, b]$, we first need to find an antiderivative $F(x)$ of $f(x)$ (any antiderivative will do), and then compute $F(b) - F(a)$. This number is then equal to $\int_a^b f(x)\,dx$. The table of indefinite integrals in the previous subsection will help us to find such antiderivatives. Note that an indefinite integral is a function, whereas a definite integral is simply a number.

Using the FTC (part II) to evaluate definite integrals

EXAMPLE 10   Evaluate $\int_{-1}^2 (x^2 - 3x)\,dx$.

Solution   Note that $f(x) = x^2 - 3x$ is continuous on $[-1, 2]$. We need to find an antiderivative of $f(x) = x^2 - 3x$; for instance, $F(x) = \frac{1}{3}x^3 - \frac{3}{2}x^2$ is an antiderivative of $f(x)$, since $F'(x) = f(x)$. We then must evaluate $F(2) - F(-1)$:

$$F(2) = \frac{1}{3}2^3 - \frac{3}{2}2^2 = \frac{8}{3} - 6 = -\frac{10}{3}$$

$$F(-1) = \frac{1}{3}(-1)^3 - \frac{3}{2}(-1)^2 = -\frac{1}{3} - \frac{3}{2} = -\frac{11}{6}$$

We find $F(2) - F(-1) = -\frac{10}{3} - \left(-\frac{11}{6}\right) = -\frac{9}{6} = -\frac{3}{2}$ and, therefore,

$$\int_{-1}^2 (x^2 - 3x)\,dx = -\frac{3}{2}$$

EXAMPLE 11   Evaluate $\int_0^\pi \sin x\,dx$.

Solution   Note that $\sin x$ is continuous on $[0, \pi]$. Since $F(x) = -\cos x$ is an antiderivative of $\sin x$,

$$\int_0^\pi \sin x\,dx = F(\pi) - F(0) = -\cos \pi - (-\cos 0) = -(-1) + 1 = 2$$

In the preceding two examples, we chose the simplest antiderivative, that is, the one where the constant was equal to 0. We could have chosen $C \neq 0$, and the answers would have been the same. We illustrate this in the following example.

EXAMPLE 12   Evaluate

$$\int_1^5 \frac{1}{x}\, dx$$

Solution   Note that $\frac{1}{x}$ is continuous on $[1, 5]$. Now, $\ln|x|$ is an antiderivative of $\frac{1}{x}$; the general antiderivative of $\frac{1}{x}$ is therefore $\ln|x| + C$. We use this antiderivative to evaluate the integral, and find

$$\int_1^5 \frac{1}{x}\, dx = (\ln|5| + C) - (\ln|1| + C) = \ln 5$$

since $\ln 1 = 0$ and the constant C cancelled out.   █

It follows from the calculation in the previous example that the value of the constant C is unimportant; it simply cancelled out. Therefore, we can always choose $C = 0$.

We introduce additional notation. If $F(x)$ is an antiderivative of $f(x)$, we write

$$\int_a^b f(x)\, dx = F(x)\big]_a^b = F(b) - F(a)$$

For instance,

$$\int_1^5 \frac{1}{x}\, dx = \ln|x|\,]_1^5 = \ln|5| - \ln|1|$$

The notation $F(x)]_a^b$ indicates that we evaluate the antiderivative $F(x)$ at b and a, respectively, and compute the difference $F(b) - F(a)$.

EXAMPLE 13   Evaluate

$$\int_0^3 2xe^{x^2}\, dx$$

Solution   Observe that $2xe^{x^2}$ is continuous on $[0, 3]$, and that $F(x) = e^{x^2}$ is an antiderivative of $f(x) = 2xe^{x^2}$ since applying the chain rule, we find

$$F'(x) = e^{x^2}\left(\frac{d}{dx}x^2\right) = e^{x^2}2x$$

Therefore,

$$\int_0^3 2xe^{x^2}\, dx = e^{x^2}\Big]_0^3 = e^9 - e^0 = e^9 - 1$$ █

EXAMPLE 14   Evaluate

$$\int_1^4 \frac{2x^2 - 3x + \sqrt{x}}{\sqrt{x}}\, dx$$

Solution   The integrand is continuous on $[1, 4]$. We first simplify the integrand:

$$f(x) = \frac{2x^2 - 3x + \sqrt{x}}{\sqrt{x}} = 2x^{3/2} - 3\sqrt{x} + 1$$

An antiderivative of $f(x)$ is, therefore,

$$F(x) = 2 \cdot \frac{2}{5}x^{5/2} - 3 \cdot \frac{2}{3}x^{3/2} + x = \frac{4}{5}x^{5/2} - 2x^{3/2} + x$$

which can be checked by differentiating $F(x)$. We can now evaluate the integral:

$$\int_1^4 \frac{2x^2 - 3x + \sqrt{x}}{\sqrt{x}}\,dx = \frac{4}{5}x^{5/2} - 2x^{3/2} + x \Big]_1^4$$

$$= \left(\frac{4}{5} \cdot 4^{5/2} - 2 \cdot 4^{3/2} + 4\right) - \left(\frac{4}{5} \cdot 1^{5/2} - 2 \cdot 1^{3/2} + 1\right)$$

$$= \left(\frac{4}{5} \cdot 32 - (2)(8) + 4\right) - \left(\frac{4}{5} - 2 + 1\right)$$

$$= \frac{68}{5} - \left(-\frac{1}{5}\right) = \frac{69}{5}$$

Finding an integrand

I EXAMPLE 15   Suppose that

$$\int_0^x f(t)\,dt = \cos(2x) + a$$

where a is a constant. Find $f(x)$ and a.

Solution   We solve this problem in two steps. First, we use the FTC, part I, to conclude that

$$\frac{d}{dx}\int_0^x f(t)\,dt = f(x)$$

Hence,

$$f(x) = \frac{d}{dx}[\cos(2x) + a] = -2\sin(2x)$$

In the second step, we use the FTC, part II, to determine a. Namely,

$$\int_0^x (-2\sin(2t))\,dt = \cos(2t)]_0^x$$
$$= \cos(2x) - \cos(0) = \cos(2x) - 1$$

We conclude that $a = -1$.

You might wonder why we always check that the integrand is continuous on the interval between the lower and the upper limit of integration. The following is an example of what can go wrong when the integrand is discontinuous. Try to evaluate

$$\int_{-2}^1 \frac{1}{x^2}\,dx$$

An antiderivative of $f(x)$ is $F(x) = -\frac{1}{x}$. We find $F(1) = -1$ and $F(-2) = \frac{1}{2}$. When we compute $F(1) - F(-2)$, we get $-\frac{3}{2}$. This is obviously not $\int_{-2}^1 \frac{1}{x^2}\,dx$ since $f(x) = \frac{1}{x^2}$ is positive on $[-2, 1]$ (see Figure 5.26) and, therefore, the integral of $f(x)$ between -2 and 1 should not be negative. The function $f(x)$ is discontinuous at $x = 0$ (it has a vertical asymptote at $x = 0$). The second part of the fundamental theorem of calculus therefore cannot be applied. We will learn how to deal with such discontinuities (see Section 6.4). In any case, before you evaluate an integral, always check whether the integrand is continuous between the limits of integration.

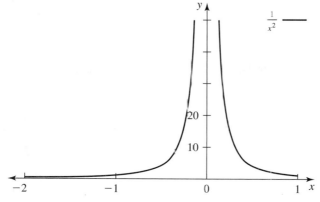

◀ **Figure 5.26**
The graph of $y = \frac{1}{x^2}$ between $x = -2$ and $x = 1$. The function is discontinuous at $x = 0$.

◆ 5.2 PROBLEMS

[5.2.1]

In Problems 1–10, find $\frac{dy}{dx}$.

1. $y = \int_0^x \sqrt{2 + u}\, du$, $x > 0$

2. $y = \int_1^x u e^{-u^2}\, du$

3. $y = \int_{-2}^x \frac{1}{u + 3}\, du$, $x > -2$

4. $y = \int_{\pi/2}^x \sin(u^2 + 1)\, du$

5. $y = \int_0^{x^2} \sqrt{t}\, dt$, $x > 0$

6. $y = \int_2^{\ln x} e^{-t}\, dt$, $x > 0$

7. $y = \int_{-x}^x \tan u\, du$, $0 < x < \frac{\pi}{4}$

8. $y = \int_{x^2}^{x^3} \ln t\, dt$, $x > 0$

9. $y = \int_x^5 \frac{1}{u^2}\, du$, $x > 0$

10. $y = \int_{x^2}^1 \sec t\, dt$, $-1 < x < 1$

[5.2.2]

In Problems 11–36, compute the indefinite integrals.

11. $\int (3x^2 - 2x)\, dx$

12. $\int \left(\frac{1}{2}x^5 + 2x^3 - 1 \right)\, dx$

13. $\int \frac{2x^2 - x}{\sqrt{x}}\, dx$

14. $\int x^2 \sqrt{x}\, dx$

15. $\int (x^{7/2} + x^{2/7})\, dx$

16. $\int \left(\sqrt{x} + \frac{1}{\sqrt{x}} \right)\, dx$

17. $\int (x - 1)(x + 1)\, dx$

18. $\int (x - 1)^2\, dx$

19. $\int e^{2x}\, dx$

20. $\int 3e^{-x}\, dx$

21. $\int x e^{-x^2/2}\, dx$

22. $\int e^x (1 - e^{-x})\, dx$

23. $\int \sin(2x)\, dx$

24. $\int \cos(3x)\, dx$

25. $\int \frac{\sin x}{1 - \sin^2 x}\, dx$

26. $\int \sec^2(3x)\, dx$

27. $\int (\sec^2 x + \tan x)\, dx$

28. $\int (\cot x - \csc^2 x)\, dx$

29. $\int \frac{4}{1 + x^2}\, dx$

30. $\int \frac{1}{x - 1}\, dx$

31. $\int \frac{x - 1}{x}\, dx$

32. $\int \frac{x + 3}{x^2 - 9}\, dx$

33. $\int \frac{5x^2}{x^2 + 1}\, dx$

34. $\int 3^x\, dx$

35. $\int (x^2 + 2^x)\, dx$

36. $\int (\sqrt{x} + \sqrt{e^x})\, dx$

[5.2.3]

In Problems 37–62, evaluate the definite integrals.

37. $\int_2^4 (3 - 2x)\, dx$

38. $\int_{-1}^3 (2x^2 - 1)\, dx$

39. $\int_0^1 (x^2 - \sqrt{x})\, dx$

40. $\int_1^2 x^{5/2}\, dx$

41. $\displaystyle\int_1^8 x^{-2/3}\,dx$

42. $\displaystyle\int_4^9 \frac{1+\sqrt{x}}{\sqrt{x}}\,dx$

43. $\displaystyle\int_0^2 (2t-1)(t+3)\,dt$

44. $\displaystyle\int_{-1}^2 (4-3t)^2\,dt$

45. $\displaystyle\int_0^{\pi/4} \sin(2x)\,dx$

46. $\displaystyle\int_{-\pi/3}^{\pi/3} 2\cos\left(\frac{x}{2}\right)\,dx$

47. $\displaystyle\int_0^{\pi/8} \sec^2(2x)\,dx$

48. $\displaystyle\int_{-\pi/4}^{\pi/4} \tan x\,dx$

49. $\displaystyle\int_0^1 \frac{1}{1+x^2}\,dx$

50. $\displaystyle\int_{-\sqrt{3}}^{-1} \frac{4}{1+x^2}\,dx$

51. $\displaystyle\int_0^{1/2} \frac{1}{\sqrt{1-x^2}}\,dx$

52. $\displaystyle\int_{-1/2}^{1/2} \frac{3}{\sqrt{1-x^2}}\,dx$

53. $\displaystyle\int_0^{\pi/6} \tan(2x)\,dx$

54. $\displaystyle\int_{\pi/20}^{\pi/15} \sec(5x)\tan(5x)\,dx$

55. $\displaystyle\int_{-1}^0 e^{3x}\,dx$

56. $\displaystyle\int_0^2 2te^{t^2}\,dt$

57. $\displaystyle\int_{-1}^1 |x|\,dx$

58. $\displaystyle\int_{-1}^1 e^{-|s|}\,ds$

59. $\displaystyle\int_1^e \frac{1}{x}\,dx$

60. $\displaystyle\int_2^3 \frac{1}{z+1}\,dz$

61. $\displaystyle\int_{-2}^{-1} \frac{1}{1-u}\,du$

62. $\displaystyle\int_2^3 \frac{2}{t-1}\,dt$

63. Use l'Hospital's Rule to compute
$$\lim_{x\to 0} \frac{1}{x^2}\int_0^x \sin t\,dt$$

64. Use l'Hospital's Rule to compute
$$\lim_{h\to 0} \frac{1}{h}\int_0^h e^x\,dx$$

65. Suppose that
$$\int_0^x f(t)\,dt = 2x^2$$
Find $f(x)$.

66. Suppose that
$$\int_0^x f(t)\,dt = \frac{1}{2}\tan(2x)$$
Find $f(x)$.

◆ 5.3 APPLICATIONS OF INTEGRATION

In this section, we will discuss a number of applications of integrals. In the first application, we will revisit the interpretation of integrals as areas; the second interprets integrals as cumulative (or net) change; the third will allow us to compute averages using integrals; and, finally, we will use integrals to compute volumes. In each application, you will see that integrals can be interpreted as "sums of many small increments."

5.3.1 Areas

The first application is already familiar to us. If f is a nonnegative, continuous function on $[a, b]$, then

$$A = \int_a^b f(x)\,dx$$

represents the area of the region bounded by the graph of $f(x)$ between a and b, the vertical lines $x = a$ and $x = b$, and the x-axis between a and b. In all of the preceding examples, one of the boundaries of the region whose area we wanted to know has been the x-axis. We will now discuss how to find the geometric area between two arbitrary curves. We emphasize that we want to compute geometric areas; that is, the areas we compute in this subsection will always be positive.

Suppose that $f(x)$ and $g(x)$ are continuous functions on $[a, b]$. We wish to find the area between the graphs of f and g. We start with a simple example (see Figure 5.27). We assume for the moment that both f and g are nonnegative on $[a, b]$

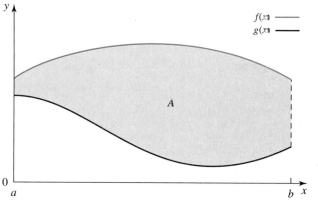

◄ **Figure 5.27**
Computing the area between
two curves.

and that $f(x) \geq g(x)$ on $[a, b]$. We see from Figure 5.27 that

$$A = \begin{bmatrix} \text{area between} \\ f \text{ and } x\text{-axis} \end{bmatrix} - \begin{bmatrix} \text{area between} \\ g \text{ and } x\text{-axis} \end{bmatrix}$$

$$= \int_a^b f(x)\,dx - \int_a^b g(x)\,dx$$

Using property (4) of Subsection 5.1.3, we can write this as

$$A = \int_a^b [f(x) - g(x)]\,dx$$

We obtained this formula under the assumption that both f and g are nonnegative on $[a, b]$; we now show that this assumption is not necessary. To do so, we derive a formula for the area between two curves from Riemann sums.

We assume that f and g are continuous on $[a, b]$ and that $f(x) \geq g(x)$ for all $x \in [a, b]$. We approximate the area between the two curves by rectangles: We divide $[a, b]$ into n equal subintervals, each of length Δx, that is, we set $a = x_0 < x_1 < x_2 < \cdots < x_n = b$ with $x_k - x_{k-1} = (b-a)/n$. The kth subinterval is thus between x_{k-1} and x_k. We choose left endpoints to compute the heights of the approximating rectangles. From Figure 5.28, we see that the height of the kth rectangle is equal to

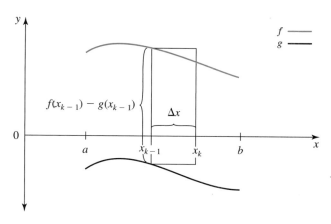

◄ **Figure 5.28**
The kth rectangle is between
x_{k-1} and x_k.

$f(x_{k-1}) - g(x_{k-1})$. We therefore find

$$A = \lim_{n \to \infty} \sum_{k=1}^{n} [f(x_{k-1}) - g(x_{k-1})]\Delta x$$

$$= \int_a^b [f(x) - g(x)]\,dx$$

If f and g are continuous on $[a, b]$, with $f(x) \geq g(x)$ for all $x \in [a, b]$, then the area of the region between the curves $y = f(x)$ and $y = g(x)$ from a to b is equal to

$$\text{Area} = \int_a^b [f(x) - g(x)]\,dx$$

Before looking at a number of examples, we point out once more that this area formula always yields a nonnegative number, since it computes the geometric area. To contrast this with the concept of a signed area, let us consider a function $h(x)$ on $[a, b]$, with $h(x) \leq 0$ for all $x \in [a, b]$ (see Figure 5.29). The definite integral $\int_a^b h(x)\,dx$ represents a signed area, and is negative in this case; more precisely, $\int_a^b h(x)\,dx$ is the negative of the geometric area of the region between the x-axis and the graph of $h(x)$ from $x = a$ to $x = b$. That this is consistent with our definition of area can be seen as follows. The region of interest is bounded by the two curves $y = 0$ and $y = h(x)$. Since $h(x) \leq 0$ for $x \in [a, b]$, the area formula yields

$$\text{Area} = \int_a^b [0 - h(x)]\,dx = -\int_a^b h(x)\,dx$$

which is a positive number.

When you compute the area of the region between two curves, you should always graph the bounding curves. This will show you how to set up the appropriate integral(s).

▮EXAMPLE I   Find the area between the curves $y = \sec^2 x$ and $y = \cos x$, from $x = 0$ to $x = \pi/4$.

Solution   We first graph the bounding curves, as shown in Figure 5.30. We see that $\sec^2 x \geq \cos x$ for $x \in \left[0, \frac{\pi}{4}\right]$. (This should also be clear from the fact that

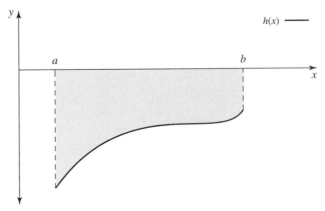

◀ Figure 5.29

The definite integral $\int_a^b h(x)\,dx$ represents a signed area; it is negative here.

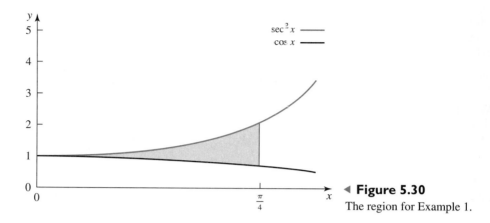

◀ **Figure 5.30**
The region for Example 1.

$\sec x \ge 1$ and $\cos x \le 1$ for $x \in \left[0, \frac{\pi}{4}\right]$.) Therefore,

$$\text{Area} = \int_0^{\pi/4} [\sec^2 x - \cos x]\, dx$$

$$= \tan x - \sin x\big]_0^{\pi/4}$$

$$= \left(\tan \frac{\pi}{4} - \sin \frac{\pi}{4}\right) - (\tan 0 - \sin 0)$$

$$= \left(1 - \frac{1}{2}\sqrt{2}\right) - (0 - 0) = 1 - \frac{1}{2}\sqrt{2}$$

▮

EXAMPLE 2 Find the area of the region enclosed by $y = (x-1)^2 - 1$ and $y = -x + 2$.

Solution The bounding curves are graphed in Figure 5.31. To find the points where the two curves intersect, we solve

$$(x-1)^2 - 1 = -x + 2$$
$$x^2 - 2x + 1 - 1 = -x + 2$$
$$x^2 - x - 2 = 0$$
$$(x+1)(x-2) = 0$$

Therefore,

$$x = -1 \qquad \text{and} \qquad x = 2$$

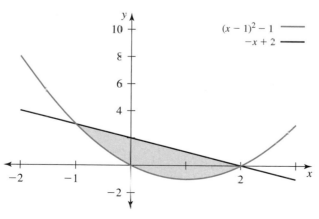

◀ **Figure 5.31**
The region for Example 2.

are the x-coordinates of the points of intersection. Since $-x + 2 \geq (x-1)^2 - 1$ for $x \in [-1, 2]$, the area of the enclosed region is

$$
\begin{aligned}
\text{Area} &= \int_{-1}^{2} [(-x + 2) - ((x-1)^2 - 1)] \, dx \\
&= \int_{-1}^{2} [-x + 2 - x^2 + 2x - 1 + 1] \, dx \\
&= \int_{-1}^{2} [-x^2 + x + 2] \, dx = -\frac{1}{3}x^3 + \frac{1}{2}x^2 + 2x \Big]_{-1}^{2} \\
&= \left(-\frac{1}{3}(2)^3 + \frac{1}{2}(2)^2 + (2)(2) \right) - \left(-\frac{1}{3}(-1)^3 + \frac{1}{2}(-1)^2 + (2)(-1) \right) \\
&= -\frac{8}{3} + 2 + 4 - \frac{1}{3} - \frac{1}{2} + 2 = \frac{9}{2}
\end{aligned}
$$

I EXAMPLE 3 Find the area of the region bounded by $y = \sqrt{x}$, $y = x - 2$, and the x-axis.

Solution We first graph the bounding curves in Figure 5.32. We see that $y = \sqrt{x}$ and $y = x - 2$ intersect. To find the point of intersection, we solve

$$ x - 2 = \sqrt{x} $$

Squaring both sides, we find

$$ (x - 2)^2 = (\sqrt{x})^2 \qquad \text{and thus} \qquad x^2 - 4x + 4 = x $$

or

$$ x^2 - 5x + 4 = 0 $$

We can factor the equation,

$$ (x - 4)(x - 1) = 0 $$

which yields the solutions $x = 4$ and $x = 1$. Since squaring an equation can introduce extraneous solutions, we need to check whether the solutions satisfy $x - 2 = \sqrt{x}$. When $x = 4$, we find $4 - 2 = \sqrt{4}$, which is a true statement; when $x = 1$, we find $1 - 2 = \sqrt{1}$, which is a false statement. Hence, $x = 4$ is the only solution.

The graph of $y = x - 2$ intersects the x-axis at $x = 2$. To compute the area, we need to split the integral into two parts, since the lower bounding curve is composed of two parts: namely, the x-axis from $x = 0$ to $x = 2$ and the line $y = x - 2$ from

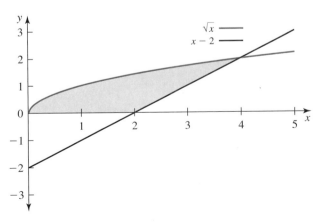

◄ **Figure 5.32**
The region for Example 3.

$x = 2$ to $x = 4$. We find

$$\text{Area} = \int_0^2 \sqrt{x}\, dx + \int_2^4 [\sqrt{x} - (x - 2)]\, dx$$

$$= \left[\frac{2}{3}x^{3/2}\right]_0^{\,2} + \left[\frac{2}{3}x^{3/2} - \frac{1}{2}x^2 + 2x\right]_2^4$$

$$= \frac{2}{3}\cdot 2^{3/2} + \frac{2}{3}\cdot 4^{3/2} - \frac{1}{2}\cdot 16 + 8 - \frac{2}{3}\cdot 2^{3/2} + 2 - 4 = \frac{10}{3} \qquad ▌$$

In the preceding example, we needed to split the integral into two parts since the lower boundary of the area was composed of two different curves that determined the heights of the approximating rectangles in the Riemann integral. Recall that the rectangles are obtained by partitioning the x-axis. It is sometimes more convenient to partition the y-axis. We illustrate this in Figure 5.33 for the region of Example 3, where we partition the interval $[0, 2]$ on the y-axis into n equal subintervals, each of length Δy. We need to express the boundary curves as functions of y. In the case of Example 3, the right boundary curve is $x = f(y) = y + 2$ and the left boundary curve is $x = g(y) = y^2$.

We set $y_0 = 0 < y_1 < y_2 < \cdots < y_n = 2$ where $y_k - y_{k-1} = \Delta y$ for $k = 1, 2, \ldots, n$. As shown in Figure 5.33, the kth rectangle has width $y_k - y_{k-1}$ and height $f(y_k) - g(y_k)$. Therefore, the area of the kth rectangle is

$$[f(y_k) - g(y_k)]\Delta y$$

If we sum this from $k = 1$ to $k = n$ and let $\Delta y \to 0$, we find

$$\lim_{\Delta y \to 0} \sum_{k=1}^n [f(y_k) - g(y_k)]\Delta y = \int_0^2 [f(y) - g(y)]\, dy$$

Since each rectangle is bounded by the same two curves, there is no need to split the integral. Using $f(y) = y + 2$ and $g(y) = y^2$, we find, for the total area,

$$\text{Area} = \int_0^2 (y + 2 - y^2)\, dy$$

$$= \frac{1}{2}y^2 + 2y - \frac{1}{3}y^3 \Big]_0^2 = 2 + 4 - \frac{8}{3} = \frac{10}{3}$$

which is the same as the result in Example 3.

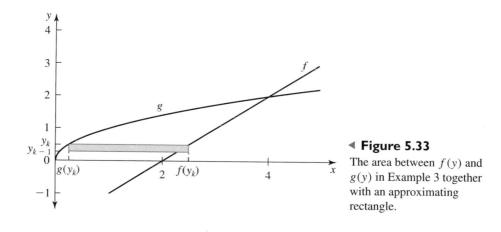

◀ **Figure 5.33**

The area between $f(y)$ and $g(y)$ in Example 3 together with an approximating rectangle.

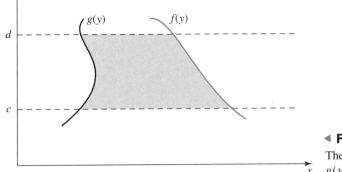

◄ **Figure 5.34**
The area between $f(y)$ and $g(y)$.

To summarize the general case, as illustrated in Figure 5.34: Suppose that a region is bounded by $x = f(y)$ and $x = g(y)$, with $g(y) \le f(y)$ for $c \le y \le d$; that is, $f(y)$ is to the right of $g(y)$ for all $y \in [c, d]$. Then the area of the shaded region is given by the following formula.

$$\text{Area} = \int_c^d [f(y) - g(y)]\,dy$$

5.3.2 Cumulative Change

Consider a population whose size at time t, $t \ge 0$, is $N(t)$ and whose dynamics are given by the initial value problem

$$\frac{dN}{dt} = f(t) \quad \text{with } N(0) = N_0$$

for some $f(t)$ that is continuous for $t \ge 0$. To solve this initial value problem, we must find an antiderivative of $f(t)$, and determine the constant so that $N(0) = N_0$. Now, using (5.8), we find that

$$N(t) = \int_0^t f(u)\,du + C$$

is the general antiderivative of $N(t)$. We choose 0 as the lower limit of integration for convenience, because it gives a simple expression for the constant when we take the initial condition into account. Namely, with

$$N(0) = \int_0^0 f(u)\,du + C = C,$$

it follows that $C = N(0) = N_0$. Therefore,

$$N(t) = N_0 + \int_0^t f(u)\,du$$

or

$$N(t) - N_0 = \int_0^t f(u)\,du \tag{5.14}$$

Since $f(u) = dN/du$ and $N(0) = N_0$, we can write

$$N(t) - N(0) = \int_0^t \frac{dN}{du}\,du$$

which allows us to interpret the definite integral $\int_0^t \frac{dN}{du}\, du$ as the **net or cumulative change** in population size between times 0 and t, since it is a "sum" of instantaneous changes accumulated over time. That is,

$$\begin{bmatrix} \text{cumulative} \\ \text{change in } [0,\, t] \end{bmatrix} = \int_0^t \begin{bmatrix} \text{instantaneous rate of} \\ \text{change at time } u \end{bmatrix} du$$

We present another example in which we can interpret an integral as the cumulative change of a quantity. Recall that velocity is the instantaneous rate of change of distance. That is, if a particle moves along a straight line, and we denote by $s(t)$ the location of the particle at time t, with $s(0) = s_0$, and by $v(t)$ the velocity at time t, then $s(t)$ and $v(t)$ are related via

$$\frac{ds}{dt} = v(t) \quad \text{for } t > 0 \text{ with } s(0) = s_0$$

To solve this initial value problem, we must find an antiderivative of $v(t)$ that satisfies $s(0) = s_0$. That is,

$$s(t) = \int_0^t v(u)\, du + C$$

with

$$s_0 = s(0) = \int_0^0 v(u)\, du + C = 0 + C$$

which implies that $C = s_0$. Hence,

$$s(t) = s_0 + \int_0^t v(u)\, du$$

or, with $s(0) = s_0$

$$s(t) - s(0) = \int_0^t v(u)\, du$$

Again, the cumulative change in distance, $s(t) - s(0)$, can be represented as a "sum" of instantaneous changes.

5.3.3 Average Values

The concentration of soil nitrogen was measured in gm^{-3} along a transect in moist tundra measured every meter, and yielded the following data.

distance from origin [m]	1	2	3	4	5
concentration [gm^{-3}]	589.3	602.7	618.5	667.2	641.2

distance from origin [m]	6	7	8	9	10
concentration [gm^{-3}]	658.3	672.8	661.2	652.3	669.8

If we denote the concentration at distance x from the origin by $c(x)$, then the average concentration, denoted by $\bar{c}$ (read: "c bar"), is the arithmetic average

$$\bar{c} = \frac{1}{10} \sum_{k=1}^{10} c(k) = 643.3 \text{ gm}^{-3}$$

More generally, suppose that we wish to find the average concentration between two points a and b along a transect. We divide $[a, b]$ into n subintervals of equal lengths $\Delta x = \frac{b-a}{n}$, and measure the concentration at the right endpoint of each subinterval. If the concentration at location x_k is denoted by $c(x_k)$, then the average concentration $\bar{c}$ is

$$\bar{c} = \frac{1}{n} \sum_{k=1}^{n} c(x_k)$$

Since $\Delta x = \frac{b-a}{n}$, we can write $n = \frac{b-a}{\Delta x}$. Hence,

$$\bar{c} = \frac{1}{b-a} \sum_{k=1}^{n} c(x_k) \Delta x$$

If we let the number of subintervals grow ($n \to \infty$), then the length of each subinterval goes to 0 ($\Delta x \to 0$), and

$$\bar{c} = \frac{1}{b-a} \lim_{n \to \infty} \sum_{k=1}^{n} c(x_k) \Delta x$$

$$= \frac{1}{b-a} \int_{a}^{b} c(x)\, dx$$

That is, the average concentration can be expressed as an integral over $c(x)$ between a and b, divided by the length of the interval $[a, b]$.

Assume that $f(x)$ is a continuous function on $[a, b]$. The average value of f on the interval $[a, b]$ is

$$f_{\text{avg}} = \frac{1}{b-a} \int_{a}^{b} f(x)\, dx$$

▮ EXAMPLE 4   Find the average value of $f(x) = 4 - x^2$ on the interval $[-2, 2]$.

Solution   We use the formula for computing average values.

$$f_{\text{avg}} = \frac{1}{2 - (-2)} \int_{-2}^{2} (4 - x^2)\, dx = \frac{1}{4} \left[4x - \frac{1}{3}x^3 \right]_{-2}^{2}$$

$$= \frac{1}{4} \left[8 - \frac{8}{3} + 8 - \frac{8}{3} \right] = \frac{1}{4} \cdot \frac{32}{3} = \frac{8}{3}$$ ▮

The following theorem says a bit more about the value of f_{avg}.

The Mean Value Theorem for Definite Integrals   Assume that $f(x)$ is a continuous function on $[a, b]$. Then there exists a number $c \in [a, b]$ such that

$$f(c)(b - a) = \int_{a}^{b} f(x)\, dx$$

That is, when we compute the average value of a function that is continuous on $[a, b]$, there exists a number c such that $f(c) = f_{\text{avg}}$. This should be fairly obvious when you look at the graph of a function f and at f_{avg}. For simplicity, let's assume that $f(x) \geq 0$. Since $\int_{a}^{b} f(x)\, dx$ is then equal to the area between the graph of $f(x)$ and the x-axis, and $f_{\text{avg}}(b - a)$ is equal to the area of the rectangle with height f_{avg} and width $b - a$, and since the two areas are equal, the horizontal line $y = f_{\text{avg}}$ must

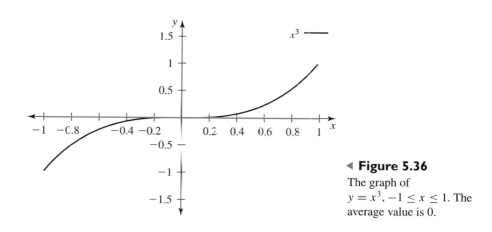

▸ **Figure 5.35**
An illustration of the average value of a function:
$\int_a^b f(x)\,dx = f_{avg}(b - a)$.

intersect the graph of $f(x)$ at some point in the interval $[a, b]$ (see Figure 5.35). The x-coordinate of this point of intersection is then the value c in the mean value theorem for definite integrals. A similar argument can be made when we do not assume that $f(x)$ is positive. In this case, "area" is replaced by "signed area." The proof of this theorem is short, and we supply it for completeness at the end of this subsection.

▌**EXAMPLE 5**   Find the average value of $f(x) = x^3$ on the interval $[-1, 1]$, and determine x, where $f(x)$ is average.

Solution

$$f_{avg} = \frac{1}{2}\int_{-1}^{1} x^3\,dx = \frac{1}{2}\left[\frac{1}{4}x^4\right]_{-1}^{1} = \frac{1}{8}(1^4 - (-1)^4) = 0$$

The graph of $y = x^3$ (see Figure 5.36) is symmetric about the origin, which explains why the average value is 0. Since $x^3 = 0$ for $x = 0$, the function $f(x)$ takes on its average value at $x = 0$. ▌

▌**EXAMPLE 6**   The speed of water in a channel varies considerably with depth. Due to friction, the velocity reaches zero at the bottom and along the sides of the channel; the velocity is highest near the surface of the water. The average velocity of a stream is of interest in characterizing rivers. One way to obtain this average value would be to measure stream velocity at various depths along a vertical transect and then average these

▸ **Figure 5.36**
The graph of
$y = x^3,\, -1 \le x \le 1$. The average value is 0.

values. In practice, however, a much simpler method is employed; namely, the speed measured at 60% of the depth from the surface is very close to the average speed. Explain why it is possible that the measurement at just one depth would yield the average stream velocity.

Solution   Assuming that the velocity profile of the stream along a vertical transect is a continuous function of depth, the MVT for definite integrals guarantees that there exists a depth h at which the velocity is equal to the average stream velocity.

The MVT only gives the existence of such a depth; it does not tell us where the velocity is equal to the average velocity. It is surprising and fortunate that the depth where the velocity is average, is quite universal; that is, it does not depend much on the specifics of the river. (The 0.6 depth rule is derived in Problems 3–5 in Section 5.5.)   ▮

Proof of the Mean Value Theorem for Definite Integrals   Since $f(x)$ is continuous on $[a, b]$, we can apply the extreme value theorem to conclude that f attains an absolute maximum and an absolute minimum in $[a, b]$. If we denote the absolute maximum by M and the absolute minimum by m, then

$$m \leq f(x) \leq M \quad \text{for all } x \in [a, b]$$

and f takes on both m and M for some values in $[a, b]$. We therefore find

$$m(b - a) \leq \int_a^b f(x)\, dx \leq M(b - a)$$

or

$$m \leq \frac{1}{b - a} \int_a^b f(x)\, dx \leq M$$

We set $I = \frac{1}{b-a} \int_a^b f(x)\, dx$; then, $m \leq I \leq M$.

Using the facts that $f(x)$ takes on all values between m and M in the interval $[a, b]$, and that I is a number between m and M, it follows from the intermediate value theorem that there must be a number $c \in [a, b]$ such that $f(c) = I$; that is,

$$f(c) = \frac{1}{b - a} \int_a^b f(x)\, dx \qquad \blacklozenge$$

5.3.4 The Volume of a Solid (Optional)

For certain regular solids, such as a circular cylinder, we know formulas from geometry for computing the volume. To compute the volume of a less regularly shaped solid, we will use an approach that is very similar to that for computing areas of irregularly shaped regions; there, we approximated the area by rectangles whose areas were easy to compute using simple formulas from geometry.

We begin with the volume of a right cylinder; the volume is simply the base area times the height. The base can be an arbitrarily shaped region (see Figure 5.37 for example).

If we denote the base area by A and the height of the cylinder by h, then the volume of the cylinder is

$$V = Ah$$

As an example, consider the circular cylinder whose base is a disk. If the disk has radius r and the cylinder has height h, then the volume of the circular cylinder is $\pi r^2 h$. We will use cylinders to approximate volumes of more complicated solids.

▲ **Figure 5.37**

A right cylinder with an irregularly shaped base.

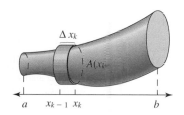

▲ **Figure 5.38**

The volume of an irregularly shaped solid using the disk method.

Suppose that we wish to compute the volume of the solid in Figure 5.38. We can slice the solid into small slabs by cutting it perpendicular to the x axis; that is, we choose points $x_0 = a < x_1 < x_2 < \cdots < x_n = b$ that partition the interval $[a, b]$ perpendicular to the x axis through $x_k, k = 1, 2, \ldots, n - 1$. The intersection of such a plane and the solid is called a cross-section. We denote the area of the cross-section at x_k by $A(x_k)$. By cutting the solid along these planes, we obtain slices, just like when cutting bread. We will approximate the volume of a slice between x_{k-1} and x_k by the volume of a cylinder with base area equal to that of the slice at x_k, and height $\Delta x_k = x_k - x_{k-1}$. The volume of the slice between x_{k-1} and x_k is then approximately

$$A(x_k)\Delta x_k$$

Adding the volumes of all the slices gives us an approximation for the total volume of the solid, that is,

$$V \approx \sum_{k=1}^{n} A(x_k)\Delta x_k$$

By making the partition of $[a, b]$ finer, we can improve the approximation.

Recall that we used $\|P\|$ (norm of P) as a measure for how fine the partition is, that is, $\|P\| = \max_{k=1,2,\ldots,n} \Delta x_k$. This suggests that we should define the volume of the solid as the limit of our approximation as $\|P\| \to 0$. We summarize this in the following.

> The volume of a solid of integrable cross-sectional area $A(x)$ between a and b is
>
> $$\int_a^b A(x)\,dx$$

▮**EXAMPLE 7** Find the volume of the sphere of radius r centered at the origin.

Solution   The cross-section at x is perpendicular to the x-axis (see Figure 5.39). It is a disk of radius $y = \sqrt{r^2 - x^2}$, whose area is

$$A(x) = \pi y^2 = \pi(r^2 - x^2)$$

Since the solid is between $-r$ and r, we find

$$\text{Volume} = \int_{-r}^{r} \pi(r^2 - x^2)\,dx$$

and evaluating this integral yields

$$= \pi\left[r^2 x - \frac{1}{3}x^3\right]_{-r}^{r}$$

$$= \pi\left[\left(r^3 - \frac{1}{3}r^3\right) - \left(-r^3 + \frac{1}{3}r^3\right)\right] = \pi\left(\frac{2}{3}r^3 + \frac{2}{3}r^3\right) = \frac{4}{3}\pi r^3$$

This agrees with the formula for a sphere that we know from geometry. ▮

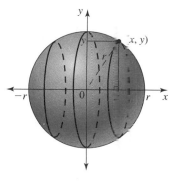

▲ **Figure 5.39**

The volume of a sphere.

The cross-sections of the sphere in the last example were of a particularly simple form, namely disks. We can think of a sphere as a **solid of revolution**, that is, a solid that is obtained by revolving a curve about the x-axis (or the y-axis). In

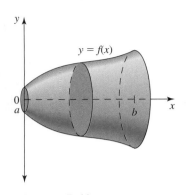

▲ Figure 5.40
The solid of rotation when rotating $f(x)$ about the x axis.

this case, we rotate the curve $y = \sqrt{r^2 - x^2}$, $-r \le x \le r$, about the x-axis, which creates circular cross-sections.

We can use other curves $y = f(x)$, rotate them about the x-axis, and obtain solids in this way. This is illustrated in Figure 5.40, where we rotate the graph of $y = f(x)$, $a \le x \le b$, about the x-axis. A cross-section through x perpendicular to the x-axis is then a disk with radius $f(x)$; hence, its cross-sectional area is $A(x) = \pi[f(x)]^2$. If we use the formula $\int_a^b A(x)\,dx$ to compute the volume of the solid, we find that the **volume** of the solid of revolution is

$$V = \int_a^b \pi[f(x)]^2\,dx \tag{5.15}$$

Computing volumes using (5.15) is called the **disk method**.

▌ EXAMPLE 8   Compute the volume of the solid obtained by rotating $y = x^2$, $0 \le x \le 2$, about the x-axis.

Solution   We illustrate this in Figure 5.41. When we rotate the graph about the x-axis, we find that the cross-section at x is a disk with radius $y = x^2$. Thus the volume is

$$V = \int_0^2 \pi[f(x)]^2\,dx$$

$$= \int_0^2 \pi x^4\,dx$$

$$= \frac{\pi}{5}x^5\Big]_0^2 = \frac{32}{5}\pi$$

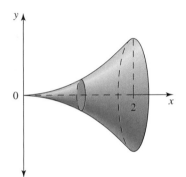

▲ Figure 5.41
The solid of rotation for Example 8.

▌ EXAMPLE 9   Rotate the area bounded by the curves $y = \sqrt{x}$ and $y = x/2$ about the x-axis, and compute the volume of the solid of rotation.

Solution   The curves $y = \sqrt{x}$ and $y = x/2$ are graphed in Figure 5.42, together with a vertical bar to indicate the cross-section. We see that the curves intersect at $x = 0$ and $x = 4$. We can compute the volume of this solid of rotation by first rotating $y = \sqrt{x}$, $0 \le x \le 4$, about the x-axis, computing the volume of this solid, and then subtracting the volume of the solid obtained by rotating $y = x/2$, $0 \le x \le 4$. We

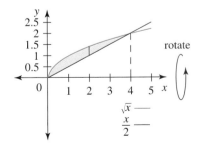

◀ Figure 5.42
The plane region for Example 9.

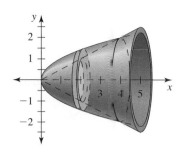

▲ **Figure 5.43**

The solid of rotation in Example 9.

find

$$V = \int_0^4 \pi(\sqrt{x})^2\,dx - \int_0^4 \pi\left(\frac{1}{2}x\right)^2 dx$$

$$= \pi\frac{1}{2}x^2\bigg]_0^4 - \pi\frac{1}{12}x^3\bigg]_0^4$$

$$= 8\pi - \frac{16}{3}\pi = \frac{8}{3}\pi$$

Looking at Figure 5.43, the cross-sectional area is that of a washer. In this case, the disk method is also referred to as the washer method. ▮

We can also rotate a curve about the y-axis, as illustrated in the following example.

▮ **EXAMPLE 10**   Rotate the area bounded by $y = 2$, $x = 0$, $y = 0$, and $y = \ln x$ about the y-axis, and compute the volume of the resulting solid.

Solution   When we rotate about the y-axis, the cross-sections are perpendicular to the y-axis. At $y = \ln x$, the radius of the cross-sectional disk is x (see Figure 5.44). The solid is shown in Figure 5.45. Since we "sum" the slices along the y-axis, we must integrate with respect to y; since $y = \ln x$, we find $x = e^y$. Therefore, the cross-sectional area at y is $A(y) = \pi(e^y)^2$, and the volume is

$$V = \int_0^2 \pi(e^y)^2\,dy = \int_0^2 \pi e^{2y}\,dy$$

$$= \pi\frac{1}{2}e^{2y}\bigg]_0^2 = \frac{\pi}{2}(e^4 - 1)$$ ▮

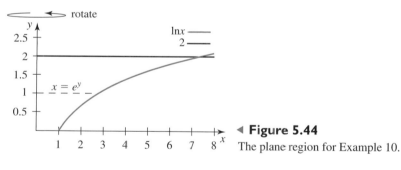

◀ **Figure 5.44**
The plane region for Example 10.

▲ **Figure 5.45**
The solid of rotation for Example 10.

◆ 5.3 PROBLEMS

[5.3.1]

Find the areas of the regions bounded by the lines and curves in Problems 1–12.

1. $y = e^x$, $y = -x$, $x = 0$, $x = 2$
2. $y = \sin x$, $y = 1$, $x = 0$, $x = \frac{\pi}{2}$
3. $y = x^2 - 1$, $y = x + 1$
4. $y = x^2$, $y = 2 - x^4$
5. $y = x^2 + 1$, $y = 4x - 2$ (in the first quadrant)
6. $y = x^2$, $y = 2 - x$, $y = 0$ (in the first quadrant)
7. $y = x^2$, $y = \frac{1}{x}$, $y = 4$ (in the first quadrant)
8. $y = \sin x$, $y = \cos x$ from $x = 0$ to $x = \frac{\pi}{2}$
9. $y = \frac{1}{x}$, $y = 1$ from $x = \frac{1}{2}$ to $x = \frac{3}{2}$
10. $y = x^2$, $y = (x - 2)^2$, $y = 0$ from $x = 0$ to $x = 2$
11. $y = x^2$, $y = x^3$ from $x = 0$ to $x = 2$
12. $y = e^{-x}$, $y = x + 1$ from $x = -1$ to $x = 1$

In Problems 13–16, find the areas of the regions bounded by the lines and curves by expressing x as a function of y and integrating with respect to y.

13. $y = x^2$, $y = (x - 2)^2$, $y = 0$ from $x = 0$ to $x = 2$
14. $y = x$, $yx = 1$, $y = \frac{1}{2}$ (in the first quadrant)
15. $x = (y - 1)^2 + 3$, $x = 1 - (y - 1)^2$ from $y = 0$ to $y = 2$ (in the first quadrant)
16. $x = (y - 1)^2 - 1$, $x = (y - 1)^2 + 1$ from $y = 0$ to $y = 2$

[5.3.2]

17. Consider a population whose size at time t is $N(t)$ and whose dynamics are given by the initial value problem

$$\frac{dN}{dt} = e^{-t}$$

with $N(0) = 100$.
(a) Find $N(t)$ by solving the initial value problem.
(b) Compute the cumulative change in population size between $t = 0$ and $t = 5$.
(c) Express the cumulative change in population size between time 0 and time t as an integral. Provide a geometric interpretation of this quantity.
18. Suppose that a change in biomass $B(t)$ at time t during the interval $[0, 12]$ follows the equation

$$\frac{d}{dt} B(t) = \cos\left(\frac{\pi}{6} t\right)$$

for $0 \le t \le 12$.
(a) Graph $\frac{dB}{dt}$ as a function of t.
(b) Suppose that $B(0) = B_0$. Express the cumulative change of biomass during the time interval $[0, t]$ as an integral. Give a geometric interpretation. What is the value of the biomass at the end of the interval $[0, 12]$, compared to the beginning of the study at time 0? How are these two quantities related to the cumulative change of the biomass during the interval $[0, 12]$.

19. A particle moves along the x-axis with velocity

$$v(t) = -(t - 2)^2 + 1$$

for $0 \le t \le 5$. Assume that the particle is at the origin at time 0.
(a) Graph $v(t)$ as a function of t.
(b) Use the graph of $v(t)$ to determine when the particle moves to the left and when it moves to the right.
(c) Find the location $s(t)$ of the particle at time t for $0 \le t \le 5$. Give a geometric interpretation of $s(t)$ in terms of the graph of $v(t)$.
(d) Graph $s(t)$ and find the leftmost and rightmost positions.
20. Recall that the acceleration $a(t)$ of a particle moving along a straight line is the instantaneous rate of change of the velocity $v(t)$; that is,

$$a(t) = \frac{d}{dt} v(t)$$

Assume that $a(t) = 32\text{ft}/\text{sec}^2$. Express the cumulative change in velocity during the time interval $[0, t]$ as a definite integral, and compute the integral.
21. If $\frac{dl}{dt}$ represents the growth rate of an organism at time t (measured in months), explain what

$$\int_2^7 \frac{dl}{dt} \, dt$$

represents.
22. If $\frac{dw}{dx}$ represents the rate of change of the weight of an organism of age x, explain what

$$\int_3^5 \frac{dw}{dx} \, dx$$

means.
23. If $\frac{dB}{dt}$ represents the rate of change of biomass at time t, explain what

$$\int_1^6 \frac{dB}{dt} \, dt$$

means.
24. Let $N(t)$ denote the size of a population at time t, and assume that

$$\frac{dN}{dt} = f(t)$$

Express the cumulative change of the population size in the interval $[2, 4]$ as an integral.

[5.3.3]

25. Let $f(x) = x^2 - 2$. Compute the average value of $f(x)$ over the interval $[0, 2]$.
26. Let $g(t) = e^{-t}$. Compute the average value of $g(t)$ over the interval $[-1, 1]$.
27. Suppose that the temperature T (measured in Fahrenheit) varies over a 24-hour period according to

$$T(t) = 68 + \sin\left(\frac{\pi}{12} t\right)$$

for $0 \le t \le 24$.

(a) Graph the temperature T as a function of time t.

(b) Find the average temperature and explain your answer graphically.

28. Suppose that the concentration of soil nitrogen (measured in gm^{-3}) along a transect in moist tundra yields data points that follow a straight line with equation

$$y = 673.8 - 34.7x$$

for $0 \leq x \leq 10$, where x is the distance to the beginning of the transect. What is the average concentration of soil nitrogen along this transect?

29. Let $f(x) = \tan x$. Give a geometric argument to explain why the average value of $f(x)$ over $[-1, 1]$ is equal to 0.

30. Suppose that you drive from St. Paul to Duluth and you average 50 mph. Explain why there must be a time during your trip at which your speed is exactly 50 mph.

31. Let $f(x) = 2x, 0 \leq x \leq 2$. Use a geometric argument to find the average value of f over the interval $[0, 2]$, and find x such that $f(x)$ is equal to the average value.

32. A particle moves along the x-axis with velocity

$$v(t) = -(t - 3)^2 + 5$$

for $0 \leq t \leq 6$.

(a) Graph $v(t)$ as a function of t for $0 \leq t \leq 6$.

(b) Find the average velocity of this particle during the time interval $[0, 6]$.

(c) Find a time $t^* \in [0, 6]$ such that the velocity at time t^* is equal to the average velocity during the time interval $[0, 6]$. Is it clear that such a point exists? Is there more than one such point in this case? Use your graph in (a) to explain how you would find t^* graphically.

[5.3.4]

33. Find the volume of a right-circular cone with base radius r and height h.

34. Find the volume of a pyramid with square base of side length a and height h.

In Problems 35–40, find the volumes of the solids obtained by rotating the region bounded by the given curves about the x-axis. In each case, sketch the region and a typical disk element.

35. $y = 4 - x^2, y = 0, x = 0$ (in the first quadrant)

36. $y = \sqrt{x}, y = 0, x = 4$

37. $y = \sqrt{\sin x}, 0 \leq x \leq \pi, y = 0$

38. $y = e^x, y = 0, x = 0, x = \ln 2$

39. $y = \sec x, -\frac{\pi}{3} \leq x \leq \frac{\pi}{3}, y = 0$

40. $y = \sqrt{4 - x^2}, -2 \leq x \leq 2, y = 0$

In Problems 41–46, find the volumes of the solids obtained by rotating the region bounded by the given curves about the x-axis. In each case, sketch the region together with a typical disk element.

41. $y = x^2, y = x, 0 \leq x \leq 1$

42. $y = 1 - x^2, y = 1 + x^2, 0 \leq x \leq 1$

43. $y = e^x, y = e^{-x}, 0 \leq x \leq 2$

44. $y = \sqrt{1 - x^2}, y = 1, x = 1$ (in the first quadrant)

45. $y = \sqrt{\cos x}, y = 1, x = \frac{\pi}{2}$

46. $y = \frac{1}{x}, x = 0, y = 1, y = 2$ (in the first quadrant)

In Problems 47–52, find the volumes of the solids obtained by rotating the region bounded by the given curves about the y-axis. In each case, sketch the region together with a typical disk element.

47. $y = \sqrt{x}, y = 2, x = 0$

48. $y = x^2, y = 4, x = 0$ (in the first quadrant)

49. $y = \ln(x + 1), y = \ln 3, x = 0$

50. $y = \sqrt{x}, y = x, 0 \leq x \leq 1$

51. $y = x^2, y = \sqrt{x}, 0 \leq x \leq 1$

52. $y = \frac{1}{x}, x = 0, y = \frac{1}{2}, y = 1$

◆ 5.4 KEY TERMS

Chapter 5 Review: Topics Discuss the following definitions and concepts:

1. Area

2. Summation notation

3. Algebraic rules for sums

4. A partition of an interval and the norm of a partition

5. Riemann sum

6. Definite integral

7. Riemann integrable

8. Geometric interpretation of definite integrals

9. The constant multiple and sum rule for integrals

10. The definite integral over a union of intervals

11. Comparison results for definite integrals

12. The fundamental theorem of calculus, part I

13. Leibniz's rule

14. Antiderivatives and indefinite integrals

15. The fundamental theorem of calculus, part II

16. Evaluating definite integrals using the FTC, part II

17. Computing the area between curves using definite integrals

18. Cumulative change and definite integrals

19. The mean value theorem for definite integrals

20. The volume of a solid and definite integrals

◆ 5.5 REVIEW PROBLEMS

1. (Discharge of a river)

When studying the flow of water in an open channel, such as a river in its bed, the amount of water passing through a cross-section per second, the discharge (Q), is of interest. The following formula is used to compute the discharge:

$$Q = \int_0^B \bar{v}(b)h(b)\,db \qquad (5.16)$$

where b is the distance from one bank of the river to the point where the depth of the river $h(b)$ and the average velocity $\bar{v}(b)$ of the vertical velocity profile of the river at b were measured. The total width of the cross section is B (see Figure 5.46).

▲ **Figure 5.46**

The river for Problem 1.

To evaluate the integral in (5.16), we would need to know $\bar{v}(b)$ and $h(b)$ at every location b along the cross-section. In practice, the cross-section is divided into a finite number of subintervals, and measurements of $\bar{v}$ and h are taken at the right endpoints, say, of each subinterval. The following table contains an example of such measurements. The location 0 corresponds to the left bank and the location $B = 16$ to the right bank of the river. The units of the location and of h are meters, and of $\bar{v}$, meter per second.

location	h	$\bar{v}$
0	0	0
1	0.28	0.172
3	0.76	0.213
5	1.34	0.230
7	1.57	0.256
9	1.42	0.241
11	1.21	0.206
13	0.83	0.187
15	0.42	0.116
16	0	0

Approximate the integral in (5.16) by a Riemann sum, using the locations in the table, and find the approximate discharge using the data from the table.

2. Suppose that you grow plants in several study plots and wish to measure the response of total biomass to the respective treatment in each plot. One way to measure this would be to determine the average specific growth rate of the biomass for each plot over the course of the growing season.

We denote by $B(t)$ the biomass in a given plot at time t. Then the specific growth rate of the biomass at time t is given

by

$$\frac{1}{B(t)}\frac{dB}{dt}$$

(a) Explain why

$$\frac{1}{t}\int_0^t \frac{1}{B(s)}\frac{dB(s)}{ds}\,ds$$

is a way to express the average specific growth rate over the time interval $[0, t]$.

(b) Use the chain rule to show that

$$\frac{1}{B(t)}\frac{dB}{dt} = \frac{d}{dt}(\ln B(t))$$

(c) Use the results in (a) and (b) to show that the average specific growth rate of $B(s)$ over the interval $[0, t]$ is given by

$$\frac{1}{t}\int_0^t \frac{d}{ds}(\ln(B(s)))\,ds = \frac{1}{t}\ln\frac{B(t)}{B(0)}$$

provided that $B(s) > 0$ for $s \in [0, t]$.

(d) Explain the measurements that you would need to take if you wanted to determine the average specific growth rate of biomass in a given plot over the time interval $[0, t]$.

Problems 3–6 discuss stream velocity profiles and provide a justification for the two measurement methods described in the following.

(Adapted from Herschy 1995) The speed of water in a channel varies considerably with depth. Due to friction, the velocity reaches zero at the bottom and along the sides of the channel. The velocity is highest near the surface of the stream. To find the average velocity for the vertical velocity profile, two methods are frequently employed in practice:

1. The 0.6 depth method: The speed is measured at 0.6 of the depth from the surface, and this value is taken as the average speed.

2. The 0.2 and 0.8 depth method: The speed is measured at 0.2 and 0.8 of the depth from the surface, and the average of the two readings is taken as the average speed.

The theoretical velocity distribution of water flowing in an open channel is given approximately by

$$v(d) = \left(\frac{D-d}{a}\right)^{1/c} \qquad (5.17)$$

where $v(d)$ is the velocity at depth d below the water surface, c is a constant varying from 5 for coarse beds to 7 for smooth beds, D is the total depth of the channel, and a is a constant that is equal to the distance above the bottom of the channel at which the velocity has unit value.

3. (a) Sketch the graph of $v(d)$ as a function of d for $D = 3$ m and $a = 1$ m for (i) $c = 5$ and (ii) $c = 7$.

(b) Show that the velocity is equal to 0 at the bottom ($d = D$) and is maximal at the surface ($d = 0$).

4. (a) Show by integration that the average velocity in the vertical profile ($\bar{v}$) is given by

$$\bar{v} = \frac{c}{c+1}\left(\frac{D}{a}\right)^{1/c}$$ (5.18)

(b) What fraction of the maximum speed is the average speed $\bar{v}$?

(c) If you knew that the maximum speed occurred at the surface of the river (as predicted in the approximate formula for $v(d)$), how could you find $\bar{v}$? (In practice, the maximum speed may occur quite a bit below the surface due to friction between the water on the surface and the atmosphere. Therefore, the speed at the surface would not provide an accurate measurement of the maximum speed.)

5. Explain why the depth d_1, at which $v = \bar{v}$, is given by the equation

$$\bar{v} = \left(\frac{D-d_1}{a}\right)^{1/c}$$ (5.19)

We can find d_1 by equating (5.18) and (5.19). Show that

$$\frac{d}{D} = 1 - \left(\frac{c}{c+1}\right)^c$$

and that d_1/D is approximately 0.6 for values of c between 5 and 7, thus resulting in the rule

$$\bar{v} \approx v_{0.6}$$

where $v_{0.6}$ is the velocity at depth $0.6D$. (*Hint*: Graph $1 - (c/(c+1))^c$ as a function of c for $c \in [5, 7]$, and investigate the range of this function.)

6. We denote by $v_{0.2}$ the velocity at depth $0.2D$. We will now find the depth d_2 so that

$$\bar{v} = \frac{1}{2}(v_{0.2} + v_{d_2})$$

(a) Show that d_2 satisfies

$$\frac{1}{2}\left[\left(\frac{D-0.2D}{a}\right)^{1/c} + \left(\frac{D-d_2}{a}\right)^{1/c}\right]$$

$$= \frac{c}{c+1}\left(\frac{D}{a}\right)^{1/c}$$

(*Hint*: Use (5.17) and (5.18).)

(b) Show that

$$\frac{d_2}{D} = 1 - \left[\frac{2c}{c+1} - (0.8)^{1/c}\right]^c$$

and confirm that d_2/D is approximately 0.8 for values of c between 5 and 7, thus resulting in the rule

$$\bar{v} \approx \frac{1}{2}(v_{0.2} + v_{0.8})$$

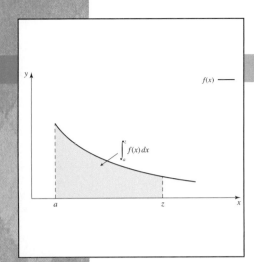

Integration Techniques and Computational Methods

In the first two sections of this chapter, we will learn two important integration techniques that are essentially differentiation rules applied backwards. (Because of the connection between differentiation and integration, it should not come as a surprise that some integration techniques are closely related to differentiation rules.) The first technique, called the substitution rule, is the chain rule backwards; the second, called integration by parts, is the product rule backwards. The third section is devoted to using these integration techniques, and the fourth deals with improper integrals, which are integrals where the integrand goes to infinity somewhere over the integration interval or the integration interval is unbounded. Finally, we devote a section to numerical integration, another to the use of tables to evaluate more complicated integrals, and a section to the approximation of functions by polynomials.

◆ 6.1 THE SUBSTITUTION RULE

6.1.1 Indefinite Integrals

The substitution rule is the chain rule in integral form. We therefore begin by recalling the chain rule. Suppose that we wish to differentiate

$$f(x) = \sin(3x^2 + 1)$$

This is clearly a situation in which we need to use the chain rule. We set

$$f(u) = \sin u \qquad \text{and} \qquad u = 3x^2 + 1$$

and find that $f'(u) = \cos u$. To differentiate the inner function $u = 3x^2 + 1$, we use Leibniz notation:

$$\frac{du}{dx} = 6x \qquad \text{or} \qquad du = 6x\,dx$$

Using the chain rule, we obtain

$$\frac{d}{dx}\sin(3x^2 + 1) = \frac{df}{du}\frac{du}{dx} = (\cos u)(6x) = \cos(3x^2 + 1) \cdot 6x$$

Reversing this, we get

$$\int \underbrace{\cos(3x^2 + 1)}_{\cos u} \cdot \underbrace{6x\,dx}_{du} = \int \cos u\,du = \underbrace{\sin u}_{\sin(3x^2+1)} + C = \sin(3x^2 + 1) + C$$

Here, we substituted u for $3x^2 + 1$ and used $du = 6x\,dx$. This substitution simplified the integrand.

To see the general principle behind this, we write $u = g(x)$ (and hence $du = g'(x)\,dx$). Our integral in the preceding is of the form

$$\int f[g(x)]g'(x)\,dx$$

If we denote by $F(x)$ an antiderivative of $f(x)$ (that is, $F'(x) = f(x)$), then using the chain rule to differentiate $F[g(x)]$, we find

$$\frac{d}{dx}F[g(x)] = F'[g(x)]g'(x) = f[g(x)]g'(x)$$

which shows that $F[g(x)]$ is an antiderivative of $f[g(x)]g'(x)$. We can therefore write

$$\int f[g(x)]g'(x)\,dx = F[g(x)] + C \qquad (6.1)$$

If we set $u = g(x)$, then the right-hand side of (6.1) can be written as $F(u) + C$. But since $F(u)$ is an antiderivative of $f(u)$, it also has the representation

$$\int f(u)\,du = F(u) + C \qquad (6.2)$$

which shows that the left-hand side of (6.1) is the same as the left-hand side of (6.2). Equating the left-hand sides of (6.1) and (6.2) results in the substitution rule.

SUBSTITUTION RULE FOR INDEFINITE INTEGRALS

If $u = g(x)$, then

$$\int f[g(x)]g'(x)\,dx = \int f(u)\,du$$

We present a number of examples that will illustrate how to use this rule, and when the rule can be successfully applied. The first two examples are straightforward.

∎ **EXAMPLE 1**   **Using Substitution**

Evaluate

$$\int (2x + 1)e^{x^2+x}\,dx$$

Solution   The expression $2x + 1$ is the derivative of $x^2 + x$, which is the inner function of e^{x^2+x}. This suggests the following substitution:

$$u = x^2 + x \qquad \text{with } \frac{du}{dx} = 2x + 1 \text{ or } du = (2x + 1)\,dx$$

Hence,

$$\int \underbrace{e^{x^2+x}}_{e^u}\underbrace{(2x + 1)\,dx}_{du} = \int e^u\,du = e^u + C = e^{x^2+x} + C$$

In the last step, we substituted $x^2 + x$ back for u since we want the final result in terms of x. ▮

▮ **EXAMPLE 2**   **Using Substitution**

Evaluate

$$\int \frac{1}{x \ln x} \, dx$$

Solution   We see that $1/x$ is the derivative of $\ln x$. We try

$$u = \ln x \qquad \text{with } \frac{du}{dx} = \frac{1}{x} \text{ or } du = \frac{1}{x} \, dx$$

Then

$$\int \frac{1}{\ln x} \frac{1}{x} \, dx = \int \frac{1}{u} \, du = \ln |u| + C = \ln | \ln x | + C \qquad ▮$$

Examples 1 and 2 illustrate types of integrals that are frequently encountered, and we display them in the following for ease of reference.

$$\int g'(x) e^{g(x)} \, dx = e^{g(x)} + C$$

$$\int \frac{g'(x)}{g(x)} \, dx = \ln |g(x)| + C$$

In the previous two examples, the derivative of the inner function appeared exactly in the integrand. This will not always be the case, as shown in the following example where the derivative of the inner function appears in the integrand only up to a multiplicative constant.

▮ **EXAMPLE 3**   **Multiplicative Constant**

Evaluate

$$\int 4x \sqrt{x^2 + 1} \, dx$$

Solution   If we set $u = x^2 + 1$, then

$$\frac{du}{dx} = 2x \qquad \text{or} \qquad \frac{du}{2} = x \, dx$$

The integrand contains the derivative of the inner function up to a multiplicative constant. But this is good enough:

$$\int 4x \sqrt{x^2 + 1} \, dx = \int 4 \underbrace{\sqrt{x^2 + 1}}_{\sqrt{u}} \underbrace{x \, dx}_{\frac{du}{2}} = \int 4\sqrt{u} \, \frac{du}{2}$$

$$= \int 2\sqrt{u} \, du = 2 \cdot \frac{2}{3} u^{3/2} + C = \frac{4}{3}(x^2 + 1)^{3/2} + C \qquad ▮$$

▮ **EXAMPLE 4**   **Rewriting the Integrand**

Evaluate

$$\int \tan x \, dx$$

Solution   The trick here is to rewrite $\tan x = \frac{\sin x}{\cos x}$ and realize that the derivative of the function in the denominator is the function in the numerator (up to a minus sign). The integral is therefore of the type $\int [g'(x)/g(x)]\,dx$, which we discussed before. We use the substitution

$$u = \cos x \qquad \text{with } \frac{du}{dx} = -\sin x \text{ or } -du = \sin x\,dx$$

Then

$$\int \tan x\,dx = \int \frac{\sin x}{\cos x}\,dx = \int \underbrace{\frac{1}{\cos x}}_{\frac{1}{u}}\underbrace{\sin x\,dx}_{-du}$$

$$= -\int \frac{1}{u}\,du = -\ln|u| + C = -\ln|\cos x| + C$$

In Table 5.1 of Section 5.2, we listed $\int \tan x = \ln|\sec x| + C$. This is the same result that we have just obtained, since $-\ln|\cos x| = \ln|\cos x|^{-1} = \ln|\sec x|$. ▌

In Problem 43, you will evaluate $\int \cot x\,dx$, where the same trick as in Example 4 is used. That is, both $\int \tan x\,dx$ and $\int \cot x\,dx$ are special cases of $\int [g'(x)/g(x)]\,dx$. We collect both integrals in the following.

$$\int \tan x\,dx = -\ln|\cos x| + C$$

$$\int \cot x\,dx = \ln|\sin x| + C$$

It is not always obvious that substitution will be successful, as in the following example.

▌EXAMPLE 5   **Substitution and Square Roots**

Evaluate

$$\int x\sqrt{2x-1}\,dx$$

Solution   Obviously, x is not the derivative of $2x - 1$, so this does not seem to fit our scheme. But watch what we do: Set

$$u = 2x - 1 \quad \text{with } \frac{du}{dx} = 2 \text{ or } dx = \frac{du}{2}$$

Since $u = 2x - 1$, we find $x = \frac{1}{2}(u + 1)$. Making all the substitutions, we find

$$\int x\sqrt{2x-1}\,dx = \int \frac{1}{2}(u+1)\sqrt{u}\,\frac{du}{2} = \frac{1}{4}\int \left(u^{3/2} + u^{1/2}\right)\,du$$

$$= \frac{1}{4}\left(\frac{2}{5}u^{5/2} + \frac{2}{3}u^{3/2}\right) + C$$

$$= \frac{1}{10}(2x-1)^{5/2} + \frac{1}{6}(2x-1)^{3/2} + C \qquad ▌$$

Functions in the integrand are not always explicitly given, as in the following example.

▮**EXAMPLE 6**   Assume that $g(x)$ is a differentiable function whose derivative $g'(x)$ is continuous. Evaluate

$$\int g'(x) \cos[g(x)] \, dx$$

Solution   If we set

$$u = g(x) \quad \text{with } \frac{du}{dx} = g'(x) \text{ or } du = g'(x) \, dx$$

then

$$\int g'(x) \cos[g(x)] dx = \int \cos u \, du = \sin u + C$$
$$= \sin[g(x)] + C \qquad \blacksquare$$

6.1.2 Definite Integrals

When we evaluate a definite integral, the FTC (part II) says that we must find an antiderivative of the integrand and then evaluate the antiderivative at the limits of integration. When we use the substitution $u = g(x)$ to find an antiderivative of an integrand, the antiderivative will be given in terms of u at first. To complete the calculation, we can proceed in two ways. 1. We can substitute $g(x)$ for u in the antiderivative and then evaluate the antiderivative at the limits of integration in terms of x. 2. We can leave the antiderivative in terms of u and change the limits of integration according to $u = g(x)$.

We illustrate these two ways in the following example where we wish to evaluate

$$\int_0^4 2x \sqrt{x^2 + 1} \, dx$$

First Way   To find an antiderivative of $f(x) = 2x\sqrt{x^2 + 1}$, we choose the substitution

$$u = x^2 + 1 \quad \text{with } \frac{du}{dx} = 2x \text{ or } du = 2x \, dx$$

Then

$$\int 2x \sqrt{x^2 + 1} \, dx = \int \sqrt{u} \, du = \frac{2}{3} u^{3/2} + C = \frac{2}{3}(x^2 + 1)^{3/2} + C$$

and $F(x) = \frac{2}{3}(x^2 + 1)^{3/2}$ is an antiderivative of $2x\sqrt{x^2 + 1}$. Using the FTC, part II, we can now compute the definite integral:

$$\int_0^4 2x \sqrt{x^2 + 1} \, dx = F(4) - F(0)$$
$$= \frac{2}{3}(17)^{3/2} - \frac{2}{3}(1)^{3/2} = \frac{2}{3}[(17)^{3/2} - 1]$$

The region corresponding to the definite integral before substitution is shown in Figure 6.1.

Second Way   We change the limits of integration along with the substitution. That is, we set

$$u = x^2 + 1 \quad \text{with } \frac{du}{dx} = 2x \text{ or } du = 2x \, dx$$

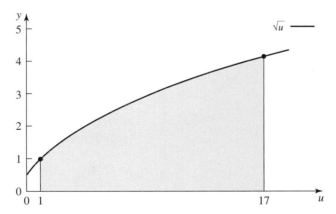

◀ Figure 6.1

The region corresponding to $\int_0^4 2x\sqrt{x^2+1}\,dx$ before substitution has an area of $\frac{2}{3}[(17)^{3/2}-1]$.

as before, and note that

$$\text{if } x = 0 \quad \text{then } u = 1$$
$$\text{if } x = 4 \quad \text{then } u = 17$$

Hence

$$\int_0^4 2x\sqrt{x^2+1}\,dx = \int_1^{17} \sqrt{u}\,du = \frac{2}{3}u^{3/2}\Big]_1^{17} = \frac{2}{3}[(17)^{3/2}-1]$$

After substitution, the integrand is $\sqrt{u}$, and the limits of integration are $u = 1$ and $u = 17$. The region corresponding to the definite integral after substitution is shown in Figure 6.2.

The second way is often more convenient, and we summarize the procedure in the following.

SUBSTITUTION RULE FOR DEFINITE INTEGRALS

If $u = g(x)$, then

$$\int_a^b f[g(x)]g'(x)\,dx = \int_{g(a)}^{g(b)} f(u)\,du$$

▲ Figure 6.2

The definite integral $\int_0^4 2x\sqrt{x^2+1}\,dx$ becomes $\int_1^{17} \sqrt{u}\,du$ after the substitution $u = x^2 + 1$. The region corresponding to $\int_1^{17} \sqrt{u}\,du$ has an area of $\frac{2}{3}[(17)^{3/2}-1]$.

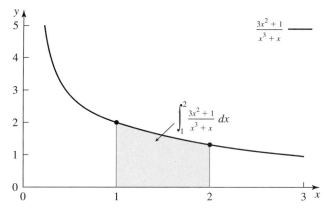

◀ **Figure 6.3**
The region corresponding to
the definite integral in
Example 7.

▮ **EXAMPLE 7**   **Definite Integral**

Compute

$$\int_1^2 \frac{3x^2 + 1}{x^3 + x}\, dx$$

Solution   (The region corresponding to the definite integral is shown in Figure 6.3.)
The integrand is of the form $\frac{g'(x)}{g(x)}$. We set

$$u = x^3 + x \quad \text{with} \quad \frac{du}{dx} = 3x^2 + 1 \text{ or } du = (3x^2 + 1)\, dx$$

and change the limits of integration:

$$\text{if } x = 1 \quad \text{then } u = 2$$
$$\text{if } x = 2 \quad \text{then } u = 10$$

Therefore,

$$\int_1^2 \frac{3x^2 + 1}{x^3 + x}\, dx = \int_2^{10} \frac{1}{u}\, du = \ln |u|]_2^{10} = \ln 10 - \ln 2 = \ln \frac{10}{2} = \ln 5 \qquad ▮$$

▮ **EXAMPLE 8**   **Substitution Function is Decreasing**

Compute

$$\int_{1/2}^1 \frac{1}{x^2} e^{1/x}\, dx$$

Solution   (The region corresponding to the definite integral is shown in Figure 6.4.)
Since $-1/x^2$ is the derivative of $1/x$, we set

$$u = \frac{1}{x} \quad \text{with} \quad \frac{du}{dx} = -\frac{1}{x^2} \text{ or } -du = \frac{1}{x^2}\, dx$$

and change the limits of integration

$$\text{if } x = \frac{1}{2} \quad \text{then } u = 2$$
$$\text{if } x = 1 \quad \text{then } u = 1$$

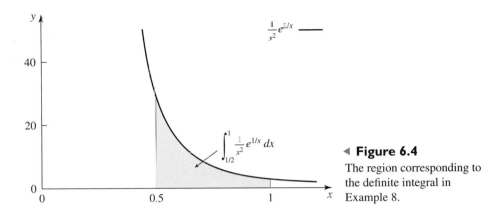

◀ Figure 6.4
The region corresponding to the definite integral in Example 8.

Therefore,

$$\int_{1/2}^{1} \frac{1}{x^2} e^{1/x}\, dx = -\int_{2}^{1} e^{u}\, du = \int_{1}^{2} e^{u}\, du = e^{u}\Big]_{1}^{2} = e^2 - e$$

Note that since $\frac{1}{x}$ is a decreasing function for $x > 0$, the lower limit is greater than the upper limit of integration after the substitution. When we reversed the order of integration in the second step, we removed the negative sign.

However, you need not reverse the order of integration. Instead, you can compute directly

$$-\int_{2}^{1} e^{u}\, du = -e^{u}\Big]_{2}^{1} = -(e - e^2) = e^2 - e$$ ■

┃EXAMPLE 9   **Trigonometric Substitution**

Compute

$$\int_{0}^{\pi/6} \cos x\, e^{\sin x}\, dx$$

Solution   (The region corresponding to the definite integral is shown in Figure 6.5.) The integrand is of the form $g'(x)e^{g(x)}$, which suggests the substitution

$$u = \sin x \quad \text{with} \quad \frac{du}{dx} = \cos x \quad \text{or } du = \cos x\, dx$$

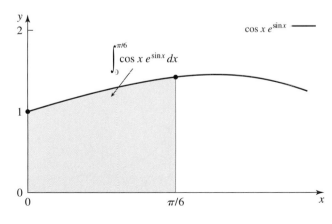

◀ Figure 6.5
The region corresponding to the definite integral in Example 9.

and change the limits of integration:

$$\text{if } x = 0 \quad \text{then } u = \sin 0 = 0$$

$$\text{if } x = \frac{\pi}{6} \quad \text{then } u = \sin\frac{\pi}{6} = \frac{1}{2}$$

Therefore,

$$\int_0^{\pi/6} \cos x\, e^{\sin x}\, dx = \int_0^{1/2} e^u\, du = e^u]_0^{1/2} = e^{1/2} - 1$$

▌ **EXAMPLE 10**   **Rational Function**

Compute

$$\int_4^9 \frac{2}{x - 3}\, dx$$

Solution   (The region corresponding to the definite integral is shown in Figure 6.6.)
We set

$$u = x - 3 \quad \text{with } \frac{du}{dx} = 1 \text{ or } du = dx$$

and change the limits of integration:

$$\text{if } x = 4 \quad \text{then } u = 1$$
$$\text{if } x = 9 \quad \text{then } u = 6$$

Therefore,

$$\int_4^9 \frac{2}{x - 3}\, dx = \int_1^6 \frac{2}{u}\, du = 2\ln|u|]_1^6 = 2(\ln 6 - \ln 1) = 2\ln 6$$

One can easily spend a great deal of time on integration techniques. The problems can get very involved; to solve them all, one would need a big bag full of tricks. There are excellent software programs (such as Mathematica and Matlab) that can integrate symbolically. These programs do not render integration techniques useless, but it has become less important to acquire a large number of tricks.

So far, we have only learned one technique: substitution. Unless you can immediately recognize an antiderivative, substitution is the only method you can try at this point.

As we proceed, you will learn other techniques. An additional complication then will be to recognize which technique to use. If you don't see right away what to

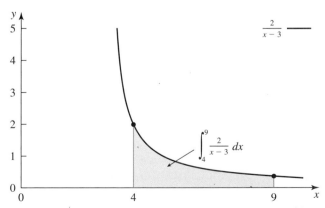

◀ **Figure 6.6**
The region corresponding to the definite integral in Example 10.

do, just try something. Don't always expect the first attempt to succeed. With practice, you will see much more quickly whether or not your approach will succeed. If your attempt does not seem to work, try to determine the reason. That way, failed attempts can be quite useful for gaining experience in integration.

◆ 6.1 PROBLEMS

[6.1.1]

In Problems 1–8, evaluate the indefinite integral by making the given substitution.

1. $\int 2x\sqrt{x^2+3}\,dx$, with $u = x^2 + 3$.

2. $\int 5\cos(3x)\,dx$, with $u = 3x$.

3. $\int 7x^2\sin(4x^3)\,dx$ with $u = 4x^3$.

4. $\int e^{2x+3}\,dx$, with $u = 2x + 3$.

5. $\int xe^{-x^2/2}\,dx$, with $u = -x^2/2$.

6. $\int \dfrac{x+2}{x^2+4x}\,dx$, with $u = x^2 + 4x$.

7. $\int \dfrac{3}{x+4}\,dx$, with $u = x + 4$.

8. $\int \dfrac{1}{5-x}\,dx$, with $u = 5 - x$.

In Problems 9–20, use substitution to evaluate the indefinite integrals.

9. $\int \sqrt{x+3}\,dx$

10. $\int \sqrt{4-x}\,dx$

11. $\int (4x-3)\sqrt{2x^2-3x+2}\,dx$

12. $\int (x^2-2x)\sqrt{x^3-3x^2+3}\,dx$

13. $\int \sin\left(\dfrac{3\pi}{2}x+\dfrac{\pi}{4}\right)\,dx$

14. $\int \cos(2x-1)\,dx$

15. $\int \tan x \sec^2 x\,dx$

16. $\int \sin^3 x \cos x\,dx$

17. $\int \dfrac{(\ln x)^2}{x}\,dx$

18. $\int \dfrac{dx}{(x-3)\ln(x-3)}$

19. $\int x\sqrt{5+x}\,dx$

20. $\int \sqrt{1+\ln x}\,\dfrac{\ln x}{x}\,dx$

In Problems 21–26, a, b, and c are constants, and $g(x)$ is a continuous function whose derivative $g'(x)$ is also continuous. Use substitution to evaluate the indefinite integrals.

21. $\int \dfrac{2cx+b}{ax^2+bx+c}\,dx$

22. $\int \dfrac{1}{ax+b}\,dx$

23. $\int g'(x)[g(x)]^n\,dx$

24. $\int g'(x)\sin[g(x)]\,dx$

25. $\int g'(x)e^{-g(x)}\,dx$

26. $\int \dfrac{g'(x)}{[g(x)]^2+1}\,dx$

[6.1.2]

In Problems 27–42, use substitution to evaluate the definite integrals.

27. $\int_0^3 x\sqrt{x^2+1}\,dx$

28. $\int_1^2 x^2\sqrt{x^3+2}\,dx$

29. $\int_2^3 \dfrac{2x+3}{(x^2+3x)^3}\,dx$

30. $\int_0^2 \dfrac{2x}{\sqrt{4x^2+3}}\,dx$

31. $\int_2^5 (x-2)e^{-(x-2)^2/2}\,dx$

32. $\int_{\ln 4}^{\ln 7} \dfrac{e^x}{(e^x-3)^2}\,dx$

33. $\int_0^{\pi/3} \sin x \cos x\,dx$

34. $\int_{-\pi/6}^{\pi/6} \sin^2 x \cos x\,dx$

35. $\int_0^{\pi/4} \tan x \sec^2 x\,dx$

36. $\int_0^{\pi/3} \dfrac{\sin x}{\cos^2 x}\,dx$

37. $\int_5^c \dfrac{x}{x-3}\,dx$

38. $\int_0^2 \dfrac{x}{x+2}\,dx$

39. $\int_e^{e^2} \dfrac{dx}{x(\ln x)^2}$

40. $\int_1^2 \dfrac{x\,dx}{(x^2+1)\ln(x^2+1)}$

41. $\int_1^9 \dfrac{1}{\sqrt{x}}e^{-\sqrt{x}}\,dx$

42. $\int_0^2 x\sqrt{4-x^2}\,dx$

43. Use the fact that

$$\cot x = \frac{\cos x}{\sin x}$$

to evaluate

$$\int \cot x\,dx$$

◆ 6.2 INTEGRATION BY PARTS

As mentioned at the beginning of this chapter, integration by parts is the product rule in integral form. Let $u = u(x)$ and $v = v(x)$ be differentiable functions, then

$$(uv)' = u'v + uv'$$

or, after rearranging,

$$uv' = (uv)' - u'v$$

Integrating both sides, we find

$$\int uv' \, dx = \int (uv)' \, dx - \int u'v \, dx$$

Since uv is an antiderivative of $(uv)'$,

$$\int (uv)' \, dx = uv + C$$

We therefore find

$$\int uv' \, dx = uv - \int u'v \, dx$$

(Note that the constant C can be absorbed into the indefinite integral on the right-hand side.) Since $u' = du/dx$ and $v' = dv/dx$, we can write this in short form:

$$\int u \, dv = uv - \int v \, du$$

We summarize this result in the following.

INTEGRATION BY PARTS RULE

If $u(x)$ and $v(x)$ are differentiable functions, then

$$\int u(x)v'(x) \, dx = u(x)v(x) - \int u'(x)v(x) \, dx$$

or, in short form,

$$\int u \, dv = uv - \int v \, du$$

You are probably wondering how this will help, given that we traded one integral for another one. Here is a first example.

❙ EXAMPLE 1   **Integration by Parts**

Evaluate

$$\int x \sin x \, dx$$

Solution   The integrand $x \sin x$ is a product of two functions, one of which will be designated as u, the other as v'. Since applying the integration by parts rule will result in another integral of the form $\int u'v \, dx$, we must choose u and v' so that $u'v$ is of a simpler form. This suggests the following choices:

$$u = x \qquad \text{and} \qquad v' = \sin x$$

Since $v = -\cos x$ and $u' = 1$, the integral $\int u'v\,dx$ is then of the form $-\int \cos x\,dx$, which is indeed simpler. We obtain

$$\int x \sin x\,dx = (-\cos x)(x) - \int (-\cos x)(1)\,dx$$

$$= -x \cos x + \int \cos x\,dx$$

$$= -x \cos x + \sin x + C$$

If we had chosen $v' = x$ and $u = \sin x$, then we would have had $v = \frac{1}{2}x^2$ and $u' = \cos x$. The integral $\int u'v\,dx$ would have been of the form $\int \frac{1}{2}x^2 \cos x\,dx$, which is even more complicated than $\int x \sin x\,dx$.

If we use the short form $\int u\,dv = uv - \int v\,du$, we would write

$$u = x \qquad \text{and} \qquad dv = \sin x\,dx$$

Then

$$du = dx \qquad \text{and} \qquad v = -\cos x$$

and

$$\int \underbrace{x}_{u}\ \underbrace{\sin x\,dx}_{dv} = \underbrace{x}_{u}\ \underbrace{(-\cos x)}_{v} - \int \underbrace{(-\cos x)}_{v}\ \underbrace{dx}_{du}$$

$$= -x \cos x + \sin x + C \qquad\qquad ▌$$

▌ EXAMPLE 2   Integration by Parts

Evaluate

$$\int x \ln x\,dx$$

Solution   Since we do not know an antiderivative of $\ln x$, we try

$$u = \ln x \qquad \text{and} \qquad v' = x$$

Then

$$u' = \frac{1}{x} \qquad \text{and} \qquad v = \frac{1}{2}x^2$$

and

$$\int x \ln x\,dx = \frac{1}{2}x^2 \ln x - \int \frac{1}{2}x^2 \cdot \frac{1}{x}\,dx = \frac{1}{2}x^2 \ln x - \frac{1}{4}x^2 + C \qquad\qquad ▌$$

Before we present a few more useful tricks, we show how to evaluate definite integrals with this method.

▌ EXAMPLE 3   Definite Integral

Compute

$$\int_0^1 xe^{-x}\,dx$$

Solution   The region representing the definite integral is shown in Figure 6.7. We set

$$u = x \qquad \text{and} \qquad dv = e^{-x}\,dx$$

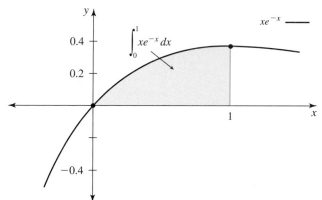

◀ **Figure 6.7**
The region corresponding to the definite integral in Example 3.

Then

$$du = dx \qquad \text{and} \qquad v = -e^{-x}$$

Therefore,

$$\int_0^1 xe^{-x} \, dx = -xe^{-x}\big]_0^1 - \int_0^1 (-e^{-x}) \, dx$$

$$= -1e^{-1} - (-0e^{-0}) + \int_0^1 e^{-x} \, dx$$

$$= -e^{-1} + [-e^{-x}]_0^1 = -e^{-1} + (-e^{-1} - (-e^{-0}))$$

$$= -e^{-1} - e^{-1} + 1 = 1 - 2e^{-1}$$

▮

In the next two examples, we demonstrate a trick that is sometimes useful when integrating by parts. The technique is called "multiplying by 1."

▮**EXAMPLE 4**   **Multiplying by 1**

Evaluate

$$\int \ln x \, dx$$

Solution   The integrand $\ln x$ is not a product of two functions, but we can write it as $(1)(\ln x)$ and set

$$u = \ln x \qquad \text{and} \qquad v' = 1$$

Then

$$u' = \frac{1}{x} \qquad \text{and} \qquad v = x$$

We find

$$\int \ln x \, dx = \int (1)(\ln x) \, dx = x \ln x - \int x \frac{1}{x} \, dx$$

$$= x \ln x - \int 1 \, dx = x \ln x - x + C$$

Our choices for u and v' might surprise you, as we said that our goal was to make the integral look simpler, which often means that we try to reduce the power of functions of the form x^n. In this case, however, integrating 1 and differentiating $\ln x$ yielded a simpler integral. In fact, if we had chosen $u' = \ln x$ and $v = 1$, we would not have been able to carry out the integration by parts since we would have needed the antiderivative of $\ln x$ to compute uv and $\int uv' \, dx$.

If you prefer the short form notation, you don't need to multiply by 1 since

$$u = \ln x \qquad \text{and} \qquad dv = dx$$

with

$$du = \frac{1}{x}dx \qquad \text{and} \qquad v = x$$

produces immediately

$$\int \ln x \, dx = x \ln x - \int x \frac{1}{x} dx = x \ln x - \int dx$$
$$= x \ln x - x + C$$

▮

▮ EXAMPLE 5   Multiplying by 1

Evaluate

$$\int \tan^{-1} x \, dx$$

Solution   We write $\tan^{-1} x = (1)(\tan^{-1} x)$, and

$$u = \tan^{-1} x \qquad \text{and} \qquad v' = 1$$

Then

$$u' = \frac{1}{x^2 + 1} \qquad \text{and} \qquad v = x$$

We find

$$\int \tan^{-1} x \, dx = x \tan^{-1} x - \int \frac{x}{x^2 + 1} dx$$

We need to use substitution to evaluate the integral on the right-hand side. With

$$w = x^2 + 1 \qquad \text{and} \qquad \frac{dw}{dx} = 2x$$

we obtain

$$\int \frac{x}{x^2 + 1} dx = \frac{1}{2} \int \frac{dw}{w} = \frac{1}{2} \ln|w| + C = \frac{1}{2} \ln(x^2 + 1) + C$$

Hence,

$$\int \tan^{-1} x \, dx = x \tan^{-1} x - \frac{1}{2} \ln(x^2 + 1) + C$$

▮

▮ EXAMPLE 6   Using Integration by Parts Repeatedly

Compute

$$\int_0^1 x^2 e^x \, dx$$

Solution   When you evaluate a definite integral using integration by parts, it is often easier to evaluate the indefinite integral first and then to use the FTC, part II, to compute the definite integral. To evaluate $\int x^2 e^x \, dx$, we set

$$u = x^2 \qquad \text{and} \qquad v' = e^x$$

Then

$$u' = 2x \qquad \text{and} \qquad v = e^x$$

Therefore

$$\int x^2 e^x \, dx = x^2 e^x - \int 2x e^x \, dx \tag{6.3}$$

To evaluate the integral $\int x e^x \, dx$ we must use integration by parts for a second time. We set

$$u = x \qquad \text{and} \qquad v' = e^x$$

Then

$$u' = 1 \qquad \text{and} \qquad v = e^x$$

Therefore,

$$\int x e^x \, dx = x e^x - \int e^x \, dx = x e^x - e^x + C \tag{6.4}$$

Combining (6.3) and (6.4), we find

$$\int x^2 e^x \, dx = x^2 e^x - 2[x e^x - e^x + C]$$
$$= x^2 e^x - 2x e^x + 2e^x - 2C$$

After evaluating the indefinite integral, we can now compute the definite integral. We set $F(x) = x^2 e^x - 2x e^x + 2e^x$. Then

$$\int_0^1 x^2 e^x \, dx = F(1) - F(0)$$
$$= (e - 2e + 2e) - (0 - 0 + 2) = e - 2 \qquad \blacksquare$$

▮ **EXAMPLE 7**   **Using Integration by Parts Repeatedly**

Evaluate

$$\int e^x \cos x \, dx$$

Solution   You can check that it does not matter which of the functions you call u and which v'. We set

$$u = \cos x \qquad \text{and} \qquad v' = e^x$$

Then

$$u' = -\sin x \qquad \text{and} \qquad v = e^x$$

Therefore,

$$\int e^x \cos x \, dx = e^x \cos x + \int e^x \sin x \, dx \tag{6.5}$$

We apply the integration by parts rule a second time, with

$$u = \sin x \qquad \text{and} \qquad v' = e^x$$

Then

$$u' = \cos x \qquad \text{and} \qquad v = e^x$$

Therefore,

$$\int e^x \sin x \, dx = e^x \sin x - \int e^x \cos x \, dx \tag{6.6}$$

Combining (6.5) and (6.6) yields

$$\int e^x \cos x \, dx = e^x \cos x + e^x \sin x - \int e^x \cos x \, dx$$

We see that the integral $\int e^x \cos x \, dx$ appears on both sides. Rearranging the equation, we obtain

$$2 \int e^x \cos x \, dx = e^x \cos x + e^x \sin x + C_1$$

or

$$\int e^x \cos x \, dx = \frac{1}{2} e^x (\cos x + \sin x) + C$$

with $C = C_1/2$. (Note that we introduced the constant C_1 (and C) in the final answer.) ∎

We conclude this section with a piece of practical advice: In integrals of the form $\int P(x) \sin(ax) \, dx$, $\int P(x) \cos(ax) \, dx$, and $\int P(x) e^{ax} \, dx$, where $P(x)$ is a polynomial and a is a constant, the polynomial $P(x)$ should be considered as u and the expressions $\sin(ax)$, $\cos(ax)$ and e^{ax} as v'. If an integral contains functions $\ln x$, $\tan^{-1} x$, or $\sin^{-1} x$, then these functions are usually treated as u.

◆ 6.2 PROBLEMS

In Problems 1–18, use integration by parts to evaluate the integrals.

1. $\displaystyle\int x \cos x \, dx$

2. $\displaystyle\int x \sin(2x) \, dx$

3. $\displaystyle\int 2x e^{-x} \, dx$

4. $\displaystyle\int 2x^2 e^{-x} \, dx$

5. $\displaystyle\int x \ln x \, dx$

6. $\displaystyle\int x^2 \ln x \, dx$

7. $\displaystyle\int x \sec^2 x \, dx$

8. $\displaystyle\int x \csc^2 x \, dx$

9. $\displaystyle\int_0^{\pi/3} x \sin x \, dx$

10. $\displaystyle\int_1^2 \ln x \, dx$

11. $\displaystyle\int_1^4 \ln \sqrt{x} \, dx$

12. $\displaystyle\int_1^4 \sqrt{x} \ln \sqrt{x} \, dx$

13. $\displaystyle\int_0^3 x^2 e^{-x} \, dx$

14. $\displaystyle\int_0^{\pi/3} e^x \sin x \, dx$

15. $\displaystyle\int e^{-3x} \cos\left(\frac{\pi}{2} x\right) dx$

16. $\displaystyle\int e^{-2x} \sin\left(\frac{x}{2}\right) dx$

17. $\displaystyle\int \sin(\ln x) \, dx$

18. $\displaystyle\int \cos(\ln x) \, dx$

19. Evaluating the integral

$$\int \cos^2 x \, dx$$

requires two steps.

First, write

$$\cos^2 x = (\cos x)(\cos x)$$

and apply integration by parts to show that

$$\int \cos^2 x \, dx = \sin x \cos x + \int \sin^2 x \, dx$$

Then, use $\sin^2 x + \cos^2 x = 1$ to replace $\sin^2 x$ in the integral on the right-hand side, and complete the integration of $\int \cos^2 x \, dx$.

20. Evaluating the integral

$$\int \sin^2 x \, dx$$

requires two steps.
First, write

$$\sin^2 x = (\sin x)(\sin x)$$

and apply integration by parts to show that

$$\int \sin^2 x \, dx = -\sin x \cos x + \int \cos^2 x \, dx$$

Then, use $\sin^2 x + \cos^2 x = 1$ to replace $\cos^2 x$ in the integral on the right-hand side, and complete the integration of $\int \sin^2 x \, dx$.

21. Evaluating the integral

$$\int \arcsin x \, dx$$

requires two steps.
(a) Write

$$\arcsin x = 1 \cdot \arcsin x$$

and apply integration by parts once to show that

$$\int \arcsin x \, dx = x \arcsin x - \int \frac{x}{\sqrt{1 - x^2}} \, dx$$

(b) Use substitution to compute

$$\int \frac{x}{\sqrt{1 - x^2}} \, dx \qquad (6.7)$$

and combine your result in (a) and (6.7) to complete the computation of $\int \arcsin x \, dx$.

22. (a) Use integration by parts to show that for $x > 0$

$$\int \frac{1}{x} \ln x \, dx = (\ln x)^2 - \int \frac{1}{x} \ln x \, dx$$

(b) Use your result in (a) to evaluate

$$\int \frac{1}{x} \ln x \, dx$$

23. (a) Use integration by parts to show that

$$\int x^n e^x \, dx = x^n e^x - n \int x^{n-1} e^x \, dx$$

Such formulas are called **reduction formulas** since they reduce the exponent of x by 1 each time they are applied.

(b) Apply the reduction formula in (a) repeatedly to compute

$$\int x^3 e^x \, dx$$

24. (a) Use integration by parts to verify the validity of the reduction formula

$$\int x^n e^{ax} \, dx = \frac{1}{a} x^n e^{ax} - \frac{n}{a} \int x^{n-1} e^{ax} \, dx$$

where a is a constant not equal to 0.

(b) Apply the reduction formula in (a) to compute

$$\int x^2 e^{-3x} \, dx$$

25. (a) Use integration by parts to verify the validity of the reduction formula

$$\int (\ln x)^n \, dx = x(\ln x)^n - n \int (\ln x)^{n-1} \, dx$$

(b) Apply the reduction formula in (a) repeatedly to compute

$$\int (\ln x)^3 \, dx$$

In Problems 26–31, first make an appropriate substitution, and then use integration by parts to evaluate the indefinite integrals.

26. $\int \cos \sqrt{x} \, dx$ **27.** $\int \sin \sqrt{x} \, dx$

28. $\int x^3 e^{-x^2/2} \, dx$ **29.** $\int x^5 e^{x^2} \, dx$

30. $\int e^{\sqrt{x}} \, dx$ **31.** $\int \ln(\sqrt{x} + 1) \, dx$

◆ 6.3 PRACTICING INTEGRATION AND PARTIAL FRACTIONS

In the preceding two sections, we learned the two main integration techniques, substitution and integration by parts. One of the major difficulties in integration is deciding which rule to use. This takes a lot of practice. This section is devoted to practicing integration; in addition, we will learn how to integrate some rational functions.

6.3.1 Practicing Integration

EXAMPLE 1 Find

$$\int \tan x \sec^2 x e^{\tan x} \, dx$$

Solution Since $\frac{d}{dx} \tan x = \sec^2 x$, we try the substitution $w = \tan x$. Then $dw = \sec^2 x \, dx$ and

$$\int \tan x \sec^2 x e^{\tan x} \, dx = \int w e^w \, dw$$

To continue, we need to use integration by parts, with

$$u = w \qquad \text{and} \qquad v' = e^w$$

Then

$$u' = 1 \qquad \text{and} \qquad v = e^w$$

and

$$\int we^w\,dw = we^w - \int e^w\,dw = we^w - e^w + C$$

With $w = \tan x$, we therefore find

$$\int \tan x\,\sec^2 x e^{\tan x}\,dx = e^{\tan x}(\tan x - 1) + C$$ ∎

Frequently, we must perform algebraic manipulations of the integrand before we can integrate.

∎ EXAMPLE 2 · Find

$$\int_0^{\sqrt{3}} \frac{1}{9 + x^2}\,dx$$

Solution   The region corresponding to the definite integral is shown in Figure 6.8. The integrand should remind you of the function $\frac{1}{1+u^2}$, whose antiderivative is $\tan^{-1} u$. However, we have a 9 in the denominator. To get a 1 there, we factor 9 in the denominator, and find

$$\frac{1}{9 + x^2} = \frac{1}{9\left(1 + \frac{x^2}{9}\right)} = \frac{1}{9\left(1 + (\frac{x}{3})^2\right)}$$

The last expression now suggests that we should try the substitution

$$u = \frac{x}{3} \qquad \text{with} \qquad dx = 3\,du$$

Since we wish to evaluate a definite integral, we must change the limits of integration as well. We find that $x = 0$ corresponds to $u = 0$ and $x = \sqrt{3}$ corresponds to $u = \frac{1}{3}\sqrt{3}$. We end up with

$$\int_0^{\sqrt{3}} \frac{1}{9 + x^2}\,dx = \frac{1}{9}\int_0^{\sqrt{3}} \frac{1}{1 + (\frac{x}{3})^2}\,dx = \frac{1}{9}\int_0^{\frac{1}{3}\sqrt{3}} \frac{3}{1 + u^2}\,du$$

$$= \frac{1}{3}\int_0^{\frac{1}{3}\sqrt{3}} \frac{1}{1 + u^2}\,du = \frac{1}{3}\tan^{-1} u\Big]_0^{\frac{1}{3}\sqrt{3}}$$

$$= \frac{1}{3}\left[\tan^{-1}\left(\frac{1}{3}\sqrt{3}\right) - \tan^{-1} 0\right] = \frac{1}{3}\left(\frac{\pi}{6} - 0\right) = \frac{\pi}{18}$$ ∎

The next example shows that simplifying the integrand first can help greatly.

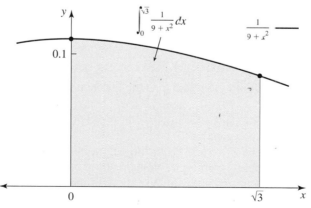

◄ **Figure 6.8**
The region corresponding to the definite integral in Example 2.

I EXAMPLE 3   Find

$$\int x^{1/2} \ln(x^{1/2}e^x)\, dx$$

Solution   Before we try any of our techniques, let's simplify the logarithm first. We find

$$\ln(x^{1/2}e^x) = \ln x^{1/2} + \ln e^x = \frac{1}{2}\ln x + x$$

The integral now becomes

$$\int x^{1/2} \ln(x^{1/2}e^x)\, dx = \int x^{1/2}\left(\frac{1}{2}\ln x + x\right) dx$$

$$= \frac{1}{2}\int x^{1/2}\ln x\, dx + \int x^{3/2}\, dx$$

We can evaluate the first integral using integration by parts, with

$$u = \ln x \qquad \text{and} \qquad v' = x^{1/2}$$

Then

$$u' = \frac{1}{x} \qquad \text{and} \qquad v = \frac{2}{3}x^{3/2}$$

We find for the first integral

$$\int x^{1/2}\ln x\, dx = \frac{2}{3}x^{3/2}\ln x - \int \frac{1}{x}\cdot\frac{2}{3}x^{3/2}\, dx$$

$$= \frac{2}{3}x^{3/2}\ln x - \frac{2}{3}\int x^{1/2}\, dx$$

$$= \frac{2}{3}x^{3/2}\ln x - \frac{2}{3}\cdot\frac{2}{3}x^{3/2} + C$$

$$= \frac{2}{3}x^{3/2}\left(\ln x - \frac{2}{3}\right) + C$$

The other integral is straightforward:

$$\int x^{3/2}\, dx = \frac{2}{5}x^{5/2} + C$$

Combining our results, we find

$$\int x^{1/2}\ln(x^{1/2}e^x)\, dx = \frac{1}{2}\int x^{1/2}\ln\, dx + \int x^{3/2}dx$$

$$= \frac{1}{3}x^{3/2}\left(\ln x - \frac{2}{3}\right) + \frac{2}{5}x^{5/2} + C \qquad\blacksquare$$

6.3.2 Rational Functions and Partial Fractions

A rational function f is the quotient of two polynomials. That is,

$$f(x) = \frac{P(x)}{Q(x)} \tag{6.8}$$

where $P(x)$ and $Q(x)$ are polynomials. To integrate rational functions, we use an algebraic technique, called method of **partial fractions**, to write $f(x)$ as a sum of a polynomial and simpler rational functions. Such a sum is called a **partial fraction decomposition**. These simpler rational functions, which can be integrated with the methods we have learned, are of the form

$$\frac{A}{(ax+b)^n} \quad \text{or} \quad \frac{Bx+C}{(ax^2+bx+c)^n} \qquad (6.9)$$

where A, B, C, a, b, and c are constants, and n is a positive integer. In this form, the quadratic polynomial ax^2+bx+c can no longer be factored into a product of two linear functions with real coefficients. Such polynomials are called **irreducible**. (In other words, if ax^2+bx+c is irreducible, then $ax^2+bx+c = 0$ has no real roots.)

If the degree of $P(x)$ in (4.8) is greater than or equal to the degree of $Q(x)$, then the first step in the partial fraction decomposition is to use long division to write $f(x)$ as a sum of a polynomial and a rational function, where the degree of the polynomial in the numerator is less than the degree of the polynomial in the denominator (such rational functions are called **proper**). We illustrate this step in the following two examples.

▮ **EXAMPLE 4**   **Long Division before Integration**

Find

$$\int \frac{x}{x+2}\,dx$$

Solution   The degree of the numerator is equal to the degree of the denominator; using long division or writing the integrand in the form

$$\frac{x}{x+2} = \frac{x+2-2}{x-2} = 1 - \frac{2}{x+2}$$

results in a polynomial of degree 0 and a proper rational function. We can integrate the integrand in this new form:

$$\int \frac{x}{x+2}\,dx = \int \left(1 - \frac{2}{x+2}\right)\,dx = x - 2\ln|x+2| + C \qquad \blacksquare$$

▮ **EXAMPLE 5**   **Long Division before Integration**

Find

$$\int \frac{3x^3 - 7x^2 + 17x - 3}{x^2 - 2x + 5}\,dx$$

Solution   Since the degree of the numerator is higher than the degree of the denominator, we use long division to simplify the integrand.

$$
\begin{array}{r}
3x - 1 \\
x^2 - 2x + 5 \overline{) 3x^3 - 7x^2 + 17x - 3} \\
\underline{3x^3 - 6x + 15x} \\
-x^2 + 2x - 3 \\
\underline{-x^2 + 2x - 5} \\
2
\end{array}
$$

That is,

$$\frac{3x^3 - 7x^2 + 17x - 3}{x^2 - 2x + 5} = 3x - 1 + \frac{2}{x^2 - 2x + 5}$$

Now, $x^2 - 2x + 5 = 0$ does not have real solutions, which means that it is irreducible. Instead, we will complete the square. That is,

$$x^2 - 2x + 5 = (x^2 - 2x + 1) + 4 = (x - 1)^2 + 4$$

The integral we wish to evaluate is therefore

$$\int \left(3x - 1 + \frac{2}{(x-1)^2 + 4}\right) dx = \int (3x - 1)\, dx + 2 \int \frac{1}{(x-1)^2 + 4}\, dx$$

The first integral is easy; we find

$$\int (3x - 1)\, dx = \frac{3}{2}x^2 - x + C$$

To evaluate the second integral, we use a similar trick as in Example 2 of this section. Namely, we factor 4 in the denominator, and find

$$\int \frac{1}{(x-1)^2 + 4}\, dx = \frac{1}{4} \int \frac{1}{1 + \left(\frac{x-1}{2}\right)^2}\, dx$$

Setting

$$u = \frac{x - 1}{2} \qquad \text{with} \qquad dx = 2\, du$$

yields

$$\frac{1}{4} \int \frac{1}{1 + \left(\frac{x-1}{2}\right)^2}\, dx = \frac{1}{4} \int \frac{2\, du}{1 + u^2}$$

$$= \frac{1}{2} \tan^{-1} u + C = \frac{1}{2} \tan^{-1}\left(\frac{x-1}{2}\right) + C$$

Hence,

$$\int \frac{3x^3 - 7x^2 + 17x - 3}{x^2 - 2x + 5}\, dx = \frac{3}{2}x^2 - x + \tan^{-1}\left(\frac{x-1}{2}\right) + C \qquad \blacksquare$$

Assume now that the rational function is proper. Unless the integrand is already of one of the types in (6.9), we need to decompose it further. A result in algebra tells us that every polynomial can be written as a product of linear and irreducible quadratic factors. This is the key to the partial fraction decomposition. We factor the denominator into a product of linear and irreducible quadratic factors. The linear factors then correspond to a sum of rational functions given as follows:

If the linear factor $ax + b$ is contained n times in the factorization of the denominator of a proper rational function, then the partial fraction decomposition contains terms of the form

$$\frac{A_1}{ax + b} + \frac{A_2}{(ax + b)^2} + \cdots + \frac{A_n}{(ax + b)^n}$$

where $A_1, A_2, \ldots, A_n$ are constants.

EXAMPLE 6   Distinct Linear Factors

Find

$$\int \frac{1}{x(x - 1)}\, dx$$

Solution   The integrand is a rational function whose denominator is a product of two distinct linear functions. We claim that the integrand can be written in the form

$$\frac{1}{x(x-1)} = \frac{A}{x} + \frac{B}{x-1} \tag{6.10}$$

where A and B are constants that we need to determine. To find A and B, we write the right-hand side of (6.10) with a common denominator. That is,

$$\frac{A}{x} + \frac{B}{x-1} = \frac{A(x-1) + Bx}{x(x-1)} = \frac{(A+B)x - A}{x(x-1)}$$

Since this must be equal to $\frac{1}{x(x-1)}$, we conclude that

$$(A+B)x - A = 1$$

Therefore,

$$A + B = 0 \quad \text{and} \quad -A = 1$$

This yields $A = -1$ and $B = -A = 1$. We thus find

$$\frac{1}{x(x-1)} = -\frac{1}{x} + \frac{1}{x-1}$$

The integral can now be written as a sum of two rational functions, which can be integrated immediately:

$$\int \frac{1}{x(x-1)}\, dx = \int \left[\frac{1}{x-1} - \frac{1}{x} \right] dx = \int \frac{1}{x-1}\, dx - \int \frac{1}{x}\, dx$$

$$= \ln|x-1| - \ln|x| + C = \ln\left| \frac{x-1}{x} \right| + C \qquad \blacksquare$$

▌EXAMPLE 7   Repeated Linear Factors

Evaluate

$$\int \frac{dx}{x^2(x+1)}$$

Solution   The integrand is a proper rational function whose denominator is a product of three linear functions, namely, x, x (again), and $x + 1$. We claim that we can write the integrand in the form

$$\frac{1}{x^2(x+1)} = \frac{A}{x} + \frac{B}{x^2} + \frac{C}{x+1} \tag{6.11}$$

where A, B, and C are constants. As in the previous example, we find A, B, and C by writing the right-hand side of (6.11) with a common denominator. This yields

$$\frac{A}{x} + \frac{B}{x^2} + \frac{C}{x+1} = \frac{Ax(x+1) + B(x+1) + Cx^2}{x^2(x+1)}$$

$$= \frac{Ax^2 + Ax - Bx + B + Cx^2}{x^2(x+1)}$$

$$= \frac{(A+C)x^2 + (A+B)x + B}{x^2(x+1)}$$

Comparing the last expression with the left-hand side of (6.11), we conclude that

$$A + C = 0, \quad A + B = 0, \quad \text{and } B = 1$$

This implies $A = -1$ and $C = 1$. Therefore,

$$\frac{1}{x^2(x+1)} = -\frac{1}{x} + \frac{1}{x^2} + \frac{1}{x+1}$$

and

$$\int \frac{1}{x^2(x+1)}\, dx = \int \left(-\frac{1}{x} + \frac{1}{x^2} + \frac{1}{x+1} \right) dx$$

$$= -\ln|x| - \frac{1}{x} + \ln|x+1| + C \qquad ▌$$

Irreducible quadratic factors in the denominator of a proper rational functions are dealt with as follows in the partial fraction decomposition.

> If the irreducible quadratic factor $ax^2 + bx + c$ is contained n times in the factorization of the denominator of a proper rational function, then the partial fraction decomposition contains terms of the form
>
> $$\frac{B_1 x + C_1}{ax^2 + bx + c} + \frac{B_2 x + C_2}{(ax^2 + bx + c)^2} + \cdots + \frac{B_n x + C_n}{(ax^2 + bx + c)^n}$$

▌**EXAMPLE 8** **Distinct Irreducible Quadratic Factors**

Evaluate

$$\int \frac{2x^3 - x^2 + 2x - 2}{(x^2 + 2)(x^2 + 1)}\, dx$$

Solution   The rational function in the integrand is proper. The denominator is already factored, each factor is an irreducible quadratic polynomial, and the two factors are distinct. We can therefore write the integrand as

$$\frac{2x^3 - x^2 + 2x - 2}{(x^2 + 2)(x^2 + 1)} = \frac{Ax + B}{x^2 + 2} + \frac{Cx + D}{x^2 + 1}$$

$$= \frac{(Ax + B)(x^2 + 1) + (Cx + D)(x^2 + 2)}{(x^2 + 2)(x^2 + 1)}$$

$$= \frac{Ax^3 + Ax + Bx^2 + B + Cx^3 + 2Cx + Dx^2 + 2D}{(x^2 + 2)(x^2 + 1)}$$

$$= \frac{(A + C)x^3 + (B + D)x^2 + (A + 2C)x + B + 2D}{(x^2 + 2)(x^2 + 1)}$$

Comparing the last expression with the integrand, we find

$$A + C = 2, \quad B + D = -1, \quad A + 2C = 2, \quad \text{and } B + 2D = -2$$

which yields $C = 0$ (write $A + 2C = 2$ as $A + C + C = 2$ and use $A + C = 2$) and $D = -1$ (write $B + 2D = -2$ as $B + D + D = -2$ and use $B + D = -1$). Then, $A = 2$ and $B = 0$. Therefore,

$$\frac{2x^3 - x^2 + 2x - 2}{(x^2 + 2)(x^2 + 1)} = \frac{2x}{x^2 + 2} - \frac{1}{x^2 + 1}$$

and

$$\int \frac{2x^3 - x^2 + 2x - 2}{(x^2 + 2)(x^2 + 1)} \, dx = \int \frac{2x}{x^2 + 2} \, dx - \int \frac{1}{x^2 + 1} \, dx$$

$$= \int \frac{du}{u} - \tan^{-1} x + C = \ln |u| - \tan^{-1} x + C$$

$$= \ln |x^2 + 2| - \tan^{-1} x + C$$

where we used the substitution $u = x^2 + 2$. ▮

▮ EXAMPLE 9   Repeated Irreducible Quadratic Factors

Evaluate

$$\int \frac{x^2 + x + 1}{(x^2 + 1)^2} \, dx$$

Solution   The rational function $\frac{x^2+x+1}{(x^2+1)^2}$ is proper since the numerator is a polynomial of degree 2 and the denominator is of degree 4 (because $(x^2+1)^2 = x^4+2x^2+1$). The denominator contains the irreducible quadratic factor x^2+1 twice. We can therefore write the integrand as

$$\frac{x^2 + x + 1}{(x^2 + 1)^2} = \frac{Ax + B}{x^2 + 1} + \frac{Cx + D}{(x^2 + 1)^2} = \frac{(Ax + B)(x^2 + 1) + Cx + D}{(x^2 + 1)^2}$$

$$= \frac{Ax^3 + Ax + Bx^2 + B + Cx + D}{(x^2 + 1)^2}$$

$$= \frac{Ax^3 + Bx^2 + (A + C)x + (B + D)}{(x^2 + 1)^2}$$

Comparing the last expression with the integrand, we conclude

$$A = 0, \quad B = 1, \quad A + C = 1, \quad \text{and } B + D = 1$$

which implies $C = 1$ and $D = 0$. Therefore

$$\frac{x^2 + x + 1}{(x^2 + 1)^2} = \frac{1}{x^2 + 1} + \frac{x}{(x^2 + 1)^2}$$

and

$$\int \frac{x^2 + x + 1}{(x^2 + 1)^2} \, dx = \int \frac{1}{x^2 + 1} \, dx + \int \frac{x}{(x^2 + 1)^2} \, dx$$

The first integral on the right-hand side is $\tan^{-1} x + C$. To evaluate the second integral on the right-hand side, we use substitution: $u = x^2 + 1$ with $du/2 = x \, dx$. This yields

$$\int \frac{x}{(x^2 + 1)^2} \, dx = \frac{1}{2} \int \frac{du}{u^2} = -\frac{1}{2u} - C = -\frac{1}{2(x^2 + 1)} + C$$

Combining these two results, we find

$$\int \frac{x^2 + x + 1}{(x^2 + 1)^2} \, dx = \tan^{-1} x - \frac{1}{2(x^2 + 1)} + C \qquad ▮$$

We conclude this section by providing a summary of the two most important cases, namely, when the integrand is a rational function for which the denominator is a polynomial of degree 2 and either a product of two distinct linear factors or an irreducible quadratic polynomial.

The first step is to make sure that the degree of the numerator is less than the degree of the denominator. If not, use long division to simplify the integrand.

We will now assume that the degree of the numerator is strictly less than the degree of the denominator (that is, the integrand is a proper rational function). We write the rational function $f(x)$ as

$$f(x) = \frac{P(x)}{Q(x)}$$

with $Q(x) = ax^2 + bx + c, a \neq 0$, and $P(x) = rx + s$. Either $Q(x)$ can be factored into two linear factors, or it is irreducible (that is, does not have real roots).

Case 1   Q(x) is a product of two distinct linear factors.   In this case, we write

$$Q(x) = a(x - x_1)(x - x_2)$$

where x_1 and x_2 are the roots of $Q(x)$. We then use the method of partial fractions to simplify the rational function:

$$\frac{P(x)}{Q(x)} = \frac{rx + s}{ax^2 + bx + c} = \frac{1}{a}\left[\frac{A}{x - x_1} + \frac{B}{x - x_2}\right]$$

The constants A and B must be determined, as in Example 6.

Case 2   Q(x) is an irreducible quadratic polynomial.   In this case,

$$Q(x) = ax^2 + bx + c \quad \text{with } b^2 - 4ac < 0$$

and we must complete the square as in Example 5. This then leads to integrals of the form

$$\int \frac{dx}{x^2 + 1} \qquad \text{or} \qquad \int \frac{x}{x^2 + 1}\,dx$$

The first integral is $\tan^{-1} x + C$, whereas the second integral can be evaluated using substitution.

◆ 6.3 PROBLEMS

[6.3.1]

In Problems 1–12, use either substitution or integration by parts to evaluate each integral.

1. $\displaystyle\int xe^{-2x}\,dx$

2. $\displaystyle\int xe^{-2x^2}\,dx$

3. $\displaystyle\int \frac{1}{\tan x}\,dx$

4. $\displaystyle\int \frac{1}{\csc x \sec x}\,dx$

5. $\displaystyle\int 2x \sin(x^2)\,dx$

6. $\displaystyle\int 2x^2 \sin x\,dx$

7. $\displaystyle\int \frac{1}{16 + x^2}\,dx$

8. $\displaystyle\int \frac{1}{x^2 + 5}\,dx$

9. $\displaystyle\int \frac{x}{x + 3}\,dx$

10. $\displaystyle\int \frac{1}{x^2 + 3}\,dx$

11. $\displaystyle\int \frac{x}{x^2 + 3}\,dx$

12. $\displaystyle\int \frac{x + 2}{x^2 + 2}\,dx$

13. The integral

$$\int \ln x\,dx$$

can be evaluated in two ways.

(a) Write $\ln x = 1 \cdot \ln x$ and use integration by parts to evaluate the integral.

(b) Use the substitution $u = \ln x$ and integration by parts to evaluate the integral.

14. Use an appropriate substitution followed by integration by parts to evaluate

$$\int x^3 e^{-x^2/2}\,dx$$

15. Use an appropriate substitution to evaluate

$$\int x(x - 2)^{1/4}\,dx$$

16. Use an appropriate substitution after simplifying the integrand to evaluate

$$\int \frac{\sin^2 x - \cos^2 x}{(\sin x - \cos x)^2} \, dx$$

In Problems 17–22, evaluate each definite integral.

17. $\int_1^4 e^{\sqrt{x}} \, dx$

18. $\int_1^2 \ln(x^2 e^x) \, dx$

19. $\int_{-1}^0 \frac{2}{1 + x^2} \, dx$

20. $\int_1^2 x^2 \ln x \, dx$

21. $\int_0^{\pi/4} e^x \sin x \, dx$

22. $\int_0^{\pi/6} (1 + \tan^2 x) \, dx$

[6.3.2]

23. (a) Find A and B so that

$$\frac{1}{x(x - 2)} = \frac{A}{x} + \frac{B}{x - 2}$$

(b) Use the partial fraction decomposition in (a) to evaluate

$$\int \frac{1}{x(x - 2)} \, dx$$

24. (a) Find A and B so that

$$\frac{1}{(x - 1)(x + 2)} = \frac{A}{x - 1} + \frac{B}{x + 2}$$

(b) Use the partial fraction decomposition in (a) to evaluate

$$\int \frac{1}{(x - 1)(x + 2)} \, dx$$

25. Complete the square in the denominator to evaluate

$$\int \frac{1}{x^2 - 4x + 13} \, dx$$

26. Complete the square in the denominator to evaluate

$$\int \frac{1}{x^2 + 2x + 5} \, dx$$

In Problems 27–36, evaluate each integral.

27. $\int \frac{1}{(x - 3)(x + 2)} \, dx$

28. $\int \frac{2x - 1}{(x + 4)(x + 1)} \, dx$

29. $\int \frac{1}{x^2 - 9} \, dx$

30. $\int \frac{1}{x^2 + 9} \, dx$

31. $\int \frac{1}{x^2 - x - 2} \, dx$

32. $\int \frac{1}{x^2 - x + 2} \, dx$

33. $\int \frac{x^2 + 1}{x^2 + 3x + 2} \, dx$

34. $\int \frac{x^3 + 1}{x^2 + 3} \, dx$

35. $\int \frac{x^2 + 4}{x^2 - 4} \, dx$

36. $\int \frac{x^4 + 3}{x^2 - 4x + 3} \, dx$

In Problems 37–44, evaluate each definite integral.

37. $\int_3^5 \frac{x - 1}{x} \, dx$

38. $\int_3^5 \frac{x}{x - 1} \, dx$

39. $\int_0^1 \frac{x}{x^2 + 1} \, dx$

40. $\int_1^2 \frac{x^2 + 1}{x} \, dx$

41. $\int_2^3 \frac{1}{1 - x} \, dx$

42. $\int_2^3 \frac{1}{1 - x^2} \, dx$

43. $\int_0^1 \tan^{-1} x \, dx$

44. $\int_0^1 x \tan^{-1} x \, dx$

In Problems 45–52, evaluate each integral.

45. $\int \frac{1}{(x + 1)^2 x} \, dx$

46. $\int \frac{1}{x^2(x - 1)^2} \, dx$

47. $\int \frac{4}{(1 - x)(1 + x)^2} \, dx$

48. $\int \frac{2x^2 + 2x - 1}{x^3(x - 3)} \, dx$

49. $\int \frac{1}{(x^2 + 9)^2} \, dx$

50. $\int \frac{1}{(x^2 + x + 1)^2} \, dx$

51. $\int \frac{1}{x^2(x^2 + 1)} \, dx$

52. $\int \frac{1}{(x + 1)^2(x^2 + 1)} \, dx$

◆ 6.4 IMPROPER INTEGRALS

In this section we discuss definite integrals of two types with the following characteristics:

1. One or both limits of integration are infinite, that is, the integration interval is unbounded; or

2. The integrand becomes infinite at one or more points of the interval of integration.

We call such integrals **improper integrals**.

6.4.1 Type 1: Unbounded Intervals

Suppose that we wanted to compute the area of the unbounded region below the graph of $f(x) = e^{-x}$ and above the x-axis for $x \geq 0$ (see Figure 6.9). How would

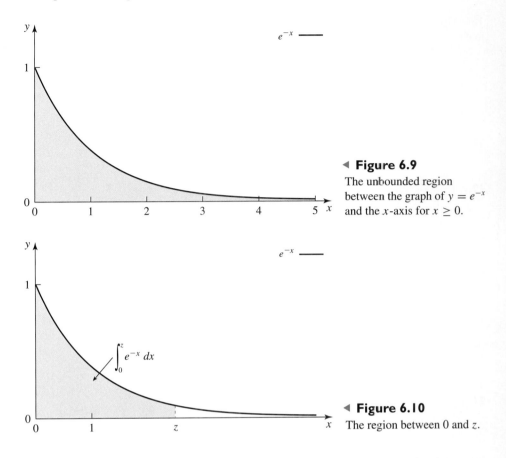

◀ **Figure 6.9**

The unbounded region between the graph of $y = e^{-x}$ and the x-axis for $x \geq 0$.

◀ **Figure 6.10**

The region between 0 and z.

we proceed? We know how to find the area of a region between 0 and z, namely,

$$A(z) = \int_0^z e^{-x} dx = -e^{-x}\Big]_0^z = 1 - e^{-z}$$

This is the shaded area in Figure 6.10. If we now let z tend to infinity, we may regard the limiting value (if it exists) as the area of the unbounded region below the graph of $f(x) = e^{-x}$ and above the x-axis for $x \geq 0$ (see Figure 6.9):

$$A = \lim_{z \to \infty} A(z) = \lim_{z \to \infty} (1 - e^{-z}) = 1$$

We therefore define (see Figures 6.11 and 6.12).

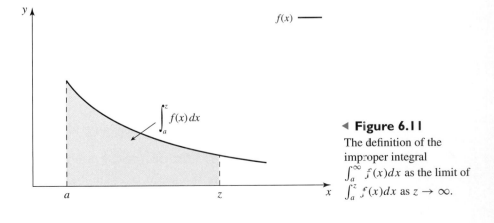

◀ **Figure 6.11**

The definition of the improper integral $\int_a^\infty f(x)dx$ as the limit of $\int_a^z f(x)dx$ as $z \to \infty$.

13. $\displaystyle\int_0^{\pi/2} \frac{\cos x}{\sqrt{\sin x}}\, dx$

14. $\displaystyle\int_{-2}^0 \frac{dx}{(x+1)^{1/3}}$

15. $\displaystyle\int_{-1}^1 \ln|x|\, dx$

16. $\displaystyle\int_0^2 \frac{dx}{(x-1)^{2/5}}$

In Problems 17–28, determine whether each integral is convergent. If the integral is convergent, compute its value.

17. $\displaystyle\int_1^\infty \frac{1}{x^3}\, dx$

18. $\displaystyle\int_1^\infty \frac{1}{x^{1/3}}\, dx$

19. $\displaystyle\int_0^4 \frac{1}{x^4}\, dx$

20. $\displaystyle\int_0^4 \frac{1}{x^{1/4}}\, dx$

21. $\displaystyle\int_0^2 \frac{1}{(x-1)^{1/3}}\, dx$

22. $\displaystyle\int_0^2 \frac{1}{(x-1)^4}\, dx$

23. $\displaystyle\int_0^\infty \frac{1}{\sqrt{x+1}}\, dx$

24. $\displaystyle\int_{-1}^0 \frac{1}{\sqrt{x+1}}\, dx$

25. $\displaystyle\int_e^\infty \frac{dx}{x \ln x}$

26. $\displaystyle\int_1^e \frac{dx}{x \ln x}$

27. $\displaystyle\int_{-2}^2 \frac{2x\, dx}{(x^2-1)^{1/3}}$

28. $\displaystyle\int_{-\infty}^1 \frac{3}{1+x^2}\, dx$

29. Determine whether

$$\int_{-\infty}^\infty \frac{1}{x^2-1}\, dx$$

is convergent. *Hint:* Use the partial fraction decomposition

$$\frac{1}{x^2-1} = \frac{1}{2}\left(\frac{1}{x-1} - \frac{1}{x+1}\right).$$

30. Though the antiderivative of $f(x) = e^{-x^2/2}$ cannot be expressed with elementary functions, it is known that

$$\int_{-\infty}^\infty e^{-x^2/2}\, dx = \sqrt{2\pi}$$

Use this fact to show that

$$\int_{-\infty}^\infty x^2 e^{-x^2/2}\, dx = \sqrt{2\pi}$$

(*Hint:* Write the integrand as

$$x \cdot (xe^{-x^2/2})$$

and use integration by parts.)

31. Determine the constant c so that

$$\int_0^\infty ce^{-3x}\, dx = 1$$

32. Determine the constant c so that

$$\int_{-\infty}^\infty \frac{c}{1+x^2}\, dx = 1$$

33. In this problem, we investigate the integral

$$\int_1^\infty \frac{1}{x^p}\, dx$$

for $0 < p < \infty$.

(a) For $z > 1$, set

$$A(z) = \int_1^z \frac{1}{x^p}\, dx$$

and show that

$$A(z) = \frac{1}{1-p}(z^{-p+1} - 1)$$

for $p \neq 1$ and

$$A(z) = \ln z$$

for $p = 1$.

(b) Use your results in (a) to show that for $0 < p \leq 1$

$$\lim_{z\to\infty} A(z) = \infty$$

(c) Use your results in (a) to show that for $p > 1$

$$\lim_{z\to\infty} A(z) = \frac{1}{p-1}$$

34. In this problem, we investigate the integral

$$\int_0^1 \frac{1}{x^p}\, dx$$

for $0 < p < \infty$.

(a) Compute

$$\int \frac{1}{x^p}\, dx$$

for $0 < p < \infty$. (*Hint:* Treat the case where $p = 1$ separately.)

(b) Use your result in (a) to compute

$$\int_c^1 \frac{1}{x^p}\, dx$$

for $0 < c < 1$.

(c) Use your result in (b) to show that

$$\int_0^1 \frac{1}{x^p}\, dx = \frac{1}{1-p}$$

for $0 < p < 1$.

(d) Show that

$$\int_0^1 \frac{1}{x^p}\, dx$$

is divergent for $p \geq 1$.

[6.4.3]

35. (a) Show that

$$0 \leq e^{-x^2} \leq e^{-x}$$

for $x \geq 1$.

(b) Use your result in (a) to show that

$$\int_1^\infty e^{-x^2}\,dx$$

is convergent.

36. Show that

$$0 \le \frac{1}{\sqrt{1+x^4}} \le \frac{1}{x^2}$$

for $x > 0$.

(b) Use your result in (a) to show that

$$\int_1^\infty \frac{1}{\sqrt{1+x^4}}\,dx$$

is convergent.

37. (a) Show that

$$\frac{1}{\sqrt{1+x^2}} \ge \frac{1}{2x} > 0$$

for $x \ge 1$.

(b) Use your result in (a) to show that

$$\int_1^\infty \frac{1}{\sqrt{1+x^2}}\,dx$$

is divergent.

38. (a) Show that

$$\frac{1}{\sqrt{x+\ln x}} \ge \frac{1}{\sqrt{2x}} > 0$$

for $x \ge 1$.

(b) Use your result in (a) to show that

$$\int_1^\infty \frac{1}{\sqrt{x+\ln x}}\,dx$$

is divergent.

In Problems 39–42, find a comparison function for each integrand and determine whether the integral is convergent.

39. $\displaystyle\int_{-\infty}^\infty e^{-x^2/2}\,dx$ **40.** $\displaystyle\int_1^\infty \frac{1}{\sqrt{1+x^6}}\,dx$

41. $\displaystyle\int_1^\infty \frac{1}{\sqrt{1+x}}\,dx$ **42.** $\displaystyle\int_{-\infty}^\infty \frac{1}{e^x + e^{-x}}\,dx$

43. (a) Show that

$$\lim_{x\to\infty} \frac{\ln x}{\sqrt{x}} = 0$$

(b) Use your result in (a) to show that

$$2\ln x \le \sqrt{x} \qquad (6.14)$$

for sufficiently large x. Use a graphing calculator to determine just how large x must be for (6.14) to hold.

(c) Use your result in (b) to show that

$$\int_0^\infty e^{-\sqrt{x}}\,dx \qquad (6.15)$$

converges. Use a graphing calculator to sketch the function $f(x) = e^{-\sqrt{x}}$ together with its comparison function(s), and use your graph to explain how you showed that the integral in (6.15) is convergent.

44. (a) Show that

$$\lim_{x\to\infty} \frac{\ln x}{x} = 0$$

(b) Use your result in (a) to show that for any $c > 0$

$$cx \ge \ln x$$

for sufficiently large x.

(c) Use your result in (b) to show that for any $p > 0$,

$$x^p e^{-x} \le e^{-x/2}$$

provided that x is sufficiently large.

(d) Use your result in (c) to show that for any $p > 0$

$$\int_0^\infty x^p e^{-x}\,dx$$

is convergent.

◆ 6.5 NUMERICAL INTEGRATION

There are integrals that are impossible to evaluate exactly, such as

$$\int_0^4 e^{-x^2}\,dx$$

In such situations, numerical approximations are needed.

One way to numerically approximate an integral should be obvious from our initial approach to the area problem—namely, approximating areas by rectangles, that is, the Riemann sum approximation. Recall that for f continuous,

$$\int_a^b f(x)\,dx = \lim_{\|P\|\to 0} \sum_{k=1}^n f(c_k)\Delta x_k$$

where $P = [x_0, x_1, \ldots, x_n]$ is a partition of $[a, b]$ with $x_0 = a$ and $x_n = b$. The number c_k is in $[x_{k-1}, x_k]$, and $\Delta x_k = x_k - x_{k-1}$ for $1 \le k \le n$.

◀ **Figure 6.32**
The midpoint rule.

In the following, we will assume that we partition the interval $[a, b]$ into n equal subintervals, that is, each subinterval is of length

$$\Delta x = \frac{b - a}{n}$$

We assume that the function f is continuous on $[a, b]$. We will discuss two methods, the midpoint rule and the trapezoidal rule.

6.5.1 The Midpoint Rule

This is the Riemann sum approximation, where we choose the midpoint of each subinterval for the point c_k. The midpoint of the interval $[x_{k-1}, x_k]$ is

$$c_k = \frac{x_{k-1} + x_k}{2}$$

The rule is defined as follows (also see Figure 6.32).

MIDPOINT RULE

Suppose that $[x_0, x_1, \ldots, x_n]$ is a partition of $[a, b]$ into n subintervals of equal lengths. We approximate

$$\int_a^b f(x)\, dx$$

by

$$M_n = \frac{b - a}{n} \sum_{k=1}^{n} f(c_k)$$

where $c_k = \frac{x_{k-1}+x_k}{2}$ is the midpoint of $[x_{k-1}, x_k]$.

In the following example, we choose an integral that we can evaluate exactly, so that we can see how close the approximation is.

▮ **EXAMPLE 1**   **Midpoint Rule**

Use the midpoint rule with $n = 4$ to approximate

$$\int_0^1 x^2\, dx$$

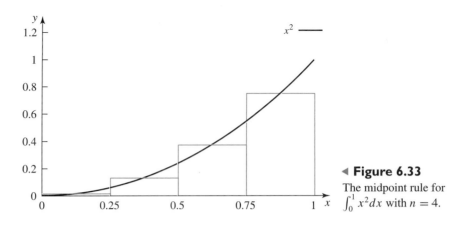

◀ **Figure 6.33**
The midpoint rule for $\int_0^1 x^2 dx$ with $n = 4$.

Solution   For $n = 4$, we find $\Delta x = \frac{b-a}{2} = \frac{1}{4}$ and we obtain four subintervals, each of length $\frac{1}{4}$ (see Figure 6.33).

subinterval $[x_{k-1}, x_k]$	midpoint c_k	$f(c_k)$	$f(c_k)\Delta x$
$\left[0, \dfrac{1}{4}\right]$	$\dfrac{1}{8}$	$\dfrac{1}{64}$	$\dfrac{1}{64}\dfrac{1}{4}$
$\left[\dfrac{1}{4}, \dfrac{1}{2}\right]$	$\dfrac{3}{8}$	$\dfrac{9}{64}$	$\dfrac{9}{64}\dfrac{1}{4}$
$\left[\dfrac{1}{2}, \dfrac{3}{4}\right]$	$\dfrac{5}{8}$	$\dfrac{25}{64}$	$\dfrac{25}{64}\dfrac{1}{4}$
$\left[\dfrac{3}{4}, 1\right]$	$\dfrac{7}{8}$	$\dfrac{49}{64}$	$\dfrac{49}{64}\dfrac{1}{4}$

We find the approximation

$$M_4 = \frac{b-a}{4} \sum_{k=1}^{4} f(c_k) = \frac{1}{4}\left(\frac{1}{64} + \frac{9}{64} + \frac{25}{64} + \frac{49}{64}\right) = \frac{1}{4}\frac{84}{64} = \frac{21}{64} = 0.3281$$

We know that $\int_0^1 x^2 \, dx = \frac{1}{3}$. Hence, the error is

$$\left|\int_0^1 x^2 dx - M_4\right| = 0.0052$$

Larger values of n improve the approximation.   ∎

Instead of memorizing the formula for the midpoint rule, it is easier to keep a picture in mind. We illustrate this in the following example.

▌EXAMPLE 2   Midpoint Rule

Use the midpoint rule with $n = 5$ to approximate

$$\int_1^2 \frac{1}{x}\, dx$$

Solution   The graph of $f(x) = \frac{1}{x}$ together with the five approximating rectangles is shown in Figure 6.34.

With $n = 5$, the partition of $[1, 2]$ is given by $P = [1, 1.2, 1.4, 1.6, 1.8, 2]$ and the midpoints are 1.1, 1.3, 1.5, 1.7, and 1.9. Since the width of each rectangle

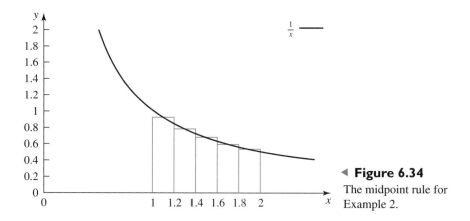

◀ **Figure 6.34**
The midpoint rule for
Example 2.

is 0.2 and $f(x) = \frac{1}{x}$, the area of the first rectangle is $(0.2)\frac{1}{1.1}$, the area of the second rectangle is $(0.2)\frac{1}{1.3}$, and so on. We thus find

$$M_5 = (0.2)\left[\frac{1}{1.1} + \frac{1}{1.3} + \frac{1}{1.5} + \frac{1}{1.7} - \frac{1}{1.9}\right] = 0.6919$$

Note that we factored 0.2, the width of each rectangle, since it is a common factor of the areas of the five rectangles.

We know that

$$\int_1^2 \frac{1}{x}\,dx = \ln x\Big]_1^2 = \ln 2 - \ln 1 = \ln 2$$

Hence the error in the approximation is

$$\left|\int_1^2 \frac{1}{x}\,dx - M_5\right| = |\ln 2 - 0.6919| = 0.0012$$ ▮

We typically use an approximation when we cannot evaluate the integral exactly. Thus, we cannot use the exact value to determine how close the approximation is. Fortunately, there are results that allow us to obtain upper bounds for the error.

ERROR BOUND FOR THE MIDPOINT RULE

Suppose that $|f''(x)| \le K$ for all $x \in [a, b]$. Then the error in the midpoint rule is at most

$$\left|\int_a^b f(x)\,dx - M_n\right| \le K\frac{(b-a)^3}{24n^2}$$

Let's check this bound for our two examples. In the first example, $f(x) = x^2$, and therefore $f''(x) = 2$. Hence with $n = 4$, the error is at most

$$2\frac{(1-0)^3}{24(4^2)} = 0.0052$$

This is in fact the error that we obtained.

In the second example, $f(x) = \frac{1}{x}$. Since $f'(x) = -\frac{1}{x^2}$ and $f''(x) = \frac{2}{x^3}$, we find that

$$|f''(x)| = \left|\frac{2}{x^3}\right| \le 2 \quad \text{for } 1 \le x \le 2$$

Hence, with $n = 5$, the error is at most

$$2\frac{(2 - 1)^3}{24(5^2)} = 0.0033$$

The actual error was in fact smaller, only 0.0012.

The actual error can differ quite a bit from the theoretical error bound, which is the worst-case scenario. The actual error might be quite a bit smaller, but will not be larger than the theoretical error bound. The advantage of such an error bound is that it allows us to find the number of subintervals required to obtain a certain accuracy. For instance, if we want to numerically approximate $\int_0^1 x^2\, dx$ so that the error was at most 10^{-4}, then we must choose n so that

$$K\frac{(b - a)^3}{24n^2} \le 10^{-4}$$

$$2\frac{1}{24n^2} \le 10^{-4}$$

$$\frac{1}{12}10^4 \le n^2$$

$$28.9 \le n$$

That is, $n = 29$ would suffice to produce an error of at most 10^{-4}.

Finding a value for K in the estimate is not always easy. A graph of $f''(x)$ over the interval of interest can facilitate finding a bound on the second derivative. One need not find the best possible bound. For instance, if we wanted to integrate $f(x) = e^x$ over the interval $[1, 2]$, we would need to find a bound on $f''(x) = e^x$ over the interval $[1, 2]$. Since $|e^x| \le e^2$ over the interval $[1, 2]$, we could use $K = 9$, for instance (see Figure 6.35). This is not the best possible bound, but a number that we can find without using a calculator. (The best possible bound would be $K = e^2$.)

6.5.2 The Trapezoidal Rule

Instead of rectangles, we use trapezoids to approximate integrals in this method, as is illustrated in Figure 6.36. We assume again that f is a continuous function on $[a, b]$, and divide $[a, b]$ into n equal subintervals. But this time we approximate the function $f(x)$ by a polygon $P(x)$. To obtain the polygon $P(x)$, we connect the points $(x_k, f(x_k))$, $k = 0, 1, 2, \ldots, n$ by straight lines, as shown in Figure 6.36. The integral $\int_a^b P(x)\, dx$ is then the approximation to $\int_a^b f(x)\, dx$.

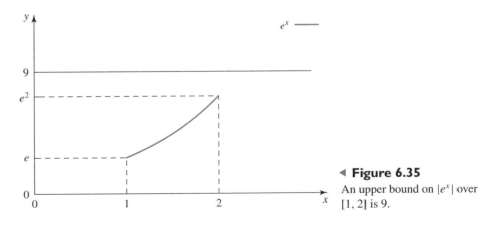

◀ **Figure 6.35**

An upper bound on $|e^x|$ over $[1, 2]$ is 9.

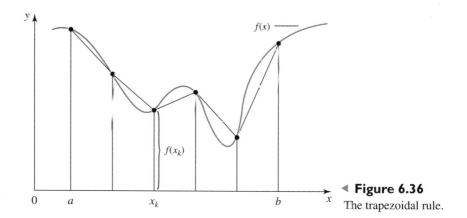

◀ **Figure 6.36**
The trapezoidal rule.

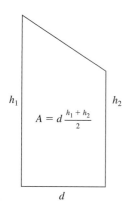

▲ **Figure 6.37**
The area of a trapezoid.

We see from Figure 6.36 that this amounts to adding up (signed) areas of trapezoids. Recall from planar geometry that the area of the trapezoid in Figure 6.37 is

$$A = d \frac{h_1 + h_2}{2}$$

The width of each trapezoid in Figure 6.36 is $d = \frac{b-a}{n}$. Adding up the areas of the trapezoids in Figure 6.36 then yields

$$T_n = \frac{b - a}{n} \left[\frac{f(x_0) + f(x_1)}{2} + \frac{f(x_1) + f(x_2)}{2} + \frac{f(x_2) + f(x_3)}{2} \right.$$
$$\left. + \cdots + \frac{f(x_{n-2}) + f(x_{n-1})}{2} + \frac{f(x_{n-1}) + f(x_n)}{2} \right]$$
$$= \frac{b - a}{n} \left[\frac{f(x_0)}{2} + f(x_1) + f(x_2) + \cdots + f(x_{n-1}) + \frac{f(x_n)}{2} \right]$$

TRAPEZOIDAL RULE

Suppose that $P = [x_0, x_1, x_2, \ldots, x_n]$ is a partition of $[a, b]$ into n subintervals of equal lengths. We approximate

$$\int_a^b f(x)\,dx$$

by

$$T_n = \frac{b - a}{n} \left[\frac{f(x_0)}{2} + f(x_1) + f(x_2) + \cdots + f(x_{n-1}) + \frac{f(x_n)}{2} \right]$$

EXAMPLE 3   **Trapezoidal Rule**

Use the trapezoidal rule with $n = 4$ to approximate

$$\int_0^1 x^2\,dx$$

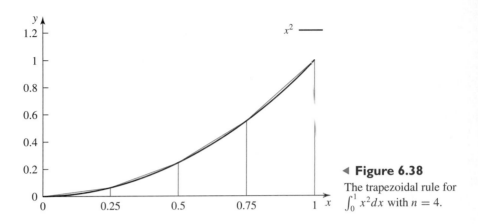

◀ **Figure 6.38**
The trapezoidal rule for $\int_0^1 x^2 dx$ with $n = 4$.

Solution   As before, since $n = 4$, there are four subintervals, each of length $\frac{1}{4}$ (see Figure 6.38). We find the following

k	x_k	$f(x_k)$
0	0	0
1	$\frac{1}{4}$	$\frac{1}{16}$
2	$\frac{1}{2}$	$\frac{1}{4}$
3	$\frac{3}{4}$	$\frac{9}{16}$
4	1	1

The approximation is

$$T_4 = \frac{1}{4}\left[\frac{0}{2} + \frac{1}{16} + \frac{1}{4} + \frac{9}{16} + \frac{1}{2}\right] = 0.34375$$

Since we know that $\int_0^1 x^2\, dx = \frac{1}{3}$, we can compute the error:

$$\left|\int_0^1 x^2\, dx - T_4\right| = 0.0104$$

I EXAMPLE 4   **Trapezoidal Rule**

Use the trapezoidal rule with $n = 5$ to approximate

$$\int_1^2 \frac{1}{x}\, dx$$

Solution   This is illustrated in Figure 6.39. With $n = 5$, the partition of $[1, 2]$ is given by $P = [1.0, 1.2, 1.4, 1.6, 1.8, 2.0]$. The base of each trapezoid has length 0.2. Hence

$$T_5 = (0.2)\left[\frac{1}{2}\cdot\frac{1}{1.0} + \frac{1}{1.2} + \frac{1}{1.4} + \frac{1}{1.6} + \frac{1}{1.8} + \frac{1}{2}\cdot\frac{1}{2.0}\right] = 0.69563$$

◀ **Figure 6.39**
The trapezoidal rule for
$\int_1^2 \frac{1}{x}\,dx$ with $n = 5$.

Since we know from Example 2 that $\int_1^2 \frac{1}{x}\,dx = \ln 2$, we can compute the error:

$$\left| \int_1^2 \frac{1}{x}\,dx - T_5 \right| = 0.00249$$

There is also a theoretical error bound for the trapezoidal rule.

ERROR BOUND FOR THE TRAPEZOIDAL RULE

Suppose that $|f''(x)| \le K$ for all $x \in [a, b]$. Then the error in the trapezoidal rule is at most

$$\left| \int_a^b f(x)\,dx - T_n \right| \le K \frac{(b-a)^3}{12n^2}$$

In Example 3, since $f(x) = x^2$, we find $f''(x) = 2$ and hence $K = 2$. The error is therefore bounded by

$$2\frac{1}{12(4^2)} = 0.0104$$

which is the same as the actual error.

In Example 4, since $f(x) = 1/x$, we find $|f''(x)| = 2/x^3 \le 2$ for $1 \le x \le 2$ (as in Example 2). Hence, with $n = 5$, the error bound is at most

$$2\frac{(2-1)^3}{12(5^2)} = 0.0067$$

The actual error was in fact smaller, only 0.00249. As in the midpoint rule, the theoretical and the actual error can differ (and typically will).

◆ **6.5 PROBLEMS** _____

[6.5.1], [6.5.2]

In Problems 1–4, use the midpoint rule to approximate each integral with the specified value of n.

In Problems 5–8, use the midpoint rule to approximate each integral with the specified value of n. Compare your approximation with the exact value.

1. $\displaystyle\int_1^2 x^2\,dx, n = 4$

2. $\displaystyle\int_{-1}^0 x^3\,dx, n = 5$

3. $\displaystyle\int_0^1 e^{-x}\,dx, n = 3$

4. $\displaystyle\int_0^{\pi/2} \sin x\,dx, n = 4$

5. $\displaystyle\int_2^4 \frac{1}{x}\,dx, n = 4$

6. $\displaystyle\int_{-1}^1 (e^x - 1)\,dx, n = 4$

7. $\displaystyle\int_0^4 \sqrt{x}\,dx, n = 4$

8. $\displaystyle\int_1^3 \frac{1}{\sqrt{x}}\,dx, n = 5$

In Problems 9–12, use the trapezoidal rule to approximate each integral with the specified value of n.

9. $\displaystyle\int_1^2 x^2\,dx,\ n = 4$

10. $\displaystyle\int_{-1}^0 x^3\,dx,\ n = 5$

11. $\displaystyle\int_0^1 e^{-x}\,dx,\ n = 3$

12. $\displaystyle\int_0^{\pi/2} \sin x\,dx,\ n = 4$

In Problems 13–16, use the trapezoidal rule to approximate each integral with the specified value of n. Compare your approximation with the exact value.

13. $\displaystyle\int_1^3 x^3\,dx,\ n = 5$

14. $\displaystyle\int_{-1}^1 (1 - e^{-x})\,dx,\ n = 4$

15. $\displaystyle\int_0^2 \sqrt{x}\,dx,\ n = 4$

16. $\displaystyle\int_1^2 \frac{1}{x}\,dx,\ n = 5$

17. How large should n be so that the midpoint rule approximation of

$$\int_0^2 x^2\,dx$$

is accurate to within 10^{-4}?

In Problems 18–24, use the theoretical error bound to determine how large n should be. (Hint: In each case, find the second derivative of the integrand, graph it, and use a graphing calculator to find an upper bound on $|f''(x)|$.)

18. How large should n be so that the midpoint rule approximation of

$$\int_1^2 \frac{1}{x}\,dx$$

is accurate to within 10^{-3}?

19. How large should n be so that the midpoint rule approximation of

$$\int_0^2 e^{-x^2/2}\,dx$$

is accurate to within 10^{-4}?

20. How large should n be so that the midpoint rule approximation of

$$\int_2^8 \frac{1}{\ln t}\,dt$$

is accurate to within 10^{-3}?

21. How large should n be so that the trapezoidal rule approximation of

$$\int_0^1 e^{-x}\,dx$$

is accurate to within 10^{-5}?

22. How large should n be so that the trapezoidal rule approximation of

$$\int_0^2 \sin x\,dx$$

is accurate to within 10^{-4}?

23. How large should n be so that the trapezoidal rule approximation of

$$\int_1^2 \frac{e^t}{t}\,dt$$

is accurate to within 10^{-4}?

24. How large should n be so that the trapezoidal rule approximation of

$$\int_1^2 \frac{\cos x}{x}\,dx$$

is accurate to within 10^{-3}?

25. **(a)** Show graphically, that for $n = 5$, the trapezoidal rule overestimates and the midpoint rule underestimates

$$\int_0^1 x^3\,dx$$

In each case, compute the approximate value of the integral and compare it to the exact value.

(b) The result in (a) has to do with the fact that $y = x^3$ is concave up on $[0, 1]$. To generalize this, we assume that the function $f(x)$ is continuous, nonnegative, and concave up on the interval $[a, b]$. Denote by M_n the midpoint rule and by T_n the trapezoidal rule approximation of $\int_a^b f(x)\,dx$. Explain in words why

$$M_n \le \int_a^b f(x)\,dx \le T_n$$

(c) If we assume that $f(x)$ is continuous, nonnegative and concave *down* on $[a, b]$, then

$$M_n \ge \int_a^b f(x)\,dx \ge T_n$$

Explain why this is so. Use this result to give an upper and a lower bound on

$$\int_0^1 \sqrt{x}\,dx$$

when $n = 4$ in the approximation.

◆ 6.6 TABLES OF INTEGRALS

Tables of indefinite integrals are useful aids when evaluating integrals. Often, it is still necessary to bring the integrand of interest into a form that is listed in the table; and there are many integrals that simply cannot be evaluated exactly and must be evaluated numerically. We will give a very brief list of indefinite integrals, and explain how to use such tables.

I. Basic Functions

1. $\displaystyle\int x^n dx = \frac{1}{n+1}x^{n+1} + C, \ n \neq -1$

2. $\displaystyle\int \frac{1}{x} dx = \ln|x| + C$ 　　　　　**3.** $\displaystyle\int e^x dx = e^x + C$

4. $\displaystyle\int \ln x \, dx = x \ln x - x + C$ 　　　　**5.** $\displaystyle\int \sin x \, dx = -\cos x + C$

6. $\displaystyle\int \cos x \, dx = \sin x + C$ 　　　　　**7.** $\displaystyle\int \tan x \, dx = -\ln|\cos x| + C$

II. Rational Functions

8. $\displaystyle\int \frac{1}{ax+b} dx = \frac{1}{a} \ln|ax+b| + C$

9. $\displaystyle\int \frac{x}{ax+b} dx = \frac{x}{a} - \frac{b}{a^2} \ln|ax+b| + C$

10. $\displaystyle\int \frac{x}{(ax+b)^2} dx = \frac{b}{a^2(ax+b)} + \frac{1}{a^2} \ln|ax+b| + C$

11. $\displaystyle\int \frac{x}{ax^2+bx+c} dx = \frac{1}{2a} \ln|ax^2+bx+c| - \frac{b}{2a} \int \frac{1}{ax^2+bx+c} dx$

12. $\displaystyle\int \frac{1}{a^2+x^2} dx = \frac{1}{a} \arctan\frac{x}{a} + C$

13. $\displaystyle\int \frac{1}{a^2-x^2} dx = \frac{1}{2a} \ln\left|\frac{x+a}{x-a}\right| + C$

III. Integrands Involving $\sqrt{a^2+x^2}$, $\sqrt{a^2-x^2}$, or $\sqrt{x^2-a^2}$

14. $\displaystyle\int \frac{1}{\sqrt{a^2-x^2}} dx = \arcsin\frac{x}{a} + C$

15. $\displaystyle\int \frac{1}{\sqrt{x^2 \pm a^2}} dx = \ln|x + \sqrt{x^2 \pm a^2}| + C$

16. $\displaystyle\int \sqrt{a^2 \pm x^2} \, dx = \frac{1}{2}\left(x\sqrt{a^2 \pm x^2} + a^2 \int \frac{1}{\sqrt{a^2 \pm x^2}} dx\right)$

17. $\displaystyle\int \sqrt{x^2 - a^2} \, dx = \frac{1}{2}\left(x\sqrt{x^2 - a^2} - a^2 \int \frac{1}{\sqrt{x^2 - a^2}} dx\right)$

IV. Integrands Involving Trigonometric Functions

18. $\displaystyle\int \sin(ax)\,dx = -\frac{1}{a}\cos(ax) + C$

19. $\displaystyle\int \sin^2(ax)\,dx = \frac{1}{2}x - \frac{1}{4a}\sin(2ax) + C$

20. $\displaystyle\int \sin(ax)\sin(bx)\,dx = \frac{\sin(a-b)x}{2(a-b)} - \frac{\sin(a+b)x}{2(a+b)} + C,\ \text{for}\ a^2 \neq b^2$

21. $\displaystyle\int \cos(ax)\,dx = \frac{1}{a}\sin(ax) + C$

22. $\displaystyle\int \cos^2(ax)\,dx = \frac{1}{2}x + \frac{1}{4a}\sin(2ax) + C$

23. $\displaystyle\int \cos(ax)\cos(bx)\,dx = \frac{\sin(a-b)x}{2(a-b)} + \frac{\sin(a+b)x}{2(a+b)} + C,\ \text{for}\ a^2 \neq b^2$

24. $\displaystyle\int \sin(ax)\cos(ax)\,dx = \frac{1}{2a}\sin^2(ax) + C$

25. $\displaystyle\int \sin(ax)\cos(bx)\,dx = -\frac{\cos(a+b)x}{2(a+b)} - \frac{\cos(a-b)x}{2(a-b)} + C,\ \text{for}\ a^2 \neq b^2$

V. Integrands Involving Exponential Functions

26. $\displaystyle\int e^{ax}\,dx = \frac{1}{a}e^{ax} + C$ 　　　　**27.** $\displaystyle\int xe^{ax}\,dx = \frac{e^{ax}}{a^2}(ax - 1) + C$

28. $\displaystyle\int x^n e^{ax}\,dx = \frac{1}{a}x^n e^{ax} - \frac{n}{a}\int x^{n-1}e^{ax}\,dx$

29. $\displaystyle\int e^{ax}\sin(bx)\,dx = \frac{e^{ax}}{a^2 + b^2}(a\sin(bx) - b\cos(bx)) + C$

30. $\displaystyle\int e^{ax}\cos(bx)\,dx = \frac{e^{ax}}{a^2 + b^2}(a\cos(bx) + b\sin(bx)) + C$

VI. Integrands Involving Logarithmic Functions

31. $\displaystyle\int \ln x\,dx = x\ln x - x + C$

32. $\displaystyle\int (\ln x)^2\,dx = x(\ln x)^2 - 2x\ln x + 2x + C$

33. $\displaystyle\int x^m \ln x\,dx = x^{m+1}\left[\frac{\ln x}{m+1} - \frac{1}{(m+1)^2}\right] + C,\ m \neq -1$

34. $\displaystyle\int \frac{\ln x}{x}\,dx = \frac{(\ln x)^2}{2} + C$ 　　　　**35.** $\displaystyle\int \frac{1}{x\ln x}\,dx = \ln(\ln x) + C$

36. $\displaystyle\int \sin(\ln x)\,dx = \frac{x}{2}(\sin(\ln x) - \cos(\ln x)) + C$

37. $\displaystyle\int \cos(\ln x)\,dx = \frac{x}{2}(\sin(\ln x) + \cos(\ln x)) + C$

We will now illustrate how to use the table. We begin with examples that fit one of the listed integrals exactly.

EXAMPLE 1   Square Root

Find

$$\int \sqrt{3 - x^2} \, dx$$

Solution   The integrand involves $\sqrt{a^2 - x^2}$ and is of the form III.16 with $a^2 = 3$. Hence,

$$\int \sqrt{3 - x^2} \, dx = \frac{1}{2} \left(x\sqrt{3 - x^2} + 3 \int \frac{1}{\sqrt{3 - x^2}} \, dx \right)$$

To evaluate $\int \frac{1}{\sqrt{3-x^2}} \, dx$, we use III.14 with $a^2 = 3$, and find

$$\int \frac{1}{\sqrt{3 - x^2}} \, dx = \arcsin \frac{x}{\sqrt{3}} + C$$

Hence,

$$\int \sqrt{3 - x^2} \, dx = \frac{1}{2} \left(x\sqrt{3 - x^2} + 3 \arcsin \frac{x}{\sqrt{3}} \right) + C \qquad \blacksquare$$

EXAMPLE 2   Trigonometric Function

Find

$$\int \sin(3x) \cos(4x) \, dx$$

Solution   The integrand involves trigonometric functions, and we can find it in IV.25 with $a = 3$ and $b = 4$. Hence,

$$\int \sin(3x) \cos(4x) \, dx = -\frac{\cos(7x)}{14} - \frac{\cos(-x)}{(2)(-1)} + C$$

Since $\cos(-x) = \cos x$, this simplifies to

$$\int \sin(3x) \cos(4x) \, dx = -\frac{\cos(7x)}{14} + \frac{\cos x}{2} + C \qquad \blacksquare$$

EXAMPLE 3   Exponential Function

Find

$$\int x^2 e^{3x} \, dx$$

Solution   This is of the form V.28 with $n = 2$ and $a = 3$. Hence

$$\int x^2 e^{3x} \, dx = \frac{1}{3} x^2 e^{3x} - \frac{2}{3} \int x e^{3x} \, dx$$

We now use V.27 to continue the evaluation of the integral, and find

$$\int x e^{3x} \, dx = \frac{e^{3x}}{9} (3x - 1) + C$$

Hence,

$$\int x^2 e^{3x} \, dx = \frac{1}{3} x^2 e^{3x} - \frac{2}{3} \left[\frac{e^{3x}}{9} (3x - 1) \right] + C \qquad \blacksquare$$

Thus far, each of our examples matched exactly one of the integrals in our table. Often, this will not be the case, and the integrand must be manipulated until it matches one of the integrals in the table. These manipulations include expansions, long division, completion of the square, and substitution. We give a few examples to illustrate this.

I EXAMPLE 4   **Exponential Function**

Find

$$\int e^{2x} \sin(3x - 1) \, dx$$

Solution   This looks similar to V.29. If we use the substitution

$$u = 3x - 1 \quad \text{with } dx = \frac{1}{3} \, du \text{ and } 2x = \frac{2}{3}(u + 1)$$

then the integrand can be transformed so that it matches V.29 exactly:

$$\int e^{2x} \sin(3x - 1) \, dx = \int e^{2(u+1)/3} (\sin u) \frac{1}{3} \, du$$

$$= \frac{e^{2/3}}{3} \int e^{2u/3} \sin u \, du$$

$$= \frac{e^{2/3}}{3} \frac{e^{2u/3}}{\frac{4}{9} + 1} \left[\frac{2}{3} \sin u - \cos u \right] + C$$

$$= \frac{e^{2/3}}{3 \cdot \frac{13}{9}} e^{2(3x-1)/3} \left[\frac{2}{3} \sin(3x - 1) - \cos(3x - 1) \right] + C$$

$$= \frac{3}{13} e^{2x} \left[\frac{2}{3} \sin(3x - 1) - \cos(3x - 1) \right] + C \qquad \blacksquare$$

I EXAMPLE 5   **Rational Function**

Find

$$\int \frac{x^2}{9 + x^2} \, dx$$

Solution   The integrand is a rational function; we can use long division to simplify it:

$$\frac{x^2}{9 + x^2} = 1 - \frac{9}{9 + x^2}$$

Then, using II.12 with $a = 3$, we find

$$\int \frac{x^2}{9 + x^2} \, dx = \int dx - 9 \int \frac{1}{9 + x^2} \, dx$$

$$= x - 9 \left(\frac{1}{3} \arctan \frac{x}{3} \right) + C$$

$$= x - 3 \arctan \frac{x}{3} + C \qquad \blacksquare$$

▎**EXAMPLE 6**   **Rational Function**

Find

$$\int \frac{1}{x^2 - 2x - 3} \, dx$$

Solution   The first step is to complete the square in the denominator:

$$\frac{1}{x^2 - 2x - 3} = \frac{1}{(x^2 - 2x + 1) - 1 - 3}$$

$$= \frac{1}{(x - 1)^2 - 4}$$

Hence, using the substitution $u = x - 1$ with $du = dx$, we find

$$\int \frac{dx}{(x - 1)^2 - 4} = \int \frac{du}{u^2 - 4} = -\int \frac{du}{4 - u^2}$$

which is of the form II.13 with $a = 2$. Therefore,

$$\int \frac{1}{x^2 - 2x - 3} \, dx = -\int \frac{du}{4 - u^2} = -\frac{1}{4} \ln \left| \frac{u + 2}{u - 2} \right| + C$$

$$= -\frac{1}{4} \ln \left| \frac{x + 1}{x - 3} \right| + C$$

◆ **6.6 PROBLEMS**

In Problems 1–8, use the table to compute each integral.

1. $\displaystyle\int \frac{x}{2x - 3} \, dx$

2. $\displaystyle\int \frac{dx}{16 + x^2}$

3. $\displaystyle\int \sqrt{x^2 - 16} \, dx$

4. $\displaystyle\int \sin(2x) \cos(2x) \, dx$

5. $\displaystyle\int_0^1 x^3 e^{-x} \, dx$

6. $\displaystyle\int_0^{\pi/4} e^{-x} \cos(2x) \, dx$

7. $\displaystyle\int_1^e x^2 \ln x \, dx$

8. $\displaystyle\int_e^{e^2} \frac{dx}{x \ln x}$

In Problems 9–22, use the table to compute each integral, after manipulating the integrand in a suitable way.

9. $\displaystyle\int_0^{\pi/6} e^x \cos\left(x - \frac{\pi}{6}\right) dx$ **10.** $\displaystyle\int_1^2 x \ln(2x - 1) \, dx$

11. $\displaystyle\int (x^2 - 1)e^{-x/2} \, dx$

12. $\displaystyle\int (x + 1)^2 e^{-2x} \, dx$

13. $\displaystyle\int \cos^2(5x - 3) \, dx$

14. $\displaystyle\int \frac{x^2}{4x^2 + 4x + 1} \, dx$

15. $\displaystyle\int \sqrt{9 + 4x^2} \, dx$

16. $\displaystyle\int \frac{1}{\sqrt{16 - 9x^2}} \, dx$

17. $\displaystyle\int e^{2x+1} \sin\left(\frac{\pi}{2}x\right) dx$

18. $\displaystyle\int (x - 1)^2 e^{2x} \, dx$

19. $\displaystyle\int_2^4 \frac{1}{x \ln \sqrt{x}} \, dx$

20. $\displaystyle\int_1^e (x + 2)^2 \ln x \, dx$

21. $\displaystyle\int \cos(\ln(3x)) \, dx$

22. $\displaystyle\int \frac{3}{x^2 - 4x + 8} \, dx$

◆ **6.7 THE TAYLOR APPROXIMATION**

In many ways, polynomials are the easiest functions to work with. In this section, we will learn how to approximate functions by polynomials. We will see that the approximation typically improves when we use higher-degree polynomials. Typically, we can only do this locally, that is, close to one particular point.

6.7.1 Taylor Polynomials

In Section 3.8, we discussed how to linearize a function about a given point. This led us to the linear or tangent approximation. We found the following.

> The linear approximation of $f(x)$ at $x = a$:
> $$L(x) = f(a) + f'(a)(x - a)$$

As an example, we look at

$$f(x) = e^x$$

and approximate this function by its linearization at $x = 0$. We find

$$L(x) = f(0) + f'(0)x = 1 + x \tag{6.16}$$

since $f'(x) = e^x$ and $f(0) = f'(0) = 1$. To see how close the approximation is, we graph both $f(x)$ and $L(x)$ in the same coordinate system (the result is shown in Figure 6.40). The approximation is quite good as long as x is close to 0; it gets gradually worse as we move away from 0. This should not be surprising, since $f(x)$ and $L(x)$ have in common only that $f(0) = L(0)$ and that $f'(0) = L'(0)$.

To improve the approximation, we may wish to use an approximating function whose higher-order derivatives also agree with those of $f(x)$ at $x = 0$. The function $L(x)$ is a polynomial of degree 1. To improve the approximation, we will continue to work with polynomials but require that the function and its first n derivatives at $x = 0$ agree with those of the polynomial. To be able to match up the first n derivatives, the polynomial must be of degree n. (If the degree of the polynomial is $k < n$, then all derivatives of degree $k + 1$ or higher are equal to 0.) A polynomial of degree n can be written as

$$P_n(x) = a_0 + a_1 x + a_2 x^2 + \cdots + a_n x^n \tag{6.17}$$

If we want to approximate $f(x)$ at $x = 0$, then we require

$$\begin{aligned}
f(0) &= P_n(0) \\
f'(0) &= P_n'(0) \\
f''(0) &= P_n''(0) \\
&\vdots \\
f^{(n)}(0) &= P_n^{(n)}(0)
\end{aligned} \tag{6.18}$$

Now

$$P_n(0) = a_0 + a_1 x + \cdots + a_n x^n \big|_{x=0} = a_0$$
$$P_n'(0) = a_1 + 2a_2 x + 3a_3 x^2 + \cdots + n a_n x^{n-1} \big|_{x=0} = a_1$$

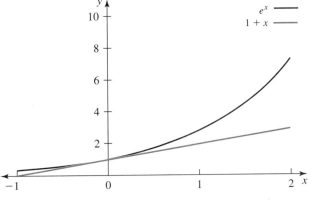

$$e^x \, \underline{\quad\quad}$$
$$1 + x \, \underline{\quad\quad}$$

◄ Figure 6.40
The graph of $y = e^x$ and its linear approximation at 0.

$$P_n''(0) = 2a_2 + (3)(2)a_3 x + \cdots + n(n-1)a_n x^{n-2}\big|_{x=0} = 2a_2$$

$$P_n'''(0) = (3)(2)a_3 + (4)(3)(2)a_4 x + \cdots + n(n-1)(n-2)a_n x^{n-3}\big|_{x=0}$$
$$= (3)(2)a_3$$
$$\vdots$$
$$P_n^{(n)}(0) = n(n-1)(n-2)\cdots(3)(2)(1)a_n\big|_{x=0}$$
$$= n(n-1)(n-2)\cdots(3)(2)(1)a_n$$

We introduce the notation

$$k! = k(k-1)(k-2)\cdots(3)(2)(1)$$

where $k!$ is read "k factorial." Solving these equations for a_k, $k = 0, 1, 2, \ldots, n$ and using $f^{(k)}(0) = P^{(k)}(0)$, $k = 0, 1, 2, \ldots, n$, we find

$$a_0 = P_n(0) = f(0)$$
$$a_1 = P_n'(0) = f'(0)$$
$$a_2 = \frac{1}{2}P_n''(0) = \frac{1}{2!}f''(0)$$
$$a_3 = \frac{1}{3\cdot 2}P_n'''(0) = \frac{1}{3!}f'''(0) \tag{6.19}$$
$$\vdots$$
$$a_n = \frac{1}{n(n-1)\cdots 2\cdot 1}P_n^{(n)}(0) = \frac{1}{n!}f^{(n)}(0)$$

A polynomial of degree n of the form (4.17), whose coefficients satisfy (4.19), is called a **Taylor polynomial** of degree n. We summarize this definition in the following.

> **Definition**   The Taylor polynomial of degree n about $x = 0$ for the function $f(x)$ is given by
>
> $$P_n(x) = f(0) + f'(0)x + \frac{f''(0)}{2!}x^2 + \frac{f'''(0)}{3!}x^3$$
> $$+ \frac{f^{(4)}(0)}{4!}x^4 + \cdots + \frac{f^{(n)}(0)}{n!}x^n$$

EXAMPLE I   Compute the Taylor polynomial of degree 3 about $x = 0$ for the function $f(x) = e^x$.

Solution   To find the Taylor polynomial of degree 3, we need the first three derivatives of $f(x)$ at $x = 0$. We find

$$f(x) = e^x \quad \text{so} \quad f(0) = 1$$
$$f'(x) = e^x \quad \text{so} \quad f'(0) = 1$$
$$f''(x) = e^x \quad \text{so} \quad f''(0) = 1$$
$$f'''(x) = e^x \quad \text{so} \quad f'''(0) = 1$$

Therefore,

$$P_3(x) = 1 + x + \frac{x^2}{2!} + \frac{x^3}{3!} = 1 + x + \frac{x^2}{2} + \frac{x^3}{6}$$

since $2! = (2)(1) = 2$ and $3! = (3)(2)(1) = 6$.

Our claim was that this polynomial would provide a better approximation to e^x than the linearization $1 + x$. We check this by evaluating e^x, $L(x)$ and $P_3(x)$ at a few values. The results are summarized in the following table.

x	e^x	$1+x$	$1+x+\dfrac{x^2}{2}+\dfrac{x^3}{6}$
-1	0.36788	0	0.3333
-0.1	0.90484	0.9	0.9048
0	1	1	1.0000
0.1	1.1052	1.1	1.1052
1	2.7183	2	2.6667

We see from the table that the third-degree Taylor polynomial provides a better approximation. Indeed, for x sufficiently close to 0, the values of $f(x)$ and $P_3(x)$ are very close. For instance,

$$f(0.1) = 1.105170918 \qquad \text{and} \qquad P_3(0.1) = 1.105166667$$

That is, their first five digits are identical. The error of approximation is

$$|f(0.1) - P_3(0.1)| = 4.25 \times 10^{-6}$$

which is quite small.

In Figure 6.41, we display the graphs of $f(x)$ and the Taylor polynomials $P_1(x)$, $P_2(x)$, and $P_3(x)$. We see from the graphs that the approximation is only good as long as x is close to 0. We also see that increasing the degree of the Taylor polynomial generally improves the approximation. ▐

When we look at Example 1, we find that the successive Taylor polynomials for $f(x) = e^x$ about $x = 0$ are

$$P_0(x) = 1$$

$$P_1(x) = 1 + x$$

$$P_2(x) = 1 + x + \frac{x^2}{2!}$$

$$P_3(x) = 1 + x + \frac{x^2}{2!} + \frac{x^3}{3!}$$

The first thing we notice is that $P_1(x)$ is the linear approximation $L(x)$ that we found in (4.16).

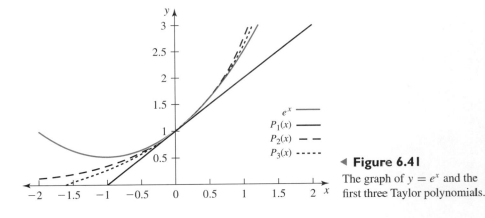

◀ **Figure 6.41**

The graph of $y = e^x$ and the first three Taylor polynomials.

The next thing we notice is that there is a pattern, and we might be tempted to guess the form of $P_n(x)$ for an arbitrary n. Our guess would be

$$P_n(x) = 1 + x + \frac{x^2}{2!} + \frac{x^3}{3!} + \cdots + \frac{x^n}{n!}$$

and this is indeed the case.

You might wonder why we bother to find an approximation for the function $f(x) = e^x$. To compute $f(1) = e$, for instance, it seems a lot easier to simply use a calculator and get the exact value directly. However, since $f(x) = e^x$ is not an algebraic function, the values of $f(x)$ cannot be found exactly using only the basic algebraic operations. Taylor polynomials are one way to evaluate such functions on a computer.

We will give additional examples in which we can find the Taylor polynomial of degree n.

EXAMPLE 2   Compute the Taylor polynomial of degree n about $x = 0$ for the function $f(x) = \sin x$.

Solution   We begin by computing successive derivatives of $f(x) = \sin x$ at $x = 0$.

$$
\begin{aligned}
f(x) &= \sin x && \text{and} && f(0) = 0 \\
f'(x) &= \cos x && \text{and} && f'(0) = 1 \\
f''(x) &= -\sin x && \text{and} && f''(0) = 0 \\
f'''(x) &= -\cos x && \text{and} && f'''(0) = -1 \\
f^{(4)}(x) &= \sin x && \text{and} && f^{(4)}(0) = 0
\end{aligned}
$$

Since $f^{(4)}(x) = f(x)$, we find that $f^{(5)}(x) = f'(x)$, $f^{(6)}(x) = f''(x)$, and so on. We also conclude that all even derivatives are equal to 0 at $x = 0$, and that the odd derivatives alternate between 1 and -1 at $x = 0$. We find

$$P_1(x) = P_2(x) = x$$

$$P_3(x) = P_4(x) = x - \frac{x^3}{3!}$$

$$P_5(x) = P_6(x) = x - \frac{x^3}{3!} + \frac{x^5}{5!}$$

$$P_7(x) = P_8(x) = x - \frac{x^3}{3!} + \frac{x^5}{5!} - \frac{x^7}{7!}$$

and so on. To find the Taylor polynomial of degree n, we must find out how to write the last term. Note that the sign in front of successive terms alternates between plus and minus. To account for this alternating sign, we introduce the factor

$$(-1)^n = \begin{cases} 1 & \text{if } n \text{ is even} \\ -1 & \text{if } n \text{ is odd} \end{cases}$$

An odd number can be written as $2n + 1$ for an integer n. For a term of the form $\pm\frac{x^k}{k!}$ with k odd, we write

$$(-1)^n \frac{x^{2n+1}}{(2n+1)!} \tag{6.20}$$

where n is an integer. Inserting successive values of n into (6.20), we find the following.

n	$(-1)^n \dfrac{x^{2n+1}}{(2n+1)!}$
0	x
1	$-\dfrac{x^3}{3!}$
2	$\dfrac{x^5}{5!}$
3	$-\dfrac{x^7}{7!}$

We see from the table that the term (6.20) produces successive terms in the Taylor polynomial for $f(x) = \sin x$. The Taylor polynomial of degree $2n + 1$ is

$$P_{2n+1}(x) = x - \frac{x^3}{3!} + \frac{x^5}{5!} - \frac{x^7}{7!} + \cdots + (-1)^n \frac{x^{2n+1}}{(2n+1)!}$$

▮

EXAMPLE 3   Compute the Taylor polynomial of degree n about $x = 0$ for the function $f(x) = \frac{1}{1-x}$, $x \neq 1$.

Solution   We begin by computing successive derivatives of $f(x) = \frac{1}{1-x}$ at $x = 0$.

$$f(x) = \frac{1}{1-x} \qquad \text{so} \qquad f(0) = 1$$

$$f'(x) = \frac{1}{(1-x)^2} \qquad \text{so} \qquad f'(0) = 1$$

$$f''(x) = \frac{2}{(1-x)^3} \qquad \text{so} \qquad f''(0) = 2 = 2!$$

$$f'''(x) = \frac{(2)(3)}{(1-x)^4} \qquad \text{so} \qquad f'''(0) = (2)(3) = 3!$$

$$f^{(4)}(x) = \frac{(2)(3)(4)}{(1-x)^5} \qquad \text{so} \qquad f^{(4)}(0) = (2)(3)(4) = 4!$$

and so on. Continuing in this way, we find

$$f^{(k)}(x) = \frac{(2)(3)(4) \cdots (k)}{(1-x)^{k+1}} \qquad \text{so } f^{(k)}(0) = k!$$

We obtain for the Taylor polynomial of degree n about $x = 0$

$$P_n(x) = 1 + x + \frac{2!}{2!}x^2 + \frac{3!}{3!}x^3 + \frac{4!}{4!}x^4 + \frac{5!}{5!}x^5 + \cdots + \frac{n!}{n!}x^n$$

$$= 1 + x + x^2 + x^3 + \cdots + x^n$$

▮

6.7.2 The Taylor Polynomial about $x = a$

Thus far, we have only considered Taylor polynomials about $x = 0$. Because Taylor polynomials, typically, are good approximations only close to the point of approximation, it is useful to have approximations about points other than $x = 0$. We have already done this for linear approximations. For instance, the tangent line approximation of $f(x)$ at $x = a$ is

$$L(x) = f(a) + f'(a)(x - a) \tag{6.21}$$

Note that $L(a) = f(a)$ and $L'(a) = f'(a)$. That is, the linear approximation and the original function, together with their first derivatives, agree at $x = a$. If we want to approximate $f(x)$ at $x = a$ by a polynomial of degree n, we then require that the polynomial and the original function, together with their first n derivatives, agree at $x = a$. This leads us to a polynomial of the form

$$P_n(x) = c_0 + c_1(x - a) + c_2(x - a)^2 + \cdots + c_n(x - a)^n \tag{6.22}$$

Comparing (6.22) and (6.21), we conclude that $c_0 = f(a)$ and $c_1 = f'(a)$. To find the remaining coefficients, we proceed as in the case $a = 0$. That is, we differentiate $f(x)$ and $P_n(x)$, and require that their first n derivatives agree at $x = a$. We then arrive at the following formula.

The Taylor polynomial of degree n about $x = a$ for the function $f(x)$ is given by

$$P_n(x) = f(a) + f'(a)(x - a) + \frac{f''(a)}{2!}(x - a)^2$$
$$+ \frac{f'''(a)}{3!}(x - a)^3 + \cdots + \frac{f^{(n)}(a)}{n!}(x - a)^n$$

▮ EXAMPLE 4   Find the Taylor polynomial of degree 3 for

$$f(x) = \ln x$$

at $x = 1$.

Handwritten annotation:
$$\frac{1}{(1-x)^2} = (1-x)^{-2}$$
$$-2(1-x)^{-3}(-1)$$
$$\frac{1}{2(1-x)^3}$$

Solution   We need to evaluate $f(x)$ and its first three derivatives at $x = 1$. We find

$$f(x) = \ln x \quad \text{so} \quad f(1) = 0$$

$$f'(x) = \frac{1}{x} \quad \text{so} \quad f'(1) = 1$$

$$f''(x) = -\frac{1}{x^2} \quad \text{so} \quad f''(1) = -1$$

$$f'''(x) = \frac{2}{x^3} \quad \text{so} \quad f'''(1) = 2$$

Using the definition of the Taylor polynomial, we get

$$P_3(x) = 0 + (1)(x - 1) + \frac{(-1)}{2!}(x - 1)^2 + \frac{2}{3!}(x - 1)^3$$

$$= (x - 1) - \frac{1}{2}(x - 1)^2 + \frac{1}{3}(x - 1)^3$$

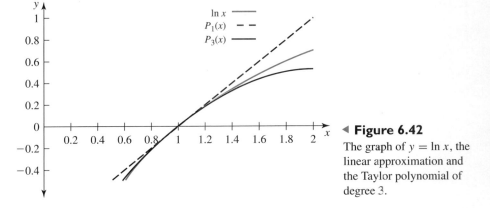

◄ **Figure 6.42**
The graph of $y = \ln x$, the linear approximation and the Taylor polynomial of degree 3.

Figure 6.42 shows $f(x)$, the linear approximation $P_1(x) = x - 1$, and $P_3(x)$. We see that the approximation is good when x is close to 1, and that the approximation $P_3(x)$ is better than the linear approximation.

6.7.3 How Accurate is the Approximation? (Optional)

We saw in Example 1 that the approximation improved when the degree of the polynomial was higher. We will now investigate how accurate the Taylor approximation is. We can assess the accuracy of the approximation directly for the function in Example 3.

In Example 3, we showed that the Taylor polynomial of degree n about $x = 0$ for $f(x) = \frac{1}{1-x}$, $x \neq 1$, is

$$P_n(x) = 1 + x + x^2 + x^3 + \cdots + x^{n-1} + x^n \tag{6.23}$$

There is a nice trick to find an expression for the sum (6.23). Note that

$$x P_n(x) = x + x^2 + x^3 + \cdots + x^n + x^{n+1} \tag{6.24}$$

Subtracting (6.24) from (6.23), we find

$$P_n(x) - x P_n(x) = 1 + x + x^2 + \cdots + x^n - x - x^2 - \cdots - x^n - x^{n+1}$$
$$= 1 - x^{n+1}$$

that is,

$$(1 - x) P_n(x) = 1 - x^{n+1}$$

or

$$P_n(x) = \frac{1 - x^{n+1}}{1 - x} = \frac{1}{1-x} - \frac{x^{n+1}}{1-x}$$

provided $x \neq 1$. We therefore conclude that

$$|f(x) - P_n(x)| = \left| \frac{x^{n+1}}{1-x} \right|$$

We can interpret the term $x^{n+1}/(1-x)$ as the error of approximation. Since

$$\lim_{n \to \infty} \left| \frac{x^{n+1}}{1-x} \right| = \begin{cases} \infty & \text{if } |x| > 1 \\ 0 & \text{if } |x| < 1 \end{cases}$$

we conclude that the error of approximation can be made small only when $|x| < 1$. For $|x| > 1$, the error of approximation increases with increasing n. (When $x = 1$, the function $f(x)$ is not defined.)

In general, it is not straightforward to obtain error estimates. In its general form, the error is given as an integral. We motivate this by looking at the error terms for $P_0(x)$ and $P_1(x)$.

Using the FTC, part II, we find

$$f(x) - f(a) = \int_a^x f'(t)\, dt$$

or

$$f(x) = f(a) + \int_a^x f'(t)\, dt$$

Since $f(a) = P_0(x)$, we can interpret $\int_a^x f'(t)\, dt$ as the error term in the Taylor approximation of $f(x)$ about $x = a$ when $n = 0$.

We can use integration by parts to obtain the next higher approximation:

$$\int_a^x f'(t)\, dt = \int_a^x 1 \cdot f'(t)\, dt$$

We set $u' = 1$ with $u = -(x - t)$, and $v = f'(t)$ with $v' = f''(t)$. (Writing $u = -(x - t)$ turns out to be a more convenient antiderivative of $u' = 1$ than $u = t$, as you will see shortly.) We find

$$\int_a^x f'(t)\, dt = -(x - t)f'(t)]_a^x + \int_a^x (x - t)f''(t)\, dt$$

$$= (x - a)f'(a) + \int_a^x (x - t)f''(t)\, dt$$

That is,

$$f(x) = f(a) + f'(a)(x - a) + \int_a^x (x - t)f''(t)\, dt$$

The expression $f(a) + f'(a)(x - a)$ is the linear approximation $P_1(x)$; the integral can then be considered as the error term.

Continuing in this way, we find the general formula.

TAYLOR'S FORMULA

Suppose that $f : I \to \mathbf{R}$, where I is an interval, $a \in I$, and f and its first $n + 1$ derivatives are continuous at $a \in I$. Then, for $x, a \in I$,

$$f(x) = f(a) + \frac{f'(a)}{1!}(x - a) + \frac{f''(a)}{2!}(x - a)^2$$

$$+ \cdots + \frac{f^{(n)}(a)}{n!}(x - a)^n + R_{n+1}(x)$$

where

$$R_{n+1}(x) = \frac{1}{n!} \int_a^x (x - t)^n f^{(n+1)}(t)\, dt$$

We will now examine the error term in Taylor's formula more closely. The error term is given in integral form, and it is often difficult (or impossible) to evaluate the integral. We will first look at the case $n = 0$; that is, we approximate $f(x)$ by the constant function $f(a)$. The error term is then $R_{n+1}(x)$ when $n = 0$, that is,

$$R_1(x) = \int_a^x f'(t)\, dt$$

Using the mean value theorem for integrals, we can find a value c in the interval between a and x, such that

$$\int_a^x f'(t)\,dt = f'(c)(x - a)$$

That is, we find

$$R_1(x) = f'(c)(x - a)$$

for some number c between a and x. Though we don't know the value of c, this form is quite useful. That is, if we set

$$K = \left[\begin{array}{c} \text{largest value of } |f'(t)| \\ \text{for } t \text{ between } a \text{ and } x \end{array} \right]$$

then

$$|R_1(x)| \le K|x - a|$$

Before we give the corresponding results for $R_{n+1}(x)$, we look at one example that illustrates how to find K.

I EXAMPLE 5   Estimate the error in the approximation of $f(x) = e^x$ by $P_0(x)$ about $x = 0$ on the interval $[0, 1]$.

Solution   Since $f(0) = 1$, we find

$$P_0(x) = 1$$

and

$$f(x) = 1 + R_1(x)$$

with

$$R_1(x) = \int_0^x f'(t)\,dt = xf'(c)$$

for some c between 0 and x. Since $f'(t) = e^t$, the largest value of $|f'(t)|$ in the interval $[0, 1]$ occurs when $t = 1$, namely $|f'(1)| = e$. Since we want to find an approximation of $f(x) = e^x$ for $x \in [0, 1]$, we should not use e in our error estimate, as e is one of the values we want to estimate. Instead, we use $|f'(t)| \le 3$ for $t \in [0, 1]$. We thus have

$$|R_1(x)| \le 3x \quad \text{for } x \in [0, 1]$$ **I**

The error term for general n can be dealt with in a similar fashion, so that one finds the following.

> There exists a c between a and x such that the error term in Taylor's formula is of the form
> $$R_{n+1}(x) = \frac{f^{(n+1)}(c)}{(n + 1)!}(x - a)^{n+1}$$

As in the case where $n = 0$, this form of the error term is quite useful. Though we don't know what c is, we can try to estimate $f^{(n+1)}(c)$ between a and x as before, when $n = 0$. Let

$$K = \left[\begin{array}{c} \text{largest value of } |f^{(n+1)}(t)| \\ \text{for } t \text{ between } a \text{ and } x \end{array} \right]$$

Then

$$|R_{n+1}(x)| \le \frac{K|x - a|^{n+1}}{(n + 1)!}$$

We will use this in the following example to determine in advance what degree of Taylor polynomial will allow us to achieve a given accuracy.

▮ **EXAMPLE 6**   Suppose that $f(x) = e^x$. What degree of Taylor polynomial about $x = 0$ will allow us to approximate $f(1)$ so that the error is less than 10^{-5}?

Solution   In Example 1, we found that for any $n \geq 1$

$$f^{(n+1)}(t) = e^t$$

We need to find out how large $f^{(n+1)}(t)$ can get for $t \in [0, 1]$. We find

$$|f^{(n+1)}(t)| = e^t \leq e \quad \text{for } 0 \leq t \leq 1$$

As in Example 5, instead of using e as a bound, we use a slightly larger value, namely 3. Therefore

$$|R_{n+1}(1)| \leq \frac{3|1|^{n-1}}{(n+1)!} = \frac{3}{(n+1)!} \tag{6.25}$$

We want the error to be less than 10^{-5}, that is. we want

$$|R_{n-1}(1)| < 10^{-5}$$

Inserting different values of n shows that

$$\frac{3}{8!} = 7.44 \times 10^{-5} \quad \text{and} \quad \frac{3}{9!} = 8.27 \times 10^{-6}$$

That is, when $n = 8$,

$$|R_{n+1}(1)| = |R_9(1)| \leq 8.27 \times 10^{-6} < 10^{-5}$$

Because the estimate of the error is greater than 10^{-5} when $n = 7$, we conclude that a polynomial of degree 8 would certainly give us the desired accuracy, whereas a polynomial of degree 7 might not. We can easily check this; we find

$$1 + 1 + \frac{1}{2!} + \frac{1}{3!} + \frac{1}{4!} + \cdots - \frac{1}{7!} = 2.71825396825$$

$$1 + 1 + \frac{1}{2!} + \frac{1}{3!} + \frac{1}{4!} + \cdots - \frac{1}{8!} = 2.71827876984$$

Comparing this to $e = 2.71828182846\ldots$, we see that the error is equal to 2.79×10^{-5} when $n = 7$ and 3.06×10^{-6} when $n = 8$. The error that we computed using (6.25) is a "worst-case scenario"; that is, the true error can be (and typically is) smaller than the error bound.   ▮

We have already seen one example where a Taylor polynomial was useful only for values close to the point at which we approximated the function, regardless of n, the degree of the polynomial. There are situations in which the error in the approximation cannot be made small for *any* value close to the point of approximation, regardless of n, the degree of the polynomial. One such example is the continuous function

$$f(x) = \begin{cases} e^{-1/x} & \text{for } x > 0 \\ 0 & \text{for } x \leq 0 \end{cases}$$

which is used, for instance, to describe the height of a tree depending on age. One can show that $f^{(k)}(0) = 0$ for *all* $k \geq 1$, which implies that a Taylor polynomial of degree n about $x = 0$ is

$$P_n(x) = 0$$

for all n. This clearly shows that it will not help to increase n; the approximation just will not improve.

When one uses Taylor polynomials to approximate functions, it is important to know for which values of x the approximation can be made arbitrarily close by choosing n large.

In the following, we list a few of the most important functions, together with their Taylor polynomials about $x = 0$ and the range of x values where the approximation can be made arbitrarily close by choosing n large enough.

$$e^x = 1 + x + \frac{x^2}{2!} + \frac{x^3}{3!} + \cdots + \frac{x^n}{n!} + R_{n+1}(x), \quad -\infty < x < \infty$$

$$\sin x = x - \frac{x^3}{3!} + \frac{x^5}{5!} - \frac{x^7}{7!} + \frac{x^9}{9!} - + \cdots + (-1)^n \frac{x^{2n+1}}{(2n+1)!}$$
$$+ R_{2n+2}(x), \quad -\infty < x < \infty$$

$$\cos x = 1 - \frac{x^2}{2!} + \frac{x^4}{4!} - \frac{x^6}{6!} + \frac{x^8}{8!} - + \cdots + (-1)^n \frac{x^{2n}}{(2n)!}$$
$$+ R_{2n+1}(x), \quad -\infty < x < \infty$$

$$\ln(1+x) = x - \frac{x^2}{2} + \frac{x^3}{3} - \frac{x^4}{4} + \frac{x^5}{5} - + \cdots + (-1)^{n+1} \frac{x^n}{n}$$
$$+ R_{n+1}(x), \quad -1 < x \le 1$$

$$\frac{1}{1-x} = 1 + x + x^2 + x^3 + x^4 + \cdots + x^n + R_{n+1}(x), \quad -1 < x < 1$$

◆ 6.7 PROBLEMS

[6.7.1]

In Problems 1–5, find the linear approximation of $f(x)$ at $x = 0$.

1. $f(x) = e^{2x}$

2. $f(x) = \sin x$

3. $f(x) = \dfrac{1}{1-x}$

4. $f(x) = x^2$

5. $f(x) = \ln(2 + x^2)$

In Problems 6–10, compute the Taylor polynomials of degree n about $a = 0$ for the indicated functions.

6. $f(x) = \dfrac{1}{1+x}, n = 4$

7. $f(x) = \cos x, n = 5$

8. $f(x) = e^{3x}, n = 3$

9. $f(x) = x^5, n = 6$

10. $f(x) = \sqrt{1+x}, n = 3$

In Problems 11–16, compute the Taylor polynomials of degree n about $x = 0$ for the indicated functions, and compare the value of the functions at the indicated point to the value of the corresponding Taylor polynomial.

11. $f(x) = \sqrt{2+x}, n = 3, x = 0.1$

12. $f(x) = \dfrac{1}{1-x}, n = 3, x = 0.1$

13. $f(x) = \sin x, n = 5, x = 1$

14. $f(x) = e^{-x}, n = 5, x = 0.3$

15. $f(x) = \tan x, n = 2, x = 0.1$

16. $f(x) = \ln(1 + x), n = 4, x = 0.2$

17. **(a)** Find the Taylor polynomial of degree 3 about $x = 0$ for $f(x) = \sin x$.

(b) Use your result in (a) to explain why

$$\lim_{x \to 0} \frac{\sin x}{x} = 1$$

18. **(a)** Find the Taylor polynomial of degree 2 about $x = 0$ for $f(x) = \cos x$.

(b) Use your result in (a) to explain why

$$\lim_{x \to 0} \frac{\cos x - 1}{x} = 0$$

[6.7.2]

In Problems 19–23, compute the Taylor polynomials of degree n about $x = a$, and compare the value of the approximation to the value of the function at the given point.

19. $f(x) = \sqrt{x}, a = 1, n = 3; x = 2$

20. $f(x) = \ln x, a = 1, n = 3; x = 2$

21. $f(x) = \cos x, a = \frac{\pi}{6}, n = 3; x = \frac{\pi}{7}$

22. $f(x) = x^{1/3}, a = -1, n = 3; x = -0.9$

23. $f(x) = e^x, a = 2, n = 3; x = 2.1$

24. Show that
$$T^4 \approx T_a^4 + 4T_a^3(T - T_a)$$
for T close to T_a.

25. Show that
$$rN\left(1 - \frac{N}{K}\right) \approx rN$$
for N close to 0 (where r and K are positive constants).

26. (a) Show that
$$f(R) = \frac{aR}{k + R} \approx \frac{a}{k}R$$
for R close to 0 (where a and k are positive constants).

(b) Show that
$$f(R) = \frac{aR}{k + R} \approx \frac{a}{2} + \frac{a}{4k}(R - k)$$
for R close to k (where a and k are positive constants).

[6.7.3]

In Problems 27–30, use the following form of the error term
$$R_{n+1}(x) = \frac{f^{(n+1)}(c)}{(n + 1)!}x^{n+1}$$
where c is between 0 and x, to determine in advance the degree of Taylor polynomial at a = 0 that would achieve the indicated accuracy in the given interval [0, x]. (Do not compute the Taylor polynomial.)

27. $f(x) = e^x$, $x = 2$, error $< 10^{-3}$

28. $f(x) = \sin x$, $x = 1$, error $< 10^{-2}$

29. $f(x) = 1/(1 + x)$, $x = 0.2$, error $< 10^{-2}$

30. $f(x) = \ln(1 + x)$, $x = 0.1$, error $< 10^{-3}$

31. Let $f(x) = e^{-1/x}$ for $x > 0$ and $f(x) = 0$ for $x = 0$. Compute a Taylor polynomial of degree 2 at $x = 0$, and determine how large the error is.

32. One can show that the Taylor polynomial for $f(x) = (1 + x)^\alpha$ about $x = 0$, with α a constant, converges for $x \in (-1, 1)$. Show that
$$(1 + x)^\alpha = 1 + \alpha x + \frac{\alpha(\alpha - 1)}{2!}x^2$$
$$+ \frac{\alpha(\alpha - 1)(\alpha - 2)}{3!}x^3 \cdots$$

33. One can show that the Taylor polynomial for $f(x) = \tan^{-1} x$ about $x = 0$ converges for $|x| \leq 1$.

(a) Show that the following is true:
$$\tan^{-1} x = x - \frac{x^3}{3} + \frac{x^5}{5} - \frac{x^7}{7} \pm \cdots$$

(b) Explain why the following holds:
$$\frac{\pi}{4} = 1 - \frac{1}{3} + \frac{1}{5} - \frac{1}{7} \pm \cdots$$

(This expression converges very slowly, as you would see if you used it to approximate π.)

◆ 6.8 KEY TERMS

Chapter 6 Review: Topics   Discuss the following definitions and concepts:

1. The substitution rule for indefinite integrals
2. The substitution rule for definite integrals
3. Integration by parts
4. The "Multiplying by 1" trick
5. Partial fraction decomposition
6. Partial fraction method
7. Proper rational function
8. Irreducible quadratic factor
9. Improper integral
10. Integration when the interval is unbounded
11. Integration when the integrand is discontinuous
12. Convergence and divergence of improper integrals
13. Comparison results for improper integrals
14. Numerical integration: midpoint and trapezoidal rule
15. Error bounds for the midpoint and the trapezoidal rule
16. Using tables of integrals for integration
17. Linear approximation
18. Taylor polynomial of degree n
19. Taylor's formula

◆ 6.9 REVIEW PROBLEMS

In Problems 1–30, evaluate the given indefinite integrals.

1. $\displaystyle\int x^2(1 - x^3)^2 \, dx$

2. $\displaystyle\int \frac{\cos x}{1 + \sin^2 x} \, dx$

3. $\displaystyle\int 4xe^{-x^2} \, dx$

4. $\displaystyle\int \frac{x \ln(1 + x^2)}{1 + x^2} \, dx$

5. $\displaystyle\int (1 + \sqrt{x})^{1/2} \, dx$

6. $\displaystyle\int z\sqrt{3x + 1} \, dx$

7. $\displaystyle\int x \sec^2(3x^2) \, dx$

8. $\displaystyle\int \tan x \sec^2 x \, dx$

9. $\displaystyle\int x \ln x \, dx$

10. $\displaystyle\int x^3 \ln x^2 \, dx$

11. $\displaystyle\int \sec^2 x \ln(\tan x) \, dx$

12. $\displaystyle\int \sqrt{x} \ln \sqrt{x} \, dx$

13. $\displaystyle\int \frac{1}{4+x^2}\,dx$

14. $\displaystyle\int \frac{1}{4-x^2}\,dx$

15. $\displaystyle\int \tan x\,dx$

16. $\displaystyle\int \tan^{-1} x\,dx$

17. $\displaystyle\int \sin x\,e^{2x}\,dx$

18. $\displaystyle\int x\sin x\,dx$

19. $\displaystyle\int \sqrt{e^x}\,dx$

20. $\displaystyle\int \ln\sqrt{x}\,dx$

21. $\displaystyle\int \sin^2 x\,dx$

22. $\displaystyle\int \sin x\cos x\,e^{\sin x}\,dx$

23. $\displaystyle\int \frac{1}{x(x-1)}\,dx$

24. $\displaystyle\int \frac{1}{(x+1)(x-2)}\,dx$

25. $\displaystyle\int \frac{x}{x+5}\,dx$

26. $\displaystyle\int \frac{x}{x^2+5}\,dx$

27. $\displaystyle\int \frac{1}{x+5}\,dx$

28. $\displaystyle\int \frac{1}{x^2+5}\,dx$

29. $\displaystyle\int \frac{(x+1)^2}{x-1}\,dx$

30. $\displaystyle\int \frac{2x+1}{\sqrt{1-x^2}}\,dx$

In Problems 31–50, evaluate the given definite integrals.

31. $\displaystyle\int_1^3 \frac{x^2+1}{x}\,dx$

32. $\displaystyle\int_0^{\pi/2} x\sin x\,dx$

33. $\displaystyle\int_0^1 x e^{-x^2/2}\,dx$

34. $\displaystyle\int_1^2 \ln x\,dx$

35. $\displaystyle\int_0^2 \frac{1}{4+x^2}\,dx$

36. $\displaystyle\int_0^{1/2} \frac{2}{\sqrt{1-x^2}}\,dx$

37. $\displaystyle\int_2^6 \frac{1}{\sqrt{x-2}}\,dx$

38. $\displaystyle\int_0^2 \frac{1}{x-2}\,dx$

39. $\displaystyle\int_0^\infty \frac{1}{9+x^2}\,dx$

40. $\displaystyle\int_0^\infty \frac{1}{x^2+3}\,dx$

41. $\displaystyle\int_0^\infty \frac{1}{x+3}\,dx$

42. $\displaystyle\int_0^\infty \frac{1}{(x+3)^2}\,dx$

43. $\displaystyle\int_0^1 \frac{1}{x^2}\,dx$

44. $\displaystyle\int_1^\infty \frac{1}{x^2}\,dx$

45. $\displaystyle\int_0^1 \frac{1}{\sqrt{x}}\,dx$

46. $\displaystyle\int_1^\infty \frac{1}{\sqrt{x}}\,dx$

47. $\displaystyle\int_0^1 x\ln x\,dx$

48. $\displaystyle\int_0^1 x2^x\,dx$

49. $\displaystyle\int_0^{\pi/4} e^{\cos x}\sin x\,dx$

50. $\displaystyle\int_0^{\pi/4} x\sin(2x)\,dx$

In Problems 51–54, use (a) the midpoint rule, and (b) the trapezoidal rule to approximate each integral with the specified value of n.

51. $\displaystyle\int_0^2 (x^2-1)\,dx,\ n=4$

52. $\displaystyle\int_{-1}^1 (x^3-1)\,dx,\ n=4$

53. $\displaystyle\int_0^1 e^{-x}\,dx,\ n=5$

54. $\displaystyle\int_0^{\pi/4} \sin(4x)\,dx,\ n=4$

In Problems 55–58, find the Taylor polynomial of degree n about x = a for each function.

55. $f(x)=\sin(2x),\ a=0,\ n=3$

56. $f(x)=e^{-x^2/2},\ a=0,\ n=3$

57. $f(x)=\ln x,\ a=1,\ n=3$

58. $f(x)=\dfrac{1}{x-3},\ a=4,\ n=4$

59. (*Cost of gene substitution*)
(Adapted from Roughgarden 1996) Suppose that an advantageous mutation arises in a population. Initially, the gene carrying this mutation is at a low frequency. As the gene carrying the mutation spreads through the population, the average fitness of the population increases. We denote by $f_{avg}(t)$ the average fitness of the population at time t, by $f_{avg}(0)$ the average fitness of the population at time 0 (when the mutation arose), and by K the final value of the average fitness after the mutation has spread through the population. Haldane (1957) suggested to measure the *cost of evolution* (now known as the *cost of gene substitution*) by the cumulative difference between the current and the final fitness, that is,

$$\int_0^\infty (K - f_{avg}(t))\,dt$$

In Figure 6.43 shade the region whose area is equal to the cost of gene substitution.

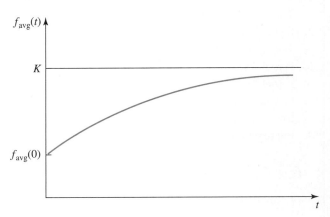

▲ **Figure 6.43**
The cost of gene substitution. See Problem 59.

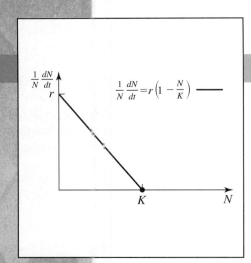

Differential Equations

In the simplest model of population growth, the number of births and deaths at any time are proportional to the population size at that time. How would we express such a situation mathematically? Let $N(t)$ denote the size of a population at time t, $t \geq 0$. The change in population size during the time interval $[t, t + \Delta t]$, denoted by $\Delta N = N(t + \Delta t) - N(t)$, is

$$\Delta N = N(t + \Delta t) - N(t) = \begin{bmatrix} \text{no. of births} - \text{no. of deaths} \\ \text{in } [t, t + \Delta t] \end{bmatrix}$$

If Δt is small, the population size will not change much during $[t, t + \Delta t]$, and the change ΔN is approximately proportional to $N(t)\Delta t$, where the error in the approximation goes to 0 as $\Delta t \to 0$. Introducing the proportionality factor r, the change in population size over the time interval $[t, t + \Delta t]$ satisfies

$$\frac{\Delta N}{\Delta t} \approx r N(t) \tag{7.1}$$

If we let Δt go to 0, the left-hand side of (7.1) becomes the derivative of $N(t)$ (if it exists) and is equal to the right-hand side (since the error vanishes in the limit). Thus

$$\frac{dN}{dt} = r N(t), \ t \geq 0 \tag{7.2}$$

(We introduced this equation in Section 3.6.) Equation (7.2) contains the derivative of a function. Such equations are called **differential equations**. Modeling biological situations frequently leads to differential equations.

Differential equations can contain any order derivative, for example,

$$\frac{d^2 y}{dx^2} + \frac{dy}{dx} = xy$$

is a differential equation that contains the first and second derivative of the function $y = y(x)$. The differential equation (7.2) contains only the first derivative; such an equation is called a **first order** differential equation.

Throughout this chapter, we will restrict ourselves to first order differential equations of the form

$$\frac{dy}{dx} = f(x)g(y) \tag{7.3}$$

where the right-hand side of (7.3) is the product of two functions, one depends only on x, the other only on y. Such equations are called **separable differential equations** (the reason for this name will become clear shortly). This type of differential equation includes two special cases, namely,

$$\frac{dy}{dx} = f(x) \tag{7.4}$$

and

$$\frac{dy}{dx} = g(y) \tag{7.5}$$

We discussed differential equations of the form (7.4) in Section 4.7. Differential equations of the form (7.5) include (7.2), and are frequently used in biological models.

◆ 7.1 SOLVING DIFFERENTIAL EQUATIONS

Let's return to the growth model in (7.2). To be concrete, we choose $r = 2$, which results in the equation

$$\frac{dN}{dt} = 2N(t), \ t \geq 0 \tag{7.6}$$

We might be interested in finding a function $N(t)$ that satisfies (7.6). A function that satisfies a differential equation, is called a **solution** of the differential equation. We claim that with $N_0 = N(0)$,

$$N(t) = N_0 e^{2t}, \ t \geq 0$$

is a solution of (7.6). To check this, we differentiate $N(t)$:

$$\frac{dN}{dt} = 2 \underbrace{N_0 e^{2t}}_{N(t)} = 2N(t)$$

At any point $(t, N(t))$ of the graph of $N(t)$, the slope is equal to $2N(t)$ (see Figure 7.1).

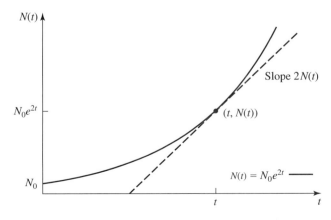

◀ Figure 7.1

The solution $N(t) = N_0 e^{2t}$: The slope at $(t, N(t))$ is $2N(t)$.

A first order differential equation tells us what the derivative of a function is. In order to find a solution, we must therefore integrate. (Since it is not always possible to integrate a function, it is not always possible to write the solution of a differential equation in explicit form.)

We begin with a general method for solving differential equations of the form (7.3)

$$\frac{dy}{dx} = f(x)g(y) \tag{7.7}$$

We divide both sides of (7.7) by $g(y)$:

$$\frac{1}{g(y)}\frac{dy}{dx} = f(x) \tag{7.8}$$

Now, if $y = u(x)$ is a solution of (7.8), then $u(x)$ satisfies

$$\frac{1}{g[u(x)]}u'(x) = f(x)$$

and, therefore,

$$\int \frac{1}{g[u(x)]}u'(x)\,dx = \int f(x)\,dx$$

But, since $g[u(x)] = g(y)$ and $u'(x)\,dx = dy$, this is the same as

$$\int \frac{1}{g(y)}\,dy = \int f(x)\,dx \tag{7.9}$$

This suggests the following procedure of solving separable differential equations. Namely, we separate the variables x and y so that one side of the equation depends only on y and the other side only on x. To do so, we treat dy and dx as if they were regular numbers: Writing (7.7) formally as

$$\frac{dy}{g(y)} = f(x)\,dx$$

and then integrating both sides:

$$\int \frac{dy}{g(y)} = \int f(x)\,dx$$

yields equation (7.9). The method of separating the variables x and y works since the right-hand side of (7.7) is of the special form $f(x)g(y)$, and it gave this type of differential equation its name. A note of caution: We simply divided (7.7) by $g(y)$. One must be careful, since $g(y)$ might be 0 for some values of y, in which case we could not divide. We will address this problem in Subsection 7.1.2. In the next two subsections, we discuss how to solve differential equations of the form (7.4) and (7.5); we give an example of the general type (7.3) in Subsection 7.1.3.

7.1.1 Pure Time Differential Equations

In many applications, the independent variable represents time. If the rate of change of a function depends only on time, we call the resulting differential equation a **pure time differential equation**, that is, such a differential equation is of the form

$$\frac{dy}{dx} = f(x), \; x \in I \tag{7.10}$$

where I is an interval, and x represents time. We discussed such equations in Section 4.7, where we found that the solution is of the form

$$y(x) = \int_{x_0}^{x} f(u)\,du + C \tag{7.11}$$

The constant C comes from finding the general antiderivative of $f(x)$; the number x_0 is in the interval I. To determine C, we must phrase the problem as an initial value problem; we assume $y(x_0) = y_0$, then the solution is

$$y(x) = y_0 + \int_{x_0}^{x} f(u)\,du$$

To solve (7.10) formally, we separate variables; we write the differential equation (7.10) in the form

$$dy = f(x)\,dx$$

and then integrate both sides:

$$\int dy = \int f(x)\,dx$$

or

$$y(x) = \int f(x)\,dx$$

which is the same as (7.11).

I EXAMPLE I   Suppose that the volume $V(t)$ of a cell at time t changes according to

$$\frac{dV}{dt} = \sin t \quad \text{with } V(0) = 3$$

Find $V(t)$.

Solution   Since

$$V(t) = V(0) + \int_{0}^{t} \sin u\,du$$

we find

$$V(t) = 3 + [-\cos u]_{0}^{t}$$
$$= 3 + (-\cos t + \cos 0)$$
$$= 4 - \cos t$$

since $\cos 0 = 1$ (see Figure 7.2).   ∎

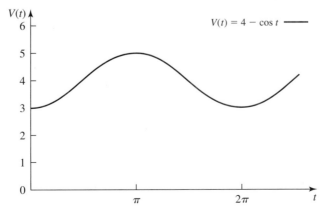

$V(t) = 4 - \cos t$ ——

◀ **Figure 7.2**
The solution $V(t) = 4 - \cos t$ in Example 1.

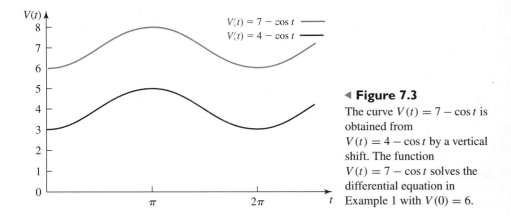

◀ **Figure 7.3**
The curve $V(t) = 7 - \cos t$ is obtained from
$V(t) = 4 - \cos t$ by a vertical shift. The function
$V(t) = 7 - \cos t$ solves the differential equation in
Example 1 with $V(0) = 6$.

If we changed the initial condition in Example 1 to some other value, the graph of the new solution would be obtained from the old solution by simply shifting the solution vertically so that it would satisfy the new initial condition (see Figure 7.3). (This was discussed in Section 4.7.)

7.1.2 Autonomous Differential Equations

Many of the differential equations that model biological situations, such as (7.2), are of the form

$$\frac{dy}{dx} = g(y) \tag{7.12}$$

where the right-hand side of (7.12) does not explicitly depend on x. Such equations are called **autonomous differential equations**.

To interpret the biological meaning of autonomous, let's return to the growth model

$$\frac{dN}{dt} = 2N(t), \ t \geq 0 \tag{7.13}$$

Assume we conduct an experiment, where we follow a population over time, which satisfies (7.13) with $N(0) = 20$. Then the size of the population at time t is given by

$$N(t) = 20e^{2t}, \ t \geq 0$$

If we repeat the experiment at time $t = 10$ (say) with the exact same initial population size, then, everything else being equal, the population evolves in exactly the same way as the one starting at $t = 0$. The graph of the solution is merely shifted to the new starting point $(10, 20)$ (see Figure 7.4).

When we look at the slopes of the tangent lines at points where $N = 30$ (say), we see that both curves have the same slope, namely, $dN/dt = 2N = (2)(30) = 60$. We can draw such line segments representing tangent lines to solution curves at other points as well. Such a graph is called a **direction field**. The line segments are tangential to the graphs of the solutions (see Figure 7.5). At each point $(t, N(t))$, the slope of the tangent line is given by (7.13). We see that along horizontal lines, all line segments have the same slope, confirming that any solution starting at $(t, 20)$, $t > 0$, follows the same curve as the one starting at $(0, 20)$, but shifted horizontally to the new starting point.

This means that a population starting with $N = 20$ follows the same trajectory, regardless of when we start the experiment. This makes sense biologically:

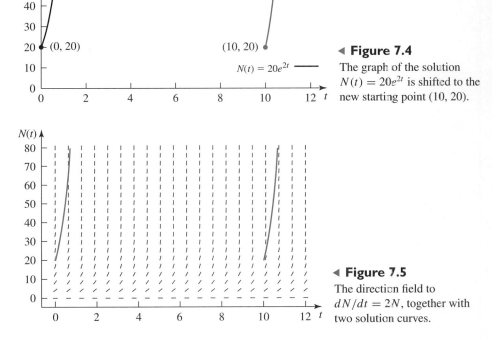

◄ Figure 7.4

The graph of the solution $N(t) = 20e^{2t}$ is shifted to the new starting point $(10, 20)$.

◄ Figure 7.5

The direction field to $dN/dt = 2N$, together with two solution curves.

If the conditions don't change, the experiment should be replicable. If the growth conditions for the population do change over time, we would not be able to use an autonomous differential equation to describe the growth of the population; we would need to explicitly include the time dependence in the dynamics.

Formally, we can solve (7.12) by separation of variables, namely, dividing both sides of (7.12) by $g(y)$ and multiplying both sides of (7.12) by dx, we obtain

$$\frac{dy}{g(y)} = dx$$

Integrating both sides gives

$$\int \frac{dy}{g(y)} = \int dx$$

We will discuss two cases: $g(y) = k(y - a)$, and $g(y) = k(y - a)(y - b)$. The growth model (7.2) is an example of the first case; the logistic equation, which we saw in Section 2.3 for the first time, is an example of the second case.

Case I: $g(y) = k(y - a)$ We wish to solve

$$\frac{dy}{dx} = k(y - a) \tag{7.14}$$

where k and a are constants. We assume $k \neq 0$. Separating variables yields

$$\frac{dy}{y - a} = k\,dx \tag{7.15}$$

It is somewhat arbitrary whether one leaves k on the right-hand side or moves it to the left-hand side. Leaving it on the right-hand side is more convenient. Integrating both sides of (7.15) results in

$$\int \frac{dy}{y-a} = \int k\,dx$$

or

$$\ln|y-a| = kx + C_1$$

Exponentiating both sides yields

$$|y-a| = e^{kx+C_1}$$

or

$$|y-a| = e^{C_1}e^{kx}$$

Removing the absolute value signs, we find

$$y-a = \pm e^{C_1}e^{kx}$$

We set $C = \pm e^{C_1}$ and write the solution as

$$y = Ce^{kx} + a \qquad (7.16)$$

Here we must repeat the note of caution before Subsection 7.1.1: To obtain (7.15), we divided by $y-a$. We are only allowed to do this as long as $y \neq a$. If $y = a$, then $dy/dx = 0$ and the constant function $y = a$ is a solution of (7.14). We *lost* this solution when we divided (7.14) by $y-a$. Note that the constant C in (7.16) is different from 0 since $C = \pm e^{C_1}$ and $e^{C_1} \neq 0$ for any real number C_1. But we can combine the constant solution $y = a$ and the solution in (7.16) by allowing C to be equal to 0 in (7.16).

Before we turn to biological applications, we will give an example where we see how to solve such a differential equation in a concrete situation.

▮ **EXAMPLE 2**   Solve

$$\frac{dy}{dx} = 2 - 3y \quad \text{where } y = 1 \text{ when } x = 1$$

Solution   Instead of trying to identify the constants C, k, and a in equation (7.16), it is actually easier to solve the equation directly. We separate variables and then integrate, which results in

$$\int \frac{dy}{2-3y} = \int dx$$

Since an antiderivative of $\frac{1}{2-3y}$ is $-\frac{1}{3}\ln|2-3y|$, we find

$$-\frac{1}{3}\ln|2-3y| = x + C_1$$

Solving for y yields

$$\ln|2-3y| = -3x - 3C_1$$
$$|2-3y| = e^{-3x-3C_1}$$
$$2-3y = \pm e^{-3C_1}e^{-3x}$$

Setting $C = \pm e^{-3C_1}$, we find

$$2-3y = Ce^{-3x}$$

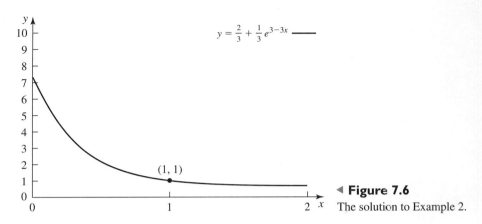

◀ **Figure 7.6**
The solution to Example 2.

To determine C, we use the fact that $y = 1$ when $x = 1$. That is,

$$2 - 3 = Ce^{-3} \quad \text{or} \quad C = -e^3$$

Hence,

$$2 - 3y = -e^3 e^{-3x}$$

or

$$y = \frac{2}{3} + \frac{1}{3}e^{3-3x}$$

(see Figure 7.6).

We will now turn to two biological applications that are covered by Case 1.

▌ **EXAMPLE 3**   **Exponential Population Growth**

This model is given by (7.2) and was introduced in Section 3.6. We denote by $N(t)$ the population size at time $t \geq 0$ and assume $N(0) = N_0 > 0$. The change in population size is described by the initial value problem

$$\frac{dN}{dt} = rN \quad \text{where } N(0) = N_0 \tag{7.17}$$

The parameter r is called the intrinsic rate of growth, and is the per capita rate of growth since

$$r = \frac{1}{N}\frac{dN}{dt}$$

When $r > 0$, this represents a growing population. When $r < 0$, the size of the population decreases. (Note that the per capita growth rate, r, is independent of the population size.)

We can either solve (7.17) directly as in Example 2, or use (7.14). It is typically easier to solve such equations directly than to memorize their solutions. But we should also learn how to use (7.14).

If we compare (7.17) with (7.14), we find $k = r$ and $a = 0$. Hence, using (7.16), we find the solution

$$N(t) = Ce^{rt}$$

Since $N(0) = N_0 = C$, we write the solution as

$$N(t) = N_0 e^{rt} \tag{7.18}$$

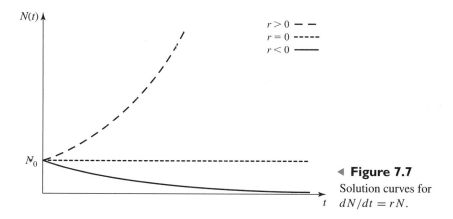

◀ **Figure 7.7**
Solution curves for
$dN/dt = rN$.

Equation (7.18) shows that the population grows exponentially when $r > 0$. When $r < 0$, the population size decreases exponentially fast.

We show solution curves of $N(t) = N_0 e^{rt}$ for $r > 0$, $r = 0$, and $r < 0$ in Figure 7.7. Exponential growth (or decay) is one of the most important growth equations in biology. You should therefore memorize both the differential equation (7.17) and its solution (7.18), together with the graphs in Figure 7.7, and know that (7.17) describes a situation where the per capita growth rate (or intrinsic rate of growth) is a constant.

When $r > 0$, the population size grows without bound ($\lim_{t \to \infty} N(t) = \infty$). This is found, for instance, when individuals are not limited by food availability or by competition. If we start a bacterial colony on a nutrient-rich substrate by inoculating the substrate with a few bacteria, then the bacteria, initially, can grow and divide unrestricted. Subsequently, when the substrate becomes more crowded and the food source depleted, the growth will be restricted, and a different differential equation will be required to describe this situation. (We will discuss this in Case 2.) Exponential decay in a population can be seen when the death rate exceeds the birth rate (for instance, in cases of starvation).   ▮

The type of growth in Example 3 is sometimes referred to as Malthusian growth, named after Thomas Malthus (1766–1834), a British clergyman and economist. Malthus wrote about the consequences of unrestricted growth on the welfare of humans. He observed that exponential growth of the human population would lead to starvation, because food supply would remain constant.

Recall from Section 3.6 that when $r < 0$, (7.17) has the same form as the differential equation that describes radioactive decay. $N(t)$ would then denote the amount of radioactive material left at time t. We will revisit this application in Problem 20.

EXAMPLE 4   **Restricted Growth: von Bertalanffy Equation**

This example describes the simplest form of restricted growth, and can be used to describe the growth of fish. We denote by $L(t)$ the length of the fish at age t and assume $L(0) = L_0$. Then

$$\frac{dL}{dt} = k(A - L) \tag{7.19}$$

where A is a positive constant that we will interpret in the following. We assume $L_0 < A$ and explain this restriction subsequently. The constant k is also positive; the equation says that the growth rate dL/dt is proportional to $A - L$, so k should be interpreted as a constant of proportionality. We see that the growth rate dL/dt

decreases linearly with length as long as $L < A$, and that the growth stops (that is, $dL/dt = 0$) when $L = A$. We could compare (7.19) with (7.14) and then use (7.16) to write the solution. Instead of memorizing the solution, it is easier to solve (7.19) using separation of variables. Separating the variables and integrating, we find

$$\int \frac{dL}{A - L} = \int k\,dt$$

Hence,

$$-\ln|A - L| = kt + C_1$$

or, after multiplying this equation by -1 and exponentiating,

$$|A - L| = e^{-C_1} e^{-kt}$$

or

$$A - L = Ce^{-kt}$$

with $C = \pm e^{-C_1}$. Since $L(0) = L_0$, we find

$$A - L_0 = C$$

Hence, the solution is given by

$$A - L(t) = (A - L_0)e^{-kt}$$

or

$$L(t) = A\left[1 - \left(1 - \frac{L_0}{A}\right)e^{-kt}\right] \tag{7.20}$$

(see Figure 7.8).

This is the von Bertalanffy equation that we encountered previously. Since

$$\lim_{t \to \infty} L(t) = A$$

we see that the parameter A denotes the **asymptotic length** of the fish. Mathematically, there are no restrictions on L_0; biologically, however, we require that $0 < L_0 < A$. Otherwise, the growth rate could be negative, which would mean that the fish shrinks in size. Note that A is an asymptotic length that is never reached, since there is no finite age T with $L(T) = A$ if $L(0) < A$.

Returning to (7.15), we can now interpret this differential equation. Namely, the growth rate is proportional to the difference in the current length and the asymptotic length, with k representing the constant of proportionality. Since the difference between the current length and the asymptotic length is decreasing over time, this

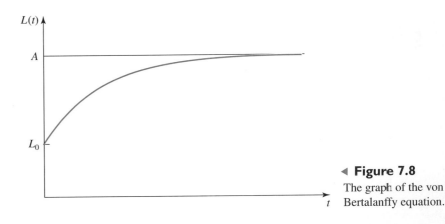

◀ Figure 7.8
The graph of the von Bertalanffy equation.

also shows that the growth rate decreases over time. That is, juveniles grow at a faster rate than adults. However, the growth rate is always positive, that is, fish grow throughout their lives, which is indeed the case. ▮

Case 2: $g(y) = k(y - a)(y - b)$ We now turn to differential equations of the form

$$\frac{dy}{dx} = k(y - a)(y - b) \tag{7.21}$$

where k, a, and b are constant. We assume $k \neq 0$. Separating variables and integrating both sides yields

$$\int \frac{dy}{(y - a)(y - b)} = \int k \, dx \tag{7.22}$$

When $a = b$, we must find an antiderivative of $\frac{1}{(y-a)^2}$, which is $-\frac{1}{y-a}$. In this case, we find

$$-\frac{1}{y - a} = kx + C$$

or

$$y = a - \frac{1}{kx + C}$$

The constant C can then be determined from the initial condition.

To find the solution when $a \neq b$, we must use the partial fraction method that we introduced in Section 6.3. Let's first do an example.

▮ **EXAMPLE 5**   Solve

$$\frac{dy}{dx} = 2(y - 1)(y + 2) \quad \text{with } y = 2 \text{ when } x = 0$$

Solution   Separation of variables yields

$$\int \frac{dy}{(y - 1)(y + 2)} = \int 2 \, dx$$

We use the partial fraction method to integrate the left hand side.

$$\frac{1}{(y - 1)(y + 2)} = \frac{A}{y - 1} + \frac{B}{y + 2}$$
$$= \frac{A(y + 2) + B(y - 1)}{(y - 1)(y + 2)}$$
$$= \frac{(A + B)y + 2A - B}{(y - 1)(y + 2)}$$

Comparing the last term to the integrand, we find

$$A + B = 0 \quad \text{and} \quad 2A - B = 1$$

which yields

$$A = -B \quad \text{and} \quad 2A - B = 3A = 1$$

and thus

$$A = \frac{1}{3} \quad \text{and} \quad B = -\frac{1}{3}$$

Using the partial fraction decomposition, we must integrate

$$\frac{1}{3} \int \left(\frac{1}{y-1} - \frac{1}{y+2} \right) dy = \int 2 \, dx$$

which yields

$$\frac{1}{3}[\ln|y-1| - \ln|y+2|] = 2x + C_1$$

Simplifying this results in

$$\ln \left| \frac{y-1}{y+2} \right| = 6x + 3C_1$$

$$\left| \frac{y-1}{y+2} \right| = e^{3C_1} e^{6x}$$

$$\frac{y-1}{y+2} = \pm e^{3C_1} e^{6x}$$

$$\frac{y-1}{y+2} = C e^{6x}$$

Using the initial condition $y = 2$ when $x = 0$, we find

$$\frac{1}{4} = C$$

The solution is therefore

$$\frac{y-1}{y+2} = \frac{1}{4} e^{6x}$$

If we want the solution in the form $y = f(x)$, we must solve for y. We find

$$y - 1 = (y+2)\frac{1}{4} e^{6x}$$

$$y \left(1 - \frac{1}{4} e^{6x} \right) = \frac{1}{2} e^{6x} + 1$$

$$y = \frac{\frac{1}{2} e^{6x} + 1}{1 - \frac{1}{4} e^{6x}}$$

$$y = \frac{2 e^{6x} + 4}{4 - e^{6x}}$$

(see Figure 7.9).

�N=$y = \frac{2e^{6x}+4}{4-e^{6x}}$

◀ **Figure 7.9**
The solution for Example 5.

We return now to (7.22) when $a \neq b$, and use the partial fraction method to simplify the integral on the left-hand side of (7.22). We decompose the integrand on the left-hand side of (7.22) into a sum of simpler rational functions that we know how to integrate; namely, we write the integrand in the form

$$\frac{1}{(y-a)(y-b)} = \frac{A}{y-a} + \frac{B}{y-b} \tag{7.23}$$

where A and B are constants that we must find. We do the following algebraic manipulation on the right-hand side of (7.23):

$$\begin{aligned} \frac{A}{y-a} + \frac{B}{y-b} &= \frac{A(y-b) + B(y-a)}{(y-a)(y-b)} \\ &= \frac{y(A+B) - (Ab+Ba)}{(y-a)(y-b)} \end{aligned} \tag{7.24}$$

Comparing the last expression in (7.24) to the left-hand side of (7.23), we conclude that the constants A and B must satisfy

$$A + B = 0 \qquad \text{and} \qquad Ab + Ba = -1$$

Substituting $B = -A$ from the first equation into the second, we find

$$Ab - Aa = -1 \qquad \text{or} \qquad A(b-a) = -1 \qquad \text{or} \qquad A = \frac{1}{a-b}$$

and, therefore,

$$B = -\frac{1}{a-b}$$

That is,

$$\frac{1}{(y-a)(y-b)} = \frac{1}{a-b}\left[\frac{1}{y-a} - \frac{1}{y-b}\right] \tag{7.25}$$

Equation (7.25) allows us to evaluate

$$\int \frac{dy}{(y-a)(y-b)}$$

which is the left-hand side of (7.22). Namely,

$$\begin{aligned} \int \frac{dy}{(y-a)(y-b)} &= \frac{1}{a-b} \int \left(\frac{1}{y-a} - \frac{1}{y-b}\right) dy \\ &= \frac{1}{a-b} \left[\ln|y-a| - \ln|y-b|\right] + C_1 \end{aligned}$$

Integrating the right-hand side of (7.22) as well, we find

$$\frac{1}{a-b}\left[\ln|y-a| - \ln|y-b|\right] = kx + C_2$$

or

$$\ln\left|\frac{y-a}{y-b}\right| = k(a-b)x + C_2(a-b)$$

Exponentiating this yields

$$\left|\frac{y-a}{y-b}\right| = e^{C_2(a-b)}e^{k(a-b)x}$$

or

$$\frac{y-a}{y-b} = \pm e^{C_2(a-b)} e^{k(a-b)x}$$

Defining $C = \pm e^{C_2(a-b)}$, we find as the solution of (7.21)

$$\frac{y-a}{y-b} = C e^{k(a-b)x}$$

We can solve this for y, namely

$$y = a + (y-b)C e^{k(a-b)x}$$
$$y(1 - C e^{k(a-b)x}) = a - bC e^{k(a-b)x}$$

or

$$y = \frac{a - bC e^{k(a-b)x}}{1 - C e^{k(a-b)x}} \qquad (7.26)$$

The constant C can then be determined from the initial condition.

We will now turn to an example that illustrates one of the most important applications of this case, namely, density dependent growth. In Example 3, we considered unrestricted growth. It had the unrealistic feature that if the intrinsic rate of growth is positive, the size of a population grows without bound. Frequently, the per capita growth rate decreases as the populations size increases. The growth rate thus depends on the population density. The simplest such model for which the growth rate is density dependent, is the logistic equation. In the logistic equation, the size of the population is restricted and cannot grow without bound.

This equation was originally developed around 1835, by Pierre-François Verhulst who used the term "logistic" for the equation. His work was completely forgotten until 1920 when Raymond Pearl and Lowell J. Reed published a series of papers on population growth. They used the same equation as Verhulst. After discovering Verhulst's work, Pearl and Reed adopted the name logistic for the equation. Pearl and Reed (1920) used the logistic equation to predict future growth of the US population based on census data from 1790 to 1920. Their equation would have predicted about 185 million people in the US for the year 2000, which is a gross underestimate of the actual population size (over 260 million people). Though the equation does not seem to fit actual populations very well, it is a useful model for analyzing growth under limiting resources.

EXAMPLE 6   **The Logistic Equation**

The logistic equation describes the change in size of a population for which growth is density dependent. If we denote by $N(t)$ the population size at time t, then the change in growth is given by the initial value problem

$$\frac{dN}{dt} = rN\left(1 - \frac{N}{K}\right) \quad \text{with } N(0) = N_0 \qquad (7.27)$$

where r and K are positive constants. This is the simplest way of incorporating density dependence in the growth rate; namely, the per capita rate of growth is given by

$$\frac{1}{N}\frac{dN}{dt} = r\left(1 - \frac{N}{K}\right)$$

that is, it decreases linearly with population size (see Figure 7.10).

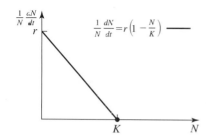

The per capita growth rate in the logistic equation is a linearly decreasing function of population size.

We can interpret the parameter r as follows: r describes the per capita growth rate in the absence of competition, that is, when the population size is close to 0. Therefore, one way to measure r is to grow the organism at a very low density, that is, when N is much smaller than K, so that the per capita growth rate is close to r. (In Problem 42, we discuss a different way to find r.)

The quantity K is called the **carrying capacity**. Looking at Figure 7.10, we see that the per capita growth rate is 0 when the population size is at the carrying capacity. Since the per capita growth rate is positive below K and negative above K, the size of the population will increase below K and decrease above K. The number K thus determines the population size that can be supported by the environment. We will see from the solution of (7.27) that starting from any positive initial population size the size of the population will eventually reach K (that is, $\lim_{t \to \infty} N(t) = K$ if $N(0) > 0$).

Before we discuss (7.27) further, we will solve it. We write the right-hand side of (7.27) as

$$g(N) = -\frac{r}{K}(N - 0)(N - K) \tag{7.28}$$

Comparing (7.28) with the right-hand side of (7.21), we find

$$k = -\frac{r}{K} \qquad a = 0 \qquad b = K \tag{7.29}$$

The solution of (7.27) is given in (7.26). Using (7.29), the solution of (7.27) is therefore

$$N(t) = \frac{0 - KCe^{-(r/K)(0-K)t}}{1 - Ce^{-(r/K)(0-K)t}}$$

or

$$N(t) = \frac{-CKe^{rt}}{1 - Ce^{rt}} = \frac{CK}{C - e^{-rt}}$$

$$= \frac{K}{1 - e^{-rt}/C} \tag{7.30}$$

Since $N(0) = N_0$, we find

$$N_0 = \frac{CK}{C - 1}$$

Solving this for C,

$$N_0(C - 1) = CK \qquad \text{or} \qquad C(N_0 - K) = N_0$$

thus

$$C = \frac{N_0}{N_0 - K} \tag{7.31}$$

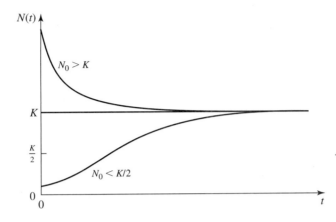

◀ **Figure 7.11**

Solution curves for the logistic equation for different initial values N_0.

Substituting (7.31) into (7.30), we find

$$N(t) = \frac{K}{1 - \frac{N_0 - K}{N_0}e^{-rt}}$$

or

$$N(t) = \frac{K}{1 + \left(\frac{K}{N_0} - 1\right)e^{-rt}} \tag{7.32}$$

We see from (7.32) that

$$\lim_{t \to \infty} N(t) = K$$

We plot (7.32) for $N_0 > K$ and $N_0 < K$ in Figure 7.11. We see from Figure 7.11 that the solution $N(t)$ approaches K, the carrying capacity, as $t \to \infty$. When $N_0 > K$, then $N(t)$ approaches K from above; when $N_0 < K$, then $N(t)$ approaches K from below.

One can show that if $0 < N_0 < \frac{K}{2}$, then the solution curve is S-shaped, as seen in Figure 7.11. This is characteristic of populations that show this type of density dependent growth. At low densities, the growth is almost like unrestricted growth. At higher densities, the growth is restricted and the curve bends around and eventually levels off at the carrying capacity. If the population size is initially greater than the carrying capacity, the population size decreases and becomes asymptotically (that is, when $t \to \infty$) equal to the carrying capacity K.

Returning to (7.27), we see that if $N = K$, then $dN/dt = 0$, that is, the population size does not change. K is therefore called an equilibrium. In particular, if $N(0) = K$, then $N(t) = K$ for all $t \geq 0$. (The constant solution $N(t) = K$ is contained in (7.32) if we choose $N_0 = K$, even though when we derived (7.32), we needed to exclude $N = K$ to avoid division by 0 when separating variables.) The constant solution $N(t) = 0$ is also an equilibrium, although not a very interesting one. We call it the trivial equilibrium. If $N(0) = 0$, then nothing happens, that is, $N(t)$ stays equal to 0 for all later times. This makes sense; if there aren't any individuals to begin with, then there won't be any later on.

7.1.3 Allometric Growth

Solving differential equations of the form (7.3) gets complicated quickly. Before we show the role such differential equations play in biology, we will give an example in which we solve a differential equation of this type.

▮ EXAMPLE 7   Solve

$$\frac{dy}{dx} = \frac{y+1}{x} \quad \text{with } y_0 = 0 \text{ when } x_0 = 1$$

Solution   To solve this differential equation, we separate variables first and then integrate. We find

$$\int \frac{dy}{y+1} = \int \frac{dx}{x}$$

Carrying out the integration on both sides, we obtain

$$\ln|y+1| = \ln|x| + C_1 \tag{7.33}$$

Solving for y, we find

$$|y+1| = e^{C_1}|x|$$
$$y+1 = \pm e^{C_1} x$$

Setting $C = \pm e^{C_1}$, we obtain

$$y = Cx - 1$$

Using the initial condition $y_0 = 0$ when $x_0 = 1$ allows us to determine C:

$$0 = C - 1 \quad \text{or} \quad C = 1$$

The solution is therefore

$$y = x - 1 \qquad\qquad ▮$$

We will now turn to a biological application, namely allometric growth. We have seen a number of allometric relationships throughout the earlier chapters of this book. These are typically relationships between sizes of parts of an organism (for instance, skull length and body length, or leaf area and stem diameter). We denote by $L_1(t)$ and $L_2(t)$ the sizes of two organs of an individual of age t. We say that L_1 and L_2 are related through an allometric law if their specific growth rates are proportional, that is, if

$$\frac{1}{L_1}\frac{dL_1}{dt} = k\frac{1}{L_2}\frac{dL_2}{dt} \tag{7.34}$$

If the constant k in (7.34) is equal to 1, then the growth is called isometric, otherwise it is called allometric. Cancelling dt on both sides of (7.34) and integrating, we find

$$\int \frac{dL_1}{L_1} = k \int \frac{dL_2}{L_2}$$

or

$$\ln|L_1| = k \ln|L_2| + C_1 \tag{7.35}$$

Solving for L_1, we obtain

$$L_1 = CL_2^k \tag{7.36}$$

where $C = \pm e^{C_1}$. (Since L_1 and L_2 are typically positive, the constant C will typically be positive.)

▮ EXAMPLE 8   In a study of 45 species of unicellular algae, the relationship between cell volume (V) and cell biomass (B) was found to be

$$B \propto V^{0.794}$$

Find a differential equation that relates the relative growth rates of cell biomass and volume.

Solution   The relationship between cell biomass and volume can be expressed as

$$B(V) = CV^{0.794} \qquad (7.37)$$

where C is the constant of proportionality. Comparing (7.37) with (7.36), we see that $k = 0.794$. It therefore follows from (7.34) that

$$\frac{1}{B}\frac{dB}{dt} = (0.794)\frac{1}{V}\frac{dV}{dt} \qquad (7.38)$$

We can also get (7.38) by differentiating (7.37) with respect to V, that is,

$$\frac{dB}{dV} = C(0.794)V^{0.794-1}$$

Equation (7.37) allows us to eliminate C. Solving (7.37) for C, we find $C = BV^{-0.794}$ and, therefore,

$$\frac{dB}{dV} = BV^{-0.794}(0.794)V^{0.794-1}$$

$$= (0.794)BV^{-1}$$

Rearranging terms yields

$$\frac{dB}{B} = (0.794)\frac{dV}{V}$$

Dividing both sides by dt, we get

$$\frac{1}{B}\frac{dB}{dt} = (0.794)\frac{1}{V}\frac{dV}{dt}$$

which is the same as (7.38).

◆ 7.1 PROBLEMS

[7.1.1]

In Problems 1–8, solve each pure time differential equation.

1. $\dfrac{dy}{dx} = x + \sin x$ where $y_0 = 0$ for $x_0 = 0$

2. $\dfrac{dy}{dx} = e^{-x}$ where $y_0 = 10$ for $x_0 = 0$

3. $\dfrac{dy}{dx} = \dfrac{1}{x}$ where $y_0 = 0$ for $x_0 = 1$

4. $\dfrac{dy}{dx} = \dfrac{1}{1+x^2}$ where $y_0 = 1$ when $x_0 = 0$

5. $\dfrac{dx}{dt} = \dfrac{1}{1-t}$ where $x(0) = 2$

6. $\dfrac{dx}{dt} = \cos(t-3)$ where $x(3) = 1$

7. $\dfrac{ds}{dt} = \sqrt{3t+1}$ where $s(0) = 1$

8. $\dfrac{dh}{dt} = 5 - 16t^2$ where $h(3) = -11$

9. Suppose that the volume $V(t)$ of a cell at time t changes according to

$$\frac{dV}{dt} = \cos t \quad \text{with } V(0) = 5$$

Find $V(t)$.

10. Suppose that the amount of phosphorus in a lake at time t, denoted by $P(t)$, follows the equation

$$\frac{dP}{dt} = 3t + 1 \quad \text{with } P(0) = 0$$

Find the amount of phosphorus at time $t = 10$.

[7.1.2]

In Problems 11–16, solve the given autonomous differential equations.

11. $\dfrac{dy}{dx} = 3y$ where $y_0 = 2$ for $x_0 = 0$

12. $\dfrac{dy}{dx} = 2(1 - y)$ where $y_0 = 2$ for $x_0 = 0$

13. $\dfrac{dx}{dt} = -2x$ where $x(1) = 5$

14. $\dfrac{dx}{dt} = 1 - 3x$ where $x(-1) = -2$

15. $\dfrac{dh}{ds} = 2h + 1$ where $h(0) = 4$

16. $\dfrac{dN}{dt} = 5 - \dfrac{1}{2}N$ where $N(2) = 3$

17. Suppose that a population, whose size at time t is denoted by $N(t)$, grows according to

$$\frac{dN}{dt} = 0.3N(t) \quad \text{with } N(0) = 20$$

Solve this differential equation, and find the size of the population at time $t = 5$.

18. Suppose that you follow the size of a population over time. When you plot the size of the population versus time on a semi-log plot (that is, the horizontal axis, representing time, is on a linear scale; the vertical axis, representing the size of the population, is on a logarithmic scale), you find that your data fits a straight line that intercepts the vertical axis at 1 (log scale) and has slope -0.43. Find a differential equation that relates the growth rate of the population at time t to the size of the population at time t.

19. Suppose that a population, whose size at time t is denoted by $N(t)$, grows according to

$$\frac{1}{N}\frac{dN}{dt} = r \tag{7.39}$$

where r is a constant.

(a) Solve (7.39).

(b) Transform your solution in (a) appropriately so that the resulting graph is a straight line. How can you determine the constant r from your graph?

(c) Suppose now that you followed a population over time that evolved according to (7.39). Describe how you would determine r from your data.

20. Assume that $W(t)$ denotes the amount of radioactive material at time t. Radioactive decay is then described by the differential equation

$$\frac{dW}{dt} = -\lambda W(t) \quad \text{with } W(0) = W_0 \tag{7.40}$$

where λ is a positive constant, called the decay constant.

(a) Solve (7.40).

(b) Assume that $W(0) = 123\,\text{gr}$ and $W(5) = 20\,\text{gr}$ and that time is measured in minutes. Find the decay constant λ and determine the half-life of the radioactive substance.

21. Suppose that a population, whose size at time t is $N(t)$, grows according to

$$\frac{dN}{dt} = \frac{1}{100}N^2 \quad \text{with } N(0) = 10 \tag{7.41}$$

(a) Solve (7.41).

(b) Graph $N(t)$ as a function of t for $0 \leq t < 10$. What happens as $t \to 10$? Explain in words what this means.

22. Denote by $L(t)$ the length of a fish at time t and assume that the fish grows according to the von Bertalanffy equation

$$\frac{dL}{dt} = k(34 - L(t)) \quad \text{with } L(0) = 2 \tag{7.42}$$

(a) Solve (7.42).

(b) Use your solution in (a) to determine k under the assumption that $L(4) = 10$. Sketch the graph of $L(t)$ for this value of k.

(c) Find the length of the fish when $t = 10$.

(d) Find the asymptotic length of the fish, that is, find $\lim_{t \to \infty} L(t)$.

23. Denote by $L(t)$ the length of a certain fish at time t and assume that this fish grows according to the von Bertalanffy equation

$$\frac{dL}{dt} = k(L_\infty - L(t)) \quad \text{with } L(0) = 1 \tag{7.43}$$

where k and L_∞ are positive constants. A study showed that the asymptotic length is equal to 123 in. and that it takes this fish 27 months to reach half its asymptotic length.

(a) Use this information to determine the constants k and L_∞ in (7.43). (*Hint*: Solve (7.43).)

(b) Determine the length of the fish after ten months.

(c) How long will it take until the fish reaches 90% of its asymptotic length?

24. Let $N(t)$ denote the size of a population at time t. Assume that the population exhibits exponential growth.

(a) If you plot $\log N(t)$ versus t, what do you get?

(b) Find a differential equation that describes the growth of this population and sketch possible solution curves.

25. Use the partial fraction method to solve

$$\frac{dy}{dx} = y(1 + y)$$

where $y_0 = 2$ for $x_0 = 0$.

26. Use the partial fraction method to solve

$$\frac{dy}{dx} = y(1 - y)$$

where $y_0 = 2$ for $x_0 = 0$.

27. Use the partial fraction method to solve

$$\frac{dy}{dx} = y(y - 5)$$

where $y_0 = 1$ for $x_0 = 0$.

28. Use the partial fraction method to solve

$$\frac{dy}{dx} = (y - 1)(y - 2)$$

where $y_0 = 0$ for $x_0 = 0$.

29. Use the partial fraction method to solve

$$\frac{dy}{dx} = 2y(3 - y)$$

where $y_0 = 5$ for $x_0 = 1$.

30. Use the partial fraction method to solve

$$\frac{dy}{dt} = \frac{1}{2}y^2 - 2y$$

where $y_0 = -3$ for $t_0 = 0$.

In Problems 31–34, solve the given differential equations.

31. $\dfrac{dy}{dx} = y(1 + y)$

32. $\dfrac{dy}{dx} = (1 + y)^2$

33. $\dfrac{dy}{dx} = (1 + y)^3$

34. $\dfrac{dy}{dx} = (3 - y)(2 + y)$

35. (a) Use the partial fraction method to show that

$$\int \frac{du}{u^2 - a^2} = \frac{1}{2a} \ln\left|\frac{u - a}{u + a}\right| + C$$

(b) Use your result in (a) to find a solution of

$$\frac{dy}{dx} = y^2 - 4$$

that passes through (i) $(0, 0)$, (ii) $(0, 2)$, and (iii) $(0, 4)$.

36. Find a solution of

$$\frac{dy}{dx} = y^2 + 4$$

that passes through $(0, 2)$.

37. Suppose that the size of a population at time t is denoted by $N(t)$ and that $N(t)$ satisfies the differential equation

$$\frac{dN}{dt} = 0.34N\left(1 - \frac{N}{200}\right) \quad \text{with } N(0) = 50$$

Solve this differential equation, and determine the size of the population in the long run, that is, find $\lim_{t\to\infty} N(t)$.

38. Assume that the size of a population, denoted by $N(t)$, evolves according to the logistic equation. Find the intrinsic rate of growth if the carrying capacity is 100, $N(0) = 10$, and $N(1) = 20$.

39. Suppose that $N(t)$ denotes the size of a population at time t, and that

$$\frac{dN}{dt} = 1.5N\left(1 - \frac{N}{50}\right)$$

(a) Solve this differential equation when $N(0) = 10$.

(b) Solve this differential equation when $N(0) = 90$.

(c) Graph your solutions in (a) and (b) in the same coordinate system.

(d) Find $\lim_{t\to\infty} N(t)$ for your solutions in (a) and (b).

40. Suppose that the size of a population, denoted by $N(t)$, satisfies

$$\frac{dN}{dt} = 0.7N\left(1 - \frac{N}{35}\right) \tag{7.44}$$

(a) Determine all equilibria by solving $dN/dt = 0$.

(b) Solve (7.44) for (i) $N(0) = 10$, (ii) $N(0) = 35$, (iii) $N(0) = 50$, and (iv) $N(0) = 0$. Find $\lim_{t\to\infty} N(t)$ for each of the four initial conditions.

(c) Compare your answer in (a) to the limiting values you found in (b).

41. Let $N(t)$ denote the size of a population at time t. Assume that the population evolves according to the logistic equation. Furthermore, assume that the intrinsic growth rate is 5 and that the carrying capacity is 30.

(a) Find a differential equation that describes the growth of this population.

(b) Without solving the differential equation in (a), sketch solution curves of $N(t)$ as a function of t when (i) $N(0) = 10$, (ii) $N(0) = 20$, and (iii) $N(0) = 40$.

42. Logistic growth is described by the differential equation

$$\frac{dN}{dt} = rN\left(1 - \frac{N}{K}\right)$$

The solution of this differential equation with initial condition $N(0) = N_0$ is given by

$$N(t) = \frac{K}{1 + \left(\frac{K}{N_0} - 1\right)e^{-rt}} \tag{7.45}$$

(a) Show that

$$r = \frac{1}{t}\ln\left(\frac{K - N_0}{N_0}\right) + \frac{1}{t}\ln\left(\frac{N(t)}{K - N(t)}\right) \tag{7.46}$$

by solving (7.45) for r.

(b) The expression for r in (7.46) can be used to estimate r. Suppose we follow a population that grows according to the logistic equation, and find that $N(0) = 10$, $N(5) = 22$, $N(100) = 30$, and $N(200) = 30$. Estimate r.

43. (*Selection at a single locus*) We consider one locus with two alleles, A_1 and A_2, in a randomly mating diploid population. That is, each individual in the population is either of type A_1A_1, A_1A_2, or A_2A_2. We denote by $p(t)$ the frequency of the A_1 allele and by $q(t)$ the frequency of the A_2 allele in the population a time t. Note that $p(t) + q(t) = 1$. We denote the fitness of the A_iA_j type by w_{ij} and assume that $w_{11} = 1$, $w_{12} = 1 - s/2$, and $w_{22} = 1 - s$, where s is a nonnegative constant less than or equal to 1. That is, the fitness of the heterozygote A_1A_2 is halfway between the fitness of the two homozygotes, and the type A_1A_1 is the fittest. If s is small, one can show that approximately

$$\frac{dp}{dt} = \frac{1}{2}sp(1 - p) \quad \text{with } p(0) = p_0 \tag{7.47}$$

(a) Use separation of variables and the partial fraction method to find the solution of (7.47).

(b) Suppose $p_0 = 0.1$ and $s = 0.01$, how long will take until $p(t) = 0.5$?

(c) Find $\lim_{t\to\infty} p(t)$. Explain in words what this limit means.

[7.1.3]

In Problems 44–52, solve each differential equation with the given initial condition.

44. $\dfrac{dy}{dx} = 2\dfrac{y}{x}$ with $y_0 = 1$ if $x_0 = 1$

45. $\dfrac{dy}{dx} = \dfrac{x + 1}{y}$ with $y_0 = 2$ if $x_0 = 0$

46. $\dfrac{dy}{dx} = \dfrac{y}{x+1}$ with $y_0 = 1$ if $x_0 = 0$

47. $\dfrac{dy}{dx} = (y+1)e^{-x}$ with $y_0 = 2$ if $x_0 = 0$

48. $\dfrac{dy}{dx} = x^2 y^2$ with $y_0 = 1$ if $x_0 = 1$

49. $\dfrac{dy}{dx} = \dfrac{y+1}{x-1}$ with $y_0 = 5$ if $x_0 = 2$

50. $\dfrac{du}{dt} = \dfrac{\sin t}{u^2 + 1}$ with $u_0 = 3$ if $t_0 = 0$

51. $\dfrac{dr}{dt} = re^{-t}$ with $r_0 = 1$ if $t_0 = 0$

52. $\dfrac{dx}{dy} = \dfrac{1}{2}\dfrac{x}{y}$ with $x_0 = 2$ if $y_0 = 3$

53. (Adapted from Reiss 1989) In a case study by Taylor et al. (1980), where the maximal rate of oxygen consumption (in ml s^{-1}) for nine species of wild African mammals was plotted against body mass (in kg) on a log-log plot, it was found that the data points fall on a straight line with slope approximately equal to 0.3. Find a differential equation that relates maximal oxygen consumption and body mass.

54. Consider the following differential equation, which is of importance in population genetics:

$$a(x)g(x) - \frac{1}{2}\frac{d}{dx}[b(x)g(x)] = 0$$

where $b(x) > 0$.

(a) Define $y = b(x)g(x)$, and show that y satisfies

$$\frac{a(x)}{b(x)}y - \frac{1}{2}\frac{dy}{dx} = 0 \qquad (7.48)$$

(b) Separate variables in (7.48), and show that if $y > 0$, then

$$y = C\exp\left[2\int \frac{a(x)}{b(x)}\,dx\right]$$

◆ 7.2 EQUILIBRIA AND THEIR STABILITY

In Example 6 of Section 7.1, we discussed the logistic equation

$$\frac{dN}{dt} = rN\left(1 - \frac{N}{K}\right) \qquad (7.49)$$

If $N(0) = K$, then $N(t) = K$ for *all* $t > 0$, since $rN(1 - N/K) = 0$ when $N = K$, implying that $dN/dt = 0$. The constant solution $N = K$ is called an **equilibrium**. If a system is in equilibrium initially, it will stay there for all later times. The constant solution $N = 0$ is another equilibrium of (7.49) since $rN(1 - N/K) = 0$ when $N = 0$. This follows also from the biological interpretation of (7.49): If the population size is 0 initially, it will remain so for all later times since individuals are not created spontaneously.

In this section, we will consider autonomous differential equations of the form

$$\frac{dy}{dx} = g(y) \qquad (7.50)$$

where we will typically think of x as time. In addition to finding equilibria, we are also interested in the *long-term behavior* of the solution $y = y(x)$, that is in the behavior of $y(x)$ as $x \to \infty$. If we can solve (7.50) (as in (7.49), where we concluded that $N(t) \to K$, the carrying capacity, as $t \to \infty$, provided $N(0) > 0$), we can study the solution directly to obtain information about the long-term behavior. But what should we do if we cannot solve (7.50)?

It turns out that candidates for describing the long-term behavior are certain constant solutions $y = \hat{y}$ (read: "y hat") that satisfy $g(\hat{y}) = 0$.

If $\hat{y}$ satisfies

$$g(\hat{y}) = 0$$

then $\hat{y}$ is an equilibrium of

$$\frac{dy}{dx} = g(y)$$

Of course, $g(y) = 0$ need not have a solution. However, in this section, $g(y)$ will always have real roots.

We will explain the meaning of equilibria in more detail before we discuss specific examples. The basic property of equilibria is that if, initially (say, at $x = 0$), $y(0) = \hat{y}$ and $\hat{y}$ is an equilibrium, then $y(x) = \hat{y}$ for all $x > 0$.

A physical analogue of equilibria is provided in Figure 7.12. On the left side, a ball rests on top of a hill, on the right side, a ball rests at the bottom of a valley. In either case, the ball is in equilibrium because it does not move.

Of great interest is the **stability** of equilibria. What we mean by this is best explained using our example of a ball on a hill versus a ball in the valley. If we perturb the ball by a small amount, that is, if we move it out of its equilibrium slightly, the ball on the left side will roll down the hill and not return to the top, whereas the ball on the right side will return to the bottom of the valley. We call the situation on the left side unstable and the situation on the right side stable.

The analogue of stability for equilibria of differential equations is as follows. Suppose that $\hat{y}$ is an equilibrium of $\frac{dy}{dx} = g(y)$, that is, $g(\hat{y}) = 0$. We say $\hat{y}$ is **locally stable** if the solution returns to $\hat{y}$ after a small perturbation; if it does not return, we say $\hat{y}$ is **unstable**. These concepts will be developed in the following subsection, in which we will discuss a graphical and an analytical method for analyzing stability of equilibria. This is then followed by a number of applications.

▲ **Figure 7.12**

Stability illustrated with a ball on a hill and in a valley.

7.2.1 A First Look at Stability

Graphical Approach   Suppose that $g(y)$ is of the form given in Figure 7.13. To find the equilibria of (7.50), we set $g(y) = 0$. We see from Figure 7.13 that for our choice of $g(y)$, the equilibria are at $y = 0$, y_1, and y_2. Equilibria are characterized by the property that a system in an equilibrium state stays there for all later times (unless some external force disturbs the system). This does *not* imply that the system will necessarily reach a particular equilibrium when starting from some initial value that is different from the equilibrium, nor does it imply that it will return to an equilibrium after a small perturbation.

Whether or not the system will return to an equilibrium after a small perturbation depends on the local stability of the equilibrium. By this we mean the following. Suppose that the system is in equilibrium, which we denote by $\hat{y}$. We apply a small perturbation to the system, so that after the perturbation the new state of the

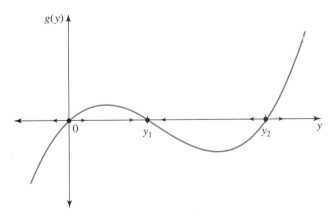

▲ **Figure 7.13**

The function $g(y)$.

system is

$$y = \hat{y} + z \tag{7.51}$$

where z is small and may be positive or negative. We explain what can happen using the function $g(y)$ in Figure 7.13.

Suppose that $\hat{y} = 0$ and we subject this equilibrium to a small perturbation. If the new value $y = \hat{y} + z = z > 0$ for z small, then $dy/dx > 0$, that is, y will increase. If the new value $y = \hat{y} + z = z < 0$ for z small, then $dy/dx < 0$ and y will decrease. In either case, the system will not return to 0. We say that $\hat{y} = 0$ is unstable.

We now turn to the equilibrium $\hat{y} = y_1$. We perturb the equilibrium to $y = y_1 + z$ for z small. Looking at Figure 7.13, if $z > 0$, then $dy/dx < 0$ and hence y decreases. If $z < 0$, then $dy/dx > 0$ and y increases. That is, the system will return to the equilibrium y_1 after a small perturbation. We say that y_1 is locally stable. The attribute "locally" refers to the fact that the system will return to y_1 if the perturbation is sufficiently small. It does not say anything about what happens when the perturbation is large. For instance, if the perturbation is large and the new value is less than 0, then $dy/dx < 0$ and the system will not return to y_1.

Analyzing the equilibrium y_2 in the same way shows that it behaves like the equilibrium $\hat{y} = 0$. That is, y_2 is an unstable equilibrium.

We see from the preceding discussion that it is not necessarily the case that the system will reach an equilibrium value. We see from Figure 7.13 that if initially $y(0) \in (0, y_2)$, then $y(x)$ will approach y_1. If $y(0) < 0$, $y(x) \to -\infty$, and if $y > y_2$, then $y(x) \to \infty$.

■ EXAMPLE ■ Let $N(t)$ denote the size of a population at time t that evolves according to the logistic equation

$$\frac{dN}{dt} = 2N \left(1 - \frac{N}{100} \right) \quad \text{for } N \geq 0$$

Find the equilibria and analyze their stability.

Solution To find the equilibria, we set $\frac{dN}{dt} = 0$. That is,

$$2N \left(1 - \frac{N}{100} \right) = 0$$

We see that either

$$N_1 = 0 \quad \text{or} \quad N_2 = 100$$

To analyze the stability, we draw the graph of dN/dt versus N in Figure 7.14. Note that $N \geq 0$, since N represents the size of a population. To perturb the trivial equilibrium $N_1 = 0$, we therefore choose only values that are slightly bigger than 0. We see from the graph that such small perturbations result in $dN/dt > 0$, and hence N increases. This implies that $N_1 = 0$ is an unstable equilibrium.

To perturb the nontrivial equilibrium $N_2 = 100$, we can either increase or decrease the population size a bit. If we decrease it, then $dN/dt > 0$ and the population size will increase. If we increase it, then $dN/dt < 0$ and the population size will decrease. That is, if the system is subjected to a small perturbation about the nontrivial equilibrium $N_2 = 100$, the population size will return to the equilibrium value of $N_2 = 100$. This implies that $N_2 = 100$ is a locally stable equilibrium. ■

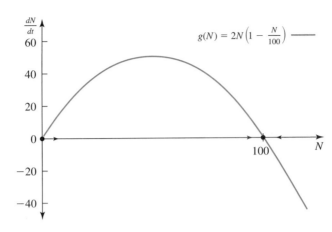

◀ **Figure 7.14**
The graph of dN/dt versus N in Example 1.

Analytical Approach   We assume that $\hat{y}$ is an equilibrium

$$\frac{dy}{dx} = g(y)$$

(thus $\hat{y}$ satisfies $g(\hat{y}) = 0$). We consider a small perturbation about the equilibrium $\hat{y}$, which we express as

$$y = \hat{y} + z$$

where z is small and may be either positive or negative. Then

$$\frac{dy}{dx} = \frac{d}{dx}(\hat{y} + z) = \frac{dz}{dx}$$

since $d\hat{y}/dx = 0$ ($\hat{y}$ is a constant). We find

$$\frac{dz}{dx} = g(\hat{y} + z)$$

If z is sufficiently small, we can approximate $g(\hat{y} + z)$ by its linear approximation. The linear approximation of $g(y)$ about $\hat{y}$ is given by

$$L(y) = g(\hat{y}) + (y - \hat{y})g'(\hat{y})$$

Since $g(\hat{y}) = 0$,

$$L(y) = (y - \hat{y})g'(\hat{y})$$

Therefore, the linear approximation of $g(\hat{y} + z)$ is given by

$$L(\hat{y} + z) = (\hat{y} + z - \hat{y})g'(\hat{y}) = zg'(\hat{y})$$

and we obtain

$$\frac{dz}{dx} \approx zg'(\hat{y}) \tag{7.52}$$

provided that z is small. The right-hand side of (7.52) is linear in z, since $g'(\hat{y})$ is just a constant. If we set

$$\lambda = g'(\hat{y})$$

then

$$\frac{dz}{dx} \approx \lambda z$$

has the approximate solution

$$z(x) = z(0)e^{\lambda x} \tag{7.53}$$

Returning to (7.51), namely $y(x) = \hat{y} + z(x)$, we see from (7.53) that

$$\lim_{x \to \infty} z(x) = 0 \quad \text{if } \lambda < 0$$

that is, if $\lambda < 0$, the system returns to the equilibrium $\hat{y}$ after a small perturbation $z(0)$. This means that $\hat{y}$ is locally stable if $\lambda < 0$. On the other hand, if $\lambda > 0$, then $z(x)$ does not go to 0 as $x \to \infty$, that is, the system will not return to the equilibrium $\hat{y}$ after a small perturbation. This means that $\hat{y}$ is unstable. The value $\lambda = g'(\hat{y})$ is called an eigenvalue; it is the slope of the tangent line of $g(y)$ at $\hat{y}$.

STABILITY CRITERION

Consider the differential equation

$$\frac{dy}{dx} = g(y)$$

where $g(y)$ is a differentiable function. Assume that $\hat{y}$ is an equilibrium, that is, $g(\hat{y}) = 0$. Then

$\hat{y}$ is locally stable if $g'(\hat{y}) < 0$
$\hat{y}$ is unstable if $g'(\hat{y}) > 0$

We wish to tie this analytical approach in with the more informal graphical analysis in the beginning of this section. Looking at Figure 7.13, we find that the slope of the tangent line at $y = 0$ is positive, that is, $g'(0) > 0$; similarly, we find $g'(y_1) < 0$ and $g'(y_2) > 0$. Hence, the equilibria 0 and y_2 are unstable and the equilibrium y_1 is locally stable, as found in the graphical analysis.

The analytical approach is more powerful than the graphical approach: in addition to determining whether an equilibrium is locally stable or unstable, it allows us to say something about how quickly a solution returns to an equilibrium after a small perturbation. This follows from (7.53). There, we found that the perturbation has the approximate solution

$$z(x) = z(0)e^{\lambda x}$$

If $\lambda > 0$, then the larger λ, the faster the solution moves away from the equilibrium. If $\lambda < 0$, then the smaller λ, the faster the solution will return to the equilibrium after a small perturbation (see Figure 7.15).

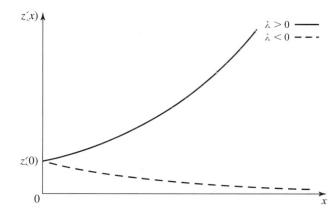

◀ **Figure 7.15**
The graph of $z(x) = z(0)e^{\lambda x}$ for $\lambda > 0$ and $\lambda < 0$.

We wish to emphasize that the above derivation for the analytical stability criterion was based on linearizing $g(y)$ about the equilibrium $\hat{y}$. Since the linearization is only close for values close to $\hat{y}$ (unless $g(y)$ is linear), the perturbations about the equilibrium must be small, and hence the stability analysis is always local, that is, within close vicinity of the equilibrium. When $g(y)$ is linear, the analysis is exact; in particular, we can compute exactly how quickly a solution returns to a locally stable equilibrium (or moves away from an unstable equilibrium) after a perturbation.

▮EXAMPLE 2   **$g(y)$ is linear**

Show that the differential equation

$$\frac{dy}{dx} = 1 - y \tag{7.54}$$

has a stable equilibrium at $\hat{y} = 1$, and determine how quickly a solution starting at $y(0) = y_0 \neq 1$ will reach $\hat{y} = 1$.

Solution   Since $g(y) = 0$ for $y = 1$, $\hat{y} = 1$ is an equilibrium. Differentiating $g(y) = 1 - y$, we find $g'(y) = -1$, which is negative regardless of y. Therefore, $\hat{y} = 1$ is a locally stable equilibrium.

We can solve (7.54) exactly using separation of variables:

$$\int \frac{dy}{1-y} = \int dx$$
$$-\ln|1-y| = x + C_1$$
$$1 - y = \pm e^{-C_1} e^{-x}$$
$$1 - y = Ce^{-x}$$

where we set $\pm e^{-C_1} = C$. Solving for y yields

$$y(x) = 1 - Ce^{-x}$$

If we set $y(0) = y_0$, then $y_0 = 1 - C$, and the solution is

$$y(x) = 1 - (1 - y_0)e^{-x}$$

We see that for any initial value y_0,

$$\lim_{x \to \infty} y(x) = 1$$

and it takes an infinite amount of time to return to the equilibrium $\hat{y} = 1$ if $y_0 \neq 1$.

A common approach to determine how quickly a solution approaches an equilibrium is to compute the time it takes to reduce the initial deviation from the equilibrium, $y_0 - 1$, to a fraction e^{-1}; that is, we wish to determine the number x_R such that

$$y(x_R) - 1 = e^{-1}(y_0 - 1)$$

Since $y(x_R) - 1 = -(1 - y_0)e^{-x_r}$, we find

$$-(1 - y_0)e^{-x_R} = e^{-1}(y_0 - 1)$$
$$e^{-x_R} = e^{-1}$$
$$x_R = 1$$

(see Figure 7.16). It takes one unit of time to reduce the initial deviation to a fraction e^{-1}; this does not depend on how large the initial deviation is.   ▮

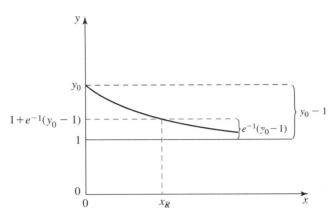

◀ **Figure 7.16**
An illustration of the time x_R in Example 2.

Using Both Approaches

▮ EXAMPLE 3   Suppose that

$$\frac{dy}{dx} = y(4 - y)$$

Find the equilibria of this differential equation and discuss their stability, using both the graphical and the analytical approach.

Solution   To find the equilibria, we set $dy/dx = 0$. That is,

$$y(4 - y) = 0$$

which yields

$$y_1 = 0 \qquad \text{and} \qquad y_2 = 4$$

If we set $g(y) = y(4 - y)$ and graph $g(y)$ (see Figure 7.17), we see that $y_1 = 0$ is unstable, since if we perturb $y_1 = 0$ to $y = z > 0$ where z is small, then $g(y) = dy/dx > 0$, and if we perturb $y_1 = 0$ to $y = z < 0$ where z is small, then $g(y) = dy/dx < 0$. That is, in either case, y will not return to 0. On the other hand, y_2 is a locally stable equilibrium. A small perturbation to the right results in $dy/dx < 0$, whereas a small perturbation to the left results in $dy/dx > 0$, in either case, the solution y will return to $y_2 = 4$.

If we use the analytical approach, we need to compute the eigenvalues. The eigenvalue associated with $y_1 = 0$ is

$$\lambda_1 = g'(0) = 4 - 2y|_{y=0} = 4 > 0$$

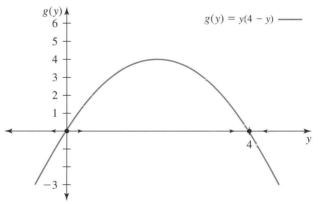

◀ **Figure 7.17**
The graph of $g(y)$ in Example 3.

which implies that y_1 is unstable. The eigenvalue associated with $y_2 = 4$ is

$$\lambda_2 = g'(4) = 4 - 2y|_{y=4} = -4 < 0$$

which implies that $y_2 = 4$ is locally stable. ▮

7.2.2 Single Compartment or Pool

This example is adapted from DeAngelis (1992). Compartment models are frequently used to model the flow of matter (nutrients, energy, and so forth). The simplest such model consists of one compartment, for instance, a fixed volume V of water (such as a tank or lake) containing a solute (such as phosphorus) of initial concentration C_0. Assume that water enters the compartment at a constant rate q and leaves the compartment at the same rate (see Figure 7.18). (Having the same input and output rate keeps the volume of the pool constant.) We will investigate the effects of different input concentrations of the solute on the concentration of the solute in the pool.

We denote by $C(t)$ the concentration of the solute in the compartment at time t. Then the total mass of the solute is $C(t)V$ where V is the volume of the compartment. For instance, if the concentration of the solute is 2 grams per liter and the volume of the compartment is 10 liters, then the total mass of the solute in the compartment is 2 gl^{-1} times 10 l, which is equal to 20 g.

If C_I is the concentration of the incoming solute and q is the rate at which water enters, then qC_I, the **input loading**, is the rate at which mass enters. For instance, if the concentration of the solute is 5 gl^{-1} and the rate at which water enters is 0.1 ls^{-1}, then the input loading, that is, the rate at which mass enters, is 5 gl^{-1} times 0.1 ls^{-1}, which is equal to 0.5 gs^{-1}.

If we assume that the water in the compartment is well mixed, so that the outflowing water has the same concentration as the water in the compartment, namely $C(t)$ at time t, then $qC(t)$ is the rate at which mass leaves the compartment at time t.

These different processes can be schematically illustrated in a flow diagram, as shown in Figure 7.19. Because mass is conserved in the system, we can use the **conservation of mass law** to derive an equation that describes the flow of matter in this system:

$$\begin{bmatrix} \text{rate of change} \\ \text{of mass of} \\ \text{solute in pool} \end{bmatrix} = \begin{bmatrix} \text{rate at which} \\ \text{mass enters} \end{bmatrix} - \begin{bmatrix} \text{rate at which} \\ \text{mass leaves} \end{bmatrix}$$

or, writing C for $C(t)$ (and being careful not to confuse C with a constant),

$$\frac{d}{dt}(CV) = qC_I - qC \tag{7.55}$$

Since V is constant, we can write (7.55) as

$$\frac{dC}{dt} = \frac{q}{V}(C_I - C) \tag{7.56}$$

This is a linear differential equation of the type discussed in Subsection 7.1.2, Case 1. It can be solved using separation of variables. We skip the steps, to concentrate on the discussion of the system. If $C(0) = C_0$, then the solution is

$$C(t) = C_I \left[1 - \left(1 - \frac{C_0}{C_I} \right) e^{-(q/V)t} \right] \tag{7.57}$$

Solution curves for different values of C_0 are shown in Figure 7.20. We see from (7.56) that C_I is the only equilibrium. Looking at the solution $C(t)$ in (7.57), we see

▲ Figure 7.18

Input and output rates are the same the water in the tank remains at the same level.

▲ Figure 7.19

Flow diagram for the single compartment model.

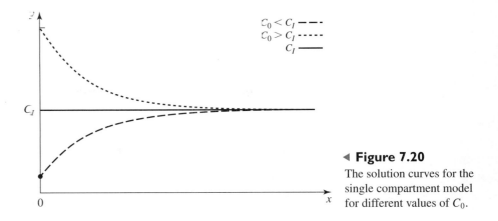

▶ **Figure 7.20**
The solution curves for the single compartment model for different values of C_0.

that

$$\lim_{t \to \infty} C(t) = C_I$$

regardless of the initial concentration C_0 in the compartment. This shows that C_I is **globally stable**. Global stability implies that no matter how much we perturb the equilibrium, the system will return to the equilibrium.

We can also obtain this result using the eigenvalue method previously discussed. We write

$$C(t) = C_I + z(t)$$

Then, as before,

$$\frac{dC(t)}{dt} = \frac{d}{dt}[C_I + z(t)] = \frac{dz(t)}{dt}$$

that is,

$$\frac{dz}{dt} = -\frac{q}{V} z(t) \tag{7.58}$$

since $C_I - C(t) = -z(t)$. Note that (7.58) is exact, that is, we need not use a linear approximation for the right-hand side of (7.56), since it is already linear. We see from (7.58) that $-q/V$ is the eigenvalue associated with the equilibrium C_I. Since the eigenvalue is negative, it follows that C_I is locally stable. We can obtain more information here, since (7.58) is exact with solution

$$z(t) = z(0)e^{-(q/V)t} \tag{7.59}$$

which shows that for *any* perturbation $z(0)$, $z(t) \to 0$ as $t \to \infty$. That is, the system returns to the equilibrium C_I (that is, $C(t) \to C_I$ as $t \to \infty$) regardless of how much we perturb the system. The reason that the eigenvalue method allows us to show global stability lies in the fact that the differential equation (7.55) is *linear*. In other cases, the eigenvalue method only allows us to obtain local stability, because we must first linearize and, consequently, the linearized differential equation is only an approximation.

Equation (7.59) shows that the system returns to the equilibrium C_I exponentially fast and that the eigenvalue $-(q/V)$ determines the time scale, that is, how quickly the system reaches the equilibrium. The larger q/V, the faster the system recovers from a perturbation. In this context, we can define a return time to equilibrium, denoted by T_R. By convention, the return time to equilibrium, T_R, is defined as the amount of time it takes to reduce the initial difference $C_0 - C_I$ to a fraction

e^{-1}, that is,

$$C(T_R) - C_I = e^{-1}(C_0 - C_I) \tag{7.60}$$

Using (7.57), we see that

$$C(T_R) = C_I \left[1 - \left(1 - \frac{C_0}{C_I} \right) e^{-(q/V)T_R} \right]$$

which we can rewrite as

$$C(T_R) - C_I = (C_0 - C_I)e^{-(q/V)T_R} \tag{7.61}$$

Equating (7.60) and (7.61), we find

$$e^{-1}(C_0 - C_I) = (C_0 - C_I)e^{-(q/V)T_R}$$

which yields

$$1 = \frac{q}{V}T_R \qquad \text{or} \qquad T_R = \frac{V}{q}$$

That is, the return time increases with the volume V and decreases with the flow rate q. This should be intuitively clear; the larger the volume and the smaller the input rate, the longer it takes to return to equilibrium. One can show that T_R can also be interpreted as the mean residence time of a molecule of the solute, that is, T_R is the average time a molecule of the solute spends in the compartment before leaving, when the system is in equilibrium.

7.2.3 Levins's Model

The ecological importance of spatially structured populations was pointed out by Andrewartha and Birch (1954) based on studies of insect populations. They observed that local populations become frequently extinct and subsequently recolonized. Fifteen years later, in 1969, Richard Levins introduced the concept of metapopulations (Levins 1969). This was a major theoretical advance; this concept provided a theoretical framework for studying spatially structured populations. Over the last ten years, the use of spatially structured population models has been firmly established in population biology.

A metapopulation is a collection of subpopulations. Each subpopulation occupies a patch, and different patches are linked via migration of individuals between patches (see Figure 7.21). In this setting, one only keeps track of what proportion of patches are occupied by subpopulations. Subpopulations go extinct at a constant rate, denoted by m (m stands for mortality). Vacant patches can be colonized at a rate that is proportional to the fraction of occupied patches; the constant of proportionality is denoted by c (c stands for colonization rate). If we denote by $p(t)$ the fraction of patches that are occupied at time t, then writing $p = p(t)$,

$$\frac{dp}{dt} = cp(1 - p) - mp \tag{7.62}$$

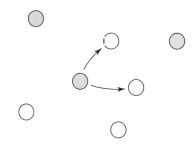

▲ **Figure 7.21**

A schematic description of a metapopulation model. The shaded patches are occupied; arrows indicate migration events.

The first term on the right hand-side describes the colonization process. Note that an increase in the fraction of occupied patches occurs only if a vacant patch becomes occupied, hence the product $p(1 - p)$ in the first term on the right-hand side. The minus sign in front of m shows that an extinction event decreases the fraction of occupied patches.

We will not solve (7.62). Instead, we will analyze its equilibria. We set

$$cp(1 - p) - mp = 0$$

After factoring cp, we obtain

$$cp\left(1 - \frac{m}{c} - p\right) = 0$$

which has the two solutions

$$p_1 = 0 \quad \text{and} \quad p_2 = 1 - \frac{m}{c}$$

We call the solution $p_1 = 0$ a trivial solution because it corresponds to the situation in which all patches are vacant. Since individuals are not created spontaneously, a vacant patch can be recolonized only through migration from other occupied patches. Therefore, once a metapopulation is extinct, it stays extinct. The other equilibrium $p_2 = 1 - m/c$ is only relevant when $p_2 \in (0, 1]$, because p represents a fraction that is a number between 0 and 1. Since m and c are both positive, it follows immediately that $p_2 < 1$ for all choices of m and c. To see when $p_2 > 0$, we check

$$1 - \frac{m}{c} > 0$$

which holds when

$$m < c$$

That is, the nontrivial equilibrium $p_2 = 1 - m/c$ is in $(0, 1]$ if the extinction rate m is less than the colonization rate c. If $m \geq c$, then there is only one equilibrium in $[0, 1]$, namely $p_1 = 0$. We illustrate this in Figures 7.22 and 7.23; looking at the figures, it is easy to analyze the stability of the equilibria.

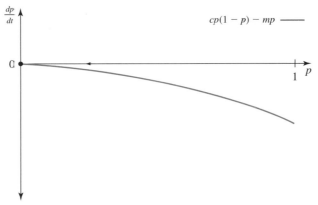

▮ **Figure 7.22**
The case $m \geq c$.

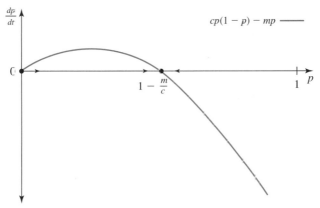

◀ **Figure 7.23**
The case $m < c$.

Case 1: $m \geq c$ There is only the trivial equilibrium $p_1 = 0$. For any $p \in (0, 1]$, we see that $dp/dt < 0$, hence the fraction of occupied patches declines. The equilibrium is locally and globally stable.

Case 2: $m < c$ There are two equilibria, namely 0 and $1 - m/c$. The trivial equilibrium $p_1 = 0$ is unstable, since if we perturb $p_1 = 0$ to some value in $(0, 1 - m/c)$, then $dp/dt > 0$, which implies that $p(t)$ increases. The system will therefore not return to 0.

The other equilibrium, $p_2 = 1 - m/c$, is locally stable. After a small perturbation of this equilibrium, to the right of p_2, $dp/dt < 0$, and to the left of p_2, $dp/dt > 0$; therefore, the system will return to p_2.

We can also use the eigenvalue approach to analyze the stability of the equilibria. In addition, this will allow us to obtain information on how quickly the system returns to the stable equilibrium. We set

$$g(p) = cp(1 - p) - mp$$

To linearize this function about the equilibrium values, we must find

$$g'(p) = c - 2cp - m$$

Now, if $p_1 = 0$, then

$$g'(0) = c - m$$

whereas, if $p_2 = 1 - m/c$, then

$$g'\left(1 - \frac{m}{c}\right) = c - 2c\left(1 - \frac{m}{c}\right) - m = c - 2c + 2m - m = m - c$$

From this we see that $c - m$ is the eigenvalue corresponding to $p_1 = 0$ and $m - c$ is the eigenvalue corresponding to $p_2 = 1 - m/c$. We find that

$$\text{if } c - m < 0 \quad \text{then } p_1 = 0 \text{ is locally stable}$$
$$\text{if } m - c < 0 \quad \text{then } p_2 = 1 - \tfrac{m}{c} \text{ is locally stable}$$

To summarize our results, if $m \geq c$, then $p_1 = 0$ is the only equilibrium in $[0, 1]$ and $p_1 = 0$ is locally stable. (In fact, the graphical analysis showed that $p_1 = 0$ is globally stable.) If $m < c$, then there are two equilibria in $[0, 1]$. The equilibrium $p_1 = 0$ is now unstable and the equilibrium $p_2 = 1 - m/c$ is locally stable.

7.2.4 The Allee Effect

In a sexually reproducing species, individuals may experience a disproportionately low recruitment rate when the population density falls below a certain level, due to lack of suitable mates. This is called an Allee effect (see Allee 1931). A simple extension of the logistic equation incorporates this effect. We denote by $N(t)$ the size of a population at time t; then, writing $N = N(t)$,

$$\frac{dN}{dt} = rN(N - a)\left(1 - \frac{N}{K}\right) \tag{7.63}$$

where r, a, and K are positive constants. We assume that $0 < a < K$. We will see that, as in the logistic equation, K denotes the carrying capacity. The constant a is a threshold population size, below which the recruitment rate is negative, meaning that the population will shrink and ultimately go to extinction.

The equilibria of (7.63) are given by $\hat{N} = 0$, a, and K. We set

$$g(N) = rN(N - a)\left(1 - \frac{N}{K}\right) = r\left(N^2 + \frac{a}{K}N^2 - \frac{N^3}{K} - aN\right)$$

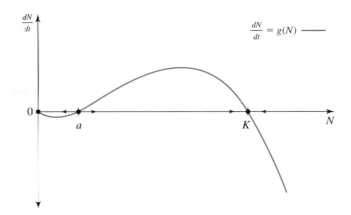

▲ **Figure 7.24**
The graph of $g(N)$ for the Allee effect.

A graph of $g(N)$ is shown in Figure 7.24. Differentiating $g(N)$ yields

$$g'(N) = r\left(2N + \frac{2a}{K}N - \frac{3N^2}{K} - a\right) = \frac{r}{K}(2NK + 2aN - 3N^2 - aK)$$

We can compute the eigenvalue $g'(\hat{N})$ associated with the equilibrium $\hat{N}$.

$$\text{if } \hat{N} = 0 \quad \text{then } g'(0) = \frac{r}{K}(-aK) < 0$$

$$\text{if } \hat{N} = a \quad \text{then } g'(a) = \frac{r}{K}a(K - a) > 0$$

$$\text{if } \hat{N} = K \quad \text{then } g'(K) = \frac{r}{K}K(a - K) < 0$$

As we continue, you should compare the results from the eigenvalue method with the graph of $g(N)$.

Since $g'(0) < 0$, it follows that $\hat{N} = 0$ is locally stable. Likewise, since $g'(K) < 0$, it follows that $\hat{N} = K$ is locally stable. The equilibrium $\hat{N} = a$ is unstable since $g'(a) > 0$. This is also evident from Figure 7.24. This is an example where both stable equilibria are locally but not globally stable.

We see from Figure 7.24 that if $0 \leq N(0) < a$, then $N(t) \to 0$ as $t \to \infty$. If $a < N(0) \leq K$ or $N(0) \geq K$, then $N(t) \to K$ as $t \to \infty$. To interpret our results, we see that if the initial population $N(0)$ is too small (that is, $N(0) < a$), then the population goes extinct. If the initial population is large enough (that is, $N(0) > a$), the population persists. That is, the parameter a is a threshold level. The recruitment rate is only large enough when the population size exceeds this level.

◆ 7.2 PROBLEMS

[7.2.1]

1. Suppose that

$$\frac{dy}{dx} = y(2 - y)$$

(a) Find the equilibria of this differential equation.

(b) Graph dy/dx as a function of y, and use your graph to discuss the stability of the equilibria.

(c) Compute the eigenvalues associated with each equilibrium and discuss the stability of the equilibria.

2. Suppose that

$$\frac{dy}{dx} = (2 - y)(3 - y)$$

(a) Find the equilibria of this differential equation.

(b) Graph dy/dx as a function of y, and use your graph to discuss the stability of the equilibria.

(c) Compute the eigenvalues associated with each equilibrium and discuss the stability of the equilibria.

3. Suppose that

$$\frac{dy}{dx} = y(y-1)(y-2)$$

(a) Find the equilibria of this differential equation.

(b) Graph dy/dx as a function of y, and use your graph to discuss the stability of the equilibria.

(c) Compute the eigenvalues associated with each equilibrium and discuss the stability of the equilibria.

4. Suppose that

$$\frac{dy}{dx} = y(1-y)(y-2)$$

(a) Find the equilibria of this differential equation.

(b) Graph dy/dx as a function of y, and use your graph to discuss the stability of the equilibria.

(c) Compute the eigenvalues associated with each equilibrium and discuss the stability of the equilibria.

5. (*Logistic equation*) Assume that the size of a population evolves according to the logistic equation with intrinsic rate of growth $r = 1.5$. Assume that the carrying capacity $K = 100$.

(a) Find the differential equation that describes the rate of growth of this population.

(b) Find all equilibria, and discuss the stability of the equilibria using the graphical approach.

(c) Find the eigenvalues associated with the equilibria, and use the eigenvalues to determine stability of the equilibria. Compare your answers with your results in (b).

6. (*A simple model of predation*) Suppose that $N(t)$ denotes the size of a population at time t. The population evolves according to the logistic equation but, in addition, predation reduces the size of the population so that the rate of change is given by

$$\frac{dN}{dt} = N\left(1 - \frac{N}{50}\right) - \frac{9N}{5+N} \qquad (7.64)$$

The first term on the right-hand side describes the logistic growth, the second term describes the effect of predation.

(a) Set

$$g(N) = N\left(1 - \frac{N}{50}\right) - \frac{9N}{5+N}$$

and graph $g(N)$.

(b) Find all equilibria of (7.64).

(c) Use your graph in (a) to determine the stability of the equilibria found in (b).

(d) Use the method of eigenvalues to determine the stability of the equilibria found in (b).

7. (*Logistic equation*) Assume that the size of a population evolves according to the logistic equation with intrinsic rate of growth $r = 2$. Assume that $N(0) = 10$.

(a) Determine the carrying capacity K if the population grows fastest when the population size is 1000. (*Hint*: Show that the graph of dN/dt as a function of N has a maximum at $K/2$.)

(b) If $N(0) = 10$, how long will it take the population size to reach 1000?

(c) Find $\lim_{t\to\infty} N(t)$.

8. (*Logistic equation*) The logistic curve $N(t)$ is an S-shaped curve that satisfies

$$\frac{dN}{dt} = rN\left(1 - \frac{N}{K}\right) \qquad \text{with } N(0) = N_0 \qquad (7.65)$$

Assume that $N_0 < K$.

(a) Use the differential equation (7.65) to show that the inflection point of the logistic curve is at exactly half the saturation value of the curve. (*Hint*: do not solve (7.65), instead differentiate the right-hand side of (7.65) with respect to t.)

(b) The solution $N(t)$ of (7.65) can be defined for all $t \in \mathbf{R}$. Show that $N(t)$ is symmetric about the inflection point and that $N(0) = N_0$. That is, first use the solution of (7.65) that is given in (7.32), and find the time t_0 so that $N(t_0) = K/2$, that is, the inflection point is at $t = t_0$. Compute $N(t_0 + h)$ and $N(t_0 - h)$ for $h > 0$, and show that

$$N(t_0 + h) - \frac{K}{2} = \frac{K}{2} - N(t_0 - h)$$

Use a sketch of the graph of $N(t)$ to explain why this shows that $N(t)$ is symmetric about the inflection point $(t_0, N(t_0))$.

9. Suppose that a fish population evolves according to the logistic equation, and that a fixed number of fish per unit time are removed. That is,

$$\frac{dN}{dt} = rN\left(1 - \frac{N}{K}\right) - H$$

Assume in the following that $r = 2$ and $K = 1000$.

(a) Find possible equilibria, and discuss their stability when $H = 100$.

(b) What is the maximal harvesting rate that maintains a positive population size?

10. Suppose that a fish population evolves according to a logistic equation, and that fish are harvested at a rate proportional to the population size. If $N(t)$ denotes the population size at time t, then

$$\frac{dN}{dt} = rN\left(1 - \frac{N}{K}\right) - hN$$

Assume that $r = 2$ and $K = 1000$.

(a) Find possible equilibria and discuss their stability when $h = 0.1$, using the graphical approach, and find the maximal harvesting rate that maintains a positive population size.

(b) Show that if $h < r = 2$, then there is a nontrivial equilibrium. Find the equilibrium.

(c) Analyze the stability of the equilibrium found in (b) using (i) the eigenvalue approach and (ii) the graphical approach.

[7.2.2]

11. Assume the single compartment model defined in Subsection 7.2.2: if $C(t)$ is the concentration of the solute at time t, then dC/dt is given by (7.56), that is,

$$\frac{dC}{dt} = \frac{q}{V}(C_I - C)$$

where q, V, and C_I are defined as in Subsection 7.2.2. Use the graphical approach to discuss the stability of the equilibrium $\hat{C} = C_I$.

12. Assume the single compartment model defined in Subsection 7.2.2, that is, denote by $C(t)$ the concentration of the solute at time t and assume that

$$\frac{dC}{dt} = 3(20 - C(t)) \quad \text{for } t \geq 0 \qquad (7.66)$$

(a) Solve (7.66) when $C(0) = 5$.

(b) Find $\lim_{t \to \infty} C(t)$.

(c) Use your answer in (a) to determine t so that $C(t) = 10$.

13. Assume the single compartment model defined in Subsection 7.2.2, that is, denote by $C(t)$ the concentration of the solute at time t and assume that the concentration of the incoming solute is $3\,\mathrm{gl}^{-1}$ and the rate at which the solution enters is $0.2\,\mathrm{ls}^{-1}$. Furthermore, assume that the volume of the compartment $V = 400\,\mathrm{l}$.

(a) Find the differential equation for the rate of change of the concentration at time t.

(b) Solve the differential equation in (a) when $C(0) = 0$ and find $\lim_{t \to \infty} C(t)$.

(c) Find all equilibria of the differential equation and discuss their stability.

14. Suppose that a tank holds $1000\,\mathrm{l}$ of water, and $2\,\mathrm{kg}$ of salt is poured into the tank.

(a) Compute the concentration of salt in gl^{-1}.

(b) Assume now that you want to reduce the salt concentration. One method would be to remove a certain amount of the salt water from the tank and then replace it by pure water. How much salt water do you have to replace by pure water to obtain a salt concentration of $1\,\mathrm{gl}^{-1}$?

(c) Another method to reduce the salt concentration would be to hook up an overflow pipe and pump pure water into the tank. This way, the salt concentration would be gradually reduced. Assume that you have two pumps, one that pumps water at a rate of $1\,\mathrm{ls}^{-1}$, the other at a rate of $2\,\mathrm{ls}^{-1}$. For each pump, find out how long it would take to reduce the salt concentration from the original concentration to $1\,\mathrm{gl}^{-1}$, and how much pure water is needed in each case. (Note that the rate at which water enters the tank is equal to the rate at which water leaves the tank.) Compare the amount of water needed using the pumps with the amount of water needed in part (b).

15. Assume the single compartment model we introduced in Subsection 7.2.2. Denote by $C(t)$ the concentration at time t and assume

$$\frac{dC}{dt} = 0.37(254 - C(t)) \quad \text{for } t \geq 0$$

(a) Find the equilibrium concentration.

(b) Assume that the concentration is suddenly increased from the equilibrium concentration to 400. Find the return time to equilibrium, denoted by T_R, which is the amount of time until the initial difference is reduced to a fraction e^{-1}.

(c) Repeat (b) for the case when the concentration is suddenly increased from the equilibrium concentration to 800.

(d) Are the values for T_R computed in (b) and (c) different?

16. Assume the compartment model as in Subsection 7.2.2. Suppose that the equilibrium concentration is C_I, and the initial concentration is C_0. Express the time it takes until the initial deviation $C_0 - C_I$ is reduced to a fraction p in terms of T_R.

17. Assume the compartment model as in Subsection 7.2.2. Suppose that the equilibrium concentration is C_I. The time T_R has an integral representation which can be generalized to systems with more than one compartment. Show that

$$T_R = \int_0^\infty \frac{C(t) - C_I}{C(0) - C_I}\, dt$$

(*Hint*: Use (7.57) to show that

$$\frac{C(t) - C_I}{C(0) - C_I} = e^{-(q/V)t}$$

and integrate both sides with respect to t from 0 to ∞.)

18. Use the compartment model defined in Subsection 7.2.2 to investigate how the size of a lake influences nutrient dynamics in the lake after a perturbation. Mary Lake and Elizabeth Lake are two fictitious lakes in the North Woods that are used as experimental lakes to study nutrient dynamics. Mary Lake has a volume of $5.8 \times 10^3\,\mathrm{m}^3$, Elizabeth Lake has twice this volume, namely $13.6 \times 10^3\,\mathrm{m}^3$. Both lakes have the same inflow/outflow rate $q = 170\,\mathrm{ls}^{-1}$. Because both lakes share the same drainage area, the concentration C_I of the incoming solute is the same for both lakes, namely $C_I = 0.7\,\mathrm{mg\,l}^{-1}$. Assume that in the beginning of the experiment both lakes are in equilibrium, that is, the concentration of the solute in both lakes is $0.7\,\mathrm{mg\,l}^{-1}$. Your experiment consists of increasing the concentration of the solute by 10% in each lake at time 0 and then watching how the concentration of the solute in each lake changes with time. Assume the single compartment model to make predictions on how the concentration of the solute will evolve. (Note that $1\,\mathrm{m}^3$ of water corresponds to 1000 liters of water.)

(a) Find the initial concentration C_0 of the solute in each lake at time 0 (that is, immediately after the 10% increase in solute concentration).

(b) Use equation (7.57) to determine how the concentration of the solute changes over time in each lake. Graph your results.

(c) Which lake returns to equilibrium faster? Compute the return time to equilibrium, T_R, for each lake, and explain how this is related to the eigenvalues corresponding to the equilibrium concentration C_I for each lake.

19. Use the compartment model defined in Subsection 7.2.2 to investigate the effect of an increase in the input concentration C_I on the nutrient concentration in a lake. Suppose a lake in a pristine environment has an equilibrium phosphorus concentration of $0.3\,\mathrm{mg}^-$. The volume V of the lake is $12.3 \times 10^6\,\mathrm{m}^3$

and the inflow/outflow rate q is equal to $220\,\mathrm{ls}^{-1}$. Conversion of land in the drainage area of the lake to agricultural use has increased the input concentration from $0.3\,\mathrm{mg}\,\mathrm{l}^{-1}$ to $1.1\,\mathrm{mg}\,\mathrm{l}^{-1}$. Assume that this increase happened instantaneously. Compute the return time to the new equilibrium, denoted by T_R, in days, and find the nutrient concentration in the lake T_R units of time after the change in input concentration. (Note that $1\,\mathrm{m}^3$ of water corresponds to about $1000\,\mathrm{l}$ of water.)

[7.2.3]

20. (*Levins's model*) Denote by $p = p(t)$ the fraction of occupied patches in a metapopulation model, and assume that

$$\frac{dp}{dt} = 2p(1-p) - p \quad \text{for } t \geq 0 \qquad (7.67)$$

(a) Set $g(p) = 2p(1-p) - p$. Graph $g(p)$ for $p \in [0, 1]$.

(b) Find all equilibria of (7.67) that are in $[0, 1]$, and determine their stability using your graph in (a).

(c) Use the eigenvalue approach to analyze the stability of the equilibria that you found in (b).

21. (*Levins's model*) Denote by $p = p(t)$ the fraction of occupied patches in a metapopulation model, and assume that

$$\frac{dp}{dt} = 0.5p(1-p) - 1.5p \quad \text{for } t \geq 0 \qquad (7.68)$$

(a) Set $g(p) = 0.5p(1-p) - 1.5p$. Graph $g(p)$ for $p \in [0, 1]$.

(b) Find all equilibria of (7.68) that are in $[0, 1]$, and determine their stability using your graph in (a).

(c) Use the eigenvalue approach to analyze the stability of the equilibria that you found in (b).

22. (*A metapopulation model with density-dependent extinction*) Denote by $p = p(t)$ the fraction of occupied patches in a metapopulation model, and assume that

$$\frac{dp}{dt} = cp(1-p) - p^2 \quad \text{for } t \geq 0 \qquad (7.69)$$

where $c > 0$. The term p^2 describes density-dependent extinction of patches, namely the per patch extinction rate is p and a fraction p of patches are occupied, resulting in an extinction rate of p^2. The colonization of vacant patches is the same as in Levins's model.

(a) Set $g(p) = cp(1-p) - p^2$ and sketch the graph of $g(p)$.

(b) Find all equilibria of (7.69) in $[0, 1]$, and determine their stability.

(c) Is there a nontrivial equilibrium when $c > 0$? Contrast your findings with the corresponding results in Levins's model.

23. (*Habitat destruction*) In Subsection 7.2.3, we introduced Levins's model. To study the effects of habitat destruction on a single species, we modify equation (7.62) in the following way. We assume that a fraction D of patches is permanently destroyed. Consequently, only patches that are vacant and undestroyed can be successfully colonized. These patches have frequency $1 - p(t) - D$ if $p(t)$ denotes the fraction of occupied patches at time t. Then

$$\frac{dp}{dt} = cp(1 - p - D) - mp \qquad (7.70)$$

(a) Explain in words the meaning of the different terms in (7.70).

(b) Show that there are two possible equilibria, the trivial equilibrium $p_1 = 0$ and the nontrivial equilibrium $p_2 = 1 - D - \frac{m}{c}$. Sketch the graph of p_2 as a function of D.

(c) Assume that $m < c$, such that the nontrivial equilibrium is stable when $D = 0$. Find a condition for D such that the nontrivial equilibrium is between 0 and 1, and investigate the stability of both the nontrivial equilibrium and the trivial equilibrium under this condition.

(d) Assume the condition that you derived in (c), that is, the nontrivial equilibrium is between 0 and 1. Show that when the system is in equilibrium, the fraction of patches that are vacant and undestroyed, that is, the sites that are *available* for colonization, is independent of D. Show that the **effective colonization rate** in equilibrium, that is, c times the fraction of available patches, is equal to the extinction rate. This shows that the effective birth rate of new colonies balances their extinction rate at equilibrium.

[7.2.4]

24. (*Allee effect*) Denote by $N(t)$ the size of a population at time t, and assume that

$$\frac{dN}{dt} = 2N(N - 10)\left(1 - \frac{N}{100}\right) \quad \text{for } t \geq 0 \qquad (7.71)$$

(a) Find all equilibria of (7.71).

(b) Use the eigenvalue approach to determine the stability of the equilibria found in (a).

(c) Set

$$g(N) = 2N(N - 10)\left(1 - \frac{N}{100}\right)$$

for $N \geq 0$ and graph $g(N)$. Identify the equilibria of (7.71) on your graph, and use your graph to determine the stability of the equilibria. Compare your results with your findings in (b). Use your graph to give a graphical interpretation of the eigenvalues associated with the equilibria.

25. (*Allee effect*) Denote by $N(t)$ the size of a population at time t, and assume that

$$\frac{dN}{dt} = 0.3N(N - 17)\left(1 - \frac{N}{200}\right) \quad \text{for } t \geq 0 \qquad (7.72)$$

(a) Find all equilibria of (7.72).

(b) Use the eigenvalue approach to determine the stability of the equilibria found in (a).

(c) Set

$$g(N) = 0.3N(N - 17)\left(1 - \frac{N}{200}\right)$$

for $N \geq 0$ and graph $g(N)$. Identify the equilibria of (7.72) on your graph, and use your graph to determine the stability of the equilibria. Compare your results with your findings in (b). Use your graph to give a graphical interpretation of the eigenvalues associated with the equilibria.

◆ 7.3 SYSTEMS OF AUTONOMOUS EQUATIONS (OPTIONAL)

In the preceding two sections, we discussed models that could be described by a single differential equation. If we wish to describe models in which several quantities interact, such as a competition model in which various species interact, more than one differential equation is needed. We call this then a system of differential equations. We will restrict ourselves again to autonomous systems, that is, systems whose dynamics do not depend explicitly on the independent variable (which is typically time).

This section is a preview of Chapter 11, where we will discuss systems of differential equations in detail. A thorough analysis of systems of differential equations requires a fair amount of theory, which we will develop in Chapters 9 and 10. Since we are not yet equipped with the right tools to analyze systems of differential equations, this section will be rather informal. As with movie previews, you will not know the full story after you finished this sections, but reading it will (hopefully) convince you that systems of differential equations provide a rich tool for modeling biological systems.

7.3.1 A Simple Epidemic Model

We begin our discussion of systems of autonomous differential equations with a classical model of an infectious disease, the Kermack-McKendrick model (1927; 1932; 1933). We consider a population of fixed size N that, at time t, can be divided into three classes: the susceptibles, $S(t)$, who can get infected; the infectives, $I(t)$, who are infected and can transmit the disease; and the removed class, $R(t)$, who are immune to the disease. The flow among these classes can be described by

$$S \longrightarrow I \longrightarrow R$$

We assume that the infection spreads according to the mass action law that we encountered in the discussion of chemical reactions. Each susceptible becomes infected at a rate that is proportional to the number of infectives I. Each infected individual recovers at a constant rate. A gain in the class of infectives is a simultaneous loss in the class of susceptibles. Likewise, a gain in the class of recovered individuals is a loss in the class of infectives. We can therefore describe the dynamics by

$$\frac{dS}{dt} = -bSI \tag{7.73}$$

$$\frac{dI}{dt} = bSI - aI \tag{7.74}$$

$$\frac{dR}{dt} = aI \tag{7.75}$$

Note that

$$\frac{dS}{dt} + \frac{dI}{dt} + \frac{dR}{dt} = -bSI + bSI - aI + aI = 0$$

Since

$$\frac{dS}{dt} + \frac{dI}{dt} + \frac{dR}{dt} = \frac{d}{dt}(S + I + R)$$

it follows that $S(t) + I(t) + R(t)$ is a constant that we can identify as the population size N. To analyze the system, we assume that at time 0

$$S(0) > 0 \qquad\qquad I(0) > 0 \qquad\qquad R(0) = 0$$

A question of interest is whether the infection will spread. We say that the infection spreads if

$$I(t) > I(0) \quad \text{for some } t > 0$$

Equation (7.74) allows us to answer this question. If

$$\frac{dI}{dt} = I(bS - a) > 0 \quad \text{at } t = 0$$

then $I(t)$ increases at the beginning and hence the infection can spread. This condition can be written as

$$\frac{bS(0)}{a} > 1$$

The value $bS(0)/a$ is called the **basic reproductive rate** of the infection and is typically denoted by R_0 (not to be confused with the number of recovered individuals at time 0, $R(0)$). The quantity R_0 is of great importance in epidemiology, because R_0 tells us whether an infection can spread. It is the key to understanding why vaccination programs work. It explains why it is not necessary to vaccinate every one against an infectious disease: as long as the number of susceptibles is reduced below a certain threshold, the infection will not spread. The theroretical threshold, based on the Kermack-McKendrick model, is a/b. In practice, there are additional factors that influence whether or not an infection will spread (such as spatial proximity of infected individuals). But the basic conclusion is the same: as long as the number of susceptibles is below a certain threshold, the infection will not spread.

To find out how the infection progresses over time, we divide (7.73) by (7.75). For $I > 0$, we obtain

$$\frac{dS/dt}{dR/dt} = \frac{dS}{dR} = -\frac{bSI}{aI} = -\frac{b}{a}S$$

That is,

$$\frac{dS}{dR} = -\frac{b}{a}S$$

Separating variables and integrating yields

$$\int \frac{dS}{S} = -\frac{b}{a} \int dR$$

Since $S(t) > 0$, we find

$$\ln S(t) = -\frac{b}{a}R(t) + C$$

With $R(0) = 0$, we obtain

$$\ln S(0) = C$$

hence,

$$S(t) = S(0)e^{-(b/a)R(t)} \tag{7.76}$$

Since $R(t)$ is nondecreasing (no one can leave the removed class), it follows from (7.76) that $S(t)$ is nonincreasing (see Figure 7.25).

Letting $t \to \infty$, we find

$$S(\infty) = \lim_{t \to \infty} S(t) = \lim_{t \to \infty} S(0)e^{-(b/a)R(t)}$$
$$= S(0)e^{-(b/a)R(\infty)}$$

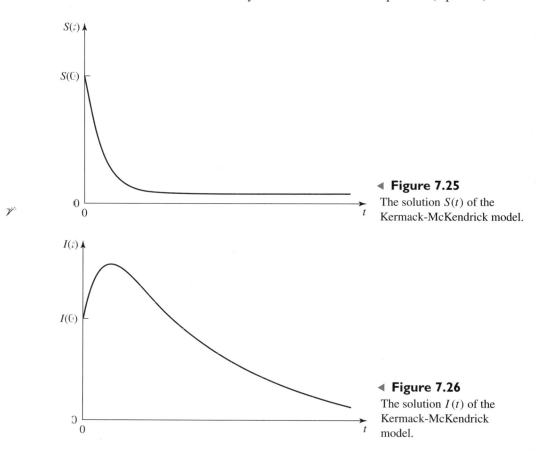

◀ **Figure 7.25**
The solution $S(t)$ of the Kermack-McKendrick model.

◀ **Figure 7.26**
The solution $I(t)$ of the Kermack-McKendrick model.

Since $R(\infty) \le N$, it follows that

$$S(\infty) \ge S(0)e^{-(b/a)N} > 0$$

That is, not everyone becomes infected. As long as the number of susceptibles is greater than a/b, $dI/dt > 0$, and the number of infected individuals increases. Since the populations size is constant, the infection "uses up" susceptibles, and there will be a time, where the number of susceptibles falls below a/b (see Figure 7.26). From then on, dI/dt is negative, and the number of infected individuals begins to decline. The infection will eventually cease, that is,

$$\lim_{t \to \infty} I(t) = 0$$

Since not everyone becomes infected ($S(\infty) > 0$), when the infection finally comes to an end, it is because it runs out of infectives, not because of a lack of susceptibles.

7.3.2 A Compartment Model

In Subsection 7.2.2, we introduced a single-compartment model that led to an autonomous differential equation with one dependent variable. In this subsection, we introduce a model with two compartments that describes the interaction of an autotroph[1] and its nutrient pool. (The following discussion is partially adapted from DeAngelis 1992.) A schematic description of the interactions can be found in Figure 7.27.

▲ **Figure 7.27**
A schematic description of the interaction between autotrophs and the nutrient pool.

---

[1] An autotroph is an organism that can manufacture organic compounds entirely from inorganic components. Examples include most chlorophyll-containing plants and blue-green algae.

We assume that the nutrient pool has an external source and denote the input rate of nutrients by N_I. That is, N_I denotes the total mass per unit time that flows into the system. Nutrients may be washed out, and we assume that the output rate is proportional to the total mass of nutrients in the compartment, N, with proportionality constant a.

The autotroph feeds on the nutrients in the nutrient pool and its growth rate is proportional to the quantity of autotroph biomass, X. The specific rate of growth, $r(N)$, depends on the amount of available nutrients.

Autotrophs can leave the autotroph compartment in two ways. At rate c, they get completely lost (for instance, through harvesting or by being washed out of their habitat). At rate m, the biomass of the autotroph gets recycled back into the nutrient pool (for instance, after the death of an organism).

This system is then described by the following set of differential equations.

$$\frac{dN}{dt} = N_I - aN - r(N)X + mX \tag{7.77}$$

$$\frac{dX}{dt} = r(N)X - (m+c)X \tag{7.78}$$

You should compare the set of differential equations to the schematic description of the model in Figure 7.27. In particular, you should pay attention to the direction of the arrows. For instance, the term $r(N)X$ shows up in both equations: because this term corresponds to the nutrient uptake by the autotrophs, it appears as a loss to the nutrient pool, which is indicated by the minus sign in front of the term in equation (7.77). The same term appears with a plus sign in equation (7.78), because the uptake of nutrients by the autotroph results in an increase of autotroph biomass. Note that the arrow labelled $r(N)X$ goes from the nutrient pool to the autotroph pool. Hence it represents a loss in the nutrient pool (a minus sign in (7.77) in front of $r(N)X$) and a gain in the autotroph pool (a plus sign in (7.78) in front of $r(N)X$).

It is important to learn how to go from the schematic description to the set of differential equations and back. A schematic description quickly summarizes the flow of matter (such as nutrients or energy), whereas the set of differential equations is indispensable if we want to analyze the system.

To keep the discussion concrete, we assume that the function $r(N)$ is linear, that is,

$$r(N) = bN$$

for some constant $b > 0$.

As in Section 7.2, we can introduce the concept of equilibria. An equilibrium for the system given by (7.77) and (7.78) is characterized by simultaneously requiring that

$$\frac{dN}{dt} = 0 \qquad \text{and} \qquad \frac{dX}{dt} = 0 \tag{7.79}$$

since when the rates of change of both quantities are equal to 0, the values for N and X no longer change.

There is a graphical method for finding equilibria, by plotting the **zero-isoclines**. These curves are obtained by setting $dN/dt = 0$ and $dX/dt = 0$. The curves for which $dX/dt = 0$ are given by

$$N = \frac{m+c}{b} \qquad \text{and} \qquad X = 0$$

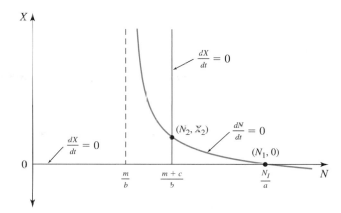

▲ **Figure 7.28**
The zero isoclines in the $N - X$ plane.

The curve for which $dN/dt = 0$, is given by

$$X = \frac{N_I - aN}{bN - m}$$

We plot the three curves in the N-X plane, as illustrated in Figure 7.28.

The zero-isocline for N intersects the horizontal line $X = 0$ at $(N_1, 0)$, where N_1 satisfies

$$0 = \frac{N_I - aN_1}{bN_1 - m}$$

and thus $N_1 = N_I/a$. We call this a trivial equilibrium since it corresponds to the case in which there is no autotroph in the system (the system reduces to the single-compartment model we discussed in Section 7.2).

Depending on where the vertical zero-isocline for X is located, there can be another equilibrium (N_2, X_2) for which both N_2 and X_2 are positive. Namely, looking at Figure 7.28, when

$$\frac{m}{b} < \frac{m + c}{b} < \frac{N_I}{a} \tag{7.80}$$

the two zero-isoclines $N = (m + c)/b$ and $X = (N_I - aN)/(bN - m)$ intersect in the first quadrant. We call this equilibrium a nontrivial equilibrium.

Since $(m + c)/b > m/b$ when $c > 0$, the vertical zero-isocline $dX/dt = 0$ is always to the right of the vertical asymptote $N = m/b$. The first inequality in (7.80) therefore always holds. If we solve the second inequality

$$\frac{m + c}{b} < \frac{N_I}{a}$$

for b, we find

$$b > \frac{a}{N_I}(m + c)$$

That is, the growth parameter b must exceed a certain threshold in order for the autotroph to survive. This should be intuitively clear, since m and c are the rates at which the autotroph pool is depleted. This depletion must be balanced by an increase in biomass of the autotroph.

Suppose that the ratio m/b is fixed. Then the smaller c, the further to the left is the vertical zero-isocline $dX/dt = 0$ and, consequently, the larger the equilibrium value X_2. This should also be intuitively clear, since losing fewer autotrophs should result in a higher equilibrium value.

To find the nontrivial equilibrium (N_2, X_2), we need to solve the following system of equations:

$$0 = N_1 - aN_2 - bN_2X_2 + mX_2 \qquad (7.81)$$

$$0 = bN_2X_2 - (m + c)X_2 \qquad (7.82)$$

where we now assume that $X_2 \neq 0$. Equation (7.82) yields

$$N_2 = \frac{m + c}{b}$$

Using this in (7.81), we find

$$0 = N_1 - a\frac{m + c}{b} - (m + c)X_2 + mX_2$$

or

$$X_2 = \frac{1}{c}\left(N_1 - a\frac{m + c}{b}\right)$$

It can be shown that if the nontrivial equilibrium exists, it is locally stable. In Chapter 11, we will learn two methods for analyzing the stability of a system of differential equations, a graphical and an analytical one; these are extensions of the methods we developed in Section 7.1.

7.3.3 A Hierarchical Competition Model

We now extend the metapopulation model (Levins's model), introduced in Subsection 7.2.3, to a multispecies setting. We recall Levins's model, but use a somewhat different interpretation. In Levins's model, the dynamics of subpopulations was described. We will now view it as an occupancy model of sites for single individuals; that is, we consider the habitat as divided into patches that are now so small that at most one individual occupies a patch. Using this interpretation, we denote by $p(t)$ the fraction of sites that are occupied by a single individual at time t. Then

$$\frac{dp}{dt} = cp(1 - p) - mp$$

where m is the extinction rate (the death rate of an individual), and c is the colonization rate (the rate at which individuals send out propagules). Tilman (1994) extended this model to a system in which species are ranked according to their competitiveness. The fraction of sites that are occupied by species i at time t is denoted by $p_i(t)$. We assume that species 1 is the best competitor, species 2 the next best, and so on. A superior competitor can invade a site that is occupied by an inferior competitor. The inferior competitor is displaced upon invasion of the superior competitor. The dynamics are then given by

$$\frac{dp_1}{dt} = c_1p_1(1 - p_1) - m_1p_1$$

$$\frac{dp_2}{dt} = c_2p_2(1 - p_1 - p_2) - m_2p_2 - c_1p_1p_2$$

$$\frac{dp_3}{dt} = c_3 p_3 (1 - p_1 - p_2 - p_3) - m_3 p_3 - c_1 p_1 p_3 - c_2 p_2 p_3$$

$$\vdots$$

$$\frac{dp_i}{dt} = c_i \left(1 - \sum_{j=1}^{i} p_j \right) - m_i p_i - \sum_{j=1}^{i-1} c_j p_j p_i$$

$$\vdots$$

The first term describes the colonization by species i of sites that are either occupied by an inferior competitor or vacant; the second term describes the extinction of sites that are occupied by species i; and the remainder of the term(s) describe the competitive displacement by superior competitors.

To simplify the discussion, we assume that there are only two species and that $m_1 = m_2 = 1$. The equations are then given by

$$\frac{dp_1}{dt} = c_1 p_1 (1 - p_1) - p_1$$

$$\frac{dp_2}{dt} = c_2 p_2 (1 - p_1 - p_2) - p_2 - c_1 p_1 p_2$$

The hierarchical structure of the model makes it easy to find possible equilibria. Setting $dp_1/dt = 0$, we find

$$0 = \hat{p}_1 [c_1 (1 - \hat{p}_1) - 1]$$

which, aside from the trivial equilibrium 0, gives

$$\hat{p}_1 = 1 - \frac{1}{c_1}$$

We saw in Subsection 7.2.2 that if $1 < c_1$, $\hat{p}_1 \in (0, 1)$ and the nontrivial equilibrium is stable. Because we are interested in coexistence of species, we assume that $c_1 > 1$ in the following, so that species 1 can survive.

Setting $dp_2/dt = 0$ with $\hat{p}_1 = 1 - 1/c_1$ allows us to find $\hat{p}_2$.

$$0 = \hat{p}_2 [c_2 (1 - \hat{p}_1 - \hat{p}_2) - 1 - c_1 \hat{p}_1]$$

which, aside from the trivial equilibrium 0, yields

$$\hat{p}_2 = (1 - \hat{p}_1) - \frac{1}{c_2} - \frac{c_1}{c_2} \hat{p}_1$$

$$= \frac{1}{c_1} - \frac{1}{c_2} - \frac{c_1}{c_2} + \frac{1}{c_2}$$

$$= \frac{1}{c_1} - \frac{c_1}{c_2}$$

Note that the sum of the two nontrivial equilibria $\hat{p}_1 + \hat{p}_2 = 1 - \frac{c_1}{c_2} < 1$. That is, if both nontrivial equilibria are in $(0, 1)$, then the two species can coexist. We see that if

$$0 < \frac{1}{c_1} - \frac{c_1}{c_2} < 1$$

then $\hat{p}_2 \in (0, 1)$. The second inequality $\frac{1}{c_1} - \frac{c_1}{c_2} < 1$ always holds, because we assumed that $c_1 > 1$. To satisfy the first inequality, we need

$$c_2 > c_1^2$$

So far, we only know that there is a nontrivial equilibrium if $c_1 > 1$ and $c_2 > c_1^2$, but this does not tell us anything about stability. Even though we cannot analyze stability directly, we can take an approach that is very common in the ecological literature. We determine whether species 2 can invade a monoculture equilibrium of species 1. Why does this help us? First, note that species 1 is unaffected by the presence of species 2. As long as species 1 can survive in the absence of species 2, it can also survive in its presence. Therefore, we need only worry about species 2. If species 2 can invade the monoculture equilibrium of species 1, then this says that if species 2 is at a low density, it will be able to increase its density, and will therefore be able to coexist with species 1. The invasion criterion is then

$$\frac{dp_2}{dt}\bigg|_{p_1=\hat{p}_1} > 0 \quad \text{when } p_2 \text{ is small}$$

This works as follows.

$$\frac{dp_2}{dt} = p_2[c_2(1 - p_1 - p_2) - 1 - c_1 p_1]$$

We assume that $p_1 = \hat{p}_1 = 1 - 1/c_1$ and that p_2 is very small. If p_2 is small, then $1 - \hat{p}_1 - p_2 \approx 1 - \hat{p}_1$. Hence

$$\frac{dp_2}{dt} \approx p_2\left[c_2\frac{1}{c_1} - 1 - c_1 + 1\right] = p_2\left[\frac{c_2}{c_1} - c_1\right] > 0$$

if

$$\frac{c_2}{c_1} - c_1 > 0 \quad \text{or} \quad c_2 > c_1^2$$

Since $dp_2/dt > 0$ when species 1 is in equilibrium and species 2 has a low abundance, this means that species 2 can invade. We conclude that species 1 and 2 can coexist when $c_2 > c_1^2$.

This mechanism of coexistence is referred to as the **competition-colonization tradeoff**. That is, the weaker competitor (species 2) can compensate for inferior competitiveness by being a superior colonizer $\left(c_2 > c_1^2\right)$.

◆ 7.3 PROBLEMS

[7.3.1]

In Problems 1–4, we will investigate the classical Kermack-McKendrick model for the spread of an infectious disease in a population of fixed size N. (This model was introduced in Subsection 7.3.1, and you should refer to this subsection when working out the problems.) If $S(t)$ denotes the number of susceptibles at time t, $I(t)$ the number of infectives at time t, and $R(t)$ the number of immune individuals at time t, then

$$\frac{dS}{dt} = -bSI$$

$$\frac{dI}{dt} = bSI - aI$$

and $R(t) = N - S(t) - I(t)$.

1. Determine in each of the following cases whether or not the disease can spread. (*Hint*: Compute R_0.)

(a) $S(0) = 1000, a = 200, b = 0.3$

(b) $S(0) = 1000, a = 200, b = 0.1$

2. Assume that $a = 100$ and $b = 0.2$. The **critical number** of susceptibles $S_c(0)$ at time 0 for the spread of a disease that is introduced into the population at time 0, is defined as the minimum number of susceptibles for which the disease can spread. Find $S_c(0)$.

3. Suppose that $a = 100$, $b = 0.01$, and $N = 10,000$. Can the disease spread if at time 0 there is one infected individual?

4. (a) Show that when $I > 0$,

$$\frac{dI}{dS} = \frac{a}{b}\frac{1}{S} - 1 \tag{7.83}$$

and solve (7.83) when $R(0) = 0$, $I(0) = i_0$, and $S(0) = S_0$.

(b) Since $I(t)$ gives the number of infectives at time t and $dI/dt = bSI - aI$, if $S(0) > a/b$, then $dI/dt > 0$ at time $t = 0$. Since $\lim_{t\to\infty} I(t) = 0$, there is a time $t > 0$ at which $I(t)$ is maximal. Show that the number of susceptibles when $I(t)$ is maximal is given by $S = a/b$.

(c) In part (a), you expressed $I(t)$ as a function of $S(t)$. Use your result in (b) to show that the maximal number of infectives, I_{max}, is given by

$$I_{max} = N - \frac{a}{b} + \frac{a}{b}\ln\left(\frac{a/b}{S_0}\right)$$

(d) Use your result in (c) to show that I_{max} is a decreasing function of the parameter a/b for $a/b < S_0$ (that is, in the case in which the infection can spread). Use this to explain how a and b determine the severity (as measured by I_{max}) of the disease. Does this make sense?

[7.3.2]

5. Assume the compartment model of Subsection 7.3.2, with $a = 5, b = 0.02, m = 1$, and $c = 1$.

(a) Find the system of differential equations that corresponds to these values.

(b) Determine which values of N_I result in a nontrivial equilibrium, and find the equilibrium values for both the autotroph and the nutrient pool

6. Assume the compartment model of Subsection 7.2.3, with $a = 1, b = 0.01, m = 2, c = 1$, and $N_I = 500$.

(a) Find the system of differential equations that corresponds to these values.

(b) Plot the zero-isoclines corresponding to this system.

(c) Use your graph in (b) to determine whether this system has a nontrivial equilibrium.

7. Assume the compartment model of Subsection 7.3.2, with $a = 1, b = 0.01, m = 2, c = 1$ and $N_I = 200$.

(a) Find the system of differential equations that corresponds to these values.

(b) Plot the zero-isoclines corresponding to this system.

(c) Use your graph in (b) to determine whether this system has a nontrivial equilibrium.

[7.3.3]

8. Assume the hierarchical competition model with two species, introduced in Subsection 7.3.3. Specifically, assume

$$\frac{dp_1}{dt} = 2p_1(1 - p_1) - p_1$$

$$\frac{dp_2}{dt} = 5p_2(1 - p_1 - p_2) - p_2 - 2p_1 p_2$$

(a) Find all equilibria.

(b) Determine whether species 2 can invade a monoculture of species 1. (Assume that species 1 is in equilibrium.)

9. Assume the hierarchical competition model with two species, introduced in Subsection 7.3.3. Specifically, assume

$$\frac{dp_1}{dt} = 2p_1(1 - p_1) - p_1$$

$$\frac{dp_2}{dt} = 3p_2(1 - p_1 - p_2) - p_2 - 2p_1 p_2$$

(a) Find all equilibria.

(b) Determine whether species 2 can invade a monoculture equilibrium of species 1.

10. Assume the hierarchical competition model with two species, introduced in Subsection 7.3.3. Specifically, assume

$$\frac{dp_1}{dt} = 2p_1(1 - p_1) - p_1$$

$$\frac{dp_2}{dt} = 6p_2(1 - p_1 - p_2) - p_2 - 2p_1 p_2$$

(a) Use the zero-isocline approach to find all equilibria graphically.

(b) Determine the numerical values of all equilibria.

11. Assume the hierarchical competition model with two species, introduced in Subsection 7.3.3. Specifically, assume

$$\frac{dp_1}{dt} = 3p_1(1 - p_1) - p_1$$

$$\frac{dp_2}{dt} = 5p_2(1 - p_1 - p_2) - p_2 - 3p_1 p_2$$

(a) Use the zero-isocline approach to find all equilibria graphically.

(b) Determine the numerical values of all equilibria.

12. (Adapted from Crawley 1997) Denote by V plant biomass, and by N herbivore number. The plant-herbivore interaction is modeled as

$$\frac{dV}{dt} = aV\left(1 - \frac{V}{K}\right) - bVN$$

$$\frac{dN}{dt} = cVN - dN$$

(a) Suppose the herbivore number is equal to 0. What differential equation describes the dynamics of the plant biomass? Can you explain the resulting equation? Determine the plant biomass equilibrium in the absence of herbivores.

(b) Now assume that herbivores are present. Describe the effect of herbivores on plant biomass, that is, explain the term $-bVN$ in the first equation. Describe the dynamics of the herbivores, that is, how their population size increases and what contributes to decreases in population size.

(c) Determine the equilibria by solving

$$\frac{dV}{dt} = 0 \quad \text{and} \quad \frac{dN}{dt} = 0$$

Also determine the equilibria also graphically. Explain why this model implies that "plant abundance is determined solely by attributes of the herbivore," as stated in Crawley (1997).

◆ 7.4 KEY TERMS

Chapter 7 Review Topics   Discuss the following definitions and concepts:

1. Differential equation
2. Separable differential equation
3. Solution of a differential equation
4. Pure time differential equation
5. Autonomous differential equation
6. Direction field
7. Exponential growth

8. Von Bertalanffy equation

9. Logistic equation

10. Allometric growth

11. Equilibrium

12. Stability

13. Eigenvalue

14. Single compartment model

15. Levins's model

16. Allee effect

17. Kermack-McKendrick model

18. Zero isocline

19. Hierarchical competition model

◆ 7.5 REVIEW PROBLEMS

1. (*Newton's law of cooling*) Suppose that an object has temperature T and is brought into a room that is kept at a constant temperature T_a. Newton's law of cooling states that the rate of temperature change of the object is proportional to the difference between the temperature of the object and the surrounding medium.

(a) Denote by $T(t)$ the temperature at time t, and explain why

$$\frac{dT}{dt} = k(T - T_a)$$

is the differential equation that expresses Newton's law of cooling.

(b) Suppose that it takes the object 20 min to cool from 30°C to 28°C in a room whose temperature is 21°C. How long will it take the object to cool to 25°C, if it is at 30°C when it is brought into the room? (*Hint*: Solve the differential equation in (a) with the initial condition $T(0) = 30$°C and with $T_a = 21$°C. Use $T(20) = 28$°C to determine the constant k.)

2. (Adapted from Cain et al. 1995) In this problem we discuss a model for clonal growth in the white clover *Trifolium repens*. *T. repens* is a widespread perennial clonal plant species. It spreads through stolon growth (a stolon is a horizontal stem). By mapping the shape of a clone over time, Cain et al. estimated stolon elongation and die-back rates as follows. Denote by $S(t)$ the stolon length of the clone at time t. They observed that the change in stolon length was proportional to stolon length, that is,

$$\frac{dS}{dt} \propto S$$

Introducing the proportionality constant r, called the net growth rate, we find

$$\frac{dS}{dt} = rS \tag{7.84}$$

(a) Suppose that S_f and S_0 are the final and the initial stolon lengths, respectively, and T denotes the time period of observation. Use (7.84) to show that r, the net growth rate, can be estimated from

$$r = \frac{1}{T} \ln \frac{S_f}{S_0}$$

(*Hint*: Solve the differential equation (7.84) with initial condition $S(0) = S_0$, and use the fact that $S(T) = S_f$.)

(b) The net growth rate r is the difference between the stolon elongation rate b and the stolon die-back rate m, that is,

$$r = b - m$$

Let B be the total amount of stolon elongation and D be the total amount of stolon die-back over the observation period of length T. Show that

$$B = \int_0^T bS(t)\,dt = \frac{bS_0}{r}(e^{rT} - 1)$$

$$D = \int_0^T mS(t)\,dt = \frac{mS_0}{r}(e^{rT} - 1)$$

(c) Show that $B - D = S_f - S_0$, and rearrange the equations for B and D in (b) so that you could estimate b and m from r, B, and D; that is, show that

$$b = \frac{rB}{S_f - S_0} = \frac{rB}{B - D}$$

$$m = \frac{rD}{S_f - S_0} = \frac{rD}{B - D}$$

(d) Explain how B and r can be estimated if S_f, S_0 and D are known from field measurements. Use your result in (c) to explain how you would then find estimates for b and m.

3. (*Diversification of life*) (Adapted from Benton 1997 and Walker 1985) Several models have been proposed to explain the diversification of life during geological periods. According to Benton (1997),

> The diversification of marine families in the past 600 million years (Myr) appears to have followed two or three logistic curves, with equilibrium levels that lasted for up to 200 Myr. In contrast, continental organisms clearly show an exponential pattern of diversification, and although it is not clear whether the empirical diversification patterns are real or are artefacts of a poor fossil record, the latter explanation seems unlikely.

In the following, we will investigate three models for diversification. They are analogous to models for population growth; however, the quantities involved have a different interpretation. We denote by $N(t)$ the diversification function, which counts the number of taxa as a function of time, and by r the intrinsic rate of diversification.

(a) (*Exponential model*) This model is described by

$$\frac{dN}{dt} = r_e N \tag{7.85}$$

Solve (7.85) with the initial condition $N(0)$ at time 0, and show that r_e can be estimated from

$$r_e = \frac{1}{t} \ln \left[\frac{N(t)}{N(0)} \right] \qquad (7.86)$$

(*Hint*: To find (7.86), solve for r in the solution of (7.85).)

(b) (*Logistic growth*) This model is described by

$$\frac{dN}{dt} = r_l N \left(1 - \frac{N}{K} \right) \qquad (7.87)$$

where K is the equilibrium value. Solve (7.87) with the initial condition $N(0)$ at time 0, and show that r_l can be estimated from

$$r_l = \frac{1}{t} \ln \left[\frac{K - N(0)}{N(0)} \right] + \frac{1}{t} \ln \left[\frac{N(t)}{K - N(t)} \right] \qquad (7.88)$$

for $N(t) < K$.

(c) Assume that $N(0) = 1$ and $N(10) = 1000$. Estimate r_e and r_l for both $K = 1001$ and $K = 10,000$.

(d) Use your answer in (c) to explain the following quote from Stanley (1979):

> ... there must be a general tendency for calculated values of $[r]$ to represent underestimates of exponential rates, because some radiation will have followed distinctly sigmoid paths during the interval evaluated...

(e) Explain why the exponential model is a good approximation to the logistic model when N/K is small compared to 1.

4. (*A simple model for photosynthesis of individual leaves*) (Adapted from Horn 1971) Photosynthesis is a complex mechanism; the following model is a very simplified caricature. Suppose that a leaf contains a number of traps that can capture light. If a trap captures light, it becomes energized. The energy in the trap can then be used to produce sugar, which causes the energized trap to become unenergized. The number of traps that can become energized is proportional to the number of unenergized traps and the light intensity. Denote by T the total number of traps (unenergized and energized) in a leaf, by I the light intensity, and by x the number of energized traps. Then the following differential equation describes how the number of energized traps changes over time:

$$\frac{dx}{dt} = k_1 (T - x) I - k_2 x$$

where k_1 and k_2 are positive constants. Find all equilibria and study their stability using the eigenvalue approach.

5. (*Gompertz growth model*) This model is sometimes used to study the growth of a population for which the per capita growth rate is density dependent. Denote by $N(t)$ the size of a population at time t; then for $N \geq 0$,

$$\frac{dN}{dt} = kN(\ln K - \ln N) \quad \text{with } N(0) = N_0 \qquad (7.89)$$

(a) Show that

$$N(t) = K \exp \left[-\left(\ln \frac{K}{N_0} \right) e^{-kt} \right]$$

is a solution of (7.89). To do this, differentiate $N(t)$ with respect to t and show that it can be written in the form (7.89). Don't forget to show that $N(0) = N_0$. Use a graphing calculator to sketch the graph of $N(t)$ for $N_0 = 100$, $k = 2$, and $K = 1000$. The function $N(t)$ is called the Gompertz growth curve.

(b) Use l'Hospital's rule to show that

$$\lim_{N \to 0} N \ln N = 0$$

and use this to show that $\lim_{N \to 0} dN/dt = 0$. Are there any other values of N where $dN/dt = 0$?

(c) Sketch the graph of dN/dt as a function of N for $k = 2$ and $K = 1000$. Find the equilibria and discuss their stability using your graph. Explain the meaning of K.

6. (*Island biogeography*) Preston (1962) and MacArthur and Wilson (1963) investigated the effect of area on species diversity in oceanic islands. It is assumed that species can immigrate to an island from a species pool of size P, and that species on the island can go extinct. We denote the immigration rate by $I(S)$ and the extinction rate by $E(S)$, where S is the number of species on the island. Then the change in species diversity over time is

$$\frac{dS}{dt} = I(S) - E(S) \qquad (7.90)$$

The simplest functional forms for $I(S)$ and $E(S)$, for a fixed island, are

$$I(S) = c \left(1 - \frac{S}{P} \right) \qquad (7.91)$$

$$E(S) = m \frac{S}{P} \qquad (7.92)$$

where c, m, and P are positive constants.

(a) Find the equilibrium species diversity $\hat{S}$ of (7.90) with $I(S)$ and $E(S)$ given in (7.91) and (7.92).

(b) It is reasonable to assume that the extinction rate is a decreasing function of island size. That is, if A denotes the area of the island, we assume that m is a function of the island size A with $dm/dA < 0$. Furthermore, we assume that the immigration rate I does not depend on the size of the island. Use these assumptions to investigate how the equilibrium species diversity changes with island size.

(c) Assume that $S(0) = S_0$. Solve (7.90) with $I(S)$ and $E(S)$ as given in (7.91) and (7.92), respectively.

(d) Assume that $S_0 = 0$. That is, the island is initially void of species. The time constant T for the system is defined as

$$S(T) = (1 - e^{-1})\hat{S}$$

Show that under the assumption $S_0 = 0$,

$$T = \frac{P}{c + m}$$

(e) Use the assumptions in (b) and your answer in (d) to investigate the effect of island size on the time constant T; that is, determine whether $T(A)$ is an increasing or decreasing function of A.

7. (*Chemostat*) A chemostat is an apparatus for growing bacteria in a medium in which all nutrients but one are available in excess. One nutrient, whose concentration can be controlled, is held at a concentration that limits the growth of bacteria. The growth chamber of the chemostat is continually flushed by adding nutrients dissolved in liquid at a constant rate, and allowing the liquid in the growth chamber, which contains bacteria, to leave the growth chamber at the same rate. If X denotes the number of bacteria in the growth chamber, then the growth dynamics of the bacteria are given by

$$\frac{dX}{dt} = r(N)X - qX \tag{7.93}$$

where $r(N)$ is the growth rate depending on the nutrient concentration N and q is the input and output flow rate. The equation for the nutrient flow is given by

$$\frac{dN}{dt} = qN_0 - qN - r(N)X \tag{7.94}$$

Note that equation (7.73) is (7.78) with $m = 0$, $N_I = qN_0$, and $a = e = q$, and (7.94) is (7.77) with $m = 0$.

(a) Explain in words the meaning of the terms in (7.93) and (7.94)

(b) Assume that $r(N)$ is given by the Monod growth function

$$r(N) = b\frac{N}{k + N}$$

where k and b are positive constants. Draw the zero isoclines in the N-X plane and explain how to find equilibria $(\hat{N}, \hat{X})$ graphically.

(c) Show that a nontrivial equilibrium (an equilibrium for which $\hat{N}$ and $\hat{X}$ are both positive) satisfies

$$r(\hat{N}) - q = 0 \tag{7.95}$$
$$qN_0 - q\hat{N} - r(\hat{N})\hat{X} = 0 \tag{7.96}$$

Show that (7.95) has a positive solution $\hat{N}$ if $q < b$, and find an expression for $\hat{N}$. Use this and (7.96) to find $\hat{X}$.

(d) Assume that $q < b$. Use your results in (c) to show that $\hat{X} > 0$ if $\hat{N} < N_0$ and $\hat{N} < N_0$ if $q < bN_0/(k + N_0)$. Furthermore, show that $\hat{N}$ is an increasing function of q for $q < b$.

(e) Use your results in (d) to explain why the following is true. As we increase the flow rate q from 0 to $bN_0/(k + N_0)$, the nutrient concentration $\hat{N}$ increases until it reaches the value N_0 and the number of bacteria decreases to 0.

8. (Adapted from Nee and May 1992 and Tilman 1994) In Subsection 7.3.3, we introduced a hierarchical competition model. We will use this model to investigate the effects of habitat destruction on coexistence. We assume that a fraction D of the

sites is permanently destroyed. Furthermore, we restrict our discussion to two species and assume that species 1 is the superior and species 2 the inferior competitor. In the case in which both species have the same mortality ($m_1 = m_2$) which we set equal to 1, the dynamics are described by

$$\frac{dp_1}{dt} = c_1 p_1(1 - p_1 - D) - p_1 \tag{7.97}$$
$$\frac{dp_2}{dt} = c_2 p_2(1 - p_1 - p_2 - D) - p_2 - c_1 p_1 p_2 \tag{7.98}$$

where p_i, $i = 1, 2$, is the fraction of sites occupied by species i.

(a) Explain in words the meanings of the different terms in (7.97) and (7.98).

(b) Show that

$$\hat{p}_1 = 1 - \frac{1}{c_1} - D$$

is an equilibrium for species 1 that is in $(0, 1)$ and is stable if $D < 1 - 1/c_1$ and $c_1 > 1$.

(c) Assume that $c_1 > 1$ and $D < 1 - 1/c_1$. Show that species 2 can invade the nontrivial equilibrium of species 1 (computed in (b)) if

$$c_2 > c_1^2(1 - D)$$

(d) Assume that $c_1 = 2$ and $c_2 = 5$. Then species 1 can survive as long as $D < 1/2$. Show that the fraction of sites that are occupied by species 1 is then

$$\hat{p}_1 = \begin{cases} \frac{1}{2} - D & \text{for } 0 \leq D \leq \frac{1}{2} \\ 0 & \text{for } \frac{1}{2} \leq D \leq 1 \end{cases}$$

Show that

$$\hat{p}_2 = \frac{1}{10} + \frac{2}{5}D \quad \text{for} \quad 0 \leq D \leq \frac{1}{2}$$

For $D > 1/2$, species 1 can no longer persist. Explain why the dynamics for species 2 reduce in this case to

$$\frac{dp_2}{dt} = 5p_2(1 - p_2 - D) - p_2$$

Show that the nontrivial equilibrium is of the form

$$\hat{p}_2 = 1 - \frac{1}{5} - D \quad \text{for } \frac{1}{2} \leq D \leq 1 - \frac{1}{5}$$

Plot $\hat{p}_1$ and $\hat{p}_2$ as functions of D in the same coordinate system. What happens for $D > 1 - 1/5$? Use the plot to explain in words how each species is affected by habitat destruction.

(e) Repeat (d) for $c_1 = 2$ and $c_2 = 3$.

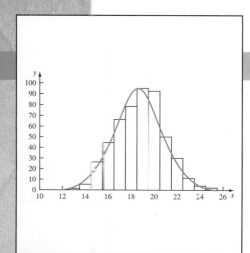

8

Applications of Integration

In this chapter, we collect a number of applications of integration. The first, rectification of curves, is of historical interest, and is another example that stresses the meaning of integrals as limiting sums. The second topic, treated in Section 8.2, deals with densities and histograms. Section 8.3 discusses average values and Section 8.4 the normal distribution. The last section introduces age-structured populations; integrals appear quite naturally in this situation.

Section 8.1 is independent of the rest of the chapter. Section 8.2 is needed for Sections 8.3-5. Sections 8.3-4 and Section 8.5 are independent.

◆ 8.1 RECTIFICATION OF CURVES

In this section, we will study how to compute the lengths of curves in the plane. This is yet another example in which integrals appear as we add up a large number of small increments. The first curve whose length was determined was the semicubical parabola $y^2 = x^3$; this was done by William Neile (1637–1670) in 1657. Interestingly enough, only about twenty years earlier, Descartes asserted that there was no rigorous way to determine the exact length of a curve. It turns out that the exact formula is in essence an infinitesimal version of the Pythagorean theorem.

How would we rectify a curve (that is, determine its length) with a ruler? We could approximate the curve by short line segments, measure the length of each segment and add up these measurements, as illustrated in Figure 8.1. By choosing smaller line segments, our approximation would improve. This is precisely the method that we will employ to find an exact formula.

To find the exact formula, assume that the curve whose length we want to find is given by a function $y = f(x)$, $a \leq x \leq b$, which has a continuous first derivative on (a, b). We partition the interval $[a, b]$ into subintervals using the partition $P = [x_0, x_1, x_2, \ldots, x_n]$, where $a = x_0 < x_1 < x_2 < \cdots < x_n = b$, and approximate the curve by a polygon that consists of the straight line segments connecting neighboring points on the curve, as shown in Figure 8.1. A typical line segment connecting the points P_{k-1} and P_k is shown in Figure 8.2. Using the Pythagorean theorem, we can find its length. We set $\Delta x_k = x_k - x_{k-1}$ and $\Delta y_k = y_k - y_{k-1}$. Then

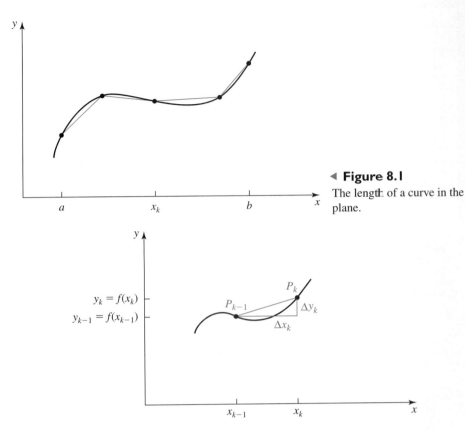

◀ **Figure 8.1**
The length of a curve in the plane.

▲ **Figure 8.2**
A typical line segment.

the length of the line segment is given by

$$\sqrt{(\Delta x_k)^2 + (\Delta y_k)^2}$$

The length of the polygon using the partition P is then

$$L_P = \sum_{k=1}^{n} \sqrt{(\Delta x_k)^2 + (\Delta y_k)^2} \tag{8.1}$$

By choosing finer and finer partitions, the length of the polygon will become a better and better approximation of the length of the corresponding curve. However, before we can take this limit, we need to work on the sum.

The difference Δy_k is equal to $f(x_k) - f(x_{k-1})$. The mean value theorem guarantees that there is a number $c_k = [x_{k-1}, x_k]$ such that

$$f'(c_k) = \frac{f(x_k) - f(x_{k-1})}{x_k - x_{k-1}}$$

Since $\Delta x_k = x_k - x_{k-1}$, we find

$$\Delta y_k = f'(c_k)(x_k - x_{k-1}) = f'(c_k)\Delta x_k \tag{8.2}$$

Replacing Δy_k in (8.1) by (8.2), we find for the length of the polygon

$$L_P = \sum_{k=1}^{n} \sqrt{(\Delta x_k)^2 + [f'(c_k)\Delta x_k]^2}$$

$$= \sum_{k=1}^{n} \sqrt{1 + [f'(c_k)]^2}\ \Delta x_k$$

This form allows us to take the limit $\|P\| \to 0$. We find

$$L = \lim_{\|P\|\to 0} \sum_{k=1}^{n} \sqrt{1 + [f'(c_k)]^2}\, \Delta x_k$$

$$= \int_a^b \sqrt{1 + [f'(x)]^2}\, dx$$

If $f(x)$ is differentiable on $[a, b]$ and $f'(x)$ is continuous on (a, b), then the length of the curve $y = f(x)$ from a to b is given by

$$L = \int_a^b \sqrt{1 + [f'(x)]^2}\, dx$$

As mentioned, the first curve that was rectified was given by $y = f(x) = x^{3/2}$. We will choose this function for our first example.

▮EXAMPLE 1   Determine the length of the curve given by the graph of $y = f(x) = x^{3/2}$ between $a = 5/9$ and $b = 21/9$.

Solution   To determine the length of the curve, we need to find $f'(x)$ first.

$$f'(x) = \frac{3}{2}x^{1/2}$$

Then,

$$L = \int_{5/9}^{21/9} \sqrt{1 + \left[\frac{3}{2}x^{1/2}\right]^2}\, dx = \int_{5/9}^{21/9} \sqrt{1 + \frac{9}{4}x}\, dx$$

$$= \left[\frac{4}{9} \cdot \frac{2}{3}\left(1 + \frac{9}{4}x\right)^{3/2}\right]_{5/9}^{21/9} = \frac{8}{27}\left[\left(1 + \frac{9}{4}\cdot\frac{21}{9}\right)^{3/2} - \left(1 + \frac{9}{4}\cdot\frac{5}{9}\right)^{3/2}\right]$$

$$= \frac{8}{27}\left[\left(\frac{5}{2}\right)^3 - \left(\frac{3}{2}\right)^3\right] = \frac{8}{27}\left(\frac{125}{8} - \frac{27}{8}\right) = \frac{98}{27}$$

The length of the curve is thus $98/27$.   ▮

Before we show one more example, we discuss the formula in more detail; namely, using

$$\frac{dy}{dx} = f'(x)$$

we can write for the length

$$L = \int_a^b \sqrt{1 + \left(\frac{dy}{dx}\right)^2}\, dx$$

Treating dy and dx like regular numbers, we can rewrite this as

$$L = \int_a^b \sqrt{(dx)^2 + (dy)^2}$$

We call the expression $\sqrt{(dx)^2 + (dy)^2}$ the **arc length differential**, and denote it by ds. We can think of ds as a typical infinitesimal line segment. The Pythagorean theorem in this infinitesimal form then reads $(ds)^2 = (dx)^2 + (dy)^2$. "Adding up" these line segments (that is, computing $\int_a^b ds$), then yields the length of the curve.

Due to the somewhat complicated form of the integrand in the computation of the length, we quickly run into computational problems when we actually try to compute the integral. Sometimes, however, the integrand simplifies enough for easy computation. The following is such an example, where the integrand turns out to be a perfect square.

▮ **EXAMPLE 2**   Determine the length of

$$f(x) = \frac{1}{4}x^2 - \frac{1}{2}\ln x \quad \text{from } x = 1 \text{ to } x = e$$

Solution   Differentiating $f(x)$, we find

$$f'(x) = \frac{x}{2} - \frac{1}{2x}$$

The length of the curve is then given by

$$L = \int_1^e \sqrt{1 + \left(\frac{x}{2} - \frac{1}{2x}\right)^2}\, dx = \int_1^e \sqrt{1 + \left(\frac{x^2}{4} - \frac{1}{2} + \frac{1}{4x^2}\right)}\, dx$$

$$= \int_1^e \sqrt{\frac{x^2}{4} + \frac{1}{2} + \frac{1}{4x^2}}\, dx$$

We notice that the expression under the square root is a perfect square, namely,

$$\frac{x^2}{4} + \frac{1}{2} + \frac{1}{4x^2} = \left(\frac{x}{2} + \frac{1}{2x}\right)^2$$

Hence, the integral for the length simplifies to

$$L = \int_1^e \sqrt{\left(\frac{x}{2} + \frac{1}{2x}\right)^2}\, dx$$

$$= \int_1^e \left(\frac{x}{2} + \frac{1}{2x}\right) dx = \frac{1}{4}x^2 + \frac{1}{2}\ln x \Big]_1^e$$

$$= \frac{1}{4}e^2 + \frac{1}{2} - \frac{1}{4} = \frac{1}{4}(e^2 + 1)$$

▮

◆ 8.1 PROBLEMS

1. Find the length of the straight line
$$y = 2x$$
from $x = 0$ to $x = 2$ by the following methods.
(a) using planar geometry
(b) using the integral formula for the lengths of curves, derived in this section

2. Find the length of the straight line
$$y = mx$$
from $x = 0$ to $x = a$, where m and a are positive constants, by the following methods.
(a) using planar geometry
(b) using the integral formula for the lengths of curves, derived in this section

3. Find the length of the curve
$$y^2 = x^3$$
from $x = 1$ to $x = 4$.

4. Find the length of the curve
$$2y^2 = 3x^3$$
from $x = 0$ to $x = 1$.

5. Find the length of the curve
$$y = \frac{x^3}{6} + \frac{1}{2x}$$
from $x = 1$ to $x = 3$.

6. Find the length of the curve

$$y = \frac{x^4}{4} + \frac{1}{8x^2}$$

from $x = 2$ to $x = 4$.

In Problems 7–16, set up but do not evaluate the integrals for the lengths of the following curves.

7. $y = x^2, -1 \le x \le 1$

8. $\sin x, 0 \le x \le \frac{\pi}{2}$

9. $y = e^{-x}, 0 \le x \le 1$

10. $y = \ln x, 1 \le x \le e$

11. Find the length of the quarter circle

$$y = \sqrt{1 - x^2}$$

for $0 \le x \le 1$ by the following methods.

(a) using a formula from geometry

(b) using the integral formula from this section

12. Use the formula $\tan^2 x + 1 = \sec^2 x$ to find the length of the curve

$$y = -\ln(\cos x)$$

between $x = 1$ and $x = \pi/3$.

13. A cable that hangs between two poles at $x = -M$ and $x = M$ takes the shape of a catenary, with equation

$$y = \frac{1}{2a}(e^{ax} + e^{-ax})$$

where a is a positive constant. Compute the length of the cable when $a = 1$ and $M = \ln 2$.

14. Show that if

$$f(x) = \frac{e^x + e^{-x}}{2}$$

then the length of the curve of $f(x)$ between $x = 0$ and $x = a$ for any $a > 0$ is given by $f'(a)$.

◆ 8.2 DENSITIES AND HISTOGRAMS

The following example is adapted from de Roos (1996). The waterflea *Daphnia pulex* feeds on the alga *Chlamydomonas rheinhardii*. An important component of the feeding behavior of *Daphnia* is the strong dependence of the amount of food consumed on the sizes of individuals. To model the feeding behavior of *Daphnia*, we need to describe the size structure of the population. One way to do this would be to determine the fraction of *Daphnia* whose size is less than or equal to x, for all possible values of x. If the population size is large, then this fraction can be closely approximated by a continuous function. We denote this function by $F(x)$. Since $F(x)$ represents a fraction, its values are in the interval $[0, 1]$. Furthermore, $F(x)$ is nondecreasing since *Daphnia* whose length is less than or equal to x_1 are also contained in the group of *Daphnia* whose length is less than or equal to x_2, when $x_1 \le x_2$. We assume that the individual size lies in the interval $[l_1, l_2]$; it follows that $F(x) = 0$ for $x \le l_1$, and $F(x) = 1$ for $x \ge l_2$ (see Figure 8.3).

If the function $F(x)$ is differentiable in the open interval (l_1, l_2), we can represent $F(x)$ in integral form as

$$F(x) = \int_{l_1}^{x} f(u)\,du, \quad l_1 \le x \le l_2$$

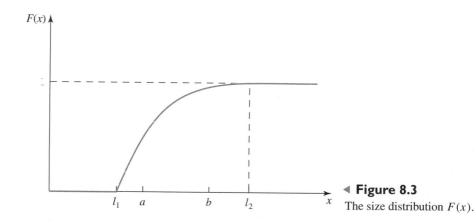

◀ **Figure 8.3**
The size distribution $F(x)$.

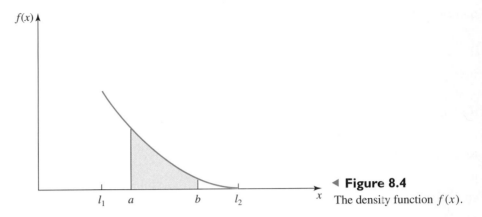

◀ **Figure 8.4**
The density function $f(x)$.

where $f(x) = F'(x)$ for $l_1 < x < l_2$ according to the FTC (part I). The function $f(x)$ is called a **density function**.

The fraction of the population with size between a and b, for $l_1 \leq a \leq b \leq l_2$, is then

$$F(b) - F(a) = \int_a^b f(x)\, dx \tag{8.3}$$

Geometrically, this means that fractions are represented by area (see Figure 8.4).

Since a fraction is a number between 0 and 1, it follows that $f(x) \geq 0$ for $l_1 \leq x \leq l_2$. Furthermore, the total are between the graph of $f(x)$ and the x-axis is equal to 1, since this area represents the entire population.

We can interpret $F(b) - F(a)$ in (8.3) also in the following way, namely, if we sampled from this population, that is, if we picked one individual at random from this population, then the chance that this observation would fall into the interval $[a, b]$ is $F(b) - F(a)$.

It is important to realize that in (8.3), fractions are represented by area; in particular, the *integral* of $f(x)$ — not $f(x)$ itself — has this physical interpretation. The density function $f(x)$ does not have an immediate physical interpretation. Later in the section, we will explain how one can find $f(x)$ empirically.

In our example, we defined $f(x)$ only for $x \in (l_1, l_2)$. We can extend its definition to all of **R** by setting $f(x) = 0$ for $x \notin (l_1, l_2)$. We then write

$$F(x) = \int_{-\infty}^x f(u)\, du \quad \text{for } x \in \mathbf{R}$$

The two properties, $f(x) \geq 0$ and $\int_{-\infty}^\infty f(x)\, dx$, characterize a density function and we will use it as its definition.

A function $f(x)$ is called a density function if

$$f(x) \geq 0 \qquad \text{and} \qquad \int_{-\infty}^\infty f(x)\, dx = 1$$

The first condition is needed because for any $a < b$, we want $\int_a^b f(x)\, dx$ to represent the fraction of the population for which x is between a and b, and this quantity is always nonnegative. The second condition says that the fraction of the population for which x is between $-\infty$ and $+\infty$ is equal to $1 = 100\%$, and this makes sense because it covers the entire possible range of the population.

In the biology literature, density functions are not always normalized, that is, the area under the density function is not always equal to 1. For instance, in our

Daphnia example, the integral $\int_a^b f(x)\,dx$ may represent the *number* of *Daphnia* whose length is between a and b instead of the fraction. In this case, $\int_{-\infty}^{\infty} f(x)\,dx$ would represent the total number of *Daphnia*. In this book, however, density functions will always be normalized so that the total area under the graph is equal to 1.

The first example will help us to become more familiar with density functions.

▮ EXAMPLE 1 The exponential function

$$f(r) = \begin{cases} ae^{-ar} & \text{for } r > 0 \\ 0 & \text{for } r \le 0 \end{cases}$$

where $a > 0$ is a constant, is frequently used to model seed dispersal. The function $f(r)$ is a density function, and $\int_a^b f(r)\,dr$ describes the fraction of seeds dispersed between distances a and b from the source at 0.

a. Show that $f(r)$ is a density function.

b. Show that the fraction of seeds that are dispersed distance R or more declines exponentially with R.

c. Find R such that 60% of the seeds are dispersed within distance R of the source. How does R depend on a?

Solution

a. To show that $f(r)$ is a density function, we need to show that $f(r) \ge 0$ for all $r \in \mathbf{R}$ and that $\int_{-\infty}^{\infty} f(r)dr = 1$. Since $a > 0$ and $e^{-ar} > 0$, it follows immediately that $f(r) \ge 0$ for $r > 0$. Combining this with $f(r) = 0$ for $r \le 0$, we find $f(r) \ge 0$ for all $r \in \mathbf{R}$. To check the second criterion, we need to carry out the integration. Since the function $f(r)$ is a piecewise defined function, we need to split up the integral into two parts:

$$\int_{-\infty}^{\infty} f(r)\,dr = \int_{-\infty}^{0} f(r)\,dr + \int_0^{\infty} f(r)\,dr$$

$$= \int_{-\infty}^{0} 0\,dr + \int_0^{\infty} ae^{-ar}\,dr$$

The term $\int_{-\infty}^{0} 0\,dr$ is equal to 0. An antiderivative of ae^{-ar} is $-e^{-ar}$. Hence,

$$\int_0^{\infty} ae^{-ar}\,dr = \lim_{z \to \infty} \int_0^z ae^{-ar}\,dr = \lim_{z \to \infty} [-e^{-ar}]_0^z$$

$$= \lim_{z \to \infty} [-e^{-az} - (-1)] = 1$$

since $\lim_{z \to \infty} e^{-az} = 0$.

b. Let $G(R)$ denote the fraction of seeds that are dispersed distance R or more. Then

$$G(R) = \int_R^{\infty} f(r)\,dr = \int_R^{\infty} ae^{-ar}\,dr = \lim_{z \to \infty} \int_R^z ae^{-ar}\,dr$$

$$= \lim_{z \to \infty} [-e^{-ar}]_R^z = \lim_{z \to \infty} (-e^{-az} + e^{-aR}) = e^{-aR}$$

This shows that $G(R)$ declines exponentially with R.

c. The number R satisfies

$$0.6 = \int_0^R ae^{-ar}\,dr$$

Carrying out the integration, we find

$$0.6 = [-e^{-ar}]_0^R = 1 - e^{-aR}$$

To find R, we need to solve

$$e^{-aR} = 0.4$$

$$-aR = \ln 0.4$$

$$R = -\frac{\ln 0.4}{a} = \frac{1}{a} \ln \frac{5}{2}$$

where we used in the last step the fact that $-\ln 0.4 = \ln \frac{1}{0.4} = \ln \frac{5}{2}$. We see that $R \propto 1/a$ (that is, the bigger a, the smaller R), which means that seeds are dispersed more closely to the source for larger values of a.   ❚

We will now try to find a way to empirically determine $f(x)$. We set

$$F(x) = \int_{-\infty}^x f(t)\,dt$$

It follows from the fundamental theorem of calculus, part I, that

$$F'(x) = f(x)$$

Using the definition of derivatives, we can write this as

$$\lim_{\Delta x \to 0} \frac{F(x + \Delta x) - F(x)}{\Delta x} = f(x)$$

We can rewrite the numerator on the left-hand side:

$$F(x + \Delta x) - F(x) = \int_{-\infty}^{x+\Delta x} f(t)\,dt - \int_{-\infty}^x f(t)\,dt$$

$$= \int_x^{x+\Delta x} f(t)\,dt$$

If Δx is sufficiently small, then $f(t)$ will not vary much over the interval $[x, x+\Delta x)$. We will therefore approximate $f(t)$ by $f(x)$ over the interval $[x, x + \Delta x)$. Hence,

$$\int_x^{x+\Delta x} f(t)\,dt \approx f(x)\Delta x \tag{8.4}$$

for Δx sufficiently small. That is, for Δx sufficiently small, we can think of $f(x)\Delta x$ as approximately representing the fraction of individuals whose quantity of interest, such as length in the *Daphnia* example, is in the interval $[x, x + \Delta x)$. This is an important interpretation which will help us to determine density functions empirically. This is of course the Riemann sum approximation that we discussed in Section 5.1. For Δx small, just one rectangle gives a pretty good approximation, as illustrated in Figure 8.5.

To determine the density $f(x)$ empirically, we will use the approximation (8.4). We take a sufficiently large sample from the population and measure the quantity of interest of each individual in the sample. We partition the interval over which the quantity of interest varies into subintervals of length Δx_i. For each subinterval, we count the number of sample points that fall into the respective subintervals. To display the data graphically, we use a **histogram**, which consists of rectangles whose widths are equal to the lengths of the corresponding subintervals and whose *areas* are equal to the number of sample points that fall into the corresponding subintervals.

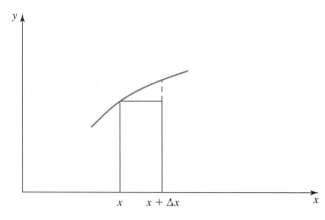

◀ **Figure 8.5**

The area of the rectangle is $f(x)\Delta x$. It is a good approximation of $\int_x^{x+\Delta x} f(t)\,dt$ which is the area under the curve between x and $x + \Delta x$.

This is analogous to approximating areas under curves by rectangles. We illustrate this procedure in the following example.

Brachiopods form a marine invertebrate phylum whose soft body parts are enclosed in shells. They were the dominant seabed shelled animals in the Paleozoic, but suffered greatly during the Permian-Triassic mass extinction.[1] They are still present today (with approximately 120 genera), and occupy a diverse range of habitats—but they are no longer the dominant seabed shelled animal, their place having been taken by bivalve mollusks. (The brachiopod story is described in Ward 1992.)

The brachiopod *Dielasma* is common in Permian reef deposits in the north of England. The following table (adapted from Benton and Harper 1997) represents measurements of the sagittal length of the shell measured in mm (see Figure 8.6).

▲ **Figure 8.6**

The sagittal length of a shell.

length	5	7	9	11	13	15	17	19	21	23	25
frequency	3	28	12	2	4	4	6	6	5	3	1

The length measurements are divided into classes $[0, 2)$, $[2, 4)$, $[4, 6)$, ..., $[24, 26)$. The midpoint of each subinterval represents the size class. For instance, length 11 in the table below corresponds to the size class of lengths between 10 mm and 12 mm. The number below each size class, the frequency, represents the number of brachiopods in the sample whose lengths fell into the corresponding size class. For instance, there were two brachiopods in the sample whose lengths fell into the size class $[10, 12)$.

To display this data set graphically, we use a histogram, as shown in Figure 8.7. The horizontal axis shows the midpoints of each size class. The graph consists of rectangles whose widths are equal to the length of the corresponding size class, and whose *area* is equal to the number of specimens in the corresponding class. It is very important to note that a histogram represents numbers by area, not by height. For instance, the number of specimens in size class $[8, 10)$ is equal to 12. This size class is represented by 9, the midpoint of the interval $[8, 10)$. Because the width of the size class is 2, the height of the rectangle must be equal to 6 units so that the area of the corresponding rectangle is equal to 12 units. In our example, all size classes are of the same length; of course, this need not be the case in general.

---

[1]The Permian geological period lasted from 286 to 248 million years ago; the Triassic followed the Permian and lasted from 248 to 213 million years ago. The Permian-Triassic mass extinction is believed to have been the most severe mass extinction of life that has ever occurred.

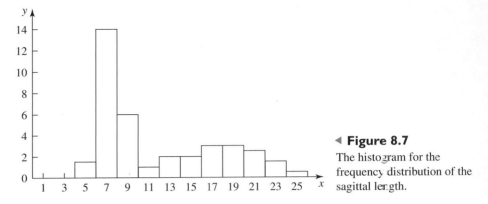

◀ **Figure 8.7**

The histogram for the frequency distribution of the sagittal length.

Displaying data graphically has certain advantages. For instance, we see immediately that the size distribution is biased towards smaller sizes, because the rectangles in the histogram that correspond to smaller shell lengths have larger areas. This bias towards smaller sizes might indicate, for instance, that brachiopods suffered a high juvenile mortality.

It is often convenient to scale the vertical axis of the histogram so that the total area of the histogram is equal to 1. One of the advantages of this is that the histogram then does not directly reflect the sample size, as only proportions are represented. This makes it easier to compare histograms from different samples. For instance, if someone else had obtained a different sample of this type of brachiopod in the same location, and if both samples are representative of its length distribution, then both histograms should look similar.

If we scale the total area of the histogram to 1, then the area of each rectangle in the histogram represents the fraction of the sample in the corresponding class. To obtain the fraction of sample points in a certain class, we divide the number of sample points in this class by the total sample size. In our example, the sample size was 74. The fraction in the sample in size class [8, 10), for instance, would then be $12/74 = 0.16$ or 16%.

The choice of the widths of the classes in the histogram is somewhat arbitrary. The goal is to obtain an informative graph. Typically, the larger the sample size, the smaller the widths of the classes can be.

The outline of the normalized histogram (that is, where the total area is equal to 1) can be used as an approximation to the density function $f(x)$ (see Figure 8.8). Typically, the larger the sample size, the smaller the widths of the size classes, and the better the approximation of $f(x)$.

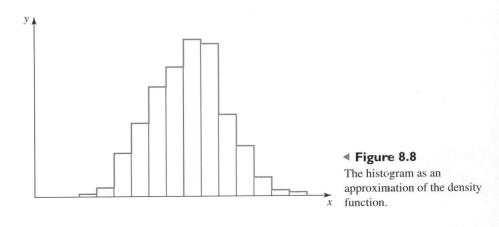

◀ **Figure 8.8**

The histogram as an approximation of the density function.

◆ 8.2 PROBLEMS

In Problems 1–3, show that the given functions are density functions.

1. $f(x) = \begin{cases} \frac{1}{3} & \text{for } -1 < x < 2 \\ 0 & \text{otherwise} \end{cases}$

2. $f(x) = \begin{cases} 2e^{-2x} & \text{for } x > 0 \\ 0 & \text{otherwise} \end{cases}$

3. For $a > 1$,
$$f(x) = \begin{cases} (a-1)x^{-a} & \text{for } x > 1 \\ 0 & \text{otherwise} \end{cases}$$

4. Suppose that
$$f(x) = \begin{cases} c(2 - |x|) & \text{for } -2 < x < 2 \\ 0 & \text{otherwise} \end{cases}$$

Find c so that $f(x)$ is a density function.

5. Suppose that
$$f(x) = \begin{cases} cx^{-1/2}e^{-ax^{1/2}} & \text{for } x > 0 \\ 0 & \text{otherwise} \end{cases}$$

where a is a positive constant. Find c so that $f(x)$ is a density function.

6. Suppose that
$$f(x) = \begin{cases} c\frac{1-x}{1+x} & \text{for } 0 < x < 1 \\ 0 & \text{otherwise} \end{cases}$$

is a density function that describes seed dispersal. That is, for $0 < a < b$,
$$\int_a^b f(x)\,dx$$

is the fraction of seeds dispersed between distance a and b from a source at 0.

(a) Determine c and graph $f(x)$.

(b) Use a graphing calculator to find R such that 90% of the seeds are dispersed within distance R of the source.

(c) Explain in words what $f(x)\,dx$ represents.

(d) Explain in words how you would determine $f(x)$ in an experiment.

7. Suppose that seeds are dispersed according to the density function

$$f(x) = \begin{cases} ae^{-ax} & \text{for } x > 0 \\ 0 & \text{otherwise} \end{cases} \qquad (8.5)$$

where x denotes the distance to the source at 0 and a is a positive constant. Define
$$F(r) = \int_0^r f(x)\,dx$$

(a) Explain why $F(r)$ represents the fraction of seeds that fall within distance r of the source at 0. Find $F(r)$. Show that a graph of
$$g(r) = -\ln(1 - F(r))$$

results in a straight line. How is the slope of the straight line related to the parameter a?

(b) Suppose that a fictitious plant, *Dispersa exponentialis*, disperses its seeds according to $f(r)$ as given in (8.5). To estimate a, you determine the fraction $F(r)$ of seeds that fall within distance r of the plant. You obtain the following data set.

r	$F(r)$
1	0.14
2	0.26
4	0.45
7	0.65
10	0.78
15	0.89
20	0.95

Use your results in (a) to transform your data set so that the transformed data are linearly related. Graph the corresponding straight line and determine a. If someone showed you this graph and claimed that it was real data, would you believe them?

8. We consider a one dimensional habitat (for concreteness, the real line **R**) in which plants grow. These plants produce seeds that are dispersed along this one-dimensional habitat. The number of seeds that fall into a particular interval on the line depends on the plant density in the neighborhood, and on the likelihood that seeds from a source some distance away will land in this interval. The following steps will result in a model for seed density.

(a) Suppose that $n(y)\,dy$ represents the number of plants in $[y, y + dy)$. Explain the meaning of
$$N_{[a,b]} = \int_a^b n(y)\,dy$$

(b) Assume that $k(z) \geq 0$ and that $k(z)$ is symmetric about $z = 0$. Let $k(z)\,dz$ be the fraction of seeds that land in $[z, z+dz)$ if the source is at 0. Explain why
$$\int_{-\infty}^\infty k(z)\,dz = 1$$

(c) Suppose that a plant is located at y and disperses its seeds according to the function $k(z)$ defined in (b). Explain why $k(x - y)\,dx$ represents the fraction of seeds dispersed by the plant at y that land in $[x, x + dx)$.

(d) Assume now that S is the total number of seeds produced per plant, and that the dispersal follows the function $k(z)$ defined in (b). Explain the meaning of
$$s(x) = S \int_{-\infty}^\infty k(x - y)n(y)\,dy$$

9. The following table contains the number of leaves per basil plant in a sample of size 25.

19	21	20	13	18
14	17	14	17	17
13	15	12	15	17
15	16	18	17	14
14	14	13	20	13

6	2	3	7	9	5
2	3	6	6	7	9
1	0	4	6	7	1
6	6	7	9	8	2
7	5	6	1	7	6

Use the data to construct a histogram by grouping the data into (a) 10 equal size classes, and (b) 5 equal size classes. Scale both histograms so that the total area is equal to 1.

10. The following table contains the litter size of mice in a sample of size 30.

Use the data to construct a histogram. Choose appropriate size classes, and then scale the histogram so that the total area is equal to 1.

◆ 8.3 AVERAGE VALUES

In this section we will revisit average values. Recall that in Subsection 5.3.3 we introduced the integral as an average value. We will use results from the previous section to provide another way to compute average values; to illustrate this, we look at the following example.

On a winter day somewhere in southern Minnesota, the following temperature readings T_k (in Fahrenheit) at hour k were obtained.

k	0	1	2	3	4	5	6	7	8	9	10	11	12
T_k	6	6	6	5	5	5	5	5	8	10	12	12	12

k	13	14	15	16	17	18	19	20	21	22	23
T_k	12	12	12	10	8	8	8	5	5	3	3

The average temperature on that day was

$$\bar{T} = \frac{1}{24}(6 + 6 + 6 + 5 + 5 + 5 + 5 + 5 + 8 + 10 + 12$$

$$+12 + 12 + 12 + 12 + 12 + 10 + 8 + 8 + 8 + 5 + 5 + 3 + 3)$$

$$= \frac{183}{24} = 7.625$$

Rearranging these values according to size, we find

$$\bar{T} = \frac{1}{24}[(3 + 3) + (5 + 5 + 5 + 5 + 5 + 5 + 5) + (6 + 6 + 6)$$

$$+ (8 + 8 + 8 + 8) + (10 + 10) + (12 + 12 + 12 + 12 + 12 + 12)]$$

$$= \frac{1}{24}[(3)(2) + (5)(7) + (6)(3) + (8)(4) + (10)(2) + (12)(6)]$$

$$= 3 \cdot \frac{2}{24} + 5 \cdot \frac{7}{24} + 6 \cdot \frac{3}{24} + 8 \cdot \frac{4}{24} + 10 \cdot \frac{2}{24} + 12 \cdot \frac{6}{24}$$

$$= \sum [\text{temperature}] \times [\text{relative frequency of temperature}]$$

$$= \frac{183}{24} = 7.625$$

This suggests another way of computing average values; depending on whether the quantity takes on discrete values or ranges over a continuous interval, this involves either summation or integration. We will treat the two cases separately.

8.3.1 The Discrete Case

The discrete case deals with quantities that take on only discrete values, such as the number of leaves per plant. If we collect data for such a quantity, we might ask what the **relative frequency** of each value is. The following table contains the number of leaves per basil plant in a sample of size 25.

16	15	13	16	16
14	16	15	18	17
16	18	16	13	16
16	16	15	15	16
15	18	16	16	15

To find the **relative frequency distribution**, we must count how often each value occurs and then divide by the sample size, which is 25 in this case. We summarize this in the following table.

no. of leaves	13	14	15	16	17	18
relative frequency	$\frac{2}{25}$	$\frac{1}{25}$	$\frac{6}{25}$	$\frac{12}{25}$	$\frac{1}{25}$	$\frac{3}{25}$

Relative frequencies are simply percentages; for instance, the relative frequency of the group with thirteen leaves is $2/25 = 8\%$ of the total sample.

The relative frequency distribution has the property that if we sum all of the frequencies, they add up to 1, that is,

$$\frac{2}{25} + \frac{1}{25} + \frac{6}{25} + \frac{12}{25} - \frac{1}{25} + \frac{3}{25} = 1$$

This should also be clear from the way we obtained the relative frequency distribution.

Suppose that the quantity of interest takes on value x_k, $k = 1, 2, \ldots, m$, with relative frequency p_k. Then the average value or **mean** of this quantity (denoted by $\bar{x}$), is, according to the introductory example,

$$\bar{x} = \sum_{k=1}^{m} x_k p_k$$

To find the average value for our example, we must compute

$$\text{avg. no. of leaves} = 13 \cdot \frac{2}{25} + 14 \cdot \frac{1}{25} + 15 \cdot \frac{6}{25} + 16 \cdot \frac{12}{25} + 17 \cdot \frac{1}{25} + 18 \cdot \frac{3}{25}$$

$$= 393 \cdot \frac{1}{25} = 15.72$$

Note that though the number of leaves per plant is an integer value, the average number of leaves per plant is not. You would actually lose valuable information if you rounded the average number to the closest integer.

To emphasize this point, consider the average number of lifetime births expected by women, 18 to 34 years old, in 1992. (The following data is taken from the U.S. Bureau of the Census 1994.) The number of lifetime births expected by a woman who is not a high school graduate is 2.393; whereas the corresponding number for a woman with a graduate or professional degree is 1.990. If we rounded these numbers to the closest integer (after all, the actual number of children per woman is an integer), they would be the same, namely 2; we would no longer see the difference between these groups. These numbers are to be interpreted in terms of large populations. Multiplying each number by the number of women in the respective groups, would give us an estimate of the expected number of children.

8.3.2 The Continuous Case

The length of *Daphnia* ranges over a continuous interval; the density function $f(x)$, defined in Section 8.2, describes its size distribution. How would we find the average length of *Daphnia* in our population? One way would be to measure the length of each individual, and average all the values. Another way would be to find the relative frequency distribution by **discretizing** (as in the brachiopod example), that is, we would determine a histogram and then proceed as in the discrete case. This procedure will give us a formula for computing the average value in the continuous case.

Suppose that we divide the interval $[l_1, l_2]$ into k subintervals of equal widths $\Delta l = (l_2 - l_1)/k$. We count the relative frequency of lengths in each subinterval. We can use the resulting histogram to find an approximate value for the average length (see Figure 8.9), namely, if we denote by l_i the midpoint of the ith subinterval, and by $n(i)$ the relative frequency of *Daphnia* in the ith subinterval, then

$$\text{approximate average length} = \sum_{i=1}^{k} l_i n(i)$$

Since

$$n(i) \approx f(l_i)\Delta l$$

where $f(l_i)$ is the density function evaluated at l_i, we find that

$$\text{approximate average length} = \sum_{i=1}^{k} l_i f(l_i)\Delta l$$

If we let the number of subintervals tend to infinity ($\Delta l \to 0$), then the sum converges to the true average length. That is,

$$\text{average length} = \int_{l_1}^{l_2} x f(x)\,dx$$

where $f(x)$ is the density function of the distribution of sizes of *Daphnia*.

For the general case, we consider a quantity X that ranges over a continuous interval. We say that X has a **continuous distribution** with density $f(x)$ if $\int_a^b f(x)\,dx$ represents the fraction of the population for which X is in the interval $[a, b]$.

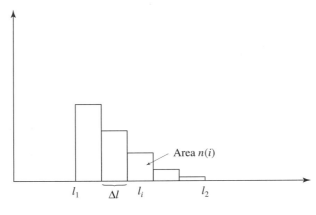

Area $n(i)$

l_1 Δl l_i l_2

◀ **Figure 8.9**
The histogram can be used to approximate the average length. (The total area of the histogram is 1.)

> Suppose that a quantity X has a continuous distribution with density function $f(x)$. The average value or mean of X is then given by
>
> $$\text{average value of } X = \int_{-\infty}^{\infty} xf(x)\,dx$$
>
> provided that the integral exists.

In the next two examples, we will use this formula to compute average values.

EXAMPLE 1 In Example 1 of Section 8.2, we considered the following density function,

$$f(r) = \begin{cases} ae^{-ar} & \text{for } r > 0 \\ 0 & \text{for } r \leq 0 \end{cases}$$

where a is a positive constant. This density is used to describe seed dispersal. Find the average dispersal distance.

Solution We use the preceding formula for average value and find

$$\text{average dispersal distance} = \int_{-\infty}^{\infty} rf(r)\,dr = \int_{0}^{\infty} rae^{-ar}\,dr$$

since $f(r) = 0$ for $r \leq 0$. To evaluate this integral, we must integrate by parts. We find

$$\text{average dispersal distance} = \int_{0}^{\infty} rae^{-ar}\,dr$$

$$= \lim_{z \to \infty} [r(-e^{-ar})]_0^z + \lim_{z \to \infty} \int_0^z e^{-ar}\,dr$$

The first expression is equal to

$$\lim_{z \to \infty} [-ze^{-az} + 0] = \lim_{z \to \infty} \frac{z}{e^{az}}$$

This is of the form $\frac{\infty}{\infty}$. Using l'Hospital's rule, we find

$$\lim_{z \to \infty} \frac{z}{e^{az}} = \lim_{z \to \infty} \frac{1}{ae^{az}} = 0$$

since $a > 0$. The second expression is

$$\lim_{z \to \infty} \int_0^z e^{-ar}\,dr = \lim_{z \to \infty} \left[-\frac{1}{a}e^{-ar} \right]_0^z = \frac{1}{a}$$

We therefore find that

$$\text{average dispersal distance} = \frac{1}{a} \qquad\blacksquare$$

EXAMPLE 2 Consider the density function

$$f(x) = \begin{cases} 1 & \text{for } 0 < x < 1 \\ 0 & \text{otherwise} \end{cases}$$

A quantity X that is distributed according to $f(x)$ is said to be uniformly distributed. To explain the name of the distribution, observe that for $0 < a < b < 1$, the fraction of the population with values in $[a, b]$ is equal to

$$\int_a^b f(x)\,dx = b - a$$

implying that intervals of equal lengths in $(0, 1)$ contain the same fraction of the population.

To find the average value or mean of X, we compute

$$\text{mean of } X = \int_{-\infty}^{\infty} xf(x)\,dx = \int_{0}^{1} x\,dx = \frac{1}{2}x^2\Big]_{0}^{1} = \frac{1}{2}$$

which is the midpoint of the interval $(0, 1)$.

In both Examples 1 and 2, the density function $f(x)$ was equal to 0 over part of the real line. This possibility must be considered when computing average values. That is, when evaluating $\int_{-\infty}^{\infty} xf(x)\,dx$, the limits of integration must be chosen appropriately, to match the domain of the density function where the function is different from 0. For instance, in Example 2, $f(x) = 0$ for $x \leq 0$ and $x \geq 1$. Hence, when evaluating $\int_{-\infty}^{\infty} xf(x)\,dx$, both $\int_{-\infty}^{0} xf(x)\,dx$ and $\int_{1}^{\infty} xf(x)\,dx$ were equal to 0, and $\int_{-\infty}^{\infty} xf(x)\,dx$ was reduced to $\int_{0}^{1} xf(x)\,dx$.

◆ 8.3 PROBLEMS

[8.3.1]

1. The following table contains the number of leaves per basil plant in a sample of size 25.

19	21	20	13	18
14	17	14	17	17
13	15	12	15	17
15	16	18	17	14
14	14	13	20	13

(a) Find the relative frequency distribution.

(b) Compute the average value by (i) averaging the values in the table directly, and (ii) using the relative frequency distribution obtained in (a).

2. The following table contains the number of aphids per plant in a sample of size 30.

15	27	13	2	0	16
26	0	2	1	17	15
21	13	5	0	19	25
12	11	0	16	22	1
28	9	0	0	1	17

(a) Find the relative frequency distribution.

(b) Compute the average value by (i) averaging the values in the table directly, and (ii) using the relative frequency distribution obtained in (a).

3. The following table contains exam scores of 25 students.

7	8	8	3	2
5	6	9	10	6
8	8	7	6	9
10	4	4	8	6
9	10	5	5	8

(a) Find the relative frequency distribution.

(b) Compute the average value by (i) averaging the values in the table directly, and (ii) using the relative frequency distribution obtained in (a).

4. The following table contains the number of flower heads per plant in a sample of size 20.

15	17	19	18	15
17	18	15	14	19
17	15	15	18	19
20	17	14	17	18

(a) Find the relative frequency distribution.

(b) Compute the average value by (i) averaging the values in the table directly, and (ii) using the relative frequency distribution obtained in (a).

[8.3.2]

5. Suppose that a quantity X is distributed according to the density function

$$f(x) = \begin{cases} \frac{1}{3} & \text{for } -1 < x < 2 \\ 0 & \text{otherwise} \end{cases}$$

Find the mean of X.

6. Suppose that a quantity X is distributed according to the density function

$$f(x) = \begin{cases} 2e^{-2x} & \text{for } x > 0 \\ 0 & \text{otherwise} \end{cases}$$

Find the mean of X.

7. Suppose that a quantity X is distributed according to the density function

$$f(x) = \begin{cases} (a-1)x^{-a} & \text{for } x > 1 \\ 0 & \text{otherwise} \end{cases}$$

where a is a constant greater than 2. Find the mean of X.

8. Suppose that a quantity X is distributed according to the density function

$$f(x) = \begin{cases} \frac{1}{4}(2 - |x|) & \text{for } -2 < x < 2 \\ 0 & \text{otherwise} \end{cases}$$

Find the mean of X.

9. Suppose that a quantity X is distributed according to the density function

$$f(x) = \frac{c}{1 + x^2}, \quad x \in \mathbf{R}$$

where c is a constant chosen so that $f(x)$ is a density function.

(a) Find the constant c.

(b) Does the mean of X exist?

10. Suppose that the distribution of a nonnegative quantity X is described by the density function $f(x)$.

(a) Define

$$G(x) = \int_x^\infty f(t)\, dt$$

and explain the meaning of $G(x)$.

(b) Assume that

$$\lim_{x \to \infty} xG(x) = 0$$

Use integration by parts to show that

$$\text{Mean of X} = \int_0^\infty G(x)\, dx \qquad (8.6)$$

(Hint: Start at the right-hand side of (8.6) and use the "multiplying by 1" trick, followed by integration by parts.)

11. Suppose a quantity X is distributed according to the density function $f(x)$. Assume that $f(x)$ is an even function and that $\int_{-\infty}^\infty xf(x)\, dx$ exists. Find the mean of X.

12. In Example 2, we introduced the density function of a quantity that was uniformly distributed over the interval $[0, 1]$. Suppose now that X is uniformly distributed over the interval $[0, L]$.

(a) Find its density function.

(b) Find the mean of X.

In Problems 13–15, suppose that a quantity X is distributed according to the density function $f(x)$. The **median** *is defined as the number m so that*

$$\int_{-\infty}^m f(x)\, dx = \int_m^\infty f(x)\, dx = \frac{1}{2}$$

13. Sketch the graph of a density function and locate the median on the graph.

14. Suppose that X is distributed according to the density function

$$f(x) = \begin{cases} 3e^{-3x} & \text{for } x > 0 \\ 0 & \text{otherwise} \end{cases}$$

(a) Compute the mean of X.

(b) Find the median.

(c) Graph $f(x)$ and label the mean and the median on the graph.

(d) Use the graph in (c) to explain why the median is smaller than the mean.

15. The median income in the U.S. is smaller than the average income. Can you think of a reason why? (*Hint*: How are the mean and the median affected by increases in the income bracket of the top 20% of incomes?)

16. Sketch the graph of a density function of a quantity X for which the mean is equal to the median.

◆ 8.4 THE NORMAL DISTRIBUTION

Quantitative genetics is concerned with metric characters, such as plant height, litter size, body weight, and so on. Such characters are called **quantitative characters**. There are many quantitative characters whose frequency distributions follow a bell-shaped curve. For instance, counting the bristles on some particular part of the abdomen (fifth sternite) of a strain of *Drosophila melanogaster*, Mackay (1984) found that the number of bristles varied according to a bell-shaped curve. (This is shown in Figure 8.10, which is adapted from Hartl and Clark 1989.)

The smooth curve in Figure 8.10 that is fitted to the histogram is proportional to the density function of a **normal distribution** (the curve is not scaled so that the area under the curve is equal to 1). The density function of the normal distribution is described by just two parameters, called μ and σ, which can be estimated from data. The parameter μ can be any real number; the parameter σ is a positive real number (we will interpret μ and σ subsequently). The density function is given in the following.

A normal distribution with parameters μ and σ has density function

$$f(x) = \frac{1}{\sigma\sqrt{2\pi}} e^{-(x-\mu)^2/2\sigma^2}, \qquad -\infty < x < \infty$$

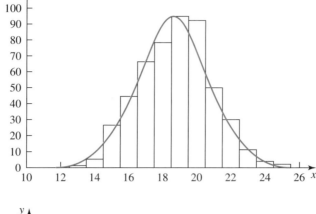

◀ **Figure 8.10**

The number of abdominal bristles.

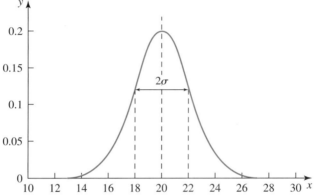

◀ **Figure 8.11**

The graph of the normal density with $\mu = 20$ and $\sigma = 2$. The maximum is at $\mu = 20$; the two inflection points are at 18 and 22, respectively.

The parameter μ is called the mean and the parameter σ is called the **standard deviation**. In Problem 1, we will investigate the shape of the density function of the normal distribution (a graph is shown in Figure 8.11). We collect the properties here.

1. $f(x)$ is symmetric about $x = \mu$.

2. The maximum of $f(x)$ is at $x = \mu$.

3. The inflection points of $f(x)$ are at $x = \mu - \sigma$ and $x = \mu + \sigma$.

Since $f(x)$ is a density function,

$$f(x) \geq 0 \quad \text{and} \quad \int_{-\infty}^{\infty} f(x)\, dx = 1$$

With the tools we have so far, we cannot show that the density function is normalized to 1. What we can show, however, is that the mean μ is the average value that we defined in this section; that is,

$$\mu = \int_{-\infty}^{\infty} x f(x)\, dx$$

(We will do the calculation in Problem 2.) Furthermore, if a quantity X is normally distributed with parameters μ and σ, then

$$\begin{bmatrix} \text{fraction of population} \\ \text{with values in } [a,\, b] \end{bmatrix} = \int_{a}^{b} \frac{1}{\sigma\sqrt{2\pi}} e^{-(x-\mu)^2/2\sigma^2}\, dx$$

It is not possible to evaluate this integral using elementary functions; it can only be evaluated numerically. There are tables for the normal distribution with parameters

◀ **Figure 8.12**
The density of the normal distribution with mean μ and standard deviation σ.

$\mu = 0$ and $\sigma = 1$ that list values for

$$F(x) = \int_{-\infty}^{x} \frac{1}{\sqrt{2\pi}} e^{-z^2/2} \, dz$$

(A table for $F(x)$ is reproduced in the Appendix B.)

At the moment, we will only need the following three values. The area $A(k)$ under the density function of the normal distribution with mean μ and standard deviation σ between $\mu - k\sigma$ and $\mu + k\sigma$ for $k = 1, 2$, and 3 is

k	$A(k)$
1	68%
2	95%
3	99%

(see Figure 8.12). That is, if a certain quantity X of a population is normally distributed with mean μ and standard deviation σ, then 68% of the population fall within one standard deviation of the mean, 95% fall within two standard deviations of the mean, and 99% fall within three standard deviations of the mean. This property also explains the meaning of the standard deviation: it tells us how spread out the distribution of the quantity of interest is.

These percentages can also be interpreted in the following way. Suppose that a quantity X in a population is normally distributed with mean μ and standard deviation σ. If we sampled from this population, that is, if we picked one individual at random from this population, then there is a 68% chance that the observation would fall within one standard deviation of the mean. We can therefore say that the probability that X is in the interval $[\mu - \sigma, \mu + \sigma]$ is equal to 0.68. We write

$$P(X \in [\mu - \sigma, \mu + \sigma]) = 0.68$$

where P stands for probability. Likewise,

$$P(X \in [\mu - 2\sigma, \mu + 2\sigma]) = 0.95 \qquad \text{and} \qquad P(X \in [\mu - 3\sigma, \mu + 3\sigma]) = 0.99$$

Finding Probabilities Using the Mean and the Standard Deviation

EXAMPLE 1 Assume that a certain quantitative character X is normally distributed with mean $\mu = 4$ and standard deviation $\sigma = 1.5$. We make one observation; find an interval centered at the mean such that there is a 95% chance (99% chance) that the observation will fall into this interval.

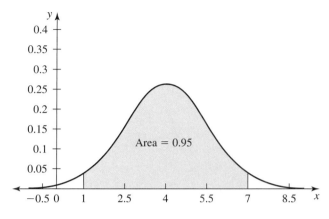

◀ **Figure 8.13**
The normal density with mean 4 and standard deviation 1.5 in Example 1: 95% of the observations fall into the interval [1, 7].

Solution   Since 95% corresponds to a range within two standard deviations of the mean (see Figure 8.13), the resulting interval is

$$[4 - (2)(1.5), 4 + (2)(1.5)] = [1, 7]$$

We can therefore write

$$P(X \in [1, 7]) = 0.95$$

Similarly, 99% corresponds to a range within three standard deviations of the mean, resulting in an interval of the form

$$[4 - (3)(1.5), 4 + (3)(1.5)] = [-0.5, 8.5]$$

We can therefore write

$$P(X \in [-0.5, 8.5]) = 0.99$$

▮

EXAMPLE 2   Assume that a certain quantitative character X is normally distributed with mean $\mu = 3$ and standard deviation $\sigma = 2$. We take a sample of size 1. What is the chance that we observe a value greater than 9?

Solution   Since $9 = 3 + (3)(2)$, we find that we want to know the chance that the observation is three standard deviations above the mean (see Figure 8.14). Now, 99% of the population is within three standard deviations of the mean; therefore, 1% is outside of the interval $[\mu - 3\sigma, \mu + 3\sigma]$. Since the density function of the normal distribution is symmetric about the mean, it follows that the area to the left of $\mu - 3\sigma$

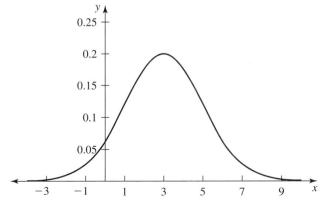

◀ **Figure 8.14**
The normal density with mean 3 and standard deviation 2 in Example 2: 0.5% of the observations are greater than 9.

and the area to the right of $\mu + 3\sigma$ are the same. Hence, there is a $(1\%)/2 = 0.5\%$ chance that the observation is above 9. We can therefore write

$$P(X > 9) = 0.005$$
■

EXAMPLE 3   Assume that a certain quantitative character X is normally distributed with parameters μ and σ. What is the probability that an observation lies below $\mu + \sigma$?

Solution   Since 68% of the population fall within one standard deviation of the mean, and since the density curve is symmetric about the mean, it follows that 34% of the population fall into the interval $[\mu, \mu + \sigma]$. Furthermore, because of the symmetry of the density function, 50% of the population are below the mean. Hence, $50\% + 34\% = 84\%$ of the population lie below $\mu + \sigma$, as illustrated in Figure 8.15. We can therefore write

$$P(X < \mu + \sigma) = 0.84$$
■

Finding Probabilities Using the Table   The table for a normal distribution with mean 0 and standard deviation 1 (see Appendix B) can be used to compute probabilities when the distribution is normal with mean μ and standard deviation σ.

We begin by explaining how to use the table for the normal distribution with mean 0 and standard deviation 1, called the **standard normal distribution**, whose density is given by

$$f(u) = \frac{1}{\sqrt{2\pi}} e^{-u^2/2} \quad \text{for } -\infty < u < \infty$$

The table lists values for

$$F(z) = \int_{-\infty}^{z} \frac{1}{\sqrt{2\pi}} e^{-u^2/2} \, du$$

Geometrically, $F(z)$ is the area to the left of the line $x = z$ under the graph of the density function, as illustrated in Figure 8.16.

Probabilistically, $F(z)$ is equal to the probability that an observation is to the left of z. For instance, when $z = 1$, we find $F(1) = 0.8413$; that is, the probability that an observation has a value less than or equal to 1, is equal to 0.8413. It also means that 84.13% of the population has a value less than or equal to 1.

As you can see, the table does not provide entries for negative values of z. To compute such values, we take advantage of the symmetries of the density function. For instance, if we wish to compute $F(-1)$, we see from the graph of the density

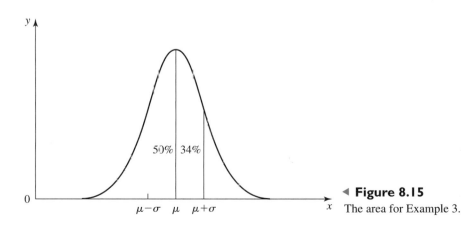

◄ **Figure 8.15**
The area for Example 3.

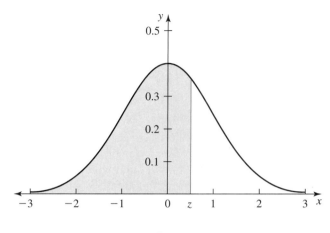

◀ **Figure 8.16**
The area to the left of $x = z$ under the graph of the normal density is $F(z)$, which is listed in the table for the normal distribution.

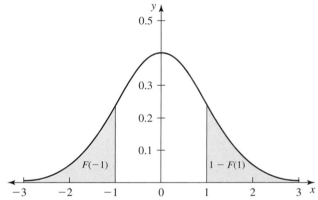

◀ **Figure 8.17**
The areas used to compute $F(-1)$.

function that the area to the left of -1 is the same as the area to the right of 1 (see Figure 8.17). We can thus write

$$F(-1) = \int_{-\infty}^{-1} \frac{1}{\sqrt{2\pi}} e^{-u^2/2}\, du = \int_{1}^{\infty} \frac{1}{\sqrt{2\pi}} e^{-u^2/2}\, du$$

$$= \int_{-\infty}^{\infty} \frac{1}{\sqrt{2\pi}} e^{-u^2/2}\, du - \int_{-\infty}^{1} \frac{1}{\sqrt{2\pi}} e^{-u^2/2}\, du$$

$$= 1 - F(1) = 1 - 0.8413 = 0.1587$$

Here we used the fact that the total area is equal to 1.

We can use the table to compute areas under the graph of a normal density with arbitrary mean μ and standard deviation σ; that is, this table will allow us to find the value of

$$\int_{a}^{b} \frac{1}{\sigma\sqrt{2\pi}} e^{-(x-\mu)^2/2\sigma^2}\, dx$$

where $-\infty < a < b < \infty$. If we use the substitution

$$u = \frac{x - \mu}{\sigma} \quad \text{with} \quad \frac{du}{dx} = \frac{1}{\sigma}$$

we find

$$\int_{a}^{b} \frac{1}{\sigma\sqrt{2\pi}} e^{-(x-\mu)^2/2\sigma^2}\, dx = \int_{(a-\mu)/\sigma}^{(b-\mu)/\sigma} \frac{1}{\sqrt{2\pi}} e^{-u^2/2}\, du$$

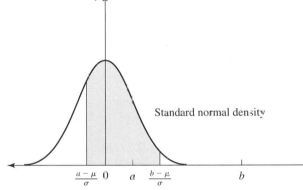

◀ **Figure 8.18**
The area under the normal density with mean μ and standard deviation σ between a and b is the same as the area under the standard normal density between $(a - \mu)/\sigma$ and $(b - \mu)/\sigma$.

We recognize the right-hand side as the area under the *standard normal density* between $(a - \mu)/\sigma$ and $(b - \mu)/\sigma$. That is, the area under the normal density with mean μ and standard deviation σ between a and b is the same as the area under the standard normal density between $(a - \mu)/\sigma$ and $(b - \mu)/\sigma$ (see Figure 8.18). We illustrate this in the following example.

EXAMPLE 4   Suppose that a quantity X is normally distributed with mean 3 and standard deviation 2. Find the fraction of the population that falls into the interval [2, 5], that is, find $P(X \in [2, 5])$.

Solution   To answer this question we must compute

$$\int_2^5 \frac{1}{2\sqrt{2\pi}} e^{(x-3)^2/8} \, dx \qquad (8.7)$$

Using the transformation $u = (x - 3)/2$, we find that when

$$x = 2 \quad \text{then } u = \frac{2-3}{2} = -\frac{1}{2}$$

and when

$$x = 5 \quad \text{then } u = \frac{5-3}{2} = 1$$

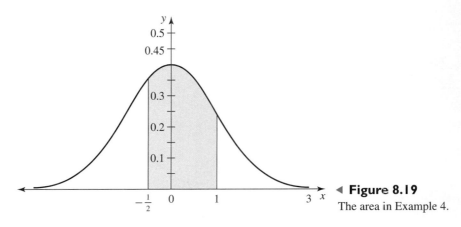

◀ **Figure 8.19**
The area in Example 4.

hence, the integral in (8.7) is equal to

$$\int_{-1/2}^{1} \frac{1}{\sqrt{2\pi}} e^{-u^2/2}\, du = \int_{-\infty}^{1} \frac{1}{\sqrt{2\pi}} e^{-u^2/2}\, du - \int_{-\infty}^{-1/2} \frac{1}{\sqrt{2\pi}} e^{-u^2/2}\, du$$

$$= F(1) - F\left(-\frac{1}{2}\right) = F(1) - \left(1 - F\left(\frac{1}{2}\right)\right)$$

$$= F(1) + F\left(\frac{1}{2}\right) - 1 = 0.8413 + 0.6915 - 1 = 0.5328$$

Hence, $P(X \in [2, 5]) = 0.5328$.

Instead of writing out these integrals, it is easier to determine what we need to compute when we sketch the relevant area under the standard normal curve, as done in Figure 8.19. We see from Figure 8.19 that we need to compute $F(1) - F\left(-\frac{1}{2}\right)$. Since $F\left(-\frac{1}{2}\right) = 1 - F\left(\frac{1}{2}\right)$, it follows that we need to find $F(1) - 1 + F\left(\frac{1}{2}\right)$, which we computed above. **I**

EXAMPLE 5   Suppose that a quantitative character X is normally distributed with mean 2 and standard deviation 1/2. Find x such that 30% of the population is above x.

Solution   We need to find x such that

$$\int_{x}^{\infty} \frac{1}{\sqrt{2\pi}/2} e^{-(u-2)^2/(2(1/2)^2)}\, du = 0.3$$

Using the transformation $z = (x - \mu)/\sigma$, and denoting by Z a quantity that is normally distributed with mean 0 and standard deviation 1, we find that

$$P(X > x) = P\left(\frac{X - \mu}{\sigma} > \frac{x - \mu}{\sigma}\right) = P\left(Z > \frac{x - 2}{1/2}\right)$$

$$= P(Z > 2(x - 2)) = 0.3$$

Now, $P(Z > z) = 1 - F(z) = 0.3$; hence, $F(z) = 0.7$. We find $F(0.52) = 0.6985 \approx 0.7$. Hence,

$$2(x - 2) = 0.52 \qquad \text{or} \qquad x = 2.26$$

That is, $P(X > 2.26) = 0.3$. Therefore, the value of x we are seeking is 2.26. **I**

A Notes On Samples   To obtain information about a quantity, such as size or bristle number, we cannot survey the entire population. Instead, we take a sample from the population and find the distribution of the quantity of interest in the sample. We must choose the sample so that it is representative of the population; this is a difficult problem, which we cannot discuss here. Even if we assume that the sample is representative of the population, it will still be the case that samples differ.

▮ EXAMPLE 6   The numbers in the following table represent two samples, each from the same population. They represent values for a quantity that is normally distributed, with mean $\mu = 0$ and standard deviation $\sigma = 1$.

Sample 1				
−1.633	0.542	0.250	−0.166	0.032
1.114	0.882	1.265	−0.202	0.151
1.151	−1.210	−0.927	0.425	0.290
−1.939	0.891	−0.227	0.602	0.873
0.385	−0.649	−0.577	0.237	−0.289

Sample 2				
−0.157	0.693	1.710	0.800	−0.265
1.492	−0.713	0.821	−0.031	−0.780
−0.042	1.615	−1.440	−0.989	−0.580
0.289	−0.904	0.259	−0.600	−1.635
0.721	−1.117	0.635	0.592	−1.362

Both samples are obtained from a table of random numbers that are normally distributed with mean 0 and standard deviation 1 (Beyer 1991).

a. Count the number of observations in each sample that fall below the mean $\mu = 0$, and compare the number to what you would expect based on properties of the normal distribution.

b. Count the number of observations in each sample that fall within one standard deviation of the mean, and compare the number to what you would expect based on properties of the normal distribution.

Solution

a. Since the mean is equal to 0, to find the number of observations that are below the mean, we simply count the number of observations that are negative. In the first sample, 10 observations are below the mean; in the second sample, 14 observations are below the mean. We expect that half of the sample points are below the mean. Since each sample is of size 25, we expect about 12 or 13 sample points to be below the mean.

b. Since the standard deviation is 1, we count the number of observations that fall into the interval $[-1, 1]$. In the first sample, there are 19 such observations; in the second sample, there are 18 such observations. To compare this with the theoretical value, note that 68% of the population falls within one standard deviation of the mean. Since the sample size is equal to 25 and $(0.68)(25) = 19.04$, we expect about 19 observations to fall into the interval $[-1, 1]$.   ▮

The preceding example illustrates an important point. Even if random samples are taken from the same population, they are not identical. For instance, in the preceding example we expect that half of the observations will be below the mean. In the first sample, less than half of the observations will be below the mean; whereas in the second sample, more than half of the observations are below the mean.

As the sample size increases, however, it will reflect the population increasingly better (at least, in all practical situations). That is, in order to determine the distribution of a quantitative character, such as the number of bristles in *D. melanogaster*, you would take a sample and find, for instance, the histogram associated with the quantity of interest. If the sample is large enough, the histogram will reflect the population distribution quite well. But if you repeat the experiment, you should not expect the two histograms to be exactly the same. If the sample size is large enough, however, they will be close.

◆ 8.4 PROBLEMS

1. Denote by

$$f(x) = \frac{1}{\sigma\sqrt{2\pi}} e^{-(x-\mu)^2/2\sigma^2}$$

for $-\infty < x < \infty$, the density of a normal distribution with mean μ and standard deviation σ.

(a) Show that $f(x)$ is symmetric about $x = \mu$.

(b) Show that the maximum of $f(x)$ is at $x = \mu$.

(c) Show that the inflection points of $f(x)$ are at $x = \mu - \sigma$ and $x = \mu + \sigma$.

(d) Graph $f(x)$ for $\mu = 2$ and $\sigma = 1$.

2. Suppose that $f(x)$ is the density function of a normal distribution with mean μ and standard deviation σ. Show that

$$\mu = \int_{-\infty}^{\infty} xf(x)\, dx$$

is the mean of this distribution. (*Hint:* Use substitution.)

3. Suppose that a quantitative character is normally distributed with mean $\mu = 12.8$ and standard deviation $\sigma = 2.7$. Find an interval centered at the mean such that 95% (99%) of the population falls into the interval.

4. Suppose a quantitative character is normally distributed with mean $\mu = 15.4$ and standard deviation $\sigma = 3.1$. Find an interval centered at the mean such that 95% (99%) of the population falls into this interval.

In Problems 5–10, assume that a quantitative character is normally distributed with mean μ and standard deviation σ. Determine what fraction of the population falls into the given intervals.

5. $[\mu, \infty)$ **6.** $[\mu - 2\sigma, \mu + \sigma]$

7. $(-\infty, \mu + 3\sigma]$ **8.** $[\mu + \sigma, \mu + 2\sigma]$

9. $(-\infty, \mu - 2\sigma]$ **10.** $[\mu - 3\sigma, \mu]$

11. Suppose that X is normally distributed with mean $\mu = 3$ and standard deviation $\sigma = 2$. Use the table in the appendix to find the following.

(a) $P(X \le 4)$ **(b)** $P(2 \le X \le 4)$

(c) $P(X > 5)$ **(d)** $P(X \le 0)$

12. Suppose that X is normally distributed with mean $\mu = -1$ and standard deviation $\sigma = 1$. Use the table in the appendix to find the following.

(a) $P(X > 0)$ **(b)** $P(0 < X < 1)$

(c) $P(-1.5 < X < 2.5)$ **(d)** $P(X > 1.5)$

13. Suppose that X is normally distributed with mean $\mu = 1$ and standard deviation $\sigma = 2$. Use the table in the appendix to find x such that the following hold.

(a) $P(X \le x) = 0.9$ **(b)** $P(X > x) = 0.4$

(c) $P(X \le x) = 0.4$ **(d)** $P(|X - 1| < x) = 0.5$

14. Suppose that X is normally distributed with mean $\mu = -2$ and standard deviation $\sigma = 1$. Use the table in the appendix to find x such that the following hold.

(a) $P(X \ge x) = 0.8$ **(b)** $P(X < 2x + 1) = 0.5$

(c) $P(X \le x) = 0.1$ **(d)** $P(|X - 2| > x) = 0.4$

15. Assume that the mathematics score X on the Scholastic Aptitude Test (SAT) is normally distributed with mean 500 and standard deviation 100.

(a) Find the probability that an individual's score exceeds 700.

(b) Find the math SAT score so that 10% of the students who took the test have that score or greater.

16. In a study of *Drosophila melanogaster* by Mackey (1984), the number of bristles on the fifth abdominal sternite in males was shown to follow a normal distribution with mean 18.7 and standard deviation 2.1.

(a) What percentage of the male population has fewer than 17 abdominal bristles?

(b) Find an interval centered at the mean so that 90% of the male population have bristle numbers that fall into this interval.

17. Suppose the weight of an animal is normally distributed with mean 3720 gr and standard deviation 527 gr. What percentage of the population has a weight that exceeds 5000 gr?

18. Suppose the height of an adult animal is normally distributed with mean 17.2 in. Find the standard deviation if 10% of the animals have a height that exceeds 19 in.

◆ 8.5 AGE-STRUCTURED POPULATIONS (OPTIONAL)

Integrals appear quite naturally in structured populations, as we saw in Section 8.2 when we encountered a size-structured model. We will now discuss the case of age structure, using ideas from Section 8.2.

We consider a population in which reproduction takes place continually over time; that is, there are no breeding seasons. Suppose that we wish to take a census of the number of newborns over time. We denote by $B_{[a,b]}$ the total number of newborns during the interval $[a, b]$. It is convenient to express this in integral form, writing

$$B_{[a,b]} = \int_a^b B(t) \, dt$$

The function $B(t)$ is called the **total birth rate** at time t and we can interpret $B(t) \, dt$ as the total number of newborns in the infinitesimal time interval $[t, t + dt)$. The contribution of females in the population to the total birth rate depends on their age. We can express this in integral form, namely

$$B(t) = \int_0^D \begin{bmatrix} \text{total birth rate at time } t \\ \text{of females of age } x \end{bmatrix} dx$$

where D is the maximum attainable age.

The following discussion can be somewhat simplified by choosing t large enough so that none of the females who were present at time 0 are still reproducing at time t.

To find the total birth rate at time t of females of age x, we must find the number of females who were born at time $t - x$ and survive until age x. Furthermore, we need to know the average birth rate of a female of age x. We denote by $l(x)$ the fraction of females who survive until age x and by $m(x)$ the average birth rate per female of age x. We assume that $l(x)$ and $m(x)$ depend only on x and not on t.

Since the number of females that were born in the interval $[t - x, t - x + dt)$ is $B(t - x) \, dt$, and the proportion of females who survive until age x is $l(x)$, we find that

$$\begin{bmatrix} \text{number of females of age } x \\ \text{in the interval } [t, t + dt) \end{bmatrix} = [B(t - x) \, dt] \, l(x)$$

Multiplying this by the average number of births per female whose age is in $[x, x + dx)$, namely $m(x) \, dx$, yields

$$\begin{bmatrix} \text{number of births during } [t, t + dt) \\ \text{to females with age in } [x, x + dx) \end{bmatrix} = [B(t - x) \, dt] \, l(x) \, [m(x) \, dx]$$

Dividing this by dt gives the rate of births at time t to females in the age class $[x, x + dx)$. Summing (that is, integrating) over all age classes then gives the total birth rate at time t for t large enough:

$$B(t) = \int_0^D B(t - x) l(x) m(x) \, dx \tag{8.8}$$

Equation (8.8) is an integral equation. This type of equation is typical for structured populations. The theory that goes along with these equations is not easy to follow, and we will not develop it here. The important part for us is to understand how such equations arise and to be able to interpret them when we see them.

Populations that evolve according to the age-structured model just described typically converge to an equilibrium over time, in which the age structure of the population is stable and the population shows a constant rate of change. This means that for sufficiently large t, $B(t)$ grows like e^{rt}, where the constant r is the intrinsic rate of growth or the Malthusian parameter that we have encountered in other population models. This parameter is of interest because it allows us to predict future population sizes. If we set $B(t) = e^{rt} B(0)$, then we obtain an equation that can be used to determine r from $l(x)$ and $m(x)$. (We will show subsequently how both $l(x)$ and $m(x)$ can be obtained experimentally.)

If $B(t) = e^{rt} B(0)$, then $B(t - x) = e^{r(t-x)} B(0)$. Plugging this into equation (8.8), we find

$$e^{rt} B(0) = \int_0^D e^{r(t-x)} B(0) l(x) m(x) \, dx$$

Dividing both sides by $e^{rt} B(0)$, we obtain

$$1 = \int_0^D e^{-rx} l(x) m(x) \, dx \tag{8.9}$$

This is called Euler's equation, and it is fundamental to the study of age-structured populations. If we know $l(x)$ and $m(x)$, then r is determined by Euler's equation. In practice, r is found by numerically solving equation (8.9). To illustrate how this is done in principle, we choose simple functions for $l(x)$ and $m(x)$ in the following calculation so that the integration can be carried out without resorting to numerical methods. (Solving Euler's equation will still require numerical methods.) We assume

$$l(x) = 1 \quad \text{for } 0 \leq x \leq 10$$
$$m(x) = m \quad \text{for } 2 \leq x \leq 5$$

Both $l(x)$ and $m(x)$ are otherwise equal to 0. This is certainly a rather unrealistic example; it means that all newborns survive until age 10 and then die. Furthermore, they reproduce at a constant rate between ages 2 and 5 but not outside of this range. (Evaluating the integral becomes rather difficult for realistic choices of $l(x)$ and $m(x)$.)

$$1 = \int_2^5 e^{-rx} \cdot 1 \cdot m \, dx$$
$$= \left[-\frac{m}{r} e^{-rx} \right]_2^5$$
$$= \frac{m}{r} (e^{-2r} - e^{-5r})$$

The equation

$$1 = \frac{m}{r} (e^{-2r} - e^{-5r})$$

can only be solved numerically. The following table shows pairs of m and r that satisfy this equation.

m	0.5	1	5	20
r	0.117	0.325	0.847	1.34

This table was obtained by setting

$$f(r) = \frac{m}{r} (e^{-2r} - e^{-5r}) - 1$$

for a given value of m and solving $f(r) = 0$ numerically. Recall that we learned two methods to solve equations of the form $f(r) = 0$, the bisection method and the Newton-Raphson method. We can also use a graphing calculator to solve $f(r) = 0$. Graph the function $f(r)$ for a given value of m; the graphing calculator can find roots of functions within the graphing menu. (You may need to consult the calculator handbook to find out how.)

After deriving Euler's equation, we simply said that if we know $l(x)$ and $m(x)$, then r can be computed. We will now discuss how one can find $l(x)$ and $m(x)$ experimentally.

We suppose that our population is described by a continuous time model. Measuring $l(x)$ and $m(x)$ is done by discretizing, a tool we have used previously. We partition the age interval into subintervals. As you can probably guess, making the subintervals smaller gives a better approximation, provided that the sample size is large enough.

To simplify the argument, we choose all subintervals of equal lengths. Suppose that the maximum age is D, then the age interval is $[0, D]$. We divide this into n subintervals of equal length $\Delta x_k = D/n$. We represent each subinterval $[x_{k-1}, x_k)$ by its midpoint $c_k = (x_{k-1} + x_k)/2$. Then, using Riemann sums, we find

$$B(t) = \lim_{n \to \infty} \sum_{k=1}^{n} B(t - c_k) l(c_k) m(c_k) \Delta x_k$$

With $\Delta x_k = D/n$, we can approximate $B(t)$ by

$$B(t) \approx \sum_{k=1}^{n} B(t - c_k) l(c_k) m(c_k) \frac{D}{n} \tag{8.10}$$

To find the intrinsic rate of change, r, we discretize Euler's equation,

$$1 = \int_0^D e^{-rx} l(x) m(x)\, dx$$

$$= \lim_{n \to \infty} \sum_{k=1}^{n} e^{-rc_k} l(c_k) m(c_k) \Delta x_k$$

The number $l(c_k)$ is the fraction of individuals that survive to age c_k. We denote this quantity by l_k. The product $m(c_k) \Delta x_k$ is approximately the number of offspring per female produced during a time interval of length Δx_k, in which the female is of age c_k in the middle of the time interval. We denote the product $m(c_k) \Delta x_k$ by m_k. We therefore obtain the approximate Euler's equation

$$1 \approx \sum_{k=1}^{n} e^{-rc_k} l_k m_k$$

We will now discuss how to find l_k and m_k. Of course, the smaller Δx_k, the better the approximation. In the following example, Δx_k is not small, so the approximation is probably not good, but the point is to simply illustrate the procedure.

In sufficiently short-lived organisms, we can follow a **cohort**; that is, we start with a group of newborns at time 0 and follow them through time, taking censuses at certain times. The following table contains data from a fictitious population that evolves according to a continuous-time model. The population consists of 10,000 organisms whose maximum age is 10 months, for whom reproduction occurs continually over time. We divide the interval $[0, 10]$ into five subintervals, each of length 2. We count the number of individuals alive at the midpoint c_i of each subinterval and count the number of female offspring during each of the subintervals.

time interval	[0,2)	[2,4)	[4,6)	[6,8)	[8,10]
no. of individuals alive	8,000	1,000	600	300	0
no. of births during interval	0	0	9000	1500	0

We can compute the values of l_k and m_k from this table. The quantity l_k is the proportion of the original population that is still alive at the midpoint c_k of the corresponding subinterval. The quantity m_k is the number of female offspring produced per surviving female in the corresponding subinterval. We find the following.

k	c_k	l_k	m_k	$l_k m_k$
1	1	0.8	0	0
2	3	0.10	0	0
3	5	0.06	15	0.9
4	7	0.03	5	0.15
5	9	0.00	0	0

Here is a verbal explanation of both tables. We start with a population of 10,000 female newborns at time 0. 8000 females are alive at time $t = 1$, the midpoint of $[0, 2)$; this is 80% of the original population, hence $l_1 = 0.8$. There are no offspring of the individuals in the population during $[0, 2)$. The next time we take a census (time $t = 3$), there are 1000 females left; this is 10% of the original population, which gives $l_2 = 0.1$. None of the females has started reproducing. At time $t = 5$, there are 600 females left; this is 6% of the original population, which gives $l_3 = 0.06$. We count the number of female offspring produced during this time interval and find 9000 female offspring. To find m_k, we divide 9000 by 600. This is certainly an approximation, because the number of females is declining over time. We find $m_3 = 15$. The last column contains the product $l_k m_k$. The entries for stages 4 and 5 are obtained similarly.

We will now show how to find r, the intrinsic rate of growth. Our intervals were each of length 2, and the midpoints c_i are listed in the preceding. The approximate Euler's equation is then

$$1 \approx e^{-5r}(0.9) + e^{-7r}(0.15)$$

As before, we can find r by solving this equation numerically, and we find $r = 0.0092$.

◆ 8.5 PROBLEMS

1. Let $B_{[0,s]}$ be the total number of newborns during the time interval $[0, s]$. If $B(s) = e^{3s} - 1, s \geq 0$, find the total birth rate at time t, $t \in [0, s]$.

2. Assume that the total birth rate at time t is $B(t) = 3t$, $0 \leq t \leq 5$. Find the total number of newborns during the time interval $[0, 5]$.

3. Assume that the total birth rate at time t of females of age x is $x(5 - x), 0 \leq x \leq 5$. Find the total birth rate at time t.

4. Assume that the total birth rate at time t of females of age x is given by a function $g(x), x \geq 0$.

(a) If D is the maximum attainable age, what can you say about $g(x)$ for $x > D$?

(b) Can $g(x)$ be negative anywhere in the interval $[0, D]$?

(c) Assume that juveniles and old females cannot reproduce. How is this reflected in the graph of $f(x)$?

(d) Interpret the total birth rate at time t geometrically in terms of the function $g(x)$.

5. Consider a population of total size N in which the age of individuals is between 0 and D. Denote by $N_{[a,b]}$ the number of individuals whose age is in the interval $[a, b]$.

(a) Explain the meaning of $N(x)\,dx$ in the integral representation of

$$N_{[a,b]} = \int_a^b N(x)\,dx$$

(b) Use the integral representation in (a) to express the total number of individuals in the population.

6. Consider a population of total size N in which the age of individuals is between 0 and D. Denote by $N_{[a,b]}$ the number of individuals whose age is in the interval $[a, b]$.

(a) Explain why $N_{[0,x]}$ must be a nondecreasing function of x for $x \geq 0$.

(b) Explain the total size of the population in terms of $N_{[0,x]}$.

(c) Assume that $N_{[0,x]}$ has the integral interpretation

$$N_{[0,x]} = \int_0^x N(s)\,ds$$

and $N_{[0,x]}$ is concave down. (i) Can $N(s)$ be ever negative? (ii) What can you say about the sign of $N'(s)$?

7. Suppose that

$$l(x) = \begin{cases} 1 & \text{for } 0 \leq x \leq 2 \\ 0 & \text{otherwise} \end{cases}$$

and

$$m(x) = \begin{cases} m & \text{for } 0 \leq x \leq 2 \\ 0 & \text{otherwise} \end{cases}$$

where m is a positive constant. The functions $l(x)$ and $m(x)$ are as introduced in Section 8.3.

(a) Find Euler's equation.

(b) For which values of m is the population size increasing?

8. Suppose that

$$l(x) = \begin{cases} 1 - \frac{1}{10}x & \text{for } 0 \leq x \leq 10 \\ 0 & \text{otherwise} \end{cases}$$

and

$$m(x) = \begin{cases} 3 & \text{for } 1 \leq x \leq 6 \\ 0 & \text{otherwise} \end{cases}$$

where the functions $l(x)$ and $m(x)$ are as introduced in Section 8.3.

(a) Graph $l(x)$ and $m(x)$, and explain in words what $l(x)$ and $m(x)$ mean in this case.

(b) Suppose that at $t = 0$ there are 1000 newborn females. What percentage of this cohort is left after 2, 5, and 7 years?

(c) The quantity

$$R = \int_0^D l(x)m(x)\,dx$$

denotes the reproductive value or the average number of female offspring produced by a female during her lifetime. Find R.

(d) Use Euler's equation to find r, the intrinsic rate of growth, numerically.

9. Suppose that we follow a cohort over a period of 8 months. At time 0 there are 5000 individuals. We divide the 8 months into four equal subinterval and take measurements at the midpoints of the respective subintervals (that is, at times $t = 1, 3, 5,$ and 7), and find that the number of individuals left at these times are 4000, 2000, 200, and 0, respectively. We also count the number of offspring in each subinterval and find 0, 6000, 400, and 0, respectively. Use these data to obtain an approximation for r, the intrinsic rate of growth.

10. Suppose that we follow a cohort over a period of 10 months. At time 0 there are 4000 individuals. We divide the 10 months into five equal subinterval and take measurements at the midpoints of the respective subintervals (that is, at times $t = 1, 3, 5, 7,$ and 9), and find that the number of individuals left at these times are 3000, 1800, 500, 100, and 0, respectively. We also count the number of offspring in each subinterval and find 0, 1000, 3000, 200, and 0, respectively. Use these data to obtain an approximation for r, the intrinsic rate of growth.

◆ 8.6 KEY TERMS

Chapter 8 Review Topics Discuss the following definitions and concepts:

1. Rectification of curves
2. Length of a curve
3. Arc length differential
4. Density function
5. Histogram
6. Average value
7. Relative frequency
8. Normal distribution
9. Mean and standard deviation
10. Age-structured model
11. Euler's equation

◆ 8.7 REVIEW PROBLEMS

1. The scarlet monkey flower (*Mimulus cardinalis*) is pollinated by hummingbirds. The following table contains the number of pollinator visits per plant over a period of three hours in a sample of 25 plants.

1	3	0	0	4
2	1	1	0	1
5	2	1	0	1
3	1	0	0	1
1	0	2	4	0

(a) Construct a histogram from this data.

(b) Find the frequency distribution.

(c) Compute the average number of pollinator visits per plant for this sample.

2. Suppose that the amount of time T that a butterfly spends feeding on the nectar of a flower is a random variable whose density is given by

$$f(x) = 3e^{-3x} \quad \text{for } x > 0$$

(a) Show that $f(x)$ is a density function.

(b) Compute the mean of T.

3. Suppose that the height of a certain plant is normally distributed with mean 12.3 in. and standard deviation 2.1 in. Find the probability that the height of a randomly selected plant falls into the interval [9, 13].

4. Suppose that the cholesterol level in a certain section of the population is normally distributed with mean 187 mg/dl and standard deviation 37 mg/dl. Find the probability that the cholesterol level of a randomly selected individual in this population is greater than 200.

5. (*Truncation selection*) Suppose that $f(x)$ is the density function of a trait that is normally distributed with mean μ and standard deviation σ. Truncation selection allows breeding only between individuals whose trait exceeds value T.

(a) The density of the population after truncation is

$$f_S(x) = \begin{cases} \frac{f(x)}{B} & \text{for } x > T \\ 0 & \text{for } x \leq T \end{cases}$$

Express the normalizing constant B in terms of $f(x)$.

(b) The average value of the trait in the population after truncation is

$$\mu_S = \int_{-\infty}^{\infty} x f_S(x)\, dx$$

Show that

$$\mu_S = \frac{1}{B} \int_T^{\infty} x f(x)\, dx$$

(c) Assume that $\mu = 0$ and $\sigma = 1$. Show that

$$\int_T^{\infty} x e^{-x^2/2} dx = e^{-T^2/2}$$

and use this to derive the relationship

$$\mu_S = \frac{f(T)}{B}$$

Use the graph of $f(x)$ to give a geometric interpretation of μ_S (that is, find the quantities $f(T)$ and B on the graph).

(d) For arbitrary values of μ and σ, show that

$$\frac{\mu_S - \mu}{\sigma^2} = \frac{f(T)}{B}$$

6. In a study of *Drosophila melanogaster* by Mackey (1984), the number of bristles on the fifth abdominal sternite in males was shown to follow a normal distribution with mean 18.7 and standard deviation 2.1. Find the probability that a randomly selected male from this population has more than 22 abdominal bristles.

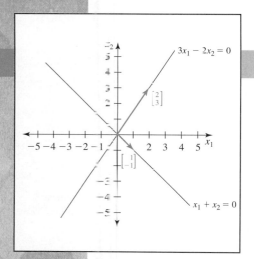

Linear Algebra and Analytic Geometry

◆ 9.1 LINEAR SYSTEMS

Two different species of insects are reared together in a laboratory cage. They are supplied with two different types of food each day. Each individual of species 1 consumes 5 units of food A and 3 units of food B, whereas each individual of species 2 consumes 2 units of food A and 4 units of food B, on average, per day. Each day, 900 units of food A and 960 units of food B are supplied. How many of each species can be reared together?

To solve such a problem, we will set up a system of equations. If

$$x = \text{number of individuals of species 1}$$

$$y = \text{number of individuals of species 2}$$

then the following two equations must be satisfied:

food A: $5x + 2y = 900$

food B: $3x + 4y = 960$

We refer to these two equations as a system of two linear equations in two variables. This section is devoted to finding solutions to such systems.

9.1.1 Graphical Solution

In this subsection, we restrict ourselves to systems of two linear equations in two variables. Recall that the standard form of a linear equation in two variables is

$$Ax + By = C$$

where A, B, and C are constants, A and B are not both equal to 0, and x and y are the two variables; its graph is a straight line (see Figure 9.1). Any point (x, y) on this straight line satisfies (or solves) the equation $Ax + By = C$. We can extend this to more than one equation, in which case we talk about a **system** of linear equations.

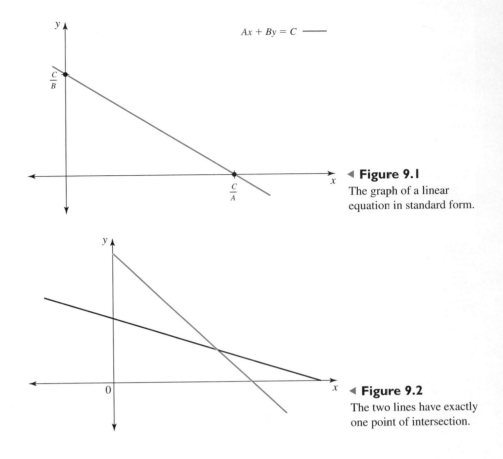

◀ Figure 9.1
The graph of a linear
equation in standard form.

◀ Figure 9.2
The two lines have exactly
one point of intersection.

In the case of two linear equations in two variables, the system is of the form

$$Ax + By = C$$
$$Dx + Ey = F$$

(9.1)

where A, B, C, D, E, and F are constants, and x and y are the two variables. (We require that A and B (D and E, respectively) are not both equal to 0.) When we say that we "solve" (9.1) for x and y, we mean that we find an ordered pair (x, y) that satisfies *each* equation of the system (9.1). Because each equation in (9.1) describes a straight line, we are therefore asking for the point of intersection of these two lines. The following three cases are possible.

1. The two lines have exactly one point of intersection. In this case, the system (9.1) has exactly one solution, as illustrated in Figure 9.2.

2. The two lines are parallel and do not intersect. In this case, the system (9.1) has no solution, as illustrated in Figure 9.3.

3. The two lines are parallel and intersect (that is, they are identical). In this case, the system (9.1) has infinitely many solutions, namely each point on the line, as illustrated in Figure 9.4.

Exactly One Solution

▮ **EXAMPLE 1**   Find the solution of

$$2x + 3y = 6$$
$$2x + y = 4$$

(9.2)

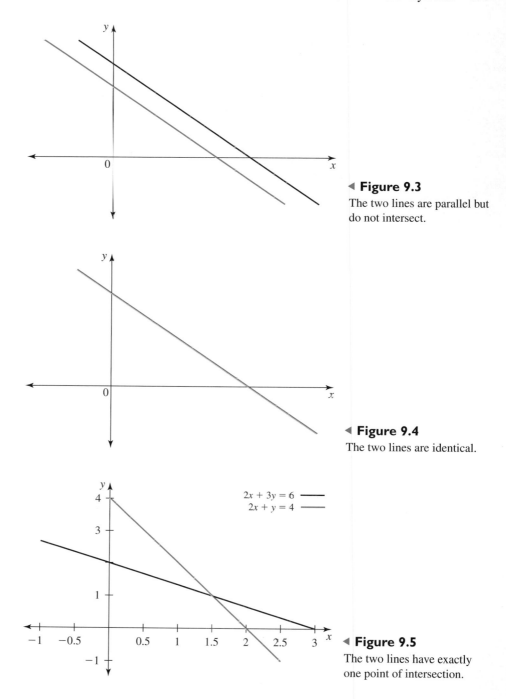

◀ **Figure 9.3**
The two lines are parallel but do not intersect.

◀ **Figure 9.4**
The two lines are identical.

◀ **Figure 9.5**
The two lines have exactly one point of intersection.

Solution   The line corresponding to $2x + 3y = 6$ has y-intercept $(0, 2)$ and x-intercept $(3, 0)$; the line corresponding to $2x + y = 4$ has y-intercept $(0, 4)$ and x-intercept $(2, 0)$ (see Figure 9.5). To find the solution of the linear system (9.2), we need to find the point of intersection of the two lines. Solving each equation for y produces the new set of equations

$$y = 2 - \frac{2}{3}x$$
$$y = 4 - 2x$$

Setting the right-hand sides equal to each other, we find

$$2 - \frac{2}{3}x = 4 - 2x$$

$$\frac{4}{3}x = 2$$

$$x = \frac{3}{2}$$

To find y, we substitute the value of x back into one of the two original equations, say $2x + y = 4$ (it does not matter which one you choose). Then,

$$y = 4 - 2x = 4 - (2)\left(\frac{3}{2}\right) = 1$$

The solution is therefore the point $(3/2, 1)$.

Looking at Figure 9.5, we see that the two lines have exactly one point of intersection. This is the case since the two lines have different slopes, namely $-2/3$ and -2, respectively. The point of intersection corresponds to the solution of (9.2). The system (9.2) therefore has exactly one solution, namely $(3/2, 1)$. ▌

No Solution

▌**EXAMPLE 2**   Solve

$$\begin{aligned} 2x + \ y &= 4 \\ 4x + 2y &= 6 \end{aligned} \tag{9.3}$$

Solution   The line corresponding to $2x + y = 4$ has y-intercept $(0, 4)$ and x-intercept $(2, 0)$; the line corresponding to $4x + 2y = 6$ has y-intercept $(0, 3)$ and x-intercept $(3/2, 0)$ (see Figure 9.6). Since

$$2x + y = 4 \iff y = 4 - 2x$$

$$4x + 2y = 6 \iff y = 3 - 2x$$

we immediately see that both lines have the same slope, namely -2, but different y-intercepts, namely 4 and 3, respectively. This implies that the two lines are parallel and do not intersect.

Let's see what happens when we solve the system (9.3). We equate the two equations $y = 4 - 2x$ and $y = 3 - 2x$:

$$4 - 2x = 3 - 2x$$

$$4 = 3$$

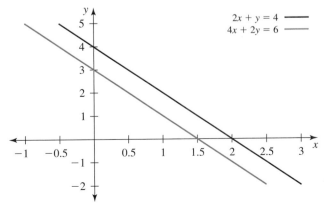

▲ **Figure 9.6**

The two lines are parallel but do not intersect.

which is obviously wrong; that is, there is no point (x, y) that satisfies both equations in (9.3) simultaneously.

Looking at Figure 9.6, we see that the two lines are parallel. Since parallel lines that are not identical do not intersect, it should be clear that (9.3) has no solution. In this case, we write the solution as $\emptyset$, the symbol for the empty set. ▌

Infinitely Many Solutions

▌**EXAMPLE 3**   Solve

$$2x + \;\; y = 4$$
$$4x + 2y = 8$$

(9.4)

Solution   If we divide the second equation by 2, we find that both equations are identical, namely $2x + y = 4$. That is, both equations describe the same line with x-intercept $(2, 0)$ and y-intercept $(0, 4)$, as shown in Figure 9.7. Every point (x, y) on this line is therefore a solution of (9.4). To see what such a solution looks like, we solve (9.4) algebraically. Using the same procedure as in Examples 1 and 2, we first solve each equation for y. This produces

$$2x + y = 4 \iff y = 4 - 2x$$

$$4x + 2y = 8 \iff y = 4 - 2x$$

Equating the two equations $y = 4 - 2x$ and $y = 4 - 2x$ yields

$$4 - 2x = 4 - 2x$$

$$0 = 0$$

This is a true statement for any value of x, which implies that any value of x is a solution. We introduce a new variable, namely t, to denote the x-coordinate. That is, $x = t$ for $t \in \mathbf{R}$, is the x-coordinate of the solution. To find the corresponding y-coordinate, we compute

$$y = 4 - 2x = 4 - 2t$$

The solution is therefore given by the set of points

$$\{(t, 4 - 2t) : t \in \mathbf{R}\}$$

This shows that the system (9.4) has infinitely many solutions, as expected from the graphical considerations (Figure 9.7 shows that the two lines representing (9.4) are identical). ▌

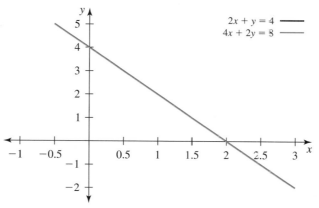

◀ **Figure 9.7**
The two lines are identical.

In Example 3, we introduced a new variable, namely t, to describe the set of solutions. We call t a **dummy variable**; it stands for any real number. Introducing a dummy variable is a convenient way to describe the set of solutions when there are infinitely many.

A Solution Method   The graphical illustration is only useful when we restrict ourselves to systems in two unknowns. Also, when we solved each system algebraically, we solved the equations for y and then equated the two resulting equations to find x; this method does not work when we solve larger systems. In the next subsection, we will discuss systems of m linear equations in n variables. We will therefore need to develop a systematic way of solving systems of linear equations that will work for any size system. The basic strategy will be to transform the system of linear equations into a new system of equations that has the same solutions as the original. The new system is called an **equivalent system**. It will be of a simpler form, so that we can solve for the unknown variables one by one and thus arrive at a solution. We illustrate this in the following example. We will label all equations with labels of the form (R_i), where i is an integer and the letter R stands for "row." This will allow us to keep track of our computations.

▮ **EXAMPLE 4**   Solve

$$3x + 2y = 8 \quad (R_1)$$
$$2x + 4y = 5 \quad (R_2)$$

Solution   There are two basic operations that transform a system of linear equations into an equivalent system. (1) We can multiply an equation by a nonzero number, and (2) can add one equation to the other.

Our goal will be to eliminate x in the second equation. If we multiply the first equation by 2 and the second equation by -3, we obtain

$$2(R_1) \quad\quad 6x + 4y = 16$$
$$-3(R_2) \quad -6x - 12y = -15$$

If we add the two equations, we find

$$-8y = 1$$

We thus eliminated x from the second equation. We can replace the original system of equations by a new (equivalent) system by leaving the first equation unchanged and replacing the second equation by $-8y = 1$. We obtain

$$(R_1) \quad 3x + 2y = 8 \quad (R_3)$$
$$2(R_1) - 3(R_2) \quad\quad -8y = 1 \quad (R_4)$$

We can now successively solve the system. It follows from equation (R_4) that

$$y = -\frac{1}{8}$$

Substituting this into equation (R_3), we find

$$3x + (2)\left(-\frac{1}{8}\right) = 8$$

$$3x = 8 + \frac{1}{4}$$

$$x = \frac{11}{4}$$

The solution is therefore $(11/4, -1/8)$.   ▮

When we look at equations (R_3) and (R_4), the left hand side has the shape of a triangle

$$*x + *y = *$$
$$*y = *$$

This is called **upper triangular** form. In the preceding example, we reduced the system of linear equations to upper triangular form and then solved it by back-substitution. As we will see in the next subsection, this method can be generalized to larger systems of equations.

9.1.2 Solving Systems of Linear Equations

In this subsection, we will extend the solution method of Example 4 to systems of m equations in n variables. We write such systems in the following form:

$$
\begin{aligned}
a_{11}x_1 + a_{12}x_2 + \cdots + a_{1n}x_n &= b_1 \\
a_{21}x_1 + a_{22}x_2 + \cdots + a_{2n}x_n &= b_2 \\
&\cdots\cdots\cdots\cdots\cdots\cdots\cdots\cdots \\
a_{m1}x_1 + a_{m2}x_2 + \cdots + a_{mn}x_n &= b_m
\end{aligned}
\tag{9.5}
$$

The variables are now $x_1, x_2, \ldots, x_n$. The coefficients a_{ij} on the left hand side have two subscripts. The first subscript indicates the equation and the second subscript to which x_j it belongs. For instance, you would find a_{43} in the fourth equation as the coefficient of x_3. Using double subscripts is a convenient way of labeling the coefficients. The subscript of the coefficients b_i indicate the equation.

We will transform this system into an equivalent system in upper triangular form. (Recall that "equivalent" means that the new system has the same solutions as the old system.) To do so we will use the following three basic operations:

1. Multiplying an equation by a nonzero constant
2. Adding one equation to another
3. Rearranging the order of the equations

This method is also called **Gaussian elimination**.

As in the case of a linear system with two equations in two variables, the general system (9.5) may have

1. Exactly one solution
2. No solutions
3. Infinitely many solutions

When a system has no solutions, we say that the system is **inconsistent**.

Exactly One Solution

∎ EXAMPLE 5   Solve

$$
\begin{aligned}
3x + 5y - z &= 10 \quad (R_1) \\
2x - y + 3z &= 9 \quad (R_2) \\
4x + 2y - 3z &= -1 \quad (R_3)
\end{aligned}
$$

Solution   Our goal is to reduce this system to upper triangular form. The first step is to eliminate the elements below $3x$ in the first column, that is, we wish to eliminate $2x$ from the second equation and $4x$ from the third equation. We leave the first equation unchanged. Multiplying the first equation by 2 and the second equation by -3 and then adding the two equations yields a new equation that replaces the second equation. In the new equation the term involving x has been eliminated. That is, we add

$$
\begin{array}{ll}
2(R_1) & 6x+ 10y- 2z = 20 \\
-3(R_2) & -6x+ 3y- 9z = -27
\end{array}
$$

which yields

$$13y - 11z = -7$$

We transform the third equation by multiplying the second equation by 2 and the third equation by -1 and then adding the two equations. In the resulting equation, the term involving x has been eliminated. That is, we add

$$
\begin{array}{ll}
2(R_2) & 4x- 2y+ 6z = 18 \\
-(R_3) & -4x- 2y+ 3z = 1
\end{array}
$$

which yields

$$-4y + 9z = 19$$

We thus transform the original set of equations into the following equivalent set of equations.

$$
\begin{array}{llll}
& (R_1) & 3x+ 5y- z = 10 & (R_4) \\
2(R_1) - 3(R_2) & & 13y- 11z = -7 & (R_5) \\
2(R_2) - (R_3) & & -4y+ 9z = 19 & (R_6)
\end{array}
$$

Our next step is to eliminate y from (R_6). To do this, we multiply (R_5) by 4 and (R_6) by 13, and then add the two equations. That is, we add

$$
\begin{array}{ll}
4(R_5) & 52y- 44z = -28 \\
13(R_6) & -52y+ 117z = 247
\end{array}
$$

which yields

$$73z = 219$$

We leave the first two equations unchanged. Our new (equivalent) system of equations is thus

$$
\begin{array}{llll}
(R_4) & 3x+ 5y- z = 10 & (R_7) \\
(R_5) & 13y- 11z = -7 & (R_8) \\
4(R_5) + 13(R_6) & 73z = 219 & (R_9)
\end{array} \tag{9.6}
$$

The systems of equations is now in upper triangular form, as can be seen from its shape. It is now easy to find the solution by back-substitution. Equation (R_9) yields

$$z = \frac{219}{73} = 3$$

Substituting this into (R_8) and solving for y yields

$$y = \frac{1}{13}(-7 + (11)(3)) = \frac{26}{13} = 2$$

Substituting both y and z into (R_7), we find

$$x = \frac{1}{3}(10 - 5y + z)$$

$$= \frac{1}{3}(10 - (5)(2) + 3) = 1$$

Hence the solution is $x = 1$, $y = 2$, and $z = 3$. ▌

A comment on the various steps in Example 5: We eliminated the terms involving x in (R_2) and (R_3); typically, we have a choice in how to do this. We used equations (R_1) and (R_2) to replace (R_2), and used (R_2) and (R_3) to replace (R_3). We could have used (R_2) and (R_3) to replace (R_2), but then we could not have used (R_2) and (R_3) again to replace (R_3). This would have resulted in two identical equations. To replace (R_3), we would then have had to choose (R_1) and (R_3). When we replaced (R_6), we had to use (R_5) and (R_6). If we had used (R_6) and (R_4), we would have introduced the variable x into the resulting equation and then the system would not have been reduced to upper triangular form.

No Solution

▮ EXAMPLE 6   Solve

$$2x - y + z = 3 \quad (R_1)$$
$$4x - 4y + 3z = 2 \quad (R_2)$$
$$2x - 3y + 2z = 1 \quad (R_3)$$

Solution   As in Example 5, we try to reduce the system to upper triangular form. We begin by eliminating terms involving x from the second and third equation. We leave the first equation unchanged. We replace the second equation by the result of subtracting the second equation from the first, after multiplying the first equation by 2. We replace the third equation by the result of subtracting the third equation from the first. We obtain

$$\begin{array}{lll} (R_1) & 2x - y + z = 3 & (R_4) \\ 2(R_1) - (R_2) & 2y - z = 4 & (R_5) \\ (R_1) - (R_3) & 2y - z = 2 & (R_6) \end{array}$$

To obtain (R_5), we also could have computed $(R_2) - 2(R_1)$; we would have found $-2y + z = -4$. Similarly, for (R_6), we could have computed $(R_3) - (R_1)$. As long as we do not use the same pair of equations twice, we have some freedom in how we reduce the system to upper triangular form.

Now, let's continue with our calculations. To eliminate terms involving y from Equation (R_6), we replace (R_6) by the difference between (R_5) and (R_6). We keep the first two equations. Then,

$$\begin{array}{lll} (R_4) & 2x - y + z = 3 \\ (R_5) & 2y - z = 4 \\ (R_5) - (R_6) & 0 = 2 \end{array}$$

The last equation, $0 = 2$, is a wrong statement, which means that this system does not have a solution. We write the solution as $\emptyset$, the symbol denoting the empty set. ▮

Infinitely Many Solutions

▮ EXAMPLE 7   Solve

$$x - 3y + z = 4 \quad (R_1)$$
$$x - 2y + 3z = 6 \quad (R_2)$$
$$2x - 6y + 2z = 8 \quad (R_3)$$

Solution   We proceed as in the preceding examples. The first step is to eliminate terms involving x from the second and third equation. We leave the first equation unchanged. We replace the second equation by the difference between the first and the second equation. We replace the third equation by the difference between the first and the third equation after dividing the third equation by 2. We find

$$(R_1) \quad x - 3y + z = 4 \quad (R_4)$$
$$(R_2) - (R_1) \qquad y + 2z = 2 \quad (R_5)$$
$$(R_1) - \tfrac{1}{2}(R_3) \qquad 0z = 0 \quad (R_6)$$

The third equation, $0z = 0$, is a correct statement. It means that we can substitute any number for z and still obtain a correct result. We introduce the dummy variable t, and set

$$z = t \quad \text{for } t \in \mathbf{R}$$

If we substitute this into equation (R_5), we find

$$y = 2 - 2t$$

If we substitute y and z into equation (R_4), we find

$$x = 4 + 3y - z = 4 + 3(2 - 2t) - t$$
$$= 4 + 6 - 6t - t = 10 - 7t$$

The solution is therefore the set

$$\{(x, y, z) : x = 10 - 7t, \ y = 2 - 2t, \ z = t, \ \text{for } t \in \mathbf{R}\} \qquad ▮$$

A Shorthand Notation   When we transform a system of linear equations, we make changes only to the coefficients of the variables. It is therefore convenient to introduce notation that will simply keep track of all the coefficients. This motivates the following definition.

> **Definition**   A **matrix** is a rectangular array of numbers
>
> $$A = \begin{bmatrix} a_{11} & a_{12} & \dots & a_{1n} \\ a_{21} & a_{22} & \dots & a_{2n} \\ \dots & \dots & \dots & \dots \\ a_{m1} & a_{m2} & \dots & a_{mn} \end{bmatrix}$$
>
> The elements a_{ij} of the matrix A are called **entries**. If the matrix has m rows and n columns, it is called an $m \times n$ matrix.

We can think of the matrix A in the definition as the **coefficient matrix** of the linear system (9.5). The entries a_{ij} of the matrix have two subscripts: the entry a_{ij} is located in the ith row and the jth column. You can think of this numbering system as the street address of the entries, using streets and avenues. Suppose that avenues go north-south and streets go east-west; finding the corner of 2nd Street and 3rd Avenue would be like finding element a_{23} in a matrix.

If a matrix has the same number of rows as columns, it is called a **square matrix**. An $m \times 1$ matrix is called a **column vector** and a $1 \times n$ matrix is called a **row vector**. Examples follow of a 3×3 square matrix, a 3×1 column vector and a 1×4 row vector, which also explain their names.

$$\begin{bmatrix} 1 & 3 & 0 \\ 0 & 1 & -1 \\ 5 & 4 & 3 \end{bmatrix}, \qquad \begin{bmatrix} 2 \\ 7 \\ 4 \end{bmatrix}, \qquad [1 \quad 3 \quad 0 \quad 5]$$

If A is a square matrix, then the **diagonal line** of A consists of the elements $a_{11}, a_{22}, \ldots, a_{nn}$. In the above 3×3 square matrix, the diagonal line thus consists of the elements 1, 1, and 3.

To solve systems of linear equations using matrices, we introduce the **augmented matrix**. The augmented matrix is the coefficient matrix of the linear system (9.5), augmented by an additional column representing the right hand side of (9.5). The augmented matrix representing the linear system (9.5) is therefore

$$
\left[
\begin{array}{cccc|c}
a_{11} & a_{12} & \cdots & a_{1n} & b_1 \\
a_{21} & a_{22} & \cdots & a_{2n} & b_2 \\
& \cdots\cdots\cdots & & & \cdots \\
a_{m1} & a_{m2} & \cdots & a_{mn} & b_m
\end{array}
\right]
$$

So, how can we use augmented matrices when solving systems of linear equations?

▮ **EXAMPLE 8**   Find the augmented matrix for the system given in Example 5, and solve the system using the augmented matrix.

Solution   The augmented matrix is given by

$$
\begin{array}{c}
\\
\\
\\
\end{array}
\left[
\begin{array}{ccc|c}
3 & 5 & -1 & 10 \\
2 & -1 & 3 & 9 \\
4 & 2 & -3 & -1
\end{array}
\right]
\begin{array}{c}
(R_1) \\
(R_2) \\
(R_3)
\end{array}
$$

Our goal is to transform the augmented matrix into a simpler form, also called upper triangular form, in which all the entries below the diagonal line are zero. We use the same basic transformations as before, and you should compare the following to the steps in Example 5. We indicate on the left side of the matrix the transformation that we used, and relabel the rows on the right side of the matrix.

$$
\begin{array}{c}
(R_1) \\
2(R_1) - 3(R_2) \\
2(R_2) - (R_3)
\end{array}
\left[
\begin{array}{ccc|c}
3 & 5 & -1 & 10 \\
0 & 13 & -11 & -7 \\
0 & -4 & 9 & 19
\end{array}
\right]
\begin{array}{c}
(R_4) \\
(R_5) \\
(R_6)
\end{array}
$$

That is, we copied row 1 which we now call row 4; to get the second row, we multiplied the first row by 2 and the second row by 3 and then took the difference $(2(R_1) - 3(R_2))$, calling the resulting row (R_5), and so on.

The next step is to eliminate the -4 in (R_6). We find

$$
\begin{array}{c}
(R_4) \\
(R_5) \\
4(R_5) + 13(R_6)
\end{array}
\left[
\begin{array}{ccc|c}
3 & 5 & -1 & 10 \\
0 & 13 & -11 & -7 \\
0 & 0 & 73 & 219
\end{array}
\right]
$$

This is now the augmented matrix for the system of linear equations in (9.6), and we can proceed as in Example 5 to solve the system by back-substitution.   ▮

So far, all of the systems that we have considered have had the same number of equations as variables. This need not be the case. When a system has fewer equations than variables, we say that it is **underdetermined**. Though underdetermined systems can be inconsistent (have no solutions), they frequently have infinitely many solutions. When a system has more equations than variables, we say that it is **overdetermined**. Overdetermined systems are frequently inconsistent.

In Example 9, we will look at an underdetermined system, and in Example 10, an overdetermined system.

▎**EXAMPLE 9**　Solve the following underdetermined system.

$$2x + 2y - z = 1 \quad (R_1)$$
$$2x - y + z = 2 \quad (R_2)$$

Note that the system has fewer equations than variables. We use the augmented matrix to solve the system.

$$\begin{bmatrix} 2 & 2 & -1 & | & 1 \\ 2 & -1 & 1 & | & 2 \end{bmatrix} \begin{matrix} (R_1) \\ (R_2) \end{matrix}$$

Transforming the augmented matrix into upper triangular form, we find

$$\begin{matrix} (R_1) \\ (R_1) - (R_2) \end{matrix} \begin{bmatrix} 2 & 2 & -1 & | & 1 \\ 0 & 3 & -2 & | & -1 \end{bmatrix}$$

Translating this back into a system of equations, we find

$$2x + 2y - z = 1$$
$$3y - 2z = -1$$

We see that

$$y = -\frac{1}{3} + \frac{2}{3}z$$

and

$$2x = 1 - 2y + z = 1 - 2\left(-\frac{1}{3} + \frac{2}{3}z\right) + z = \frac{5}{3} - \frac{1}{3}z$$

or

$$x = \frac{5}{6} - \frac{1}{6}z$$

We use a dummy variable again, and set $z = t$, $t \in \mathbf{R}$; then, $x = \frac{5}{6} - \frac{1}{6}t$ and $y = -\frac{1}{3} + \frac{2}{3}t$. The solution can then be written as

$$\left\{ (x, y, z) : x = \frac{5}{6} - \frac{1}{6}t, \, y = -\frac{1}{3} + \frac{2}{3}t, z = t, t \in \mathbf{R} \right\}$$ ▎

▎**EXAMPLE 10**　Solve the following overdetermined system.

$$2x - y = 1 \quad (R_1)$$
$$x + y = 2 \quad (R_2)$$
$$x - y = 3 \quad (R_3)$$

Solution　Note that the system has more equations than variables. To solve it, we write

$$\begin{matrix} (R_1) \\ 2(R_2) - (R_1) \\ (R_2) - (R_3) \end{matrix} \quad \begin{matrix} 2x - y = 1 \quad (R_4) \\ 3y = 3 \quad (R_5) \\ 2y = -1 \quad (R_6) \end{matrix}$$

It follows from (R_6) that $y = -\frac{1}{2}$, and from (R_5) that $y = 1$. Since $1 \neq -\frac{1}{2}$, there cannot be a solution. The system is inconsistent, and the solution set is the empty set. In this example, we can illustrate that this system is inconsistent if we graph the three straight lines represented by (R_1), (R_2), and (R_3), as in Figure 9.8.

In order for the system to have solutions, the three lines would need to intersect in one point (or be identical). They don't, so there is no solution. ▎

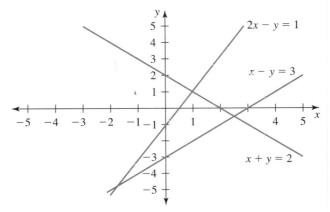

◀ **Figure 9.8**
The straight lines in
Example 10 do not intersect
in one point.

◆ 9.1 PROBLEMS

[9.1.1]

In Problems 1–4, solve each linear system of equations. In addition, for each system, graph the two lines corresponding to the two equations in a single coordinate system, and explain your solution using your graph.

1. $x - y = 1$
$x - 2y = -2$

2. $x + 3y = 6$
$x - 4y = -4$

3. $x - 3y = 6$
$y = 3 + \frac{1}{3}x$

4. $2x + y = \frac{1}{3}$
$6x + 3y = 1$

5. Determine c such that

$$2x - 3y = 5$$
$$4x - 6y = c$$

has (a) infinitely many solutions, and (b) no solutions. (c) Is it possible to choose a number for c so that the system has exactly one solution? Explain your answer.

6. Show that the solution of

$$a_{11}x_1 + a_{12}x_2 = b_1$$
$$a_{21}x_1 + a_{22}x_2 = b_2$$

is given by

$$x_1 = \frac{a_{22}b_1 - a_{12}b_2}{a_{11}a_{22} - a_{21}a_{12}}$$

and

$$x_2 = \frac{-a_{21}b_1 + a_{11}b_2}{a_{11}a_{22} - a_{21}a_{12}}$$

7. Zach wants to buy fish and plants for his aquarium. Each fish costs \$2.30, each plant costs \$1.70. He buys a total of 11 items and spends a total of \$21.70. Set up a system of linear equations that will allow you to determine how many fish and how many plants Zach bought, and solve the system.

8. Laboratory mice are fed with a mixture of two foods that contain two essential nutrients. Food 1 contains 3 units of nutrient A and 2 units of nutrient B per ounce; food 2 contains 4 units of nutrient A and 5 units of nutrient B per ounce.

(a) In what proportion should you mix the food if the mice require the nutrients A and B in equal amounts?

(b) Assume now that the mice require the nutrients A and B in the ratio 1:2. Is it possible to satisfy their dietary needs with the two foods available?

9. Show that if

$$a_{11}a_{22} - a_{21}a_{12} \neq 0$$

then the system

$$a_{11}x_1 + a_{12}x_2 = 0$$
$$a_{21}x_1 + a_{22}x_2 = 0$$

has exactly one solution, namely $x_1 = 0$ and $x_2 = 0$.

[9.1.2]

In Problems 10–14, solve each system of linear equations.

10. $2x - 3y + z = -1$
$x + y - 2z = -3$
$3x - 2y + z = 2$

11. $5x - y + 2z = 6$
$x + 2y - z = -1$
$3x + 2y - 2z = 1$

12. $x + 4y - 3z = -13$
$2x - 3y + 5z = 18$
$3x + y - 2z = 1$

13. $-2x + 4y - z = -1$
$x + 7y + 2z = -4$
$3x - 2y + 3z = -3$

14. $2x - y + 3z = 3$
$2x + y + 4z = 4$
$2x - 3y + 2z = 2$

In Problems 15–18, find the augmented matrix and use it to solve the system of linear equations.

15. $-x - 2y + 3z = -9$
$2x + y - z = 5$
$4x - 3y + 5z = -9$

16. $3x - 2y + z = 4$
$4x + y - 2z = -12$
$2x - 3y + z = 7$

17. $y + x = 3$
$z - y = -1$
$x + z = 2$

18. $2x - z = 1$
$y + 3z = x - 1$
$x + z = y - 3$

In Problems 19–22, determine whether each system is over-determined or underdetermined, then solve each system.

19. $x - 2y + z = 3$
$2x - 3y + z = 8$

20. $x - y \quad = 2$
$x + y + z = 3$

21. $2x - y = 3$
$x - y = 4$
$3x - y = 1$

22. $4y - 3z = 6$
$2y + z = 1$
$y + z = 0$

23. SplendidLawn sells three types of lawn fertilizer, SL 24-4-8, SL 21-7-12 and SL 17-0-0. The three numbers refer to the percentages of nitrogen, phosphate, and potassium, in that order, of the contents. (For instance, 100 gr of SL 24-4-8 contains 24 gr of actual nitrogen.) Suppose that each year your lawn requires 500 gr of actual nitrogen, 100 gr of actual phosphate, and 180 gr of actual potassium per 1000 square feet. How much of each of the three types of fertilizer should you apply per 1000 square feet per year?

24. Three different species of insects are reared together in a laboratory cage. They are supplied with two different types of food each day. Each individual of species 1 consumes 3 units of food *A* and 5 units of food *B*; each individual of species 2 consumes 2 units of food *A* and 3 units of food *B*; and individual of species 3 consumes 1 unit of food *A* and 2 units of food *B*. Each day, 500 units of food *A* and 900 units of food *B* are supplied. How many individuals of each species can be reared together?

◆ 9.2 MATRICES

We introduced matrices in the previous section; in this section, we will learn various matrix operations.

9.2.1 Basic Matrix Operations

We recall the definition of a matrix. An $m \times n$ matrix *A* is a rectangular array of numbers with *m* rows and *n* columns. We write this as

$$A = \begin{bmatrix} a_{11} & a_{12} & \cdots & a_{1n} \\ a_{21} & a_{22} & \cdots & a_{2n} \\ \cdots\cdots\cdots\cdots\cdots\cdots \\ a_{m1} & a_{m2} & \cdots & a_{mn} \end{bmatrix} = [a_{ij}]_{\substack{1 \le i \le m \\ 1 \le j \le n}}$$

We will also use the shorthand notation $A = [a_{ij}]$ if the size of the matrix is clear. We list a few simple definitions that do not need much explanation.

> **Definition**   Suppose that $A = [a_{ij}]$ and $B = [b_{ij}]$ are two $m \times n$ matrices. Then
>
> $$A = B$$
>
> if and only if, for all $1 \le i \le m$ and $1 \le j \le n$,
>
> $$a_{ij} = b_{ij}$$

This definition says that we can compare matrices of the same size, and they are equal if all their corresponding entries are equal. The next definition shows how to add matrices.

> **Definition**   Suppose that $A = [a_{ij}]$ and $B = [b_{ij}]$ are two $m \times n$ matrices. Then
>
> $$C = A + B$$
>
> is an $m \times n$ matrix with entries
>
> $$c_{ij} = a_{ij} + b_{ij} \quad \text{for } 1 \le i \le m, 1 \le j \le n$$

Note that addition is only defined for matrices of equal sizes. Matrix addition satisfies the following two properties.

1. $A + B = B + A$
2. $(A + B) + C = A + (B + C)$

The matrix with all its entries equal to zero is called the **zero matrix** and denoted by **0**. The following holds:

$$A + \mathbf{0} = A$$

We can multiply matrices by numbers. Such numbers are called **scalars**.

> **Definition**   Suppose that $A = [a_{ij}]$ is an $m \times n$ matrix and c is a scalar. Then cA is an $m \times n$ matrix with entries ca_{ij} for $1 \le i \le m$ and $1 \le j \le n$.

▌**EXAMPLE 1**   Find $A + 2B - 3C$, if

$$A = \begin{bmatrix} 2 & 3 \\ 1 & 0 \end{bmatrix}, \quad B = \begin{bmatrix} 0 & 1 \\ -1 & -3 \end{bmatrix}, \quad C = \begin{bmatrix} 1 & 0 \\ 0 & 3 \end{bmatrix}$$

Solution

$$A + 2B - 3C = \begin{bmatrix} 2 & 3 \\ 1 & 0 \end{bmatrix} + 2\begin{bmatrix} 0 & 1 \\ -1 & -3 \end{bmatrix} - 3\begin{bmatrix} 1 & 0 \\ 0 & 3 \end{bmatrix}$$

$$= \begin{bmatrix} 2 & 3 \\ 1 & 0 \end{bmatrix} + \begin{bmatrix} 0 & 2 \\ -2 & -6 \end{bmatrix} + \begin{bmatrix} -3 & 0 \\ 0 & -9 \end{bmatrix}$$

$$= \begin{bmatrix} -1 & 5 \\ -1 & -15 \end{bmatrix}$$
▌

▌**EXAMPLE 2**   Let

$$A = \begin{bmatrix} 1 & 3 \\ 0 & 4 \end{bmatrix}, \quad B = \begin{bmatrix} 5 & 3 \\ -2 & 1 \end{bmatrix}, \quad C = \begin{bmatrix} -1 & -3 \\ 0 & -4 \end{bmatrix}$$

Show that $A + B = B - C$.

Solution   One way is to simply compute $A + B$ and $B - C$ and then compare the results. We find

$$A + B = \begin{bmatrix} 1 & 3 \\ 0 & 4 \end{bmatrix} + \begin{bmatrix} 5 & 3 \\ -2 & 1 \end{bmatrix} = \begin{bmatrix} 6 & 6 \\ -2 & 5 \end{bmatrix}$$

and

$$B - C = \begin{bmatrix} 5 & 3 \\ -2 & 1 \end{bmatrix} - \begin{bmatrix} -1 & -3 \\ 0 & -4 \end{bmatrix} = \begin{bmatrix} 6 & 6 \\ -2 & 5 \end{bmatrix}$$

Comparing $A + B$ with $B - C$, we find that they are equal since all their entries are the same.

Another way to reach this conclusion is to subtract B from both sides of the assertion $A + B = B - C$, which results in $A = -C$. Adding C to both sides, we find $A + C = 0$. That is, to show the assertion $A + B = B - C$, we can also show that $A + C = 0$. Computing $A + C$, we obtain

$$A + C = \begin{bmatrix} 1 & 3 \\ 0 & 4 \end{bmatrix} + \begin{bmatrix} -1 & -3 \\ 0 & -4 \end{bmatrix} = \begin{bmatrix} 0 & 0 \\ 0 & 0 \end{bmatrix} = 0$$

which shows that $A + C = 0$, and implies that the assertion is true.

The operation that interchanges rows and columns is called **transposition**.

> **Definition**   Suppose that $A = [a_{ij}]$ is an $m \times n$ matrix. The **transpose** of A, denoted by A', is an $n \times m$ matrix with entries
>
> $$a'_{ij} = a_{ji}$$

To understand how this operation works, we look at the following example.

I EXAMPLE 3   Transpose the following matrices.

$$A = \begin{bmatrix} 1 & 2 & 3 \\ 4 & 5 & 6 \end{bmatrix}, \quad B = \begin{bmatrix} 1 \\ 2 \end{bmatrix}, \quad C = [3 \quad 4]$$

Solution   To find the transpose, we need to interchange rows and columns. Since A is a 2×3 matrix, its transpose A' is a 3×2 matrix, namely

$$A' = \begin{bmatrix} 1 & 4 \\ 2 & 5 \\ 3 & 6 \end{bmatrix}$$

To check this with our definition, note that

$$a'_{11} = a_{11} = 1, \quad a'_{12} = a_{21} = 4, \quad a'_{21} = a_{12} = 2, \dots$$

The matrix B is a 2×1 matrix, which is a column vector Its transpose is a 1×2 matrix, which is a row vector. We find

$$B' = [1 \quad 2]$$

Similarly, C is a 1×2 row vector; its transpose is then a 2×1 column vector. We find

$$C' = \begin{bmatrix} 3 \\ 4 \end{bmatrix}$$

In the preceding example, we saw that the transpose of a row vector is a column vector, and vice versa. When we need to write a column vector in text, such as $X = \begin{bmatrix} 1 \\ 2 \\ 3 \end{bmatrix}$, we can instead write $X = [1 \quad 2 \quad 3]'$ which, written as the transpose of a row vector, is really a column vector. A large column vector written as the transpose of a row vector is more legible in text.

9.2.2 Matrix Multiplication

The multiplication of two matrices is more complicated. We give the definition first.

Definition   Suppose that $A = [a_{ij}]$ is an $m \times l$ matrix and $B = [b_{ij}]$ is an $l \times n$ matrix. Then

$$C = AB$$

is an $m \times n$ matrix with

$$c_{ij} = a_{i1}b_{1j} + a_{i2}b_{2j} + \cdots + a_{il}b_{lj} = \sum_{k=1}^{l} a_{ik}b_{kj}$$

for $1 \leq i \leq m$ and $1 \leq j \leq n$.

To explain this in words, note that c_{ij} is the entry in C that is located in the ith row and the jth column. To obtain it, we multiply the entries of the ith row of A with the entries of the jth column of B as indicated in the definition. For the product AB to be defined, the number of columns in A must equal the number of rows in B. The definition looks rather formidable, let's see how it actually works.

EXAMPLE 4   **Matrix multiplication**

Compute AB when

$$A = \begin{bmatrix} 1 & 2 & 3 \\ -1 & 0 & 4 \end{bmatrix} \quad \text{and} \quad B = \begin{bmatrix} 1 & 2 & 3 & -3 \\ 0 & -1 & 4 & 0 \\ -1 & 0 & -2 & 1 \end{bmatrix}$$

Solution   First, note that A is a 2×3 matrix and B is a 3×4 matrix. That is, A has 3 columns and B has 3 rows. Therefore, the product AB is defined and AB is a 2×4 matrix. We write the product as

$$C = AB = \begin{bmatrix} 1 & 2 & 3 \\ -1 & 0 & 4 \end{bmatrix} \cdot \begin{bmatrix} 1 & 2 & 3 & -3 \\ 0 & -1 & 4 & 0 \\ -1 & 0 & -2 & 1 \end{bmatrix}$$

For instance, to find c_{11}, we multiply the first row of A with the first column of B as follows:

$$c_{11} = \begin{bmatrix} 1 & 2 & 3 \end{bmatrix} \begin{bmatrix} 1 \\ 0 \\ -1 \end{bmatrix} = (1)(1) + (2)(0) + (3)(-1) = -2$$

You can see from this calculation that the number of columns of A must be equal to the number of rows of B. Otherwise, we would run out of numbers when multiplying the corresponding entries.

To find c_{23}, we multiply the second row of A with the third column of B:

$$c_{23} = \begin{bmatrix} -1 & 0 & 4 \end{bmatrix} \begin{bmatrix} 3 \\ 4 \\ -2 \end{bmatrix} = (-1)(3) + (0)(4) + (4)(-2) = -11$$

The other entries are obtained in a similar way, and we find

$$C = \begin{bmatrix} -2 & 0 & 5 & 0 \\ -5 & -2 & -11 & 7 \end{bmatrix}$$

∎

When multiplying matrices, the order is important. For instance, BA is not defined for the matrices in Example 4. From this, it should be clear that, typically, $AB \neq BA$. Even if both AB and BA exist, the resulting matrices are usually not the same. They can differ in both the number of rows and columns, and in the actual entries. We give a couple of examples to illustrate this important point.

EXAMPLE 5   Order is important

Suppose that

$$A = [\,2 \quad 1 \quad -1\,] \qquad \text{and} \qquad B = \begin{bmatrix} 1 \\ -1 \\ 0 \end{bmatrix}$$

Show that both AB and BA are defined but $AB \neq BA$.

Solution   Note that A is a 1×3 matrix and B is a 3×1 matrix. The product AB is defined since the number of columns of A is the same as the number of rows of B (namely 3); the product AB is a 1×1 matrix. When we carry out the matrix multiplication, we find

$$AB = [\,2 \quad 1 \quad -1\,] \begin{bmatrix} 1 \\ -1 \\ 0 \end{bmatrix} = [2 - 1 + 0] = [1] = 1$$

We see from this that a 1×1 matrix is simply a number.

The product BA is also defined, since the number of columns of B is the same as the number of rows of A (namely 1); the product BA is a 3×3 matrix. When we carry out the matrix multiplication, we find

$$BA = \begin{bmatrix} 1 \\ -1 \\ 0 \end{bmatrix} [\,2 \quad 1 \quad -1\,] = \begin{bmatrix} 2 & 1 & -1 \\ -2 & -1 & 1 \\ 0 & 0 & 0 \end{bmatrix}$$

When we compare AB and BA, we immediately see that $AB \neq BA$; the two matrices are not even of the same size.   ∎

EXAMPLE 6   Order is important

Suppose that

$$A = \begin{bmatrix} 2 & -1 \\ 4 & -2 \end{bmatrix} \qquad \text{and} \qquad B = \begin{bmatrix} 1 & -1 \\ 2 & -2 \end{bmatrix}$$

Show that both AB and BA are defined but $AB \neq BA$.

Solution   Both A and B are 2×2 matrices. We conclude that both AB and BA are 2×2 matrices as well. We find

$$AB = \begin{bmatrix} 0 & 0 \\ 0 & 0 \end{bmatrix} \qquad \text{and} \qquad BA = \begin{bmatrix} -2 & 1 \\ -4 & 2 \end{bmatrix}$$

Since the entries in AB are not the same as in BA, it follows that $AB \neq BA$.   ∎

The product AB in Example 6 resulted in a matrix with all entries equal to 0. Earlier, we called this matrix the zero matrix and denoted it by **0**. This example shows that a product of two matrices can be the zero matrix without either matrix being the zero matrix.

We list the following properties of matrix multiplication. We assume that all matrices are of appropriate sizes so that all matrix multiplications are defined.

1. $(A + B)C = AC + BC$
2. $A(B + C) = AB + AC$
3. $(AB)C = A(BC)$
4. $A\mathbf{0} = \mathbf{0}A = \mathbf{0}$

(We will practice applying these properties in the problem section.)

Next, we note that if A is a square matrix (that is, if it has the same number of rows as columns), we can define powers of A; namely, if k is a positive integer, then

$$A^k = A^{k-1}A = AA^{k-1} = \underbrace{AA\ldots A}_{k \text{ factors}}$$

For instance, $A^2 = AA$, $A^3 = AAA$, and so on.

▌**EXAMPLE 7**   **Powers of matrices**

Find A^3 if

$$A = \begin{bmatrix} 1 & 2 \\ 3 & 4 \end{bmatrix}$$

Solution   To find A^3, we first need to compute A^2. We obtain

$$A^2 = AA = \begin{bmatrix} 1 & 2 \\ 3 & 4 \end{bmatrix}\begin{bmatrix} 1 & 2 \\ 3 & 4 \end{bmatrix} = \begin{bmatrix} 7 & 10 \\ 15 & 22 \end{bmatrix}$$

To compute A^3 now, we compute either A^2A or AA^2. Both will yield A^3. This is a case in which the order of multiplication does not matter, since $A^3 = (AA)A = A(AA) = AAA$. We do it both ways.

$$A^3 = A^2A = \begin{bmatrix} 7 & 10 \\ 15 & 22 \end{bmatrix}\begin{bmatrix} 1 & 2 \\ 3 & 4 \end{bmatrix} = \begin{bmatrix} 37 & 54 \\ 81 & 118 \end{bmatrix}$$

and

$$A^3 = AA^2 = \begin{bmatrix} 1 & 2 \\ 3 & 4 \end{bmatrix}\begin{bmatrix} 7 & 10 \\ 15 & 22 \end{bmatrix} = \begin{bmatrix} 37 & 54 \\ 81 & 118 \end{bmatrix}$$

An important square matrix is the **identity matrix**, denoted by I_n. The identity matrix is an $n \times n$ matrix with 1s on its diagonal line and 0s elsewhere, that is,

$$I_n = \begin{bmatrix} 1 & 0 & 0 & \cdots & 0 \\ 0 & 1 & 0 & \cdots & 0 \\ & & \cdots & & \\ 0 & 0 & 0 & \cdots & 1 \end{bmatrix}$$

For instance,

$$I_1 = [1], \qquad I_2 = \begin{bmatrix} 1 & 0 \\ 0 & 1 \end{bmatrix}, \qquad I_3 = \begin{bmatrix} 1 & 0 & 0 \\ 0 & 1 & 0 \\ 0 & 0 & 1 \end{bmatrix}$$

(We frequently write I instead of I_n if the size of I_n is clear from the context.) The identity matrix has an important property: if A is an $m \times n$ matrix, then

$$AI_n = I_m A = A$$

It follows that $I^k = I$ for any positive integer k.

The identity matrix serves the same purpose as the number "1" in the multiplication of numbers. If a and x are numbers, then you know that $ax = x$ can be written as $(a - 1)x = 0$ or $x(a - 1) = 0$. There is a similar procedure for matrices, but we must be more careful, since the order in which we multiply matrices is important.

∎ **EXAMPLE 8**   **Matrix equation**

Let A be a 2×2 matrix and X be a 2×1 matrix. Show that $AX = X$ can be written as $(A - I_2)X = \mathbf{0}$.

Solution   We find that

$$AX = X \qquad \text{is equivalent to} \qquad AX - X = \mathbf{0}$$

To factor X, we multiply X from the left by I_2. This yields

$$AX - I_2 X = \mathbf{0}$$

We can now factor X and find

$$(A - I_2)X = \mathbf{0}$$

An important observation: we multiply X from the left by $A - I_2$. We are not allowed to reverse the order (the product $X(A - I_2)$ is not even defined). ∎

There is a close connection between systems of linear equations and matrices. The system of linear equations

$$a_{11}x_1 + a_{12}x_2 + \cdots + a_{1n}x_n = b_1$$
$$a_{21}x_1 + a_{22}x_2 + \cdots + a_{2n}x_n = b_2$$
$$\cdots\cdots\cdots\cdots\cdots\cdots\cdots\cdots\cdots\cdots\cdots$$
$$a_{m1}x_1 + a_{m2}x_2 + \cdots + a_{mn}x_n = b_m$$

can be written in matrix form

$$AX = B$$

with

$$A = \begin{bmatrix} a_{11} & \cdots & a_{1n} \\ a_{21} & \cdots & a_{2n} \\ \cdots\cdots\cdots\cdots \\ a_{m1} & \cdots & a_{mn} \end{bmatrix}, \quad X = \begin{bmatrix} x_1 \\ x_2 \\ \vdots \\ x_n \end{bmatrix}, \quad B = \begin{bmatrix} b_1 \\ b_2 \\ \vdots \\ b_m \end{bmatrix}$$

∎ **EXAMPLE 9**   **Matrix representation of linear systems**

Write

$$2x_1 + 3x_2 + 4x_3 = 1$$
$$-x_1 + 5x_2 - 6x_3 = 7$$

in matrix form.

Solution   When we write this system in matrix form, we find

$$\begin{bmatrix} 2 & 3 & 4 \\ -1 & 5 & -6 \end{bmatrix} \begin{bmatrix} x_1 \\ x_2 \\ x_3 \end{bmatrix} = \begin{bmatrix} 1 \\ 7 \end{bmatrix}$$

We see that A is a 2×3 matrix and X is a 3×1 matrix. We therefore expect (based on the rules of multiplication) that B is a 2×1 matrix, which it is indeed. ∎

We will use matrix representation to solve systems of linear equations in the next subsection.

9.2.3 Inverse Matrices

In this subsection, we will learn how to solve systems of linear equations when they are written in matrix form,

$$AX = B \tag{9.7}$$

We will restrict ourselves to the case when A is a square matrix, and start with a simple example. When we solve

$$5x = 10$$

for x, we simply divide both sides by 5 or, equivalently, multiply both sides by $\frac{1}{5} = 5^{-1}$. We find

$$5^{-1} \cdot 5x = 5^{-1} \cdot 10$$
$$x = 2$$

To solve (9.7), we therefore need an operation that is analogous to division or multiplication by the "reciprocal" of A. We will define a matrix A^{-1}, called the **inverse matrix** of A, which will allow us to write the solution of (9.7) as

$$X = A^{-1}B$$

As with division by 0, it will not always be possible to do this.

> **Definition**   Suppose that $A = [a_{ij}]$ is an $n \times n$ square matrix. If there exists an $n \times n$ square matrix B such that
>
> $$AB = BA = I_n$$
>
> then B is called the inverse matrix of A and denoted by A^{-1}.

If A has an inverse matrix, A is called **invertible** or **nonsingular**; if A does not have an inverse, then A is called **singular**. If A is invertible, its inverse matrix is unique, that is, if B and C are both inverse matrices of A, then $B = C$. (In other words, if you and your friend compute the inverse of A, you both should get the exact same answer.) To see why, assume that B and C are both inverse matrices of A. Using $BA = I_n$ and $AC = I_n$, we find that

$$B = BI_n = B(AC) = (BA)C = I_nC = C$$

The following hold for inverse matrices.

1. If A is an invertible $n \times n$ matrix, then

$$(A^{-1})^{-1} = A$$

2. If A and B are invertible $n \times n$ matrices, then

$$(AB)^{-1} = B^{-1}A^{-1}$$

The first statement says that if we apply the inverse operation twice, we get the original matrix back. This is familiar from dealing with real numbers: Take the number 2, its inverse is $2^{-1} = 1/2$. If we take the inverse of $1/2$, we get 2 again. To see why this is true in general, assume that A is an $n \times n$ matrix with inverse matrix A^{-1}. Then

$$AA^{-1} = A^{-1}A = I_n$$

But this equation also says that A is the inverse matrix of A^{-1}, that is, $A = (A^{-1})^{-1}$.

To see why the second statement is true, we must check that

$$(AB)(B^{-1}A^{-1}) = (B^{-1}A^{-1})(AB) = I_n$$

We compute

$$(AB)(B^{-1}A^{-1}) = A(BB^{-1})A^{-1} = AI_nA^{-1} = AA^{-1} = I_n$$

and

$$(B^{-1}A^{-1})(AB) = B^{-1}(A^{-1}A)B = B^{-1}I_nB = B^{-1}B = I_n$$

In the next example, we will see how we can check whether two matrices are inverses of each other. The matrices will be of size 3×3. It is somewhat cumbersome to find the inverse of a matrix of size 3×3 or larger (unless you use a calculator or computer software). Here, we start out with the two matrices, and simply check whether they are inverse matrices of each other.

▮ EXAMPLE 10   Checking whether matrices are inverses of each other

Show that the inverse of

$$A = \begin{bmatrix} 3 & 5 & -1 \\ 2 & -1 & 3 \\ 4 & 2 & -3 \end{bmatrix} \quad \text{is} \quad B = \begin{bmatrix} -\frac{3}{73} & \frac{13}{73} & \frac{14}{73} \\ \frac{18}{73} & -\frac{5}{73} & -\frac{11}{73} \\ \frac{8}{73} & \frac{14}{73} & -\frac{13}{73} \end{bmatrix}$$

Solution   Carrying out the matrix multiplications AB and BA, we see that

$$AB = I_3 \quad \text{and} \quad BA = I_3$$

Therefore, $B = A^{-1}$.   ▮

In this book, we will only need to invert 2×2 matrices. In the next example, we will show one way to do this.

▮ EXAMPLE 11   Finding an inverse

Find the inverse of

$$A = \begin{bmatrix} 2 & 5 \\ 1 & 3 \end{bmatrix}$$

Solution   We need to find a matrix

$$B = \begin{bmatrix} b_{11} & b_{12} \\ b_{21} & b_{22} \end{bmatrix}$$

such that $AB = I_2$. We will then check that $BA = I_2$ and, hence, that B is the inverse of A. We must solve

$$\begin{bmatrix} 2 & 5 \\ 1 & 3 \end{bmatrix} \begin{bmatrix} b_{11} & b_{12} \\ b_{21} & b_{22} \end{bmatrix} = \begin{bmatrix} 1 & 0 \\ 0 & 1 \end{bmatrix}$$

or

$$\begin{array}{ccc} 2b_{11} + 5b_{21} = 1 & & 2b_{12} + 5b_{22} = 0 \\ & \text{and} & \\ b_{11} + 3b_{21} = 0 & & b_{12} + 3b_{22} = 1 \end{array} \tag{9.8}$$

The left set of equations has the solution

$$b_{11} = 3 \quad \text{and} \quad b_{21} = -1$$

The right set of equations has the solution

$$b_{12} = -5 \quad \text{and} \quad b_{22} = 2$$

Hence

$$B = \begin{bmatrix} 3 & -5 \\ -1 & 2 \end{bmatrix}$$

You can verify that both $AB = I_2$ and $BA = I_2$, hence $B = A^{-1}$.   **I**

In the optional Subsection 9.2.4, we will present a method that can be employed for larger matrices. It is essentially the same method as in Example 11, but uses augmented matrices, and is computationally quite intensive. To invert larger matrices, I strongly recommend using a graphing calculator or computer software.

At the end of Subsection 9.2.2, we mentioned the connection between systems of linear equations and matrix multiplication. A system of linear equations can be written as

$$AX = B \tag{9.9}$$

If we consider a linear system of n equations in n unknowns, then A is an $n \times n$ matrix. If A is invertible, then multiplying (9.9) by A^{-1} from the left, we find

$$A^{-1}AX = A^{-1}B$$

Since $A^{-1}A = I_n$ and $I_nX = X$, it follows that

$$X = A^{-1}B$$

We will use this to repeat the solution of Example 5 of Section 9.1.

EXAMPLE 12   **Using inverse matrices to solve linear systems**

Solve

$$\begin{array}{rcr} 3x + 5y - z &=& 10 \\ 2x - y + 3z &=& 9 \\ 4x + 2y - 3z &=& -1 \end{array}$$

Solution   The coefficient matrix of A is

$$A = \begin{bmatrix} 3 & 5 & -1 \\ 2 & -1 & 3 \\ 4 & 2 & -3 \end{bmatrix}$$

which we encountered in Example 10, where we saw that

$$A^{-1} = \begin{bmatrix} -\frac{3}{73} & \frac{13}{73} & \frac{14}{73} \\ \frac{18}{73} & -\frac{5}{73} & -\frac{11}{73} \\ \frac{8}{73} & \frac{14}{73} & -\frac{13}{73} \end{bmatrix}$$

We set

$$B = \begin{bmatrix} 10 \\ 9 \\ -1 \end{bmatrix}$$

and compute $A^{-1}B$. We find

$$A^{-1} \begin{bmatrix} 10 \\ 9 \\ -1 \end{bmatrix} = \begin{bmatrix} 1 \\ 2 \\ 3 \end{bmatrix}$$

that is, $x = 1$, $y = 2$, and $z = 3$, as we found in Example 5 of Section 9.1.   ▮

We now have two methods for solving systems of linear equations when the number of equations is equal to the number of variables. We can either reduce the system to upper triangular form and then use back-substitution, or we can write the system in matrix form $AX = B$, find A^{-1}, and then compute $X = A^{-1}B$. Of course, the second method only works when A^{-1} exists. If A^{-1} does not exist, the system has either no solution or infinitely many solutions. But when A^{-1} exists, then $AX = B$ has exactly one solution.

There is a simple criterion for checking whether a 2×2 matrix is invertible. Deriving this criterion will also provide us with a formula for the inverse of an invertible 2×2 matrix. We set

$$A = \begin{bmatrix} a_{11} & a_{12} \\ a_{21} & a_{22} \end{bmatrix} \quad \text{and} \quad B = \begin{bmatrix} b_{11} & b_{12} \\ b_{21} & b_{22} \end{bmatrix}$$

We wish to write the entries of B in terms of the entries of A, such that $AB = BA = I$ or $B = A^{-1}$. From

$$\begin{bmatrix} a_{11} & a_{12} \\ a_{21} & a_{22} \end{bmatrix} \begin{bmatrix} b_{11} & b_{12} \\ b_{21} & b_{22} \end{bmatrix} = \begin{bmatrix} 1 & 0 \\ 0 & 1 \end{bmatrix}$$

we obtain the following set of equations

$$\begin{aligned} a_{11}b_{11} + a_{12}b_{21} &= 1 \\ a_{21}b_{11} + a_{22}b_{21} &= 0 \end{aligned} \tag{9.10}$$

$$\begin{aligned} a_{11}b_{12} + a_{12}b_{22} &= 0 \\ a_{21}b_{12} + a_{22}b_{22} &= 1 \end{aligned} \tag{9.11}$$

We solve the system of linear equations (9.10) for b_{11} and b_{21}:

$$\begin{aligned} a_{11}b_{11} + a_{12}b_{21} &= 1 \quad (R_1) \\ a_{21}b_{11} + a_{22}b_{21} &= 0 \quad (R_2) \end{aligned}$$

If we compute $a_{11}(R_2) - a_{21}(R_1)$, we find

$$a_{11}a_{22}b_{21} - a_{12}a_{21}b_{21} = -a_{21}$$

Solving this for b_{21}, we first factor b_{21} on the left-hand side. This yields

$$(a_{11}a_{22} - a_{12}a_{21})b_{21} = -a_{21}$$

If $a_{11}a_{22} - a_{12}a_{21} \neq 0$, we can isolate b_{21}. We find

$$b_{21} = -\frac{a_{21}}{a_{11}a_{22} - a_{12}a_{21}}$$

Substituting b_{21} into (R_1) and solving for b_{11}, we find, for $a_{11} \neq 0$,

$$b_{11} = \frac{1}{a_{11}}[1 - a_{12}b_{21}]$$

$$= \frac{1}{a_{11}}\left[1 - \frac{-a_{12}a_{21}}{a_{11}a_{22} - a_{12}a_{21}}\right]$$

$$= \frac{1}{a_{11}}\frac{a_{11}a_{22}}{a_{11}a_{22} - a_{12}a_{21}}$$

$$= \frac{a_{22}}{a_{11}a_{22} - a_{12}a_{21}}$$

(When $a_{11} = 0$, the same final result holds.) When we solve (9.11), we find

$$b_{12} = -\frac{a_{12}}{a_{11}a_{22} - a_{12}a_{21}}$$

$$b_{22} = \frac{a_{11}}{a_{11}a_{22} - a_{12}a_{21}}$$

Our calculations yield two important results.

If

$$A = \begin{bmatrix} a_{11} & a_{12} \\ a_{21} & a_{22} \end{bmatrix}$$

and $a_{11}a_{22} - a_{12}a_{21} \neq 0$, then A is invertible, and

$$A^{-1} = \frac{1}{a_{11}a_{22} - a_{12}a_{21}}\begin{bmatrix} a_{22} & -a_{12} \\ -a_{21} & a_{11} \end{bmatrix}$$

The expression $a_{11}a_{22} - a_{12}a_{21}$ is called the **determinant** of A and is denoted by $\det A$. We summarize this in the following.

Definition   If

$$A = \begin{bmatrix} a_{11} & a_{12} \\ a_{21} & a_{22} \end{bmatrix}$$

then the determinant of A is defined as

$$\det A = a_{11}a_{22} - a_{12}a_{21}$$

Looking back now at the formula for A^{-1}, where A is a 2×2 matrix whose determinant is nonzero, to find the inverse of A, we divide A by the determinant of A, switch the diagonal elements of A and change the sign of the off-diagonal elements. If the determinant is equal to 0, then the inverse of A does not exist.

The determinant can be defined for any $n \times n$ matrix. We will not give the general formula, which is computationally somewhat complicated for $n \geq 3$. Graphing calculators or more sophisticated computer software can compute determinants. But we mention the following result that allows us to determine whether or not an $n \times n$ matrix has an inverse.

> **Theorem**   Suppose that A is an $n \times n$ matrix. A is nonsingular if and only if $\det A \neq 0$.

▌ **EXAMPLE 13**   **Using the determinant to find inverse matrices**

Determine which of the following matrices is invertible. In each case, compute the inverse if it exists.

$$\textbf{a.} \qquad A = \begin{bmatrix} 3 & 5 \\ 2 & 4 \end{bmatrix} \qquad\qquad \textbf{b.} \qquad B = \begin{bmatrix} 2 & 6 \\ 1 & 3 \end{bmatrix}$$

Solution

a. To check whether A is invertible, we compute the determinant of A.

$$\det A = (3)(4) - (2)(5) = 2 \neq 0$$

Hence, A is invertible and we find

$$A^{-1} = \frac{1}{2} \begin{bmatrix} 4 & -5 \\ -2 & 3 \end{bmatrix} = \begin{bmatrix} 2 & -\frac{5}{2} \\ -1 & \frac{3}{2} \end{bmatrix}$$

b. Since

$$\det B = (2)(3) - (1)(6) = 0$$

B is not invertible.

We mentioned that if A is invertible, then

$$AX = B$$

has exactly one solution, namely $X = A^{-1}B$. Of particular importance is the case when $B = \mathbf{0}$. That is, assume that A is a 2×2 matrix and $B = \begin{bmatrix} 0 \\ 0 \end{bmatrix}$, then $X = \begin{bmatrix} 0 \\ 0 \end{bmatrix}$ is a solution of

$$AX = \begin{bmatrix} 0 \\ 0 \end{bmatrix}$$

We call the solution $\begin{bmatrix} 0 \\ 0 \end{bmatrix}$ a **trivial** solution. It is the only solution of $AX = \mathbf{0}$ when A is invertible.   If $AX = \mathbf{0}$ has a solution $X \neq \begin{bmatrix} 0 \\ 0 \end{bmatrix}$, then X is called a **nontrivial** solution. In order to get a nontrivial solution for

$$AX = \mathbf{0},$$

A must be singular. We formulate this as a theorem in the more general case when *A* is an $n \times n$ matrix; we will need this result repeatedly.

> **Theorem**   Suppose that *A* is an $n \times n$ matrix, and *X* and **0** are $n \times 1$ matrices. The equation
>
> $$AX = 0$$
>
> has a nontrivial solution if and only if *A* is singular.

▮ EXAMPLE 14   Nontrivial solutions

Let

$$A = \begin{bmatrix} a & 6 \\ 3 & 2 \end{bmatrix}$$

Determine *a* so that $AX = 0$ has a nontrivial solution, and find a nontrivial solution.

Solution   To determine when $AX = 0$ has a nontrivial solution, we must find conditions for which *A* is singular or, equivalently, for which $\det A = 0$.

Since $\det A = 2a - 18$, it follows that *A* is singular if $2a - 18 = 0$ or $a = 9$. Therefore, if $a = 9$, $AX = 0$ has a nontrivial solution. To compute a nontrivial solution, we must solve

$$\begin{bmatrix} 9 & 6 \\ 3 & 2 \end{bmatrix}\begin{bmatrix} x_1 \\ x_2 \end{bmatrix} = \begin{bmatrix} 0 \\ 0 \end{bmatrix}$$

which can be written as a system of two linear equations

$$9x_1 + 6x_2 = 0$$
$$3x_1 + 2x_2 = 0$$

We see that the first equation is three times the second equation. We can therefore simplify the system and obtain

$$3x_1 + 2x_2 = 0$$
$$0x_2 = 0$$

This shows that the system has infinitely many solutions, namely

$$\left\{ (x_1, x_2) : x_2 = t \text{ and } x_1 = -\frac{2}{3}t, \text{ for } t \in \mathbf{R} \right\}$$

In particular, the system has nontrivial solutions, for instance, choosing $t = 3$, we find $x_1 = -2$ and $x_2 = 3$.   ▮

9.2.4 Computing Inverse Matrices (Optional)

We saw in the preceding subsection how to invert 2×2 matrices by solving two systems of linear equations. This method becomes computationally quite intensive when we try to invert larger matrices. Inverting an $n \times n$ matrix would result in n linear systems, each system consisting of n equations in n unknowns. By solving these n systems simultaneously, we can speed up the process quite a bit.

When we look back at equation (9.8) of Example 11 of Subsection 9.2.3,

$$2b_{11} + 5b_{21} = 1 \qquad\qquad 2b_{12} + 5b_{22} = 0$$
$$b_{11} + 3b_{21} = 0 \quad \text{and} \quad b_{12} + 3b_{22} = 1$$

and rewrite both systems as augmented matrices, namely

$$\begin{bmatrix} 2 & 5 & | & 1 \\ 1 & 3 & | & 0 \end{bmatrix} \quad \text{and} \quad \begin{bmatrix} 2 & 5 & | & 0 \\ 1 & 3 & | & 1 \end{bmatrix}$$

we see that each augmented matrix has the same matrix A on its left side. To solve each system, we must perform row transformations until we can read off the solutions. Reading off the solutions is particularly easy when the matrix on the left is transformed into the identity matrix. We do this for the augmented matrix

$$\begin{bmatrix} 2 & 5 & | & 1 \\ 1 & 3 & | & 0 \end{bmatrix} \quad \begin{matrix} (R_1) \\ (R_2) \end{matrix}$$

We divide the first row by 2 in order to get a 1 in the first position. In the second row we want a 0 in the first position; to get this we subtract $2(R_2)$ from (R_1). That is,

$$\begin{matrix} \frac{1}{2}(R_1) \\ (R_1) - 2(R_2) \end{matrix} \quad \begin{bmatrix} 1 & \frac{5}{2} & | & \frac{1}{2} \\ 0 & -1 & | & 1 \end{bmatrix} \quad \begin{matrix} (R_3) \\ (R_4) \end{matrix}$$

By adding $\frac{5}{2}(R_4)$ to (R_3), we get a 0 in the second position in the first row. Multiplying (R_4) by -1, we get a 1 in the second position in the second row. That is, we obtain the identity matrix on the left side.

$$\begin{matrix} (R_3) + \frac{5}{2}(R_4) \\ (-1)(R_4) \end{matrix} \quad \begin{bmatrix} 1 & 0 & | & 3 \\ 0 & 1 & | & -1 \end{bmatrix}$$

Rewriting this as a system of linear equations, we see that

$$b_{11} = 3$$
$$b_{21} = -1$$

To find b_{12} and b_{22}, we need to transform the augmented matrix

$$\begin{bmatrix} 2 & 5 & | & 0 \\ 1 & 3 & | & 1 \end{bmatrix}$$

so that the matrix on the left is the identity matrix. This involves the exact same transformations, since the coefficient matrix is the same for both systems of linear equations. Now comes the trick that speeds up the calculation. Instead of doing each augmented matrix separately, we do them simultaneously. Namely, we write

$$\begin{bmatrix} 2 & 5 & | & 1 & 0 \\ 1 & 3 & | & 0 & 1 \end{bmatrix}$$

and then make row transformations until this is of the form

$$\begin{bmatrix} 1 & 0 & | & b_{11} & b_{12} \\ 0 & 1 & | & b_{21} & b_{22} \end{bmatrix}$$

It then follows that

$$A^{-1} = \begin{bmatrix} b_{11} & b_{12} \\ b_{21} & b_{22} \end{bmatrix}$$

Let's do the calculation:

$$\left[\begin{array}{cc|cc} 2 & 5 & 1 & 0 \\ 1 & 3 & 0 & 1 \end{array}\right] \quad \begin{array}{c} (R_1) \\ (R_2) \end{array}$$

$$\begin{array}{c} \frac{1}{2}(R_1) \\ (R_1) - 2(R_2) \end{array} \left[\begin{array}{cc|cc} 1 & \frac{5}{2} & \frac{1}{2} & 0 \\ 0 & -1 & 1 & -2 \end{array}\right] \quad \begin{array}{c} (R_3) \\ (R_4) \end{array}$$

$$\begin{array}{c} (R_3) + \frac{5}{2}(R_4) \\ (-1)(R_4) \end{array} \left[\begin{array}{cc|cc} 1 & 0 & 3 & -5 \\ 0 & 1 & -1 & 2 \end{array}\right]$$

We recognize the column

$$\left[\begin{array}{c} 3 \\ -1 \end{array}\right]$$

which gives

$$b_{11} = 3 \qquad \text{and} \qquad b_{21} = -1$$

The column on the right,

$$\left[\begin{array}{c} -5 \\ 2 \end{array}\right]$$

corresponds to the solution

$$b_{12} = -5 \qquad \text{and} \qquad b_{22} = 2$$

We conclude that

$$A^{-1} = \left[\begin{array}{cc} 3 & -5 \\ -1 & 2 \end{array}\right]$$

as we saw in Example 11.

This method works for larger matrices as well. It is quite efficient, considering that we must solve a large number of linear equations. We illustrate the technique on a 3×3 matrix.

▮ EXAMPLE 15 Find the inverse (if it exists) of

$$A = \left[\begin{array}{ccc} 1 & -1 & -1 \\ 2 & -1 & 1 \\ -1 & 1 & -1 \end{array}\right]$$

Solution   We write

$$\left[\begin{array}{ccc|ccc} 1 & -1 & -1 & 1 & 0 & 0 \\ 2 & -1 & 1 & 0 & 1 & 0 \\ -1 & 1 & -1 & 0 & 0 & 1 \end{array}\right] \quad \begin{array}{c} (R_1) \\ (R_2) \\ (R_3) \end{array}$$

and perform appropriate row transformations in order to get the identity matrix on the left side. The matrix on the right side is then the inverse matrix. The first step is to get 0s below the 1 in the first column.

$$\begin{array}{c} (R_1) \\ 2(R_1) - (R_2) \\ (R_1) + (R_3) \end{array} \left[\begin{array}{ccc|ccc} 1 & -1 & -1 & 1 & 0 & 0 \\ 0 & -1 & -3 & 2 & -1 & 0 \\ 0 & 0 & -2 & 1 & 0 & 1 \end{array}\right] \quad \begin{array}{c} (R_4) \\ (R_5) \\ (R_6) \end{array}$$

Now, we turn to the second column. We want a 1 in the middle and 0s on top and bottom of the second column.

$$
\begin{matrix}
(R_4) - (R_5) \\
(-1)(R_5) \\
(R_6)
\end{matrix}
\quad
\left[
\begin{array}{ccc|ccc}
1 & 0 & 2 & -1 & 1 & 0 \\
0 & 1 & 3 & -2 & 1 & 0 \\
0 & 0 & -2 & 1 & 0 & 1
\end{array}
\right]
\quad
\begin{matrix}
(R_7) \\
(R_8) \\
(R_9)
\end{matrix}
$$

Next, we turn to the third column. We need 0s in the first two rows and a 1 in the third row. Note that we can only use (R_9) to do the transformations.

$$
\begin{matrix}
(R_7) + (R_9) \\
2(R_8) + 3(R_9) \\
-\frac{1}{2}(R_9)
\end{matrix}
\quad
\left[
\begin{array}{ccc|ccc}
1 & 0 & 0 & 0 & 1 & 1 \\
0 & 2 & 0 & -1 & 2 & 3 \\
0 & 0 & 1 & -\frac{1}{2} & 0 & -\frac{1}{2}
\end{array}
\right]
\quad
\begin{matrix}
(R_{10}) \\
(R_{11}) \\
(R_{12})
\end{matrix}
$$

Finally, we need to get a 1 in the second row and second column.

$$
\begin{matrix}
(R_{10}) \\
\frac{1}{2}(R_{11}) \\
(R_{12})
\end{matrix}
\quad
\left[
\begin{array}{ccc|ccc}
1 & 0 & 0 & 0 & 1 & 1 \\
0 & 1 & 0 & -\frac{1}{2} & 1 & \frac{3}{2} \\
0 & 0 & 1 & -\frac{1}{2} & 0 & -\frac{1}{2}
\end{array}
\right]
$$

We succeeded in getting the identity matrix on the left side. Therefore,

$$
A^{-1} =
\begin{bmatrix}
0 & 1 & 1 \\
-\frac{1}{2} & 1 & \frac{3}{2} \\
-\frac{1}{2} & 0 & -\frac{1}{2}
\end{bmatrix}
$$

This method is reasonably quick as long as the matrices are not too big. If you cannot get the identity matrix on the left side, then the matrix does not have an inverse, as illustrated in the next example.

▮ **EXAMPLE 16** Find the inverse (if it exists) of

$$
A =
\begin{bmatrix}
2 & 6 \\
1 & 3
\end{bmatrix}
$$

Solution Since $\det A = (2)(3) - (1)(6) = 0$, we already know that A is singular, that is, does not have an inverse. But let's see what happens if we try to find the inverse.

$$
\left[
\begin{array}{cc|cc}
2 & 6 & 1 & 0 \\
1 & 3 & 0 & 1
\end{array}
\right]
\quad
\begin{matrix}
(R_1) \\
(R_2)
\end{matrix}
$$

As before, we try to get a 1 in the first row and first column and a 0 in the second row and first column. We find

$$
\begin{matrix}
\frac{1}{2}(R_1) \\
(R_1) - 2(R_2)
\end{matrix}
\quad
\left[
\begin{array}{cc|cc}
1 & 3 & \frac{1}{2} & 0 \\
0 & 0 & 1 & -2
\end{array}
\right]
$$

Since the last row in the left half consists only of 0s, we can never obtain the identity matrix on the left side. That is, our method fails to provide an inverse matrix and we conclude that A does not have an inverse.

9.2.5 An Application: The Leslie Matrix

We introduced age-structured populations in Section 8.5, where we considered a population that reproduced continually over time. In this subsection, we will discuss populations with discrete breeding seasons, such as perennial plants in the temperate zone, where reproduction is limited to a particular season of the year. Such populations are better described by discrete-time models. (The results of Section 8.5 are not needed here.)

We begin with the simplest case. We measure time in units of generations and denote the size of a population at generation t by $N(t)$, where $t = 0, 1, 2, \ldots$. The discrete-time analogue of unrestricted growth is described by

$$N(t + 1) = RN(t) \quad \text{for } t = 0, 1, 2, \ldots \tag{9.12}$$

with $N(0) = N_0$. The constant R can be interpreted as the number of individuals in the next generation per individual in the current generation. We assume that $R \geq 0$. We encountered this type of equation in Subsection 1.2.5, when we introduced exponential functions. There, we saw how to find a solution of (9.12). Namely,

$$N(1) = RN(0)$$
$$N(2) = RN(1) = R[RN(0)] = R^2 N(0)$$
$$N(3) = RN(2) = R[R^2 N(0)] = R^3 N(0)$$

Continuing in this way, we find

$$N(t) = R^t N(0)$$

That is, the unrestricted growth of (9.12) yields exponential growth of the population. Depending on the value of R, we obtain the following long-term behavior

$$\lim_{t \to \infty} N(t) = \begin{cases} \infty & \text{if } R > 1 \\ N(0) & \text{if } R = 1 \\ 0 & \text{if } 0 \leq R < 1 \end{cases}$$

(see Figure 9.9)

The population model described by (9.12) is the simplest model of population growth. The quantity R describes the relative change of the population size from generation to generation.

In many species, reproduction is highly age dependent. For instance, periodical cicadas spend 13–17 years in the nymphal stage; they only reproduce once in their life. The purple coneflower, a prairie flower, does not reproduce until it is about three years old but then continues to produce seeds throughout its life. To take the life

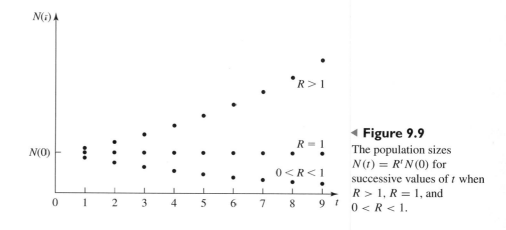

◀ **Figure 9.9**
The population sizes $N(t) = R^t N(0)$ for successive values of t when $R > 1$, $R = 1$, and $0 < R < 1$.

history into account, we will examine discrete-time, age-structured models. These models were introduced by Patrick Leslie (1945). They are formulated as matrix models, which allows one to use life tables to calculate important demographic quantities, such as the population size in each age class and the growth rate of the population.

Leslie's approach is widely used, not only in population biology but also in the life insurance industry. It is an example of a relatively simple mathematical method that allows one to use field data for demographic predictions.

We begin with a numerical example to illustrate this method. For simplicity, we will only follow the females in the population. We assume that breeding occurs once a year, and that we take a census of the population at the end of each breeding season. Individuals that are born during a particular breeding season are of age 0 at the end of this breeding season. If a zero-year-old survives until the end of the following breeding season, it will be of age 1 when we take a census at the end of that breeding season. If a one-year-old survives until the end of the next breeding season, it will be of age 2 at the end of that breeding season, and so on. In our example, we will assume that no individual lives beyond age 3, that is, there are no individuals of age 4 or older in the population. We denote by

$$N_x(t) = \text{number of females of age } x \text{ at time } t$$

where $t = 0, 1, 2, \ldots$. Then,

$$N(t) = \begin{bmatrix} N_0(t) \\ N_1(t) \\ N_2(t) \\ N_3(t) \end{bmatrix}$$

is the vector that describes the number of females in each age class at time t.

We assume that 40% of the females of age 0, 30% of the females of age 1, and 10% of the females of age 2 at time t are alive when we take the census at time $t + 1$. That is,

$$N_1(t + 1) = (0.4)N_0(t)$$

$$N_2(t + 1) = (0.3)N_1(t)$$

$$N_3(t + 1) = (0.1)N_2(t)$$

The number of newborn females at time $t + 1$ depends on the number of births of each surviving female during the breeding season. We assume that

$$N_0(t + 1) = 2N_1(t) + 1.5N_2(t)$$

which means that females reach sexual maturity when they are of age 1. The factor 2 in front of $N_1(t)$ should be interpreted as the average number of female offspring of a one-year-old female. (The factor 1.5 in front of $N_2(t)$ has a similar interpretation.) There is no contribution of three-year-olds to the newborn class.

We can summarize the dynamics in matrix form:

$$\begin{bmatrix} N_0(t + 1) \\ N_1(t + 1) \\ N_2(t + 1) \\ N_3(t + 1) \end{bmatrix} = \begin{bmatrix} 0 & 2 & 1.5 & 0 \\ 0.4 & 0 & 0 & 0 \\ 0 & 0.3 & 0 & 0 \\ 0 & 0 & 0.1 & 0 \end{bmatrix} \begin{bmatrix} N_0(t) \\ N_1(t) \\ N_2(t) \\ N_3(t) \end{bmatrix} \qquad (9.13)$$

The 4×4 matrix in this equation is called the **Leslie matrix**. We denote this matrix by L. We can write the matrix equation (9.13) in short form,

$$N(t+1) = LN(t) \tag{9.14}$$

To see how this works, let's assume that the population at time t has the following age distribution:

$$N_0(t) = 1000, \quad N_1(t) = 200, \quad N_2(t) = 100 \quad \text{and} \quad N_3(t) = 10$$

When we take a census after the following breeding season, we find that 40% of the females of age 0 at time t are alive at time $t+1$, that is, $N_1(t+1) = (0.4)(1000) = 400$; 30% of the females of age 1 at time t are alive at time $t+1$, that is, $N_2(t+1) = (0.3)(200) = 60$; and 10% of the females of age 2 at time t are alive at time $t+1$, that is, $N_3(t+1) = (0.1)(100) = 10$. The number of newborns at time $t+1$ is

$$N_0(t+1) = 2N_1(t) + 1.5N_2(t) = (2)(200) + (1.5)(100) = 550$$

We obtain the same result when we use the Leslie matrix and equation (9.13):

$$\begin{bmatrix} N_0(t+1) \\ N_1(t+1) \\ N_2(t+1) \\ N_3(t+1) \end{bmatrix} = \begin{bmatrix} 0 & 2 & 1.5 & 0 \\ 0.4 & 0 & 0 & 0 \\ 0 & 0.3 & 0 & 0 \\ 0 & 0 & 0.1 & 0 \end{bmatrix} \begin{bmatrix} 1000 \\ 200 \\ 100 \\ 10 \end{bmatrix}$$

$$= \begin{bmatrix} (2)(200) + (1.5)(100) \\ (0.4)(1000) \\ (0.3)(200) \\ (0.1)(100) \end{bmatrix} = \begin{bmatrix} 550 \\ 400 \\ 60 \\ 10 \end{bmatrix}$$

To obtain the age distribution at time $t+2$, we apply the Leslie matrix to the population vector at time $t+1$, that is,

$$\begin{bmatrix} N_0(t+2) \\ N_1(t+2) \\ N_2(t+2) \\ N_3(t+2) \end{bmatrix} = \begin{bmatrix} 0 & 2 & 1.5 & 0 \\ 0.4 & 0 & 0 & 0 \\ 0 & 0.3 & 0 & 0 \\ 0 & 0 & 0.1 & 0 \end{bmatrix} \begin{bmatrix} 550 \\ 400 \\ 60 \\ 10 \end{bmatrix} = \begin{bmatrix} 890 \\ 220 \\ 120 \\ 6 \end{bmatrix}$$

These three breeding seasons are illustrated in Figure 9.10.

From the preceding discussion, we can now find the general form of the Leslie matrix. We assume that the population is divided into $m+1$ age classes. The census

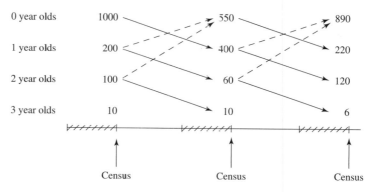

▲ **Figure 9.10**
An illustration of the three breeding seasons.

is again taken at the end of each breeding season. Then, for the survival of each age class, we find

$$N_1(t + 1) = P_0 N_0(t)$$
$$N_2(t + 1) = P_1 N_1(t)$$
$$\vdots$$
$$N_m(t + 1) = P_{m-1} N_{m-1}(t)$$

where P_i denotes the fraction of females of age i at time t that survive to time $t + 1$. Since P_i denotes a fraction, it follows that $0 \le P_i \le 1$ for $i = 0, 1, \ldots, m - 1$. The equation for the newborns is given by

$$N_0(t + 1) = F_0 N_0(t) + F_1 N_1(t) + \cdots + F_m N_m(t)$$

where F_i is the number of female offspring per female individual of age i. Writing this in matrix form, we find that the Leslie matrix is given by

$$\begin{bmatrix} F_0 & F_1 & F_2 & \cdots & F_{m-1} & F_m \\ P_0 & 0 & \cdots & \cdots & \cdots & 0 \\ 0 & P_1 & 0 & \cdots & \cdots & 0 \\ \vdots & \cdots & \cdots & \cdots & \cdots & \vdots \\ 0 & \cdots & \cdots & 0 & P_{m-1} & 0 \end{bmatrix}$$

This is a matrix in which all elements are 0, except possibly those in the first row and in the first subdiagonal below the diagonal.

EXAMPLE 17   Suppose that the Leslie matrix of a population is of the form

$$L = \begin{bmatrix} 5 & 7 & 1.5 \\ 0.2 & 0 & 0 \\ 0 & 0.4 & 0 \end{bmatrix}$$

Interpret this matrix, and determine what happens if you follow a population for two breeding seasons, starting with 1000 zero-year-old females.

Solution   The population is divided into three age classes, zero-year-olds, one-year-olds, and two-year-olds. The elements in the subdiagonal describe the survival probabilities, that is, 20% of the zero-year-olds survive until the next census and 40% of the one-year-olds survive until the next census. The maximum age is 2 years. Zero-year-olds produce an average of five females per female; one-year-olds produce an average of seven females per female; two-year-olds produce an average of 1.5 females per female.

A population that starts with 1000 zero-year-old females at time 0 has the population vector

$$N(0) = \begin{bmatrix} 1000 \\ 0 \\ 0 \end{bmatrix}$$

Using $N(t + 1) = LN(t)$, we find

$$N(1) = \begin{bmatrix} 5 & 7 & 1.5 \\ 0.2 & 0 & 0 \\ 0 & 0.4 & 0 \end{bmatrix} \begin{bmatrix} 1000 \\ 0 \\ 0 \end{bmatrix} = \begin{bmatrix} 5000 \\ 200 \\ 0 \end{bmatrix}$$

and

$$N(2) = \begin{bmatrix} 5 & 7 & 1.5 \\ 0.2 & 0 & 0 \\ 0 & 0.4 & 0 \end{bmatrix} \begin{bmatrix} 5000 \\ 200 \\ 0 \end{bmatrix} = \begin{bmatrix} 26{,}400 \\ 1000 \\ 80 \end{bmatrix}$$

▮

Stable Age Distribution   We will now investigate what happens when we run such a model for a long time. Let's assume that the Leslie matrix is given by

$$L = \begin{bmatrix} 1.5 & 2 \\ 0.08 & 0 \end{bmatrix}$$

We can compute successive population vectors using the equation

$$N(t+1) = LN(t)$$

That is,

$$N_0(t+1) = 1.5N_0(t) + 2N_1(t)$$

$$N_1(t+1) = 0.08N_0(t)$$

Suppose that we start with $N_0(0) = 100$ and $N_1(0) = 100$. Then

$$\begin{bmatrix} N_0(1) \\ N_1(1) \end{bmatrix} = \begin{bmatrix} 1.5 & 2 \\ 0.08 & 0 \end{bmatrix} \begin{bmatrix} 100 \\ 100 \end{bmatrix}$$

$$= \begin{bmatrix} 350 \\ 8 \end{bmatrix}$$

Continuing in this way, we find the population vectors at successive times, starting at time 0,

$$\begin{bmatrix} 100 \\ 100 \end{bmatrix}, \quad \begin{bmatrix} 350 \\ 8 \end{bmatrix}, \quad \begin{bmatrix} 541 \\ 28 \end{bmatrix}, \quad \begin{bmatrix} 868 \\ 43 \end{bmatrix}, \quad \begin{bmatrix} 1388 \\ 69 \end{bmatrix},$$

$$\begin{bmatrix} 2221 \\ 111 \end{bmatrix}, \quad \begin{bmatrix} 3553 \\ 178 \end{bmatrix}, \quad \begin{bmatrix} 5685 \\ 284 \end{bmatrix}, \quad \cdots$$

(In these calculations, we have rounded to the nearest integer.) The first thing that you notice is that the total population is growing (the population size at time t is $N_0(t) + N_1(t)$). But we can say a lot more. If we look at the successive ratios

$$q_0(t) = \frac{N_0(t)}{N_0(t-1)}$$

we find the following:

t	1	2	3	4	5	6	7
$q_0(t)$	3.5	1.55	1.60	1.5991	1.6001	1.5997	1.6001

That is, $q_0(t)$ seems to approach a limiting value, namely 1.6. The same happens when we look at the ratio

$$q_1(t) = \frac{N_1(t)}{N_1(t-1)}$$

We find

t	1	2	3	4	5	6	7
$q_1(t)$	0.08	3.5	1.536	1.605	1.609	1.604	1.596

Both ratios seem to approach the same value, namely 1.6. In fact, one can show that

$$\lim_{t \to \infty} \frac{N_0(t)}{N_0(t-1)} = \lim_{t \to \infty} \frac{N_1(t)}{N_1(t-1)} = 1.6$$

(We will see in the next section how to get this result.) If we look at the fraction of females of age class 0, that is,

$$p(t) = \frac{N_0(t)}{N_0(t) + N_1(t)}$$

we find

t	0	1	2	3	4	5	6	7
$p(t)$	0.5	0.9777	0.9508	0.9528	0.9526	0.9524	0.9523	0.9524

That is, this fraction also seems to converge. It looks as if about 95.2% of the population are in age class 0 when t is sufficiently large.

Though the population is increasing in size, the fraction of females in age class 0 (and hence also in age class 1) seems to converge. This constant fraction is referred to as the **stable age distribution**.

This method of finding the stable age distribution does not always work, and we will give an example in Problem 68 in which the population does not reach a stable age distribution when starting from a particular initial condition. In the next section, we will learn another method to determine the stable age distribution. But first, we examine further the stable age distribution.

If we start the population in the stable age distribution, the fraction of females in age class 0 will remain the same, namely about 95.2%, and the population will increase by a constant factor each generation, namely by a factor of 1.6. Here is a numerical illustration of these two important properties. We will learn in the next section how to find a stable age distribution; so, take it for granted that

$$N(0) = \begin{bmatrix} 2000 \\ 100 \end{bmatrix}$$

is a stable age distribution. If we start with this stable age distribution, then

$$N(1) = LN(0) = \begin{bmatrix} 1.5 & 2 \\ 0.08 & 0 \end{bmatrix} \begin{bmatrix} 2000 \\ 100 \end{bmatrix} = \begin{bmatrix} 3200 \\ 160 \end{bmatrix}$$

The fraction of females in age class 0 remains the same, namely, $3200/3360 = 2000/2100$. We compare this to

$$(1.6) \begin{bmatrix} 2000 \\ 100 \end{bmatrix} = \begin{bmatrix} 3200 \\ 160 \end{bmatrix}$$

which yields the same result as expected. That is, if N denotes a stable age distribution, then

$$LN = \lambda N \tag{9.15}$$

where $\lambda = 1.6$ and $N = [N_C, N_1]'$, with $N_0/(N_0 + N_1) \approx 95.2\%$. (We will see in the next section how to compute λ and the ratio $N_0/(N_0 + N_1)$.) Equation (9.15) is the equation for the stable age distribution. The vector N is called an **eigenvector** and the value λ is an eigenvalue. Typically, a matrix has more than one eigenvalue. In the case of the Leslie matrix, the largest eigenvalue determines how the population grows.

In the next section, we will learn how to compute eigenvalues and eigenvectors. Because we will need this in other contexts as well, we will develop the theory in a more general setting. At the end of the next section, we will return to Leslie matrices, and see how to compute the stable age distribution and the growth rate of the population.

◆ 9.2 PROBLEMS

[9.2.1], [9.2.2]

In Problems 1–6, let

$$A = \begin{bmatrix} -1 & 2 \\ 0 & -3 \end{bmatrix},$$

$$B = \begin{bmatrix} 0 & 1 \\ 2 & 4 \end{bmatrix},$$

$$C = \begin{bmatrix} 1 & -2 \\ 1 & -1 \end{bmatrix}$$

1. Find $A - B + 2C$.
2. Find $-2A + 3B$
3. Determine D so that $A + B = 2A - B + D$.
4. Show that $A + B = B + A$.
5. Show that $(A - B) + C = A + (B + C)$.
6. Show that $2(A + B) = 2A + 2B$.

In Problems 7–12 let

$$A = \begin{bmatrix} 1 & 3 & -1 \\ 2 & 4 & 1 \\ 0 & -2 & 2 \end{bmatrix},$$

$$B = \begin{bmatrix} 5 & -1 & 4 \\ 2 & 0 & 1 \\ 1 & -3 & -3 \end{bmatrix},$$

$$C = \begin{bmatrix} -2 & 0 & 4 \\ 1 & -3 & 1 \\ 0 & 0 & 2 \end{bmatrix}$$

7. Find $2A + 3B - C$.
8. Find $3C - B + \frac{1}{2}A$.
9. Determine D so that $A + B + C + D = 0$.
10. Determine D so that $A + 4B = 2(A + B) + D$.
11. Show that $A + B = B + A$.
12. Show that $(A + B) + C = A + (B + C)$.

13. Find the transpose of

$$A = \begin{bmatrix} -1 & 0 & 3 \\ 2 & 1 & -4 \end{bmatrix}$$

14. Find the transpose of

$$A = \begin{bmatrix} 1 \\ -3 \\ 4 \end{bmatrix}$$

In Problems 15–20, let

$$A = \begin{bmatrix} -1 & 0 \\ 1 & 2 \end{bmatrix},$$

$$B = \begin{bmatrix} 2 & 3 \\ -1 & 1 \end{bmatrix},$$

$$C = \begin{bmatrix} 1 & 2 \\ 0 & -1 \end{bmatrix}$$

15. Compute the following.
   (a) AB (b) BA
16. Compute ABC.
17. Show that $AC \neq CA$.
18. Show that $(AB)C = A(BC)$.
19. Show that $(A + B)C = AC + BC$.
20. Show that $A(B + C) = AB + AC$.
21. Suppose that A is a 3×4 matrix and B is a 4×2 matrix. What is the size of the product AB?
22. Suppose that A is a 2×5 matrix, B is a 1×3, matrix, C is a 5×1 matrix, and D is a 2×3 matrix. Which of the following matrix multiplications are defined? Whenever it is defined, state the size of the resulting matrix.
   (a) AB (b) AC (c) BD
23. Suppose that A is a 4×3 matrix, B is a 1×3 matrix, C is a 3×1 matrix, and D is a 4×3 matrix. Which of the following matrix multiplications are defined? Whenever it is defined, state the size of the resulting matrix.
   (a) BD' (b) $D'A$ (c) ACB

24. Let

$$A = \begin{bmatrix} 1 & 4 & -2 \end{bmatrix}$$

and

$$B = \begin{bmatrix} -1 \\ 2 \\ 3 \end{bmatrix}$$

(a) Compute AB.   **(b)** Compute BA.

25. Let

$$A = \begin{bmatrix} 1 & 3 \\ 0 & -2 \end{bmatrix}$$

and

$$B = \begin{bmatrix} 1 & 2 & 0 & -1 \\ 2 & 1 & 3 & 0 \end{bmatrix}$$

(a) Compute AB.   **(b)** Compute $B'A$.

26. Suppose that

$$A = \begin{bmatrix} 1 & -1 \\ 3 & 0 \\ 5 & 2 \end{bmatrix}$$

and

$$B = \begin{bmatrix} 2 & 4 & 1 \\ 6 & 0 & 0 \end{bmatrix}$$

Show that $(AB)' = B'A'$.

27. Let

$$A = \begin{bmatrix} 2 & 1 \\ -1 & -3 \end{bmatrix}$$

Find A^2, A^3, and A^4.

28. Let

$$I_3 = \begin{bmatrix} 1 & 0 & 0 \\ 0 & 1 & 0 \\ 0 & 0 & 1 \end{bmatrix}$$

Show that $I_3 = I_3^2 = I_3^3$.

29. Let

$$A = \begin{bmatrix} 1 & 3 \\ 0 & -2 \end{bmatrix}$$

and

$$I_2 = \begin{bmatrix} 1 & 0 \\ 0 & 1 \end{bmatrix}$$

Show that $AI_2 = I_2A = A$.

30. Let

$$A = \begin{bmatrix} 1 & 3 & 0 \\ 0 & -1 & 2 \\ -1 & -2 & 1 \end{bmatrix}$$

and

$$I_3 = \begin{bmatrix} 1 & 0 & 0 \\ 0 & 1 & 0 \\ 0 & 0 & 1 \end{bmatrix}$$

Show that $AI_3 = I_3A = A$.

31. Write the following in matrix form.

$$\begin{aligned} 2x_1 + 3x_2 - \ x_3 &= 0 \\ 2x_2 + \ x_3 &= 1 \\ x_1 \quad\quad - 2x_3 &= 2 \end{aligned}$$

32. Write the following in matrix form.

$$\begin{aligned} 2x_2 - x_1 &= x_3 \\ 4x_1 + x_3 &= 7x_2 \\ x_2 - x_1 &= x_3 \end{aligned}$$

33. Write the following in matrix form.

$$\begin{aligned} 2x_1 - 3x_2 &= 4 \\ -x_1 + x_2 &= 3 \\ 3x_1 &= 4 \end{aligned}$$

34. Write the following in matrix form.

$$\begin{aligned} x_1 - 2x_2 + x_3 &= 1 \\ -2x_1 + x_2 - 3x_3 &= 0 \end{aligned}$$

[9.2.3]

35. Show that the inverse of

$$A = \begin{bmatrix} 2 & 1 \\ 1 & 1 \end{bmatrix}$$

is

$$B = \begin{bmatrix} 1 & -1 \\ -1 & 2 \end{bmatrix}$$

36. Show that the inverse of

$$A = \begin{bmatrix} 2 & 3 & 1 \\ 5 & 2 & 3 \\ 1 & 2 & 0 \end{bmatrix}$$

is

$$B = \begin{bmatrix} -\frac{6}{5} & \frac{2}{5} & \frac{7}{5} \\ \frac{3}{5} & -\frac{1}{5} & -\frac{1}{5} \\ \frac{8}{5} & -\frac{1}{5} & -\frac{11}{5} \end{bmatrix}$$

In Problems 37–40, let

$$A = \begin{bmatrix} -1 & 1 \\ 2 & 3 \end{bmatrix},$$

$$B = \begin{bmatrix} 2 & -2 \\ 3 & 2 \end{bmatrix}$$

37. Find the inverse (if it exists) of A.

38. Find the inverse (if it exists) of B.

39. Show that $(A^{-1})^{-1} = A$.

40. Show that $(AB)^{-1} = B^{-1}A^{-1}$.

41. Find the inverse (if it exists) of

$$C = \begin{bmatrix} 2 & 4 \\ 3 & 6 \end{bmatrix}$$

42. Find the inverse (if it exists) of

$$I_3 = \begin{bmatrix} 1 & 0 & 0 \\ 0 & 1 & 0 \\ 0 & 0 & 1 \end{bmatrix}$$

43. Suppose that

$$A = \begin{bmatrix} -1 & 0 \\ 2 & -3 \end{bmatrix}$$

and

$$D = \begin{bmatrix} -2 \\ -5 \end{bmatrix}$$

Find X, so that $AX = D$, by

(a) solving the associated system of linear equations, and

(b) using the inverse of A.

44. (a) Show that if $X = AX + D$, then

$$X = (I - A)^{-1}D$$

provided that $I - A$ is invertible.

(b) Suppose now that

$$A = \begin{bmatrix} 3 & 2 \\ 0 & -1 \end{bmatrix}$$

and

$$D = \begin{bmatrix} -2 \\ 2 \end{bmatrix}$$

Compute $(I - A)^{-1}$, and use your result in (a) to compute X.

45. Use the determinant to determine whether the matrix

$$A = \begin{bmatrix} 2 & -1 \\ 1 & 3 \end{bmatrix}$$

is invertible.

46. Use the determinant to determine whether the matrix

$$A = \begin{bmatrix} -1 & 3 \\ 0 & 3 \end{bmatrix}$$

is invertible.

47. Use the determinant to determine whether the matrix

$$A = \begin{bmatrix} 4 & -1 \\ 8 & -2 \end{bmatrix}$$

is invertible.

48. Use the determinant to determine whether the matrix

$$A = \begin{bmatrix} -1 & 1 \\ -1 & 1 \end{bmatrix}$$

is invertible.

49. Suppose that

$$A = \begin{bmatrix} 2 & 4 \\ 3 & 6 \end{bmatrix}$$

(a) Compute $\det A$. Is A invertible?

(b) Suppose that

$$X = \begin{bmatrix} x \\ y \end{bmatrix}$$

and

$$B = \begin{bmatrix} b_1 \\ b_2 \end{bmatrix}$$

Write $AX = B$ as a system of linear equations.

(c) Show that if

$$B = \begin{bmatrix} 3 \\ \frac{9}{2} \end{bmatrix}$$

then

$$AX = B$$

has infinitely many solutions. Graph the two straight lines associated with the corresponding system of linear equations, and explain why it has infinitely many solutions.

(d) Find a column vector

$$B = \begin{bmatrix} b_1 \\ b_2 \end{bmatrix}$$

so that

$$AX = B$$

has no solutions.

50. Suppose that

$$A = \begin{bmatrix} a & 8 \\ 2 & 4 \end{bmatrix},$$

$$X = \begin{bmatrix} x \\ y \end{bmatrix},$$

and

$$B = \begin{bmatrix} b_1 \\ b_2 \end{bmatrix}$$

(a) Show that when $a \neq 4$, then $AX = B$ has exactly one solution.

(b) Suppose $a = 4$. Find conditions on b_1 and b_2 so that $AX = B$ has (i) infinitely many solutions, and (ii) no solutions.

(c) Explain your results in (a) and (b) graphically.

51. Use the determinant to determine whether

$$A = \begin{bmatrix} 1 & -1 \\ 0 & 2 \end{bmatrix}$$

is invertible. If it is invertible, compute its inverse. In either case, solve $AX = 0$.

52. Use the determinant to determine whether

$$B = \begin{bmatrix} 1 & 1 \\ 2 & 1 \end{bmatrix}$$

is invertible. If it is invertible, compute its inverse. In either case, solve $BX = \mathbf{0}$.

53. Use the determinant to determine whether

$$C = \begin{bmatrix} 1 & 3 \\ 1 & 3 \end{bmatrix}$$

is invertible. If it is invertible, compute its inverse. In either case, solve $CX = \mathbf{0}$.

54. Use the determinant to determine whether

$$D = \begin{bmatrix} -3 & 6 \\ -4 & 8 \end{bmatrix}$$

is invertible. If it is invertible, compute its inverse. In either case, solve $DX = \mathbf{0}$.

[9.2.4]

In Problems 55–58, find the inverse matrix to each given matrix, if it exists.

55. $A = \begin{bmatrix} 2 & -1 & -1 \\ 2 & 1 & 1 \\ -1 & 1 & -1 \end{bmatrix}$

56. $A = \begin{bmatrix} -1 & 3 & -1 \\ 2 & -2 & 3 \\ -1 & 1 & 2 \end{bmatrix}$

57. $A = \begin{bmatrix} -1 & 0 & -1 \\ 0 & -2 & 0 \\ -1 & 1 & 2 \end{bmatrix}$

58. $A = \begin{bmatrix} -1 & 0 & 2 \\ -1 & -2 & 3 \\ 0 & 2 & -1 \end{bmatrix}$

[9.2.5]

In Problems 59–62, assume that breeding occurs once a year and that a census is taken at the end of each breeding season.

59. Assume that a population is divided into three age classes, and that 20% of the females of age 0 and 70% of the females of age 1 survive until the end of the following breeding season. Furthermore, assume that females of age 1 have an average of 3.2 female offspring and females of age 2 have an average of 1.7 female offspring. If, at time 0, the population consists of 2000 females of age 0, 800 females of age 1, and 200 females of age 2, find the Leslie matrix and the age distribution at time 2.

60. Assume that a population is divided into three age classes, and that 80% of the females of age 0 and 10% of the females

of age 1 survive until the end of the following breeding season. Furthermore, assume that females of age 1 have an average of 1.6 female offspring and females of age 2 have an average of 3.9 female offspring. If, at time 0, the population consists of 1000 females of age 0, 100 females of age 1, and 20 females of age 2, find the Leslie matrix and the age distribution at time 3.

61. Assume that a population is divided into four age classes, and that 70% of the females of age 0, 50% of the females of age 1, and 10% of the females of age 2 survive until the end of the following breeding season. Furthermore, assume that females of age 2 have an average of 4.6 female offspring and females of age 3 have an average of 3.7 female offspring. If, at time 0, the population consists of 1500 females of age 0, 500 females of age 1, 250 females of age 2, and 50 females of age 3, find the Leslie matrix and the age distribution at time 2.

62. Assume that a population is divided into four age classes, and that 65% of the females of age 0, 40% of the females of age 1, and 30% of the females of age 2 survive until the end of the following breeding season. Furthermore, assume that females of age 1 have an average of 2.8 female offspring, females of age 2 have an average of 7.2 female offspring, and females of age 3 have an average of 3.7 female offspring. If, at time 0, the population consists of 1500 females of age 0, 500 females of age 1, 250 females of age 2, and 50 females of age 3, find the Leslie matrix and the age distribution at time 3.

In Problems 63–64, assume the given Leslie matrix L. Determine the number of age classes in the population, the fraction of one-year-olds that survive until the end of the following breeding season, and the average number of female offspring of a two-year-old female.

63. $L = \begin{bmatrix} 2 & 3 & 2 & 1 \\ 0.4 & 0 & 0 & 0 \\ 0 & 0.6 & 0 & 0 \\ 0 & 0 & 0.8 & 0 \end{bmatrix}$

64. $L = \begin{bmatrix} 0 & 5 & 0 \\ 0.8 & 0 & 0 \\ 0 & 0.3 & 0 \end{bmatrix}$

In Problems 65–66, assume the given Leslie matrix L. Determine the number of age classes in the population. What fraction of two-year-olds survive until the end of the following breeding season? Determine the average number of female offspring of a one-year-old female.

65. $L = \begin{bmatrix} 1 & 2.5 & 3 & 1.5 \\ 0.9 & 0 & 0 & 0 \\ 0 & 0.3 & 0 & 0 \\ 0 & 0 & 0.2 & 0 \end{bmatrix}$

66. $L = \begin{bmatrix} 0 & 4.2 & 3.7 \\ 0.7 & 0 & 0 \\ 0 & 0.1 & 0 \end{bmatrix}$

67. Assume that the Leslie matrix is of the form

$$L = \begin{bmatrix} 1.2 & 3.2 \\ 0.8 & 0 \end{bmatrix}$$

Suppose that at time $t = 0$, $N_0(0) = 100$, and $N_1(0) = 0$. Find the population vectors for $t = 0, 1, 2, \ldots, 10$. Compute the successive ratios

$$q_0(t) = \frac{N_0(t)}{N_0(t-1)}$$

and

$$q_1(t) = \frac{N_1(t)}{N_1(t-1)}$$

for $t = 1, 2, \ldots, 10$. What value do $q_0(t)$ and $q_1(t)$ approach as $t \to \infty$? (Take a guess.) Compute the fraction of females of age 0 for $t = 0, 1, \ldots, 10$. Can you find the stable age distribution?

68. Assume that a Leslie matrix is of the form

$$L = \begin{bmatrix} 0 & 5.2 \\ 0.7 & 0 \end{bmatrix}$$

Suppose that at time $t = 0$, $N_0(0) = 100$, and $N_1(0) = 0$. Find the population vectors for $t = 0, 1, 2, \ldots, 10$. Compute the successive ratios

$$q_0(t) = \frac{N_0(t)}{N_0(t-1)}$$

and

$$q_1(t) = \frac{N_1(t)}{N_1(t-1)}$$

for $t = 1, 2, \ldots, 10$. What value do $q_0(t)$ and $q_1(t)$ approach as $t \to \infty$? (Take a guess.) Compute the fraction of females of age 0 for $t = 0, 1, \ldots, 10$. Can you find the stable age distribution?

◆ 9.3 LINEAR MAPS, EIGENVECTORS, AND EIGENVALUES

In this section, we will consider maps of the form

$$\mathbf{x} \to A\mathbf{x} \tag{9.16}$$

where A is a 2×2 matrix and $\mathbf{x}$ is a 2×1 column vector (or, simply, vector). It follows that $A\mathbf{x}$ is a 2×1 vector. That is, this map takes a 2×1 vector and maps it into a 2×1 vector. This allows us to apply A again, that is, to compute $A(A\mathbf{x}) = A^2\mathbf{x}$, which is again a 2×1 vector, and so on. We will first look at vectors, then at maps $A\mathbf{x}$, and finally we will investigate iterates of the map A, that is $A^2\mathbf{x}$, $A^3\mathbf{x}$, and so on.

In the following we will denote vectors by boldface lowercase letters. It follows from the properties of matrix multiplications that the map (9.16) satisfies the following.

1. $A(\mathbf{x} + \mathbf{y}) = A\mathbf{x} + A\mathbf{y}$, and
2. $A(\lambda\mathbf{x}) = \lambda(A\mathbf{x})$, where λ is a scalar.

Because of these two properties, we say that the map $\mathbf{x} \to A\mathbf{x}$ is linear.

We saw an example of such a map in the previous section: if A is a 2×2 Leslie matrix and $\mathbf{x}$ is a population vector at time 0, then $A\mathbf{x}$ represents the population vector at time 1.

Linear maps are important in other contexts as well, and we will encounter them in Chapters 10 and 11. We restrict our discussion to 2×2 matrices, but point out that one can generalize the following theory to arbitrary $n \times n$ matrices. (These topics are covered in courses on matrix or linear algebra.)

9.3.1 Graphical Representation

Vectors   We begin with a graphical representation of vectors. We assume that $\mathbf{x}$ is a 2×1 matrix. We call $\mathbf{x}$ a column vector or simply a vector. Since a 2×1 matrix has just two components, we can represent a vector in the plane. For instance, to represent the vector

$$\mathbf{x} = \begin{bmatrix} 3 \\ 4 \end{bmatrix}$$

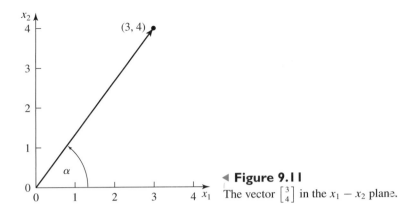

◀ **Figure 9.11**
The vector $\begin{bmatrix} 3 \\ 4 \end{bmatrix}$ in the $x_1 - x_2$ plane.

in the $x_1 - x_2$ plane, we draw an arrow from the origin $(0, 0)$ to the point $(3, 4)$, as illustrated in Figure 9.11. We see from Figure 9.11 that a vector has a length and a direction. The length of the vector $\mathbf{x} = \begin{bmatrix} 3 \\ 4 \end{bmatrix}$, denoted by $|\mathbf{x}|$, is the distance from the origin $(0, 0)$ to the point $(3, 4)$, that is,

$$\text{length of } \mathbf{x} = |\mathbf{x}| = \sqrt{9 + 16} = 5$$

We define the direction of $\mathbf{x}$ as the angle α between the positive x_1-axis and the vector $\mathbf{x}$ (measured counterclockwise), as shown in Figure 9.11. The angle α is in the interval $[0, 2\pi)$. In our example, α satisfies $\tan \alpha = 4/3$.

More generally, a vector $\begin{bmatrix} x_1 \\ x_2 \end{bmatrix}$ has length

$$|\mathbf{x}| = \sqrt{x_1^2 + x_2^2}$$

and direction α, where $\alpha \in [0, 2\pi)$ satisfies

$$\tan \alpha = \frac{x_2}{x_1}$$

The angle α is always measured counterclockwise from the positive x_1-axis.

If we denote the length of $\mathbf{x}$ by r, that is, $r = |\mathbf{x}|$, then since $x_1 = r \cos \alpha$ and $x_2 = r \sin \alpha$, the vector $\mathbf{x}$ can also be written as

$$\mathbf{x} = \begin{bmatrix} r \cos \alpha \\ r \sin \alpha \end{bmatrix}$$

We thus have an alternative way of representing vectors in the plane. We can either use the endpoint (x_1, x_2) or the length and direction (r, α). The first representation leads to our familiar Cartesian coordinate system. The second representation, using the length and direction of the corresponding vector, leads to the **polar coordinate system**. We will use both representations in the following.

Since vectors are matrices, we can use matrix addition to add vectors. For instance,

$$\begin{bmatrix} 1 \\ 2 \end{bmatrix} + \begin{bmatrix} 2 \\ 2 \end{bmatrix} = \begin{bmatrix} 3 \\ 4 \end{bmatrix}$$

This has a nice graphical interpretation (see Figure 9.12).

To add $\begin{bmatrix} 2 \\ 2 \end{bmatrix}$ to $\begin{bmatrix} 1 \\ 2 \end{bmatrix}$, we move the vector $\begin{bmatrix} 2 \\ 2 \end{bmatrix}$ to the tip of the vector $\begin{bmatrix} 1 \\ 2 \end{bmatrix}$ without changing the direction or the length of $\begin{bmatrix} 2 \\ 2 \end{bmatrix}$. The sum of the two vectors then starts at $(0, 0)$ and ends at the point where the tip of the vector that was moved, ended.

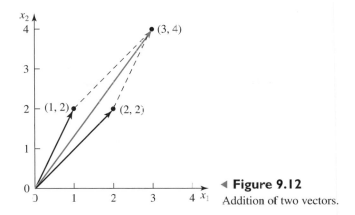

◀ **Figure 9.12**
Addition of two vectors.

This can also be described in the following way. The sum is the diagonal in the parallelogram that is formed by the two vectors $\begin{bmatrix} 1 \\ 2 \end{bmatrix}$ and $\begin{bmatrix} 2 \\ 2 \end{bmatrix}$. The rules for vector addition are therefore referred to as the **parallelogram law**.

Multiplication of a vector by a scalar is carried out componentwise. For instance, if we multiply $\begin{bmatrix} 1 \\ 2 \end{bmatrix}$ by 2, we find

$$2 \begin{bmatrix} 1 \\ 2 \end{bmatrix} = \begin{bmatrix} 2 \\ 4 \end{bmatrix}$$

This corresponds to changing the length of the vector. Since 2 is positive, the resulting vector points in the same direction as the vector $\begin{bmatrix} 1 \\ 2 \end{bmatrix}$. If we multiply $\begin{bmatrix} 1 \\ 2 \end{bmatrix}$ by -1, then the resulting vector is $\begin{bmatrix} -1 \\ -2 \end{bmatrix}$, which has the same length as the original vector but points in the opposite direction, as illustrated in Figure 9.13.

EXAMPLE 1   Let

$$\mathbf{u} = \begin{bmatrix} 2 \\ 1 \end{bmatrix}, \qquad \mathbf{v} = \begin{bmatrix} -1 \\ -3 \end{bmatrix}, \qquad \mathbf{w} = \begin{bmatrix} -2 \\ 3 \end{bmatrix}$$

Find $-\frac{1}{2}\mathbf{u}$, $\mathbf{u} + \mathbf{v}$, $\mathbf{v} + \mathbf{w}$, and $-\mathbf{v}$, and illustrate the results graphically.

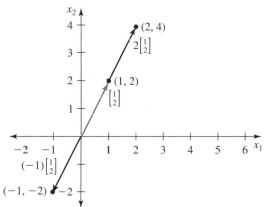

◀ **Figure 9.13**
Multiplication by a scalar.

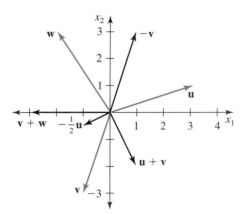

◀ **Figure 9.14**
The vectors in Example 1.

Solution

$$-\frac{1}{2}\mathbf{u} = -\frac{1}{2}\begin{bmatrix} 2 \\ 1 \end{bmatrix} = \begin{bmatrix} -1 \\ -\frac{1}{2} \end{bmatrix}$$

$$\mathbf{u} + \mathbf{v} = \begin{bmatrix} 2 \\ 1 \end{bmatrix} + \begin{bmatrix} -1 \\ -3 \end{bmatrix} = \begin{bmatrix} 1 \\ -2 \end{bmatrix}$$

$$\mathbf{v} + \mathbf{w} = \begin{bmatrix} -1 \\ -3 \end{bmatrix} + \begin{bmatrix} -2 \\ 3 \end{bmatrix} = \begin{bmatrix} -3 \\ 0 \end{bmatrix}$$

$$-\mathbf{v} = -\begin{bmatrix} -1 \\ -3 \end{bmatrix} = \begin{bmatrix} 1 \\ 3 \end{bmatrix}$$

The results are illustrated in Figure 9.14.   ▮

We summarize vector addition, and multiplication of a vector by a scalar, in the following.

Let $\mathbf{x} = \begin{bmatrix} x_1 \\ x_2 \end{bmatrix}$ and $\mathbf{y} = \begin{bmatrix} y_1 \\ y_2 \end{bmatrix}$. Then

$$\mathbf{x} + \mathbf{y} = \begin{bmatrix} x_1 + y_1 \\ x_2 + y_2 \end{bmatrix}$$

If a is a scalar, then

$$a\mathbf{x} = \begin{bmatrix} ax_1 \\ ax_2 \end{bmatrix}$$

Linear Maps   We will now turn to maps of the form

$$\mathbf{x} \rightarrow A\mathbf{x}$$

where A is a 2×2 matrix and $\mathbf{x}$ is a 2×1 vector. Since $A\mathbf{x}$ is a 2×1 vector as well, the map A takes the 2×1 vector $\mathbf{x}$ and maps it again into a 2×1 vector. We can thus represent such maps in the plane, and graphically investigate the action of A on the vector $\mathbf{x}$.

The simplest such map is the **identity map**, represented by the identity matrix I_2, that is, the map

$$\begin{bmatrix} x_1 \\ x_2 \end{bmatrix} \rightarrow \begin{bmatrix} 1 & 0 \\ 0 & 1 \end{bmatrix} \begin{bmatrix} x_1 \\ x_2 \end{bmatrix}$$

Since

$$I_2\mathbf{x} = \begin{bmatrix} 1 & 0 \\ 0 & 1 \end{bmatrix} \begin{bmatrix} x_1 \\ x_2 \end{bmatrix} = \begin{bmatrix} x_1 \\ x_2 \end{bmatrix} = \mathbf{x}$$

we conclude that the identity matrix leaves the vector $\mathbf{x}$ unchanged.

Slightly more complicated is the map

$$\begin{bmatrix} x_1 \\ x_2 \end{bmatrix} \rightarrow \begin{bmatrix} a & 0 \\ 0 & b \end{bmatrix} \begin{bmatrix} x_1 \\ x_2 \end{bmatrix} = \begin{bmatrix} ax_1 \\ bx_2 \end{bmatrix}$$

This map stretches or contracts each coordinate separately. In Figure 9.15, we show the action of the map on $\begin{bmatrix} 1 \\ 2 \end{bmatrix}$ when $a = 2$ and $b = -\frac{1}{2}$. We find

$$\begin{bmatrix} 2 & 0 \\ 0 & -\frac{1}{2} \end{bmatrix} \begin{bmatrix} 1 \\ 2 \end{bmatrix} = \begin{bmatrix} 2 \\ -1 \end{bmatrix}$$

That is, this map stretches the first coordinate by a factor of 2 and contracts the second coordinate by a factor of 1/2. The minus sign in front of $\frac{1}{2}$ corresponds to reflecting the second coordinate about the x_1-axis.

Another linear map is a **rotation**, that is, a map that rotates the $x_1 - x_2$ plane by a fixed angle. The following matrix rotates the plane counterclockwise by an angle θ.

$$R_\theta = \begin{bmatrix} \cos\theta & -\sin\theta \\ \sin\theta & \cos\theta \end{bmatrix}$$

To check that this is indeed a rotation, we investigate the action of R_θ on a vector with coordinates (x_1, x_2), as illustrated in Figure 9.16. Using the polar coordinate system, we can write this vector as

$$\begin{bmatrix} x_1 \\ x_2 \end{bmatrix} = \begin{bmatrix} r\cos\alpha \\ r\sin\alpha \end{bmatrix}$$

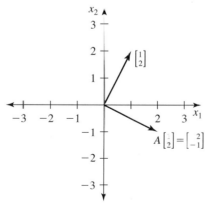

◀ **Figure 9.15**

The action of the matrix A on the vector $\begin{bmatrix} 1 \\ 2 \end{bmatrix}$.

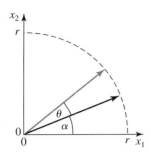

▲ Figure 9.16
The rotation of a vector.

where r is the length of the vector and α is the angle it forms with the positive x_1-axis.

Applying R_θ, we find

$$\begin{bmatrix} \cos\theta & -\sin\theta \\ \sin\theta & \cos\theta \end{bmatrix} \begin{bmatrix} r\cos\alpha \\ r\sin\alpha \end{bmatrix} = \begin{bmatrix} r(\cos\theta\cos\alpha - \sin\theta\sin\alpha) \\ r(\sin\theta\cos\alpha + \cos\theta\sin\alpha) \end{bmatrix}$$

$$= \begin{bmatrix} r\cos(\theta+\alpha) \\ r\sin(\theta+\alpha) \end{bmatrix}$$

where we used the trigonometric identities

$$\cos(\theta+\alpha) = \cos\theta\cos\alpha - \sin\theta\sin\alpha$$

$$\sin(\theta+\alpha) = \sin\theta\cos\alpha + \cos\theta\sin\alpha$$

We thus see that the resulting vector still has length r, but now the angle with the x_1-axis is $\alpha + \theta$; that is, the vector was rotated counterclockwise by the angle θ.

From this brief discussion of linear maps, we see that the map $\mathbf{x} \to A\mathbf{x}$ typically takes the vector $\mathbf{x}$ and rotates, stretches, or contracts it. For an arbitrary matrix A, vectors may be moved in a way that has no simple geometric interpretation. Take, for instance, the matrix

$$A = \begin{bmatrix} 1 & 2 \\ 3 & 2 \end{bmatrix}$$

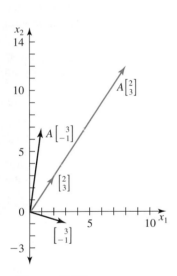

▲ Figure 9.17
The action of the matrix
$A = \begin{bmatrix} 1 & 2 \\ 3 & 2 \end{bmatrix}$ on two vectors.

If $\mathbf{x} = \begin{bmatrix} 3 \\ -1 \end{bmatrix}$, then

$$A\mathbf{x} = \begin{bmatrix} 1 & 2 \\ 3 & 2 \end{bmatrix} \begin{bmatrix} 3 \\ -1 \end{bmatrix} = \begin{bmatrix} 1 \\ 7 \end{bmatrix}$$

We can easily compute the outcome of this map, but there does not seem to be a straightforward geometric explanation (see Figure 9.17). On the other hand, if $\mathbf{x} = \begin{bmatrix} 2 \\ 3 \end{bmatrix}$, then

$$A\mathbf{x} = \begin{bmatrix} 1 & 2 \\ 3 & 2 \end{bmatrix} \begin{bmatrix} 2 \\ 3 \end{bmatrix} = \begin{bmatrix} 8 \\ 12 \end{bmatrix} = 4 \begin{bmatrix} 2 \\ 3 \end{bmatrix}$$

That is, the result of $A\mathbf{x}$ is simply a multiple of $\mathbf{x}$ (see Figure 9.17). Such a vector is called an eigenvector, and the stretching or contracting factor is called an eigenvalue. Eigenvectors and eigenvalues are the topic of the next subsection.

9.3.2 Eigenvalues and Eigenvectors

We saw in the preceding subsection that there are matrices and vectors for which the map $\mathbf{x} \to A\mathbf{x}$ takes on a particularly simple form, namely

$$A\mathbf{x} = \lambda\mathbf{x} \tag{9.17}$$

where λ is a scalar. This subsection is devoted to investigating this relationship. We will again restrict our discussion to 2×2 matrices, and begin with the following definition.

> **Definition**   Assume that A is a square matrix. A nonzero vector $\mathbf{x}$ that satisfies the equation
>
> $$A\mathbf{x} = \lambda\mathbf{x}$$
>
> is an eigenvector of the matrix A and the number λ is an eigenvalue of the matrix A.

Note that we assume that the vector $\mathbf{x}$ is different from the zero vector. (The zero vector $\mathbf{x} = \mathbf{0}$ always satisfies the equation $A\mathbf{x} = \lambda\mathbf{x}$ and thus would not be special.) The eigenvalue λ can be 0, however. We will see below that λ can even be a complex number.

The action of A on eigenvectors is of a particularly simple form; if we apply A to an eigenvector $\mathbf{x}$ (that is, compute $A\mathbf{x}$), the result is simply a constant multiple of $\mathbf{x}$. This property of an eigenvector has an important geometric interpretation when the eigenvalue λ is a real number. Namely, if we draw a straight line through the origin in the direction of an eigenvector, then any vector on this straight line will remain on this line after applying the map A (see Figure 9.18). We will discuss this in detail subsequently.

We will first discuss how to find eigenvalues and eigenvectors for 2×2 matrices. We will see that a 2×2 matrix has two eigenvalues, which are either distinct or identical. We will only discuss the case in which the two eigenvalues are distinct; the case in which the eigenvalues are identical is more complicated, and is covered in courses on linear algebra. For our purposes, it will suffice to look at the case in which the two eigenvalues are distinct.

We will show how to find eigenvalues and eigenvectors by way of example. We will use the same matrix as before, namely

$$A = \begin{bmatrix} 1 & 2 \\ 3 & 2 \end{bmatrix}$$

We are interested in finding $\mathbf{x} \neq \mathbf{0}$ and λ, such that

$$A\mathbf{x} = \lambda\mathbf{x}$$

We can rewrite this as

$$A\mathbf{x} - \lambda\mathbf{x} = \mathbf{0}$$

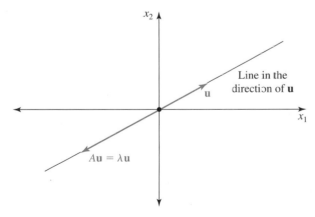

◀ **Figure 9.18**

Any vector on the line in the direction of the eigenvector will remain on the line under the map A.

In order to factor $\mathbf{x}$, we must multiply $\lambda\mathbf{x}$ by the identity matrix $I = I_2$. (Because we will be dealing only with 2×2 matrices, we simply write I instead of I_2.) This yields

$$Ax - \lambda I\mathbf{x} = \mathbf{0}$$

We can now factor $\mathbf{x}$, which results in

$$(A - \lambda I)\mathbf{x} = \mathbf{0}$$

We showed in Section 9.2 that in order to obtain a nontrivial solution ($\mathbf{x} \neq \mathbf{0}$), $A - \lambda I$ must be singular, that is,

$$\det(A - \lambda I) = 0$$

Now,

$$A - \lambda I = \begin{bmatrix} 1 - \lambda & 2 \\ 3 & 2 - \lambda \end{bmatrix}$$

and

$$\det(A - \lambda I) = (1 - \lambda)(2 - \lambda) - (2)(3)$$
$$= 2 - 3\lambda + \lambda^2 - 6 = \lambda^2 - 3\lambda - 4$$
$$= (\lambda + 1)(\lambda - 4)$$

Thus, we need to solve

$$(\lambda + 1)(\lambda - 4) = 0$$

and find

$$\lambda_1 = -1 \qquad \text{and} \qquad \lambda_2 = 4$$

These two numbers are the eigenvalues of the matrix A. Each eigenvalue will have its own eigenvector.

To compute the eigenvectors, we carry out the following calculations. If $\lambda_1 = -1$, we must find $\mathbf{x} = \begin{bmatrix} x_1 \\ x_2 \end{bmatrix}$ such that

$$A \begin{bmatrix} x_1 \\ x_2 \end{bmatrix} = (-1) \begin{bmatrix} x_1 \\ x_2 \end{bmatrix}$$

Writing this as a system of equations, we find

$$x_1 + 2x_2 = -x_1$$
$$3x_1 + 2x_2 = -x_2$$

Simplifying, we obtain

$$2x_1 + 2x_2 = 0$$
$$3x_1 + 3x_2 = 0$$

The two equations are identical, and reduce to

$$x_1 + x_2 = 0$$

We are looking for a nonzero vector that satisfies this equation. For instance,

$$\begin{bmatrix} x_1 \\ x_2 \end{bmatrix} = \begin{bmatrix} 1 \\ -1 \end{bmatrix}$$

would be a reasonable choice. To check that $\begin{bmatrix} 1 \\ -1 \end{bmatrix}$ is indeed an eigenvector corresponding to the eigenvalue $\lambda_1 = -1$, we compute

$$\begin{bmatrix} 1 & 2 \\ 3 & 2 \end{bmatrix} \begin{bmatrix} 1 \\ -1 \end{bmatrix} = \begin{bmatrix} -1 \\ 1 \end{bmatrix} = (-1) \begin{bmatrix} 1 \\ -1 \end{bmatrix}$$

that is,

$$A \begin{bmatrix} 1 \\ -1 \end{bmatrix} = (-1) \begin{bmatrix} 1 \\ -1 \end{bmatrix}$$

The vector $\begin{bmatrix} 1 \\ -1 \end{bmatrix}$ is not the only eigenvector associated with $\lambda_1 = -1$. In fact, any vector $a \begin{bmatrix} 1 \\ -1 \end{bmatrix}$, $a \neq 0$, is an eigenvector associated with the eigenvalue -1. For instance, $\begin{bmatrix} -1 \\ 1 \end{bmatrix}$ or $\begin{bmatrix} 2 \\ -2 \end{bmatrix}$ are other choices (see Figure 9.19).

To find an eigenvector associated with $\lambda_2 = 4$, we must solve

$$A \begin{bmatrix} x_1 \\ x_2 \end{bmatrix} = 4 \begin{bmatrix} x_1 \\ x_2 \end{bmatrix}$$

which yields

$$x_1 + 2x_2 = 4x_1$$
$$3x_1 + 2x_2 = 4x_2$$

Simplifying, we obtain

$$-3x_1 + 2x_2 = 0$$
$$3x_1 - 2x_2 = 0$$

For instance,

$$\begin{bmatrix} x_1 \\ x_2 \end{bmatrix} = \begin{bmatrix} 2 \\ 3 \end{bmatrix}$$

satisfies this. We see that A has two eigenvalues, namely -1 and 4. An eigenvector associated with -1 is $\begin{bmatrix} 1 \\ -1 \end{bmatrix}$, an eigenvector associated with 4 is $\begin{bmatrix} 2 \\ 3 \end{bmatrix}$ (see Figure 9.20).

We see that eigenvectors are not unique, they are determined only up to a multiplicative constant. When the eigenvalues are real, as in our example, all eigenvectors corresponding to a particular eigenvalue lie on the *same* straight line through the

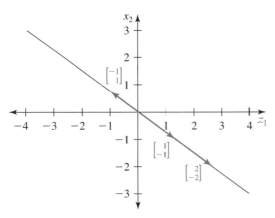

◀ **Figure 9.19**

The three vectors are all eigenvectors to the eigenvalue $\lambda_1 = -1$.

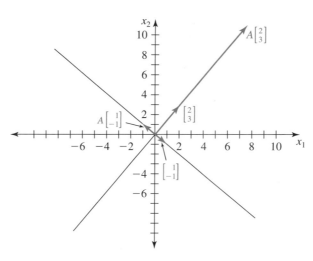

◀ **Figure 9.20**

The two eigenvectors with their corresponding lines. The images of the eigenvectors under the map A remain on their respective lines.

origin. Namely, the line represented by the vector $\begin{bmatrix} 1 \\ -1 \end{bmatrix}$ is given by

$$l_1 = \{(x_1, x_2) : x_1 + x_2 = 0\}$$

and the line represented by the vector $\begin{bmatrix} 2 \\ 3 \end{bmatrix}$ is given by

$$l_2 = \{(x_1, x_2) : 3x_1 - 2x_2 = 0\}$$

The lines l_1 and l_2 are **invariant** under the map $\mathbf{x} \to A\mathbf{x}$, in the sense that if we choose a point (x_1, x_2) on a line that is represented by an eigenvector, then since

$$A \begin{bmatrix} x_1 \\ x_2 \end{bmatrix} = \lambda \begin{bmatrix} x_1 \\ x_2 \end{bmatrix}$$

the result of the map is a point that is on the same line. We check this for the line l_1. Assume that $(x_1, x_2) \in l_1$, that is, $x_1 + x_2 = 0$; then the point $(\lambda x_1, \lambda x_2) \in l_1$ as well, since $\lambda x_1 + \lambda x_2 = \lambda(x_1 + x_2) = 0$. This is illustrated in Figure 9.21, where we draw both eigenvectors and the corresponding lines. We will refer to this as the geometric interpretation of eigenvalues and eigenvectors.

In the following example, we show the procedure for finding eigenvalues and eigenvectors once more, using a different matrix, which will also have real eigenvalues.

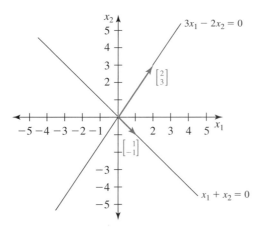

◀ **Figure 9.21**

Eigenvectors and associated lines of the matrix $\begin{bmatrix} 1 & 2 \\ 3 & 2 \end{bmatrix}$.

▮ **EXAMPLE 2**   **Finding eigenvalues and eigenvectors**

Find the eigenvalues and eigenvectors of the matrix

$$A = \begin{bmatrix} 1 & 4 \\ 1 & -2 \end{bmatrix}$$

Solution   To find the eigenvalues of A, we must solve

$$\det(A - \lambda I) = \det \begin{bmatrix} 1 - \lambda & 4 \\ 1 & -2 - \lambda \end{bmatrix}$$

$$= (1 - \lambda)(-2 - \lambda) - 4 = \lambda^2 + \lambda - 6$$

$$= (\lambda - 2)(\lambda + 3) = 0$$

We find

$$\lambda_1 = 2 \quad \text{and} \quad \lambda_2 = -3$$

To find an eigenvector associated with the eigenvalue $\lambda_1 = 2$, we must determine x_1 and x_2, so that

$$\begin{bmatrix} 1 & 4 \\ 1 & -2 \end{bmatrix} \begin{bmatrix} x_1 \\ x_2 \end{bmatrix} = 2 \begin{bmatrix} x_1 \\ x_2 \end{bmatrix}$$

Writing this as a system of linear equations, we find

$$x_1 + 4x_2 = 2x_1$$
$$x_1 - 2x_2 = 2x_2$$

We see that both equations are the same. Simplifying yields

$$-x_1 + 4x_2 = 0$$

Setting $x_2 = 1$, we find $x_1 = 4$, that is, $\begin{bmatrix} 4 \\ 1 \end{bmatrix}$ is an eigenvector associated with the eigenvalue $\lambda_1 = 2$.

To find an eigenvector associated with the eigenvalue $\lambda_2 = -3$, we must determine x_1 and x_2, so that

$$\begin{bmatrix} 1 & 4 \\ 1 & -2 \end{bmatrix} \begin{bmatrix} x_1 \\ x_2 \end{bmatrix} = -3 \begin{bmatrix} x_1 \\ x_2 \end{bmatrix}$$

Writing this as a system of linear equations, we find

$$x_1 + 4x_2 = -3x_1$$
$$x_1 - 2x_2 = -3x_2$$

We see that both equations are the same. Simplifying yields

$$x_1 + x_2 = 0$$

Setting $x_1 = 1$, we find $x_2 = -1$, that is, $\begin{bmatrix} 1 \\ -1 \end{bmatrix}$ is an eigenvector associated with the eigenvalue $\lambda_2 = -3$.

Both eigenvectors and corresponding lines are illustrated in Figure 9.22.   ▮

So far, we have only seen examples in which the eigenvalues were real and distinct. In the next example, we will see that eigenvalues can be complex. When eigenvalues are complex, we will not compute the corresponding eigenvectors, because the eigenvectors will be complex as well, which is beyond the scope of this course.

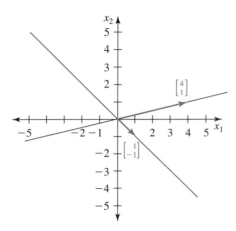

◀ **Figure 9.22**
The eigenvectors and their corresponding lines in Example 2.

▌EXAMPLE 3   **Complex eigenvalues**

Let

$$A = \begin{bmatrix} \cos 30° & -\sin 30° \\ \sin 30° & \cos 30° \end{bmatrix}$$

Describe the action of A on the vector $\begin{bmatrix} 1 \\ 0 \end{bmatrix}$. Compute the eigenvalues of A.

Solution   We recognize that A is a matrix that describes a rotation counterclockwise by $30°$. (The matrix A is the matrix R_θ for $\theta = 30°$, as defined in Subsection 9.3.1.) The vector $\begin{bmatrix} 1 \\ 0 \end{bmatrix}$ is thus rotated counterclockwise by $30°$, and

$A \begin{bmatrix} 1 \\ 0 \end{bmatrix} = \begin{bmatrix} \cos 30° \\ \sin 30° \end{bmatrix} = \begin{bmatrix} \sqrt{3}/2 \\ 1/2 \end{bmatrix}$ (see Figure 9.23).

We will now compute the eigenvalues associated with this matrix. We set

$$\det(A - \lambda I) = 0$$

Using $\cos 30° = \frac{1}{2}\sqrt{3}$ and $\sin 30° = \frac{1}{2}$, we find

$$\det \begin{bmatrix} \frac{1}{2}\sqrt{3} - \lambda & -\frac{1}{2} \\ \frac{1}{2} & \frac{1}{2}\sqrt{3} - \lambda \end{bmatrix} = 0$$

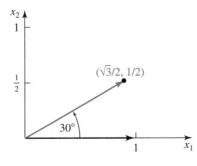

◀ **Figure 9.23**
The vector $\begin{bmatrix} 1 \\ 0 \end{bmatrix}$ is rotated counterclockwise by $30°$ in Example 3.

or

$$\left(\frac{1}{2}\sqrt{3} - \lambda\right)^2 + \frac{1}{4} = 0$$

$$\frac{3}{4} - \lambda\sqrt{3} + \lambda^2 + \frac{1}{4} = 0$$

$$\lambda^2 - \sqrt{3}\lambda + 1 = 0$$

Solving this quadratic equation, we obtain

$$\lambda_{1,2} = \frac{\sqrt{3} \pm \sqrt{3 - 4}}{2}$$

$$= \frac{1}{2}\left(\sqrt{3} \pm i\right)$$

where $i^2 = -1$. This shows that eigenvalues can be complex numbers.   ▮

In Chapter 11, we will discuss the stability of equilibria in systems of ordinary differential equations. This will lead us to investigating the eigenvalues of certain linear maps. It will be important to determine whether the eigenvalues (or more precisely, the real parts of the eigenvalues) are positive or negative. In the case where the linear map is given by a 2×2 matrix, there is a simple criterion. Consider

$$A = \begin{bmatrix} a & b \\ c & d \end{bmatrix}$$

Then the determinant of A is

$$\det A = ad - bc$$

and the **trace** of A (denoted by $\operatorname{tr} A$) is defined as

$$\operatorname{tr} A = a + d$$

That is, the trace is simply the sum of the diagonal elements of A. The trace and the determinant of a matrix are related to its eigenvalues, as follows.

The eigenvalues of A satisfiy

$$\det(A - \lambda I) = 0$$

That is,

$$\det \begin{bmatrix} a - \lambda & b \\ c & d - \lambda \end{bmatrix} = 0$$

or

$$(a - \lambda)(d - \lambda) - bc = 0$$

$$\lambda^2 - (a + d)\lambda + ad - bc = 0$$

Since $\det A = ad - bc$ and $\operatorname{tr} A = a + d$, we can write this as

$$\lambda^2 - (\operatorname{tr} A)\lambda + \det A = 0 \tag{9.18}$$

If λ_1 and λ_2 are the two solutions of (9.18), then λ_1 and λ_2 must satisfy

$$(\lambda - \lambda_1)(\lambda - \lambda_2) = 0$$

Multiplying this out, we find that λ_1 and λ_2 satisfy

$$\lambda^2 - (\lambda_1 + \lambda_2)\lambda + \lambda_1\lambda_2 = 0$$

Comparing this with (9.18), we find the following important result.

If A is a 2×2 matrix with eigenvalues λ_1 and λ_2, then

$$\text{tr}A = \lambda_1 + \lambda_2 \qquad \text{and} \qquad \det A = \lambda_1\lambda_2$$

To prepare for the next result, we make the following observations. Assume that λ_1 and λ_2 are both real and negative. Then the trace of A, which is the sum of the two eigenvalues, is negative, and the determinant of A, which is the product of the two eigenvalues, is positive.

In the case when the eigenvalues are complex conjugates, we have a similar result. Assume that λ_1 and λ_2 are complex conjugates and that their real parts are negative. One can then show that the trace of A is negative and the determinant is positive. (We will see an example of this fact below.)

That is, whenever the real parts of the two eigenvalues λ_1 and λ_2 are negative, it follows that $\text{tr}A < 0$ and $\det A > 0$. The converse of this result is also true. That is, if $\text{tr}A < 0$ and $\det A > 0$, then both eigenvalues have negative real parts. This will be a useful criterion in Chapter 11, since it will allow us to determine whether or not both eigenvalues have negative real parts without computing the eigenvalues.

Theorem   Let A be a 2×2 matrix with eigenvalues λ_1 and λ_2. Then the real parts of λ_1 and λ_2 are negative if and only if

$$\text{tr}A < 0 \qquad \text{and} \qquad \det A > 0$$

▮ **EXAMPLE 4**   **Trace and determinant**

Use the preceding theorem to show that both of the eigenvalues of

$$A = \begin{bmatrix} -1 & -3 \\ 1 & -2 \end{bmatrix}$$

have negative real parts. Then compute the eigenvalues, and use them to recompute the trace and the determinant of A.

Solution   Since

$$\text{tr}A = -1 - 2 = -3 < 0 \qquad \text{and} \qquad \det A = (-1)(-2) - (1)(-3) = 5 > 0$$

it follows from our theorem that both eigenvalues have negative real parts. To compute the eigenvalues, we solve

$$\det(A - \lambda I) = 0$$

or

$$\det \begin{bmatrix} -1 - \lambda & -3 \\ 1 & -2 - \lambda \end{bmatrix} = 0$$

Evaluating this, we find

$$(-1 - \lambda)(-2 - \lambda) - (1)(-3) = 0$$

$$\lambda^2 + 3\lambda + 5 = 0$$

that is,

$$\lambda_{1,2} = \frac{-3 \pm \sqrt{9 - 20}}{2}$$

$$= -\frac{3}{2} \pm \frac{1}{2} i \sqrt{11}$$

This also shows that the real parts of both eigenvalues are negative.

We can now use the eigenvalues to recompute the trace and the determinant of A.

$$\text{tr} A = \lambda_1 + \lambda_2 = \left(-\frac{3}{2} + \frac{1}{2} i \sqrt{11}\right) + \left(-\frac{3}{2} - \frac{1}{2} i \sqrt{11}\right) = -3$$

This is the same result that we obtained previously. Note that since λ_1 and λ_2 are complex conjugates, the imaginary parts cancel when we add λ_1 and λ_2. The determinant of A is

$$\det(A) = \lambda_1 \lambda_2 = \left(-\frac{3}{2} + \frac{1}{2} i \sqrt{11}\right) \left(-\frac{3}{2} - \frac{1}{2} i \sqrt{11}\right)$$

$$= \frac{9}{4} - \frac{1}{4} i^2 (11) = \frac{9}{4} + \frac{11}{4} = \frac{20}{4} = 5$$

which is the same result that we obtained previously. Note that since λ_1 and λ_2 are complex conjugates, we computed a product of the form $(x + iy)(x - iy) = x^2 - (iy)^2 = x^2 + y^2$, which is a real number (and positive). ▮

9.3.3 Iterated Maps (Optional)

We restrict ourselves to the case in which A is a 2×2 matrix with real eigenvalues. We saw in the preceding that in this case the eigenvectors define lines through the origin that are invariant under the map A. If the two invariant lines are *not* identical, we say that the two eigenvectors are **linearly independent** (see Figure 9.24). This can be formulated as follows in terms of eigenvectors. If we denote by $\mathbf{u}_1$ and $\mathbf{u}_2$ the two eigenvectors, then $\mathbf{u}_1$ and $\mathbf{u}_2$ are linearly independent if there does not exist a number a such that $\mathbf{u}_1 = a\mathbf{u}_2$. (Linear independence is defined more generally: any two nonzero vectors $\mathbf{x}_1$ and $\mathbf{x}_2$ are linearly independent if there does not exist a number a such that $\mathbf{x}_1 = a\mathbf{x}_2$.)

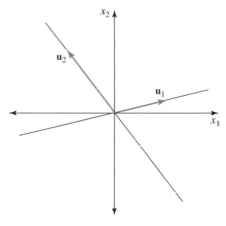

◄ **Figure 9.24**
The two eigenvectors $\mathbf{u}_1$ and $\mathbf{u}_2$ are linearly independent.

The following criterion is useful when we want to determine whether eigenvalues are linearly independent.

Let A be a 2×2 matrix with eigenvalues λ_1 and λ_2. Denote by $\mathbf{u}_1$ the eigenvector associated with λ_1 and by $\mathbf{u}_2$ the eigenvector associated with λ_2. If $\lambda_1 \neq \lambda_2$, then $\mathbf{u}_1$ and $\mathbf{u}_2$ are linearly independent.

There are also cases, in which $\mathbf{u}_1$ and $\mathbf{u}_2$ are linearly independent even though $\lambda_1 = \lambda_2$. We will, however, be concerned primarily with cases in which $\lambda_1 \neq \lambda_2$. Hence, the above criterion will suffice for our purposes. (The other cases are covered in courses on linear algebra.)

A consequence of linear independence is that we can write any vector uniquely as a **linear combination** of the two eigenvectors. That is, suppose that $\mathbf{u}_1$ and $\mathbf{u}_2$ are linearly independent eigenvectors of a 2×2 matrix; then, any 2×1 vector $\mathbf{x}$ can be written as

$$\mathbf{x} = a_1\mathbf{u}_1 + a_2\mathbf{u}_2$$

where a_1 and a_2 are uniquely determined. We will not prove this, but will examine what we can do with it.

If we apply A to $\mathbf{x}$ (written as a linear combination of the two eigenvectors of A), then using the linearity of the map A, we find

$$A\mathbf{x} = A(a_1\mathbf{u}_1 + a_2\mathbf{u}_2) = a_1 A\mathbf{u}_1 + a_2 A\mathbf{u}_2$$

Now, $\mathbf{u}_1$ and $\mathbf{u}_2$ are both eigenvectors corresponding to A. Hence, $A\mathbf{u}_1 = \lambda_1\mathbf{u}_1$ and $A\mathbf{u}_2 = \lambda_2\mathbf{u}_2$. We thus find

$$A\mathbf{x} = a_1\lambda_1\mathbf{u}_1 + a_2\lambda_2\mathbf{u}_2$$

This representation of $\mathbf{x}$ is particularly useful if we apply A repeatedly to $\mathbf{x}$. Applying A to $A\mathbf{x}$, we find

$$A^2\mathbf{x} = A(A\mathbf{x}) = A(a_1\lambda_1\mathbf{u}_1 + a_2\lambda_2\mathbf{u}_2) = a_1\lambda_1 A\mathbf{u}_1 + a_2\lambda_2 A\mathbf{u}_2$$
$$= a_1\lambda_1^2\mathbf{u}_1 + a_2\lambda_2^2\mathbf{u}_2$$

where we again used the fact that $\mathbf{u}_1$ and $\mathbf{u}_2$ are eigenvectors of the matrix A. Continuing in this way, we find that

$$A^n\mathbf{x} = a_1\lambda_1^n\mathbf{u}_1 + a_2\lambda_2^n\mathbf{u}_2$$

That is, instead of multiplying A with itself n times (which is rather time consuming), we evaluate the right-hand side, which is essentially adding two vectors (a much faster task). Are you convinced that this might be useful? If not, here is an example.

▮ EXAMPLE 5   Let

$$A = \begin{bmatrix} 1 & 2 \\ 3 & 2 \end{bmatrix}$$

Find $A^{10}\mathbf{x}$ for $\mathbf{x} = \begin{bmatrix} 4 \\ 1 \end{bmatrix}$.

Solution   We computed eigenvalues and eigenvectors for the matrix A earlier. We found

$$\lambda_1 = -1 \quad \text{with} \quad \mathbf{u}_1 = \begin{bmatrix} 1 \\ -1 \end{bmatrix}$$

and

$$\lambda_2 = 4 \quad \text{with} \quad \mathbf{u}_2 = \begin{bmatrix} 2 \\ 3 \end{bmatrix}$$

We first represent $\mathbf{x} = \begin{bmatrix} 4 \\ 1 \end{bmatrix}$ as a linear combination of $\mathbf{u}_1$ and $\mathbf{u}_2$. That is, we must find a_1 and a_2 so that

$$\begin{bmatrix} 4 \\ 1 \end{bmatrix} = a_1 \begin{bmatrix} 1 \\ -1 \end{bmatrix} + a_2 \begin{bmatrix} 2 \\ 3 \end{bmatrix}$$

Writing this as a system of linear equations yields

$$a_1 + 2a_2 = 4$$
$$-a_1 + 3a_2 = 1$$

Using the method of elimination, we obtain

$$a_1 + 2a_2 = 4$$
$$5a_2 = 5$$

Hence, $a_2 = 1$ and $a_1 = 4 - 2a_2 = 4 - 2 = 2$. That is,

$$\begin{bmatrix} 4 \\ 1 \end{bmatrix} = 2 \begin{bmatrix} 1 \\ -1 \end{bmatrix} + \begin{bmatrix} 2 \\ 3 \end{bmatrix}$$

To find $A^{10}\mathbf{x}$, we compute

$$A^{10}\mathbf{x} = A^{10} \begin{bmatrix} 4 \\ 1 \end{bmatrix} = A^{10} \left(2 \begin{bmatrix} 1 \\ -1 \end{bmatrix} + \begin{bmatrix} 2 \\ 3 \end{bmatrix} \right)$$

$$= 2A^{10} \begin{bmatrix} 1 \\ -1 \end{bmatrix} + A^{10} \begin{bmatrix} 2 \\ 3 \end{bmatrix}$$

$$= 2(-1)^{10} \begin{bmatrix} 1 \\ -1 \end{bmatrix} + 4^{10} \begin{bmatrix} 2 \\ 3 \end{bmatrix}$$

$$= 2 \begin{bmatrix} 1 \\ -1 \end{bmatrix} + 4^{10} \begin{bmatrix} 2 \\ 3 \end{bmatrix} = \begin{bmatrix} 2,097,154 \\ 3,145,726 \end{bmatrix}$$

To compute A^{10} by hand would have taken a much longer time.   ▮

The Leslie Matrix Revisited

▮ **EXAMPLE 6**   In Subsection 9.2.5, we investigated an age-structured population with Leslie matrix

$$L = \begin{bmatrix} 1.5 & 2 \\ 0.08 & 0 \end{bmatrix}$$

Find both eigenvalues and eigenvectors.

Solution   To find the eigenvalues, we must solve

$$\det(L - \lambda I) = 0$$

Since

$$L - \lambda I = \begin{bmatrix} 1.5 - \lambda & 2 \\ 0.08 & -\lambda \end{bmatrix}$$

we find

$$\det(L - \lambda I) = (1.5 - \lambda)(-\lambda) - (2)(0.08)$$
$$= -1.5\lambda + \lambda^2 - 0.16$$

Since

$$\lambda^2 - 1.5\lambda - 0.16 = (\lambda - 1.6)(\lambda + 0.1)$$

the eigenvalues are

$$\lambda_1 = 1.6 \qquad \text{and} \qquad \lambda_2 = -0.1$$

To compute the corresponding eigenvectors, we start with the larger eigenvalue $\lambda_1 = 1.6$. We need to solve

$$\begin{bmatrix} 1.5 & 2 \\ 0.08 & 0 \end{bmatrix} \begin{bmatrix} x_1 \\ x_2 \end{bmatrix} = (1.6) \begin{bmatrix} x_1 \\ x_2 \end{bmatrix}$$

or

$$1.5x_1 + 2x_2 = 1.6x_1$$
$$0.08x_1 \qquad = 1.6x_2$$

Simplifying, we find

$$-0.1x_1 + 2x_2 = 0$$
$$0.08x_1 - 1.6x_2 = 0$$

or

$$x_1 - 20x_2 = 0$$
$$x_1 - 20x_2 = 0$$

From this, it follows that

$$x_1 = 20x_2$$

For instance, $x_2 = 1$ and $x_1 = 20$ satisfies this. That is, an eigenvector corresponding to $\lambda_1 = 1.6$ is

$$\mathbf{u}_1 = \begin{bmatrix} 20 \\ 1 \end{bmatrix}$$

The eigenvector corresponding to $\lambda_2 = -0.1$ satisfies

$$\begin{bmatrix} 1.5 & 2 \\ 0.08 & 0 \end{bmatrix} \begin{bmatrix} x_1 \\ x_2 \end{bmatrix} = -0.1 \begin{bmatrix} x_1 \\ x_2 \end{bmatrix}$$

or

$$1.5x_1 + 2x_2 = -0.1x_1$$
$$0.08x_1 \qquad = -0.1x_2$$

Simplifying, we find

$$1.6x_1 + 2x_2 = 0$$
$$0.08x_1 + 0.1x_2 = 0$$

or

$$0.8x_1 + x_2 = 0$$
$$0.8x_1 + x_2 = 0$$

From this, it follows that

$$0.8x_1 = -x_2$$

For instance, $x_1 = 5$ and $x_2 = -4$ satisfies this. That is, an eigenvector corresponding to the eigenvalue $\lambda_2 = -0.1$ is

$$\mathbf{u}_2 = \begin{bmatrix} 5 \\ -4 \end{bmatrix}$$

▮

We will now show that the larger eigenvalue determines the growth rate of the population and the eigenvector corresponding to the larger eigenvalue is a stable age distribution. To illustrate this, we assume that the population vector at time 0 is

$$N(0) = \begin{bmatrix} 105 \\ 1 \end{bmatrix}$$

We can compute $N(1)$ by evaluating

$$N(1) = LN(0) = \begin{bmatrix} 1.5 & 2 \\ 0.08 & 0 \end{bmatrix} \begin{bmatrix} 105 \\ 1 \end{bmatrix} = \begin{bmatrix} 159.5 \\ 8.4 \end{bmatrix}$$

If we wanted to compute $N(t)$ for some integer value t, we would need to find

$$N(t) = L^t N(0)$$

That is, we would need to compute the tth power of L. For large t, it will be convenient to write $N(0)$ as a linear combination of the two eigenvectors and then apply L^t to this linear combination.

Looking at the eigenvalues, we see that $N(0)$, as a linear combination of the two eigenvectors, is

$$\begin{bmatrix} 105 \\ 1 \end{bmatrix} = 5 \begin{bmatrix} 20 \\ 1 \end{bmatrix} + \begin{bmatrix} 5 \\ -4 \end{bmatrix}$$

Now, if we want to compute $N(1)$, we must find

$$N(1) = LN(0) = L \begin{bmatrix} 105 \\ 1 \end{bmatrix} = L \left(5 \begin{bmatrix} 20 \\ 1 \end{bmatrix} + (-0.1) \begin{bmatrix} 5 \\ -4 \end{bmatrix} \right)$$

$$= 5L \begin{bmatrix} 20 \\ 1 \end{bmatrix} - (0.1)L \begin{bmatrix} 5 \\ -4 \end{bmatrix}$$

Since $\begin{bmatrix} 20 \\ 1 \end{bmatrix}$ is an eigenvector corresponding to $\lambda_1 = 1.6$, we find $L \begin{bmatrix} 20 \\ 1 \end{bmatrix} = 1.6 \begin{bmatrix} 20 \\ 1 \end{bmatrix}$. Likewise, since $\begin{bmatrix} 5 \\ -4 \end{bmatrix}$ is an eigenvector corresponding to $\lambda_2 = -0.1$, we find $L \begin{bmatrix} 5 \\ -4 \end{bmatrix} = -0.1 \begin{bmatrix} 5 \\ -4 \end{bmatrix}$. Hence,

$$N(1) = (5)(1.6) \begin{bmatrix} 20 \\ 1 \end{bmatrix} + (-0.1) \begin{bmatrix} 5 \\ -4 \end{bmatrix}$$

$$= \begin{bmatrix} 159.5 \\ 8.4 \end{bmatrix}$$

which, of course, is the same answer as before. Now, let's find $N(t)$. For this, we need to compute

$$N(t) = L^t N(0) = L^t \begin{bmatrix} 105 \\ 1 \end{bmatrix} = L^t \left(5 \begin{bmatrix} 20 \\ 1 \end{bmatrix} + (-0.1) \begin{bmatrix} 5 \\ -4 \end{bmatrix} \right)$$

$$= 5(1.6)^t \begin{bmatrix} 20 \\ 1 \end{bmatrix} + (-0.1)^t \begin{bmatrix} 5 \\ -4 \end{bmatrix}$$

We see from this that $(1.6)^t$ grows much faster than $(-0.1)^t$. In fact, $(-0.1)^t$ tends to 0 as $t \to \infty$. We conclude that the larger eigenvalue determines the growth rate of the population.

It turns out that the larger eigenvalue of a Leslie matrix is always real and positive, provided that a positive fraction of zero-year-old females survive, and either zero-year-old or one-year-old females have offspring. Furthermore, $\lambda_1 \geq |\lambda_2|$ and $\lambda_1 > |\lambda_2|$ if zero-year-olds can give birth. We thus find the following.

If L is a 2×2 Leslie matrix with eigenvalues λ_1 and λ_2, then the larger eigenvalue determines the growth rate of the population.

If λ_1 is the larger eigenvalue and $0 < \lambda_1 < 1$, then the population size decreases over time. If $\lambda_1 > 1$, then the population size increases over time.

We will now show that the eigenvector corresponding to the larger eigenvalue is a stable age distribution. Let λ_1 be the larger eigenvalue and $\mathbf{u}_1$ its corresponding eigenvector; then, if $N(0) = \mathbf{u}_1$, it follows that

$$N(t) = L^t N(0) = L^t \mathbf{u}_1 = \lambda_1^t \mathbf{u}_1$$

To show that $\mathbf{u}_1$ is a stable age distribution, observe that if $\mathbf{u}_1 = \begin{bmatrix} x \\ y \end{bmatrix}$, then the fraction of zero-year-olds at time t is

$$\frac{\lambda^t x}{\lambda^t x + \lambda^t y} = \frac{x}{x + y}$$

which is the same fraction of zero-year-olds at time 0. Furthermore, since $\lambda_1 > 0$, we can choose $\mathbf{u}_1$ so that both entries are positive (a condition that is needed if the entries represent population sizes). We summarize this in the following.

If L is a 2×2 Leslie matrix with eigenvalues λ_1 and λ_2, then the eigenvector corresponding to the larger eigenvalue is a stable age distribution.

For the matrix

$$L = \begin{bmatrix} 1.5 & 2 \\ 0.08 & 0 \end{bmatrix}$$

it follows that

$$\begin{bmatrix} 20 \\ 1 \end{bmatrix}$$

is a stable age distribution. In Section 9.2.5, we claimed that $\begin{bmatrix} 2000 \\ 100 \end{bmatrix}$ was a stable age distribution. In both cases, the fraction of zero-year-olds is the same, namely, $20/21 = 2000/2100$. That is, both vectors represent the same proportion of zero-year-olds in the population.

◆ 9.3 PROBLEMS

1. Let

$$A = \begin{bmatrix} 2 & 1 \\ 3 & 4 \end{bmatrix},$$

$$\mathbf{x} = \begin{bmatrix} x_1 \\ x_2 \end{bmatrix},$$

and

$$\mathbf{y} = \begin{bmatrix} y_1 \\ y_2 \end{bmatrix}.$$

(a) Show by direct calculation that $A(\mathbf{x} + \mathbf{y}) = A\mathbf{x} + A\mathbf{y}$.
(b) Show by direct calculation that $A(\lambda\mathbf{x}) = \lambda(A\mathbf{x})$.

2. Let

$$A = \begin{bmatrix} a_{11} & a_{12} \\ a_{21} & a_{22} \end{bmatrix},$$

$$\mathbf{x} = \begin{bmatrix} x_1 \\ x_2 \end{bmatrix},$$

and

$$\mathbf{y} = \begin{bmatrix} y_1 \\ y_2 \end{bmatrix}.$$

(a) Show by direct calculation that $A(\mathbf{x} + \mathbf{y}) = A\mathbf{x} + A\mathbf{y}$.
(b) Show by direct calculation that $A(\lambda\mathbf{x}) = \lambda(A\mathbf{x})$.

In Problems 3–8, represent each given vector $\mathbf{x} = \begin{bmatrix} x_1 \\ x_2 \end{bmatrix}$ in the $x_1 - x_2$ plane, and determine its length and the angle that it forms with the positive x_1-axis (measured counterclockwise).

3. $\mathbf{x} = \begin{bmatrix} 2 \\ 2 \end{bmatrix}$ **4.** $\mathbf{x} = \begin{bmatrix} -1 \\ 0 \end{bmatrix}$

5. $\mathbf{x} = \begin{bmatrix} 0 \\ 3 \end{bmatrix}$ **6.** $\mathbf{x} = \begin{bmatrix} -\sqrt{3} \\ 1 \end{bmatrix}$

7. $\mathbf{x} = \begin{bmatrix} -\sqrt{3} \\ 1 \end{bmatrix}$ **8.** $\mathbf{x} = \begin{bmatrix} 1 \\ -\sqrt{3} \end{bmatrix}$

In Problems 9–14, find $\mathbf{x} + \mathbf{y}$ for the given vectors $\mathbf{x}$ and $\mathbf{y}$. Represent $\mathbf{x}$, $\mathbf{y}$, and $\mathbf{x} - \mathbf{y}$ in the plane, and explain graphically how you add $\mathbf{x}$ and $\mathbf{y}$.

9. $\mathbf{x} = \begin{bmatrix} 1 \\ 2 \end{bmatrix}$ and $\mathbf{y} = \begin{bmatrix} 3 \\ 2 \end{bmatrix}$

10. $\mathbf{x} = \begin{bmatrix} -1 \\ 0 \end{bmatrix}$ and $\mathbf{y} = \begin{bmatrix} 1 \\ 2 \end{bmatrix}$

11. $\mathbf{x} = \begin{bmatrix} 0 \\ -2 \end{bmatrix}$ and $\mathbf{y} = \begin{bmatrix} 1 \\ -1 \end{bmatrix}$

12. $\mathbf{x} = \begin{bmatrix} -2 \\ -1 \end{bmatrix}$ and $\mathbf{y} = \begin{bmatrix} 1 \\ 2 \end{bmatrix}$

13. $\mathbf{x} = \begin{bmatrix} 1 \\ 0 \end{bmatrix}$ and $\mathbf{y} = \begin{bmatrix} -1 \\ 0 \end{bmatrix}$

14. $\mathbf{x} = \begin{bmatrix} -3 \\ -1 \end{bmatrix}$ and $\mathbf{y} = \begin{bmatrix} -2 \\ 3 \end{bmatrix}$

In Problems 15–20, compute $a\mathbf{x}$ for the given vector $\mathbf{x}$ and scalar a. Represent $\mathbf{x}$ and $a\mathbf{x}$ in the plane, and explain graphically how you obtain $a\mathbf{x}$.

15. $\mathbf{x} = \begin{bmatrix} -2 \\ 1 \end{bmatrix}$ and $a = 2$

16. $\mathbf{x} = \begin{bmatrix} 3 \\ -1 \end{bmatrix}$ and $a = -1$

17. $\mathbf{x} = \begin{bmatrix} 0 \\ -2 \end{bmatrix}$ and $a = 0.5$

18. $\mathbf{x} = \begin{bmatrix} 3 \\ -9 \end{bmatrix}$ and $a = -1/3$

19. $\mathbf{x} = \begin{bmatrix} -4 \\ 1 \end{bmatrix}$ and $a = 1/4$

20. $\mathbf{x} = \begin{bmatrix} 0.5 \\ 0.25 \end{bmatrix}$ and $a = 4$

In Problems 21–26, let

$$\mathbf{u} = \begin{bmatrix} 3 \\ 4 \end{bmatrix}, \mathbf{v} = \begin{bmatrix} -1 \\ -2 \end{bmatrix},$$

and

$$\mathbf{w} = \begin{bmatrix} 1 \\ -2 \end{bmatrix}.$$

21. Compute $\mathbf{u} - \mathbf{v}$ and illustrate the result graphically.
22. Compute $\mathbf{u} - \mathbf{v}$ and illustrate the result graphically.
23. Compute $\mathbf{w} - \mathbf{u}$ and illustrate the result graphically.
24. Compute $\mathbf{v} - \frac{1}{2}\mathbf{u}$ and illustrate the result graphically.
25. Compute $\mathbf{u} + \mathbf{v} + \mathbf{w}$ and illustrate the result graphically.
26. Compute $2\mathbf{v} - \mathbf{w}$ and illustrate the result graphically.

In Problems 27–32, give a geometric interpretation of the map $\mathbf{x} \to A\mathbf{x}$ for each given map A.

27. $A = \begin{bmatrix} 1 & 0 \\ 0 & 1 \end{bmatrix}$ **28.** $A = \begin{bmatrix} 2 & 0 \\ 0 & -1 \end{bmatrix}$

29. $A = \begin{bmatrix} 0 & -1 \\ 1 & 0 \end{bmatrix}$ **30.** $A = \begin{bmatrix} 0 & 1 \\ -1 & 0 \end{bmatrix}$

31. $A = \frac{1}{2}\begin{bmatrix} \sqrt{3} & -1 \\ 1 & \sqrt{3} \end{bmatrix}$ **32.** $A = \frac{1}{2}\begin{bmatrix} \sqrt{2} & -\sqrt{2} \\ \sqrt{2} & \sqrt{2} \end{bmatrix}$

In Problems 33–40, find the eigenvalues λ_1 and λ_2 and corresponding eigenvectors $\mathbf{v}_1$ and $\mathbf{v}_2$ for each matrix A. Determine the equations of the lines through the origin in the direction of the eigenvectors $\mathbf{v}_1$ and $\mathbf{v}_2$, and graph the lines together with the eigenvectors $\mathbf{v}_1$ and $\mathbf{v}_2$, and the vectors $A\mathbf{v}_1$ and $A\mathbf{v}_2$.

33. $A = \begin{bmatrix} 2 & 3 \\ 0 & -1 \end{bmatrix}$

34. $A = \begin{bmatrix} 0 & 0 \\ 1 & -3 \end{bmatrix}$

35. $A = \begin{bmatrix} 1 & 0 \\ 0 & -1 \end{bmatrix}$

36. $A = \begin{bmatrix} -1 & 0 \\ 0 & 2 \end{bmatrix}$

37. $A = \begin{bmatrix} -4 & 2 \\ -3 & 1 \end{bmatrix}$

38. $A = \begin{bmatrix} 3 & 6 \\ -1 & -4 \end{bmatrix}$

39. $A = \begin{bmatrix} 2 & 1 \\ 2 & 3 \end{bmatrix}$

40. $A = \begin{bmatrix} -3 & -0.5 \\ 7 & 1.5 \end{bmatrix}$

41. Let

$$A = \begin{bmatrix} 2 & 4 \\ -2 & -3 \end{bmatrix}$$

Without explicitly computing the eigenvalues of A, decide whether the real parts of both eigenvalues are negative.

42. Let

$$A = \begin{bmatrix} -2 & 3 \\ -1 & 1 \end{bmatrix}$$

Without explicitly computing the eigenvalues of A, decide whether the real parts of both eigenvalues are negative.

43. Let

$$A = \begin{bmatrix} 4 & 4 \\ -4 & -3 \end{bmatrix}$$

Without explicitly computing the eigenvalues of A, decide whether the real parts of both eigenvalues are negative.

44. Let

$$A = \begin{bmatrix} 0 & 1 \\ -2 & 1 \end{bmatrix}$$

Without explicitly computing the eigenvalues of A, decide whether the real parts of both eigenvalues are negative.

45. Let

$$A = \begin{bmatrix} 2 & -5 \\ 2 & -3 \end{bmatrix}$$

Without explicitly computing the eigenvalues of A, decide whether the real parts of both eigenvalues are negative.

46. Let

$$A = \begin{bmatrix} 2 & 5 \\ 2 & -3 \end{bmatrix}$$

Without explicitly computing the eigenvalues of A, decide whether the real parts of both eigenvalues are negative.

[9.4.3]

47. Let

$$A = \begin{bmatrix} -1 & 1 \\ 0 & 2 \end{bmatrix}$$

(a) Show that

$$\mathbf{u}_1 = \begin{bmatrix} 1 \\ 0 \end{bmatrix}$$

and

$$\mathbf{u}_2 = \begin{bmatrix} 1 \\ 3 \end{bmatrix}$$

are eigenvectors of A, and that $\mathbf{u}_1$ and $\mathbf{u}_2$ are linearly independent.

(b) Represent

$$\mathbf{x} = \begin{bmatrix} 1 \\ -3 \end{bmatrix}$$

as a linear combination of $\mathbf{u}_1$ and $\mathbf{u}_2$.

(c) Use your results in (a) and (b) to compute $A^{20}\mathbf{x}$.

48. Let

$$A = \begin{bmatrix} -2 & 1 \\ -4 & 3 \end{bmatrix}$$

(a) Show that

$$\mathbf{u}_1 = \begin{bmatrix} 1 \\ 4 \end{bmatrix}$$

and

$$\mathbf{u}_2 = \begin{bmatrix} 1 \\ 1 \end{bmatrix}$$

are eigenvectors of A, and that $\mathbf{u}_1$ and $\mathbf{u}_2$ are linearly independent.

(b) Represent

$$\mathbf{x} = \begin{bmatrix} -1 \\ 2 \end{bmatrix}$$

as a linear combination of $\mathbf{u}_1$ and $\mathbf{u}_2$.

(c) Use your results in (a) and (b) to compute $A^{10}\mathbf{x}$.

49. Let

$$A = \begin{bmatrix} -1 & 0 \\ 3 & 1 \end{bmatrix}$$

Find

$$A^{15} \begin{bmatrix} 2 \\ 0 \end{bmatrix}$$

without using a calculator.

50. Let

$$A = \begin{bmatrix} 4 & -3 \\ 2 & -1 \end{bmatrix}$$

Find

$$A^{30} \begin{bmatrix} -4 \\ -2 \end{bmatrix}$$

without using a calculator.

51. Suppose that

$$L = \begin{bmatrix} 2 & 4 \\ 0.3 & 0 \end{bmatrix}$$

is the Leslie matrix for a population with two age classes.
(a) Determine both eigenvalues.
(b) Give a biological interpretation of the larger eigenvalue.
(c) Find the stable age distribution.

52. Suppose that

$$L = \begin{bmatrix} 1 & 3 \\ 0.7 & 0 \end{bmatrix}$$

is the Leslie matrix for a population with two age classes.
(a) Determine both eigenvalues.
(b) Give a biological interpretation of the larger eigenvalue.
(c) Find the stable age distribution.

53. Suppose that

$$L = \begin{bmatrix} 7 & 3 \\ 0.1 & 0 \end{bmatrix}$$

is the Leslie matrix for a population with two age classes.

(a) Determine both eigenvalues.
(b) Give a biological interpretation of the larger eigenvalue.
(c) Find the stable age distribution.

54. Suppose that

$$L = \begin{bmatrix} 0 & 5 \\ 0.9 & 0 \end{bmatrix}$$

is the Leslie matrix for a population with two age classes.
(a) Determine both eigenvalues.
(b) Give a biological interpretation of the larger eigenvalue.
(c) Find the stable age distribution.

55. Suppose that

$$L = \begin{bmatrix} 0 & 5 \\ 0.09 & 0 \end{bmatrix}$$

is the Leslie matrix for a population with two age classes.
(a) Determine both eigenvalues.
(b) Give a biological interpretation of the larger eigenvalue.
(c) Find the stable age distribution.

◆ 9.4 ANALYTIC GEOMETRY

René Descartes (1596–1650) and Pierre de Fermat (1601–1665) are credited with the invention of analytic geometry, which combines techniques from algebra and geometry, and provides important tools for multidimensional calculus. In Chapter 10, we will discuss some aspects of multidimensional calculus; we will therefore need a few results from analytic geometry. Our first task will be to generalize points and vectors in the plane to higher dimensions. We will then introduce a product between vectors that will allow us to determine the length of a vector and the angle between two vectors. Finally, we will give a vector representation of lines and planes in three-dimensional space.

9.4.1 Points and Vectors in Higher Dimensions

We are familiar with points and vectors in the plane. To represent points, we use a Cartesian coordinate system in the plane that consists of two axes, the x_1-axis and the x_2-axis that are perpendicular to each other. Any point in the plane can be represented by an ordered pair (a_1, a_2) of real numbers, where a_1 is the x_1-coordinate and a_2 is the x_2-coordinate. Since we need two numbers to locate a point in the plane, we call the plane "two-dimensional."

The plane can thus be thought of as the set of all points (x_1, x_2) with $x_1 \in \mathbf{R}$ and $x_2 \in \mathbf{R}$. We introduce the notation $\mathbf{R}^2$ to denote the set of these points. That is, the two-dimensional plane is

$$\mathbf{R}^2 = \{(x_1, x_2) : x_1 \in \mathbf{R}, x_2 \in \mathbf{R}\}$$

To generalize this to n dimensions, we set

$$\mathbf{R}^n = \{(x_1, x_2, \ldots, x_n) : x_1 \in \mathbf{R}, x_2 \in \mathbf{R}, \ldots, x_n \in \mathbf{R}\}$$

For instance, $\mathbf{R}^3$ is the three-dimensional space. It consists of all points (x_1, x_2, x_3) with $x_i \in \mathbf{R}$ for $i = 1, 2$, and 3. In three dimensions, we use a "right-handed" Cartesian coordinate system, to represent points and vectors in $\mathbf{R}^3$. This

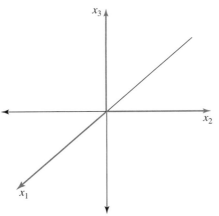

◀ **Figure 9.25**

A right-handed three-dimensional coordinate system. The axes are perpendicular. The x_1-axis points in the direction of the thumb, the x_2-axis in the direction of the index finger, and the x_3-axis in the direction of the middle finger of the right hand.

coordinate system consists of three mutually perpendicular axes oriented in a right-handed coordinate system. That is, the axes are perpendicular to each other and oriented so that the thumb of your right hand points along the positive x_1-axis, the index finger of your right hand points along the positive x_2-axis and the middle finger of your right hand points along the positive x_3-axis. This is shown in Figure 9.25, together with a point (a_1, a_2, a_3). In four and higher dimensions, we can no longer draw a coordinate system.

We introduced vectors in two dimensions in Section 9.3. A vector is a quantity that has a direction and a magnitude. We saw in Section 9.3 that a vector in two dimensions is an ordered pair that can be represented by a directed segment, as illustrated in Figure 9.26. There, the vector $\mathbf{x} = \begin{bmatrix} x_1 \\ x_2 \end{bmatrix}$ is represented by a directed segment with initial point $(0, 0)$ and endpoint (x_1, x_2). The arrow at the tip indicates the direction of the vector. We will generalize this now to n dimensions.

Definition   A vector in n-dimensional space is an ordered n-tuple

$$\mathbf{x} = \begin{bmatrix} x_1 \\ x_2 \\ \vdots \\ x_n \end{bmatrix}$$

of real numbers. The numbers $x_1, x_2, \ldots, x_n$ are called the **components** of $\mathbf{x}$.

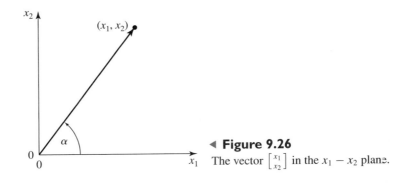

◀ **Figure 9.26**

The vector $\begin{bmatrix} x_1 \\ x_2 \end{bmatrix}$ in the $x_1 - x_2$ plane.

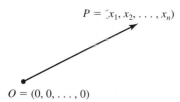

$P = (x_1, x_2, \ldots, x_n)$

$O = (0, 0, \ldots, 0)$

▲ **Figure 9.27**
A vector in n-dimensional space.

Vectors in n-dimensional space are also represented by directed segments with initial point $(0, 0, \ldots, 0)$ and endpoint $(x_1, x_2, \ldots, x_n)$. In three dimensions, we can use the three-dimensional right-handed Cartesian coordinate system to visualize vectors. In four and higher dimensions, we can no longer draw the coordinate system. However, we can still draw vectors. Namely, to draw a vector $\mathbf{x}$ whose endpoint has coordinates $(x_1, x_2, \ldots, x_n)$, we draw a directed arrow from the origin $(0, 0, \ldots, 0)$ to the point $P = (x_1, x_2, \ldots, x_n)$, as illustrated in Figure 9.27.

Next, we generalize vector addition to n dimensions. We saw in the preceding section how to add vectors in two-dimensional space. In n dimensions, vector addition is defined in a similar way.

VECTOR ADDITION

If

$$\mathbf{x} = \begin{bmatrix} x_1 \\ x_2 \\ \vdots \\ x_n \end{bmatrix} \quad \text{and} \quad \mathbf{y} = \begin{bmatrix} y_1 \\ y_2 \\ \vdots \\ y_n \end{bmatrix}$$

Then

$$\mathbf{x} + \mathbf{y} = \begin{bmatrix} x_1 + y_1 \\ x_2 + y_2 \\ \vdots \\ x_n + y_n \end{bmatrix}$$

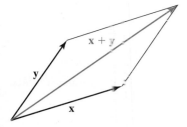

▲ **Figure 9.28**
The parallelogram for vector addition.

The geometric interpretation is the same as in two-dimensional space. This law of addition is similarly called the parallelogram law, for reasons that can be seen from Figure 9.28, where the addition of two vectors is illustrated.

We see from Figure 9.28 that vector addition can also be interpreted in the following way: To obtain $\mathbf{x} + \mathbf{y}$, we place the vector $\mathbf{y}$ at the tip of the vector $\mathbf{x}$. The sum $\mathbf{x} + \mathbf{y}$ is then the vector that starts at the same point as $\mathbf{x}$ and ends at the point where the moved vector $\mathbf{y}$ ends, as illustrated in Figure 9.29.

The multiplication of a vector by a scalar was also introduced in the preceding section. It generalizes easily to higher dimensions, and has the same geometric interpretation.

▲ **Figure 9.29**
Adding two vectors.

MULTIPLICATION OF A VECTOR BY A SCALAR

If a is a scalar and $\mathbf{x}$ is a vector in n-dimensional space, then

$$a\mathbf{x} = a \begin{bmatrix} x_1 \\ x_2 \\ \vdots \\ x_n \end{bmatrix} = \begin{bmatrix} ax_1 \\ ax_2 \\ \vdots \\ ax_n \end{bmatrix}$$

Vector Representation   A vector $\mathbf{x}$ is a directed line segment $\vec{AB}$ from the initial point A to the terminal point B. A particular representation of $\mathbf{x}$ has the origin

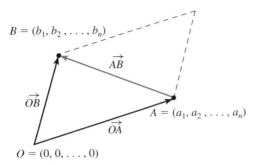

◀ **Figure 9.30**

The vector representation of a directed segment.

at its initial point and the same direction and length as the directed segment $\vec{AB}$. We can use the terminal point of $\mathbf{x}$ as the representation of $\mathbf{x}$ when its initial point is the origin. We will call this particular representation the **vector representation** of $\vec{AB}$.

We can use Figure 9.30 to find the vector representation of a directed segment $\vec{AB}$, from point A with coordinates $(a_1, a_2, \ldots, a_n)$ to point B with coordinates $(b_1, b_2, \ldots, b_n)$. Using the parallelogram law, we see that

$$\vec{OA} + \vec{AB} = \vec{OB}$$

or, solving for $\vec{AB}$,

$$\vec{AB} = \vec{OB} - \vec{OA}$$

Now, $\vec{OB}$ has the vector representation $[b_1, b_2, \ldots, b_n]'$, and $\vec{OA}$ has the vector representation $[a_1, a_2, \ldots, a_n]'$. The difference $\vec{OB} - \vec{OA}$ is then simply the difference between the two vectors, that is,

$$\vec{OB} - \vec{OA} = \begin{bmatrix} b_1 - a_1 \\ b_2 - a_2 \\ \vdots \\ b_n - a_n \end{bmatrix}$$

The vector representing the line segment $\vec{AB}$ is thus given by

$$\vec{AB} = \begin{bmatrix} b_1 - a_1 \\ b_2 - a_2 \\ \vdots \\ b_n - a_n \end{bmatrix}$$

▮EXAMPLE I Find the vector representation of $\vec{AB}$ when $A = (2, -1)$ and $B = (1, 3)$.

Solution The vector representation of $\vec{AB}$ is given by

$$\vec{AB} = \begin{bmatrix} b_1 - a_1 \\ b_2 - a_2 \end{bmatrix} = \begin{bmatrix} 1 - 2 \\ 3 - (-1) \end{bmatrix} = \begin{bmatrix} -1 \\ 4 \end{bmatrix}$$

We illustrate this graphically in Figure 9.31.

We see from Figure 9.31 that if we shift the vector $\vec{AB}$ to the origin, its tip ends at $(-1, 4)$ which confirms that the vector representation of $\vec{AB}$ is $\begin{bmatrix} -1 \\ 4 \end{bmatrix}$. ▮

Length of a Vector The length of a vector in two dimensions is computed from the Pythagorean theorem. If $\mathbf{x} = \begin{bmatrix} x_1 \\ x_2 \end{bmatrix}$, then the length of $\mathbf{x}$ is denoted by $|\mathbf{x}|$, and we

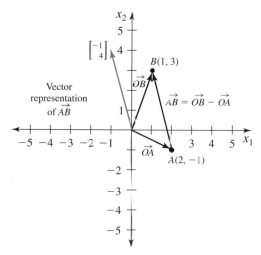

◀ **Figure 9.31**

The vector representation of the vector $\vec{AB}$.

find

$$|\mathbf{x}| = \sqrt{x_1^2 + x_2^2}$$

as illustrated in Figure 9.32. We can generalize this to n dimensions.

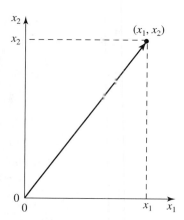

▲ **Figure 9.32**

The length of the vector $\begin{bmatrix} x_1 \\ x_2 \end{bmatrix}$.

The length of a vector $\mathbf{x} = [x_1, x_2, \ldots, x_n]'$ is

$$|\mathbf{x}| = \sqrt{x_1^2 + x_2^2 + \cdots + x_n^2}$$

▮ EXAMPLE 2

Find the length of

$$\mathbf{x} = \begin{bmatrix} 1 \\ -3 \\ 4 \end{bmatrix}$$

Solution   The length of $\mathbf{x}$ is given by

$$|\mathbf{x}| = \sqrt{x_1^2 + x_2^2 + x_3^2} = \sqrt{(1)^2 + (-3)^2 + (4)^2} = \sqrt{26}$$ ▮

If we know the length of a vector $\mathbf{x}$, we can **normalize x** to obtain a vector of length 1 in the same direction as $\mathbf{x}$ (see Figure 9.33). We call such a vector a **unit vector** in the direction of $\mathbf{x}$, and denote it by $\hat{\mathbf{x}}$, that is,

$$\hat{\mathbf{x}} = \frac{\mathbf{x}}{|\mathbf{x}|}$$

and, of course, $|\hat{\mathbf{x}}| = 1$. We summarize this in the following.

▲ **Figure 9.33**

The normalized vector $\hat{x}$ has the same direction as x; its length is 1.

$\dfrac{\mathbf{x}}{|\mathbf{x}|}$ is a vector of length 1 in the direction of $\mathbf{x}$.

▌**EXAMPLE 3**   Normalize the vector

$$\mathbf{x} = \begin{bmatrix} 3 \\ -6 \\ 6 \end{bmatrix}$$

Solution   We must first find the length of **x**.

$$|\mathbf{x}| = \sqrt{x_1^2 + x_2^2 + x_3^2} = \sqrt{(3)^2 + (-6)^2 + (6)^2} = \sqrt{81} = 9$$

The unit vector $\hat{\mathbf{x}}$ is then given by

$$\hat{\mathbf{x}} = \frac{\mathbf{x}}{|\mathbf{x}|} = \frac{1}{9} \begin{bmatrix} 3 \\ -6 \\ 6 \end{bmatrix} = \begin{bmatrix} 1/3 \\ -2/3 \\ 2/3 \end{bmatrix}$$

We check that $\hat{\mathbf{x}}$ is indeed a vector of length 1:

$$\hat{\mathbf{x}} = \sqrt{(1/3)^2 + (-2/3)^2 + (2/3)^2} = 1$$

9.4.2 The Dot Product

We will now define the dot product of two vectors, which will allow us to determine the angle between two vectors.

> **Definition**   The **scalar product** or **dot product** of two vectors $\mathbf{x} = [x_1, x_2, \ldots, x_n]'$ and $\mathbf{y} = [y_1, y_2, \ldots, y_n]'$ is the number
>
> $$\mathbf{x} \cdot \mathbf{y} = \mathbf{x}'\mathbf{y} = \sum_{i=1}^{n} x_i y_i$$

Note that the result is a scalar (hence the name scalar product). It is also called dot product because the notation uses a dot between **x** and **y**.

▌**EXAMPLE 4**   Find the dot product of

$$\mathbf{x} = \begin{bmatrix} 2 \\ 3 \\ 1 \end{bmatrix} \quad \text{and} \quad \mathbf{y} = \begin{bmatrix} -1 \\ 2 \\ 0 \end{bmatrix}$$

Solution   Using the definition of the dot product, we find

$$\mathbf{x} \cdot \mathbf{y} = \begin{bmatrix} 2 & 3 & 1 \end{bmatrix} \begin{bmatrix} -1 \\ 2 \\ 0 \end{bmatrix} = -2 + 6 + 0 = 4$$

The dot product satisfies the following two properties.

1. $\mathbf{x} \cdot \mathbf{y} = \mathbf{y} \cdot \mathbf{x}$
2. $\mathbf{x} \cdot (\mathbf{y} + \mathbf{z}) = \mathbf{x} \cdot \mathbf{y} + \mathbf{x} \cdot \mathbf{z}$

We can use the dot product to express the length of a vector. Recall that in n dimensions the length of a vector $\mathbf{x} = [x_1, x_2, \ldots, x_n]'$ is defined as

$$|\mathbf{x}| = \sqrt{x_1^2 + x_2^2 + \cdots + x_n^2}$$

If we compute the dot product of **x** with itself, we find

$$\mathbf{x} \cdot \mathbf{x} = \sum_{i=1}^{n} x_i^2$$

Hence, comparing $|\mathbf{x}|$ and $\mathbf{x} \cdot \mathbf{x}$, we see that the following holds.

$$|\mathbf{x}|^2 = \mathbf{x} \cdot \mathbf{x}$$

The length of a vector is therefore $|\mathbf{x}| = \sqrt{\mathbf{x} \cdot \mathbf{x}}$.

The Angle between Two Vectors   The other important application of the dot product is that it allows us to find the angle between two vectors. To derive this result, we need one of the trigonometric laws, the law of cosines, as illustrated in Figure 9.34.

Now, let **x** and **y** be two nonzero vectors whose initial points coincide. Then **x** − **y** is a vector that connects the endpoint of **y** to the endpoint of **x**. This follows from the parallelogram law for adding two vectors, namely $\mathbf{y} + \mathbf{x} - \mathbf{y} = \mathbf{x}$, as illustrated in Figure 9.35.

Using the law of cosines, we find

$$\mathbf{x} - \mathbf{y}|^2 = |\mathbf{x}|^2 + |\mathbf{y}|^2 - 2|\mathbf{x}||\mathbf{y}| \cos \theta$$

where θ is the angle between **x** and **y**, as illustrated in Figure 9.35. On the other hand, the length of the vector **x** − **y**, denoted $|\mathbf{x} - \mathbf{y}|$, can be computed using the dot product:

$$|\mathbf{x} - \mathbf{y}|^2 = (\mathbf{x} - \mathbf{y}) \cdot (\mathbf{x} - \mathbf{y})$$
$$= \mathbf{x} \cdot \mathbf{x} - \mathbf{x} \cdot \mathbf{y} - \mathbf{y} \cdot \mathbf{x} + \mathbf{y} \cdot \mathbf{y}$$
$$= |\mathbf{x}|^2 - 2\mathbf{x} \cdot \mathbf{y} + |\mathbf{y}|^2$$

Setting the two expressions for $|\mathbf{x} - \mathbf{y}|^2$ to be equal, we find

$$|\mathbf{x}|^2 + |\mathbf{y}|^2 - 2|\mathbf{x}||\mathbf{y}| \cos \theta = |\mathbf{x}|^2 - 2\mathbf{x} \cdot \mathbf{y} + |\mathbf{y}|^2$$

Solving this for $\mathbf{x} \cdot \mathbf{y}$, we find the following.

$$\mathbf{x} \cdot \mathbf{y} = |\mathbf{x}||\mathbf{y}| \cos \theta \qquad (9.19)$$

The significance of equation (9.19) is that it allows us to find the angle between the two nonzero vectors **x** and **y**, since

$$\theta = \cos^{-1} \left(\frac{\mathbf{x} \cdot \mathbf{y}}{|\mathbf{x}||\mathbf{y}|} \right)$$

Note that $\theta \in [0, \pi)$, since the interval $[0, \pi)$ is used to find the inverse of the cosine function.

EXAMPLE 5   Find the angle between

$$\mathbf{x} = \begin{bmatrix} 2 \\ 1 \end{bmatrix} \quad \text{and} \quad \mathbf{y} = \begin{bmatrix} 1 \\ 1 \end{bmatrix}$$

Solution   To determine the angle θ between **x** and **y**, we use

$$\mathbf{x} \cdot \mathbf{y} = |\mathbf{x}||\mathbf{y}| \cos \theta$$

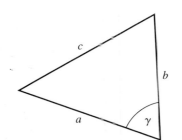

▲ **Figure 9.34**
The Law of Cosines:
$c^2 = a^2 + b^2 - 2ab \cos \gamma$.

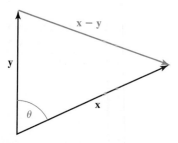

▲ **Figure 9.35**
The angle between two vectors.

We find

$$\mathbf{x} \cdot \mathbf{y} = \begin{bmatrix} 2 \\ 1 \end{bmatrix} \cdot \begin{bmatrix} 1 \\ 1 \end{bmatrix} = 2 + 1 = 3$$

$$|\mathbf{x}| = \sqrt{(2)^2 + (1)^2} = \sqrt{5}$$

$$|\mathbf{y}| = \sqrt{(1)^2 + (1)^2} = \sqrt{2}$$

Hence,

$$\cos\theta = \frac{\mathbf{x} \cdot \mathbf{y}}{|\mathbf{x}||\mathbf{y}|} = \frac{3}{\sqrt{5}\sqrt{2}} = \frac{3}{\sqrt{10}}$$

and, therefore,

$$\theta = \cos^{-1}\frac{3}{\sqrt{10}} = 18.4° \qquad \text{or} \qquad 0.3218 \qquad \blacksquare$$

We wish to single out the case in which $\theta = \pi/2$. We say that two vectors are **perpendicular** to each other if the angle between them is $\pi/2$; this is illustrated in Figure 9.36.

An important consequence of (9.19) is that it gives us a simple criterion with which to determine whether two vectors are perpendicular. Since $\cos(\pi/2) = 0$, we have the following.

▲ **Figure 9.36**

The vectors x and y are perpendicular.

Theorem

$\mathbf{x}$ and $\mathbf{y}$ are perpendicular if $\mathbf{x} \cdot \mathbf{y} = 0$.

We will now give two examples that illustrate how to use this result.

▌**EXAMPLE 6** Let $\mathbf{x} = \begin{bmatrix} 1 \\ 2 \end{bmatrix}$. Find $\mathbf{y} = \begin{bmatrix} y_1 \\ y_2 \end{bmatrix}$ so that $\mathbf{x}$ and $\mathbf{y}$ are perpendicular.

Solution   The vectors $\mathbf{x}$ and $\mathbf{y}$ are perpendicular if $\mathbf{x} \cdot \mathbf{y} = 0$. We find

$$\mathbf{x} \cdot \mathbf{y} = \begin{bmatrix} 1 \\ 2 \end{bmatrix} \cdot \begin{bmatrix} y_1 \\ y_2 \end{bmatrix} = y_1 + 2y_2$$

Setting this equal to 0, we obtain the equation

$$y_1 + 2y_2 = 0$$

Any choice of numbers (y_1, y_2) that satisfies this equation would thus give us a vector that is perpendicular to $\mathbf{x}$. For instance, if we choose $y_2 = 1$ and $y_1 = -2$, then

$$\mathbf{y} = \begin{bmatrix} -2 \\ 1 \end{bmatrix}$$

is perpendicular to $\mathbf{x}$. $\blacksquare$

▌**EXAMPLE 7**   Show that the coordinate axes in a two-dimensional Cartesian coordinate system are perpendicular .

Solution   A two-dimensional Cartesian coordinate system is illustrated in Figure 9.37. We see that the x-axis can be represented by the vector $\begin{bmatrix} 1 \\ 0 \end{bmatrix}$ and the y-axis

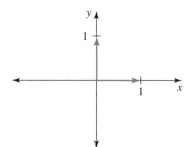

◀ **Figure 9.37**
The coordinate axes in a two dimensional Cartesian coordinate system are perpendicular.

by the vector $\begin{bmatrix} 0 \\ 1 \end{bmatrix}$. Computing the dot product between these two vectors, we find

$$\begin{bmatrix} 1 \\ 0 \end{bmatrix} \cdot \begin{bmatrix} 0 \\ 1 \end{bmatrix} = (1)(0) + (0)(1) = 0$$

and we conclude that the two vectors are perpendicular. Therefore, the x-axis and the y-axis are perpendicular. ∎

We will now use the dot product, and the result that the dot product between perpendicular vectors is zero, to obtain the equation of a line in two-dimensional space and the equation of a plane in three-dimensional space.

Lines in the Plane   We will use the dot product to write equations for lines in the plane. Suppose that we wish to find the solution of the equation of a line through the point (x_0, y_0), which is perpendicular to the vector $\mathbf{n} = \begin{bmatrix} a \\ b \end{bmatrix}$. If (x, y) is another point on the line, then the vector $\mathbf{r} - \mathbf{r}_0$ is perpendicular to $\mathbf{n}$, as illustrated in Figure 9.38.

Therefore,

$$\mathbf{n} \cdot (\mathbf{r} - \mathbf{r}_0) = 0$$

This is called the **vector equation** of a line in the plane. To obtain the **scalar equation** of this line, we set

$$\mathbf{n} = \begin{bmatrix} a \\ b \end{bmatrix}, \qquad \mathbf{r} = \begin{bmatrix} x \\ y \end{bmatrix} \qquad \text{and} \qquad \mathbf{r}_0 = \begin{bmatrix} x_0 \\ y_0 \end{bmatrix}$$

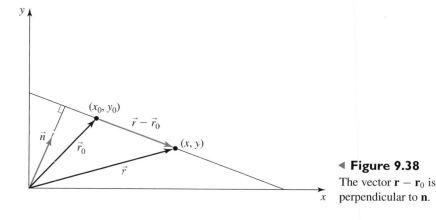

◀ **Figure 9.38**
The vector $\mathbf{r} - \mathbf{r}_0$ is perpendicular to $\mathbf{n}$.

Then

$$\mathbf{n} \cdot (\mathbf{r} - \mathbf{r}_0) = \begin{bmatrix} a \\ b \end{bmatrix} \cdot \begin{bmatrix} x - x_0 \\ y - y_0 \end{bmatrix}$$

$$= a(x - x_0) + b(y - y_0) = 0$$

That is, we find the following.

> The line through (x_0, y_0) perpendicular to $\begin{bmatrix} a \\ b \end{bmatrix}$ has the equation
>
> $$a(x - x_0) + b(y - y_0) = 0 \qquad\qquad (9.20)$$

▌EXAMPLE 8   Find an equation for the line through $(4, 3)$ perpendicular to $\begin{bmatrix} 1 \\ 2 \end{bmatrix}$.

Solution   Using (9.20), we find

$$(1)(x - 4) + (2)(y - 3) = 0$$

Simplifying yields

$$x + 2y = 10$$

Planes in Space   We can characterize planes by a point P in the plane, and the vector with initial point P that is perpendicular to all vectors in the plane whose initial points coincide with P, as illustrated in Figure 9.39. If P is the endpoint of $\mathbf{r}_0$, $\mathbf{r} - \mathbf{r}_0$ is a vector in the plane with initial point P, and if $\mathbf{n}$ is a vector that is perpendicular to the plane, then

> $$\mathbf{n} \cdot (\mathbf{r} - \mathbf{r}_0) = 0$$

This is called the vector equation of a plane. To obtain a scalar equation of a plane in three dimensional space, we set

$$\mathbf{n} = \begin{bmatrix} a \\ b \\ c \end{bmatrix}, \qquad \mathbf{r} = \begin{bmatrix} x \\ y \\ z \end{bmatrix} \qquad \text{and} \qquad \mathbf{r}_0 = \begin{bmatrix} x_0 \\ y_0 \\ z_0 \end{bmatrix}$$

Then $\mathbf{n} \cdot (\mathbf{r} - \mathbf{r}_0) = 0$ results in the scalar equation of the plane through the point (x_0, y_0, z_0) with normal vector $\mathbf{n} = \begin{bmatrix} a & b & c \end{bmatrix}'$.
    Evaluating the dot product yields

$$a(x - x_0) + b(y - y_0) + c(z - z_0) = 0$$

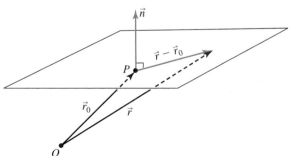

◀ **Figure 9.39**
The vector $\mathbf{r} - \mathbf{r}_0$ is perpendicular to $\mathbf{n}$.

Summarizing this result, we find the following.

> The plane through (x_0, y_0, z_0) perpendicular to $\begin{bmatrix} a, & b, & c \end{bmatrix}'$ has the equation
> $$a(x - x_0) + b(y - y_0) + c(z - z_0) = 0 \qquad (9.21)$$

▮ **EXAMPLE 9**   Find the equation of a plane in three-dimensional space through $(2, 0, 3)$ that is perpendicular to $\begin{bmatrix} -1, & 4, & 1 \end{bmatrix}'$.

Solution   Using (9.21), we find
$$(-1)(x - 2) + (4)(y - 0) + (1)(z - 3) = 0$$

Simplifying yields
$$-x + 4y + z = 1$$

▮

◆ 9.4 PROBLEMS

[9.5.1]

1. Let $x = [1, 4, -1]'$ and $y = [-2, 1, 0]'$.
(a) Find $x + y$.
(b) Find $2x$.
(c) Find $-3y$.

2. Let $x = [-4, 3, 1]'$ and $y = [0, -2, 3]'$.
(a) Find $x - y$.
(b) Find $2x + 3y$.
(c) Find $-x - 2y$

3. Let $A = (2, 3)$ and $B = (4, 1)$. Find the vector representation of $\vec{AB}$.

4. Let $A = (-1, 0)$ and $B = (2, -4)$. Find the vector representation of $\vec{AB}$.

5. Let $A = (0, 1, -3)$ and $B = (-1, -1, 2)$. Find the vector representation of $\vec{AB}$

6. Let $A = (1, 3, -2)$ and $B = (0, -1, 0)$. Find the vector representation of $\vec{AB}$

7. Find the length of $x = [1, 3]'$.

8. Find the length of $x = [-1, 4]'$.

9. Find the length of $x = [0, 1, 5]'$.

10. Find the length of $x = [-2, 3, -3]'$.

11. Normalize $[1, 3, -1]'$.

12. Normalize $[2, 0, -4]'$.

13. Normalize $[1, 0, 0]'$.

14. Normalize $[0, -2, 1, 3]'$.

[9.5.2]

15. Find the dot product of $x = [1, 2]'$ and $y = [3, -1]'$.

16. Find the dot product of $x = [-1, 2]'$ and $y = [-2, -4]'$.

17. Find the dot product of $x = [0, -1, 3]'$ and $y = [-3, 1, 1]'$.

18. Find the dot product of $x = [1, -3, 2]'$ and $y = [3, 1, -4]'$.

19. Use the dot product to compute the length of $[0, -1, 2]'$.

20. Use the dot product to compute the length of $[-1, 4, 3]'$.

21. Use the dot product to compute the length of $[1, 2, 3, 4]'$.

22. Use the dot product to compute the length of $[-1, -2, -3, -4]'$.

23. Find the angle between $x = [1, 2]'$ and $y = [3, -1]'$.

24. Find the angle between $x = [-1, 2]'$ and $y = [-2, -4]'$.

25. Find the angle between $x = [0, -1, 3]'$ and $y = [-3, 1, 1]'$.

26. Find the angle between $x = [1, -3, 2]'$ and $y = [3, 1, -4]'$.

27. Let $x = [1, -1]'$. Find y so that x and y are perpendicular.

28. Let $x = [-2, 1]'$. Find y so that x and y are perpendicular.

29. Let $x = [1, -2, 4]'$. Find y so that x and y are perpendicular.

30. Let $x = [2, 0, -1]'$. Find y so that x and y are perpendicular.

31. Find the equation for the line through $(2, 1)$ that is perpendicular to $[1, 2]'$.

32. Find the equation for the line through $(5, 2)$ that is perpendicular to $[-1, 1]'$.

33. Find the equation for the line through $(1, -2)$ that is perpendicular to $[4, 1]'$.

34. Find the equation for the line through $(0, 1)$ that is perpendicular to $[1, 0]'$.

35. Find the equation for the plane through $(1, 2, 3)$ that is perpendicular to $[0, -1, 1]'$.

36. Find the equation for the plane through $(-1, 0, -3)$ that is perpendicular to $[1, -2, 1]'$.

37. Find the equation for the plane through $(0, 0, 0)$ that is perpendicular to $[1, 0, 0]'$.

38. Find the equation for the plane through $(1, -1, 2)$ that is perpendicular to $[-1, 1, 1]'$.

◆ 9.5 KEY TERMS

Chapter 9 Review: Topics Discuss the following definitions and concepts:

1. Linear system of equations
2. Solving a linear system of equations
3. Upper triangular form
4. Gaussian elimination
5. Matrix
6. Basic matrix operations: addition, multiplication by a scalar
7. Transposition
8. Matrix multiplication
9. Identity matrix
10. Inverse matrix
11. Determinant
12. Leslie matrix
13. Stable age distribution
14. Vector
15. Parallelogram law
16. Linear map
17. Eigenvalues and eigenvectors
18. Vector addition
19. Multiplication of a vector by a scalar
20. Length of a vector
21. Dot product
22. Angle between two vectors
23. Perpendicular vectors
24. Line in the plane and in space

◆ 9.6 REVIEW PROBLEMS

1. Let
$$A = \begin{bmatrix} -1 & 1 \\ 0 & 2 \end{bmatrix}$$

(a) Find $A\mathbf{x}$ when $\mathbf{x} = \begin{bmatrix} 1 \\ 1 \end{bmatrix}$. Graph both $\mathbf{x}$ and $A\mathbf{x}$ in the same coordinate system.
(b) Find the eigenvalues λ_1 and λ_2 and the corresponding eigenvectors $\mathbf{u}_1$ and $\mathbf{u}_2$ of A.
(c) If $\mathbf{u}_i$ is the eigenvector corresponding to λ_i, find $A\mathbf{u}_i$, and explain graphically what happens when you apply A to $\mathbf{u}_i$.
(d) Write $\mathbf{x}$ as a linear combination of $\mathbf{u}_1$ and $\mathbf{u}_2$; that is, find a_1 and a_2 so that
$$\mathbf{x} = a_1\mathbf{u}_1 + a_2\mathbf{u}_2$$
Show that
$$A\mathbf{x} = a_1\lambda_1\mathbf{u}_1 + a_2\lambda_2\mathbf{u}_2$$
and illustrate this graphically.

2. Let
$$A = \begin{bmatrix} 3 & 1/2 \\ -5 & -1/2 \end{bmatrix}$$

(a) Find $A\mathbf{x}$ when $\mathbf{x} = \begin{bmatrix} 2 \\ 1 \end{bmatrix}$. Graph both $\mathbf{x}$ and $A\mathbf{x}$ in the same coordinate system.
(b) Find the eigenvalues λ_1 and λ_2 and the corresponding eigenvectors $\mathbf{u}_1$ and $\mathbf{u}_2$ of A.
(c) If $\mathbf{u}_i$ is the eigenvector corresponding to λ_i, find $A\mathbf{u}_i$, and explain graphically what happens when you apply A to $\mathbf{u}_i$.
(d) Write $\mathbf{x}$ as a linear combination of $\mathbf{u}_1$ and $\mathbf{u}_2$; that is, find a_1 and a_2 so that
$$\mathbf{x} = a_1\mathbf{u}_1 + a_2\mathbf{u}_2$$
Show that
$$A\mathbf{x} = a_1\lambda_1\mathbf{u}_1 + a_2\lambda_2\mathbf{u}_2$$
and illustrate this graphically.

3. Given the Leslie matrix
$$L = \begin{bmatrix} 1.5 & 0.875 \\ 0.5 & 0 \end{bmatrix}$$
find the growth rate of the population, and determine its stable age distribution.

4. Given the Leslie matrix
$$L = \begin{bmatrix} 0.5 & 2.99 \\ 0.25 & 0 \end{bmatrix}$$
find the growth rate of the population, and determine its stable age distribution.

5. Let
$$AB = \begin{bmatrix} 0 & 1 \\ 2 & -1 \end{bmatrix}$$
and
$$A^{-1} = \begin{bmatrix} 4 & -1 \\ 8 & -1 \end{bmatrix}$$
Find B.

6. Let
$$(AB)^{-1} = \begin{bmatrix} -1 & 3 \\ 0 & 2 \end{bmatrix}$$
and
$$B = \begin{bmatrix} 0 & -2 \\ 1 & 3 \end{bmatrix}$$
Find A.

7. Explain two different ways to solve a system of the form
$$a_{11}x_1 + a_{12}x_2 = b_1$$
$$a_{21}x_1 + a_{22}x_2 = b_2$$
when $a_{11}a_{22} - a_{12}a_{21} \neq 0$.

8. Suppose that

$$a_{11}x_1 + a_{12}x_2 = b_1$$

$$a_{21}x_1 + a_{22}x_2 = b_2$$

has infinitely many solutions. If you wrote this system in matrix form $AX = B$, could you find X by computing $A^{-1}B$?

9. Let

$$ax + 3y = 0$$

$$x - y = 0$$

How would you need to choose a so that this system had infinitely many solutions?

10. Let A be a 2×2 matrix and X and B be 2×1 matrices. Assume that $\det A = 0$. Explain how the choice of B affects the number of solutions of $AX = B$.

11. Suppose that

$$L = \begin{bmatrix} 0.5 & 2.3 \\ a & 0 \end{bmatrix}$$

is the Leslie matrix of a population with two age classes. For which values of a does this population grow?

12. Suppose that

$$L = \begin{bmatrix} 0.5 & 2.0 \\ 0.1 & 0 \end{bmatrix}$$

is the Leslie matrix of a population with two age classes.

(a) If you were to manage this population, would you need to be concerned about its long-term survival?

(b) Suppose that you can either improve the fecundity of or the survival of the zero-year-olds. But due to physiological and environmental constraints, fecundity of zero-year-olds will not exceed 1.5, and survival of zero-year-olds will not exceed 0.4. Investigate how the growth rate of the population is affected by changing either the survival or the fecundity of zero-year-olds, or both. What would be the maximum achievable growth rate?

(c) In real situations, what other factors might you need to consider when you decide on management strategies?

Multivariable Calculus

To survive in cold temperatures, species must maintain a sufficiently high metabolic rate, or regulate heat loss by covering their skin with insulating material such as feathers or fur. Humans cope with cold temperature by wearing warm clothing. There is a functional relationship that gives the lowest temperature for survival (T_e) as a function of metabolic heat production (M) and whole-body thermal conductance (g_{Hb}). The metabolic heat production depends on the type of activity; some values for humans are summarized in the following table.

Activity	M in Wm^{-2}
Sleeping	50
Working at a desk	95
Level walking at 2.5 mph	180
Level walking at 3.5 mph with a 40-lb pack	350
Data adapted from Landsberg(1969)	

The whole-body thermal conductance g_{Hb} describes how quickly heat is lost. The larger the value of g_{Hb}, the faster a body loses heat. The value of g_{Hb} depends on the type of protection, for instance, $g_{Hb} = 0.45$mol m^{-2} s^{-1} without clothing, $g_{Hb} = 0.14$mol m^{-2} s^{-1} for a wool suit, and $g_{Hb} = 0.04$mol m^{-2} s^{-1} for a warm sleeping bag. That is, the smaller g_{Hb}, the better protection from the cold the material provides. The relationship between T_e, M, and g_{Hb} is given by

$$T_e = 36 - \frac{(0.9M - 12)(g_{Hb} + 0.95)}{27.8 g_{Hb}}$$

where M is measured in Wm^{-2}, g_{Hb} is measured in mol m^{-2} s^{-1}, and T_e is measured in degree Celsius (Campbell 1986). The temperature T_e is a function of two variables, namely M and g_{Hb}; to meet the required temperature, we can change either M (by starting to move when we get cold) or g_{Hb} (by putting on more clothes when we get cold). We can plot T_e as a function of M for different values of g_{Hb}, or plot T_e as a function of g_{Hb} for different values of M, as shown in Figures 10.1 and 10.2.

Looking at Figure 10.1, we see that for a given thermal conductance (say, wearing a wool suit), we need to increase metabolic heat production as it gets colder to stay above the minimum temperature for survival. Looking at Figure 10.2, we

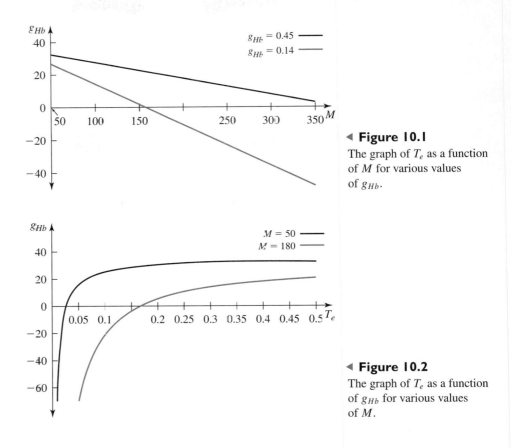

◄ **Figure 10.1**
The graph of T_e as a function of M for various values of g_{Hb}.

◄ **Figure 10.2**
The graph of T_e as a function of g_{Hb} for various values of M.

see that for a given activity (say level walking at 2.5 mph), we need to decrease thermal conductance (that is, dress warmer) as it gets colder in order to stay above the minimum temperature for survival.

Since we can vary the two variables M and g_{Hb} independently, we can think of T_e as a function of two independent variables. In this chapter, we will discuss functions of two or more independent variables, develop the theory of differential calculus for such functions, and discuss a number of applications.

◆ 10.1 FUNCTIONS OF TWO OR MORE INDEPENDENT VARIABLES

To recall the notation and terminology that we used when we considered functions of one variable, let

$$f : [0, 4] \longrightarrow \mathbf{R}$$
$$x \longrightarrow \sqrt{x}$$

The function $y = f(x)$ depends on one variable, namely x. Its domain is the set of numbers that we can use to evaluate $f(x)$, namely, the interval $[0, 4]$. Its range is the set of all possible values $y = f(x)$ for x in the domain of f. We see from Figure 10.3 that $0 \le y \le 2$ for $0 \le x \le 4$; that is, the range of $f(x)$ is the interval $[0, 2]$.

We will now consider functions for which the domain consists of pairs of real numbers (x, y) with $x, y \in \mathbf{R}$, or more generally, of n-tuples of real numbers $(x_1, x_2, \ldots, x_n)$ with $x_1, x_2, \ldots, x_n \in \mathbf{R}$. We will also call n-tuples points.

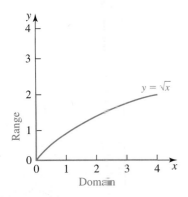

▲ **Figure 10.3**

We will use the notation $\mathbf{R}^n$ to denote the set of all n-tuples $(x_1, x_2, \ldots, x_n)$ with $x_1, x_2, \ldots, x_n \in \mathbf{R}$, that is,

$$\mathbf{R}^n = \{(x_1, x_2, \ldots, x_n) : x_1 \in \mathbf{R}, x_2 \in \mathbf{R}, \ldots, x_n \in \mathbf{R}\}$$

For instance, $\mathbf{R}^1 = \mathbf{R}$, the set of all real numbers; $\mathbf{R}^2$ is the set of all points in the plane, and so on. Note that n-tuples are *ordered*; for instance, $(2, 3) \neq (3, 2)$. The range will still be a subset of the real numbers. We will therefore call such functions "real-valued."

> **Definition**   Suppose $D \subset \mathbf{R}^n$. A **real-valued function** f on D assigns a real number to each element in D and we write
>
> $$f : D \to \mathbf{R}$$
> $$(x_1, x_2, \ldots, x_n) \to f(x_1, x_2, \ldots x_n)$$
>
> The set D is the domain of the function f and the set $\{w \in \mathbf{R} : f(x_1, x_2, \ldots, x_n) = w$ for some $(x_1, x_2, \ldots, x_n) \in D\}$, that is, the set of values taken on by f, is the range of the function f.

If a function f depends on just two independent variables, we will often denote the independent variables by x and y, and write $f(x, y)$. In the case of three variables, we will often write $f(x, y, z)$. If f is a function of more than three independent variables, it is often convenient to use subscripts to label the variables, for example, $f(x_1, x_2, x_3, x_4)$.

The first example shows how a function of more than one independent variable is evaluated at given points. You should pay attention to the fact that the domain consists of *ordered* n-tuples.

▌**EXAMPLE 1**   Evaluate the function

$$f(x, y, z) = \frac{xy}{z^2}$$

at the points $(2, 3, -1)$ and $(-1, 2, 3)$.

Solution

$$f(2, 3, -1) = \frac{(2)(3)}{(-1)^2} = 6$$

$$f(-1, 2, 3) = \frac{(-1)(2)}{(3)^2} = -\frac{2}{9}$$ ▌

The next example shows how to determine the range of a real-valued function for a given domain.

▌**EXAMPLE 2**   Let $D = \{(x, y) : 0 \le x \le 1, \ 0 \le y \le 1\}$ and set

$$f : \quad D \quad \longrightarrow \mathbf{R}$$
$$(x, y) \longrightarrow x + y$$

Graph the domain of f in the x-y plane and determine the range of f.

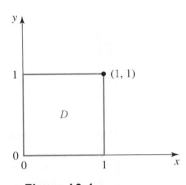

▲ Figure 10.4

The domain for the function in Example 2.

Solution   The domain of f is the set D, which consists of all points (x, y) whose x and y coordinates are between 0 and 1. This is the square shown in Figure 10.4.

To find the range of f, we need to determine what values f can take when we plug in points (x, y) from the domain D. We will argue as follows: The function $z = f(x, y)$ is smallest when $(x, y) = (0, 0)$, namely $f(0, 0) = 0$. The function $z = f(x, y)$ is largest when $(x, y) = (1, 1)$, namely $f(1, 1) = 2$. It takes on any value in between. Hence, the range of f is the set $\{z : 0 \leq z \leq 2\}$.

It is often important to find the largest possible domain, as illustrated in the next example.

❚ EXAMPLE 3   Find the largest possible domain for the function

$$f(x, y) = \sqrt{y^2 - x}$$

Solution   The largest possible domain of f is the set of all points (x, y) so that $y^2 - x \geq 0$. We can illustrate this as a set in the x-y plane, where the boundary is given by the graph of $y^2 - x = 0$. The graph of $y^2 - x = 0$ is a parabola that opens to the right. The set of points (x, y) that satisfies $y^2 - x \geq 0$ is then the shaded area in Figure 10.5. It includes the boundary curve.

The easiest way to see this is to use a test point. The boundary curve $y^2 - x = 0$ divides the plane into two regions: the set of points that lie inside the parabola and the set of points that lie outside the parabola. In one of the regions, $y^2 - x > 0$; in the other, $y^2 - x < 0$. If we use a test point from the inside of the parabola, say the point $(1, 0)$, then $y^2 - x = 0 - 1 < 0$; if we use a test point from the outside of the parabola, say the point $(-1, 0)$, then $y^2 - x = 0 - (-1) > 0$. We therefore conclude that the set of points for which $y^2 - x \geq 0$ is the outside of the parabola plus the boundary curve.

The Graph of a Function of Two Independent Variables   It is possible to graph a real-valued function of two variables. Namely, we can write $z = f(x, y)$, and then locate the point (x, y, z) in three-dimensional space. The following definition makes this precise.

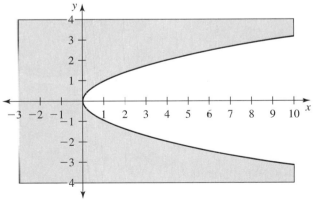

◀ Figure 10.5

The domain of $f(x, y)$ in Example 3.

> **Definition**   If f is a function of two independent variables with domain D, then the graph of f is the set of all points (x, y, z) such that $z = f(x, y)$ for $(x, y) \in D$. That is, the graph of f is the set
>
> $$S = \{(x, y, z) : z = f(x, y), \ (x, y) \in D\}$$

To locate a point (x, y, z) in three-dimensional space, we use the Cartesian coordinate system, which we introduced in Chapter 9. It consists of three mutually perpendicular axes that emanate from a common point, called the origin, which has coordinates $(0, 0, 0)$. The axes are oriented in a right-handed coordinate frame, as explained in Chapter 9. This is shown in Figure 10.6, together with a point P that has coordinates (x_0, y_0, z_0).

Figure 10.6 suggests that we can visualize the graph of $f(x, y)$, $(x, y) \in D$, as lying directly above (or below) the domain D in the x-y plane. The graph of $f(x, y)$ is therefore a **surface** in three-dimensional space, as illustrated in Figure 10.7 for $f(x, y) = 2x^2 - y^2$.

Graphing a surface in three-dimensional space is difficult. Fortunately, good computer software is now available that facilitates this task. We show a few such surfaces, generated by computer software, in Figures 10.7-10. We will make no effort to learn how to graph such functions, but will provide problems in which surfaces and functions must be matched up. To be able to do so, we will discuss the surfaces in Figures 10.7–10.

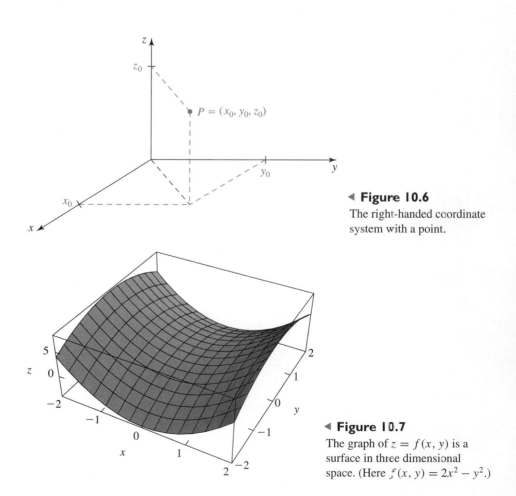

◀ **Figure 10.6**
The right-handed coordinate system with a point.

◀ **Figure 10.7**
The graph of $z = f(x, y)$ is a surface in three dimensional space. (Here $f(x, y) = 2x^2 - y^2$.)

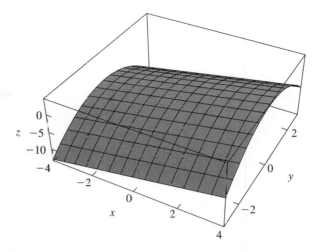

▲ **Figure 10.8**
The graph of $x - y^2$.

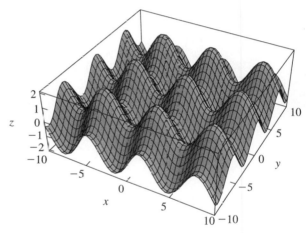

▲ **Figure 10.9**
The graph of $\sin x + \cos y$.

To understand the shape of the graph of $f(x, y) = 2x^2 - y^2$ in Figure 10.7, we can fix a value for x and then "walk" on the surface in the y-direction. Since the function is of the form "constant minus y^2" for fixed x, we expect a curve that has the shape of an upside-down parabola. Looking at the surface in Figure 10.7, this is indeed the case. On the other hand, if we fix y and walk in the direction of x, then we walk along a curve of the form "$2x^2$ minus a constant," which is a parabola. Checking Figure 10.7 confirms this.

Figure 10.8 can be analyzed in a similar way. In Figure 10.9, we recognize the waves created by the sine and cosine functions. Figure 10.10 is primarily included to show that functions of two variables can result in interesting (and complicated) shapes.

Another way to visualize functions is with **level curves** or **contour lines**. This is used, for instance, in topographical maps (see Figure 10.11). There is a subtle distinction between level curves and contour lines, in that level curves are drawn in the function domain, whereas contour lines are drawn on the surface. This distinction is not always made, and often the two terms are used interchangeably. In this text, we will only use level curves, for which we now give the precise definition.

◀ **Figure 10.10**
The graph of $xy/(1 + x^2y^2)$.

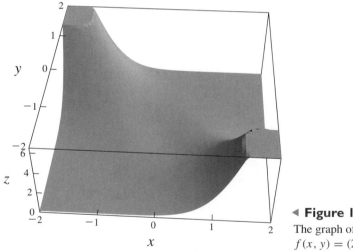

◀ **Figure 10.11**
A topographical map with level curves.

◀ **Figure 10.12**
The graph of
$f(x, y) = (2x^2 + y^2)e^{-(x+y)^2}$.

Definition   Suppose that $f : D \rightarrow \mathbf{R}$, $D \subset \mathbf{R}^2$. The level curves of f comprise the set of points (x, y) in the x-y plane where the function f has a constant value: $f(x, y) = c$.

In Figure 10.12, we graph the surface of $f(x, y) = (2x^2 + y^2)e^{-(x+y)^2}$, and in Figure 10.13, its level curves. To get an informative picture from the graph of the level curves, you should choose equidistant values for c, for instance, $c = 0, 1, 2, \ldots$

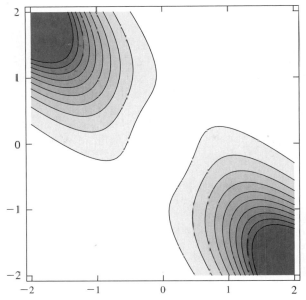

◀ **Figure 10.13**
Level curves for
$f(x, y) = (2x^2 + y^2)e^{-(x+y)^2}$.

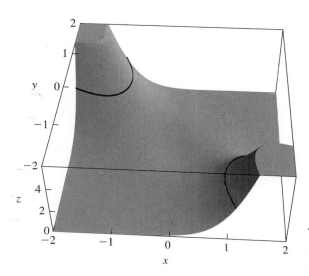

◀ **Figure 10.14**
A contour line for
$f(x, y) = (2x^2 + y^2)e^{-(x+y)^2}$.

or $c = 0, -0.1, -0.2, \ldots$, depending on the range of the function. The closer level curves are together, the steeper the surface.

If we project a level curve $f(x, y) = c$ up to its height c, the curve lies on the surface of $f(x, y)$, and traces the graph of f in a horizontal plane at height c. Such curves are then called contour lines, as illustrated in Figure 10.14.

EXAMPLE 4   Set $D = \{(x, y) : x^2 + y^2 \le 4\}$. Compare the level curves of

$$f(x, y) = 4 - x^2 - y^2 \quad \text{for } (x, y) \in D$$

and

$$g(x, y) = \sqrt{4 - x^2 - y^2} \quad \text{for } (x, y) \in D$$

Solution   The level curve $f(x, y) = c$ is the set of all points (x, y) that satisfy

$$4 - x^2 - y^2 = c \quad \text{or} \quad x^2 + y^2 = 4 - c$$

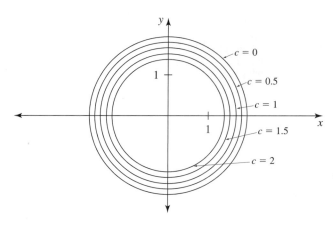

◀ **Figure 10.15**
Level curves for
$f(x, y) = 4 - x^2 - y^2$.

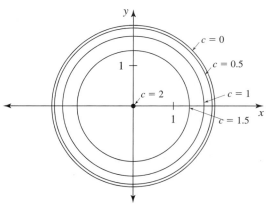

◀ **Figure 10.16**
Level curves for
$g(x, y) = \sqrt{4 - x^2 - y^2}$.

We recognize this as a circle with center at the origin and radius $\sqrt{4 - c}$, illustrated in Figure 10.15 for $c = 0, 0.5, 1, 1.5$, and 2.

The level curve $g(x, y) = c$ satisfies

$$\sqrt{4 - x^2 - y^2} = c \qquad \text{or} \qquad x^2 + y^2 = 4 - c^2$$

This is also a circle with center at the origin but the radius is $\sqrt{4 - c^2}$; it is illustrated in Figure 10.16 for $c = 0, 0.5, 1, 1.5$, and 2.

To compare the level curves of two functions, we choose the same values for c for them. Here, we chose $c = 0, 0.5, 1, 1.5$, and 2 for $f(x, y)$ and $g(x, y)$. For instance, the level curves for $c = 0$ and 1 are the same for the two functions, but when $c = 1.5$ and 2, the contour lines for $f(x, y)$ are a lot closer to the contour line at $c = 1$ than those for $g(x, y)$. We can understand this when we look at the actual surfaces.

The graph of $f(x, y)$ is obtained by rotating the parabola $z = 4 - x^2$ in the x-z plane about the z-axis; its surface is called a paraboloid (see Figure 10.17).

The graph of $g(x, y)$ is obtained by rotating the half circle $z = \sqrt{4 - x^2}$ in the x-z plane about the z-axis; its surface is the top half of a sphere (see Figure 10.18).

Because the paraboloid is steeper than the sphere for c between 1 and 2, but the reverse holds for c between 0 and 1, the contour lines of the paraboloid are closer together for c between 1 and 2, whereas the reverse holds for c between 0 and 1. ▮

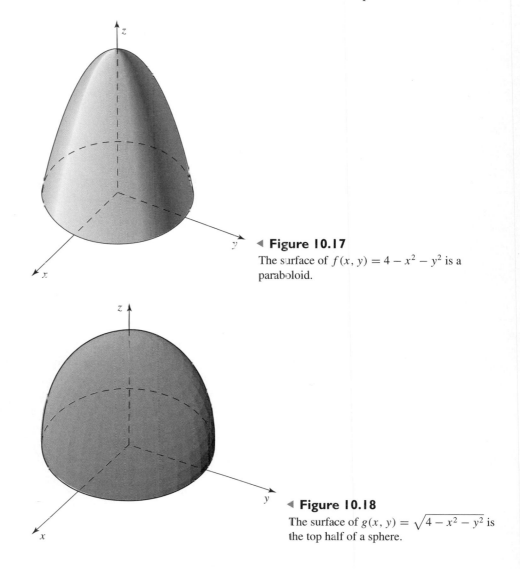

◀ **Figure 10.17**
The surface of $f(x, y) = 4 - x^2 - y^2$ is a paraboloid.

◀ **Figure 10.18**
The surface of $g(x, y) = \sqrt{4 - x^2 - y^2}$ is the top half of a sphere.

Biological Applications

EXAMPLE 5   In Figure 10.19, we show level curves for oxygen concentration (in mg/l) in Long Lake, Clear Water County (Minnesota), as a function of date and depth. For instance, on day 140 (May 20, 1998) at 10 m depth, the oxygen concentration was 12 mg/l.

The waterflea *Daphnia* needs a minimum of 3 mg/l oxygen to survive. Suppose that you went out to Long Lake on day 200 (July 19, 1998), and wanted to look for *Daphnia* in the lake. Below what depth could you be fairly sure not to find any *Daphnia*?

Solution   Because *Daphnia* needs a minimum of 3 mg/l of oxygen, we need to find the depth on July 19, 1998, below which the oxygen concentration is always less than 3 mg/l. We see that the 3 mg/l oxygen level curve goes through the point (200, −17.5). Thus, on July 19, 1998, *Daphnia* needed to stay above 17.5 m; that is, we could have been fairly sure not to have found *Daphnia* below 17.5 m. ▮

EXAMPLE 6   In Figure 10.20, we show temperature **isoclines** (called **isotherms**) for Long Lake, Clear Water County (Minnesota). Temperature isoclines are lines of equal temperature, that is, level curves. The isoclines are functions of date and depth. For

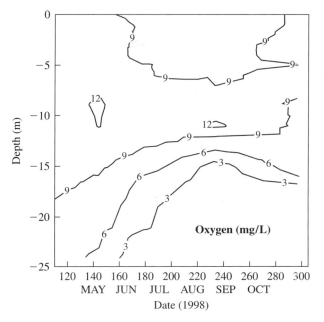

▲ **Figure 10.19**
Level curves for oxygen concentration on Long Lake, Clear Water County (Minnesota).
Courtesy of Leif Hembre.

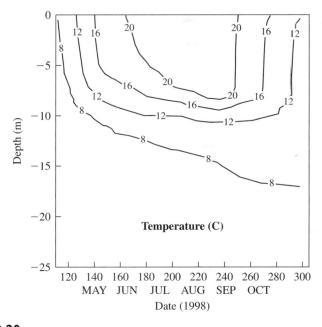

▲ **Figure 10.20**
Isotherms for Long Lake, Clear Water County (Minnesota). Courtesy of Leif Hembre.

instance, on day 130 (May 10, 1998), the temperature at 5 m was approximately 12°C. Suppose that we were interested in how temperature changes over time at a fixed depth. We would draw a horizontal line through the appropriate depth, and then read off the days and the temperature where the horizontal line intersects the isoclines. For instance, if we look at 5 meter depths, we would find that the temperature climbs from roughly 8°C in the spring to more than 20°C in the summer and then falls back to below 12°C later in the fall.

◆ **10.1 PROBLEMS**

1. Locate the following points in a three-dimensional Cartesian coordinate system.

(a) $(1, 3, 2)$

(b) $(-1, -2, 1)$

(c) $(0, 1, 2)$

(d) $(2, 0, 3)$

2. Describe in words the set of all points in $\mathbf{R}^3$ that satisfy the following.

(a) $x = 0$

(b) $y = 0$

(c) $z = 0$

(d) $z \geq 0$

(e) $y \leq 0$

In Problems 3–6, evaluate each function at the given point.

3. $f(x, y) = \dfrac{3x}{x^2 + y^2}$ at $(2, 3)$

4. $f(x, y, z) = \sqrt{x^2 - 3y + z}$ at $(3, -1, 1)$

5. $h(x, t) = \exp\left[-\dfrac{(x - 2)^2}{2t}\right]$ at $(1, 5)$

6. $g(x_1, x_2, x_3, x_4) = x_1 x_4 \sqrt{x_2 x_3}$ at $(1, 8, 2, -1)$

In Problems 7–10, find the maximal domain and the corresponding range of each function. Determine the equation of the level curves $f(x, y) = c$, together with the possible values of c.

7. $f(x, y) = x^2 - y^2$

8. $f(x, y) = \sqrt{9 - x^2 - y^2}$

9. $f(x, y) = \ln(y - x^2)$

10. $f(x, y) = \exp[-(x^2 + y^2)]$

In Problems 11–14 match each function with the appropriate graph in Figures 10.21–24.

11. $f(x, y) = 1 + x^2 + y^2$

12. $f(x, y) = \sin(x) \sin(y)$

13. $f(x, y) = y^2 - x^2$

14. $f(x, y) = 4 - x^2$

▲ **Figure 10.21**

▲ **Figure 10.22**

▲ **Figure 10.23**

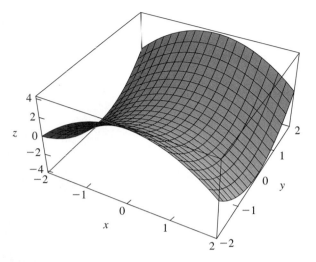

Figure 10.24

15. Let

$$f_a(x, y) = ax^2 + y^2$$

for $(x, y) \in \mathbf{R}$, where a is a positive constant.

(a) Assume that $a = 1$ and describe the level curves. The graph of $f(x, y)$ intersects both the x-z and the y-z plane; show that these two curves of intersection are parabolas.

(b) Assume that $a = 4$. Then,

$$f_4(x, y) = 4x^2 + y^2$$

and the level curves satisfy

$$4x^2 + y^2 = c$$

Use a graphing calculator to sketch the level curves for $c = 0, 1, 2, 3$, and 4. These curves are ellipses. Find the curves of intersection of $f_4(x, y)$ with the x-z and the y-z plane.

(c) Repeat (b) for $a = 1/4$.

(d) Explain in words how the surfaces of $f_a(x, y)$ change when a changes.

16. The graph in Figure 10.25 shows isotherms of a lake in the temperate climate of the northern hemisphere.

(a) Use this plot to sketch the temperature profiles in March and June. That is, plot the temperature as a function of depth for a day in March and for a day in June.

(b) Explain how it follows from your temperature plots that the lake is **homeothermic** in March, that is, has the same temperature from the surface to the bottom.

(c) Explain how it follows from your temperature plots that the lake is **stratified** in June, that is, has a warm layer on top (called the **epilimnion**), followed by a region where the temperature changes quickly (called the **metalimnion**), which is then followed by a cold layer deeper down (called the **hypolimnion**).

Figure 10.25

Isotherms for a typical lake in the northern hemisphere.

17. Figure 10.26 shows the oxygen concentration for Long Lake, Clear Water County (Minnesota). The waterflea *Daphnia* can only survive if the oxygen concentration is higher than 3 mg/l. Suppose that you wanted to sample the *Daphnia* population in 1997 on days 180, 200, and 220. Below which depths can you be fairly sure not to find any *Daphnia*?

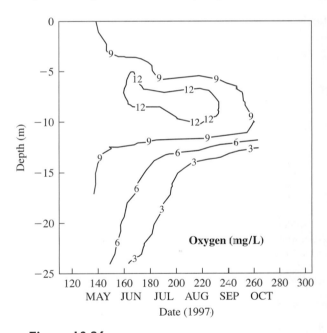

Figure 10.26

Level curves for oxygen concentration on Long Lake, Clear Water County (Minnesota). Courtesy of Leif Hembre.

18. At the beginning of this chapter, we discussed the minimum temperature required for survival as a function of metabolic heat production and whole-body thermal conductance. Suppose that you wish to go winter camping in Northern Minnesota and the predicted low temperature for the night is $-15°C$. Use the information provided in the beginning of the chapter to find the maximum value of g_{Hb} for your sleeping bag that would allow you to sleep safely.

◆ 10.2 LIMITS AND CONTINUITY

We need to extend the notion of limits and continuity to the multivariable setting. The ideas are the same as in the one-dimensional case, so we will give a very brief treatment.

10.2.1 Limits

We first recall the formal definition of limits in the one-dimensional case. For instance, to show that

$$\lim_{x \to 2} x^2 = 4$$

we must show that whenever x is close to 2 (but not equal to 2), then x^2 is close to 4. Formally, this means that we must show that for every $\epsilon > 0$ there exists $\delta > 0$ so that

$$|x^2 - 4| < \epsilon \qquad \text{whenever} \qquad 0 < |x - 2| < \delta$$

To generalize this idea to higher dimensions, we need a notion of proximity. In the one-dimensional case, we chose an interval centered at a particular point that contained all points within distance δ of the center of the interval, except for the center of the interval. In two dimensions, we replace the interval by a disk. A disk of radius δ centered at the point (x_0, y_0) is the set of all points that are within distance δ of (x_0, y_0).

As with open and closed intervals, there are open and closed disks. We have the following definition.

Definition   An **open disk** with radius r centered at $(x_0, y_0) \in \mathbf{R}^2$ is the set

$$B_r(x_0, y_0) = \{(x, y) \in \mathbf{R}^2 : \sqrt{(x - x_0)^2 + (y - y_0)^2} < r\}$$

A **closed disk** with radius r centered at $(x_0, y_0) \in \mathbf{R}^2$ is the set

$$\bar{B}_r(x_0, y_0) = \{(x, y) \in \mathbf{R}^2 : \sqrt{(x - x_0)^2 + (y - y_0)^2} \le r\}$$

An open disk is thus a disk in which the boundary line is not part of the disk; whereas a closed disk contains the boundary line. This is analogous to intervals: an open interval does not contain the two endpoints (which are the boundary of the interval), whereas a closed interval does contain them. A closed disk is shown in Figure 10.27.

When we defined $\lim_{x \to x_0} f(x)$, we emphasized that the value of f at x_0 is not important. This was reflected in the formal $\epsilon - \delta$ definition, where we required $0 < |x - x_0| < \delta$; that is, we removed x_0 from the interval $(x_0 - \delta, x_0 + \delta)$. We will need to do the same when we generalize the limit to two dimensions. We will write $B_\delta(x_0, y_0) - \{(x_0, y_0)\}$ to denote the open disk with radius δ centered at (x_0, y_0), where the center (x_0, y_0) is removed. We can now generalize the notion of limits to two dimensions.

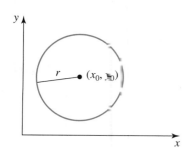

▲ **Figure 10.27**

A closed disk with radius r centered at (x_0, y_0).

Definition   The limit of $f(x, y)$ as (x, y) approaches (x_0, y_0), denoted by

$$\lim_{(x,y) \to (x_0, y_0)} f(x, y)$$

is the number L such that for every $\epsilon > 0$ there exists a $\delta > 0$ so that

$$|f(x, y) - L| < \epsilon \qquad \text{whenever} \qquad (x, y) \in B_\delta(x_0, y_0) - \{(x_0, y_0)\}$$

You should recognize that this is very similar to the definition in one dimension. Namely, we require that whenever (x, y) is close (but not equal) to (x_0, y_0), then $f(x, y)$ is close to L.

We provide one example in which we use the formal definition of limits. After this, we will give limit laws, which are extensions of the laws in the one-dimensional case. Using the formal definition is difficult, and we will not need to do so subsequently. But seeing an example might help you to understand the definition.

▌EXAMPLE I Let

$$f(x, y) = x^2 + y^2$$

and show that

$$\lim_{(x,y)\to(0,0)} f(x, y) = 0$$

Solution We must show that for every $\epsilon > 0$ there exists a $\delta > 0$ such that

$$|x^2 + y^2 - 0| < \epsilon \qquad \text{whenever} \qquad (x, y) \in B_\delta(0, 0) - \{(0, 0)\}$$

Now, $(x, y) \in B_\delta(0, 0)$ if $\sqrt{x^2 + y^2} < \delta$ or $x^2 + y^2 < \delta^2$. We need to show $|x^2 + y^2| < \epsilon$. This suggests that we should choose δ so that $\delta^2 = \epsilon$. Let's try this. We set $\delta = \sqrt{\epsilon}$ for $\epsilon > 0$, then

$$\sqrt{x^2 + y^2} < \delta = \sqrt{\epsilon}$$

implies

$$|x^2 + y^2| < \epsilon$$

But this is what we need to show. In other words, we have shown that for every $\epsilon > 0$, we can find $\delta > 0$ (namely $\delta = \sqrt{\epsilon}$) so that whenever (x, y) is close to $(0, 0)$, then $x^2 + y^2$ is close to 0. ▌

As in the one-dimensional case, the $\epsilon - \delta$ definition of limits is difficult to use. Fortunately, there are laws similar to those in the one dimensional case, which will allow us to compute limits.

LIMIT LAWS FOR THE TWO-DIMENSIONAL CASE

If a is a constant and if

$$\lim_{(x,y)\to(x_0,y_0)} f(x, y) = L_1 \qquad \text{and} \qquad \lim_{(x,y)\to(x_0,y_0)} g(x, y) = L_2$$

where L_1 and L_2 are real numbers, then the following hold.

  1. (Addition Rule)

$$\lim_{(x,y)\to(x_0,y_0)} [f(x, y) + g(x, y)] = \lim_{(x,y)\to(x_0,y_0)} f(x, y) + \lim_{(x,y)\to(x_0,y_0)} g(x, y)$$

  2. (Constant Factor Rule)

$$\lim_{(x,y)\to(x_0,y_0)} af(x, y) = a \lim_{(x,y)\to(x_0,y_0)} f(x, y)$$

  3. (Multiplication Rule)

$$\lim_{(x,y)\to(x_0,y_0)} f(x, y)g(x, y) = \left[\lim_{(x,y)\to(x_0,y_0)} f(x, y) \right] \left[\lim_{(x,y)\to(x_0,y_0)} g(x, y) \right]$$

4. (Quotient Rule)

$$\lim_{(x,y)\to(x_0,y_0)} \frac{f(x,y)}{g(x,y)} = \frac{\lim_{(x,y)\to(x_0,y_0)} f(x,y)}{\lim_{(x,y)\to(x_0,y_0)} g(x,y)}$$

provided that $L_2 \neq 0$.

Limits That Exist

EXAMPLE 2   **a.**

$$\lim_{(x,y)\to(1,2)} (x^2 y + 3x) = \left(\lim_{(x,y)\to(1,2)} x^2 y\right) + \left(3 \lim_{(x,y)\to(1,2)} x\right)$$

$$= \left(\lim_{(x,y)\to(1,2)} x^2\right)\left(\lim_{(x,y)\to(1,2)} y\right) + \left(3 \lim_{(x,y)\to(1,2)} x\right)$$

$$= (1)^2(2) + (3)(1) = 5$$

b.

$$\lim_{(x,y)\to(-1,3)} \left(2y^2 - \frac{3x}{y}\right) = (2)(3)^2 - \frac{(3)(-1)}{3} = 18 + 1 = 19$$

c.

$$\lim_{(x,y)\to(2,0)} \frac{4y + 2x}{x^2 + 2xy - 3} = \frac{(4)(0) + (2)(2)}{(2)^2 + (2)(2)(0) - 3}$$

$$= \frac{4}{4 - 3} = 4 \qquad ▌$$

We see from Example 2 that limits of polynomials and rational functions can be evaluated simply by evaluating the functions at the respective points, provided that the functions are defined at those points.

Limits That Do Not Exist   In the one-dimensional case, there were only two ways in which we could approach a number, namely, from the left or the right. If the two limits were different, we said that the limit did not exist. In two dimensions, there are many more ways that we can approach the point (x_0, y_0), namely, any curve in the x-y plane that ends up at the point (x_0, y_0). We call such curves **paths**.

> If $f(x, y)$ approaches L_1 as $(x, y) \to (x_0, y_0)$ along path C_1, and $f(x, y)$ approaches L_2 as $(x, y) \to (x_0, y_0)$ along path C_2, and if $L_1 \neq L_2$, then $\lim_{(x,y)\to(x_0,y_0)} f(x, y)$ does not exist.

EXAMPLE 3   Show that

$$\lim_{(x,y)\to(0,0)} \frac{x^2 - y^2}{x^2 + y^2}$$

does not exist.

Solution   We first let $(x, y) \to (0, 0)$ along the positive x-axis; this is the curve C_1 in Figure 10.28. On C_1, $y = 0$, $x > 0$. Then

$$\lim_{x\to 0+} \frac{x^2}{x^2} = 1$$

Next, we let $(x, y) \to (0, 0)$ along the positive y-axis; this is the curve C_2 in

▲ Figure 10.28
The paths C_1 and C_2 for Example 3.

Figure 10.28. On C_2, $x = 0$ and $y > 0$. Then

$$\lim_{y \to 0+} \frac{-y^2}{y^2} = -1$$

Since $1 \neq -1$, we conclude that the limit does not exist. ▌

Unless one has a lot of experience, it is not easy to find paths for which limits differ. We will therefore always provide the paths along which you should check the limits, in case a limit does not exist.

▌**EXAMPLE 4**   Show that

$$\lim_{(x,y) \to (0,0)} \frac{4xy}{xy + y^3}$$

does not exist.

Solution   A natural choice for paths to $(0, 0)$ are straight lines of the form $y = mx$. We assume that $m \neq 0$. If we substitute $y = mx$ in the preceding limit, then $(x, y) \to (0, 0)$ reduces to $x \to 0$ and we find

$$\lim_{x \to 0} \frac{4mx^2}{mx^2 + (mx)^3} = \lim_{x \to 0} \frac{4}{1 + m^2 x} = 4$$

That is, as long as we approach $(0, 0)$ along the straight line $y = mx, m \neq 0$, the limit is always 4, irrespective of m. Such a path, labelled C_1, is shown in Figure 10.29.

You might be tempted to say that the limit exists. But let's approach $(0, 0)$ along the parabola $x = y^2$, $y > 0$. This is the curve C_2 in Figure 10.29. Substituting y^2 for x then yields

$$\lim_{y \to 0^+} \frac{4y^3}{y^3 + y^3} = 2 \neq 4$$

Thus we found paths along which the limits differ. Therefore, the limit does not exist. ▌

The limit laws allow us to compute limits. To show that a limit does not exist, we must identify two paths along which the limits differ. This method does not work when we want to show that a limit exists. To show that a limit exists, the limits along *all* possible paths must be the same; there is no way to check all possible paths.

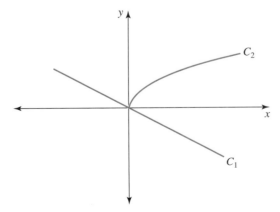

◄ Figure 10.29
The paths C_1 and C_2 for Example 4.

10.2.2 Continuity

The definition of continuity is likewise analogous to that in the one-dimensional case.

A function $f(x, y)$ is continuous at (x_0, y_0) if the following hold.

1. $f(x, y)$ is defined at (x_0, y_0)
2. $\lim\limits_{(x,y) \to (x_0, y_0)} f(x, y)$ exists
3. $\lim\limits_{(x,y) \to (x_0, y_0)} f(x, y) = f(x_0, y_0)$

EXAMPLE 5　Show that

$$f(x, y) = 2 + x^2 + y^2$$

is continuous at $(0, 0)$.

1. $f(x, y)$ is defined at $(0, 0)$, namely

$$f(0, 0) = 2$$

2. To show that the limit exists, we refer to Example 1, where we showed that

$$\lim_{(x,y) \to (0,0)} (x^2 + y^2)$$

exists. Using the limit laws, we conclude that

$$\lim_{(x,y) \to (0,0)} f(x, y) = 2 + \lim_{(x,y) \to (0,0)} (x^2 + y^2)$$

exists.

3. In Example 1, we also showed that

$$\lim_{(x,y) \to (0,0)} (x^2 + y^2) = 0$$

Therefore

$$\lim_{(x,y) \to (0,0)} (2 + x^2 + y^2) = 2 + \lim_{(x,y) \to (0,0)} (x^2 + y^2) = 2 + 0 = 2$$

Since $f(0, 0) = 2$, this shows that $f(x, y)$ is continuous at $(0, 0)$. ▮

EXAMPLE 6　Show that

$$f(x, y) = \begin{cases} \frac{x^2 - y^2}{x^2 + y^2} & \text{for } (x, y) \neq (0, 0) \\ 0 & \text{for } (x, y) = (0, 0) \end{cases}$$

is discontinuous at $(0, 0)$.

Solution　The function $f(x, y)$ is defined at $(0,0)$. Hence (1) holds. But in Example 3, we showed that

$$\lim_{(x,y) \to (0,0)} f(x, y)$$

does not exist. Therefore, (2) is violated and we conclude that $f(x, y)$ is discontinuous at $(x, y) = (0, 0)$. ▮

Composition of Functions　Using the definition of continuity and the rules for finding limits, one can show that polynomial functions of two variables (that is, functions that are sums of terms of the form $ax^n y^m$, where a is a constant and n and m are nonnegative integers) are continuous.

We can obtain a much larger class of continuous functions when we allow the composition of functions. For instance, the function $h(x, y) = e^{x^2+y^2}$ can be written as a composition of two functions. Namely, if we set $z = f(x, y) = x^2 + y^2$ and $g(z) = e^z$, then the function

$$h(x, y) = (g \circ f)(x, y) = g[f(x, y)] = e^{x^2+y^2}$$

Another example is $h(x, y) = \sqrt{y^2 - x}$. Here, $z = f(x, y) = y^2 - x$ and $g(z) = \sqrt{z}$. Then $h(x, y) = (g \circ f)(x, y)$. More generally, we set

$$f : D \to \mathbf{R}, \quad D \subset \mathbf{R}^2$$

and

$$g : I \to \mathbf{R}, \quad I \subset \mathbf{R}$$

and define the composition

$$h(x, y) = (g \circ f)(x, y) = g[f(x, y)]$$

One can then show that if f is continuous at (x_0, y_0) and g is continuous at $z = f(x_0, y_0)$, then

$$h(x, y) = (g \circ f)(x, y) = g[f(x, y)]$$

is continuous at (x_0, y_0). As an example, let's return to the function

$$h(x, y) = e^{x^2+y^2}$$

With $z = f(x, y) = x^2 + y^2$ and $g(z) = e^z$, then $h(x, y) = g[f(x, y)]$ is continuous, since $f(x, y)$ is continuous for all $(x, y) \in \mathbf{R}^2$, and $g(z)$ is continuous for all z in the range of $f(x, y)$. Likewise, $h(x, y) = \sqrt{y^2 - x}$ is continuous for all $(x, y) \in \{(x, y) : y^2 - x \geq 0\}$, since $z = f(x, y) = y^2 - x$ is continuous for all $(x, y) \in \mathbf{R}^2$, and $g(z) = \sqrt{z}$ is continuous for all $z \geq 0$.

The results of this section can be generalized to higher dimensions. The three dimensional analogue of a disk is a **sphere**. In even higher dimensions, we call these sets **balls**. For instance, an open ball with radius r centered at $(x_1, x_2, \ldots, x_n)$ is the set

$$B_r(x_1, x_2, \ldots, x_n) = \{(y_1, y_2, \ldots, y_n) \in \mathbf{R}^n :$$
$$\sqrt{(x_1 - y_1)^2 + (x_2 - y_2)^2 + \cdots + (x_n - y_n)^2} < r\}$$

The definitions of limits and continuity can then be extended by replacing (x, y) in all definitions by $(x_1, x_2, \ldots, x_n)$.

◆ 10.2 PROBLEMS

[10.2.1]

1. Draw an open disk with radius 2 centered at $(1, -1)$ in the x-y plane, and give a mathematical description of this set.

2. Draw a closed disk with radius 3 centered at $(2, 0)$ in the x-y plane, and give a mathematical description of this set.

3. Give a geometric interpretation of the set

$$A = \{(x, y) \in \mathbf{R}^2 : \sqrt{x^2 + y^2 - 4y + 4} < 3\}$$

4. Give a geometric interpretation of the set

$$A = \{(x, y) \in \mathbf{R}^2 : \sqrt{x^2 + 6x + y^2 - 2y + 10} < 2\}$$

5. Let

$$f(x, y) = 2x^2 + y^2$$

Use the $\epsilon - \delta$ definition of limits to show that

$$\lim_{(x,y) \to (0,0)} f(x, y) = 0$$

6. Let

$$f(x, y) = x^2 + 3y^2$$

Use the $\epsilon - \delta$ definition of limits to show that

$$\lim_{(x,y)\to(0,0)} f(x, y) = 0$$

In Problems 7–14, use the properties of limits to calculate the following limits.

7. $\lim_{(x,y)\to(1,0)}(x^2 - 3y^2)$

8. $\lim_{(x,y)\to(-1,1)}(2xy + 3x^2)$

9. $\lim_{(x,y)\to(0,2)}\left(4xy^2 - \frac{x+1}{y}\right)$

10. $\lim_{(x,y)\to(1,1)}(x^2 + y^2)$

11. $\lim_{(x,y)\to(0,1)}\frac{2xy-3}{x^2+y^2+1}$

12. $\lim_{(x,y)\to(-1,-2)}\frac{x^2-y^2}{2xy+2}$

13. $\lim_{(x,y)\to(2,0)}\frac{2x-4y^2}{y^2+3x}$

14. $\lim_{(x,y)\to(1,-1)}\frac{2x^2+y}{2xy+3}$

15. Show that

$$\lim_{(x,y)\to(0,0)} \frac{x^2 - 2y^2}{x^2 + y^2}$$

does not exist, by computing the limit along the positive x-axis and along the positive y-axis.

16. Show that

$$\lim_{(x,y)\to(0,0)} \frac{3x^2 - y^2}{x^2 + y^2}$$

does not exist, by computing the limit along the positive x-axis and along the positive y-axis.

17. Compute

$$\lim_{(x,y)\to(0,0)} \frac{4xy}{x^2 + y^2}$$

along the x-axis, the y-axis, and the line $y = x$. What can you conclude?

18. Compute

$$\lim_{(x,y)\to(0,0)} \frac{3xy}{x^2 + y^3}$$

along lines of the form $y = mx$, for $m \neq 0$. What can you conclude?

19. Compute

$$\lim_{(x,y)\to(0,0)} \frac{2xy}{x^3 + yx}$$

along lines of the form $y = mx$, for $m \neq 0$, and along the parabola $y = x^2$. What can you conclude?

20. Compute

$$\lim_{(x,y)\to(0,0)} \frac{3x^2y^2}{x^3 + y^6}$$

along lines of the form $y = mx$, for $m \neq 0$, and along the parabola $x = y^2$. What can you conclude?

[10.2.2]

21. Use the definition of continuity to show that

$$f(x, y) = x^2 + y^2$$

is continuous at $(0, 0)$.

22. Use the definition of continuity to show that

$$f(x, y) = \sqrt{9 + x^2 + y^2}$$

is continuous at $(0, 0)$.

23. Show that

$$f(x, y) = \begin{cases} \frac{4xy}{x^2+y^2} & \text{for } (x, y) \neq (0, 0) \\ 0 & \text{for } (x, y) = (0, 0) \end{cases}$$

is discontinuous at $(0, 0)$.

24. Show that

$$f(x, y) = \begin{cases} \frac{3xy}{x^2+y^2} & \text{for } (x, y) \neq (0, 0) \\ 0 & \text{for } (x, y) = (0, 0) \end{cases}$$

is discontinuous at $(0, 0)$.

25. Show that

$$f(x, y) = \begin{cases} \frac{5xy}{2x^3+yx} & \text{for } (x, y) \neq (0, 0) \\ 0 & \text{for } (x, y) = (0, 0) \end{cases}$$

is discontinuous at $(0, 0)$.

26. Show that

$$f(x, y) = \begin{cases} \frac{3x^2y^2}{x^3+y^6} & \text{for } (x, y) \neq (0, 0) \\ 0 & \text{for } (x, y) = (0, 0) \end{cases}$$

is discontinuous at $(0, 0)$.

27. (a) Write

$$h(x, y) = \sin(x^2 + y^2)$$

as a composition of two functions.

(b) For which values of (x, y) is $h(x, y)$ continuous?

28. (a) Write

$$h(x, y) = \sqrt{x + y}$$

as a composition of two functions.

(b) For which values of (x, y) is $h(x, y)$ continuous?

29. (a) Write

$$h(x, y) = e^{xy}$$

as a composition of two functions.

(b) For which values of (x, y) is $h(x, y)$ continuous?

30. (a) Write

$$h(x, y) = \cos(y - x)$$

as a composition of two functions.

(b) For which values of (x, y) is $h(x, y)$ continuous?

◆ 10.3 PARTIAL DERIVATIVES

10.3.1 Functions of Two Variables

Suppose that the response of an organism depends on a number of independent variables. To investigate this dependency, a common experimental design is to measure the response when changing one variable while keeping all other variables fixed. As an example, Pisek et al. (1969) measured the net assimiliation of CO_2 of *Ranunculus glacialis*, a member of the buttercup family, as a function of temperature and light intensity. They varied the temperature while keeping the light intensity constant. Repeating this experiment at different light intensities, they were able to determine how net assimiliation of CO_2 changes as a function of both temperature and light intensity.

This experimental design illustrates the idea behind partial derivatives. Suppose that we want to know how the function $f(x, y)$ changes when x and y change. Instead of changing both variables simultaneously, we might get an idea of how $f(x, y)$ depends on x and y when we change one variable while keeping the other variable fixed.

To illustrate this we look at a simple function. Let

$$f(x, y) = x^2 y$$

We want to know how $f(x, y)$ changes if we change one variable, say x, and keep the other variable, in this case y, fixed. We fix $y = y_0$, then the change of f with respect to x is simply the derivative of f with respect to x when $y = y_0$. That is,

$$\frac{d}{dx} f(x, y_0) = \frac{d}{dx} x^2 y_0 = 2x y_0$$

Such a derivative is called a partial derivative.

> **Definition**   Suppose that f is a function of two independent variables x and y. The **partial derivative** of f with respect to x is defined by
> $$\frac{\partial f(x, y)}{\partial x} = \lim_{h \to 0} \frac{f(x + h, y) - f(x, y)}{h}$$
> The partial derivative of f with respect to y is defined by
> $$\frac{\partial f(x, y)}{\partial y} = \lim_{h \to 0} \frac{f(x, y + h) - f(x, y)}{h}$$

To denote partial derivatives, we use "∂" instead of "d." We will also use the notation

$$f_x(x, y) = \frac{\partial f(x, y)}{\partial x} \qquad \text{and} \qquad f_y(x, y) = \frac{\partial f(x, y)}{\partial y}$$

In the definition of partial derivatives, you should recognize the formal definition of derivatives of Chapter 3. Namely, to compute $\partial f / \partial x$, we look at the ratio of the difference in the f values, $f(x + h, y) - f(x, y)$, and the difference in the x values, $x + h - x$. The other variable, y, is not changed. We then let h tend to 0.

To compute $\partial f(x, y) / \partial x$, we differentiate f with respect to x while treating y as a constant. When we read $\partial f(x, y) / \partial x$, we can say "the partial derivative of f of x and y with respect to x." To read $f_x(x, y)$, we say "f sub x of x and y."

Finding partial derivatives is no different from finding derivatives of functions of one variable, since by keeping all but one variable fixed, computing a partial derivative is reduced to computing a derivative of a function of one variable. You just need to keep straight which of the variables you have fixed and which one you will vary.

EXAMPLE 1   Find $\partial f/\partial x$ and $\partial f/\partial y$ when

$$f(x, y) = ye^{xy}$$

Solution   To compute $\partial f/\partial x$, we treat y as a constant, and differentiate f with respect to x using the chain rule.

$$\frac{\partial f(x, y)}{\partial x} = \frac{\partial}{\partial x}(ye^{xy}) = ye^{xy}y = y^2 e^{xy}$$

To compute $\partial f/\partial y$, we treat x as a constant, and differentiate f with respect to y using the product rule combined with the chain rule.

$$\frac{\partial f(x, y)}{\partial y} = \frac{\partial}{\partial y}(ye^{xy})$$
$$= 1 \cdot e^{xy} + ye^{xy}x$$
$$= e^{xy}(1 + xy)$$

EXAMPLE 2   Find $\partial f/\partial x$ when

$$f(x, y) = \frac{\sin(xy)}{x^2 + \cos y}$$

Solution   We treat y as a constant and use the quotient rule.

$$u = \sin(xy) \qquad\qquad\qquad v = x^2 + \cos y$$

$$\frac{\partial u}{\partial x} = y \cos(xy) \qquad\qquad\qquad \frac{\partial v}{\partial x} = 2x$$

Hence,

$$\frac{\partial f(x, y)}{\partial x} = \frac{y(x^2 + \cos y)\cos(xy) - 2x \sin(xy)}{(x^2 + \cos y)^2}$$

Geometric Interpretation   As with ordinary derivatives, partial derivatives are slopes of lines that are tangent to certain curves on the surface $z = f(x, y)$.

Let's start with the interpretation of $\partial f/\partial x$. We fix $y = y_0$; then $f(x, y_0)$ as a function of x is obtained by intersecting the surface $z = f(x, y)$ with a vertical plane that is parallel to the x-z plane and goes through $y = y_0$. The curve of intersection is the graph of $z = f(x, y_0)$, as illustrated in Figure 10.30.

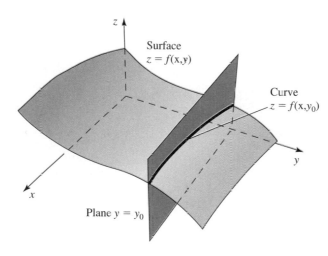

◀ **Figure 10.30**
The surface of $f(x, y)$ intersected with the plane $y = y_0$.

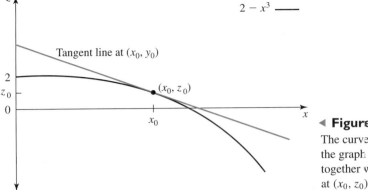

$2 - x^3$ ——

◀ **Figure 10.31**
The curve of intersection of the graph $z = f(x, z_0)$ together with its tangent line at (x_0, z_0).

We can now project this curve onto the x-z plane, which is illustrated in Figure 10.31. The curve is the graph of a function that depends only on x. As such, we can find the slope of the tangent line at any point P with coordinates (x_0, z_0), where $z_0 = f(x_0, y_0)$. The slope of the tangent line is given by the derivative of the function f in the x direction, that is, by $\partial f / \partial x$. This is the geometric meaning of $\partial f / \partial x$. We summarize this in the following.

> The partial derivative $\partial f / \partial x$ evaluated at (x_0, y_0) is the slope of the tangent line at the point (x_0, y_0, z_0) with $z_0 = f(x_0, y_0)$ to the curve $z = f(x, y_0)$.

The derivative $\partial f / \partial y$ has a similar meaning. This time we fix $x = x_0$, and intersect the surface $z = f(x, y)$ with a vertical plane that is parallel to the y-z plane and goes through $x = x_0$. The curve of intersection is the graph of $z = f(x_0, y)$, as illustrated in Figure 10.32.

The projection of this curve onto the y-z plane together with the tangent line at (y_0, z_0) is illustrated in Figure 10.33. Analogous to the interpretation of $\partial f / \partial x$, we then have the following geometric meaning of $\partial f / \partial y$.

> The partial derivative $\partial f / \partial y$ evaluated at (x_0, y_0) is the slope of the tangent line at the point (x_0, y_0, z_0) with $z_0 = f(x_0, y_0)$ to the curve $z = f(x_0, y)$.

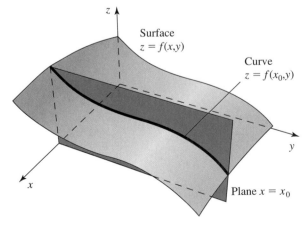

◀ **Figure 10.32**
The surface of $f(x, y)$ intersected with the plane $x = x_0$.

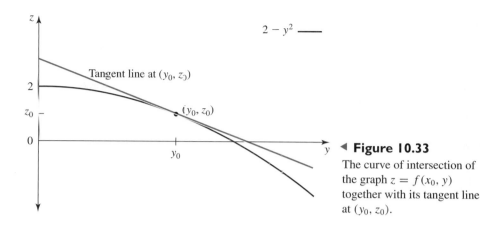

▸ **Figure 10.33**
The curve of intersection of the graph $z = f(x_0, y)$ together with its tangent line at (y_0, z_0).

EXAMPLE 3 Let

$$f(x, y) = 3 - x^3 - y^2$$

Find $f_x(1, 1)$ and $f_y(1, 1)$, and interpret your results geometrically.

Solution We have

$$f_x(x, y) = -3x^2 \qquad \text{and} \qquad f_y(x, y) = -2y$$

hence,

$$f_x(1, 1) = -3 \qquad \text{and} \qquad f_y(1, 1) = -2$$

To interpret $f_x(1, 1)$ geometrically, we fix $y = 1$. The vertical plane $y = 1$ intersects the graph of $f(x, y)$. The curve of intersection has slope -3 when $x = 1$. The projection of the curve of intersection is shown in Figure 10.31.

Similarly, to interpret $f_y(1, 1)$, we intersect the graph of $f(x, y)$ with the vertical plane $x = 1$. The tangent line at the curve of intersection has slope -2 when $y = 1$. The projection of the curve of intersection is shown in Figure 10.33. ▮

A Biological Application

EXAMPLE 4 Holling (1959) derived an expression for the number of prey items eaten by a predator, P_e, during a time interval T, as a function of prey density, N, and handling time of each prey item, T_h:

$$P_e = \frac{aNT}{1 + aT_h N} \tag{10.1}$$

where a is a positive constant (the predator attack rate). Equation (10.1) is called Holling's disc equation. (Holling came up with equation (10.1) when measuring how many sand paper discs (representing prey) a blindfolded assistant (representing the predator) could pick up during a certain time interval.) We can consider P_e as a function of N and T_h.

To determine how handling time influences the number of prey eaten, we compute

$$\frac{\partial P_e(N, T_h)}{\partial T_h} = aNT(-1)(1 + aT_h N)^{-2}aN$$

$$= -\frac{a^2 N^2 T}{(1 + aT_h N)^2} < 0$$

since $a^2N^2T > 0$ and $(1 + aT_hN)^2 > 0$. Since $\partial P_e / \partial T_h$ is negative, the number of prey items eaten decreases with increasing handling time, as you would expect.

To determine how P_e changes with N, we compute

$$\frac{\partial P_e(N, T_h)}{\partial N} = \frac{aT(1 + aT_hN) - aNTaT_h}{(1 + aT_hN)^2}$$

$$= \frac{aT}{(1 + aT_hN)^2} > 0$$

since $aT > 0$ and $(1 + aT_hN)^2 > 0$. Since $\partial P_e / \partial N$ is positive, the number of prey items eaten increases with increasing prey density, as you would expect. ▮

10.3.2 Functions of More Than Two Variables

The definition of partial derivatives extends in a straightforward way to functions of more than two variables. These are ordinary derivatives with respect to one variable while treating all other variables as constants.

EXAMPLE 5　Let f be a function of three independent variables x, y, and z:

$$f(x, y, z) = e^{yz}(x^2 + z^3)$$

Find $\partial f / \partial x$, $\partial f / \partial y$, and $\partial f / \partial z$.

Solution　To compute $\partial f / \partial x$, we treat y and z as constants and differentiate f with respect to x.

$$\frac{\partial f}{\partial x} = e^{yz}2x$$

Likewise,

$$\frac{\partial f}{\partial y} = ze^{yz}(x^2 + z^3)$$

$$\frac{\partial f}{\partial z} = ye^{yz}(x^2 + z^3) + e^{yz}3z^2$$ ▮

10.3.3 Higher Order Partial Derivatives

As in the case of functions of one variable, we can define higher order partial derivatives for functions of more than one variable. For instance, to find the second partial derivative of $f(x, y)$ with respect to x, denoted by $\partial^2 f / \partial x^2$, we compute

$$\frac{\partial^2 f}{\partial x^2} = \frac{\partial}{\partial x}\left(\frac{\partial f}{\partial x}\right)$$

We can write $\partial^2 f / \partial x^2$ also as f_{xx}.

We can also compute **mixed derivatives**. For instance,

$$\frac{\partial^2 f}{\partial x \partial y} = \frac{\partial}{\partial x}\left(\frac{\partial f}{\partial y}\right)$$

EXAMPLE 6　Set

$$f(x, y) = \sin x + xe^y$$

Find $\dfrac{\partial^2 f}{\partial x^2}$, $\dfrac{\partial^2 f}{\partial x \partial y}$, and $\dfrac{\partial^2 f}{\partial y \partial x}$.

Solution

$$\frac{\partial^2 f}{\partial x^2} = \frac{\partial}{\partial x}\left(\frac{\partial f}{\partial x}\right) = \frac{\partial}{\partial x}\left[\frac{\partial}{\partial x}(\sin x + xe^y)\right]$$

$$= \frac{\partial}{\partial x}\left[\cos x + e^y\right] = -\sin x$$

$$\frac{\partial^2 f}{\partial x \partial y} = \frac{\partial}{\partial x}\left(\frac{\partial f}{\partial y}\right) = \frac{\partial}{\partial x}\left[\frac{\partial}{\partial y}(\sin x + xe^y)\right]$$

$$= \frac{\partial}{\partial x}[0 + xe^y] = e^y$$

$$\frac{\partial^2 f}{\partial y \partial x} = \frac{\partial}{\partial y}\frac{\partial f}{\partial x} = \frac{\partial}{\partial y}\left[\frac{\partial}{\partial x}(\sin x + xe^y)\right]$$

$$= \frac{\partial}{\partial y}[\cos x + e^y] = e^y$$ **I**

In the preceding example, you probably noticed the order in which we computed $\partial^2 f/\partial x \partial y$ and $\partial^2 f/\partial y \partial x$; the results would have been the same, however, if the order had been different. The following theorem states the conditions under which the order of differentiation in mixed partial derivatives does not matter.

The Mixed Derivative Theorem   If $f(x, y)$ and its partial derivatives f_x, f_y, f_{xy}, and f_{yx} are continuous on an open disk centered at the point (x_0, y_0), then

$$f_{xy}(x_0, y_0) = f_{yx}(x_0, y_0)$$

It is straightforward to define partial derivatives of even higher order. For instance, if f is a function of two independent variables x and y, then

$$\frac{\partial^3 f}{\partial x^2 \partial y} = \frac{\partial}{\partial x}\frac{\partial^2 f}{\partial x \partial y} = \frac{\partial}{\partial x}\frac{\partial}{\partial x}\frac{\partial f}{\partial y}$$

We illustrate this with

$$f(x, y) = y^2 \sin x$$

for which

$$\frac{\partial^3 f}{\partial x^2 \partial y} = \frac{\partial}{\partial x}\frac{\partial}{\partial x}\frac{\partial}{\partial y}(y^2 \sin x)$$

$$= \frac{\partial}{\partial x}\frac{\partial}{\partial x}(2y \sin x)$$

$$= \frac{\partial}{\partial x}(2y \cos x) = -2y \sin x$$

In another example,

$$\frac{\partial^3 f}{\partial x \partial y^2} = \frac{\partial}{\partial x}\frac{\partial}{\partial y}\frac{\partial}{\partial y}(y^2 \sin x)$$

$$= \frac{\partial}{\partial x}\frac{\partial}{\partial y}(2y \sin x) = \frac{\partial}{\partial x}(2 \sin x) = 2 \cos x$$

and so on.

The mixed derivative theorem can be extended to higher order derivatives. The order of differentiation does not matter as long as the function and all its derivatives through the order in question are continuous on an open disk centered at the point at which we want to compute the derivative. For instance,

$$
\frac{\partial^3 f}{\partial y^2 \partial x} = \frac{\partial}{\partial y}\frac{\partial}{\partial y}\frac{\partial}{\partial x}(y^2 \sin x)
$$

$$
= \frac{\partial}{\partial y}\frac{\partial}{\partial y}(y^2 \cos x)
$$

$$
= \frac{\partial}{\partial y}(2y \cos x) = 2 \cos x
$$

which is the same as $\partial^3 f/(\partial x \partial y^2)$.

◆ 10.3 PROBLEMS

[10.3.1]

In Problems 1–16, find $\partial f/\partial x$ and $\partial f/\partial y$ for the given functions.

1. $f(x, y) = x^2 y + x y^2$
2. $f(x, y) = 2x\sqrt{y} - \frac{1}{xy}$
3. $f(x, y) = (xy)^{3/2} - (xy)^{2/3}$
4. $f(x, y) = x^3 y^4 - \frac{1}{x^3 y^4}$
5. $f(x, y) = \sin(x + y)$
6. $f(x, y) = \tan(x - y)$
7. $f(x, y) = \cos^2(x^2 - 2y)$
8. $f(x, y) = \sec(y^2 x - x^2)$
9. $f(x, y) = e^{\sqrt{x+y}}$
10. $f(x, y) = x^2 e^{-xy}$
11. $f(x, y) = e^x \sin(xy)$
12. $f(x, y) = e^{-y^2} \cos(x^2 + y^2)$
13. $f(x, y) = \ln(2x + y)$
14. $f(x, y) = \ln(3x^2 - y)$
15. $f(x, y) = \log_3(y^2 - x^2)$
16. $f(x, y) = \log_5(3xy)$

In Problems 17–24, find the indicated partial derivatives.

17. $f(x, y) = 3x^2 - y - 2y^2$; $f_x(1, 0)$
18. $f(x, y) = x^{1/3}y - xy^{1/3}$; $f_y(1, 1)$
19. $g(x, y) = e^{x+3y}$; $g_y(2, 1)$
20. $h(u, v) = e^u \sin(u + v)$; $h_u(1, -1)$
21. $f(x, z) = \ln(xz)$; $f_z(e, 1)$
22. $g(v, w) = \frac{w^2}{v+w}$; $g_v(1, 1)$
23. $f(x, y) = \frac{xy}{x^2+2}$; $f_x(-1, 2)$
24. $f(u, v) = e^{u^2/2} \ln(u + v)$; $f_u(2, 1)$
25. Let
$$f(x, y) = 4 - x^2 - y^2$$
Compute $f_x(1, 1)$ and $f_y(1, 1)$ and interpret these partial derivatives geometrically.
26. Let
$$f(x, y) = \sqrt{4 - x^2 - y^2}$$
Compute $f_x(1, 1)$ and $f_y(1, 1)$ and interpret these partial derivatives geometrically.
27. In Example 4, we investigated Holling's disc equation
$$P_e = \frac{aNT}{1 + aT_h N}$$
(see Example 4 for the meaning of these quantities). We will now consider P_e as a function of the predator attack rate a and the length of the time interval, T, during which the predator searches for food.

(a) Determine how the predator attack rate, a, influences the number of prey eaten per predator.

(b) Determine how the length of the time interval, T, influences the number of prey eaten per predator.

28. Suppose that the per capita growth rate of some prey at time t depends on both the prey density $H(t)$ at time t and the predator density $P(t)$ at time t. Assume the following relationship:

$$
\frac{1}{H}\frac{dH}{dt} = r\left(1 - \frac{H}{K}\right) - aP \qquad (10.2)
$$

where r, K, and a are positive constants. The right-hand side of (10.2) is a function of both prey density and predator density. Investigate how an increase in (a) prey density and (b) predator density affects the per capita growth rate of this prey species.

[10.3.2]

In Problems 29–36, find $\partial f/\partial x$, $\partial f/\partial y$, and $\partial f/\partial z$ for the given functions.

29. $f(x, y, z) = x^2 z + yz^2 - xy$

30. $f(x, y, z) = xyz$

31. $f(x, y, z) = x^2 y^2 z + \frac{x}{yz}$

32. $f(x, y, z) = \frac{xyz}{x^2 + y^2 + z^2}$

33. $f(x, y, z) = e^{x+y+z}$

34. $f(x, y, z) = e^{yz} \sin x$

35. $f(x, y, z) = \ln(x + y + z)$

36. $f(x, y, z) = y \tan(x^2 + z)$

[10.3.3]

In Problems 37–46, find the indicated partial derivatives.

37. $f(x, y) = x^2 + xy^2$; $\frac{\partial^2 f}{\partial x^2}$

38. $f(x, y) = y^3(x - y)$; $\frac{\partial^2 f}{\partial y^2}$

39. $f(x, y) = xe^y$; $\frac{\partial^2 f}{\partial x \partial y}$

40. $f(x, y) = \sin(x + y)$; $\frac{\partial^2 f}{\partial y \partial x}$

41. $f(u, w) = \tan(u + w)$; $\frac{\partial^2 f}{\partial u^2}$

42. $g(s, t) = \ln(s^2 + 3t)$; $\frac{\partial^2 g}{\partial t^2}$

43. $f(x, y) = x^3 \cos y$; $\frac{\partial^3 f}{\partial x^2 \partial y}$

44. $f(x, y) = e^{x-y}$; $\frac{\partial^3 f}{\partial y^2 \partial x}$

45. $f(x, y) = \ln(x + y)$; $\frac{\partial^3 f}{\partial x^3}$

46. $f(x, y) = \sin(3xy)$; $\frac{\partial^3 f}{\partial y^2 \partial x}$

[10.3.4]

47. The functional responses of some predators are sigmoidal; that is, the number of prey attacked per predator as a function of prey density has a sigmoidal shape. If we denote by N the prey density, by P the predator density, by T the time available for searching for prey, and by T_h the handling time of each prey item per predator, then the number of prey encounters per predator as a function of N, T, and T_h can be expressed as

$$f(N, T, T_h) = \frac{b^2 N^2 T}{1 + cN + bT_h N^2}$$

where b and c are positive constants.

(a) Investigate how an increase in the prey density N affects the function $f(N, T, T_h)$.

(b) Investigate how an increase in the time available for search, T, affects the function $f(N, T, T_h)$.

(c) Investigate how an increase in the handling time T_h affects the function $f(N, T, T_h)$.

(d) Graph $f(N, T, T_h)$ as a function of N when $T = 2.4$ hours, $T_h = 0.2$ hours, $b = 0.8$, and $c = 0.5$.

48. In this problem we will investigate how mutual interference of parasitoids affects their searching efficiency for a host. We assume that N is the host density and P is the parasitoid density. A frequently used model for host-parasitoid interactions is the **Nicholson-Bailey** model (Nicholson 1933, Bailey 1935), in which it is assumed that the number of parasitized hosts, denoted by N_a is given by

$$N_a = N[1 - e^{-bP}] \qquad (10.3)$$

where b is the searching efficiency.

(a) Show that

$$b = \frac{1}{P} \ln \frac{N}{N - N_a}$$

by solving (10.3) for b.

(b) Consider

$$b = f(P, N, N_a) = \frac{1}{P} \ln \frac{N}{N - N_a}$$

as a function of P, N, and N_a. How is the searching efficiency b affected when the parasitoid density increases?

(c) Assume now that the fraction of parasitized host depends on the host density; that is, assume that

$$N_a = g(N)$$

where $g(N)$ is a nonnegative, differentiable function. The searching efficiency b can then be written as a function of P and N, namely,

$$b = h(P, N) = \frac{1}{P} \ln \frac{N}{N - g(N)}$$

How does the searching efficiency depend on host density when $g(N)$ is a decreasing function of N? (Use the fact that $g(N) < N$.)

49. Leopold and Kriedemann (1975) measured crop growth rate of sunflowers as a function of leaf area index and percent of full sunlight. (Leaf area index is the ratio of leaf surface area to the ground area the plant covers.) They found that for a fixed level of sunlight, crop growth rate first increases, then decreases as a function of leaf area index. For a given leaf area index, crop growth rate increases with sunlight level. The leaf area index that maximizes the crop growth rate is an increasing function of sunlight. Sketch the crop growth rate as a function of leaf area index for different values of percent of full sunlight.

◆ 10.4 TANGENT PLANES, DIFFERENTIABILITY, AND LINEARIZATION

10.4.1 Functions of Two Variables

Tangent Planes Suppose that $z = f(x)$ is differentiable at $x = x_0$. Then the equation of the tangent line of $z = f(x)$ at (x_0, z_0) with $z_0 = f(x_0)$ is given by

$$z - z_0 = f'(x_0)(x - x_0) \tag{10.4}$$

This is illustrated in Figure 10.34.

We now generalize this to functions of two variables. The analogue of a tangent line is called a **tangent plane**, an example of which is shown in Figure 10.35. Let $z = f(x, y)$ be a function of two variables. We saw in the previous section that the partial derivatives $\partial f/\partial x$ and $\partial f/\partial y$, evaluated at (x_0, y_0), are the slopes of tangent lines at the point (x_0, y_0, z_0) with $z_0 = f(x_0, y_0)$, to certain curves through (x_0, y_0, z_0) on the surface $z = f(x, y)$. These two tangent lines, one in the x direction, the other in the y direction, define a unique plane. If, in addition, $f(x, y)$ has partial derivatives that are continuous on an open disk containing (x_0, y_0), then one can show that the tangent line at (x_0, y_0, z_0) to any other smooth curve on the surface $z = f(x, y)$ through (x_0, y_0, z_0) is contained in this plane. This plane is then called the tangent plane.

We will use the two original tangent lines to find the equation of the tangent plane at a point (x_0, y_0, z_0) on the surface $z = f(x, y)$. We take the curve that is obtained as the intersection of the surface $z = f(x, y)$ with the plane that is parallel to the y-z plane and contains the point (x_0, y_0, z_0) – that is, the plane $x = x_0$ – and we denote this curve by C_1. Its tangent line at (x_0, y_0, z_0) is contained in the tangent plane (see Figure 10.35). Likewise, we take the curve of intersection between $z = f(x, y)$ and the plane that is parallel to the x-z plane and contains the point (x_0, y_0, z_0) – that is, the plane $y = y_0$ – and we denote this curve by C_2. Its tangent line is contained in the tangent plane (see Figure 10.35).

In Section 9.4, we gave the general equation of a plane, namely

$$z - z_0 = A(x - x_0) + B(y - y_0) \tag{10.5}$$

We will use curves C_1 and C_2 to determine the constants A and B. C_1 satisfies the equation

$$z = f(x_0, y)$$

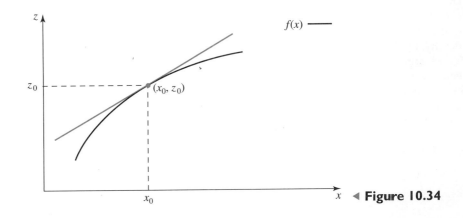

$f(x)$ ——

◄ **Figure 10.34**

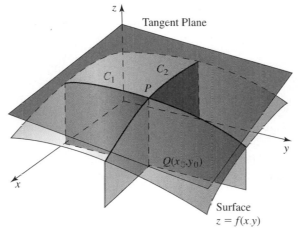

◀ **Figure 10.35**

The surface $z = f(x, y)$ and its tangent plane at $P(x_0, y_0, z_0)$.

The tangent line to C_1 at (x_0, y_0, z_0) therefore satisfies

$$z - z_0 = \frac{\partial f(x_0, y_0)}{\partial y}(y - y_0) \tag{10.6}$$

Since the tangent line is contained in the tangent plane, we set $x = x_0$ in (10.5), and obtain

$$z - z_0 = (A)(0) + B(y - y_0)$$
$$= B(y - y_0)$$

Comparing this with (10.6), we find

$$B = \frac{\partial f(x_0, y_0)}{\partial y}$$

Similarly, using C_2, we find

$$A = \frac{\partial f(x_0, y_0)}{\partial x}$$

We thus arrive at the following result.

If the tangent plane to the surface $z = f(x, y)$ at the point (x_0, y_0, z_0) exists, then that tangent plane has the equation

$$z - z_0 = \frac{\partial f(x_0, y_0)}{\partial x}(x - x_0) + \frac{\partial f(x_0, y_0)}{\partial y}(y - y_0)$$

You should observe the similarity of this equation to the equation of the tangent line in (10.4). We will see that the mere existence of the partial derivatives $\frac{\partial f(x_0, y_0)}{\partial x}$ and $\frac{\partial f(x_0, y_0)}{\partial y}$ is *not* enough to guarantee the existence of a tangent plane at (x_0, y_0); something stronger is needed. But, before we discuss the conditions under which tangent planes exist, let's look at an example.

EXAMPLE 1   It can be shown that the tangent plane to the surface

$$z = f(x, y) = 4x^2 + y^2$$

at the point $(1, 2, 8)$ exists. Find its equation.

Solution   First note that the point $(1, 2, 8)$ is contained in the surface $z = f(x, y)$, since $8 = f(1, 2)$. To find the tangent plane, we need to compute the partial derivatives of $f(x, y)$,

$$\frac{\partial f}{\partial x} = 8x \qquad \text{and} \qquad \frac{\partial f}{\partial y} = 2y$$

and evaluate them at $(x_0, y_0) = (1, 2)$,

$$\frac{\partial f(1, 2)}{\partial x} = 8 \qquad \text{and} \qquad \frac{\partial f(1, 2)}{\partial y} = 4$$

Hence, the equation of the tangent plane is

$$z - 8 = 8(x - 1) + 4(y - 2)$$

We can write this in a somewhat nicer form, namely

$$z - 8 = 8x - 8 + 4y - 8$$

or

$$8x + 4y - z = 8 \qquad\qquad ▮$$

Differentiability   We will now discuss the conditions under which tangent planes exist. For this, we need to define what it means for a function of two variables to be differentiable. To make the connection with functions of one variable clear, recall that the tangent line is used for linearly approximating $f(x)$ at $x = x_0$. The linear approximation of $f(x)$ at $x = x_0$ is given by

$$L(x) = f(x_0) + f'(x_0)(x - x_0) \tag{10.7}$$

The distance between $f(x)$ and its linear approximation at $x = x_0$ is then

$$|f(x) - L(x)| = |f(x) - f(x_0) - f'(x_0)(x - x_0)| \tag{10.8}$$

From the definition of the derivative, we know that

$$f'(x_0) = \lim_{x \to x_0} \frac{f(x) - f(x_0)}{x - x_0} \tag{10.9}$$

If we divide (10.8) by the distance between x and x_0, $x - x_0$, we find

$$\left| \frac{f(x) - L(x)}{x - x_0} \right| = \left| \frac{f(x) - f(x_0) - f'(x_0)(x - x_0)}{x - x_0} \right| \tag{10.10}$$

$$= \left| \frac{f(x) - f(x_0)}{x - x_0} - f'(x_0) \right| \tag{10.11}$$

Taking the limit $x \to x_0$ and using (10.9), we therefore find that

$$\lim_{x \to x_0} \left| \frac{f(x) - L(x)}{x - x_0} \right| = 0 \tag{10.12}$$

We could say that $f(x)$ is differentiable at $x = x_0$ if (10.12) holds.

For functions of two variables, the definition of differentiability is based on the same idea. Before we can give it, we need to address one more point. In the preceding, we divided by the distance between x and x_0. In two dimensions, the distance between two points (x, y) and (x_0, y_0) is $\sqrt{(x - x_0)^2 + (y - y_0)^2}$.

Definition   Suppose that $f(x, y)$ is a function of two independent variables and that both $\partial f/\partial x$ and $\partial f/\partial y$ are defined throughout an open disk containing (x_0, y_0). Set

$$L(x, y) = f(x_0, y_0) + \frac{\partial f(x_0, y_0)}{\partial x}(x - x_0) + \frac{\partial f(x_0, y_0)}{\partial y}(y - y_0)$$

Then $f(x, y)$ is differentiable at (x_0, y_0) if

$$\lim_{(x,y)\to(x_0,y_0)} \left| \frac{f(x, y) - L(x, y)}{\sqrt{(x - x_0)^2 + (y - y_0)^2}} \right| = 0$$

Furthermore, if $f(x, y)$ is differentiable at (x_0, y_0), then $z = L(x, y)$ defines the tangent plane to the graph of f at $(x_0, y_0, f(x_0, y_0))$. We say that $f(x, y)$ is differentiable if it is differentiable at every point of its domain.

The key idea in both the one- and the two-dimensional cases is to approximate functions by linear functions, so that the error in the approximation vanishes as we approach the point at which we approximated the function (x_0 in the one-dimensional case, and (x_0, y_0) in the two-dimensional case).

The following theorem holds, as in the one-dimensional case.

Theorem   If $f(x, y)$ is differentiable at (x_0, y_0), then f is continuous at (x_0, y_0).

Differentiability of $f(x, y)$ at (x_0, y_0) means that the function $f(x, y)$ is close to the tangent plane at (x_0, y_0) for all (x, y) close to (x_0, y_0). The mere existence of the partial derivatives $\partial f/\partial x$ and $\partial f/\partial y$ at (x_0, y_0), however, is *not* enough to guarantee differentiability (and consequently the existence of a tangent plane at a certain point). Here is a very simple example that will show what can go wrong with that assumption.

▮ **EXAMPLE 2**   Assume that

$$f(x, y) = \begin{cases} 0 & \text{if } xy \neq 0 \\ 1 & \text{if } xy = 0 \end{cases}$$

Show that $\frac{\partial f(0,0)}{\partial x}$ and $\frac{\partial f(0,0)}{\partial y}$ exist, but $f(x, y)$ is not continuous, and consequently not differentiable, at $(0, 0)$.

Solution   The graph of $f(x, y)$ is shown in Figure 10.36. To compute $\partial f/\partial x$ at $(0, 0)$, we set $y = 0$, then $f(x, 0) = 1$, and consequently

$$\frac{\partial f(0, 0)}{\partial x} = 0$$

Likewise, setting $x = 0$, we find $f(0, y) = 1$ and

$$\frac{\partial f(0, 0)}{\partial y} = 0$$

That is, both partial derivatives exist at $(0, 0)$. However, $f(x, y)$ is not continuous at $(0, 0)$. For this it is enough to show that $f(x, y)$ has different limits along two different paths as (x, y) approaches $(0, 0)$. For the first path, denoted by C_1, we

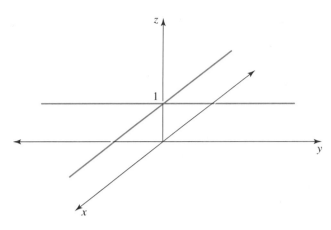

◄ **Figure 10.36**

The graph of the function

$$f(x, y) = \begin{cases} 0 \text{ if } xy \neq 0 \\ 1 \text{ if } xy = 0 \end{cases}$$

Even though $\frac{\partial f(0,0)}{\partial x}$ and $\frac{\partial f(0,0)}{\partial y}$ exist, the function is not continuous at $(0, 0)$.

choose $y = 0$. Then

$$\lim_{(x,y) \overset{C_1}{\to} (0,0)} f(x, y) = 1$$

For the second path, denoted by C_2, we choose $y = x$. Then

$$\lim_{(x,y) \overset{C_2}{\to} (0,0)} f(x, y) = 0$$

Since $1 \neq 0$, $f(x, y)$ is not continuous at $(0, 0)$. Since differentiability implies continuity, a function that is not continuous cannot be differentiable. ▮

The definition of differentiability is not easy to use if one actually wants to check whether a function is differentiable at a certain point. Fortunately, there is a simple criterion, which suffices for all practical purposes, that can be used to check whether $f(x, y)$ is differentiable. We saw in Example 2 that the mere existence of the partial derivatives $\partial f/\partial x$ and $\partial f/\partial y$ at a point (x_0, y_0) is not enough to guarantee that $f(x, y)$ is differentiable. However, if the partial derivatives are continuous on a disk centered at (x_0, y_0), then this is enough to guarantee differentiability.

SUFFICIENT CONDITION FOR DIFFERENTIABILITY

Suppose $f(x, y)$ is defined on an open disk centered at (x_0, y_0), and the partial derivatives $\partial f/\partial x$ and $\partial f/\partial y$ are continuous on an open disk centered at (x_0, y_0). Then $f(x, y)$ is differentiable at (x_0, y_0).

▮ **EXAMPLE 3**   Show that $f(x, y) = 2x^2 y - y^2$ is differentiable for all $(x, y) \in \mathbf{R}^2$.

Solution   We use the sufficient condition for differentiability. First, observe that $f(x, y)$ is defined for all $(x, y) \in \mathbf{R}^2$. The partial derivatives are given by

$$\frac{\partial f}{\partial x} = 4xy \qquad \text{and} \qquad \frac{\partial f}{\partial y} = 2x^2 - 2y$$

Since both $\partial f/\partial x$ and $\partial f/\partial y$ are polynomials, both are continuous for all $(x, y) \in \mathbf{R}^2$ and, hence, $f(x, y)$ is differentiable for all $(x, y) \in \mathbf{R}^2$. ▮

Linearization

Definition Suppose that $f(x, y)$ is differentiable at (x_0, y_0). The linearization of $f(x, y)$ at (x_0, y_0) is the function

$$L(x, y) = f(x_0, y_0) + \frac{\partial f(x_0, y_0)}{\partial x}(x - x_0) + \frac{\partial f(x_0, y_0)}{\partial y}(y - y_0)$$

The approximation

$$f(x, y) \approx L(x, y)$$

is the **standard linear approximation** or the **tangent plane approximation** of $f(x, y)$ at (x_0, y_0).

EXAMPLE 4 Find the linear approximation of

$$f(x, y) = x^2 y + 2xe^y$$

at the point $(1, 1)$.

Solution The linearization of $f(x, y)$ at $(1, 1)$ is given by

$$L(x, y) = f(1, 1) + \frac{\partial f(1, 1)}{\partial x}(x - 1) + \frac{\partial f(1, 1)}{\partial y}(y - 1)$$

Now, $f(1, 1) = 1 + 2e$,

$$\frac{\partial f}{\partial x} = 2xy + 2e^y \qquad \frac{\partial f}{\partial y} = x^2 + 2xe^y$$

hence,

$$\frac{\partial f(1, 1)}{\partial x} = 2 + 2e \qquad \frac{\partial f(1, 1)}{\partial y} = 1 + 2e$$

and we find

$$L(x, y) = 1 + 2e + (2 + 2e)(x - 1) + (1 + 2e)(y - 1)$$
$$= -2 - 2e + (2 + 2e)x + (1 + 2e)y$$

10.4.2 Vector-Valued Functions

So far, we have only considered real-valued functions,

$$f : \mathbf{R}^n \to \mathbf{R}$$

We will now extend our discussion to functions where the range is a subset of $\mathbf{R}^m$, that is,

$$f : \mathbf{R}^n \to \mathbf{R}^m$$

Such functions are **vector-valued functions** since we can represent them in a vector.

$$f : \mathbf{R}^n \to \mathbf{R}^m$$

$$(x_1, x_2, \ldots, x_n) \to \begin{pmatrix} f_1(x_1, x_2, \ldots, x_n) \\ f_2(x_1, x_2, \ldots, x_n) \\ \vdots \\ f_m(x_1, x_2, \ldots, x_n) \end{pmatrix}$$

Here, each function $f_i(x_1, \ldots, x_n)$ is a real-valued function, that is,

$$f_i : \mathbf{R}^n \to \mathbf{R}$$

$$(x_1, x_2, \ldots, x_n) \to f_i(x_1, x_2, \ldots, x_n)$$

We will encounter such vector-valued functions extensively in Chapter 11. As an example, consider a community consisting of two species. Let u and v denote the respective densities of the two species, and $f(u, v)$ and $g(u, v)$ the per capita growth rates of the two species as functions of the densities u and v. We can then write this as a map from $\mathbf{R}^2$ to $\mathbf{R}^2$, where

$$(u, v) \to \begin{pmatrix} f(u, v) \\ g(u, v) \end{pmatrix}$$

Our main task in this subsection will be to define the linearization of such vector-valued functions. We will motivate this by analogy with the cases of functions $f : \mathbf{R} \to \mathbf{R}$ and $f : \mathbf{R}^2 \to \mathbf{R}$.

We begin with differentiable functions $f : \mathbf{R} \to \mathbf{R}$. The linearization of $f : \mathbf{R} \to \mathbf{R}$ about x_0 is given by

$$L(x) = f(x_0) + f'(x_0)(x - x_0) \tag{10.13}$$

■ **EXAMPLE 5**   Find the linearization of

$$f(x) = 2 \ln x$$

at $x_0 = 1$.

Solution   The linearization of $f(x)$ at $x = 1$ is given by

$$L(x) = f(1) + f'(1)(x - 1)$$
$$= 0 + (2)(x - 1) = 2x - 2$$

since $f'(x) = 2/x$. ∎

We found the linearization of a real valued function f with domain in $\mathbf{R}^2$, that is, $f : \mathbf{R}^2 \to \mathbf{R}$, in the previous subsection. It is given by

$$L(x, y) = f(x_0, y_0) + \frac{\partial f(x_0, y_0)}{\partial x}(x - x_0) + \frac{\partial f(x_0, y_0)}{\partial y}(y - y_0)$$

We can write this in matrix notation as

$$L(x, y) = f(x_0, y_0) + \begin{bmatrix} \frac{\partial f(x_0, y_0)}{\partial x} & \frac{\partial f(x_0, y_0)}{\partial y} \end{bmatrix} \begin{bmatrix} x - x_0 \\ y - y_0 \end{bmatrix}$$

■ **EXAMPLE 6**   Find the linearization of

$$f(x, y) = \ln x + \ln y$$

at $(1, 1)$.

Solution   We find $f(1, 1) = 0$. Since $f_x(x, y) = \frac{1}{x}$ and $f_y(x, y) = \frac{1}{y}$, it follows that

$$\frac{\partial f(1, 1)}{\partial x} = 1 \quad \text{and} \quad \frac{\partial f(1, 1)}{\partial y} = 1$$

Hence,

$$L(x, y) = 0 + [1, 1] \begin{bmatrix} x - 1 \\ y - 1 \end{bmatrix}$$

$$= x - 1 + y - 1 = x + y - 2 \qquad ∎$$

We will denote vector-valued functions by boldface letters. Suppose that

$$\mathbf{h} : \mathbf{R}^2 \to \mathbf{R}^2$$

$$(x, y) \to \begin{bmatrix} f(x, y) \\ g(x, y) \end{bmatrix}$$

and assume that all first partial derivatives are continuous on a disk centered at (x_0, y_0). We can linearize each component of the function $\mathbf{h}$. The linearization of f is

$$\alpha(x, y) = f(x_0, y_0) + \frac{\partial f(x_0, y_0)}{\partial x}(x - x_0) + \frac{\partial f(x_0, y_0)}{\partial y}(y - y_0)$$

and the linearization of g is

$$\beta(x, y) = g(x_0, y_0) + \frac{\partial g(x_0, y_0)}{\partial x}(x - x_0) + \frac{\partial g(x_0, y_0)}{\partial y}(y - y_0)$$

We define the vector-valued function $L(x, y) = \begin{bmatrix} \alpha(x,y) \\ \beta(x,y) \end{bmatrix}$. The linearization of $\mathbf{h}$ can thus be written in matrix form:

$$L(x, y) = \begin{bmatrix} \alpha(x, y) \\ \beta(x, y) \end{bmatrix} = \begin{bmatrix} f(x_0, y_0) \\ g(x_0, y_0) \end{bmatrix} + \begin{bmatrix} \frac{\partial f(x_0, y_0)}{\partial x}(x - x_0) + \frac{\partial f(x_0, y_0)}{\partial y}(y - y_0) \\ \frac{\partial g(x_0, y_0)}{\partial x}(x - x_0) + \frac{\partial g(x_0, y_0)}{\partial y}(y - y_0) \end{bmatrix}$$

or

$$L(x, y) = \begin{bmatrix} f(x_0, y_0) \\ g(x_0, y_0) \end{bmatrix} + \begin{bmatrix} \frac{\partial f(x_0, y_0)}{\partial x} & \frac{\partial f(x_0, y_0)}{\partial y} \\ \frac{\partial g(x_0, y_0)}{\partial x} & \frac{\partial g(x_0, y_0)}{\partial y} \end{bmatrix} \begin{bmatrix} x - x_0 \\ y - y_0 \end{bmatrix}$$

where the matrix

$$\begin{bmatrix} \frac{\partial f(x_0, y_0)}{\partial x} & \frac{\partial f(x_0, y_0)}{\partial y} \\ \frac{\partial g(x_0, y_0)}{\partial x} & \frac{\partial g(x_0, y_0)}{\partial y} \end{bmatrix}$$

is a 2×2 matrix that is called the **Jacobi matrix** or the **derivative matrix**. We denote the Jacobi matrix by $(D\mathbf{h})$, that is,

$$(D\mathbf{h})(x_0, y_0) = \begin{bmatrix} \frac{\partial f(x_0, y_0)}{\partial x} & \frac{\partial f(x_0, y_0)}{\partial y} \\ \frac{\partial g(x_0, y_0)}{\partial x} & \frac{\partial g(x_0, y_0)}{\partial y} \end{bmatrix}$$

▮ EXAMPLE 7   Assume that

$$\mathbf{f} : \mathbf{R}^2 \to \mathbf{R}^2$$

$$(x, y) \to (u, v)$$

with

$$u(x, y) = x^2 y - y^3 \quad \text{and} \quad v(x, y) = 2x^3 y^2 + y$$

Compute the Jacobi matrix and evaluate it at $(1, 2)$.

Solution   The Jacobi matrix is

$$\begin{bmatrix} \frac{\partial u}{\partial x} & \frac{\partial u}{\partial y} \\ \frac{\partial v}{\partial x} & \frac{\partial v}{\partial y} \end{bmatrix} = \begin{bmatrix} 2xy & x^2 - 3y^2 \\ 6x^2 y^2 & 4x^3 y + 1 \end{bmatrix}$$

At $(1, 2)$ we find

$$\begin{bmatrix} 4 & -11 \\ 24 & 9 \end{bmatrix}$$

▮

▮ **EXAMPLE 8**   Let

$$\mathbf{f} : \mathbf{R}^2 \to \mathbf{R}^2$$
$$(x, y) \to (u, v)$$

with

$$u(x, y) = ye^{-x} \quad \text{and} \quad v(x, y) = \sin x + \cos y$$

Find the linear approximation to $\mathbf{f}(x, y)$ at $(0, 0)$. Compare $\mathbf{f}(0.1, -0.1)$ to its linear approximation.

Solution   We compute the Jacobi matrix first,

$$(D\mathbf{f})(x, y) = \begin{bmatrix} -ye^{-x} & e^{-x} \\ \cos x & -\sin y \end{bmatrix}$$

The linear approximation of $\mathbf{f}(x, y)$ at $(0, 0)$ is

$$L(x, y) = \begin{bmatrix} \alpha(x, y) \\ \beta(x, y) \end{bmatrix} = \begin{bmatrix} u(0, 0) \\ v(0, 0) \end{bmatrix} + (D\mathbf{f})(0, 0) \begin{bmatrix} x \\ y \end{bmatrix}$$

$$= \begin{bmatrix} 0 \\ 1 \end{bmatrix} + \begin{bmatrix} 0 & 1 \\ 1 & 0 \end{bmatrix} \begin{bmatrix} x \\ y \end{bmatrix}$$

$$= \begin{bmatrix} 0 \\ 1 \end{bmatrix} + \begin{bmatrix} y \\ x \end{bmatrix}$$

Using $(x, y) = (0.1, -0.1)$, we find

$$L(0.1, -0.1) = \begin{bmatrix} 0 \\ 1 \end{bmatrix} + \begin{bmatrix} -0.1 \\ 0.1 \end{bmatrix} = \begin{bmatrix} -0.1 \\ 1.1 \end{bmatrix}$$

and

$$\mathbf{f}(0.1, -0.1) = \begin{bmatrix} -0.1e^{-0.1} \\ \sin 0.1 + \cos(-0.1) \end{bmatrix} = \begin{bmatrix} -0.09 \\ 1.09 \end{bmatrix}$$

We see that the linear approximation is close to the actual value.   ▮

It is straightforward to generalize the Jacobi matrix to functions $\mathbf{f} : \mathbf{R}^n \to \mathbf{R}^m$. If

$$\mathbf{f}(x_1, x_2, \ldots, x_n) = \begin{bmatrix} f_1(x_1, x_2, \ldots, x_n) \\ f_2(x_1, x_2, \ldots, x_n) \\ \vdots \\ f_m(x_1, x_2, \ldots, x_n) \end{bmatrix}$$

where $f_i : \mathbf{R}^n \to \mathbf{R}$ $i = 1, 2, \ldots, n$, are real-valued functions of n independent variables. Then the Jacobi matrix is an $m \times n$ matrix of the form

$$D\mathbf{f} = \begin{bmatrix} \frac{\partial f_1}{\partial x_1} & \cdots & \frac{\partial f_1}{\partial x_n} \\ \vdots & & \vdots \\ \frac{\partial f_m}{\partial x_1} & \cdots & \frac{\partial f_m}{\partial x_n} \end{bmatrix}$$

◆ 10.4 PROBLEMS

[10.4.1]

In Problems 1–6, find the tangent plane at the indicated point (x_0, y_0, z_0).

1. $f(x, y) = 2x^2 + y^2$; $(1, 2, 7)$
2. $f(x, y) = \sin x + \cos y$; $(0, 0, 1)$
3. $f(x, y) = e^{x^2 + y^2}$; $(1, 0, e)$
4. $f(x, y) = \ln(x^2 + y^2)$; $(1, 1, \ln 2)$
5. $f(x, y) = \sin(xy)$; $(1, 0, 0)$
6. $f(x, y) = e^x \cos y$; $(0, 0, 1)$

In Problems 7–12, show that $f(x, y)$ *is differentiable at the indicated point.*

7. $f(x, y) = y^2 - x^2 y$; $(1, 1)$
8. $f(x, y) = xy - 3x^2$; $(1, 1)$
9. $f(x, y) = \cos(x + y)$; $(0, 0)$
10. $f(x, y) = e^{x - y}$; $(0, 0)$
11. $f(x, y) = x + y^2 - 2xy$; $(-1, 2)$
12. $f(x, y) = \tan(x^2 + y^2)$; $\left(\frac{\pi}{4}, -\frac{\pi}{4}\right)$

In Problems 13–18, find the linearization of $f(x, y)$ *at the indicated point* (x_0, y_0).

13. $f(x, y) = \sqrt{x} + 2y$; $(1, 0)$
14. $f(x, y) = \cos(x^2 y)$; $\left(\frac{\pi}{2}, 0\right)$
15. $f(x, y) = \tan(x + y)$; $(0, 0)$
16. $f(x, y) = e^{-x + 2y}$; $(1, 2)$
17. $f(x, y) = \ln(x^2 + y)$; $(1, 1)$
18. $f(x, y) = x^2 e^z$; $(1, 0)$

[10.4.2]

19. Find the linearization of

$$f(x) = e^x$$

at $x = 0$.

20. Find the linearization of

$$f(x) = \sin x$$

at $x = 0$.

In Problems 21–28, find the Jacobi matrix for each given function.

21. $\mathbf{f}(x, y) = \begin{bmatrix} x + y \\ x^2 - y^2 \end{bmatrix}$
22. $\mathbf{f}(x, y) = \begin{bmatrix} 2x - y \\ x^2 \end{bmatrix}$

23. $\mathbf{f}(x, y) = \begin{bmatrix} e^{x - y} \\ e^{x + y} \end{bmatrix}$
24. $\mathbf{f}(x, y) = \begin{bmatrix} (x - y)^2 \\ \sin(x + y) \end{bmatrix}$

25. $\mathbf{f}(x, y) = \begin{bmatrix} \cos(x - y) \\ \cos(x + y) \end{bmatrix}$
26. $\mathbf{f}(x, y) = \begin{bmatrix} \ln(x + y) \\ e^{x + y} \end{bmatrix}$

27. $\mathbf{f}(x, y) = \begin{bmatrix} 2x^2 y - 3y + x \\ e^x \sin y \end{bmatrix}$

28. $\mathbf{f}(x, y) = \begin{bmatrix} \sqrt{x^2 + y^2} \\ e^{-x^2} \end{bmatrix}$

In Problems 29–34, find a linear approximation to each function $f(x, y)$ *at the indicated point.*

29. $\mathbf{f}(x, y) = \begin{bmatrix} 2x^2 y \\ \frac{1}{xy} \end{bmatrix}$ at $(1, 1)$

30. $\mathbf{f}(x, y) = \begin{bmatrix} 3x - y^2 \\ 4y \end{bmatrix}$ at $(-1, -2)$

31. $\mathbf{f}(x, y) = \begin{bmatrix} e^{2x - y} \\ \ln(2x - y) \end{bmatrix}$ at $(1, 1)$

32. $\mathbf{f}(x, y) = \begin{bmatrix} e^x \sin y \\ e^{-y} \cos x \end{bmatrix}$ at $(0, 0)$

33. $\mathbf{f}(x, y) = \begin{bmatrix} \frac{x}{y} \\ \frac{y}{x} \end{bmatrix}$ at $(1, 1)$

34. $\mathbf{f}(x, y) = \begin{bmatrix} (x - y)^2 \\ xy \end{bmatrix}$ at $(-1, 1)$

◆ 10.5 MORE ABOUT DERIVATIVES (OPTIONAL)

10.5.1 The Chain Rule for Functions of Two Variables

In Section 10.3, we discussed how the net assimilation of CO_2 can change as a function of both temperature and light intensity. If we follow net assimilation of CO_2 over time, we must take into account that both temperature and light intensity depend on time. That is, if we denote by $T(t)$ the temperature at time t, by $I(t)$ the light intensity at time t, and by $N(t)$ the net assimilation of CO_2 at time t, then $N(t)$ is a function of both $T(t)$ and $I(t)$; that is, we can write

$$N(t) = f(T(t), I(t))$$

Net assimilation is thus a composite function.

To differentiate composite functions of one variable we use the chain rule. Namely, suppose that $w = f(x)$ is a function of one variable, and that x depends on t. To differentiate w with respect to t, the chain rule says that

$$\frac{dw}{dt} = \frac{dw}{dx}\frac{dx}{dt} \tag{10.14}$$

The chain rule can be extended to functions of more than one variable.

CHAIN RULE FOR FUNCTIONS OF TWO INDEPENDENT VARIABLES

If $w = f(x, y)$ is differentiable and x and y are differentiable functions of t, then w is a differentiable function of t and

$$\frac{dw}{dt} = \frac{\partial w}{\partial x}\frac{dx}{dt} + \frac{\partial w}{\partial y}\frac{dy}{dt}$$

We will not prove this, but merely outline the steps that lead to this formula. We approximate $w = f(x, y)$ at (x_0, y_0) by its linear approximation

$$L(x, y) = f(x_0, y_0) + \frac{\partial f(x_0, y_0)}{\partial x}(x - x_0) + \frac{\partial f(x_0, y_0)}{\partial y}(y - y_0)$$

If we set $\Delta x = x - x_0$, $\Delta y = y - y_0$, and $\Delta w = f(x, y) - f(x_0, y_0)$, we can approximate $\Delta w = f(x_0 + \Delta x, y_0 + \Delta y) - f(x_0, y_0)$ by its linear approximation. We find

$$\Delta w \approx \frac{\partial f(x_0, y_0)}{\partial x}\Delta x + \frac{\partial f(x_0, y_0)}{\partial y}\Delta y$$

Dividing both sides by Δt, we find

$$\frac{\Delta w}{\Delta t} \approx \frac{\partial f(x_0, y_0)}{\partial x}\frac{\Delta x}{\Delta t} + \frac{\partial f(x_0, y_0)}{\partial y}\frac{\Delta y}{\Delta t}$$

If we let $\Delta t \to 0$, then

$$\frac{\Delta w}{\Delta t} \to \frac{dw}{dt}, \qquad \frac{\Delta x}{\Delta t} \to \frac{dx}{dt}, \qquad \frac{\Delta y}{\Delta t} \to \frac{dy}{dt}$$

and we find

$$\frac{dw}{dt} = \frac{\partial f(x_0, y_0)}{\partial x}\frac{dx}{dt} + \frac{\partial f(x_0, y_0)}{\partial y}\frac{dy}{dt}$$

or, in short,

$$\frac{dw}{dt} = \frac{\partial w}{\partial x}\frac{dx}{dt} + \frac{\partial w}{\partial y}\frac{dy}{dt}$$

▮ **EXAMPLE 1**   Let

$$w = f(x, y) = x^2 y^3$$

with $x(t) = \sin t$ and $y(t) = e^{-t}$. Find the derivative of $w = f(x, y)$ with respect to t when $t = \pi/2$.

Solution　Using the chain rule, we find

$$\frac{dw}{dt} = \frac{\partial w}{\partial x}\frac{dx}{dt} + \frac{\partial w}{\partial y}\frac{dy}{dt}$$

$$= 2xy^3\frac{dx}{dt} + x^2 3y^2\frac{dy}{dt}$$

$$= 2xy^3\cos t + 3x^2 y^2(-1)e^{-t}$$

Now, $\cos\frac{\pi}{2} = 0$,

$$x\left(\frac{\pi}{2}\right) = \sin\frac{\pi}{2} = 1 \qquad \text{and} \qquad y\left(\frac{\pi}{2}\right) = e^{-\pi/2}$$

Hence,

$$\left.\frac{dw}{dt}\right|_{t=\pi/2} = (2)(1)(e^{-3\pi/2})(0) - (3)(1)^2 e^{-\pi}e^{-\pi/2} = -3e^{-3\pi/2}$$

We can check the answer directly; namely,

$$w(t) = (\sin^2 t)(e^{-3t})$$

and, therefore,

$$\left.\frac{dw}{dt}\right|_{t=\pi/2} = 2(\sin t)(\cos t)e^{-3t} - 3(\sin^2 t)e^{-3t}\Big|_{t=\pi/2} = -3e^{-3\pi/2} \qquad ▮$$

▮EXAMPLE 2　Suppose that we wish to predict the abundance of a particular plant. We suspect that the two major factors influencing the abundance of this plant are nitrogen levels and disturbance levels. Previous studies have shown that an increase in nitrogen in the soil has a negative effect on the abundance of the plant; also, an increase in disturbance due to grazing seems to have a negative effect on abundance. If both nitrogen and disturbance due to grazing increase over the next few years, how would the abundance of this species be affected?

Solution　We denote by $B(t)$ the abundance of the plant at time t. We are interested in how $B(t)$ will change over time, that is, we want to find out whether $B(t)$ will increase or decrease over time. For this, we need to compute the derivative of $B(t)$. If we assume that the abundance of the plant is primarily affected by nitrogen and disturbance levels, we can consider B as a function of N, the nitrogen level, and D, the disturbance level. Using the chain rule for functions of two independent variables, we find

$$\frac{dB}{dt} = \underbrace{\frac{\partial B}{\partial N}}_{<0}\underbrace{\frac{dN}{dt}}_{>0} - \underbrace{\frac{\partial B}{\partial D}}_{<0}\underbrace{\frac{dD}{dt}}_{>0} < 0$$

since abundance B is a decreasing function of both N and D, and both nitrogen and disturbance levels are assumed to increase over the next few years. We thus find that the abundance of the plant will likely decrease over the next few years. ▮

10.5.2 Implicit Differentiation

We discussed implicit differentiation in Section 3.4. This was a useful technique for differentiating a function $y = f(x)$ when y was given implicitly, such as in

$$x^2 y - e^{-y} = 0 \tag{10.15}$$

To find dy/dx, we differentiate both sides with respect to x, keeping in mind that y is a function of x. We obtain

$$2xy + x^2\frac{dy}{dx} - e^{-y}\left(-\frac{dy}{dx}\right) = 0$$

$$2xy + \frac{dy}{dx}(x^2 + e^{-y}) = 0$$

Solving for dy/dx, we find

$$\frac{dy}{dx} = -\frac{2xy}{x^2 + e^{-y}} \tag{10.16}$$

When we look at (10.15), we can define a function $F(x, y)$ as

$$F(x, y) = x^2y - e^{-y} \quad \text{with } F(x, y) = 0$$

$F(x, y)$ is a function of two variables. The equation $F(x, y) = 0$ defines y implicitly as a function of x.

We will turn to the general case to see why this is a useful way to look at this problem. We set

$$w = F(u, v) \quad \text{with } u(x) = x \text{ and } v(x) = y$$

This makes w a function of x, that is, $w = w(x)$. We can use the chain rule to differentiate w with respect to x, and we find

$$\frac{dw}{dx} = \frac{\partial F}{\partial u}\frac{du}{dx} + \frac{\partial F}{\partial v}\frac{dv}{dx}$$

Since $\frac{du}{dx} = 1$ and $\frac{dv}{dx} = \frac{dy}{dx}$, we obtain, with $u = x$ and $v = y$,

$$\frac{dw}{dx} = \frac{\partial F}{\partial x} + \frac{\partial F}{\partial y}\frac{dy}{dx}$$

Since y is a function of x, we can write $w(x) = F(x, y)$, with F satisfying $F(x, y) = 0$. Hence, $w(x) = 0$ for all values of x and, therefore,

$$\frac{dw}{dx} = 0$$

We thus obtain

$$0 = \frac{\partial F}{\partial x} + \frac{\partial F}{\partial y}\frac{dy}{dx}$$

which we can solve for dy/dx. We find

$$\frac{dy}{dx} = -\frac{\frac{\partial F}{\partial x}}{\frac{\partial F}{\partial y}} = -\frac{F_x}{F_y}$$

provided that $\partial F/\partial y \neq 0$. We summarize this in the following.

> Suppose that $w = F(x, y)$ is differentiable, and $F(x, y) = 0$ defines $y = g(x)$ implicitly as a differentiable function of x. Then, at any point where $F_y \neq 0$,
>
> $$\frac{dy}{dx} = -\frac{F_x}{F_y}$$

I EXAMPLE 3   Find dy/dx if $x^2y - e^{-y} = 0$.

Solution   We set $F(x, y) = x^2y - e^{-y}$. We need to find F_x and F_y:

$$F_x = \frac{\partial F}{\partial x} = 2xy$$

$$F_y = \frac{\partial F}{\partial y} = x^2 + e^{-y}$$

Then, since $F(x, y) = 0$,

$$\frac{dy}{dx} = -\frac{F_x}{F_y} = -\frac{2xy}{x^2 + e^{-y}}$$

as in (10.16). Since $x^2 + e^{-y} \neq 0$ for all $(x, y) \in \mathbf{R}^2$, dy/dx is defined for all $(x, y) \in \mathbf{R}^2$ with $F(x, y) = 0$.   **I**

The following example shows that we can use this rule to find the derivatives of inverse trigonometric functions.

I EXAMPLE 4   Find dy/dx for

$$y = \arcsin x$$

Solution   We already know the answer, since we discussed the derivative of $\arcsin x$ earlier, in the context of inverse functions. But let's see how we can use the results of this section to find the answer. Since

$$x = \sin y \quad \text{for} \quad -\frac{\pi}{2} \le y \le \frac{\pi}{2}$$

is equivalent to

$$y = \arcsin x \quad \text{for} \ -1 \le x \le 1$$

we can define a function $F(x, y)$ that satisfies $F(x, y) = 0$ and defines y implicitly as a function of x, namely,

$$F(x, y) = x - \sin y$$

Then,

$$\frac{dy}{dx} = -\frac{F_x}{F_y} = -\frac{1}{-\cos y} = \frac{1}{\cos y}$$

Since $\sin^2 y + \cos^2 y = 1$, we find

$$\cos y = \sqrt{1 - \sin^2 y} = \sqrt{1 - x^2}$$

Here we must use the fact that $x = \sin y$ is defined for $-\pi/2 \le y \le \pi/2$ and $\cos y \ge 0$ for y in this interval. Using this representation for $\cos y$, we find

$$\frac{dy}{dx} = \frac{1}{\sqrt{1 - x^2}}$$ **I**

10.5.3 Directional Derivatives and Gradient Vectors

Suppose that you are on a sloped surface, like a hillside. Depending on which direction you walk, you must either go uphill, stay at the same level, or go downhill. That is, by choosing a particular direction, you have some control over the steepness of your path. How steep your path is can be described by the slope of the tangent at your starting point in the direction of your path. This slope is given by the **directional derivative**.

We assume that $z = f(x, y)$ is a differentiable function of two independent variables. We choose a point (x_0, y_0, z_0) with $z_0 = f(x_0, y_0)$ on the surface defined by $z = f(x, y)$. To define the slope of a tangent line at (x_0, y_0, z_0), we must specify a direction in which we wish to go. We know how to deal with this problem when we go in the x or y direction. In these cases, the partial derivatives $\partial f/\partial x$ and $\partial f/\partial y$ tell us how $f(x, y)$ changes. We will now explain how we can express the slope when we choose an arbitrary direction.

The first step is to find a way to express what we mean by "going in a certain direction." We start at a point (x_0, y_0) and wish to go in the direction of a unit vector $\mathbf{u} = \begin{bmatrix} u_1 \\ u_2 \end{bmatrix}$ (recall that a unit vector has length 1). This is illustrated in Figure 10.37. From Figure 10.37, we see that

$$\mathbf{r} = \mathbf{r_0} + t\mathbf{u} \tag{10.17}$$

where $\mathbf{r_0} = \begin{bmatrix} x_0 \\ y_0 \end{bmatrix}$, $\mathbf{r} = \begin{bmatrix} x \\ y \end{bmatrix}$, and t is a real number; different values of t get us to different points on the straight line through (x_0, y_0) that points in the direction of $\mathbf{u}$. We can write (10.17) also as:

$$\begin{bmatrix} x \\ y \end{bmatrix} = \begin{bmatrix} x_0 \\ y_0 \end{bmatrix} + \begin{bmatrix} tu_1 \\ tu_2 \end{bmatrix} \quad \text{for } t \in \mathbf{R} \tag{10.18}$$

Equation (10.18) is called a **parametric equation** of a line, and t is called the **parameter**. Since $x = x_0 + tu_1$ and $y = y_0 + tu_2$, we find

$$\frac{dx}{dt} = u_1 \quad \text{and} \quad \frac{dy}{dt} = u_2 \tag{10.19}$$

We can now find out what happens to $f(x, y)$ when we start at (x_0, y_0) and go in the direction of the unit vector $\mathbf{u}$, as illustrated in Figure 10.38. For this, we use the parametric equation (10.18) of the line through (x_0, y_0) that is oriented in the direction of $\mathbf{u}$:

$$x = x_0 + tu_1 \quad \text{and} \quad y = y_0 + tu_2$$

Since $z(t) = f(x(t), y(t))$, we can use the chain rule to find out how f changes when we vary t (that is, move along the straight line),

$$\frac{dz}{dt} = \frac{\partial f}{\partial x}\frac{dx}{dt} + \frac{\partial f}{\partial y}\frac{dy}{dt} = \frac{\partial f}{\partial x}u_1 + \frac{\partial f}{\partial y}u_2 \tag{10.20}$$

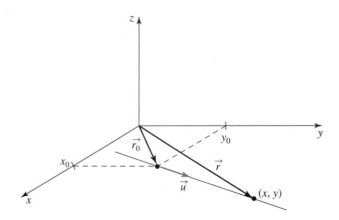

▲ **Figure 10.37**

Going in the direction of $\mathbf{u}$ from (x_0, y_0).

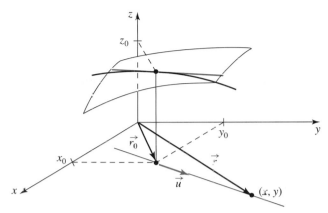

◀ **Figure 10.38**

An illustration of the directional derivative.

where we used (10.19) in the last step. Equation (10.20) can be written as a dot product:

$$\frac{dz}{dt} = \begin{bmatrix} \frac{\partial f}{\partial x} \\ \frac{\partial f}{\partial y} \end{bmatrix} \cdot \begin{bmatrix} u_1 \\ u_2 \end{bmatrix}$$

The first vector in the dot product is called the **gradient** of f.

Definition   Assume that $z = f(x, y)$ is a function of two independent variables, and $\partial f/\partial x$ and $\partial f/\partial y$ exist. Then the vector

$$\nabla f(x, y) = \begin{bmatrix} \frac{\partial f(x,y)}{\partial x} \\ \frac{\partial f(x,y)}{\partial y} \end{bmatrix}$$

is called the gradient of f at (x, y).

The notation ∇f is read "grad f" or "gradient of f." The symbol ∇ is called "del," so you can also say "del f." An alternative notation is grad f.

We can now define the derivative of f in a particular direction.

Definition   The directional derivative of $f(x, y)$ at (x_0, y_0) in the direction of the unit vector $\mathbf{u} = \begin{bmatrix} u_1 \\ u_2 \end{bmatrix}$ is

$$D_{\mathbf{u}} f(x_0, y_0) = (\nabla f(x_0, y_0)) \cdot \mathbf{u}$$

Note that in the definition of the directional derivative we assume that $\mathbf{u}$ is a *unit vector*. Choosing a unit vector (as opposed to a vector of some other length) ensures that the directional derivative of $f(x, y)$ agrees with the partial derivatives when we go along the positive x- or y-axis. That is, if $\mathbf{u} = \begin{bmatrix} 1 \\ 0 \end{bmatrix}$, then

$$D_{\mathbf{u}} f(x_0, y_0) = \begin{bmatrix} \frac{\partial f(x_0, y_0)}{\partial x} \\ \frac{\partial f(x_0, y_0)}{\partial y} \end{bmatrix} \cdot \begin{bmatrix} 1 \\ 0 \end{bmatrix} = \frac{\partial f(x_0, y_0)}{\partial x}$$

and if $\mathbf{u} = \begin{bmatrix} 0 \\ 1 \end{bmatrix}$, then

$$D_{\mathbf{u}} f(x_0, y_0) = \frac{\partial f(x_0, y_0)}{\partial y}$$

EXAMPLE 5   Compute the directional derivative of

$$f(x, y) = \sqrt{x^2 + y^2}$$

at the point $(-1, 2)$ in the direction of $(1, 1)$.

Solution   We first compute the gradient vector

$$\nabla f(x, y) = \begin{bmatrix} \dfrac{x}{\sqrt{x^2+y^2}} \\ \dfrac{y}{\sqrt{x^2+y^2}} \end{bmatrix}$$

Evaluating this at $(-1, 2)$, we find

$$\nabla f(-1, 2) = \begin{bmatrix} \dfrac{-1}{\sqrt{5}} \\ \dfrac{2}{\sqrt{5}} \end{bmatrix}$$

Since $\begin{bmatrix} 1 \\ 1 \end{bmatrix}$ is not a unit vector, we normalize it first. The vector $\begin{bmatrix} 1 \\ 1 \end{bmatrix}$ has length $\sqrt{2}$, hence

$$\mathbf{u} = \frac{1}{\sqrt{2}} \begin{bmatrix} 1 \\ 1 \end{bmatrix} = \begin{bmatrix} \dfrac{1}{\sqrt{2}} \\ \dfrac{1}{\sqrt{2}} \end{bmatrix}$$

and, hence,

$$D_{\mathbf{u}} f(-1, 2) = (\nabla f(-1, 2)) \cdot \mathbf{u}$$

$$= \begin{bmatrix} -\dfrac{1}{\sqrt{5}} \\ \dfrac{2}{\sqrt{5}} \end{bmatrix} \cdot \begin{bmatrix} \dfrac{1}{\sqrt{2}} \\ \dfrac{1}{\sqrt{2}} \end{bmatrix}$$

$$= -\frac{1}{\sqrt{10}} + \frac{2}{\sqrt{10}} = \frac{1}{\sqrt{10}} = \frac{1}{10}\sqrt{10} \qquad \blacksquare$$

Properties of the Gradient Vector   The directional derivative is a dot product. We can therefore write

$$D_{\mathbf{u}} f(x, y) = (\nabla f(x, y)) \cdot \mathbf{u}$$
$$= |\nabla f(x, y)||\mathbf{u}| \cos \theta$$

where θ is the angle between ∇f and $\mathbf{u}$ (see Figure 10.39). Since $|\mathbf{u}| = 1$ ($\mathbf{u}$ is a unit vector), we find

$$D_{\mathbf{u}} f(x, y) = |\nabla f(x, y)| \cos \theta$$

The angle θ is in the interval $[0, 2\pi)$, and $\cos \theta$ is maximal when $\theta = 0$. We therefore find that $D_{\mathbf{u}} f(x, y)$ is maximal when $\mathbf{u}$ is in the direction of ∇f.

Geometrically, the gradient vector at a point (x_0, y_0) is perpendicular to the level curve $f(x, y) = c$ that passes through this point. The level curve $f(x, y) = c$ is a curve in the x-y plane that can be described by the vector $\mathbf{r}(t) = \begin{bmatrix} x(t) \\ y(t) \end{bmatrix}$, which traces out the curve as t varies. If we differentiate $\mathbf{r}(t)$ with respect to t, we obtain

$$\frac{d}{dt}\mathbf{r}(t) = \begin{bmatrix} \dfrac{dx(t)}{dt} \\ \dfrac{dy(t)}{dt} \end{bmatrix}$$

which is the tangent vector at the point $(x(t), y(t))$. Since

$$f(x, y) = c \qquad (10.21)$$

and x and y are both functions of t, we can write $w(t) = f(x(t), y(t)) = c$ and differentiate $w(t)$ with respect to t. We find $dw/dt = 0$, since $w(t)$ is a constant.

▲ **Figure 10.39**
The angle θ between ∇f and the unit vector u.

◄ **Figure 10.40**

Using the chain rule to differentiate $f(x(t), y(t))$ yields

$$\frac{d}{dt} f(x(t), y(t)) = \frac{\partial f}{\partial x} \frac{dx}{dt} + \frac{\partial f}{\partial y} \frac{dy}{dt}$$

hence,

$$\frac{\partial f}{\partial x} \frac{dx}{dt} + \frac{\partial f}{\partial y} \frac{dy}{dt} = 0$$

or

$$\nabla f \cdot \frac{d\mathbf{r}}{dt} = 0$$

That is, the gradient of f at (x_0, y_0) is perpendicular to the level curve at (x_0, y_0), as illustrated in Figure 10.40.

We summarize these two important properties of the gradient in the following.

Suppose that $f(x, y)$ is a differentiable function. The gradient vector $\nabla f(x, y)$ has the following properties.

1. At each point (x_0, y_0), $f(x, y)$ increases most rapidly in the direction of the gradient vector $\nabla f(x_0, y_0)$.

2. The gradient vector of f at a point (x_0, y_0) is perpendicular to the level curve through (x_0, y_0).

EXAMPLE 6 Let

$$f(x, y) = x^2 y + y^2$$

In what direction does $f(x, y)$ increase most rapidly at $(1, 1)$?

Solution   The function $f(x, y)$ increases most rapidly at $(1, 1)$ in the direction of $\nabla f(1, 1)$. Since

$$\nabla f(x, y) = \begin{bmatrix} \frac{\partial f}{\partial x} \\ \frac{\partial f}{\partial y} \end{bmatrix} = \begin{bmatrix} 2xy \\ x^2 + 2y \end{bmatrix}$$

it follows that

$$\nabla f(1, 1) = \begin{bmatrix} 2 \\ 3 \end{bmatrix}$$

That is, $f(x, y)$ increases most rapidly in the direction of $\begin{bmatrix} 2 \\ 3 \end{bmatrix}$ at the point $(1, 1)$. **I**

■ EXAMPLE 7 Find a unit vector that is perpendicular to the level curve of the function

$$f(x, y) = x^2 - y^2$$

at $(1, 2)$.

Solution The gradient of f at $(1, 2)$ is perpendicular to the level curve at $(1, 2)$. The gradient of f is given as

$$\nabla f(x, y) = \begin{bmatrix} 2x \\ -2y \end{bmatrix}$$

hence,

$$\nabla f(1, 2) = \begin{bmatrix} 2 \\ -4 \end{bmatrix}$$

To normalize this vector, we divide $\nabla f(1, 2)$ by its length. Since

$$|\nabla f(1, 2)| = \sqrt{(2)^2 + (-4)^2} = \sqrt{4 + 16} = 2\sqrt{5}$$

the unit vector that is perpendicular to the level curve of $f(x, y)$ at $(1, 2)$ is

$$\mathbf{u} = \frac{1}{2\sqrt{5}} \begin{bmatrix} 2 \\ -4 \end{bmatrix} = \frac{1}{\sqrt{5}} \begin{bmatrix} 1 \\ -2 \end{bmatrix} = \begin{bmatrix} \frac{1}{5}\sqrt{5} \\ -\frac{2}{5}\sqrt{5} \end{bmatrix}$$
■

10.5.4 Diffusion

Suppose that we place a sugar cube into a glass of water without stirring the water. The sugar dissolves and the sugar molecules move about randomly in the water. If we wait long enough, the sugar concentration will eventually be uniform throughout the water. This random movement of molecules is called **diffusion**, and plays an important role in many processes of life. For instance, gas exchange in unicellular organisms and many small multicellular organisms takes place by diffusion. Diffusion is a slow process, which means that cells have to be close to the surface if they want to exchange gas by diffusion. This limits the size and shape of organisms, unless they evolve different gas exchange mechanisms. (There are large organisms, like kelp, that rely on diffusion for gas exchange, but their blades are extremely thin so that all cells are close to the surface.)

Derivation of the One-dimensional Diffusion Equation We want to understand what type of microscopic description yields a diffusion equation. We assume that molecules move along the x-axis and we denote the concentration of these molecules at x at time t by $c(x, t)$. That is, the number of molecules at time t in the interval $[x_1, x_2)$ is given by

$$N_{[x_1, x_2)}(t) = \int_{x_1}^{x_2} c(x, t)\, dx \tag{10.22}$$

Because the molecules move around, the number of molecules in a given interval changes over time. We will express this change as the difference between the net movement of molecules on the left and on the right end of the interval. The quantity that describes this net movement is called the **flux**, and is denoted by $J(x, t)$. We interpret

$$J(x, t)\Delta t = \text{ the net number of molecules crossing } x \text{ from}$$
$$\text{the left to the right during a time interval of length } \Delta t$$

That is, if we consider the change in the number of molecules in the interval $[x_0, x_0 + \Delta x)$ during the time interval $[t, t + \Delta)$, then

$$N_{[x_0, x_0 + \Delta x)}(t + \Delta t) - N_{[x_0, x_0 + \Delta x)}(t)$$
$$= J(x_0, t)\Delta t - J(x_0 + \Delta x, t)\Delta t$$

Dividing both sides by Δt, and letting $\Delta t \to 0$, we find

$$\lim_{\Delta t \to 0} \frac{N_{[x_0, x_0 + \Delta x)}(t + \Delta t) - N_{[x_0, x_0 + \Delta x)}(t)}{\Delta t}$$
$$= J(x_0, t) - J(x_0 + \Delta x, t)$$

The left-hand side is equal to

$$\frac{d}{dt} N_{[x_0, x_0 + \Delta x)}(t) \tag{10.23}$$

Using (10.22), we can write (10.23) as

$$\frac{d}{dt} \int_{x_0}^{x_0 + \Delta x} c(z, t)\, dx$$

When $c(x, t)$ is sufficiently smooth, we can interchange differentiation and integration. (We cannot justify this step here, but it is justified in courses on real analysis.) We find

$$\frac{d}{dt} \int_{x_0}^{x_0 + \Delta x} c(x, t)\, dx = \int_{x_0}^{x_0 + \Delta x} \frac{\partial c(x, t)}{\partial t}\, dx$$

Note that the "d" changed into a "∂" when we moved the derivative inside the integral. Before we moved it inside, we differentiated a function that depended only on t, but once we moved it inside, we differentiated a function that depends on two variables, namely x and t. Summarizing, we arrive at the equation

$$\int_{x_0}^{x_0 + \Delta x} \frac{\partial}{\partial t} c(x, t)\, dx = J(x_0, t) - J(x_0 + \Delta x, t) \tag{10.24}$$

To obtain the diffusion equation, we divide both sides of (10.24) by Δx, and take the limit $\Delta x \to 0$. On the left-hand side of (10.24), we find

$$\lim_{\Delta x \to 0} \frac{1}{\Delta x} \int_{x_0}^{x_0 + \Delta x} \frac{\partial}{\partial t} c(x, t)\, dx = \frac{\partial c(x_0, t)}{\partial t}$$

On the right-hand side of (10.24), we find

$$\lim_{\Delta x \to 0} \frac{J(x_0, t) - J(x_0 + \Delta x, t)}{\Delta x} = -\frac{\partial J(x_0, t)}{\partial x}$$

Putting things together, we arrive at

$$\frac{\partial c(x_0, t)}{\partial t} = -\frac{\partial J(x_0, t)}{\partial x} \tag{10.25}$$

A phenomenological law, namely Fick's law, relates the flux to the change in concentration. It holds, for instance, when molecules move around randomly in a solvent. Fick's law says that

$$J = -D \frac{\partial c}{\partial x} \tag{10.26}$$

where D is a positive constant, called the **diffusion constant**. It means that the flux is proportional to the change in concentration; the minus sign in (10.26) means that the net movement of molecules is from regions of high concentration to regions of low concentration. This agrees with our intuition: Going back to our example of sugar dissolving in water, we expect the net movement of sugar molecules to be from regions of high concentration to regions of low concentration so that ultimately the sugar concentration is uniform.

Combining (10.25) and (10.26), we arrive at the **diffusion equation**, namely

$$\frac{\partial c}{\partial t} = D\frac{\partial^2 c}{\partial x^2} \tag{10.27}$$

The diffusion approach is ubiquitous in biology. It is not only used in the description of the random movement of molecules, but in a wide array of applications, such as the change in allele frequencies due to random genetic drift, the invasion of alien species into virgin habitat, the directed movement of organisms along gradients of chemicals (chemotaxis), pattern formation, and many more. Equation (10.27) is the simplest form of an equation that incorporates diffusion. In physics, (10.27) is called the heat equation. It describes the diffusion of heat through a solid bar; in this case, $c(x, t)$ represents the temperature at x at time t.

Equation (10.27) is an example of a **partial differential equation**, which is an equation that contains partial derivatives. The theory of partial differential equations is complex, and well beyond the scope of this course. We will only be able to discuss some aspects of this equation.

Solving the Diffusion Equation   For most partial differential equations, it is not possible to find an analytical solution, such equations often can only be solved numerically, and even this is typically not an easy task. Fortunately, (10.27) is simple enough that we can find a solution. We claim that

$$c(x, t) = \frac{1}{\sqrt{4\pi Dt}} \exp\left[-\frac{x^2}{4Dt}\right] \tag{10.28}$$

is a solution of (10.27). As in the case of ordinary differential equations, one can check this by computing the appropriate derivatives. For the left-hand side of (10.27), we need the first partial derivative of $c(x, t)$ with respect to t. We find

$$\frac{\partial c(x, t)}{\partial t} = \frac{\partial}{\partial t} \frac{1}{\sqrt{4\pi Dt}} \exp\left[-\frac{x^2}{4Dt}\right]$$

$$= -\frac{1}{2}\frac{4\pi D}{(4\pi Dt)^{3/2}} \exp\left[-\frac{x^2}{4Dt}\right]$$

$$+ \frac{1}{\sqrt{4\pi Dt}} \exp\left[-\frac{x^2}{4Dt}\right] \frac{x^2}{4Dt^2}$$

$$= \exp\left[-\frac{x^2}{4Dt}\right] \left\{\frac{x^2}{4Dt^2\sqrt{4\pi Dt}} - \frac{2\pi D}{4\pi Dt\sqrt{4\pi Dt}}\right\}$$

$$= \frac{1}{2t\sqrt{4\pi Dt}} \exp\left[-\frac{x^2}{4Dt}\right] \left\{\frac{x^2}{2Dt} - 1\right\}$$

On the right-hand side of (10.27), we need the second partial derivative of $c(x, t)$ with respect to x. We find

$$\frac{\partial c(x, t)}{\partial x} = \frac{\partial}{\partial x} \frac{1}{\sqrt{4\pi Dt}} \exp\left[-\frac{x^2}{4Dt}\right]$$

$$= \frac{1}{\sqrt{4\pi Dt}} \exp\left[-\frac{x^2}{4Dt}\right]\left(-\frac{2x}{4Dt}\right)$$

and

$$\frac{\partial^2 c(x, t)}{\partial x^2} = \frac{-1}{\sqrt{4\pi Dt}} \frac{1}{2Dt} \exp\left[-\frac{x^2}{4Dt}\right]\left\{1 - x\frac{2x}{4Dt}\right\}$$

$$= \frac{1}{2Dt\sqrt{4\pi Dt}} \exp\left[-\frac{x^2}{4Dt}\right]\left\{\frac{x^2}{2Dt} - 1\right\}$$

Putting things together, we see that (10.28) satisfies (10.27).

The function in (10.28) is the Gaussian density, which we encountered in Section 8.4. Figure 10.41 shows $c(x, t)$ for $t = 1, 2$, and 4; it clearly shows that the concentration $c(x, t)$ becomes more uniform as time goes on.

Diffusion is a very slow process. The diffusion constant D measures how quickly it proceeds. In Section 8.4 we introduced the standard deviation σ for a Gaussian density, and noted that 68% of the area under the density is between one standard deviation of the mean. Comparing (10.28) with the equation for the density of the normal distribution,

$$f(x) = \frac{1}{\sigma\sqrt{2\pi}} e^{-(x-\mu)^2/2\sigma^2} \quad \text{for } -\infty < x < \infty$$

we see that $\sigma = \sqrt{2Dt}$. The larger D, the faster the concentration spreads out. However, within t units of time, molecules only spread over a region of length of order $\sqrt{t}$.

To give an idea of how slow diffusion is, here are a few examples taken from Yeargers, Shonkwiler, and Herod (1996). Oxygen in blood at 20°C has a diffusion constant of 10^{-5} cm^2/sec, which means that it takes an oxygen molecule roughly 500 seconds to cross a distance of 1 mm by diffusion alone. Ribonuclease (an enzyme

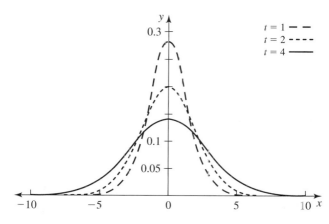

◀ **Figure 10.41**
The solution of the diffusion equation at different times.

that hydrolyses RNA) in water at $20°C$ has a diffusion constant of 1.1×10^{-6} cm²/sec, which means that ribonuclease takes roughly 4672 seconds (or 1 hr 18 min) to cross a distance of 1 mm by diffusion alone. These examples illustrate why organisms frequently rely on other active mechanisms to transport molecules.

The diffusion equation (10.27) can be generalized to higher dimensions. Namely, (10.25) becomes

$$\frac{\partial c}{\partial t} = -\nabla J \tag{10.29}$$

and (10.26) becomes

$$J = -D\nabla c \tag{10.30}$$

Combining (10.29) and (10.30), we find

$$\frac{\partial c}{\partial t} = D\nabla \cdot (\nabla c)$$

where $\nabla \cdot (\nabla c)$ is to be interpreted as a dot product. That is, if $\mathbf{x} = (x_1, x_2, x_3) \in \mathbf{R}^3$, $t \in \mathbf{R}$, then

$$\frac{\partial c}{\partial t} = D\left(\frac{\partial^2 c}{\partial x_1^2} + \frac{\partial^2 c}{\partial x_2^2} + \frac{\partial^2 c}{\partial x_3^2}\right)$$

As a shorthand notation, one defines

$$\triangle = \frac{\partial^2}{\partial x_1^2} + \frac{\partial^2}{\partial x_2^2} + \frac{\partial^2}{\partial x_3^2}$$

where $\triangle$ is called the **Laplace operator**. One then writes

$$\frac{\partial c}{\partial t} = D\triangle c$$

More generally, if $\mathbf{x} = (x_1, x_2, \ldots, x_n) \in \mathbf{R}^n$, then

$$\triangle = \frac{\partial^2}{\partial x_1^2} + \frac{\partial^2}{\partial x_2^2} + \cdots + \frac{\partial^2}{\partial x_n^2}$$

($\triangle c$ is read "Laplacian of c.")

◆ 10.5 PROBLEMS

[10.5.1]

1. Let $f(x, y) = x^2 + y^2$ with $x(t) = 3t$ and $y(t) = e^t$. Find the derivative of $w = f(x, y)$ with respect to t when $t = \ln 2$.

2. Let $f(x, y) = e^x \sin y$ with $x(t) = t$ and $y(t) = t^3$. Find the derivative of $w = f(x, y)$ with respect to t when $t = 1$.

3. Let $f(x, y) = \sqrt{x^2 + y^2}$ with $x(t) = t$ and $y(t) = \sin t$. Find the derivative of $w = f(x, y)$ with respect to t when $t = \pi/3$.

4. Let $f(x, y) = \ln(xy - x^2)$ with $x(t) = t^2$ and $y(t) = t$. Find the derivative of $w = f(x, y)$ with respect to t when $t = 5$.

5. Let $f(x, y) = \frac{1}{x} + \frac{1}{y}$ with $x(t) = \sin t$ and $y(t) = \cos t$. Find the derivative of $w = f(x, y)$ with respect to t when $t = \pi/4$.

6. Let $f(x, y) = xe^y$ with $x(t) = e^t$ and $y(t) = t^2$. Find the derivative of $w = f(x, y)$ with respect to t when $t = 0$.

7. Find $\frac{dz}{dt}$ for $z = f(x, y)$ with $x = u(t)$ and $y = v(t)$.

8. Find $\frac{dw}{dt}$ for $w = e^{f(x,y)}$ with $x = u(t)$ and $y = v(t)$.

[10.5.2]

9. Find $\frac{dy}{dx}$ if $(x^2 + y^2)e^y = 0$.

10. Find $\frac{dy}{dx}$ if $(\sin x + \cos y)x^2 = 0$.

11. Find $\frac{dy}{dx}$ if $\ln(x^2 + y^2) = 3xy$.

12. Find $\frac{dy}{dx}$ if $\cos(x^2 + y^2) = \sin(x^2 - y^2)$.

13. Find $\frac{dy}{dx}$ if $y = \arccos x$.

14. Find $\frac{dy}{dx}$ if $y = \arctan x$.

15. The growth rate r of a particular organism is affected by both the availability of food and the number of other competitors for the food source. Denote by $F(t)$ the amount of food available at time t and by $N(t)$ the number of competitors at time t. The growth rate r can then be thought of as a function of the two time-dependent variables $F(t)$ and $N(t)$. Assume that the growth rate is an increasing function of food availability and a decreasing function of the number of competitors. How is the growth rate affected if the availability of food decreases over time while the number of competitors increases?

16. Suppose that you travel along an environmental gradient, along which both temperature and precipitation increase. If the abundance of a particular plant species increases with both temperature and precipitation, would you expect to encounter this species more or less often during your journey? (Use calculus to answer this question.)

[10.5.3]

In Problems 17–24, find the gradient of each function.

17. $f(x, y) = x^2 y^2$

18. $f(x, y) = \frac{xy}{x^2 + y^2}$

19. $f(x, y) = \sqrt{x^3 - 3xy}$

20. $f(x, y) = x(x^2 - y^2)^{2/3}$

21. $f(x, y) = \exp\left[\sqrt{x^2 + y^2}\right]$

22. $f(x, y) = \tan\frac{x-y}{x+y}$

23. $f(x, y) = \ln\left(\frac{x}{y} + \frac{y}{x}\right)$

24. $f(x, y) = \cos(3x^2 - 2y^2)$

In Problems 25–30, compute the directional derivative of $f(x, y)$ at the given point in the indicated direction.

25.
$$f(x, y) = \sqrt{2x^2 + y^2}$$
at $(1, 2)$ in the direction of $\begin{bmatrix} 1 \\ 1 \end{bmatrix}$.

26.
$$f(x, y) = x^2 \sin y$$
at $(-1, 0)$ in the direction of $\begin{bmatrix} 2 \\ -1 \end{bmatrix}$.

27.
$$f(x, y) = e^{x+y}$$
at $(0, 0)$ in the direction of $\begin{bmatrix} -1 \\ -1 \end{bmatrix}$.

28.
$$f(x, y) = x^3 y^2$$
at $(2, 3)$ in the direction of $\begin{bmatrix} -2 \\ 1 \end{bmatrix}$.

29.
$$f(x, y) = 2xy^3 - 3x^2 y$$
at $(1, -1)$ in the direction of $\begin{bmatrix} 3 \\ 1 \end{bmatrix}$.

30.
$$f(x, y) = ye^{x^2}$$
at $(0, 2)$ in the direction of $\begin{bmatrix} 4 \\ -1 \end{bmatrix}$.

31. In what direction does $f(x, y) = 3xy - x^2$ increase most rapidly at $(-1, 1)$?

32. In what direction does $f(x, y) = e^x \cos y$ increase most rapidly at $(0, \pi/2)$?

33. In what direction does $f(x, y) = \sqrt{x^2 - y^2}$ increase most rapidly at $(5, 3)$?

34. In what direction does $f(x, y) = \ln(x^2 + y^2)$ increase most rapidly at $(1, 1)$?

35. Find a unit vector that is normal to the level curve of the function
$$f(x, y) = 3x + 4y$$
at the point $(-1, 1)$.

36. Find a unit vector that is normal to the level curve of the function
$$f(x, y) = x^2 + \frac{y^2}{9}$$
at the point $(1, 3)$.

37. Find a unit vector that is normal to the level curve of the function
$$f(x, y) = x^2 - y^3$$
at the point $(1, 3)$.

38. Find a unit vector that is normal to the level curve of the function
$$f(x, y) = xy$$
at the point $(2, 3)$.

39. (*Chemotaxis*) Chemotaxis is the chemically directed movement of organisms up a concentration gradient. The slime mold *Dictyostelium discoideum* exhibits this phenomenon. In this case, single-celled amoeba of this species move up the concentration gradient of a chemical called cyclic AMP. Suppose the concentration of cyclic AMP at the point (x, y) in the x-y plane is given by
$$f(x, y) = \frac{4}{xy + 1}$$
If you place an amoeba at the point $(3, 1)$ in the x-y plane, determine in which direction the amoeba will move if its movement is directed by chemotaxis.

40. Suppose an organism moves down a sloped surface along the steepest line of descent. If the surface is given by
$$f(x, y) = x^2 - y^2$$
find the direction in which the organism will move at the point $(2, 3)$.

[10.5.4]

41. Show that
$$c(x, t) = \frac{1}{\sqrt{8\pi t}} \exp\left[-\frac{x^2}{8t}\right]$$
solves
$$\frac{\partial c(x, t)}{\partial t} = 2\frac{\partial^2 c(x, t)}{\partial x^2}$$

42. Show that

$$c(x, t) = \frac{1}{\sqrt{2\pi t}} \exp\left[-\frac{x^2}{2t}\right]$$

solves

$$\frac{\partial c(x, t)}{\partial t} = \frac{1}{2}\frac{\partial^2 c(x, t)}{\partial x^2}$$

43. A solution of

$$\frac{\partial c(x, t)}{\partial t} = D\frac{\partial^2 c(x, t)}{\partial x^2}$$

is the function

$$c(x, t) = \frac{1}{\sqrt{4\pi Dt}} \exp\left[-\frac{x^2}{4Dt}\right]$$

for $x \in \mathbf{R}$ and $t > 0$.

(a) Show that $c(x, t)$ as a function of x for fixed values of $t > 0$ is (i) positive for all $x \in \mathbf{R}$, (ii) increasing for $x < 0$, decreasing for $x > 0$, (iii) has a local maximum at $x = 0$, and (iv) has inflection points at $x = \pm\sqrt{2Dt}$.

(b) Graph $c(x, t)$ as a function of x when $D = 1$, for $t = 0.01$, $t = 0.1$, and $t = 1$.

44. A solution of

$$\frac{\partial c(x, t)}{\partial t} = D\frac{\partial^2 c(x, t)}{\partial x^2}$$

is the function

$$c(x, t) = \frac{1}{\sqrt{4\pi Dt}} \exp\left[-\frac{x^2}{4Dt}\right]$$

for $x \in \mathbf{R}$ and $t > 0$.

(a) Show that a local maximum of $c(x, t)$ for fixed t occurs at $x = 0$.

(b) Show that $c(0, t)$, $t > 0$, is a decreasing function of t.

(c) Find

$$\lim_{t \to 0^+} c(x, t)$$

when $x = 0$, and when $x \neq 0$.

(d) Use the fact that

$$\int_{-\infty}^{\infty} e^{-u^2/2}\, du = \sqrt{2\pi}$$

to show that for $t > 0$,

$$\int_{-\infty}^{\infty} c(x, t)\, dx = 1$$

(e) The function $c(x, t)$ can be interpreted as the concentration of a substance diffusing in space. Explain the meaning of

$$\int_{-\infty}^{\infty} c(x, t)\, dx = 1$$

and use your results in (c) and (d) to explain why this means that initially (that is, at $t = 0$) the entire amount of the substance was released at the origin.

Mathematically, one can specify such an initial condition (in which the substance is concentrated at the origin at time 0) by the δ-function $\delta(x)$, with the property

$$\delta(x) = 0, \quad \text{for } x \neq 0$$

and

$$\int_{-\infty}^{\infty} \delta(x)\, dx = 1$$

45. The two-dimensional diffusion equation

$$\frac{\partial n(\mathbf{r}, t)}{\partial t} = D\left(\frac{\partial^2 n(\mathbf{r}, t)}{\partial x^2} + \frac{\partial^2 n(\mathbf{r}, t)}{\partial y^2}\right) \qquad (10.31)$$

where $n(\mathbf{r}, t)$, $\mathbf{r} = (x, y)$, denotes the population density at the point $\mathbf{r} = (x, y)$ in the plane at time t, can be used to describe the spread of organisms. Assume that a large number of insects are released at time 0 at the point $(0, 0)$. Furthermore, assume that at later times, the density of these insects can be described by the diffusion equation (10.31). Show that

$$n(x, y, t) = \frac{n_0}{4\pi Dt} \exp\left[-\frac{x^2 + y^2}{4Dt}\right]$$

satisfies (10.31).

◆ 10.6 KEY TERMS

Chapter 10 Review: Topics Discuss the following definitions and concepts:

1. Real-valued function
2. Function of two variables
3. Surface
4. Level curve
5. Limit
6. Limit laws
7. Continuity
8. Partial derivative
9. Geometric interpretation of a partial derivative
10. Mixed derivative theorem
11. Tangent plane
12. Differentiability
13. Differentiability and continuity
14. Sufficient condition for differentiability
15. Standard linear approximation, tangent plane approximation
16. Vector-valued function
17. Jacobi matrix, derivative matrix
18. Chain rule
19. Implicit differentiation
20. Directional derivative
21. Gradient
22. Diffusion

◆ 10.7 REVIEW PROBLEMS

1. Suppose that you conduct an experiment to measure germination success of seeds of a certain plant as a function of temperature and humidity. You find that seeds don't germinate at all when the humidity is too low, regardless of temperature; germination success is highest for intermediate values of temperature; and seeds tend to germinate better when you increase humidity levels. Use the above information to sketch a graph of germination success as a function of temperature for different levels of humidity. Also sketch the graph of germination success as a function of humidity for different temperature values.

2. Gaastra (1959) measured the effects of atmospheric CO_2 enrichment on CO_2 fixation in sugar beet leaves at various light levels. He found that increasing CO_2 at fixed light levels increases the fixation rate, and that increasing light levels at fixed atmospheric CO_2 concentration also increased fixation. If $F(A, I)$ denotes the fixation rate as a function of atmospheric CO_2 concentration (A) and light intensity (I), determine the signs of $\partial F/\partial A$ and $\partial F/\partial I$.

3. In Burke and Grime (1996), a long-term field experiment in a limestone grassland was described.

(a) One of the experiments related total area covered by *indigenous* species to fertility and disturbance gradients. The experiment was designed so that the two variables (fertility and disturbance) could be altered independently. They found that the area covered by indigenous species generally increased with the amount of fertilizer added and decreased with disturbance intensity. If $A_i(F, D)$ denotes the area covered by indigenous species as a function of the amount of fertilizer added (F) and the disturbance intensity (D), determine the signs of $\partial A_i/\partial F$ and $\partial A_i/\partial D$ for their experiment.

(b) In another experiment, they related the total area covered by *introduced* species to fertility and disturbance gradients. Let $A_e(F, D)$ denote the area covered by introduced species as a function of the amount of fertilizer added (F) and the disturbance intensity (D). They found that

$$\frac{\partial A_e}{\partial F} > 0$$

and

$$\frac{\partial A_e}{\partial D} > 0$$

Explain in words what this means.

(c) Compare the responses to fertilization and disturbance to the area covered in the two experiments.

4. Vitousek and Farrington (1997) investigated nutrient limitations in soils of different ages. In the abstract of their paper, they say that

> Walker and Syers (1976) proposed a conceptual model that describes the pattern and regulation of soil nutrient pools and availability during long-term soil and ecosystem development. Their model implies that plant production generally should be limited by N [nitrogen] on young soils and by P [phosphorus] on old soils; N

and P supply should be more or less equilibrate on intermediate aged soils.

They tested this hypothesis in fertilizer experiments along a gradient of soil age, where they measured the average diameter increment (in mm/yr) of *Metrosideros polymorpha* trees.

Denote by $D(N, P, t)$ the diameter increment (in mm/yr) as a function of the amount of nitrogen added (N), the amount of phosphorus added (P), and the age of the soil (t). Their experiments showed that

$$\frac{\partial D}{\partial t}(N, 0, t) < 0$$

and

$$\frac{\partial D}{\partial t}(0, P, t) > 0$$

for their choices of $N > 0$ and $P > 0$. Explain why their results support the Walker and Syers hypothesis.

5. Find the Jacobi matrix

$$\mathbf{f}(x, y) = \begin{bmatrix} x^2 - y \\ x^3 - y^2 \end{bmatrix}$$

6. Find a linear approximation to $\mathbf{f}(x, y)$ at the indicated point:

$$\mathbf{f}(x, y) = \begin{bmatrix} 2xy^2 \\ \frac{x}{y} \end{bmatrix}$$

at $(1, 1)$.

7. One can compute the average radius of spreading individuals at time t, denoted by r_{avg}. One finds

$$r_{avg} = \sqrt{\pi D t} \tag{10.32}$$

(a) Graph r_{avg} as a function of D for $t = 0.1, t = 1$, and $t = 5$. Describe in words how an increase in D affects the average radius of spread.

(b) Show that

$$D = \frac{(r_{avg})^2}{\pi t} \tag{10.33}$$

(c) Equation (10.33) can be used to measure D, the diffusion constant, from field data of mark-recapture experiments, taken from Kareiva (1983), as follows: Marked organisms are released from the release site, and then recaptured after a certain amount of time t from the time of release. The distance of the recaptured organisms from the release site is measured.

If N denotes the total number of recaptured organisms, d_i denotes the distance of the ith recaptured organism from the release site, and t is the time between release and recapture, use (10.33) to explain why

$$D = \frac{1}{\pi t} \left(\frac{1}{N} \sum_{i=1}^{N} d_i \right)^2$$

can be used to measure D from field data. (Note that the time between release and recapture is the same for each individual in this study.)

Systems of Differential Equations

We first encountered systems of differential equations in Section 7.3. There, the emphasis was on understanding where such systems arise, and on finding equilibria. (Section 7.3 is not a prerequisite for this chapter.) We will now give a more systematic treatment of such systems. The general form of a system of differential equations is given by

$$\frac{dx_1}{dt} = g_1(t, x_1, x_2, \ldots, x_n)$$

$$\frac{dx_2}{dt} = g_2(t, x_1, x_2, \ldots, x_n)$$

$$\vdots$$

$$\frac{dx_n}{dt} = g_n(t, x_1, x_2, \ldots, x_n)$$

(11.1)

We will first look at the case when the functions g_i are linear in the variables $x_1, x_2, \ldots, x_n$; that is, when for $i = 1, 2, \ldots, n$,

$$g_i(t, x_1, x_2, \ldots, x_n) = a_{i1}(t)x_1 + a_{i2}(t)x_2 + \cdots + a_{in}(t)x_n + f_i(t)$$

We can write the linear system in matrix form, namely

$$\frac{d\mathbf{x}}{dt} = A(t)\mathbf{x}(t) + \mathbf{f}(t)$$

(11.2)

where

$$\mathbf{x}(t) = \begin{bmatrix} x_1(t) \\ x_2(t) \\ \vdots \\ x_n(t) \end{bmatrix} \qquad A = \begin{bmatrix} a_{11}(t) & a_{12}(t) & \ldots & a_{1n}(t) \\ a_{21}(t) & a_{22}(t) & \ldots & a_{2n}(t) \\ \vdots & \vdots & \vdots & \vdots \\ a_{n1}(t) & a_{n2}(t) & \ldots & a_{nn}(t) \end{bmatrix}$$

and

$$\mathbf{f}(t) = \begin{bmatrix} f_1(t) \\ f_2(t) \\ \vdots \\ f_n(t) \end{bmatrix}$$

Equation (11.2) is called a system of **linear first order equations** (first order since only first derivatives occur). We will concentrate on the case when $\mathbf{f}(t) = \mathbf{0}$ and $A(t)$ does not depend on t. Equation (11.2) then reduces to

$$\frac{d\mathbf{x}}{dt} = A\mathbf{x}(t) \tag{11.3}$$

where

$$A = \begin{bmatrix} a_{11} & a_{12} & \cdots & a_{1n} \\ a_{21} & a_{22} & \cdots & a_{2n} \\ \vdots & \vdots & \vdots & \vdots \\ a_{n1} & a_{n2} & \cdots & a_{nn} \end{bmatrix}$$

Equation (11.3) is called a **homogeneous** linear first order system with constant coefficients. (Homogeneous refers to the fact that $\mathbf{f}(t) = \mathbf{0}$. When $\mathbf{f}(t) \neq \mathbf{0}$, the system is called **inhomogeneous**.) Since the matrix A does not depend on t, all the coefficients are constant; such a system is **autonomous** (we encountered autonomous systems in Section 7.1). Note that A is a square matrix.

In Section 11.1, we will present some of the theory for systems of the form (11.3). In Section 11.2, we will discuss some applications of linear systems. Section 11.3 is devoted to the theory of nonlinear systems, and Section 11.4 to applications of nonlinear systems.

◆ 11.1 LINEAR SYSTEMS—THEORY

In this section, we will analyze homogeneous, linear first order systems with constant coefficients, that is, systems of the form

$$\begin{aligned} \frac{dx_1}{dt} &= a_{11}x_1 + a_{12}x_2 + \cdots + a_{1n}x_n \\ \frac{dx_2}{dt} &= a_{21}x_1 + a_{22}x_2 + \cdots + a_{2n}x_n \\ &\vdots \\ \frac{dx_n}{dt} &= a_{n1}x_1 + a_{n2}x_2 - \cdots + a_{nn}x_n \end{aligned} \tag{11.4}$$

which we can write in matrix form as

$$\frac{d\mathbf{x}}{dt} = A\mathbf{x}(t)$$

where

$$\mathbf{x}(t) = \begin{bmatrix} x_1(t) \\ x_2(t) \\ \vdots \\ x_n(t) \end{bmatrix} \qquad A = \begin{bmatrix} a_{11} & a_{12} & \cdots & a_{1n} \\ a_{21} & a_{22} & \cdots & a_{2n} \\ \vdots & \vdots & \vdots & \vdots \\ a_{n1} & a_{n2} & \cdots & a_{nn} \end{bmatrix}$$

Most of the time we will restrict ourselves to the case $n = 2$.

∎ **EXAMPLE 1**   Write

$$\frac{dx_1}{dt} = 4x_1 - 2x_2$$

$$\frac{dx_2}{dt} = -3x_1 + x_2$$

in matrix notation.

Solution   We write $x_1 = x_1(t)$ and $x_2 = x_2(t)$. Using the rules for matrix multiplication, we find

$$\begin{bmatrix} \frac{dx_1}{dt} \\ \frac{dx_2}{dt} \end{bmatrix} = \begin{bmatrix} 4x_1 - 2x_2 \\ -3x_1 + x_2 \end{bmatrix} = \begin{bmatrix} 4 & -2 \\ -3 & 1 \end{bmatrix} \begin{bmatrix} x_1(t) \\ x_2(t) \end{bmatrix}$$

That is,

$$\frac{d\mathbf{x}(t)}{dt} = \begin{bmatrix} 4 & -2 \\ -3 & 1 \end{bmatrix} \mathbf{x}(t) \qquad\qquad ∎$$

First, we will be concerned with solutions of (11.4): A solution is an ordered n-tuple of functions $(x_1(t), x_2(t), \ldots, x_n(t))$ that satisfies all n equations in (11.4). Second, we will discuss equilibria: An equilibrium is a point $\hat{\mathbf{x}} = (\hat{x}_1, \hat{x}_2, \ldots, \hat{x}_n)$ so that $A\hat{\mathbf{x}} = \mathbf{0}$. We begin with a graphical approach to visualizing solutions.

11.1.1 The Direction Field

Direction fields are a way to visualize solutions of differential equations. We illustrate this concept in the following example. Consider

$$\frac{dx_1}{dt} = x_1 - 2x_2$$

$$\frac{dx_2}{dt} = x_2 \qquad\qquad (11.5)$$

The system (11.5) assigns to each point (x_1, x_2) in the x_1-x_2 plane a vector $\left[\frac{dx_1}{dt}, \frac{dx_2}{dt}\right]'$. This vector is tangent to the solution curve that passes through the point (x_1, x_2). To see why, we look at the point $(2, -1)$. The solution curve in the x_1-x_2 plane is given by the points $(x_1(t), x_2(t))$, $t \geq 0$. The slope of the tangent line at a point (x_1, x_2) is dx_2/dx_1. At the point $(2, -1)$, we find $\frac{dx_1}{dt} = x_1 - 2x_2 = 4$ and $\frac{dx_2}{dt} = x_2 = -1$. Therefore, when $(x_1, x_2) = (2, -1)$

$$\frac{dx_2}{dx_1} = \frac{dx_2/dt}{dx_1/dt} = \frac{x_2}{x_1 - 2x_2} = \frac{-1}{4}$$

which implies that the slope of the tangent line of the solution curve at this point is equal to $-1/4$. If we attach the vector $[dx_1/dt, dx_2/dt]' = [4, -1]'$ at the point $(2, -1)$, then this vector points in the direction of the tangent line with slope $-1/4$, and is therefore tangent to the solution curve (see Figure 11.1).

More generally, at each point (x_1, x_2) in the x_1-x_2 plane, we can define a vector $\left[\frac{dx_1}{dt}, \frac{dx_2}{dt}\right]'$ that is tangential to the solution curve that passes through (x_1, x_2), and that points in the direction of the solution. In our example, the vector is of the form $[x_1 - 2x_2, x_2]'$, and the slope of the solution curve that goes through (x_1, x_2) is

$$\frac{dx_2}{dx_1} = \frac{dx_2/dt}{dx_1/dt} = \frac{x_2}{x_1 - 2x_2}$$

The collection of these vectors is called a **direction** or **slope field**, and each vector of the direction field is called a **direction vector**. Since (11.5) is an autonomous

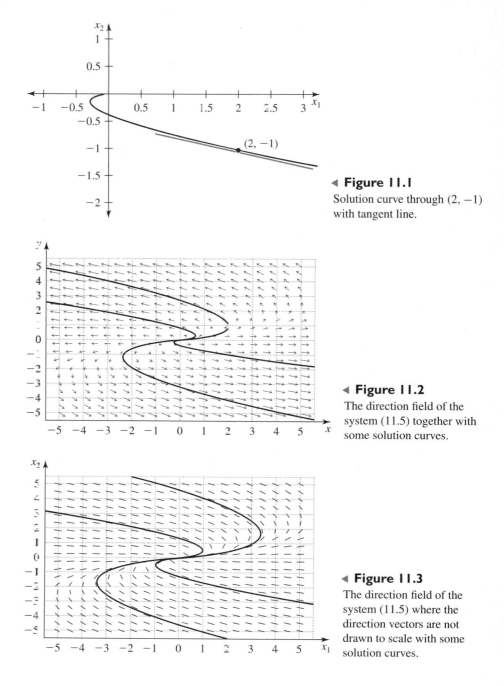

◀ **Figure 11.1**
Solution curve through $(2, -1)$ with tangent line.

◀ **Figure 11.2**
The direction field of the system (11.5) together with some solution curves.

◀ **Figure 11.3**
The direction field of the system (11.5) where the direction vectors are not drawn to scale with some solution curves.

system, the direction vector only depends on the location of the point (x_1, x_2) but not on t. This implies that the direction field is the same for all times t.

The direction field for (11.5) is shown in Figure 11.2 (which also contains four solution curves). The length of the direction vector at a given point tells us how quickly the solution curve passes through the point; it is equal to $\sqrt{(dx_1/dt)^2 + (dx_2/dt)^2}$. If we are only interested in the direction of the solution curves, we can indicate the direction by small line segments (as shown in Figure 11.3), which results in a less cluttered picture. Figure 11.3 also contains four solution curves; note that the direction vectors are always tangential to the solution curves. We can therefore use the direction field to sketch solution curves by drawing curves in such a way that the direction vectors are always tangential to the curve.

The point $(0, 0)$ is special: When we compute the direction vector at $(0, 0)$, we find

$$\begin{bmatrix} \frac{dx_1}{dt} \\ \frac{dx_2}{dt} \end{bmatrix} = \begin{bmatrix} 0 \\ 0 \end{bmatrix}$$

that is, if we start at this point, neither $x_1(t)$ nor $x_2(t)$ will change. We call such points equilibria. We will discuss the significance of these points in Subsection 11.1.3. They play a central role in the long-term behavior of solutions, that is, when $t \to \infty$.

11.1.2 Solving Linear Systems

Specific Solutions   Consider the following differential equation:

$$\frac{dx}{dt} = ax$$

This is a linear, first order differential equation with a constant coefficient (that is, it is of the form (11.4) with $n = 1$). We can find its solution simply by integrating after separating variables (as we learned in Chapter 7). The solution is of the form

$$x(t) = ce^{at}$$

where c is a constant that depends on the initial condition. We will now show that exponential functions are also solutions of systems of linear differential equations. We restrict our discussion to systems of two differential equations. Consider the system

$$\frac{dx_1}{dt} = a_{11}x_1(t) + a_{12}x_2(t) \tag{11.6}$$

$$\frac{dx_2}{dt} = a_{21}x_1(t) + a_{22}x_2(t) \tag{11.7}$$

which in matrix form is written as

$$\frac{d\mathbf{x}}{dt} = A\mathbf{x}(t) \tag{11.8}$$

The solution of (11.8) is a vector-valued function. We claim that the following vector-valued function

$$\mathbf{x}(t) = \begin{bmatrix} u_1 e^{\lambda t} \\ u_2 e^{\lambda t} \end{bmatrix} = e^{\lambda t} \begin{bmatrix} u_1 \\ u_2 \end{bmatrix} \tag{11.9}$$

where λ, u_1 and u_2 are constants, is a solution of (11.8) for an appropriate choice of values for λ, u_1, and u_2. To see how we must choose λ, u_1, and u_2, we differentiate $\mathbf{x}(t)$ in (11.9). This yields

$$\frac{d\mathbf{x}}{dt} = \begin{bmatrix} u_1 \lambda e^{\lambda t} \\ u_2 \lambda e^{\lambda t} \end{bmatrix} = \lambda e^{\lambda t} \begin{bmatrix} u_1 \\ u_2 \end{bmatrix} \tag{11.10}$$

If $\mathbf{x}(t)$ solves (11.8), then $\mathbf{x}(t)$ must satisfy (11.8). Using (11.10) for the left-hand side and (11.9) for the right-hand side of (11.8), we find

$$\underbrace{\lambda e^{\lambda t} \begin{bmatrix} u_1 \\ u_2 \end{bmatrix}}_{\frac{d\mathbf{x}(t)}{dt}} = \underbrace{A e^{\lambda t} \begin{bmatrix} u_1 \\ u_2 \end{bmatrix}}_{A\mathbf{x}(t)}$$

or, after dividing both sides by $e^{\lambda t}$,

$$\lambda \begin{bmatrix} u_1 \\ u_2 \end{bmatrix} = A \begin{bmatrix} u_1 \\ u_2 \end{bmatrix}$$

This last expression should remind you of eigenvalues and eigenvectors that we encountered in Section 9.4. Namely, if $\begin{bmatrix} u_1 \\ u_2 \end{bmatrix}$ is an eigenvector corresponding to an eigenvalue λ of A, then (11.9) is a solution of (11.8).

In this subsection, we will only look at differential equations of the form (11.8) for which the eigenvalues of A are both real and distinct. We will discuss complex eigenvalues in the next subsection (we will not discuss the case when both eigenvalues are identical). We will illustrate on the following system how to find specific solutions of a system of the form (11.8) in the case when both eigenvalues of A are real and distinct.

$$\frac{dx_1}{dt} = 2x_1 - 2x_2$$
$$\frac{dx_2}{dt} = 2x_1 - 3x_2$$

(11.11)

1. Finding Eigenvalues   The coefficient matrix of (11.11) is given by

$$A = \begin{bmatrix} 2 & -2 \\ 2 & -3 \end{bmatrix}$$

We will first determine the eigenvalues and corresponding eigenvectors of A. To find the eigenvalues of A, we must solve

$$\det(A - \lambda I) = 0$$

That is,

$$\det \begin{bmatrix} 2 - \lambda & -2 \\ 2 & -3 - \lambda \end{bmatrix} = (2 - \lambda)(-3 - \lambda) + 4$$

$$= \lambda^2 + \lambda - 2 = (\lambda - 1)(\lambda + 2) = 0$$

which has solutions

$$\lambda_1 = 1 \qquad \text{and} \qquad \lambda_2 = -2$$

2. A solution corresponding to the eigenvalue λ_1   To find the eigenvector $\mathbf{u} = \begin{bmatrix} u_1 \\ u_2 \end{bmatrix}$ corresponding to $\lambda_1 = 1$, we solve

$$A \begin{bmatrix} u_1 \\ u_2 \end{bmatrix} = \lambda_1 \begin{bmatrix} u_1 \\ u_2 \end{bmatrix} \qquad \text{with } \lambda_1 = 1$$

We find

$$\begin{bmatrix} 2u_1 - 2u_2 \\ 2u_1 - 3u_2 \end{bmatrix} = \begin{bmatrix} u_1 \\ u_2 \end{bmatrix}$$

which can be written as

$$u_1 - 2u_2 = 0$$
$$2u_1 - 4u_2 = 0$$

Both equations reduce to the same equation, namely

$$u_1 = 2u_2$$

Setting $u_2 = 1$, for instance, we find $u_1 = 2$. An eigenvector corresponding to $\lambda_1 = 1$ is then

$$\mathbf{u} = \begin{bmatrix} 2 \\ 1 \end{bmatrix}$$

We claim that

$$\mathbf{x}(t) = e^t \begin{bmatrix} 2 \\ 1 \end{bmatrix}$$

solves (11.11). Let's check. We find

$$\frac{d\mathbf{x}}{dt} = e^t \begin{bmatrix} 2 \\ 1 \end{bmatrix} = \mathbf{x}(t)$$

Since $\begin{bmatrix} 2 \\ 1 \end{bmatrix}$ is an eigenvector corresponding to $\lambda_1 = 1$, $\mathbf{x}(t)$ also satisfies

$$A\mathbf{x}(t) = Ae^t \begin{bmatrix} 2 \\ 1 \end{bmatrix} = e^t A \begin{bmatrix} 2 \\ 1 \end{bmatrix} = e^t \begin{bmatrix} 2 \\ 1 \end{bmatrix} = \mathbf{x}(t)$$

where we used the fact that

$$A \begin{bmatrix} 2 \\ 1 \end{bmatrix} = \lambda_1 \begin{bmatrix} 2 \\ 1 \end{bmatrix} = \begin{bmatrix} 2 \\ 1 \end{bmatrix}$$

since $\begin{bmatrix} 2 \\ 1 \end{bmatrix}$ is an eigenvector to $\lambda_1 = 1$. Therefore,

$$\frac{d\mathbf{x}}{dt} = A\mathbf{x}(t)$$

and we conclude that $\mathbf{x}(t) = e^t \begin{bmatrix} 2 \\ 1 \end{bmatrix}$ is indeed a solution of (11.11).

3. A solution corresponding to the eigenvalue λ_2   We can now repeat the same steps for the eigenvalue $\lambda_2 = -2$. To find the eigenvector $\mathbf{v} = \begin{bmatrix} v_1 \\ v_2 \end{bmatrix}$ corresponding to the eigenvalue $\lambda_2 = -2$, we solve

$$\begin{bmatrix} 2v_1 - 2v_2 \\ 2v_1 - 3v_2 \end{bmatrix} = -2 \begin{bmatrix} v_1 \\ v_2 \end{bmatrix}$$

which can be written as

$$4v_1 - 2v_2 = 0$$
$$2v_1 - v_2 = 0$$

The two equations reduce to the same equation, namely,

$$2v_1 = v_2$$

Setting $v_1 = 1$, we find $v_2 = 2$. An eigenvector corresponding to $\lambda_2 = -2$ is then

$$\mathbf{v} = \begin{bmatrix} 1 \\ 2 \end{bmatrix}$$

We set

$$\mathbf{x}(t) = e^{-2t} \begin{bmatrix} 1 \\ 2 \end{bmatrix}$$

and check that

$$\frac{d\mathbf{x}}{dt} = -2e^{-2t} \begin{bmatrix} 1 \\ 2 \end{bmatrix} = -2\mathbf{x}(t)$$

and

$$A\mathbf{x}(t) = Ae^{-2t} \begin{bmatrix} 1 \\ 2 \end{bmatrix} = e^{-2t}A \begin{bmatrix} 1 \\ 2 \end{bmatrix} = -2e^{-2t} \begin{bmatrix} 1 \\ 2 \end{bmatrix} = -2\mathbf{x}(t)$$

where we used the fact that

$$A \begin{bmatrix} 1 \\ 2 \end{bmatrix} = \lambda_2 \begin{bmatrix} 1 \\ 2 \end{bmatrix} = -2 \begin{bmatrix} 1 \\ 2 \end{bmatrix}$$

since $\begin{bmatrix} 1 \\ 2 \end{bmatrix}$ is an eigenvector to $\lambda_2 = -2$. This shows that $\mathbf{x}(t) = e^{-2t} \begin{bmatrix} 1 \\ 2 \end{bmatrix}$ is also a solution of (11.11).

4. The Direction Field   We wish to illustrate these two particular solutions in the corresponding direction field. Recall that an eigenvector can be used to construct a straight line through the origin in the direction of the eigenvector. In Figure 11.4, we show the direction field together with the two lines in the direction of the eigenvectors.

If $\mathbf{x}(0)$ is a point on one of the lines defined by the eigenvectors, then the solution $\mathbf{x}(t)$ will remain on that line for *all* later times. The location on the line will change with time. If the corresponding eigenvalue is positive, the solution will move away from the origin, if the eigenvalue is negative, it will move towards the origin, as can be seen from the direction of the direction vectors.

The General Solution   Suppose that $\mathbf{y}(t)$ and $\mathbf{z}(t)$ are two solutions of

$$\frac{d\mathbf{x}}{dt} = A\mathbf{x}(t) \tag{11.12}$$

that is,

$$\frac{d\mathbf{y}}{dt} = A\mathbf{y}(t) \qquad \text{and} \qquad \frac{d\mathbf{z}}{dt} = A\mathbf{z}(t)$$

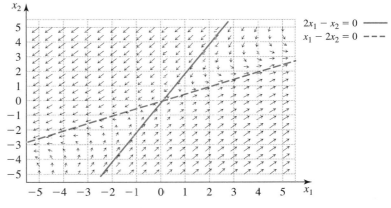

▲ **Figure 11.4**
The direction field of the system (11.11) with the lines in the direction of the eigenvectors.

Then the linear combination

$$c_1\mathbf{y}(t) + c_2\mathbf{z}(t) \tag{11.13}$$

also solves (11.12). This is not hard to see; first, note that

$$\frac{d}{dt}[c_1\mathbf{y}(t) + c_2\mathbf{z}(t)] = c_1\frac{d\mathbf{y}}{dt} + c_2\frac{d\mathbf{z}}{dt}$$

But since $\mathbf{y}(t)$ and $\mathbf{z}(t)$ are both solutions of (11.12),

$$c_1\frac{d\mathbf{y}}{dt} + c_2\frac{d\mathbf{z}}{dt} = c_1 A\mathbf{y}(t) + c_2 A\mathbf{z}(t) \tag{11.14}$$

Using the linearity properties of A, we can write the right-hand side of (11.14) as

$$A\,[c_1\mathbf{y}(t) + c_2\mathbf{z}(t)]$$

Summarizing this, we find

$$\frac{d}{dt}\underbrace{[c_1\mathbf{y}(t) + c_2\mathbf{z}(t)]}_{\mathbf{x}(t)} = A\underbrace{[c_1\mathbf{y}(t) + c_2\mathbf{z}(t)]}_{\mathbf{x}(t)}$$

which shows that the linear combination $c_1\mathbf{y}(t) + c_2\mathbf{z}(t)$ is also a solution.

Combining solutions, as in (11.13), illustrates the **superposition principle**. Since this is an important principle, we state it in the following.

SUPERPOSITION PRINCIPLE

Suppose that

$$\begin{bmatrix} \frac{dx_1}{dt} \\ \frac{dx_2}{dt} \end{bmatrix} = \begin{bmatrix} a_{11} & a_{12} \\ a_{21} & a_{22} \end{bmatrix} \begin{bmatrix} x_1(t) \\ x_2(t) \end{bmatrix} \tag{11.15}$$

If

$$\mathbf{y}(t) = \begin{bmatrix} y_1(t) \\ y_2(t) \end{bmatrix} \quad \text{and} \quad \mathbf{z}(t) = \begin{bmatrix} z_1(t) \\ z_2(t) \end{bmatrix}$$

are solutions of (11.15), then

$$\mathbf{x}(t) = c_1\mathbf{y}(t) + c_2\mathbf{z}(t)$$

is also a solution of (11.15).

We saw that $e^{\lambda t}\begin{bmatrix} w_1 \\ w_2 \end{bmatrix}$, where $\begin{bmatrix} w_1 \\ w_2 \end{bmatrix}$ is an eigenvector corresponding to the real eigenvalue λ of A, solves (11.12). If we have two real and distinct eigenvalues λ_1 and λ_2 with corresponding eigenvectors $\mathbf{u}$ and $\mathbf{v}$, then setting $\mathbf{y}(t) = e^{\lambda_1 t}\mathbf{u}$ and $\mathbf{z}(t) = e^{\lambda_2 t}\mathbf{v}$ and using the superposition principle, the following linear combination

$$\mathbf{x}(t) = c_1 e^{\lambda_1 t}\mathbf{u} + c_2 e^{\lambda_2 t}\mathbf{v} \tag{11.16}$$

where c_1 and c_2 are constants, is also a solution of (11.12). The constants c_1 and c_2 depend on the initial conditions. One can show that every solution can be written in the form (11.16); we therefore call a solution of the form (11.16) the general solution. (The situation is more complicated when A has repeated eigenvalues, that is, when $\lambda_1 = \lambda_2$. We do not give the general solution for that case here, but will discuss two such examples in Problems 25 and 26.) We summarize our findings in the following.

THE GENERAL SOLUTION

Let

$$\frac{d\mathbf{x}}{dt} = A\mathbf{x}(t) \tag{11.17}$$

where A is a 2×2 matrix with two real and distinct eigenvalues λ_1 and λ_2, and corresponding eigenvectors $\mathbf{u}$ and $\mathbf{v}$. Then

$$\mathbf{x}(t) = c_1 e^{\lambda_1 t}\mathbf{u} + c_2 e^{\lambda_2 t}\mathbf{v} \tag{11.18}$$

is the general solution of (11.17). The constants c_1 and c_2 depend on the initial condition.

We will now check that (11.18) is indeed a solution of (11.17). To do so, we differentiate (11.18) with respect to t, and find

$$\frac{d\mathbf{x}}{dt} = \lambda_1 c_1 e^{\lambda_1 t}\mathbf{u} - \lambda_2 c_2 e^{\lambda_2 t}\mathbf{v}$$

Since $\mathbf{u}$ and $\mathbf{v}$ are eigenvectors corresponding to λ_1 and λ_2, we also find

$$
\begin{aligned}
A\mathbf{x}(t) &= A(c_1 e^{\lambda_1 t}\mathbf{u} + c_2 e^{\lambda_2 t}\mathbf{v}) \\
&= c_1 e^{\lambda_1 t} A\mathbf{u} + c_2 e^{\lambda_2 t} A\mathbf{v} \\
&= c_1 e^{\lambda_1 t}\lambda_1\mathbf{u} + c_2 e^{\lambda_2 t}\lambda_2\mathbf{v}
\end{aligned}
$$

hence,

$$\frac{d\mathbf{x}}{dt} = A\mathbf{x}(t)$$

The general solution of (11.11)

$$\frac{d\mathbf{x}}{dt} = \begin{bmatrix} 2 & -2 \\ 2 & -3 \end{bmatrix} \mathbf{x}(t)$$

is

$$\mathbf{x}(t) = c_1 e^{t}\begin{bmatrix} 2 \\ 1 \end{bmatrix} + c_2 e^{-2t}\begin{bmatrix} 1 \\ 2 \end{bmatrix} \tag{11.19}$$

5. An initial condition for (11.11)   Suppose we know that at time 0,

$$\mathbf{x}(0) = \begin{bmatrix} -1 \\ 4 \end{bmatrix} \tag{11.20}$$

holds for (11.11). We can then determine the constants c_1 and c_2 in (11.19) so that $\mathbf{x}(t)$ satisfies the initial condition (11.20),

$$\mathbf{x}(0) = c_1\begin{bmatrix} 2 \\ 1 \end{bmatrix} + c_2\begin{bmatrix} 1 \\ 2 \end{bmatrix} = \begin{bmatrix} -1 \\ 4 \end{bmatrix}$$

To find c_1 and c_2, we must solve the system of linear equations

$$
\begin{aligned}
2c_1 + c_2 &= -1 \\
c_1 + 2c_2 &= 4
\end{aligned}
$$

Eliminating c_1 in the second equation yields

$$
\begin{aligned}
2c_1 + c_2 &= -1 \\
3c_2 &= 9
\end{aligned}
$$

hence, $c_2 = 3$ and, therefore,

$$2c_1 = -1 - c_2 = -1 - 3 = -4 \qquad \text{or} \qquad c_1 = -2$$

Therefore,

$$\mathbf{x}(t) = -2e^t \begin{bmatrix} 2 \\ 1 \end{bmatrix} + 3e^{-2t} \begin{bmatrix} 1 \\ 2 \end{bmatrix}$$

or, writing it out as separate equations,

$$x_1(t) = -4e^t + 3e^{-2t}$$
$$x_2(t) = -2e^t + 6e^{-2t}$$

solves the system (11.11), namely,

$$\frac{dx_1}{dt} = 2x_1 - 2x_2$$

$$\frac{dx_2}{dt} = 2x_2 - 3x_1$$

with initial condition $x_1(0) = -1$ and $x_2(0) = 4$.

We give one more example of an initial value problem that illustrates all at once the different steps that must be carried out in order to obtain a solution in the case when A has two distinct and real eigenvalues.

▌**EXAMPLE 2**   Solve

$$\frac{d\mathbf{x}}{dt} = \begin{bmatrix} 2 & -3 \\ 1 & -2 \end{bmatrix} \mathbf{x}(t) \tag{11.21}$$

with the initial condition

$$\mathbf{x}(0) = \begin{bmatrix} 3 \\ -1 \end{bmatrix} \tag{11.22}$$

Solution   The first step is to find the eigenvalues and the corresponding eigenvectors. To find the eigenvalues, we must solve

$$\det(A - \lambda I) = \det \begin{bmatrix} 2 - \lambda & -3 \\ 1 & -2 - \lambda \end{bmatrix} = (2 - \lambda)(-2 - \lambda) + 3$$

$$= \lambda^2 - 1 = 0$$

which has solutions

$$\lambda_1 = 1 \qquad \text{and} \qquad \lambda_2 = -1$$

The eigenvector $\mathbf{u}$ corresponding to the eigenvalue $\lambda_1 = 1$ satisfies

$$\begin{bmatrix} 2 & -3 \\ 1 & -2 \end{bmatrix} \begin{bmatrix} u_1 \\ u_2 \end{bmatrix} = \begin{bmatrix} u_1 \\ u_2 \end{bmatrix}$$

Rewriting this as a system of equations, we find

$$2u_1 - 3u_2 = u_1$$
$$u_1 - 2u_2 = u_2$$

These two equations reduce to the same equation, namely,

$$u_1 - 3u_2 = 0$$

If we set $u_2 = 1$, then $u_1 = 3$ and, therefore,

$$\mathbf{u} = \begin{bmatrix} 3 \\ 1 \end{bmatrix}$$

is an eigenvector corresponding to $\lambda_1 = 1$.

The eigenvector $\mathbf{v}$ corresponding to the eigenvalue $\lambda_2 = -1$ satisfies

$$\begin{bmatrix} 2 & -3 \\ 1 & -2 \end{bmatrix} \begin{bmatrix} v_1 \\ v_2 \end{bmatrix} = - \begin{bmatrix} v_1 \\ v_2 \end{bmatrix}$$

Rewriting this as a system of equations, we find

$$2v_1 - 3v_2 = -v_1$$
$$v_1 - 2v_2 = -v_2$$

These two equations reduce to the same equation, namely,

$$v_1 - v_2 = 0$$

If we set $v_1 = 1$, then $v_2 = 1$ and, therefore,

$$\mathbf{v} = \begin{bmatrix} 1 \\ 1 \end{bmatrix}$$

is an eigenvector corresponding to $\lambda_2 = -1$. The general solution of (11.21) is therefore

$$\mathbf{x}(t) = c_1 e^t \begin{bmatrix} 3 \\ 1 \end{bmatrix} + c_2 e^{-t} \begin{bmatrix} 1 \\ 1 \end{bmatrix}$$

where c_1 and c_2 are constants. The initial condition (11.22) will allow us to determine the constants c_1 and c_2, namely,

$$\mathbf{x}(0) = c_1 \begin{bmatrix} 3 \\ 1 \end{bmatrix} + c_2 \begin{bmatrix} 1 \\ 1 \end{bmatrix} = \begin{bmatrix} 3 \\ -1 \end{bmatrix}$$

That is, c_1 and c_2 satisfy

$$3c_1 + c_2 = 3$$
$$c_1 + c_2 = -1$$

We solve this system using the standard elimination method,

$$3c_1 + c_2 = 3$$
$$2c_1 = 4$$

Hence, $c_1 = 2$ and $c_2 = 3 - 3c_1 = 3 - 6 = -3$. The solution of (11.21) that satisfies the initial condition (11.22) is therefore

$$\mathbf{x}(t) = 2e^t \begin{bmatrix} 3 \\ 1 \end{bmatrix} - 3e^{-t} \begin{bmatrix} 1 \\ 1 \end{bmatrix}$$

which can also be written in the form

$$x_1(t) = 6e^t - 3e^{-t}$$
$$x_2(t) = 2e^t - 3e^{-t}$$

The direction field, with two lines in the direction of the two eigenvectors and the solution, is shown in Figure 11.5. ▮

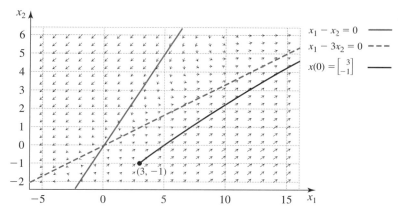

▲ **Figure 11.5**
The direction field of the system (11.21) with the lines in the direction of the eigenvectors, and the solution with initial condition $\begin{bmatrix} 3 \\ -1 \end{bmatrix}$.

You might have noticed that all initial conditions have been formulated at time $t = 0$. This is a natural choice for an initial condition; however, we could have chosen any other time, for instance, $t = 1$. Suppose that in Example 2, the initial condition had been

$$\mathbf{x}(1) = \begin{bmatrix} 2 \\ 1 \end{bmatrix} \tag{11.23}$$

The general solution would still be

$$\mathbf{x}(t) = c_1 e^t \begin{bmatrix} 3 \\ 1 \end{bmatrix} + c_2 e^{-t} \begin{bmatrix} 1 \\ 1 \end{bmatrix}$$

but the constants c_1 and c_2 would satisfy

$$3e c_1 + e^{-1} c_2 = 2$$
$$e c_1 + e^{-1} c_2 = 1$$

We solve this system using the standard elimination method,

$$3e c_1 + e^{-1} c_2 = 2$$
$$2e c_1 = 1$$

Hence, $c_1 = \frac{1}{2e}$ and $c_2 = \frac{e}{2}$. The solution satisfying (11.23) is then

$$\mathbf{x}(t) = \frac{1}{2} e^{t-1} \begin{bmatrix} 3 \\ 1 \end{bmatrix} + \frac{1}{2} e^{1-t} \begin{bmatrix} 1 \\ 1 \end{bmatrix}$$

11.1.3 Equilibria and Stability

In this subsection, we will be concerned with equilibria and stability, two concepts that we encountered in Section 7.2, where we discussed ordinary differential equations. Both concepts can be extended to systems of differential equations. We will restrict ourselves to the case

$$\frac{d\mathbf{x}}{dt} = A\mathbf{x}(t) \tag{11.24}$$

with

$$A = \begin{bmatrix} a_{11} & a_{12} \\ a_{21} & a_{22} \end{bmatrix} \qquad \text{and} \qquad \mathbf{x}(t) = \begin{bmatrix} x_1(t) \\ x_2(t) \end{bmatrix} \tag{11.25}$$

We say that a point

$$\hat{\mathbf{x}} = (\hat{x}_-, \hat{x}_2)$$

is an equilibrium of (11.24) if

$$A\hat{\mathbf{x}} = \mathbf{0}$$

that is, a point where the direction vector in the corresponding direction field has length 0. If we start a system of differential equations at an equilibrium point, it will remain there for all later times.

To find equilibria of (11.24), we must solve $A\mathbf{x} = \mathbf{0}$. We see immediately that $\hat{\mathbf{x}} = \mathbf{0}$ solves $A\mathbf{x} = \mathbf{0}$. It follows from results in Subsection 9.2.3 that if $\det A \neq 0$, then $(0, 0)$ is the only equilibrium of (11.24). If $\det A = 0$, then there will be other equilibria.

The characterizing property of an equilibrium is that if we start a system in equilibrium, it will stay there for all future times. This does *not* mean that if the system is in equilibrium and is **perturbed** by a small amount (that is, the solution is moved to a nearby point), it will return to the equilibrium. Whether or not a solution will return to an equilibrium after a small perturbation is addressed by the **stability** cf the equilibrium. We saw in the previous subsection that, in the case when A has two real and distinct eigenvalues, the solution of (11.24) is given by

$$\mathbf{x}(t) = c_1 e^{\lambda_1 t}\mathbf{u} + c_2 e^{\lambda_2 t}\mathbf{v} \tag{11.26}$$

where $\mathbf{u}$ and $\mathbf{v}$ are the eigenvectors corresponding to the eigenvalues λ_1 and λ_2 of the matrix A, and the constants c_1 and c_2 depend on the initial condition. Knowing the solution will allow us to study the behavior of the solution (11.26) as $t \to \infty$, and thus address the question of stability, at least when the eigenvalues are real.

Since A is a 2×2 matrix and all entries of A are real, the eigenvalues of A are either both real or both complex conjugate. We will treat these two cases separately. To simplify our discussion, we assume that A has two distinct eigenvalues.

The equilibria of (11.24) can be found by solving

$$A\mathbf{x} = \mathbf{0} \tag{11.27}$$

If $\det A \neq 0$, then (11.27) has only one solution, namely the trivial solution $(0, 0)$ (see Subsection 9.2.3). Since $\det A = \lambda_1 \lambda_2$, where λ_1 and λ_2 are the eigenvalues of A, we see that if λ_1 and λ_2 are both nonzero, then $\det A \neq 0$.

Case 1: A has two distinct, real and nonzero eigenvalues In this case, the equation $A\mathbf{x} = \mathbf{0}$ has only one solution, namely $(0, 0)$. That is, $(0, 0)$ is the only equilibrium. The general solution of (11.24) is given by (11.26), and we can therefore study the behavior of the solution of (11.24) directly by investigating (11.26). We are interested in determining

$$\lim_{t \to \infty} \mathbf{x}(t) = \lim_{t \to \infty} \left[c_1 e^{\lambda_1 t}\mathbf{u} + c_2 e^{\lambda_2 t}\mathbf{v} \right] \tag{11.28}$$

We see immediately that the behavior of $\mathbf{x}(t)$ is determined by $e^{\lambda_1 t}$ and $e^{\lambda_2 t}$. Recall that

$$\lim_{t \to \infty} e^{\lambda_i t} = \begin{cases} 0 & \text{if } \lambda_i < 0 \\ \infty & \text{if } \lambda_i > 0 \end{cases}$$

We distinguish the following three categories:

1. Both eigenvalues are negative: $\lambda_1 < 0$ and $\lambda_2 < 0$.
2. The eigenvalues are of opposite signs: $\lambda_1 < 0$ and $\lambda_2 > 0$ (or $\lambda_1 > 0$ and $\lambda_2 < 0$).
3. Both eigenvalues are positive: $\lambda_1 > 0$ and $\lambda_2 > 0$.

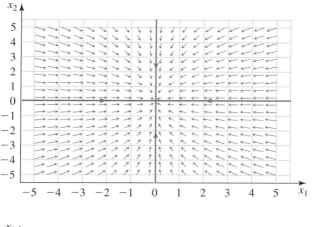

◀ **Figure 11.6**
The direction field of a linear system where both eigenvalues are negative, together with the lines in the direction of the eigenvectors.

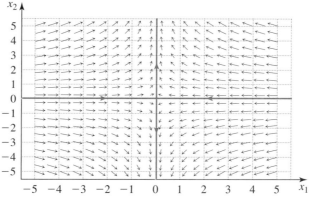

◀ **Figure 11.7**
The direction field of a linear system where both eigenvalues are of opposite sign, together with the lines in the direction of the eigenvectors.

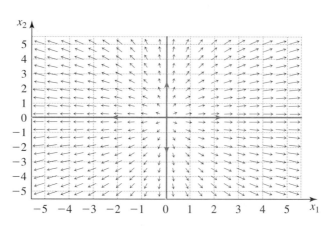

◀ **Figure 11.8**
The direction field of a linear system where both eigenvalues are positive, together with the lines in the direction of the eigenvectors.

Typical direction fields for each of the three categories are shown in Figures 11.6–11.8. We will discuss each category separately.

Category 1   When both eigenvalues are negative, we conclude from (11.28) that

$$\lim_{t \to \infty} \mathbf{x}(t) = \mathbf{0}$$

regardless of $(x_1(0), x_2(0))$. We say that the equilibrium $(0, 0)$ is **globally stable** since the solution will approach the equilibrium $(0, 0)$ regardless of the starting point (and not just from nearby points). We call $(0, 0)$ a **sink**, or a **stable node**. A direction field for this case is shown in Figure 11.6; the shape of the direction field explains why we call this equilibrium a sink.

The system of differential equations that gave rise to this direction field is

$$\frac{d\mathbf{x}}{dt} = A\mathbf{x}(t) \quad \text{with } A = \begin{bmatrix} -2 & 0 \\ 0 & -1 \end{bmatrix}$$

The eigenvalues of A are $\lambda_1 = -2$ and $\lambda_2 = -1$. Both are negative. The two straight lines in Figure 11.6 are the lines in the directions of the two eigenvectors. We see that, when starting at a point on either straight line, the solution will approach the equilibrium $(0, 0)$ along the straight line. Furthermore, we see from the direction field that, starting from any other point, the solution will approach the equilibrium $(0, 0)$, as we concluded from the general solution and the fact that both eigenvalues are negative.

Category 2   When the eigenvalues are of opposite signs, we see from (11.28) that the component of the solution associated with the negative eigenvalue goes to 0 as $t \to \infty$, and the component associated with the positive eigenvalue goes to infinity. That is, unless we start in the direction of the eigenvector associated with the negative eigenvalue, the solution will not converge to the equilibrium $(0, 0)$. We say that the equilibrium $(0, 0)$ is **unstable**, and call $(0, 0)$ a **saddle**. A direction field for this case is shown in Figure 11.7; the shape of the direction field explains why we call this equilibrium a saddle.

The system of differential equations that gave rise to this direction field is

$$\frac{d\mathbf{x}}{dt} = A\mathbf{x}(t) \quad \text{with } A = \begin{bmatrix} -2 & 0 \\ 0 & 1 \end{bmatrix}$$

The eigenvalues of A are $\lambda_1 = -2$ and $\lambda_2 = 1$. The two straight lines in Figure 11.7 are the lines in the directions of the two eigenvectors. We see that, when starting at a point on the straight line corresponding to the negative eigenvalue, the solution will approach the equilibrium $(0, 0)$ along the straight line. When starting at a point on the straight line corresponding to the positive eigenvalue, the solution will move away from the equilibrium $(0, 0)$ along the straight line. Furthermore, we see from the direction field that, starting from any other point, the solution will eventually move away from the equilibrium $(0, 0)$.

In summary, we see from the direction field that the solution can approach $(0, 0)$ from only one direction (namely, the direction of the eigenvector associated with the negative eigenvalue); it gets pushed away from $(0, 0)$ elsewhere.

Category 3   Finally, if both eigenvalues are positive, we see from (11.28) that the solution will not converge to $(0, 0)$ unless we start in $(0, 0)$. We say that the equilibrium $(0, 0)$ is unstable, and we call $(0, 0)$ a **source**, or an **unstable node**. A direction field for this case is shown in Figure 11.8; the shape of the direction field explains why we call this equilibrium a source.

The system of differential equations that gave rise to this direction field is

$$\frac{d\mathbf{x}}{dt} = A\mathbf{x}(t) \quad \text{with } A = \begin{bmatrix} 2 & 0 \\ 0 & 1 \end{bmatrix}$$

The eigenvalues of A are $\lambda_1 = 2$ and $\lambda_2 = 1$, which are both positive. The two straight lines in Figure 11.8 are the lines in the directions of the two eigenvectors. We see that, when starting at a point on either straight line, the solution will move away from the equilibrium $(0, 0)$ along the straight line. Furthermore, we see from the direction field that, starting from any other point, the solution will move away from the equilibrium $(0, 0)$.

Case 2: A has complex conjugate eigenvalues   We will not solve the system when A has complex conjugate eigenvalues. Instead, we will look at some examples to see what typical direction fields and their solutions look like. Let

$$A = \begin{bmatrix} -1 & -1 \\ 1 & 0 \end{bmatrix} \tag{11.29}$$

then,

$$\det(A - \lambda I) = \det \begin{bmatrix} -1 - \lambda & -1 \\ 1 & -\lambda \end{bmatrix} = (-1 - \lambda)(-\lambda) + 1$$

$$= \lambda^2 + \lambda + 1 = 0$$

which gives

$$\lambda_{1,2} = \frac{-1 \pm \sqrt{1 - 4}}{2} = -\frac{1}{2} \pm \frac{i}{2}\sqrt{3}$$

That is, both eigenvalues are complex and form a conjugate pair. (Note that the real parts of both eigenvalues are negative.) To see what the solutions of

$$\frac{d\mathbf{x}}{dt} = A\mathbf{x}(t)$$

look like, we graph the direction field and solutions $x_1(t)$ and $x_2(t)$ in Figures 11.9 and 11.10, respectively.

We see from the direction field that, starting from any point other than $(0, 0)$, solutions spiral into the equilibrium $(0, 0)$. For this reason, the equilibrium $(0, 0)$ is called a **stable spiral**. When we plot solutions as functions of time, they show oscillations, as illustrated in Figure 11.10. The amplitude of the oscillations decreases over time. We therefore call the oscillations **damped**.

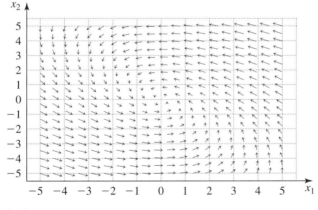

◀ **Figure 11.9**
The direction field for the system with matrix A in (11.29).

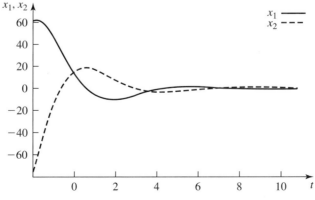

◀ **Figure 11.10**
The solutions for (11.29).

The oscillations are caused by the imaginary part of the eigenvalues; the damping of the oscillations is caused by the negative real part of the eigenvalues. Before we explain this further, we give two more examples, one in which the complex conjugate eigenvalues have positive real parts, the other in which the eigenvalues are purely imaginary.

Let

$$B = \begin{bmatrix} 1 & -1 \\ 1 & 0 \end{bmatrix}$$

then

$$\det(B - \lambda I) = \det \begin{bmatrix} 1 - \lambda & -1 \\ 1 & -\lambda \end{bmatrix} = (1 - \lambda)(-\lambda) + 1$$

$$= \lambda^2 - \lambda + 1 = 0$$

which gives

$$\lambda_{1,2} = \frac{1 \pm \sqrt{1 - 4}}{2} = \frac{1}{2} \pm \frac{i}{2}\sqrt{3}$$

That is, both eigenvalues of B are complex and form a complex conjugate pair, but their real parts are positive. To see what the solutions of

$$\frac{d\mathbf{x}}{dt} = B\mathbf{x}(t) \tag{11.30}$$

look like, we graph the direction field and solutions $x_1(t)$ and $x_2(t)$ in Figures 11.11 and 11.12, respectively.

We see from the direction field that, starting from any point other than $(0, 0)$, the solutions spiral out from the equilibrium $(0, 0)$. For this reason, the equilibrium

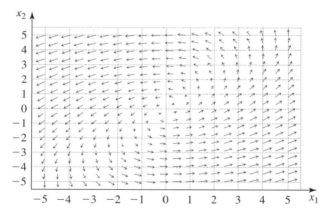

◀ **Figure 11.11**

The direction field for the system with matrix B in (11.30).

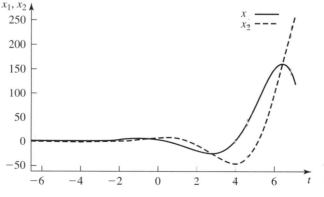

◀ **Figure 11.12**

The solutions for (11.30).

$(0, 0)$ is a **unstable spiral**. When we plot solutions as functions of time, as in Figure 11.12, we see that the solutions show oscillations as before, but this time their amplitudes are increasing. The oscillations are again caused by the imaginary parts of the eigenvalues; the increase in amplitude is caused by the positive real parts.

Finally, we look at an instance where both eigenvalues are purely imaginary. Let

$$C = \begin{bmatrix} 0 & -1 \\ 1 & 0 \end{bmatrix}$$

then

$$\det(C - \lambda I) = \det \begin{bmatrix} -\lambda & -1 \\ 1 & -\lambda \end{bmatrix}$$
$$= \lambda^2 + 1 = 0$$

which gives

$$\lambda_{1,2} = \pm i$$

That is, both eigenvalues of C are complex and form a complex conjugate pair, but they are purely imaginary (that is, their real parts are equal to 0). To see what the solutions of

$$\frac{d\mathbf{x}}{dt} = C\mathbf{x}(t) \tag{11.31}$$

look like, we graph both the direction field and solutions $x_1(t)$ and $x_2(t)$ in Figures 11.13 and 11.14.

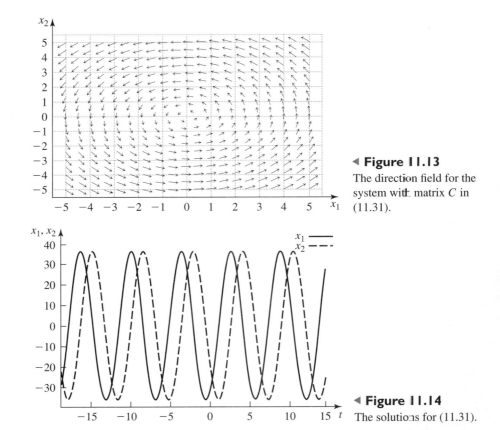

◀ **Figure 11.13**

The direction field for the system with matrix C in (11.31).

◀ **Figure 11.14**

The solutions for (11.31).

Looking at solution curves in Figure 11.14, we see that the solutions oscillate as in the previous two examples, but this time the amplitude does not change with time. Looking at the direction field in Figure 11.13, we see that solutions spiral around the equilibrium $(0, 0)$. But since the amplitude of the solutions does not change, the solutions neither approach nor move away from the equilibrium. The equilibrium $(0, 0)$ is called a **neutral spiral** or a **center**. One can show that the solutions form closed curves. (We will analyze this more closely in Example 3, where we will show this direction field with a solution curve.)

Where do the oscillations come from?   One can show that the solutions of (11.24) are given by (11.26), regardless of whether the eigenvalues are real or complex, as long as they are distinct. If the eigenvalues are complex, then the solution contains terms of the form e^z, where z is a complex number.

We can write a complex number z as

$$z = a + ib$$

where a and b are both real. Recall that the number a is called the real part of z, the number b is called the imaginary part of z.

To find out what e^z means when z is complex, we write

$$e^z = e^{a+ib} = e^a e^{ib}$$

The term e^a is a real number; therefore, we know what it means. The term e^{ib} is an exponential with a purely imaginary exponent. To find out what it means, we use the following formula, which we cannot prove here.

EULER'S FORMULA

$$e^{ib} = \cos b + i \sin b$$

Both the sine and the cosine functions show oscillations; this is the reason why systems of differential equations with complex eigenvalues have solutions that oscillate, as in the following example, where we can guess the answer.

▮ EXAMPLE 3   Solve

$$\frac{dx_1}{dt} = -x_2$$
$$\frac{dx_2}{dt} = x_1$$

(11.32)

with $x_1(0) = 1$ and $x_2(0) = 0$.

Solution   The associated matrix of coefficients is

$$A = \begin{bmatrix} 0 & -1 \\ 1 & 0 \end{bmatrix}$$

which is the same as matrix C that we saw in (11.31). We therefore know that

$$\lambda_1 = i \qquad \text{and} \qquad \lambda_2 = -i$$

We claim that

$$x_1(t) = \cos t$$
$$x_2(t) = \sin t$$

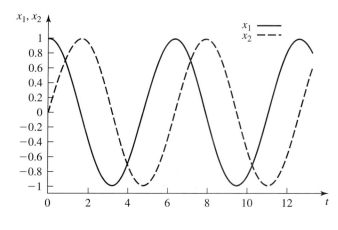

◀ **Figure 11.15**
The solution for Example 3 as functions of t with initial condition $x_1(0) = 1$ and $x_2(0) = 0$.

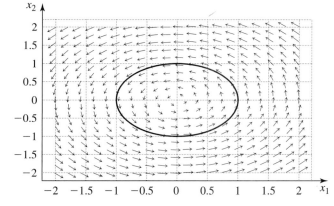

◀ **Figure 11.16**
The trajectory for Example 3 in the direction field with initial condition $x_1(0) = 1$ and $x_2(0) = 0$. The trajectory is a closed curve.

solves (11.32). Let's check. We find

$$\frac{dx_1}{dt} = -\sin t = -x_2(t)$$

and

$$\frac{dx_2}{dt} = \cos t = x_1(t)$$

In addition, we must check that our solution satisfies the initial condition. Indeed, $x_1(0) = \cos 0 = 1$ and $x_2(0) = \sin 0 = 0$. The solution for this system is a pair of real-valued functions, as claimed. The solution is shown in Figure 11.15, and the trajectory of $(x_1(t), x_2(t))$ in the x_1-x_2 plane is shown in Figure 11.16.

We see from Figure 11.15 that the solutions in this case show sustained oscillations, that is, the amplitudes of $x_1(t)$ and $x_2(t)$ do not change with time. In Figure 11.16, we see that the solution in the x_1-x_2 plane forms closed curves, indicating that the solutions are periodic. **I**

Summary We give a brief summary of the classification of the equilibrium $(0, 0)$ of

$$\frac{d\mathbf{x}}{dt} = A\mathbf{x}(t) \quad \text{with } A = \begin{bmatrix} a_{11} & a_{12} \\ a_{21} & a_{22} \end{bmatrix}$$

Recall that

$$\tau = \text{trace}(A) = a_{11} + a_{22}$$
$$\Delta = \det(A) = a_{11}a_{22} - a_{12}a_{21}$$

The eigenvalues of A are found by solving

$$\det(A - \lambda I) = \det \begin{bmatrix} a_{11} - \lambda & a_{12} \\ a_{21} & a_{22} - \lambda \end{bmatrix}$$

$$= (a_{11} - \lambda)(a_{22} - \lambda) - a_{12}a_{21}$$

$$= \lambda^2 - (a_{11} + a_{22})\lambda + a_{11}a_{22} - a_{12}a_{21}$$

$$= \lambda^2 - \tau\lambda + \Delta = 0$$

which gives

$$\lambda_1 = \frac{\tau + \sqrt{\tau^2 - 4\Delta}}{2} \quad \text{and} \quad \lambda_2 = \frac{\tau - \sqrt{\tau^2 - 4\Delta}}{2} \tag{11.33}$$

We see immediately that

if $\tau^2 > 4\Delta$ both eigenvalues are real and distinct
if $\tau^2 < 4\Delta$ eigenvalues are complex conjugates

When both eigenvalues are real and distinct, that is, when $\tau^2 > 4\Delta$, we can distinguish the following three cases:

1. $\Delta < 0, \tau \neq 0$: $\lambda_1 > 0, \lambda_2 < 0$ (saddle)
2. $\Delta > 0, \tau < 0$: $\lambda_1 < 0, \lambda_2 < 0$ (sink or stable node)
3. $\Delta > 0, \tau > 0$: $\lambda_1 > 0, \lambda_2 > 0$ (source or unstable node)

When both eigenvalues are complex conjugates, that is, when $\tau^2 < 4\Delta$, we can distinguish the following two cases:

1. $\tau < 0$: both eigenvalues have negative real parts (stable spiral)
2. $\tau > 0$: both eigenvalues have positive real parts (unstable spiral)

We can summarize this graphically in the τ-Δ plane as shown in Figure 11.17. The parabola $4\Delta = \tau^2$ is the boundary line between oscillatory and nonoscillatory behavior. The line $\tau = 0$ divides the stable and the unstable regions. The line $\Delta = 0$ separates the saddle and the node regions. The case in which the eigenvalues are identical resides on the boundary line $4\Delta = \tau^2$.

The case in which both eigenvalues are purely imaginary is on the positive part of the vertical line $\tau = 0$. We discussed this case in Example 4, where we saw that the solution shows oscillations of a constant amplitude. In this case, $(0, 0)$ is called a center.

The line $\Delta = 0$ corresponds to the case in which one eigenvalue is equal to 0. As long as the other eigenvalue is not equal to 0, both eigenvalues are again distinct

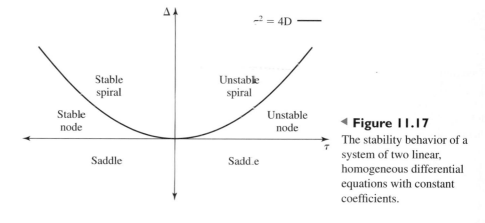

◀ **Figure 11.17**

The stability behavior of a system of two linear, homogeneous differential equations with constant coefficients.

and the solution is of the form (11.26). However, in that case, there are equilibria other than $(0, 0)$. We will discuss two such examples in Problems 57 and 58, and will also discuss an example of this case in 11.2.

◆ 11.1 PROBLEMS

[11.1.1]

In Problems 1–4, write each system of differential equations in matrix form.

1. $\dfrac{dx_1}{dt} = 2x_1 + 3x_2$

$\dfrac{dx_2}{dt} = -4x_1 + x_2$

2. $\dfrac{dx_1}{dt} = x_1 + x_2$

$\dfrac{dx_2}{dt} = -x_1$

3. $\dfrac{dx_1}{dt} = x_3$

$\dfrac{dx_2}{dt} = -x_1$

$\dfrac{dx_3}{dt} = x_1 + x_2$

4. $\dfrac{dx_1}{dt} = 2x_2 - x_3$

$\dfrac{dx_2}{dt} = -x_1$

$\dfrac{dx_3}{dt} = 5x_1 + x_3$

5. Consider

$$\frac{dx_1}{dt} = -x_1 + 2x_2$$

$$\frac{dx_2}{dt} = x_1$$

Determine the direction vectors associated with the following points in the x_1-x_2 plane, and graph the direction vectors in the x_1-x_2 plane: $(1, 0)$, $(0, 1)$, $(-1, 0)$, $(0, -1)$, $(1, 1)$, $(0, 0)$, and $(-2, 1)$.

6. Consider

$$\frac{dx_1}{dt} = 2x_1 - x_2$$

$$\frac{dx_2}{dt} = -x_2$$

Determine the direction vectors associated with the following points in the x_1-x_2 plane, and graph the direction vectors in the x_1-x_2 plane: $(2, 0)$, $(1.5, 1)$, $(1, 0)$, $(0, -1)$, $(1, 1)$, $(0, 0)$, and $(-2, -2)$.

7. Consider

$$\frac{dx_1}{dt} = x_1 + 3x_2$$

$$\frac{dx_2}{dt} = -x_1 + 2x_2$$

Determine the direction vectors associated with the following points in the x_1-x_2 plane, and graph the direction vectors in the x_1-x_2 plane: $(1, 0)$, $(0, 1)$, $(-1, 1)$, $(0, -1)$, $(-3, 1)$, $(0, 0)$, and $(-2, 1)$.

8. Consider

$$\frac{dx_1}{dt} = -x_2$$

$$\frac{dx_2}{dt} = x_1$$

Determine the direction vectors associated with the following points in the x_1-x_2 plane, and graph the direction vectors in the x_1-x_2 plane: $(1, 0)$, $(0, 1)$, $(-1, 0)$, $(0, -1)$, $(1, 1)$, $(0, 0)$, and $(-2, -2)$.

9. In Figures 11.18–11.21, direction fields are given. Each of the following systems of differential equations corresponds to exactly one of the direction fields. Match the systems to the appropriate figures.

(a) $\dfrac{dx_1}{dt} = -x_2$

$\dfrac{dx_2}{dt} = x_1$

(b) $\dfrac{dx_1}{dt} = x_1$

$\dfrac{dx_2}{dt} = x_2$

(c) $\dfrac{dx_1}{dt} = x_1 + 2x_2$

$\dfrac{dx_2}{dt} = -2x_1$

(d) $\dfrac{dx_1}{dt} = 2x_1$

$\dfrac{dx_2}{dt} = x_1 + x_2$

▲ **Figure 11.18**

▲ **Figure 11.19**

▲ **Figure 11.20**

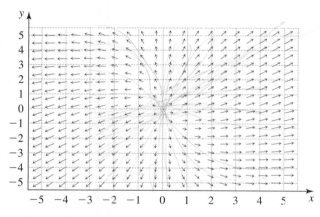

▲ **Figure 11.21**

10. The direction field of

$$\frac{dx_1}{dt} = x_1 + 3x_2$$

$$\frac{dx_2}{dt} = 2x_1 + 3x_2$$

is given in Figure 11.22. Sketch the solution curve that goes through the point $(1, 0)$.

▲ **Figure 11.22**

11. The direction field of

$$\frac{dx_1}{dt} = 2x_1 + 3x_2$$

$$\frac{dx_2}{dt} = -x_1 + x_2$$

is given in Figure 11.23. Sketch the solution curve that goes through the point $(2, -1)$.

12. The direction field of

$$\frac{dx_1}{dt} = -x_1 - x_2$$

$$\frac{dx_2}{dt} = -2x_2$$

is given in Figure 11.24. Sketch the solution curve that goes through the point $(-3, -3)$.

[11.1.2]

In Problems 13–18, find the general solution to each given system of differential equations, and sketch the lines in the direction of the eigenvectors in the given direction field. Indicate on each line the direction in which the solution would move if starting on that line.

13.
$$\begin{bmatrix} \frac{dx_1}{dt} \\ \frac{dx_2}{dt} \end{bmatrix} = \begin{bmatrix} 1 & 3 \\ 5 & 3 \end{bmatrix} \begin{bmatrix} x_1(t) \\ x_2(t) \end{bmatrix}$$

▲ **Figure 11.23**

▲ **Figure 11.24**

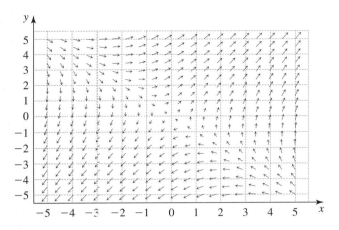

▲ **Figure 11.25**

14. $\begin{bmatrix} \frac{dx_1}{dt} \\ \frac{dx_2}{dt} \end{bmatrix} = \begin{bmatrix} 2 & 1 \\ 4 & -1 \end{bmatrix} \begin{bmatrix} x_1(t) \\ x_2(t) \end{bmatrix}$

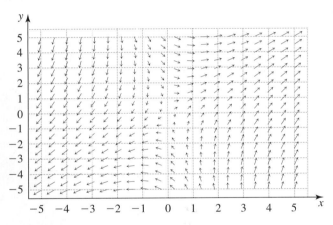

▲ **Figure 11.26**

15. $\begin{bmatrix} \frac{dx_1}{dt} \\ \frac{dx_2}{dt} \end{bmatrix} = \begin{bmatrix} -3 & 3 \\ 6 & 4 \end{bmatrix} \begin{bmatrix} x_1(t) \\ x_2(t) \end{bmatrix}$

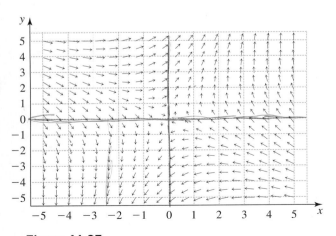

▲ **Figure 11.27**

16. $\begin{bmatrix} \frac{dx_1}{dt} \\ \frac{dx_2}{dt} \end{bmatrix} = \begin{bmatrix} -5 & 3 \\ -2 & 0 \end{bmatrix} \begin{bmatrix} x_1(t) \\ x_2(t) \end{bmatrix}$

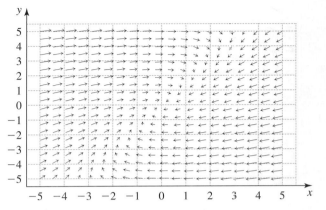

▲ **Figure 11.28**

17. $\begin{bmatrix} \frac{dx_1}{dt} \\ \frac{dx_2}{dt} \end{bmatrix} = \begin{bmatrix} -2 & 0 \\ -3 & 1 \end{bmatrix} \begin{bmatrix} x_1(t) \\ x_2(t) \end{bmatrix}$

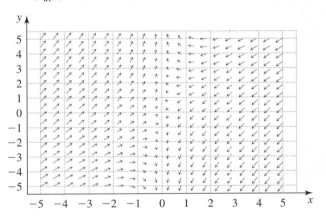

▲ **Figure 11.29**

18. $\begin{bmatrix} \frac{dx_1}{dt} \\ \frac{dx_2}{dt} \end{bmatrix} = \begin{bmatrix} 5 & 2 \\ 1 & 6 \end{bmatrix} \begin{bmatrix} x_1(t) \\ x_2(t) \end{bmatrix}$

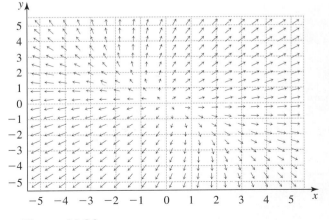

▲ **Figure 11.30**

In Problems 19–24, solve the given initial value problem.

19.

$$\begin{bmatrix} \frac{dx_1}{dt} \\ \frac{dx_2}{dt} \end{bmatrix} = \begin{bmatrix} -3 & 0 \\ 4 & 2 \end{bmatrix} \begin{bmatrix} x_1(t) \\ x_2(t) \end{bmatrix}$$

with $x_1(0) = -5$ and $x_2(0) = 5$.

20.

$$\begin{bmatrix} \frac{dx_1}{dt} \\ \frac{dx_2}{dt} \end{bmatrix} = \begin{bmatrix} 1 & 3 \\ 2 & 2 \end{bmatrix} \begin{bmatrix} x_1(t) \\ x_2(t) \end{bmatrix}$$

with $x_1(0) = 2$ and $x_2(0) = -1$.

21.

$$\begin{bmatrix} \frac{dx_1}{dt} \\ \frac{dx_2}{dt} \end{bmatrix} = \begin{bmatrix} 3 & -2 \\ 0 & 1 \end{bmatrix} \begin{bmatrix} x_1(t) \\ x_2(t) \end{bmatrix}$$

with $x_1(0) = 1$ and $x_2(0) = 1$.

22.

$$\begin{bmatrix} \frac{dx_1}{dt} \\ \frac{dx_2}{dt} \end{bmatrix} = \begin{bmatrix} -3 & 4 \\ -1 & 2 \end{bmatrix} \begin{bmatrix} x_1(t) \\ x_2(t) \end{bmatrix}$$

with $x_1(0) = 1$ and $x_2(0) = 2$.

23.

$$\begin{bmatrix} \frac{dx_1}{dt} \\ \frac{dx_2}{dt} \end{bmatrix} = \begin{bmatrix} 4 & 7 \\ 1 & -2 \end{bmatrix} \begin{bmatrix} x_1(t) \\ x_2(t) \end{bmatrix}$$

with $x_1(0) = -1$ and $x_2(0) = -2$.

24.

$$\begin{bmatrix} \frac{dx_1}{dt} \\ \frac{dx_2}{dt} \end{bmatrix} = \begin{bmatrix} 2 & 6 \\ 1 & 3 \end{bmatrix} \begin{bmatrix} x_1(t) \\ x_2(t) \end{bmatrix}$$

with $x_1(0) = -2$ and $x_2(0) = 1$.

In Problems 25 and 26, we discuss the case of repeated eigenvalues.

25. Let

$$\begin{bmatrix} \frac{dx_1}{dt} \\ \frac{dx_2}{dt} \end{bmatrix} = \begin{bmatrix} 1 & 0 \\ 0 & 1 \end{bmatrix} \begin{bmatrix} x_1(t) \\ x_2(t) \end{bmatrix} \tag{11.34}$$

(a) Show that

$$A = \begin{bmatrix} 1 & 0 \\ 0 & 1 \end{bmatrix}$$

has repeated eigenvalues, namely $\lambda_1 = \lambda_2 = 1$.
(b) Show that $\begin{bmatrix} 1 \\ 0 \end{bmatrix}$ and $\begin{bmatrix} 0 \\ 1 \end{bmatrix}$ are eigenvectors of A, and that any vector $\begin{bmatrix} c_1 \\ c_2 \end{bmatrix}$ can be written as

$$\begin{bmatrix} c_1 \\ c_2 \end{bmatrix} = c_1 \begin{bmatrix} 1 \\ 0 \end{bmatrix} + c_2 \begin{bmatrix} 0 \\ 1 \end{bmatrix}$$

(c) Show that

$$\mathbf{x}(t) = c_1 e^t \begin{bmatrix} 1 \\ 0 \end{bmatrix} + c_2 e^t \begin{bmatrix} 0 \\ 1 \end{bmatrix}$$

is a solution of (11.34) that satisfies the initial condition $x_1(0) = c_1$ and $x_2(0) = c_2$.

26. Let

$$\begin{bmatrix} \frac{dx_1}{dt} \\ \frac{dx_2}{dt} \end{bmatrix} = \begin{bmatrix} 1 & 2 \\ 0 & 1 \end{bmatrix} \begin{bmatrix} x_1(t) \\ x_2(t) \end{bmatrix} \tag{11.35}$$

(a) Show that

$$A = \begin{bmatrix} 1 & 2 \\ 0 & 1 \end{bmatrix}$$

has repeated eigenvalues, namely $\lambda_1 = \lambda_2 = 1$.
(b) Show that every eigenvector of A is of the form

$$c_1 \begin{bmatrix} 1 \\ 0 \end{bmatrix}$$

where c_1 is a real number different from 0.
(c) Show that

$$\mathbf{x}_1(t) = e^t \begin{bmatrix} 1 \\ 0 \end{bmatrix}$$

is a solution of (11.35).
(d) Show that

$$\mathbf{x}_2(t) = te^t \begin{bmatrix} 1 \\ 0 \end{bmatrix} + e^t \begin{bmatrix} 0 \\ 0.5 \end{bmatrix}$$

is a solution of (11.35).
(e) Show that

$$c_1 \mathbf{x}_1(t) + c_2 \mathbf{x}_2(t)$$

is a solution of (11.35). (It turns out that this is the general solution.)

[11.1.3]

In Problems 27–36, we consider differential equations of the form

$$\frac{d\mathbf{x}}{dt} = A\mathbf{x}(t)$$

where

$$A = \begin{bmatrix} a_{11} & a_{12} \\ a_{21} & a_{22} \end{bmatrix}$$

The eigenvalues of A will be real, distinct, and nonzero. Analyze the stability of the equilibrium $(0, 0)$, and classify the equilibrium according to sink, source, or saddle.

27. $A = \begin{bmatrix} 1 & 0 \\ 1 & -2 \end{bmatrix}$ **28.** $A = \begin{bmatrix} -2 & 4 \\ 2 & -5 \end{bmatrix}$

29. $A = \begin{bmatrix} 6 & -4 \\ -3 & 5 \end{bmatrix}$ **30.** $A = \begin{bmatrix} -1 & 3 \\ 2 & -5 \end{bmatrix}$

31. $A = \begin{bmatrix} -3 & -1 \\ 1 & -6 \end{bmatrix}$ **32.** $A = \begin{bmatrix} -3 & 1 \\ 1 & -2 \end{bmatrix}$

33. $A = \begin{bmatrix} 0 & -2 \\ -1 & 3 \end{bmatrix}$ **34.** $A = \begin{bmatrix} 1 & 2 \\ 3 & 7 \end{bmatrix}$

35. $A = \begin{bmatrix} -2 & -3 \\ 1 & 3 \end{bmatrix}$ **36.** $A = \begin{bmatrix} 4 & -1 \\ 5 & -1 \end{bmatrix}$

In Problems 37–46, we consider differential equations of the form

$$\frac{d\mathbf{x}}{dt} = A\mathbf{x}(t)$$

where

$$A = \begin{bmatrix} a_{11} & a_{12} \\ a_{21} & a_{22} \end{bmatrix}$$

The eigenvalues of A will be complex conjugates. Analyze the stability of the equilibrium $(0, 0)$, and classify the equilibrium according to stable spiral, unstable spiral, or center.

37. $A = \begin{bmatrix} 1 & 3 \\ -2 & -2 \end{bmatrix}$ 38. $A = \begin{bmatrix} 2 & -3 \\ 4 & -1 \end{bmatrix}$

39. $A = \begin{bmatrix} 4 & 5 \\ -3 & -3 \end{bmatrix}$ 40. $A = \begin{bmatrix} 2 & 2 \\ -6 & -4 \end{bmatrix}$

41. $A = \begin{bmatrix} -1 & 1 \\ -3 & 1 \end{bmatrix}$ 42. $A = \begin{bmatrix} 3 & -2 \\ 1 & 2 \end{bmatrix}$

43. $A = \begin{bmatrix} 0 & -1 \\ 1 & 0 \end{bmatrix}$ 44. $A = \begin{bmatrix} 0 & -3 \\ 2 & 2 \end{bmatrix}$

45. $A = \begin{bmatrix} 1 & 2 \\ -5 & -3 \end{bmatrix}$ 46. $A = \begin{bmatrix} 2 & -3 \\ 3 & -2 \end{bmatrix}$

In Problems 47–56, we consider differential equations of the form

$$\frac{d\mathbf{x}}{dt} = A\mathbf{x}(t)$$

where

$$A = \begin{bmatrix} a_{11} & a_{12} \\ a_{21} & a_{22} \end{bmatrix}$$

Analyze the stability of the equilibrium $(0, 0)$, and classify the equilibrium.

47. $A = \begin{bmatrix} -1 & -2 \\ 1 & 3 \end{bmatrix}$ 48. $A = \begin{bmatrix} -2 & 2 \\ -4 & 3 \end{bmatrix}$

49. $A = \begin{bmatrix} -1 & -1 \\ 5 & -3 \end{bmatrix}$ 50. $A = \begin{bmatrix} 2 & 2 \\ 2 & 3 \end{bmatrix}$

51. $A = \begin{bmatrix} 1 & 3 \\ 2 & 3 \end{bmatrix}$ 52. $A = \begin{bmatrix} -1 & 5 \\ -3 & -3 \end{bmatrix}$

53. $A = \begin{bmatrix} -2 & 3 \\ 1 & -4 \end{bmatrix}$ 54. $A = \begin{bmatrix} -2 & -7 \\ 1 & 2 \end{bmatrix}$

55. $A = \begin{bmatrix} 3 & -5 \\ 2 & -1 \end{bmatrix}$ 56. $A = \begin{bmatrix} 3 & 6 \\ -1 & -4 \end{bmatrix}$

57. The following system has two distinct real eigenvalues, but one eigenvalue is equal to 0:

$$\frac{d\mathbf{x}}{dt} = \begin{bmatrix} 4 & 8 \\ 1 & 2 \end{bmatrix} \mathbf{x}(t) \tag{11.36}$$

(a) Find both eigenvalues and associated eigenvectors.

(b) Use the general solution (11.26) to find $x_1(t)$ and $x_2(t)$.

(c) The direction field is shown in Figure 11.31. Sketch the lines corresponding to the eigenvectors. Compute dx_2/dx_1, and conclude that all direction vectors are parallel to the line in the direction of the eigenvector corresponding to the nonzero eigenvalue. Describe in words how solutions starting at different points behave.

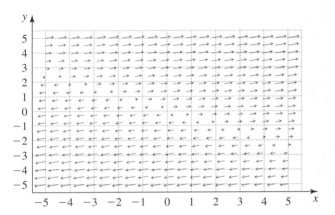

▲ **Figure 11.31**

58. The following system has two distinct real eigenvalues, but one eigenvalue is equal to 0:

$$\frac{d\mathbf{x}}{dt} = \begin{bmatrix} 2 & 4 \\ 3 & 6 \end{bmatrix} \mathbf{x}(t) \tag{11.37}$$

(a) Find both eigenvalues and associated eigenvectors.

(b) Use the general solution (11.26) to find $x_1(t)$ and $x_2(t)$.

(c) The direction field is shown in Figure 11.32. Sketch the lines corresponding to the eigenvectors. Compute dx_2/dx_1, and conclude that all direction vectors are parallel to the line in the direction of the eigenvector corresponding to the nonzero eigenvalue. Describe in words how solutions starting at different points behave.

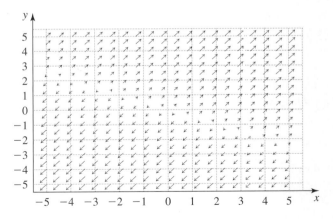

▲ **Figure 11.32**

◆ 11.2 LINEAR SYSTEMS—APPLICATIONS

11.2.1 Compartment Models

▲ **Figure 11.33**

A schematic description of a
general two compartment model.

Compartment models (which we encountered in Chapter 7) describe flow between compartments, such as nutrient flow between lakes or the flow of a radioactive tracer between different parts of an organism. In the simplest situations, the resulting model is a system of linear differential equations.

We will consider a general two-compartment model that can be described by a system of two linear differential equations. A schematic description of the model is given in Figure 11.33.

We denote by $x_1(t)$ the amount of material in compartment 1 at time t, and by $x_2(t)$ the amount of material in compartment 2 at time t. The direction of the flow of material and the rates at which material flows are shown in Figure 11.33. We see that material enters compartment 1 at the constant rate I, and moves from compartment 1 to compartment 2 at rate ax_1 if x_1 is the amount of material in compartment 1. Material in compartment 1 is lost at rate cx_1. In addition, material flows from compartment 2 to compartment 1 at rate bx_2 if x_2 is the amount of the material in compartment 2. Material in compartment 2 is lost at rate dx_2; we assume that there is no external input into compartment 2.

We describe the dynamics of $x_1(t)$ and $x_2(t)$ by the following system of differential equations:

$$\begin{aligned}\frac{dx_1}{dt} &= I - (a+c)x_1 + bx_2 \\ \frac{dx_2}{dt} &= ax_1 - (b+d)x_2\end{aligned} \tag{11.38}$$

The constants $I, a, b, c,$ and d are all nonnegative. If $I > 0$, then (11.38) is a system of inhomogeneous linear differential equations with constant coefficients. Constant input is often important in real situations, such as the flow of nutrients between soil and plants, where nutrients might be added at a constant rate. In the following discussion, however, we will set $I = 0$, since this corresponds to the situation discussed in the previous section (that is, no material is added over time). It is not difficult to guess how the system behaves when $I = 0$: Either some material is continually lost so that one or both compartments empty out, or no material is lost so that at least one compartment will contain material. We will discuss both cases.

When $I = 0$, then (11.38) reduces to the linear system

$$\frac{d\mathbf{x}}{dt} = A\mathbf{x}(t) \quad \text{with } A = \begin{bmatrix} -(a+c) & b \\ a & -(b+d) \end{bmatrix} \tag{11.39}$$

To avoid trivial situations, we assume that at least one of the parameters $a, b, c,$ or d is positive (otherwise, no material would ever move in the system).

To find the eigenvalues of A, we compute

$$\begin{aligned}\det &\begin{bmatrix} -(a+c) - \lambda & b \\ a & -(b+d) - \lambda \end{bmatrix} \\ &= (a+c+\lambda)(b+d+\lambda) - ab \\ &= \lambda^2 + (a+b+c+d)\lambda + (a+c)(b+d) - ab \\ &= \lambda^2 - \tau\lambda + \Delta = 0\end{aligned}$$

where

$$\tau = \text{tr}(A) = -(a+b+c+d) < 0$$
$$\Delta = \det(A) = (a+c)(b+d) - ab = ad + bc + cd \geq 0$$

We find

$$\lambda_{1,2} = \frac{\tau \pm \sqrt{\tau^2 - 4\Delta}}{2}$$

To check whether the eigenvalues are real, we simplify the expression under the square root,

$$\tau^2 - 4\Delta = (a + b + c + d)^2 - 4(a + c)(b + d) + 4ab$$
$$= [(a + b) - (c + d)]^2 + 4ab \geq 0$$

which implies that both eigenvalues are real.

Since $\Delta \geq 0$ and $\tau^2 - 4\Delta \geq 0$, it follows that $\sqrt{\tau^2 - 4\Delta} \leq |\tau|$. Since in addition, $\tau < 0$, we conclude that both eigenvalues are less than or equal to 0. As long as $\Delta > 0$, both eigenvalues will be strictly negative; then $(0, 0)$ will be the only equilibrium and will be a stable node.

$\Delta > \mathbf{0}$: When $\Delta = ad + bc + cd > 0$, at least one of the terms in $ad + bc + cd$ must be positive; that is, either

$$a, d > 0 \qquad \text{or} \qquad b, c > 0 \qquad \text{or} \qquad c, d > 0$$

To see what this means, we return to Figure 11.33. When a and d are both positive, the material in compartment 1 can move to compartment 2, and material in compartment 2 can leave the system. Similarly, when both b and c are positive, material from both compartments can leave the system through compartment 1. If c and d are both positive, then material can leave both compartments. It follows that in any of these three cases, all material will eventually leave the system, implying that $(0, 0)$ is a stable equilibrium.

$\Delta = \mathbf{0}$: Then

$$\lambda_1 = \tau = -(a + b + c + d) \qquad \text{and} \qquad \lambda_2 = 0$$

Since we assume that at least one of the four parameters a, b, c, or d is positive, it follows that $\lambda_1 < 0$; thus, the eigenvalues are distinct. To have $\Delta = 0$, one of the following must hold:

$$c = a = 0 \qquad \text{or} \qquad d = b = 0 \qquad \text{or} \qquad c = d = 0$$

If $c = a = 0$, then material gets stuck in compartment 1; if $d = b = 0$, then material gets stuck in compartment 2; if $c = d = 0$, no material will ever leave the system, and the amount of material present at time t is equal to the amount of material present at time 0. In other words, the total amount $x_1(t) + x_2(t)$ is constant (it does not depend on t); it is therefore called a **conserved quantity**. This follows from (11.38) as well, since in this case,

$$\frac{dx_1}{dt} = -ax_1 + bx_2$$

$$\frac{dx_2}{dt} = ax_1 - bx_2$$

Adding the two equations, we find

$$\frac{dx_1}{dt} + \frac{dx_2}{dt} = 0$$

Since

$$\frac{dx_1}{dt} + \frac{dx_2}{dt} = \frac{d}{dt}(x_1 + x_2)$$

it follows that $x_1(t) + x_2(t)$ is a constant.

Solving the system when $c = d = 0$ We can write the solution explicitly to find out the equilibrium state. To determine the solution, we must compute the eigenvectors corresponding to the eigenvalues λ_1 and λ_2.

The eigenvector **u** corresponding to $\lambda_1 = \tau$ satisfies

$$\begin{bmatrix} -a & b \\ a & -b \end{bmatrix} \begin{bmatrix} u_1 \\ u_2 \end{bmatrix} = -(a + b) \begin{bmatrix} u_1 \\ u_2 \end{bmatrix}$$

Writing this system out yields

$$-au_1 + bu_2 = -au_1 - bu_1$$
$$au_1 - bu_2 = -au_2 - bu_2$$

This simplifies to one equation, namely

$$u_1 = -u_2$$

If we set $u_1 = 1$, then $u_2 = -1$, and an eigenvector corresponding to $\lambda_1 = -(a + b)$ is

$$\mathbf{u} = \begin{bmatrix} 1 \\ -1 \end{bmatrix}$$

The eigenvector **v** corresponding to $\lambda_2 = 0$ satisfies

$$\begin{bmatrix} -a & b \\ a & -b \end{bmatrix} \begin{bmatrix} v_1 \\ v_2 \end{bmatrix} = \begin{bmatrix} 0 \\ 0 \end{bmatrix}$$

This simplifies to one equation, namely

$$-av_1 + bv_2 = 0$$

If we set $v_1 = b$, then $v_2 = a$, and an eigenvector corresponding to $\lambda_2 = 0$ is

$$\mathbf{v} = \begin{bmatrix} b \\ a \end{bmatrix}$$

The general solution when $\Delta = 0$ is, therefore,

$$\mathbf{x}(t) = c_1 e^{-(a+b)t} \begin{bmatrix} 1 \\ -1 \end{bmatrix} + c_2 \begin{bmatrix} b \\ a \end{bmatrix}$$

and

$$\lim_{t \to \infty} \mathbf{x}(t) = c_2 \begin{bmatrix} b \\ a \end{bmatrix}$$

At time 0,

$$\begin{bmatrix} x_1(0) \\ x_2(0) \end{bmatrix} = c_1 \begin{bmatrix} 1 \\ -1 \end{bmatrix} + c_2 \begin{bmatrix} b \\ a \end{bmatrix}$$

that is,

$$c_1 + bc_2 = x_1(0)$$
$$-c_1 + ac_2 = x_2(0)$$

Adding these two equations, we eliminate c_1 and find

$$(a + b)c_2 = x_1(0) + x_2(0)$$

or

$$c_2 = \frac{x_1(0) + x_2(0)}{a + b}$$

Recall that the sum $x_1(t) + x_2(t)$ is a constant for all $t \geq 0$; we set $x_1(0) + x_2(0) = K$. Then,

$$\lim_{t \to \infty} x_1(t) = K\frac{b}{a+b} \qquad \text{and} \qquad \lim_{t \to \infty} x_2(t) = K\frac{a}{a+b}$$

This means that compartment 1 will eventually contain a fraction $b/(a + b)$ of the total amount, and compartment 2 a fraction $a/(a + b)$ of the total amount. These are the relative rates at which material enters the respective compartments.

▮ EXAMPLE 1

$0.2x_1$/hour

$0.5x_2$/hour

$0.1x_1$/hour

▲ Figure 11.34

The compartment diagram for Example 1.

Find the system of differential equations corresponding to the compartment diagram shown in Figure 11.34, and analyze the stability of the equilibrium $(0, 0)$.

Solution   The compartment model is described by the following system of differential equations:

$$\frac{dx_1}{dt} = -(0.1 + 0.2)x_1 + 0.5x_2$$

$$\frac{dx_2}{dt} = 0.2x_1 - 0.5x_2$$

In matrix notation, this is

$$\frac{d\mathbf{x}}{dt} = \begin{bmatrix} -0.3 & 0.5 \\ 0.2 & -0.5 \end{bmatrix} \mathbf{x}(t)$$

To investigate the stability of $(0, 0)$, we find the eigenvalues of the matrix describing the system. That is, we solve

$$\det \begin{bmatrix} -0.3 - \lambda & 0.5 \\ 0.2 & -0.5 - \lambda \end{bmatrix} = (-0.3 - \lambda)(-0.5 - \lambda) - (0.2)(0.5)$$

$$= (0.3)(0.5) + (0.3 + 0.5)\lambda + \lambda^2 - (0.2)(0.5)$$

$$= \lambda^2 + 0.8\lambda + 0.05 = 0$$

We find

$$\lambda_{1,2} = \frac{-0.8 \pm \sqrt{0.64 - 0.2}}{2}$$

$$= \begin{cases} -0.4 + \frac{1}{2}\sqrt{0.44} \approx -0.068 \\ -0.4 - \frac{1}{2}\sqrt{0.44} \approx -0.732 \end{cases}$$

We find that both eigenvalues are negative. Therefore, the equilibrium $(0, 0)$ is a stable node. ▮

▮ EXAMPLE 2   Given the system of differential equations

$$\frac{dx_1}{dt} = -0.7x_1 + 0.2x_2$$

$$\frac{dx_2}{dt} = 0.3x_1 - 0.2x_2$$

determine the parameter for the compartment diagram in Figure 11.33.

Solution   The general compartment diagram describing a linear system with two states is shown in Figure 11.33. Comparing the diagram and the system of equations, we find that $I = 0$ and

$$a + c = 0.7$$
$$b = 0.2$$
$$a = 0.3$$
$$b + d = 0.2$$

Solving this system of equations, we conclude that

$$a = 0.3, \qquad b = 0.2, \qquad c = 0.4 \qquad \text{and} \qquad d = 0 \qquad \blacksquare$$

Compartment models are used in pharmacology to study how drug concentrations change within a body. In the simplest of such models, a drug is administered to a person in a single dose; we investigate this case in the next example.

▍EXAMPLE 3   A drug is administered to a person in a single dose. We assume that the drug does not accumulate in body tissue but is secreted through urine. We denote the drug concentration in the body at time t by $x_1(t)$ and in the urine at time t by $x_2(t)$. Initially,

$$x_1(0) = K \qquad \text{and} \qquad x_2(0) = 0$$

We describe the movement of the drug between the body and the urine by

$$\frac{dx_1}{dt} = -ax_1(t) \tag{11.40}$$

$$\frac{dx_2}{dt} = ax_1(t) \tag{11.41}$$

That is, the body's concentration of the drug decreases at the same rate as the drug accumulates in the urine. We can solve (11.40) directly, and find

$$x_1(t) = c_1 e^{-at}$$

Since $x_1(0) = K = c_1$, we find

$$x_1(t) = K e^{-at}$$

Plugging this into (11.41), we have

$$\frac{dx_2}{dt} = a K e^{-at}$$

or

$$\int dx_2 = \int a K e^{-at} \, dt$$

Hence,

$$x_2(t) = c_2 - K e^{-at}$$

With $x_2(0) = 0$, we find $0 = c_2 - K$. Hence,

$$x_2(t) = K(1 - e^{-at})$$

The constant a is called the excretion rate in this application. Finding the excretion rate is important for determining how long a drug will remain in the body. We can determine the excretion rate a by plotting $\ln(K - x_2(t))$ versus t. Namely,

$$\ln(K - x_2(t)) = \ln(K e^{-at})$$
$$= \ln K - at$$

That is, plotting $\ln(K - x_2(t))$ versus t results in a straight line with slope $-a$, which allows us to determine a. ▌

11.2.2 The Harmonic Oscillator (Optional)

We consider a particle that moves along the x-axis. We assume that the acceleration is proportional to the distance to the origin, and that the direction of the acceleration always points towards the origin. If $x(t)$ is the location of the particle at time t, then the second derivative of $x(t)$ denotes the acceleration of the particle, and we find

$$\frac{d^2 x}{dt^2} = -kx(t) \tag{11.42}$$

for some $k > 0$. This is a **second order differential equation**, because the derivative of the highest order derivative in the equation is of order 2. We will use this example to show how a second order differential equation can be transformed into a system of first order differential equations. To do so, we set

$$\frac{dx}{dt} = v(t)$$

Then

$$\frac{dv}{dt} = \frac{d^2 x}{dt^2}$$

and, hence,

$$\frac{dv}{dt} = -kx(t)$$

We thus obtain the following system of first order differential equations:

$$\frac{dx}{dt} = v$$

$$\frac{dv}{dt} = -kx$$

or

$$\frac{d}{dt} \begin{bmatrix} x \\ v \end{bmatrix} = \begin{bmatrix} 0 & 1 \\ -k & 0 \end{bmatrix} \begin{bmatrix} x \\ v \end{bmatrix}$$

If we denote the matrix by A, then

$$\text{tr} A = 0 \qquad \text{and} \qquad \det A = k > 0$$

which implies that the eigenvalues of A are complex conjugates with real parts equal to 0. We find

$$\det(A - \lambda I) = \det \begin{bmatrix} -\lambda & 1 \\ -k & -\lambda \end{bmatrix} = \lambda^2 + k = 0$$

hence,

$$\lambda_1 = i\sqrt{k} \qquad \text{and} \qquad \lambda_2 = -i\sqrt{k}$$

Thus, we expect the equilibrium $(0, 0)$ to be a neutral spiral (or center), and the solutions to exhibit oscillations whose amplitudes do not change with time.

We can solve (11.42) directly: Since

$$\frac{d}{dt} \sin(at) = a \cos(at)$$

$$\frac{d}{dt} \cos(at) = -a \sin(at)$$

we find

$$\frac{d^2}{dt^2}\sin(at) = -a^2\sin(at)$$

and

$$\frac{d^2}{dt^2}\cos(at) = -a^2\cos(at)$$

If we set $a = \sqrt{k}$, we see that $\cos(\sqrt{k}t)$ and $\sin(\sqrt{k}t)$ solve (11.42). Using the superposition principle, we therefore obtain the solution of (11.42) as

$$x(t) = c_1\sin(\sqrt{k}t) + c_2\cos(\sqrt{k}t)$$

To determine the constants c_1 and c_2, we must fix an initial condition. If we assume, for instance, that

$$x(0) = 0 \qquad \text{and} \qquad v(0) = v_0 \tag{11.43}$$

then

$$0 = c_2$$

Since $v(t) = dx/dt$, we find

$$v(t) = c_1\sqrt{k}\cos(\sqrt{k}t) - c_2\sqrt{k}\sin(\sqrt{k}t)$$

and, therefore,

$$v(0) = c_1\sqrt{k} = v_0$$

which implies that

$$c_1 = \frac{v_0}{\sqrt{k}}$$

Hence, the solution of (11.42) that satisfies the initial condition (11.43) is given by

$$x(t) = \frac{v_0}{\sqrt{k}}\sin(\sqrt{k}t)$$

The harmonic oscillator is quite important in physics. It describes, for instance, a frictionless pendulum when the displacement from the resting state is not too large.

◆ 11.2 PROBLEMS

[11.2.1]

In Problems 1–6, determine the system of differential equations corresponding to each compartment model and analyze the stability of the equilibrium $(0, 0)$. The parameters have the same meaning as in Figure 11.33.

1. $a = 0.5, b = 0.1\ c = 0.05, d = 0.02$

2. $a = 0.4, b = 0.2\ c = 0.3, d = 0$

3. $a = 2.5, b = 0.7\ c = 0, d = 0.1$

4. $a = 1.7, b = 0.6\ c = 0.1, d = 0.3$

5. $a = 0, b = 0, c = 0.1, d = 0.3$

6. $a = 0, b = 0.1, c = 0, d = 0$

In Problems 7–12, find the corresponding compartment diagram for each system of differential equations.

7. $$\frac{dx_1}{dt} = -0.4x_1 + 0.3x_2$$
$$\frac{dx_2}{dt} = 0.1x_1 - 0.5x_2$$

8. $$\frac{dx_1}{dt} = -0.4x_1 + 3x_2$$
$$\frac{dx_2}{dt} = 0.2x_1 - 3x_2$$

9. $$\frac{dx_1}{dt} = -1.2x_1$$
$$\frac{dx_2}{dt} = 0.3x_1 - 0.2x_2$$

10.
$$\frac{dx_1}{dt} = -0.2x_1 + 0.4x_2$$
$$\frac{dx_2}{dt} = 0.2x_1 - 0.4x_2$$

11.
$$\frac{dx_1}{dt} = -0.2x_1$$
$$\frac{dx_2}{dt} = -0.3x_2$$

12.
$$\frac{dx_1}{dt} = -x_1$$
$$\frac{dx_2}{dt} = x_1 - 0.5x_2$$

13. Suppose that a drug is administered to a person in a single dose, and assume that the drug does not accumulate in body tissue but is excreted through urine. Denote the drug concentration in the body at time t by $x_1(t)$ and in the urine at time t by $x_2(t)$. If $x_1(0) = 4$ mg and $x_2(0) = 0$, find $x_1(t)$ and $x_2(t)$ if

$$\frac{dx_1}{dt} = -0.3x_1(t)$$

14. Suppose that a drug is administered to a person in a single dose, and assume that the drug does not accumulate in body tissue but is excreted through urine. Denote the drug concentration in the body at time t by $x_1(t)$ and in the urine at time t by $x_2(t)$. If $x_1(0) = 6$ mg and $x_2(0) = 0$, find a system of differential equations for $x_1(t)$ and $x_2(t)$ if it takes twenty minutes for the drug to be at one-half of its initial concentration in the body.

15. A very simple two-compartment model for gap dynamics in a forest assumes that gaps are created by disturbances (wind, fire, and so forth) and that gaps revert back to forest as trees grow in the gaps. We denote by $x_1(t)$ the area occupied by gaps, and by $x_2(t)$ the area occupied by adult trees. We assume that the dynamics are given by

$$\frac{dx_1}{dt} = -0.2x_1 + 0.1x_2 \tag{11.44}$$

$$\frac{dx_2}{dt} = 0.2x_1 - 0.1x_2 \tag{11.45}$$

(a) Find the corresponding compartment diagram.

(b) Show that $x_1(t) + x_2(t)$ is a constant. Denote the constant by A and give its meaning. (Hint: Show that $\frac{d}{dt}(x_1 + x_2) = 0$.)

(c) Let $x_1(0) + x_2(0) = 20$. Use your answer in (b) to explain why this implies that $x_1(t) + x_2(t) = 20$ for all $t > 0$.

(d) Use your result in (c) to replace x_2 in (11.44) by $20 - x_1$, and show that this reduces the system (11.44) and (11.45) to

$$\frac{dx_1}{dt} = 2 - 0.3x_1 \tag{11.46}$$

with $x_1(t) + x_2(t) = 20$ for all $t \geq 0$.

(e) Solve the system (11.44) and (11.45), and determine what fraction of the forest is occupied by adult trees at time t when $x_1(0) = 2$ and $x_2(0) = 18$. What happens as $t \to \infty$?

16. A simple model for forest succession can be formulated by a three-compartment model. We assume that gaps in a forest are created by disturbances and are colonized by early successional species, which are then replaced by late successional species. We denote by $x_1(t)$ the total area occupied by gaps at time t, by $x_2(t)$ the total area occupied by early successional species at time t, and by $x_3(t)$ the total area occupied by late successional species at time t. The dynamics are given by

$$\frac{dx_1}{dt} = 0.2x_2 + x_3 - 2x_1$$

$$\frac{dx_2}{dt} = 2x_1 - 0.7x_2$$

$$\frac{dx_3}{dt} = 0.5x_2 - x_3$$

(a) Draw the corresponding compartment diagram.

(b) Show that

$$x_1(t) + x_2(t) + x_3(t) = A$$

where A is a constant, and give the meaning of A.

[11.2.2]

17. Solve

$$\frac{d^2x}{dt^2} = -4x$$

with $x(0) = 0$ and $\frac{dx(0)}{dt} = 6$.

18. Solve

$$\frac{d^2x}{dt^2} = -9x$$

with $x(0) = 0$ and $\frac{dx(0)}{dt} = 12$.

19. Transform the second order differential equation

$$\frac{d^2x}{dt^2} = 3x$$

into a system of first order differential equations.

20. Transform the second order differential equation

$$\frac{d^2x}{dt^2} = -\frac{1}{2}x$$

into a system of first order differential equations.

21. Transform the second order differential equation

$$\frac{d^2x}{dt^2} + \frac{dx}{dt} = x$$

into a system of first order differential equations.

22. Transform the second order differential equation

$$\frac{d^2x}{dt^2} - 2\frac{dx}{dt} = 3x$$

into a system of first order differential equations.

◆ 11.3 NONLINEAR AUTONOMOUS SYSTEMS—THEORY

In this section, we will develop some of the theory needed to analyze systems of differential equations of the form

$$\frac{dx_1}{dt} = f_1(x_1, x_2, \ldots, x_n)$$

$$\frac{dx_2}{dt} = f_2(x_1, x_2, \ldots, x_n)$$

$$\vdots$$

$$\frac{dx_n}{dt} = f_n(x_1, x_2, \ldots, x_n)$$

(11.47)

where $f_i : \mathbf{R}^n \to \mathbf{R}$, for $i = 1, 2, \ldots, n$. Since the functions f_i, $i = 1, 2, \ldots, n$, do not explicitly depend on t, the system (11.47) is called autonomous. Furthermore, we no longer assume that the functions f_i are linear as in Section 11.1. Using matrix notation, the system (11.47) can be written in the form

$$\frac{d\mathbf{x}}{dt} = \mathbf{f}(\mathbf{x})$$

where $\mathbf{x} = [x_1, x_2, \ldots, x_n]'$, and $\mathbf{f}(\mathbf{x})$ is a vector-valued function $\mathbf{f} : \mathbf{R}^n \to \mathbf{R}^n$ with components $f_i(x_1, x_2, \ldots, x_n) : \mathbf{R}^n \to \mathbf{R}, i = 1, 2, \ldots, n$. The function $\mathbf{f}(\mathbf{x})$ defines a direction field, just as in the linear case.

Unless the functions f_i are linear, it is typically not possible to find explicit solutions of systems of differential equations. If one wants to solve such systems, one frequently must use numerical methods. Instead of trying to find solutions, we will focus on equilibria and their stability, just as in Section 7.2.

The definition of an equilibrium (as given in Section 7.2) must be extended to systems of the form (11.47). We say that a point

$$\hat{\mathbf{x}} = (\hat{x}_1, \hat{x}_2, \ldots, \hat{x}_n)$$

is an equilibrium of (11.47) if

$$\mathbf{f}(\hat{\mathbf{x}}) = \mathbf{0}$$

An equilibrium is also called a **critical point**. As in the linear case, this is a point in the direction field at which the direction vector has length 0, implying that if we start a system of differential equations at an equilibrium point, it will stay there for all later times.

As in the linear case, a solution might not return to an equilibrium after a small perturbation; this is addressed by the stability of the equilibrium. The theory of stability for systems of nonlinear autonomous differential equations is parallel to that in Section 7.2; there is both an analytical and graphical approach that reduces to the theory in Section 7.2 when there is a single differential equation. We will restrict our discussion to systems of two equations in at most two variables.

11.3.1 Analytical Approach

A Single Autonomous Differential Equation

┃EXAMPLE 1   Find all equilibria of

$$\frac{dx}{dt} = x(1 - x)$$

(11.48)

and analyze their stability.

Solution   We developed the theory for single autonomous differential equations in Section 7.2. To find equilibria, we set

$$x(1 - x) = 0$$

which yields

$$\hat{x}_1 = 0 \quad \text{and} \quad \hat{x}_2 = 1$$

To analyze their stability, we linearize the differential equation (11.48) about each equilibrium, and compute the corresponding eigenvalue. We set

$$f(x) = x(1 - x)$$

then

$$f'(x) = 1 - 2x$$

The eigenvalue associated with the equilibrium $\hat{x}_1 = 0$ is

$$\lambda_1 = f'(0) = 1 > 0$$

which implies that $\hat{x}_1 = 0$ is unstable.

The eigenvalue associated with the equilibrium $\hat{x}_2 = 1$ is

$$\lambda_2 = f'(1) = -1 < 0$$

which implies that $\hat{x}_2 = 1$ is locally stable.

The eigenvalue corresponding to an equilibrium of the differential equation

$$\frac{dx}{dt} = f(x) \tag{11.49}$$

is the slope of the function $f(x)$ at the equilibrium value. The reason for this is discussed in detail in Section 7.2; we repeat the basic argument here. Suppose that $\hat{x}$ is an equilibrium of (11.49), that is, $f(\hat{x}) = 0$. If we perturb $\hat{x}$ slightly (that is, if we look at $\hat{x} + z$ for small $|z|$), we can find out what happens to $\hat{x} + z$ by looking at

$$\frac{dx}{dt} = \frac{d}{dt}(\hat{x} + z) = \frac{dz}{dt}$$

and

$$f(\hat{x} + z) \approx f(\hat{x}) + f'(\hat{x})(\hat{x} + z - \hat{x}) = f'(\hat{x})z$$

In the last step, we used the fact that $f(\hat{x}) = 0$. We find

$$\frac{dz}{dt} \approx f'(\hat{x})z$$

which has approximate solution

$$z(t) \approx z(0)e^{\lambda t} \quad \text{with } \lambda = f'(\hat{x})$$

Therefore, we see that if $f'(\hat{x}) < 0$, then $z(t) \to 0$ as $t \to \infty$ and, hence, $x(t) = \hat{x} + z(t) \to \hat{x}$ as $t \to \infty$; that is, the solution will return to the equilibrium $\hat{x}$ after a small perturbation. In this case, $\hat{x}$ is locally stable. If $f'(\hat{x}) > 0$, then $z(t)$ will not go to 0, which implies that $\hat{x}$ is unstable. The linearization of $f(x)$ thus tells us whether an equilibrium is locally stable or unstable. The exact same method is used for systems of differential equations.

Systems of Two Differential Equations   We consider differential equations of the form

$$\frac{dx_1}{dt} = f_1(x_1, x_2) \tag{11.50}$$

$$\frac{dx_2}{dt} = f_2(x_1, x_2) \tag{11.51}$$

or, using matrix notation,

$$\frac{d\mathbf{x}}{dt} = \mathbf{f}(\mathbf{x}) \tag{11.52}$$

where $\mathbf{x} = \mathbf{x}(t) = (x_1(t), x_2(t))$, and $\mathbf{f}(\mathbf{x})$ is a vector-valued function $\mathbf{f}(\mathbf{x}) = (f_1(\mathbf{x}), f_2(\mathbf{x}))$ with $f_i(\mathbf{x}) : \mathbf{R}^2 \to \mathbf{R}$. An equilibrium or critical point of (11.52) satisfies

$$\mathbf{f}(\mathbf{x}) = \mathbf{0}$$

Suppose that $\hat{\mathbf{x}}$ is an equilibrium (that is, $\mathbf{f}(\hat{\mathbf{x}}) = \mathbf{0}$); then, as in the case of one nonlinear equation, we look at what happens to a small perturbation of $\hat{\mathbf{x}}$. We perturb $\hat{\mathbf{x}}$, that is, we look at how $\hat{\mathbf{x}} + \mathbf{z}$ changes under the dynamics described by (11.52):

$$\frac{d}{dt}(\hat{\mathbf{x}} + \mathbf{z}) = \frac{d}{dt}\mathbf{z} = \mathbf{f}(\hat{\mathbf{x}} + \mathbf{z})$$

Since $\mathbf{z}$ is a small perturbation, we linearize $\mathbf{f}(\mathbf{x})$ about $\mathbf{x} = \hat{\mathbf{x}}$. We find

$$\mathbf{f}(\hat{\mathbf{x}} + \mathbf{z}) \approx \mathbf{f}(\hat{\mathbf{x}}) + D\mathbf{f}(\hat{\mathbf{x}})\mathbf{z} = D\mathbf{f}(\hat{\mathbf{x}})\mathbf{z}$$

where $D\mathbf{f}(\hat{\mathbf{x}})$ is the Jacobi matrix evaluated at $\hat{\mathbf{x}}$. In the last step, we used the fact that $\mathbf{f}(\hat{\mathbf{x}}) = \mathbf{0}$. Hence,

$$\frac{d\mathbf{z}}{dt} \approx D\mathbf{f}(\hat{\mathbf{x}})\mathbf{z} \tag{11.53}$$

We now have a system of linear differential equations that is a good approximation, provided that $\mathbf{z}$ is sufficiently close to $\mathbf{0}$. We learned in Section 11.1 how to analyze linear systems. We saw that eigenvalues of the matrix $D\mathbf{f}(\hat{\mathbf{x}})$ allowed us to determine the nature of the equilibrium. We will use the same approach here, but we emphasize that this is now a *local* analysis. Our statements about what happens are valid only when we start close to an equilibrium, since the linearization (11.53) is a good approximation only as long as we are sufficiently close to the point about which we linearized.

We return to our classification scheme for the linear case,

$$\frac{d\mathbf{x}}{dt} = A\mathbf{x}$$

where $\mathbf{x}(t) = (x_1(t), x_2(t))$ and A is a 2×2 matrix. We let

$$\Delta = \det A \qquad \text{and} \qquad \tau = \operatorname{tr} A$$

When we linearize a nonlinear system about an equilibrium, the matrix A is the Jacobi matrix evaluated at the equilibrium:

$$A = D\mathbf{f}(\hat{\mathbf{x}})$$

We exclude the following cases: (i) $\Delta = 0$, (ii) $\tau = 0$ and $\Delta > 0$, and (iii) $\tau^2 = 4\Delta$. Recall that when $\Delta = 0$, at least one eigenvalue is equal to 0. When $\tau = 0$ and $\Delta > 0$, the two eigenvalues are purely imaginary. When $\tau^2 = 4\Delta$,

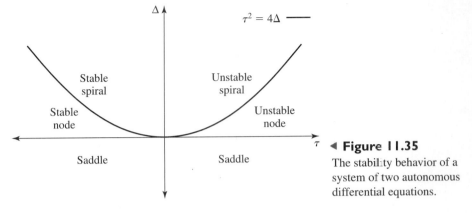

◀ **Figure 11.35**

The stability behavior of a system of two autonomous differential equations.

the two eigenvalues are identical. Except in these three cases, we can use the same classification scheme as in the linear case, as illustrated in Figure 11.35.

This is possible because the linearized vector field and the original vector field are geometrically very similar when we are close to an equilibrium point (after all, that is the idea behind linearization). This result is known as the Hartman-Grobman theorem, which says that as long as $D\mathbf{f}(\hat{\mathbf{x}})$ has no zero or purely imaginary eigenvalues, then the linearized and the original vector fields are very similar close to the equilibrium. That is,

$$\frac{d\mathbf{x}}{dt} = \mathbf{f}(\mathbf{x}) \qquad \text{and} \qquad \frac{d\mathbf{z}}{dt} = D\mathbf{f}(\hat{\mathbf{x}})\mathbf{z}$$

behave very similarly for $\mathbf{x} = \hat{\mathbf{x}} + \mathbf{z}$ with $\mathbf{z}$ close to $\mathbf{0}$.

We find the same classification scheme as in the linear case:

- The equilibrium $\hat{\mathbf{x}}$ is a node if both eigenvalues of $D\mathbf{f}(\hat{\mathbf{x}})$ are real, distinct, nonzero, and of the same sign. The node is locally stable if the eigenvalues are negative, and unstable if the eigenvalues are positive.

- The equilibrium $\hat{\mathbf{x}}$ is a saddle if both eigenvalues are real and nonzero but have opposite signs. A saddle is unstable.

- The equilibrium $\hat{\mathbf{x}}$ is a spiral if both eigenvalues are complex conjugates with nonzero real parts. The spiral is stable if the real parts of the eigenvalues are negative, and unstable if the real parts of the eigenvalues are positive.

When the two eigenvalues have zero real parts, we cannot determine the stability by linearization.

▎**EXAMPLE 2**   Consider

$$\frac{dx_1}{dt} = x_1 - 2x_1^2 - 2x_1x_2$$

$$\frac{dx_2}{dt} = 4x_2 - 5x_2^2 - 7x_1x_2 \tag{11.54}$$

a. Find all equilibria of (11.54), and

b. analyze their stability.

Solution

a. To find equilibria, we set the right-hand side of (11.54) equal to 0:

$$x_1 - 2x_1^2 - 2x_1x_2 = 0 \tag{11.55}$$

$$4x_2 - 5x_2^2 - 7x_1x_2 = 0 \tag{11.56}$$

Factoring x_1 in the first equation and x_2 in the second yields

$$x_1(1 - 2x_1 - 2x_2) = 0 \quad \text{and} \quad x_2(4 - 5x_2 - 7x_1) = 0$$

That is,

$$x_1 = 0 \quad \text{or} \quad 2x_1 + 2x_2 = 1$$

and

$$x_2 = 0 \quad \text{or} \quad 7x_1 + 5x_2 = 4$$

Combining the different solutions, we get the four cases

(i) $x_1 = 0$ and $x_2 = 0$

(ii) $x_1 = 0$ and $7x_1 + 5x_2 = 4$

(iii) $x_2 = 0$ and $2x_1 + 2x_2 = 1$

(iv) $2x_1 + 2x_2 = 1$ and $7x_1 + 5x_2 = 4$

First, we will compute the equilibria in these four cases.

Case (i)   There is nothing to compute; the equilibrium is $(\hat{x}_1, \hat{x}_2) = (0, 0)$.

Case (ii)   To find the equilibrium, we must solve the system

$$x_1 = 0$$
$$7x_1 + 5x_2 = 4$$

which has the solutions

$$x_1 = 0 \quad \text{and} \quad x_2 = \frac{4}{5}$$

Hence, the equilibrium is $(\hat{x}_1, \hat{x}_2) = \left(0, \frac{4}{5}\right)$.

Case (iii)   To find the equilibrium, we must solve the system

$$x_2 = 0$$
$$2x_1 + 2x_2 = 1$$

which has the solutions

$$x_2 = 0 \quad \text{and} \quad x_1 = \frac{1}{2}$$

Hence, the equilibrium is $(\hat{x}_1, \hat{x}_2) = \left(\frac{1}{2}, 0\right)$.

Case (iv)   To find the equilibrium, we must solve the system

$$2x_1 + 2x_2 = 1$$
$$7x_1 + 5x_2 = 4$$

We use standard elimination method: leaving the first equation, and changing the second by multiplying the first by 7 and the second by 2, and then subtracting the second equation from the first, we find that this system is equivalent to

$$2x_1 + 2x_2 = 1$$
$$4x_2 = -1$$

which has the solutions

$$x_2 = -\frac{1}{4} \quad \text{and} \quad x_1 = \frac{1}{2} - x_2 = \frac{3}{4}$$

Hence, the equilibrium is $(\hat{x}_1, \hat{x}_2) = \left(\frac{3}{4}, -\frac{1}{4}\right)$.

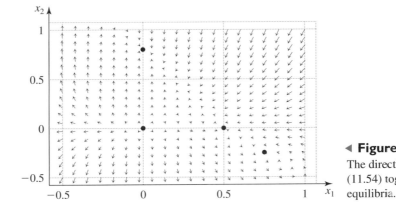

◀ **Figure 11.36**
The direction field of
(11.54) together with the
equilibria.

We can illustrate all equilibria in the direction field of (11.54), which is displayed in Figure 11.36. The equilibria are shown as dots.

b. To analyze the stability of the equilibria, we compute the Jacobi matrix

$$Df = \begin{bmatrix} \frac{\partial f_1}{\partial x_1} & \frac{\partial f_1}{\partial x_2} \\ \frac{\partial f_2}{\partial x_1} & \frac{\partial f_2}{\partial x_2} \end{bmatrix}$$

With

$$f_1(x_1, x_2) = x_1 - 2x_1^2 - 2x_1 x_2 \qquad \text{and} \qquad f_2(x_1, x_2) = 4x_2 - 5x_2^2 - 7x_1 x_2$$

we find

$$Df(x_1, x_2) = \begin{bmatrix} 1 - 4x_1 - 2x_2 & -2x_1 \\ -7x_2 & 4 - 10x_2 - 7x_1 \end{bmatrix}$$

We will now go through the four cases and analyze each equilibrium.

Case (i) The equilibrium is the point $(0, 0)$. The Jacobi matrix at $(0, 0)$ is

$$Df(0, 0) = \begin{bmatrix} 1 & 0 \\ 0 & 4 \end{bmatrix}$$

Since this matrix is in diagonal form, the eigenvalues are the diagonal elements, and we find $\lambda_1 = 1$ and $\lambda_2 = 4$. Since both eigenvalues are positive, the equilibrium is unstable. Using the same classification as in the linear case, we say that $(0, 0)$ is an unstable node.

The linearization of the direction field about $(0, 0)$ is displayed in Figure 11.37. That is, in Figure 11.37, we show the direction field of

$$\frac{d\mathbf{x}}{dt} = \begin{bmatrix} 1 & 0 \\ 0 & 4 \end{bmatrix} \mathbf{x}(t)$$

Figure 11.37 confirms that $(0, 0)$ is an unstable node (or source). You should compare Figure 11.37 to the direction field of (11.54) shown in Figure 11.36. You will find that the linearized direction field and the direction field of (11.54) close to the equilibrium $(0, 0)$ are quite similar.

Case (ii) The equilibrium is the point $\left(0, \frac{4}{5}\right)$. The Jacobi matrix at $\left(0, \frac{4}{5}\right)$ is

$$Df\left(0, \frac{4}{5}\right) = \begin{bmatrix} 1 - \frac{8}{5} & 0 \\ -\frac{28}{5} & 4 - 8 \end{bmatrix} = \begin{bmatrix} -\frac{3}{5} & 0 \\ -\frac{28}{5} & -4 \end{bmatrix}$$

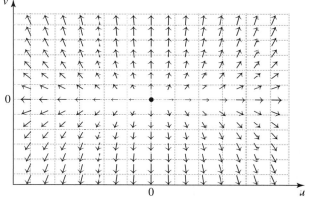

◀ **Figure 11.37**
The linearization of the direction field about $(0, 0)$.

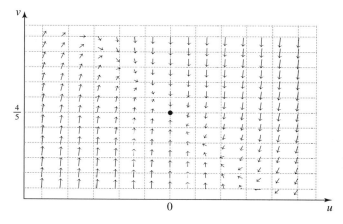

◀ **Figure 11.38**
The linearization of the direction field about $\left(0, \frac{4}{5}\right)$.

Since this matrix is in lower triangular form, the eigenvalues are simply the diagonal elements, and we find that the eigenvalues of $D\mathbf{f}(0, \frac{4}{5})$ are $\lambda_1 = -\frac{3}{5}$ and $\lambda_2 = -4$. Since both eigenvalues are negative, $(0, \frac{4}{5})$ is locally stable. Using the same classification as in the linear case, we say that $(0, \frac{4}{5})$ is a stable node.

The linearization of the direction field about $(0, \frac{4}{5})$ is displayed in Figure 11.38. Figure 11.38 confirms that $\left(0, \frac{4}{5}\right)$ is a stable node (or sink). You should compare Figure 11.38 to the direction field of (11.54) shown in Figure 11.36. You will find that the linearized direction field and the direction field of (11.54) close to the equilibrium $\left(0, \frac{4}{5}\right)$ are quite similar.

Case (iii) The equilibrium is the point $\left(\frac{1}{2}, 0\right)$. The Jacobi matrix at $\left(\frac{1}{2}, 0\right)$ is

$$D\mathbf{f}\left(\frac{1}{2}, 0\right) = \begin{bmatrix} 1 - 2 & -1 \\ 0 & 4 - \frac{7}{2} \end{bmatrix} = \begin{bmatrix} -1 & -1 \\ 0 & \frac{1}{2} \end{bmatrix}$$

Since this matrix is in upper triangular form, the eigenvalues are simply the diagonal elements, and we find that the eigenvalues of $D\mathbf{f}(\frac{1}{2}, 0)$ are $\lambda_1 = -1$ and $\lambda_2 = \frac{1}{2}$. Since one eigenvalue is positive and the other is negative, $\left(\frac{1}{2}, 0\right)$ is unstable. Using the same classification as in the linear case, we say that $\left(\frac{1}{2}, 0\right)$ is a saddle.

The linearization of the direction field about $\left(\frac{1}{2}, 0\right)$ is displayed in Figure 11.39. Figure 11.39 confirms that $\left(\frac{1}{2}, 0\right)$ is a saddle. You should compare Figure 11.39 to the direction field of (11.54) shown in Figure 11.36. You will

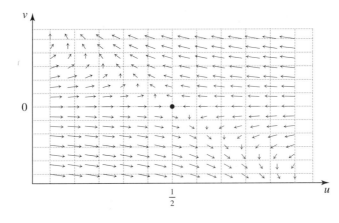

◀ **Figure 11.39**
The linearization of the
direction field about $\left(\frac{1}{2}, 0\right)$.

find that the linearized direction field and the direction field of (11.54) close
to the equilibrium $\left(\frac{1}{2}, 0\right)$ are quite similar.

Case (iv)   The Jacobi matrix at the equilibrium $\left(\frac{3}{4}, -\frac{1}{4}\right)$ is

$$\mathbf{Df}\left(\frac{3}{4}, -\frac{1}{4}\right) = \begin{bmatrix} 1 - 3 + \frac{1}{2} & -\frac{3}{2} \\ \frac{7}{4} & 4 + \frac{10}{4} - \frac{21}{4} \end{bmatrix}$$

$$= \begin{bmatrix} -\frac{3}{2} & -\frac{3}{2} \\ \frac{7}{4} & \frac{5}{4} \end{bmatrix}$$

To find the eigenvalues, we must solve

$$\det \begin{bmatrix} -\frac{3}{2} - \lambda & -\frac{3}{2} \\ \frac{7}{4} & \frac{5}{4} - \lambda \end{bmatrix} = 0$$

That is, we must solve

$$\left(-\frac{3}{2} - \lambda\right)\left(\frac{5}{4} - \lambda\right) + \left(\frac{3}{2}\right)\left(\frac{7}{4}\right) = 0$$

$$\lambda^2 + \frac{3}{2}\lambda - \frac{5}{4}\lambda - \frac{15}{8} + \frac{21}{8} = 0$$

$$\lambda^2 + \frac{1}{4}\lambda + \frac{3}{4} = 0$$

Solving this quadratic equation, we find

$$\lambda_{1,2} = \frac{-\frac{1}{4} \pm \sqrt{\frac{1}{16} - 3}}{2}$$

$$= -\frac{1}{8} \pm \frac{1}{8}\sqrt{-47} = -\frac{1}{8} \pm \frac{1}{8}i\sqrt{47}$$

That is,

$$\lambda_1 = -\frac{1}{8} + \frac{1}{8}i\sqrt{47} \qquad \text{and} \qquad \lambda_2 = -\frac{1}{8} - \frac{1}{8}i\sqrt{47}$$

The eigenvalues are complex conjugates, with negative real parts. That is,
$\left(\frac{3}{4}, -\frac{1}{4}\right)$ is locally stable, and we expect the solutions to spiral into the equilibrium when we start close to the equilibrium.

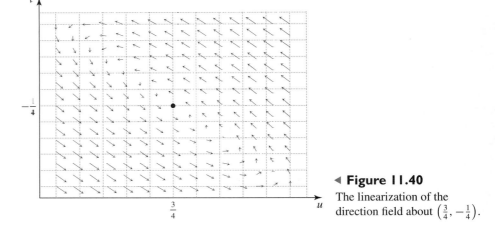

◀ **Figure 11.40**

The linearization of the direction field about $\left(\frac{3}{4}, -\frac{1}{4}\right)$.

The linearization of the direction field about $\left(\frac{3}{4}, -\frac{1}{4}\right)$ is displayed in Figure 11.40. Figure 11.40 confirms that $\left(\frac{3}{4}, -\frac{1}{4}\right)$ is a stable spiral. You should compare Figure 11.40 to the direction field of (11.54) shown in Figure 11.36. You will find that the linearized direction field and the direction field of (11.54) close to the equilibrium $\left(\frac{3}{4}, -\frac{1}{4}\right)$ are quite similar. ▪

11.3.2 Graphical Approach for 2×2 Systems

In this subsection, we will discuss a graphical approach to systems of two autonomous differential equations. Suppose that

$$\frac{dx_1}{dt} = f_1(x_1, x_2)$$

$$\frac{dx_2}{dt} = f_2(x_1, x_2)$$

which in matrix form is

$$\frac{d\mathbf{x}}{dt} = \mathbf{f}(\mathbf{x})$$

The curves

$$f_1(x_1, x_2) = 0$$
$$f_2(x_1, x_2) = 0$$

are called **zero isoclines** or **null clines**, and they represent the points in the x_1-x_2 plane where the growth rates of the respective quantities are equal to zero. This is illustrated in Figure 11.41 for a particular choice of f_1 and f_2. In the following discussion, we assume that x_1 and x_2 are nonnegative, that is, we are only interested in the first quadrant of the x_1-x_2 plane. The two curves in Figure 11.41 divide the first quadrant into four regions, and we label each region according to whether dx_i/dt is positive or negative. Of course, this depends on the functions $f_1(x_1, x_2)$ and $f_2(x_1, x_2)$. The labeling of the regions in Figure 11.41 should therefore be considered as our choice to be more concrete in the following discussion.

The point where both null clines in Figure 11.41 intersect is an equilibrium or critical point. We denote the equilibrium by $\hat{\mathbf{x}}$. We can use the graph to determine

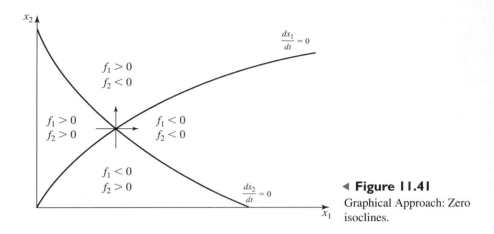

◀ **Figure 11.41**
Graphical Approach: Zero isoclines.

the signs of the entries in the Jacobi matrix

$$Df(\hat{x}) = \begin{bmatrix} a_{11} & a_{12} \\ a_{21} & a_{22} \end{bmatrix}$$

where $a_{ij} = \frac{\partial f_i}{\partial x_j}(\hat{x})$. Namely, the entry $a_{11} = \frac{\partial f_1}{\partial x_1}$ is the effect of a change of f_1 in the x_1 direction when we keep x_2 fixed. That is, we follow the horizontal arrow in Figure 11.41. The arrow goes from a region where f_1 is positive to a region where f_1 is negative; that is, f_1 is decreasing and hence $\frac{\partial f_1}{\partial x_1}(\hat{x}) < 0$. This determines the sign of a_{11} in $Df(\hat{x})$, that is,

$$Df(\hat{x}) = \begin{bmatrix} - & a_{12} \\ a_{21} & a_{22} \end{bmatrix}$$

Next, we determine $a_{12} = \frac{\partial f_1}{\partial x_2}(\hat{x})$. This time, we want to know how f_1 changes at an equilibrium when we move in the x_2 direction and keep x_1 fixed. This is the direction of the vertical arrow through the equilibrium point. Since the arrow goes from a region where f_1 is negative to a region where f_1 is positive, f_1 increases in the direction of x_2 and, therefore, $a_{12} = \frac{\partial f_1}{\partial x_2}(\hat{x}) > 0$.

The signs of the other two entries are found similarly, and we obtain

$$Df(\hat{x}) = \begin{bmatrix} - & + \\ - & - \end{bmatrix}$$

Thus, the trace of $Df(\hat{x})$ is negative and the determinant of $Df(\hat{x})$ is positive. Using the criterion stated in Subsection 9.4.2, we conclude that both eigenvalues have negative real parts and, therefore, that the equilibrium is stable.

This simple graphical approach does not always give us the signs of the real parts of the eigenvalues, as illustrated in the following example. Suppose that we arrive at the following Jacobi matrix, where the signs of the entries are

$$\begin{bmatrix} + & - \\ - & - \end{bmatrix}$$

The trace may now be positive or negative and, similarly, the determinant may be positive or negative. Therefore, we cannot conclude anything about the eigenvalues. In this case we would have to compute the eigenvalues.

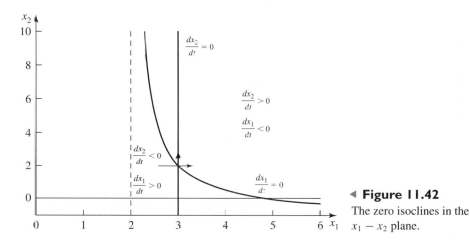

◀ **Figure 11.42**
The zero isoclines in the $x_1 - x_2$ plane.

EXAMPLE 3   Use the graphical approach to analyze the equilibrium $(3, 2)$ of

$$\frac{dx_1}{dt} = 5 - x_1 - x_1 x_2 + 2x_2$$

$$\frac{dx_2}{dt} = x_1 x_2 - 3x_2$$

Solution   First, note that $(3, 2)$ is indeed an equilibrium of this system. Now,

$$\frac{dx_1}{dt} = 0 \quad \text{for } x_2 = \frac{5 - x_1}{x_1 - 2}$$

and

$$\frac{dx_2}{dt} = 0 \quad \text{for } x_2 = 0 \text{ or } x_1 = 3$$

The zero isoclines in the x_1-x_2 plane are illustrated in Figure 11.42. The equilibrium $(3, 2)$ is given by the point where the zero isoclines $x_1 = 3$ and $x_2 = \frac{5-x_1}{x_1-2}$ intersect. The signs of $\frac{dx_1}{dt}$ and $\frac{dx_2}{dt}$ are indicated in Figure 11.42. We find

$$\mathbf{Df}(\hat{\mathbf{x}}) = \begin{bmatrix} - & - \\ + & 0 \end{bmatrix}$$

Since the trace is negative and the determinant is positive, we conclude that the equilibrium is stable.   ▮

◆ 11.3 PROBLEMS

[11.3.1]

In Problems 1–4, the point $(0, 0)$ is always an equilibrium. Investigate its stability using the analytical approach.

1. $\dfrac{dx_1}{dt} = x_1 - 2x_2 + x_1 x_2$

$\dfrac{dx_2}{dt} = -x_1 + x_2$

2. $\dfrac{dx_1}{dt} = -x_1 - x_2 + x_1^2$

$\dfrac{dx_2}{dt} = x_2 - x_1^2$

3. $\dfrac{dx_1}{dt} = x_1 + x_1^2 - 2x_1 x_2 + x_2$

$\dfrac{dx_2}{dt} = x_1$

4. $\dfrac{dx_1}{dt} = 3x_1 x_2 - x_1 + x_2$

$\dfrac{dx_2}{dt} = x_2^2 - x_1$

In Problems 5–10, find all equilibria of each system of differential equations, and use the analytical approach to determine the stability of each equilibrium.

5. $\dfrac{dx_1}{dt} = -x_1 + 2x_1(1 - x_1)$

$\dfrac{dx_2}{dt} = -x_2 + 5x_2(1 - x_1 - x_2)$

6. $\dfrac{dx_1}{dt} = -x_1 + 3x_1(1 - x_1 - x_2)$

$\dfrac{dx_2}{dt} = -x_2 + 5x_2(1 - x_1 - x_2)$

7. $\dfrac{dx_1}{dt} = 4x_1(1 - x_1) - 2x_1x_2$

$\dfrac{dx_2}{dt} = x_2(2 - x_2) - x_2$

8. $\dfrac{dx_1}{dt} = 2x_1(5 - x_1 - x_2)$

$\dfrac{dx_2}{dt} = 3x_2(7 - 3x_1 - x_2)$

9. $\dfrac{dx_1}{dt} = x_1 - x_2$

$\dfrac{dx_2}{dt} = x_1x_2 - x_2$

10. $\dfrac{dx_1}{dt} = x_1x_2 - x_2$

$\dfrac{dx_2}{dt} = x_1 + x_2$

[11.3.2]

11. Assume that

$$\dfrac{dx_1}{dt} = x_1(10 - 2x_1 - x_2)$$

$$\dfrac{dx_2}{dt} = x_2(10 - x_1 - 2x_2)$$

(a) Graph the zero isoclines.

(b) Show that $\left(\frac{10}{3}, \frac{10}{3}\right)$ is an equilibrium, and determine its stability using the analytical approach.

12. Assume that

$$\dfrac{dx_1}{dt} = x_1(10 - x_1 - 2x_2)$$

$$\dfrac{dx_2}{dt} = x_2(10 - 2x_1 - x_2)$$

(a) Graph the zero isoclines.

(b) Show that $\left(\frac{10}{3}, \frac{10}{3}\right)$ is an equilibrium, and determine its stability using the analytical approach.

In Problems 13–18, use the graphical approach for 2×2 systems to find the sign structure of the Jacobi matrix at the indicated equilibrium. If possible, determine its stability. Assume that the system of differential equations is given by

$$\dfrac{dx_1}{dt} = f_1(x_1, x_2)$$

$$\dfrac{dx_2}{dt} = f_2(x_1, x_2)$$

Furthermore, assume that x_1 and x_2 are both nonnegative. In each problem, the zero isoclines are drawn and the equilibrium we want to investigate is indicated by a dot. Assume that both x_1 and x_2 increase close to the origin.

13.

▲ **Figure 11.43**

14.

▲ **Figure 11.44**

15.

▲ **Figure 11.45**

16.

▲ **Figure 11.46**

17.

▲ **Figure 11.47**

18.

▲ **Figure 11.48**

19. Let

$$\frac{dx_1}{dt} = x_1(2 - x_1) - x_1 x_2$$

$$\frac{dx_2}{dt} = x_1 x_2 - x_2$$

(a) Graph the zero isoclines.

(b) Show that $(1, 1)$ is an equilibrium. Use the graphical approach to determine its stability.

20. Let

$$\frac{dx_1}{dt} = x_1(2 - x_1^2) - x_1 x_2$$

$$\frac{dx_2}{dt} = x_1 x_2 - x_2$$

(a) Graph the zero isoclines.

(b) Show that $(1, 1)$ is an equilibrium. Use the graphical approach to determine its stability.

◆ 11.4 NONLINEAR SYSTEMS—APPLICATIONS

11.4.1 The Lotka-Volterra Model of Interspecific Competition

Imagine two species of plants growing together in the same plot. They both use similar resources, namely light, water, and nutrients. The use of these resources by one individual reduces their availability to other individuals close by. We call this type of interaction between individuals **competition**. **Intraspecific competition** occurs between individuals of the same species and **interspecific competition** between individuals of different species. In either case, competition typically results in reduced fecundity and/or reduced survivorship. The effects of competition are often more pronounced when the number of competitors is higher, in which case we say that the effects of competition are density dependent.

In this subsection we will discuss the Lotka-Volterra model of interspecific competition, which incorporates density-dependent effects of competition in the manner described above. It is an extension of the logistic equation to the case of two species. To describe the model, we denote the population size of species 1 at time t by $N_1(t)$ and that of species 2 at time t by $N_2(t)$. Each species grows according to the logistic equation in the absence of the other species. We denote their respective carrying capacities by K_1 and K_2, and their respective intrinsic rates of growth by r_1 and r_2. We assume that K_1, K_2, r_1, and r_2 are positive. In addition, the two species may have inhibitory effects on each other. We measure the effect of species 1 on species 2 by the **competition coefficient** α_{21}; the effect of species 2 on species 1 is measured by the competition coefficient α_{12}. The Lotka-Volterra model of interspecific competition is then given by the following system of differential equations:

$$\frac{dN_1}{dt} = r_1 N_1 \left(1 - \frac{N_1}{K_1} - \alpha_{12}\frac{N_2}{K_1} \right)$$

$$\frac{dN_2}{dt} = r_2 N_2 \left(1 - \frac{N_2}{K_2} - \alpha_{21}\frac{N_1}{K_2} \right)$$

(11.57)

Let's look at the first equation. If $N_2 = 0$, the first equation reduces to the logistic equation $dN_1/dt = r_1 N_1(1 - N_1/K_1)$, as mentioned. To understand the precise meaning of the competition coefficient α_{12}, observe that N_2 individuals of species 2 have the same effect on species 1 as $\alpha_{12}N_2$ individuals of species 1. The term $\alpha_{12}N_2$ thus converts the number of N_2 individuals into "N_1-equivalents." For instance, set $\alpha_{12} = 0.2$ and assume that $N_2 = 20$; then the effect of 20 individuals of species 2 on species 1 is the same as the effect of $(0.2)(20) = 4$ individuals of species 1 on species 1, because both reduce the growth rate by the same amount. A similar interpretation can be attached to the competition coefficient α_{21} in the second equation.

This model takes a very simplistic view of competition; actual competitive interactions are more complicated. Nevertheless, it serves an important purpose. Precisely because of its simple form, it allows us to study the consequences of competition, which can give us valuable insight into complex situations.

We will analyze the model using both the graphical approach and the eigenvalue approach.

The Graphical Approach   The first step is to find the equations of the zero isoclines. To find the zero isocline for species 1, we set

$$r_1 N_1 \left(1 - \frac{N_1}{K_1} - \alpha_{12}\frac{N_2}{K_1} \right) = 0$$

The solutions are $N_1 = 0$ or

$$N_2 = \frac{K_1}{\alpha_{12}} - \frac{1}{\alpha_{12}}N_1$$

(11.58)

To find the zero isocline for species 2, we set

$$r_2 N_2 \left(1 - \frac{N_2}{K_2} - \alpha_{21}\frac{N_1}{K_2} \right) = 0$$

The solutions are $N_2 = 0$ or

$$N_2 = K_2 - \alpha_{21}N_1$$

(11.59)

The isocline $N_i = 0$, $i = 1, 2$, corresponds to the case in which species i is absent. This is biologically reasonable, because individuals of either species are not created spontaneously — once a species is absent, it remains absent.

The other two isoclines, given by (11.58) and (11.59), are of particular interest because they tell us whether both species can stably coexist. Both isoclines are straight lines in the N_1-N_2 plane, and there are four ways these isoclines can be arranged. This is illustrated in Figures 11.49–11.52, together with the corresponding direction fields. The solid dots in each figure are the equilibria. We see that there are always the equilibria $(K_1, 0)$ and $(0, K_2)$. These are the equilibria where only one species is present; they are referred to as **monoculture** equilibria. In addition, there are two cases where an equilibrium exists when both species are present; such an equilibrium is called a nontrivial equilibrium. We will now discuss each case separately.

Case 1: $K_1 > \alpha_{12}K_2$ and $K_2 < \alpha_{21}K_1$ When we look at the direction field in Figure 11.49, we see that species 1 outcompetes species 2 and drives it to extinction.

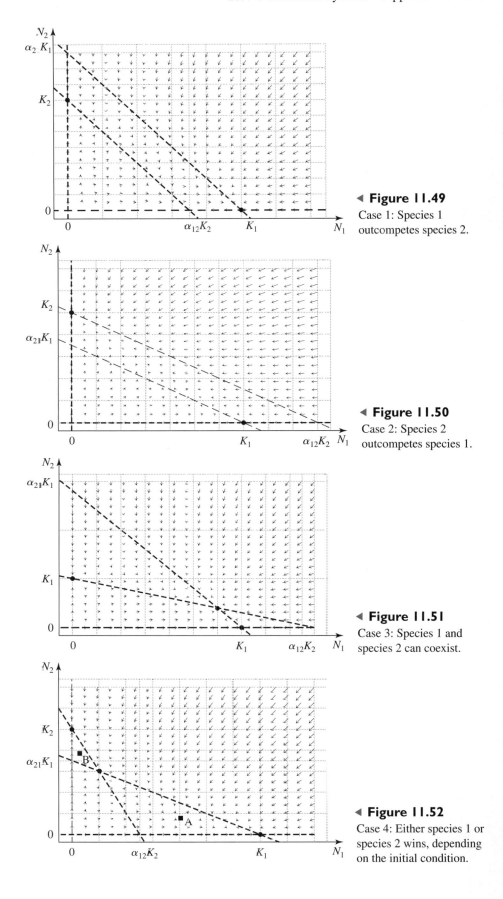

◄ **Figure 11.49**
Case 1: Species 1
outcompetes species 2.

◄ **Figure 11.50**
Case 2: Species 2
outcompetes species 1.

◄ **Figure 11.51**
Case 3: Species 1 and
species 2 can coexist.

◄ **Figure 11.52**
Case 4: Either species 1 or
species 2 wins, depending
on the initial condition.

If both species are present initially, then the abundance of species 2 declines over time, and it will eventually become extinct, whereas species 1 will reach its carrying capacity K_1. We say that species 1 outcompetes species 2, and refer to this as the case of **competitive exclusion**.

We can understand why species 1 is the better competitor by looking at the inequalities $K_1 > \alpha_{12}K_2$ and $K_2 < \alpha_{21}K_1$. The carrying capacity K_1 is the abundance of species 1 that results in a zero per capita growth rate of species 1 in the absence of species 2. Namely, if $N_1 = K_1$ and $N_2 = 0$, then $\frac{1}{N_1}\frac{dN_1}{dt} = 0$. The carrying capacity K_2 has a similar interpretation. To see the effect of species 2 on species 1, we look at the case when species 2 is at its carrying capacity K_2 and the abundance of species 1 is negligible. In this case, the per capita growth rate of species 1 is approximately equal to $r_1(1 - \alpha_{12}K_2/K_1)$, which is positive since $\alpha_{12}K_2/K_1$ is less than 1. This implies that species 2 cannot prevent species 1 from increasing its abundance when rare; we say that species 1 can invade species 2 when rare. On the other hand, since $K_2 < \alpha_{21}K_1$, species 2 cannot invade species 1. Namely, if we set $N_1 = K_1$ and assume that the abundance of species 2 is negligible, then the per capita growth rate of species 2 is approximately $r_2(1 - \alpha_{21}K_1/K_2)$, which is negative. We say that species 1 is the strong and species 2 is the weak interspecific competitor.

Case 2: $K_1 < \alpha_{12}K_2$ and $K_2 > \alpha_{21}K_1$ This is the same as case 1, but with the roles of species 1 and 2 interchanged. That is, species 2 is now the strong and species 1 is the weak interspecific competitor. We see from the direction field in Figure 11.50 that species 2 outcompetes species 1 and drives it to extinction.

Case 3: $K_1 > \alpha_{12}K_2$ and $K_2 > \alpha_{21}K_1$ The two inequalities imply that each species can invade the other species. When we look at the direction field of Figure 11.51, we see that the interior equilibrium where both species are present is stable. We say that **coexistence** is possible. The two monoculture equilibria are unstable.

Case 4: $K_1 < \alpha_{12}K_2$ and $K_2 < \alpha_{21}K_1$ In this case, neither species can invade the other. When we look at the direction field of Figure 11.52, we see that the interior equilibrium is a saddle and hence unstable. The outcome of competition depends on the initial densities. For instance, if the densities of N_1 and N_2 are given initially by the point A in Figure 11.52, then species 1 will win and species 2 will become extinct (following the direction of the direction vectors in the region that contains the point A). On the other hand, if the densities of N_1 and N_2 are given initially by the point B in Figure 11.52, then species 2 will win and species 1 will become extinct. Since the outcome of competition depends on initial abundances, we refer to this as **founder control**.

We see from the preceding analysis that the Lotka-Volterra model allows for three possible outcomes in a two-species interaction. Cases 1 and 2 show the possibility of competitive exclusion, that is, one species can competitively exclude the other species. Case 3 shows that coexistence is possible. Case 4 shows that, depending on the initial abundances, one or the other species eventually wins; this case is referred to as founder control.

Though we stated that the system (11.57) describes a highly idealized situation of competition, there is a famous example that fits the equations very well. The Russian ecologist G.F. Gause (1934) studied competition between species of the protozoan *Paramecium*. When *Paramecium aurelia* and *P. caudatum* were grown together, *P. aurelia* competitively excluded *P. caudatum*. When *P. caudatum* and *P. bursaria* were grown together, they stably coexisted. Solution curves from (11.57) were fitted to both sets of data by estimating the relevant parameters in (11.57), and an excellent fit was obtained.

We will now turn to the eigenvalue approach and analyze (11.57) again.

The Eigenvalue Approach   We will first determine all possible equilibria. Setting $dN_1/dt = 0$, we find

$$N_1 = 0 \qquad \text{or} \qquad N_1 + \alpha_{12}N_2 = K_1$$

Setting $dN_2/dt = 0$, we find

$$N_2 = 0 \qquad \text{or} \qquad \alpha_{21}N_1 + N_2 = K_2$$

There are four possible combinations:

1. The equilibrium $(\hat{N}_1, \hat{N}_2) = (0, 0)$ refers to the case when both species are absent. This is called the trivial equilibrium.
2. The equilibrium $(\hat{N}_1, \hat{N}_2) = (K_1, 0)$ refers to the case when species 2 is absent and species 1 is at its carrying capacity K_1.
3. The equilibrium $(\hat{N}_1, \hat{N}_2) = (0, K_2)$ refers to the case when species 1 is absent and species 2 is at its carrying capacity K_2.
4. The fourth equilibrium is obtained by simultaneously solving

$$N_1 + \alpha_{12}N_2 = K_1$$
$$\alpha_{21}N_1 + N_2 = K_2$$

and requiring that both solutions be positive.

To analyze the stability of these equilibria, we must compute the Jacobi matrix associated with (11.57). We find

$$D\mathbf{f}(N_1, N_2) = \begin{bmatrix} r_1 - 2\frac{r_1}{K_1}N_1 - \frac{r_1\alpha_{12}}{K_1}N_2 & -\frac{r_1\alpha_{12}}{K_1}N_1 \\ -\frac{r_2\alpha_{21}}{K_2}N_2 & r_2 - 2\frac{r_2}{K_2}N_2 - \frac{r_2\alpha_{21}}{K_2}N_1 \end{bmatrix}$$

We look at each equilibrium separately.

1. Evaluating the Jacobi matrix at the trivial equilibrium $(\hat{N}_1, \hat{N}_2) = (0, 0)$, we find

$$D\mathbf{f}(0, 0) = \begin{bmatrix} r_1 & 0 \\ 0 & r_2 \end{bmatrix}$$

which is a diagonal matrix; the eigenvalues are therefore

$$\lambda_1 = r_1 \qquad \text{and} \qquad \lambda_2 = r_2$$

Since $r_1 > 0$ and $r_2 > 0$, the equilibrium $(0, 0)$ is unstable.

2. Evaluating the Jacobi matrix at the equilibrium $(\hat{N}_1, \hat{N}_2) = (K_1, 0)$, we find

$$D\mathbf{f}(K_1, 0) = \begin{bmatrix} -r_1 & -r_1\alpha_{12} \\ 0 & r_2\left(1 - \alpha_{21}\frac{K_1}{K_2}\right) \end{bmatrix}$$

Since $D\mathbf{f}(K_1, 0)$ is an upper triangular matrix, we can immediately read off the eigenvalues. We find

$$\lambda_1 = -r_1 \qquad \text{and} \qquad \lambda_2 = r_2\left(1 - \alpha_{21}\frac{K_1}{K_2}\right)$$

Since $r_1 > 0$, it follows that $\lambda_1 < 0$. The eigenvalue $\lambda_2 < 0$ when $K_2 < \alpha_{21}K_1$. Therefore, the equilibrium

$$(K_1, 0) \text{ is } \begin{cases} \text{locally stable} & \text{for } K_2 < \alpha_{21}K_1 \\ \text{unstable} & \text{for } K_2 > \alpha_{21}K_1 \end{cases}$$

3. Evaluating the Jacobi matrix at the equilibrium $(\hat{N}_1, \hat{N}_2) = (0, K_2)$, we find

$$D\mathbf{f}(0, K_2) = \begin{bmatrix} r_1\left(1 - \alpha_{12}\frac{K_2}{K_1}\right) & 0 \\ -r_2\alpha_{21} & -r_2 \end{bmatrix}$$

Since $D\mathbf{f}(0, K_2)$ is a lower triangular matrix, we can immediately read off the eigenvalues. We find

$$\lambda_1 = r_1\left(1 - \alpha_{12}\frac{K_2}{K_1}\right) \qquad \text{and} \qquad \lambda_2 = -r_2$$

Since $r_2 > 0$, it follows that $\lambda_2 < 0$. The eigenvalue $\lambda_1 < 0$ when $K_1 < \alpha_{12}K_2$. Therefore, the equilibrium

$$(0, K_2) \text{ is } \begin{cases} \text{locally stable} & \text{for } K_1 < \alpha_{12}K_2 \\ \text{unstable} & \text{for } K_1 > \alpha_{12}K_2 \end{cases}$$

4. We stated that the fourth equilibrium can be obtained by simultaneously solving

$$N_1 + \alpha_{12}N_2 = K_1$$
$$\alpha_{21}N_1 + N_2 = K_2$$

Using the standard method of elimination, we find

$$N_1 + \alpha_{12}N_2 = K_1$$
$$(\alpha_{21}\alpha_{12} - 1)N_2 = \alpha_{21}K_1 - K_2$$

and, therefore,

$$\hat{N}_2 = \frac{\alpha_{21}K_1 - K_2}{\alpha_{21}\alpha_{12} - 1}$$

and

$$\hat{N}_1 = K_1 - \alpha_{12}\frac{\alpha_{21}K_1 - K_2}{\alpha_{21}\alpha_{12} - 1} = \frac{\alpha_{12}K_2 - K_1}{\alpha_{21}\alpha_{12} - 1}$$

We require that both $\hat{N}_1$ and $\hat{N}_2$ be positive (after all, these are population densities). That is, we require that

$$\frac{\alpha_{21}K_1 - K_2}{\alpha_{21}\alpha_{12} - 1} > 0 \qquad \text{and} \qquad \frac{\alpha_{12}K_2 - K_1}{\alpha_{21}\alpha_{12} - 1} > 0 \qquad (11.60)$$

If $\alpha_{21}\alpha_{12} > 1$, then (11.60) reduces to

$$K_2 < \alpha_{21}K_1 \qquad \text{and} \qquad K_1 < \alpha_{12}K_2 \qquad (11.61)$$

If $\alpha_{21}\alpha_{12} < 1$, then (11.60) reduces to

$$K_2 > \alpha_{21}K_1 \qquad \text{and} \qquad K_1 > \alpha_{12}K_2 \qquad (11.62)$$

It turns out that finding the eigenvalues associated with this nontrivial equilibrium is algebraically rather involved if one uses the Jacobi matrix associated with the original system of differential equations (11.57). However, if we investigate the effect of a perturbation directly, then the analysis becomes manageable. To demonstrate what we mean by this, we set

$$z_1 = N_1 - \hat{N}_1 \qquad \text{and} \qquad z_2 = N_2 - \hat{N}_2$$

then (z_1, z_2) represents the deviation from the equilibrium. First, observe that

$$\frac{dz_1}{dt} = \frac{dN_1}{dt} \quad \text{and} \quad \frac{dz_2}{dt} = \frac{dN_2}{dt}$$

Substituting $z_1 + \hat{N}_1$ for N_1 and $z_2 + \hat{N}_2$ for N_2 in (11.57) yields

$$\frac{dz_1}{dt} = r_1(z_1 + \hat{N}_1)\left(1 - \frac{z_1 + \hat{N}_1}{K_1} - \alpha_{12}\frac{z_2 + \hat{N}_2}{K_1}\right)$$

$$= r_1(z_1 + \hat{N}_1)\left(\underbrace{1 - \frac{\hat{N}_1}{K_1} - \alpha_{12}\frac{\hat{N}_2}{K_1}}_{=0} - \frac{z_1}{K_1} - \alpha_{12}\frac{z_2}{K_1}\right)$$

and

$$\frac{dz_2}{dt} = r_2(z_2 + \hat{N}_2)\left(1 - \frac{z_2 + \hat{N}_2}{K_2} - \alpha_{21}\frac{z_1 + \hat{N}_1}{K_2}\right)$$

$$= r_2(z_2 + \hat{N}_2)\left(\underbrace{1 - \frac{\hat{N}_2}{K_2} - \alpha_{21}\frac{\hat{N}_1}{K_2}}_{=0} - \frac{z_2}{K_2} - \alpha_{21}\frac{z_1}{K_2}\right)$$

Instead of analyzing the Jacobi matrix associated with (11.57) evaluated at the nontrivial equilibrium $(\hat{N}_1, \hat{N}_1)$, we investigate the equilibrium $(z_1, z_2) = (0, 0)$ of the new system

$$\frac{dz_1}{dt} = -\frac{r_1}{K_1}(z_1 + \hat{N}_1)(z_1 + \alpha_{12}z_2)$$

$$\frac{dz_2}{dt} = -\frac{r_2}{K_2}(z_2 + \hat{N}_2)(z_2 + \alpha_{21}z_1)$$

The Jacobi matrix is

$$J(z_1, z_2) = \begin{bmatrix} -\frac{r_1}{K_1}(z_1 + \alpha_{12}z_2) - \frac{r_1}{K_1}(z_1 + \hat{N}_1) & -\frac{r_1\alpha_{12}}{K_1}(z_1 + \hat{N}_1) \\ -\frac{r_2\alpha_{21}}{K_2}(z_2 + \hat{N}_2) & -\frac{r_2}{K_2}(z_2 + \alpha_{21}z_2) - \frac{r_2}{K_2}(z_2 + \hat{N}_2) \end{bmatrix}$$

Evaluating the matrix $J(z_1, z_2)$ at $(z_1, z_2) = (0, 0)$, we obtain

$$J(0, 0) = \begin{bmatrix} -\frac{r_1}{K_1}\hat{N}_1 & -\frac{r_1\alpha_{12}}{K_1}\hat{N}_1 \\ -\frac{r_2\alpha_{21}}{K_2}\hat{N}_2 & -\frac{r_2}{K_2}\hat{N}_2 \end{bmatrix}$$

Now,

$$\text{tr}(J(0, 0)) = -\frac{r_1\hat{N}_1}{K_1} - \frac{r_2\hat{N}_2}{K_2}$$

$$\det(J(0, 0)) = \frac{r_1 r_2}{K_1 K_2}\hat{N}_1\hat{N}_2(1 - \alpha_{12}\alpha_{21})$$

The nontrivial equilibrium $(\hat{N}_1, \hat{N}_2)$ satisfies $\hat{N}_1 > 0$ and $\hat{N}_2 > 0$, implying that $\text{tr}(J(0, 0)) < 0$. Since

$$\det(J(0, 0)) > 0 \quad \text{when } 1 - \alpha_{12}\alpha_{21} > 0$$

we find that $(\hat{N}_1, \hat{N}_2)$ is unstable if (11.61) holds, and is locally stable if (11.62) holds.

When we compare the conditions in (4) with those in (2) and (3), we see that the interior equilibrium is locally stable when the two boundary (monoculture) equilibria $(K_1, 0)$ and $(0, K_2)$ are unstable, and that the interior equilibrium is unstable when the two boundary equilibria are locally stable. One can also show that both eigenvalues are always real; that is, we do not expect oscillations.

Comparing the two approaches, we of course find the same results. The graphical approach gives qualitative answers only, albeit rather useful ones. The graphical approach also turns out to be easier. The eigenvalue approach allows us to obtain quantitative answers in terms of eigenvalues, which tell us something about how quickly the system returns to a stable equilibrium after a small perturbation. Such quantitative answers are important if one wants to know how the system responds to small perturbations.

11.4.2 A Predator-Prey Model

In this subsection, we will investigate a simple model for predation, defined as the consumption of prey by a predator. We assume that the prey is alive when the predator attacks it, and that the predator kills its prey and thus removes it from the population.

There are many patterns of abundance that result from predator-prey interactions; most notably, those situations in nature in which predator and prey abundances appear to be closely linked and show periodic fluctuations. A frequently cited example is the Canada lynx (*Lynx canadensis*) and snowshoe hare (*Lepus americanus*) system. The population abundances show periodic oscillations. A laboratory example of coupled predator-prey oscillations is the azuki bean weevil (*Callosbruchus chinensis*) and its parasitoid wasp (*Heterospilus prosopidis*).

The simplest predator-prey model that exhibits coupled oscillations is the Lotka-Volterra model (Lotka 1932, Volterra 1926). We describe this system in Volterra's own words:

> The first case I have considered is that of two associated species, of which one, finding sufficient food in its environment, would multiply indefinitely when left to itself, while the other would perish for lack of nourishment if left alone; but the second feeds upon the first, and so the two species can coexist together.
>
> The proportional rate of increase of the eaten species diminishes as the number of individuals of the eating species increases, while augmentation of the eating species increases with the increase of the number of individuals of the eaten species.

We will now translate this verbal explanation of the model into a system of differential equations. If we denote by $N(t)$ the abundance of the prey and by $P(t)$ the abundance of the predator, then the model is given by the following system of differential equations:

$$\frac{dN}{dt} = rN(t) - aP(t)N(t)$$

$$\frac{dP}{dt} = baP(t)N(t) - dP(t) \tag{11.63}$$

where r, a, b, and d are positive constants. Note that the prey increases exponentially in the absence of the predator ($P = 0$). The intrinsic rate of increase of the prey in the absence of the predator is r. The constant a denotes the attack rate and the term aPN is the consumption rate of prey. Predators decline exponentially in the absence of prey; the rate of decline is given by the constant d. The consumption

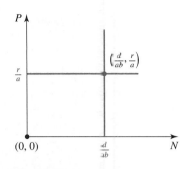

▲ **Figure 11.53**

of prey results in an increase of predator abundance. The constant b describes how efficiently predator turn prey into predator offspring.

To find possible equilibria of (11.63), we set

$$N(r - aP) = 0$$
$$P(abN - d) = 0$$

This yields the zero isoclines $N = 0$ and $P = r/a$ for $dN/dt = 0$, and $P = 0$ and $N = d/ab$ for $dP/dt = 0$. The zero isoclines are shown in Figure 11.53. The points of intersection of the zero isoclines for N and P are equilibria. We find

$$(\hat{N}, \hat{P}) = (0, 0) \qquad \text{and} \qquad (\hat{N}, \hat{P}) = \left(\frac{d}{ab}, \frac{r}{a}\right)$$

To analyze the stability of these two equilibria, we linearize (11.63). The Jacobi matrix is given by

$$\mathbf{Df}(N, P) = \begin{bmatrix} r - aP & -aN \\ baP & baN - d \end{bmatrix}$$

When $(\hat{N}, \hat{P}) = (0, 0)$, then

$$\mathbf{Df}(0, 0) = \begin{bmatrix} r & 0 \\ 0 & -d \end{bmatrix}$$

This is a diagonal matrix, hence the eigenvalues are given by the diagonal elements

$$\lambda_1 = r > 0 \qquad \text{and} \qquad \lambda_2 = -d < 0$$

We conclude that $(0, 0)$ is unstable.

When $(\hat{N}, \hat{P}) = \left(\frac{d}{ab}, \frac{r}{a}\right)$, then

$$\mathbf{Df}\left(\frac{d}{ab}, \frac{r}{a}\right) = \begin{bmatrix} 0 & -\frac{d}{b} \\ rb & 0 \end{bmatrix}$$

To determine the eigenvalues of $\mathbf{Df}\left(\frac{d}{ab}, \frac{r}{a}\right)$, we must solve

$$\det \begin{bmatrix} -\lambda & -\frac{d}{b} \\ rb & -\lambda \end{bmatrix} = \lambda^2 + rd = 0$$

Solving this, we find

$$\lambda_1 = i\sqrt{rd} \qquad \text{and} \qquad \lambda_2 = -i\sqrt{rd}$$

That is, both eigenvalues are purely imaginary. This does not allow us to determine the stability of this equilibrium by linearizing about the equilibrium, as pointed out in Section 11.3. Fortunately, however, we can solve (11.63) exactly.

To solve (11.63) exactly, we divide dP/dt by dN/dt. We find

$$\frac{dP/dt}{dN/dt} = \frac{dP}{dN} = \frac{P(abN - d)}{N(r - aP)}$$

Separating variables and integrating, we obtain

$$\int \frac{r - aP}{P} \, dP = \int \frac{abN - d}{N} \, dN$$

Carrying out the integration, we find

$$r \ln P - aP = abN - d \ln N + C$$

where C is the constant of integration. Rearranging terms and exponentiating, we find

$$\left(N^d e^{-abN}\right)\left(P^r e^{-aP}\right) = K$$

where $K = e^C$ depends on the initial condition. We define the function

$$f(N, P) = (N^d e^{-abN})(P^r e^{-aP})$$

and set

$$g(N) = N^d e^{-abN} \qquad \text{and} \qquad h(P) = P^r e^{-aP}$$

Then one can show that $g(N)$ has its absolute maximum when $N = d/ab$ and $h(P)$ has its absolute maximum when $P = r/a$. The function $f(N, P)$ therefore takes on its absolute maximum at the equilibrium point $(d/ab, r/a)$. We can therefore define level curves

$$f(N, P) = K$$

for $K \leq K_{\max}$, where $K_{\max}$ is the value of f at the equilibrium $(d/ab, r/a)$. We will not be able to demonstrate this here, but these level curves are closed curves. (We show such level curves in Figure 11.54, to convince you that these curves are indeed closed.) These level curves are solution curves.

Solutions of $N(t)$ and $P(t)$ as functions of time corresponding to the two closed curves in Figure 11.54 are shown in Figures 11.55 and 11.56. When we plot $N(t)$ and $P(t)$ versus t, we see that the closed trajectories in the N-P plane correspond to periodic solutions for the predator and the prey. The amplitudes of the oscillations depend on the initial condition. Note that the amplitudes in the two figures are different.

These closed trajectories are not stable under perturbations. That is, if a small perturbation changes the value of N or P, the solution will follow a *different* closed trajectory. This property is a major drawback of the model. It implies that if a natural population actually followed this simple model, its abundance would not exhibit regular cycles, because external factors would constantly shift the population to different trajectories. If a natural population exhibits regular cycles, we would expect these cycles to be stable, that is, the population would return to the same cycle after a small perturbation. Such cycles are called **stable limit cycles**, and can be obtained by modifying the original Lotka-Volterra model to include a nonlinear predator response to prey abundance.

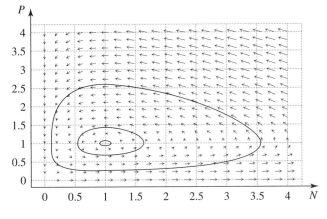

◀ **Figure 11.54**

Solutions for (11.63) in the $N - P$ plane.

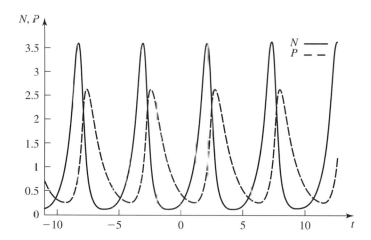

▲ **Figure 11.55**
Solutions for (11.63) as functions of time.

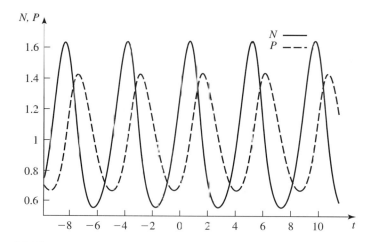

▲ **Figure 11.56**
Solutions for (11.63) as functions of time.

11.4.3 The Community Matrix

In this subsection we will consider a fairly general multispecies population model, which was initiated by Levins (1970) and further developed by May (1975). The goal is to determine how interactions between pairs of species influence the stability of the equilibria of an assemblage of species.

We assume an assemblage of two species in which the abundance of species i at time t is given by $N_i(t)$. (At the end of this subsection we will generalize this to an assemblage of m species.) Suppose that the following set of differential equations describes the dynamics of this two-species assemblage.

$$\frac{dN_1}{dt} = f_1(N_1(t), N_2(t))$$
$$\frac{dN_2}{dt} = f_2(N_1(t), N_2(t))$$

(11.64)

or, in matrix notation,

$$\frac{d\mathbf{N}}{dt} = \mathbf{f}(\mathbf{N}(t))$$

(11.65)

To determine possible equilibria, we must solve the system of equations

$$f_1(\hat{N}_1, \hat{N}_2) = 0$$
$$f_2(\hat{N}_1, \hat{N}_2) = 0$$

(11.66)

Assuming that (11.66) has a solution, which we denote by $\hat{\mathbf{N}} = (\hat{N}_1, \hat{N}_2)$, we can now proceed to determine its stability. To do so, we must evaluate the Jacobi matrix associated with the system (11.64) at the equilibrium $(\hat{N}_1, \hat{N}_2)$. The Jacobi matrix is given as

$$D\mathbf{f}(\hat{\mathbf{N}}) = \begin{bmatrix} \frac{\partial f_1(\hat{N}_1, \hat{N}_2)}{\partial N_1} & \frac{\partial f_1(\hat{N}_1, \hat{N}_2)}{\partial N_2} \\ \frac{\partial f_2(\hat{N}_1, \hat{N}_2)}{\partial N_1} & \frac{\partial f_2(\hat{N}_1, \hat{N}_2)}{\partial N_2} \end{bmatrix}$$

This 2×2 matrix is called the **community matrix**. Its elements

$$a_{ij} = \frac{\partial f_i(\hat{N}_1, \hat{N}_2)}{\partial N_j}$$

describe the effect of species j on species i at equilibrium, because the partial derivative $\frac{\partial f_i}{\partial N_j}$ tells us how the function f_i, which describes the growth of species i, changes when the abundance of species j changes.

The diagonal elements $a_{ii} = \frac{\partial f_i}{\partial N_i}$ measure the effect species i has on itself, whereas the off-diagonal elements $a_{ij} = \frac{\partial f_i}{\partial N_j}$, $i \neq j$, measure the effect species j has on species i. The signs of the elements a_{ij} thus tell us something about the pairwise effects the species in this assemblage have on each other at equilibrium.

Assume that an equilibrium of (11.64) is given by $(\hat{N}_1, \hat{N}_2)$ with $\hat{N}_1 > 0$ and $\hat{N}_2 > 0$, that is, both species have positive densities at the equilibrium. The quantity

$$a_{ij} = \frac{\partial f_i(\hat{N}_1, \hat{N}_2)}{\partial N_j}$$

describes the net effect of species j on species i at equilibrium. The quantity a_{ij} can be negative, 0, or positive. If $a_{ij} < 0$, then the growth rate of species i is decreased if species j increases its abundance; we therefore say that species j has a negative or inhibitory effect on species i. If $a_{ij} = 0$, then changes in the abundance of species j have no effect on the growth rate of species i. If $a_{ij} > 0$, then the growth rate of species i is increased if species j increases its abundance; we therefore say that species j has a positive or facilitory effect on species i.

We look at the possible combinations of the pair (a_{21}, a_{12}). This pair describes the interactions *between* the two species in the assemblage. The following table lists all possible combinations.

		a_{12}		
		+	0	−
	+	++	+0	+−
a_{21}	0	0+	00	0−
	−	−+	−0	−−

To interpret this table, take the pair $(0, 0)$, for instance. The pair $(0, 0)$ represents the case in which neither species has an effect on the other species at equilibrium. Let's look at another pair, for instance, $(0+)$. In this case, species 1 has no effect on species 2, but species 2 has a positive effect on species 1.

The case (00) is the simplest, and we will discuss it first. The community matrix in this case is

$$
\begin{bmatrix} a_{11} & 0 \\ 0 & a_{22} \end{bmatrix}
$$

Since the community matrix is in diagonal form, its eigenvalues are the diagonal elements a_{11} and a_{22}. Hence the equilibrium $(\hat{N}_1, \hat{N}_2)$ is stable, provided that both a_{11} and a_{22} are negative. That is, if neither species has an effect on the other species, a nontrivial equilibrium, in which both species coexist, only exists if they each have a negative effect on themselves, which means that each species regulates its own population size.

Following May (1975), the remaining eight combinations in the table can be categorized into five biologically distinct types of interactions.

1. **Mutualism or symbiosis (++):** Each species has a positive effect on the other.
2. **Competition (−−):** Each species has a negative effect on the other.
3. **Commensalism (+0):** One species benefits from the interaction, whereas the other is unaffected.
4. **Amensalism (−0):** One species is negatively affected, whereas the other is unaffected.
5. **Predation (+−):** One species benefits, whereas the other is negatively affected.

We will now discuss the stability of the nontrivial equilibrium $(\hat{N}_1, \hat{N}_2)$ in all five cases. Recall that we assumed that this equilibrium exists, and that both $\hat{N}_1 > 0$ and $\hat{N}_2 > 0$. The community matrix (the Jacobi matrix of (11.64) at equilibrium) is of the form

$$
A = \begin{bmatrix} a_{11} & a_{12} \\ a_{21} & a_{22} \end{bmatrix}
$$

and both eigenvalues of A have negative real parts if and only if

$$
\operatorname{tr} A = a_{11} + a_{22} < 0 \qquad \text{and} \qquad \det A = a_{11}a_{22} - a_{12}a_{21} > 0
$$

We assume in the following that

$$
a_{11} < 0 \qquad \text{and} \qquad a_{22} < 0 \tag{11.67}
$$

so that the first condition $\operatorname{tr} A < 0$ is automatically satisfied. This has the same interpretation as in the case (00).

We will now go through all five cases, and determine under which conditions the nontrivial equilibrium is stable.

Mutualism We assume (11.67). The sign structure of the community matrix at equilibrium in the case of mutualism is then of the form

$$
A = \begin{bmatrix} - & + \\ + & - \end{bmatrix}
$$

Since

$$
\det A = \underbrace{a_{11}a_{22}}_{> 0} - \underbrace{a_{12}a_{21}}_{> 0}
$$

the determinant of A may be either positive or negative. If $a_{12}a_{21}$ is sufficiently small compared to $a_{11}a_{22}$, then $\det A > 0$, and the equilibrium $(\hat{N}_1, \hat{N}_2)$ is locally

stable. In other words, if the positive effects of the species on each other are sufficiently counteracted by their own population control (represented by a_{11} and a_{22}),
the equilibrium is locally stable.

Competition   We assume (11.67). The sign structure of the community matrix at
equilibrium in the case of competition is then of the form

$$A = \begin{bmatrix} - & - \\ - & - \end{bmatrix}$$

Since

$$\det A = \underbrace{a_{11}a_{22}}_{>0} - \underbrace{a_{12}a_{21}}_{>0}$$

the determinant of A may be either positive or negative. Now, the equilibrium
$(\hat{N}_1, \hat{N}_2)$ is locally stable if the negative effects each species has on the other are
smaller than the effects each species has on itself. (In this case, $\det A > 0$ and, hence,
the equilibrium is locally stable.)

Commensalism and Amensalism   We assume (11.67). The sign structure of the
community matrices at equilibrium are then of the form

$$A = \begin{bmatrix} - & 0 \\ + & - \end{bmatrix} \quad \text{or} \quad A = \begin{bmatrix} - & 0 \\ - & - \end{bmatrix}$$

In either case, the determinant is positive; therefore, the equilibrium is locally stable.

Predation   We assume (11.67). The sign structure of the community matrix at
equilibrium in the case of predation is then of the form

$$A = \begin{bmatrix} - & - \\ + & - \end{bmatrix}$$

and $\det A$ is always positive. That is, provided a nontrivial equilibrium exists, it is
locally stable. From the biological meaning of this case, it should be clear that if the
species that serves as the food source is in short supply, the predator cannot survive
and, hence, there would be no nontrivial equilibrium.

The multispecies case   We will now briefly look at the case in which more than
two species form an assemblage. Say there are m species. We denote the density of
species i at time t by $N_i(t)$. The dynamics are described by the following system of
differential equations.

$$\frac{dN_1}{dt} = f_1(N_1(t), N_2(t), \ldots, N_m(t))$$

$$\frac{dN_2}{dt} = f_2(N_1(t), N_2(t), \ldots, N_m(t))$$

$$\vdots$$

$$\frac{dN_m}{dt} = f_m(N_1(t), N_2(t), \ldots, N_m(t))$$

The equilibria are found by solving the following system of equations.

$$f_1(\hat{N}_1, \hat{N}_2, \ldots, \hat{N}_m) = 0$$
$$f_2(\hat{N}_1, \hat{N}_2, \ldots, \hat{N}_m) = 0$$

$$\vdots$$

$$f_m(\hat{N}_1, \hat{N}_2, \ldots, \hat{N}_m) = 0$$

If we assume that $\hat{\mathbf{N}} = (\hat{N}_1, \hat{N}_2, \ldots, \hat{N}_m)$ is an equilibrium, then the community matrix at equilibrium is the Jacobi matrix at equilibrium, which is given by

$$D\mathbf{f}(\hat{\mathbf{N}}) = \begin{bmatrix} \frac{\partial f_1}{\partial N_1} & \cdots & \frac{\partial f}{\partial N_n} \\ \vdots & & \vdots \\ \frac{\partial f_m}{\partial N_1} & \cdots & \frac{\partial f_n}{\partial N_m} \end{bmatrix}$$

As in the case of two species, the element

$$a_{ij} = \frac{\partial f_i(\hat{\mathbf{N}})}{\partial N_j}$$

describes the effect of species j on species i at equilibrium.

11.4.4 A Mathematical Model for Neuron Activity

The nervous system of an organism is a communication network that allows for rapid transmission of information between cells. It consists of nerve cells, called **neurons**. A typical neuron has a cell body that contains the cell nucleus and nerve fibers. Nerve fibers that receive information are called **dendrites**, those that transport information are called **axons**, which provide links to other neurons via **synapses**. A typical vertebrate neuron is shown in Figure 11.57.

Neurons respond to electrical stimuli; this response is exploited when studying neurons. When the cell body of an isolated neuron is stimulated with a very mild electrical shock, the neuron shows no response; increasing the intensity of the electrical shock beyond a certain threshold will trigger a response, namely an impulse that travels along the axon. Increasing the intensity of the electrical shock further does not change the response. This impulse is thus an all-or-none response.

A.L. Hodgkin and A.F. Huxley studied the giant axon of a squid experimentally, and developed a mathematical model for neuron activity. Their work appeared in a series of papers in 1952. It is an excellent example of how experimental and theoretical research can be combined to gain a thorough understanding of a natural system. In 1963, Hodgkin and Huxley were awarded the Nobel prize in physiology/medicine for their work on neurons.

We begin with a brief explanation of how a neuron works. The main players in the functioning of a neuron are sodium (Na^+) and potassium (K^+) ions. The cell membrane of a neuron is impermeable to sodium and potassium ions when the cell is in a resting state. In a typical neuron in its resting state, the concentration of Na^+ in the interior of the cell is about one-tenth of the extracellular concentration of Na^+; the concentration of K^+ in the interior of the cell is about thirty times the extracellular concentration of K^+. When the neuron is in its resting state, the interior of the cell is negatively charged (at -70 mV) relative to the exterior of the cell.

When a nerve cell becomes stimulated, its surface becomes permeable to Na^+ ions, which rush into the cell through sodium channels in the surface. This results in a reversal of polarization at the points where Na^+ ions entered the cell. The surface inside the cell is now positively charged relative to the outside of the cell, and becomes permeable to K^+ ions which rush outside through potassium channels. Because the potassium ions (K^+) are positively charged and move from the inside to the outside of the cell, the polarization at the surface of the cell is again reversed, and is now below the polarization of the resting cell. To restore the original concentration of Na^+ and K^+ (and thus the original polarization), energy must be expended to run the so-called sodium and potassium pumps in the surface of the cell to pump the excess Na^+ from the interior to the exterior of the cell and to pump K^+ from the exterior to the interior of the cell.

▲ **Figure 11.57**
A typical vertebrate neuron.

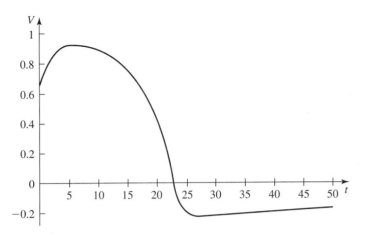

▲ Figure 11.58
The action potential.

To trigger such a reaction, the intensity of the stimulus must be above a certain threshold. The described reaction occurs locally on the surface of the cell. This large local change in polarization triggers the same reaction in the neighborhood, which allows the reaction to propagate along the nerve cell—thus creating the observed impulse that travels along the cell. This localized large change in polarization, which is then reversed to the original polarization, is called an **action potential**. An example of an action potential is illustrated in Figure 11.58.

Hodgkin and Huxley measured sodium and potassium conductance and fitted curves to their data. The sodium curve is fitted by a cubic function and the potassium curve by a quartic function. They then developed a model for the action potential. The model consists of a system of four autonomous differential equations. It is a phenomenological model, that is, the equations are based on fitting curves to experimental data for the various components of the model. There is one equation that describes the change of voltage on the cell surface, two equations that describe the sodium channel, and one equation that describes the potassium channel. We will not give the details of this system, as it is far too complicated to analyze. (It is typically solved numerically.) Instead, we will present a simplified version of this model, which was developed by Fitzhugh (1961) and Nagumo et al. (1962).

The Fitzhugh-Nagumo model is based on the fact that the time scales of the two channels are quite different. The sodium channel works on a much faster time scale than the potassium channel. This fact led Fitzhugh and Nagumo to assume that the sodium channel is essentially always in equilibrium, which allowed them to reduce the four equations of the Hodgkin and Huxley model to two equations. The Fitzhugh-Nagumo model is thus an approximation to the Hodgkin and Huxley model. It retains the essential features of the action potential, but is much easier to analyze.

The Fitzhugh-Nagumo model is described by two variables. One variable, denoted by V, describes the potential of the cell surface. The other variable, denoted by w, models the sodium and the potassium channels. The equations are given by

$$\frac{dV}{dt} = -V(V-a)(V-1) - w$$

$$\frac{dw}{dt} = b(V - cw)$$

(11.68)

where a, b, and c are constants that satisfy $0 < a < 1$, $b > 0$, and $c > 0$.

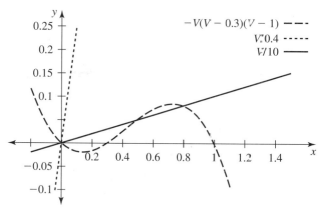

◀ **Figure 11.59**
The zero isoclines of (11.66).

We will analyze the system graphically. The zero isoclines of (11.68) are given by

$$w = -V(V - a)(V - 1) \qquad \text{and} \qquad w = \frac{1}{c}V$$

The important feature of this model is that the zero isocline $dV/dt = 0$ is the graph of a cubic function in the V-w plane. The zero isoclines are illustrated in Figure 11.59.

We see that if c is small, there is just one equilibrium, namely $(0, 0)$; whereas when c is sufficiently large, the line $w = V/c$ intersects the graph of $dV/dt = 0$ three times.

The threshold phenomenon and the action potential are observed when c is small. In this case, there is just one equilibrium, namely $(0, 0)$, which corresponds to the resting state.

We can analyze the stability of $(0, 0)$ by linearizing the system about this equilibrium. We find

$$\mathbf{Df}(V, w) = \begin{bmatrix} -3V^2 + 2V + 2aV - a & -1 \\ b & -bc \end{bmatrix}$$

Hence,

$$\mathbf{Df}(0, 0) = \begin{bmatrix} -a & -1 \\ b & -bc \end{bmatrix}$$

To find the eigenvalues, we compute

$$\det \begin{bmatrix} -a - \lambda & -1 \\ b & -bc - \lambda \end{bmatrix} = (-a - \lambda)(-bc - \lambda) + b = 0$$

That is, we must solve

$$\lambda^2 + (a + bc)\lambda + b(ac + 1) = 0$$

which has solutions

$$\lambda_{1,2} = \frac{-(a + bc) \pm \sqrt{(a + bc)^2 - 4b(ac + 1)}}{2}$$

$$= \frac{-(a + bc) \pm \sqrt{(a - bc)^2 - 4b}}{2}$$

Since a, b, and c are positive constants, it follows that the expression under the square root, namely $(a + bc)^2 - 4b(ac + 1)$, is smaller than $(a + bc)^2$. It therefore

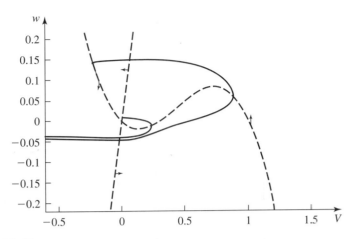

▲ **Figure 11.60**

Solution curves for the Fitzhugh-Nagumo model.

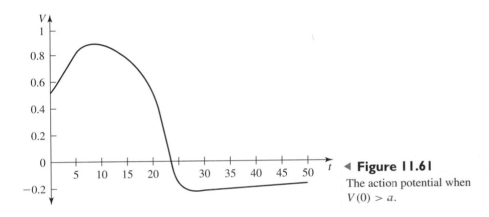

◀ **Figure 11.61**

The action potential when $V(0) > a$.

follows that both λ_1 and λ_2 have negative real parts, which implies that $(0, 0)$ is locally stable. As long as $(a - bc)^2 > 4b$, the equilibrium is a stable sink. When $(a - bc)^2 < 4b$, the eigenvalues are complex conjugates and $(0, 0)$ becomes a stable spiral.

The system mimics the action potential when both eigenvalues are real and negative. If we apply a weak stimulus, that is, increase V to a value less than a, V will quickly return to 0. However, if we apply a strong enough stimulus, that is, $V \in (a, 1)$, the trajectory will move away from the equilibrium point, as shown in Figure 11.60. A plot of voltage versus time reveals that if the stimulus is too weak, $V(t)$ will quickly return to the equilibrium, whereas if the stimulus is large enough, the solution curve of $V(t)$ resembles the action potential. In Figures 11.61 and 11.62, we present two solution curves $V(t)$. In either case, $w(0) = 0$. In Figure 11.61, $V(0) = 0.5 > a$, and we observe an action potential; whereas in Figure 11.62, $V(0) = 0.2 < a$, and the initial stimulus is too small and dies away quickly.

11.4.5 A Mathematical Model for Enzymatic Reactions

The goal of this subsection is threefold. First, we will learn how to model biochemical reactions; second, we will see how mathematical models can be used to understand empirical observations; third, we will introduce an important method, namely by

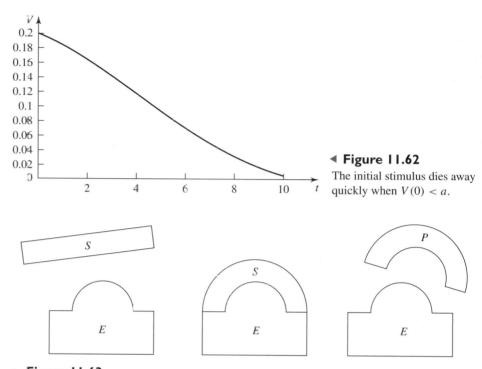

◀ **Figure 11.62**
The initial stimulus dies away quickly when $V(0) < a$.

▲ **Figure 11.63**
A schematic description of an enzymatic reaction.

making appropriate assumptions, the number of variables in a model can sometimes be reduced, which typically facilitates the analysis of the model.

The class of biochemical reactions we will study are enzymatic reactions, which are ubiquitous in the living world. Enzymes are proteins that act as catalysts in chemical reactions by reducing the required activation energy for the reaction. Enzymes are not altered by the reaction; they aid in the initial steps of the reaction and control the rate of the reaction by binding the reactants (called substrates) to the active site of the enzyme, thus forming an enzyme-substrate complex that then allows the substrates to react and to form the product. This is illustrated in Figure 11.63.

If we denote the substrate by S, the enzyme by E, and the product of the reaction by P, then an enzymatic reaction can be described by

$$E + S \underset{k_-}{\overset{k_+}{\rightleftarrows}} ES \xrightarrow{k_2} E + P \tag{11.69}$$

where k_+, k_-, and k_2 are the reaction rates in the corresponding reaction steps.

We can translate the schematic description of the reaction (11.69) into a system of differential equations. We use the following notation.

$$e(t) = [E] = \text{enzyme concentration at time } t$$
$$s(t) = [S] = \text{substrate concentration at time } t$$
$$c(t) = [ES] = \text{enzyme substrate complex concentration at time } t$$
$$p(t) = [P] = \text{product concentration at time } t$$

Using the mass action law,

$$\frac{ds}{dt} = k_- c - k_+ se$$

$$\frac{de}{dt} = (k_- + k_2)c - k_+ se$$

$$\frac{dc}{dt} = k_+ se - (k_- + k_2)c \tag{11.70}$$

$$\frac{dp}{dt} = k_2 c$$

Michaelis and Menten (1913) were instrumental in the description of enzyme kinetics, through both experimental and theoretical work. On the experimental side, they developed techniques that allowed them to measure reaction rates under controlled conditions. Their experiments showed a hyperbolic relationship between the rate of the enzymatic reaction dp/dt and the substrate concentration s, which they described as

$$\frac{dp}{dt} = \frac{v_m s}{K_m + s} \tag{11.71}$$

where v_m is the saturation constant and K_m is the half-saturation constant (that is, if $s = K_m$, then $dp/dt = v_m/2$).

On the theoretical side, they developed a mathematical model for enzyme kinetics that predicted the observed hyperbolic relationship between the substrate concentration and the initial rate at which the product is formed.

In the following, we will analyze (11.70), which is a system of four equations. It is not easy to analyze; we will simplify it, and in the end, we will obtain (11.71). We will also use (11.70) to illustrate that it is sometimes possible to reduce the number of equations in a system.

We claim that the system has a conserved quantity, that is, a quantity that does not depend on time and is therefore constant throughout the reaction. To find this quantity, note that

$$\frac{de}{dt} + \frac{dc}{dt} = 0$$

That is,

$$\frac{d}{dt}(e + c) = 0$$

which implies that

$$e(t) + c(t) = e_0 \tag{11.72}$$

where e_0 is a constant. Since $e(t) + c(t)$ is constant, we say that the sum $e(t) + c(t)$ is a conserved quantity. The advantage of this is that if we know the initial concentrations $e(0)$ and $c(0)$ and one of the two quantities $e(t)$ or $c(t)$, we immediately know the other, since $e(0) + c(0) = e(t) + c(t)$. This reduces the number of equations from four to three.

To reduce the number of equations even further, we make another assumption. Whereas the existence of a conserved quantity followed from the system of equations, and we could have obtained it without knowing the meaning of these equations, the next assumption requires a thorough understanding of the enzymatic reaction itself, and cannot be deduced from the set of equations (11.70). Briggs and Haldane (1925) who had this understanding, proposed that the rate of formation balances the rate of breakdown of the complex, that is, they assumed that

$$\frac{dc}{dt} = 0$$

This assumption yields the equation

$$0 = k_+ se - (k_- + k_2)c$$

which can be rewritten as

$$\frac{se}{c} = \frac{k_- + k_2}{k_+}$$

We denote this ratio by K_m, that is,

$$K_m = \frac{k_- + k_2}{k_+}$$

and, therefore,

$$\frac{se}{c} = K_m \tag{11.73}$$

Solving (11.72) for e, that is, $e = e_0 - c$, and plugging e into (11.73), we find

$$\frac{s(e_0 - c)}{c} = K_m$$

which we can solve for c. This yields

$$c = \frac{e_0 s}{K_m + s} \tag{11.74}$$

which allows us to rewrite the equation for the rate at which the product is formed. Since $dp/dt = k_2 c$, it follows from (11.74) that

$$\frac{dp}{dt} = \frac{k_2 e_0 s}{K_m + s} \tag{11.75}$$

We can interpret the factor $k_2 e_0$. Namely, if all of the enzyme is complexed with the substrate, then $e = 0$ and therefore $c = e_0$ (since $e + c = e_0$ is a conserved quantity). This implies that the rate at which the product is formed, $dp/dt = k_2 c$, is fastest when $c = e_0$, in which case $dp/dt = k_2 e_0$. We can therefore interpret $k_2 e_0$ as the maximum rate at which this reaction can proceed. We introduce the notation

$$v_m = k_2 e_0$$

and rewrite (11.75) as

$$\frac{dp}{dt} = \frac{v_m s}{K_m + s} \tag{11.76}$$

Equation (11.76) is known as the Michaelis-Menten law and describes the velocity of an enzymatic reaction. We see from equation (11.76) that the reaction rate dp/dt is limited by the availability of the substrate S.

Equations (11.76) and (11.71) are the same. Whereas (11.71) was derived from fitting a curve to data points that related the measured substrate concentration s to the velocity of the reaction dp/dt, (11.76) is derived from a mathematical model. The mathematical model allows us to interpret the constants v_m and K_m in terms of the enzymatic reaction, and the experiments allow us to measure v_m and K_m.

Microbial Growth in a Chemostat — An Application of Substrate-Limited Growth   A rather simplistic view of microbial growth is that microbes convert substrate through enzymatic reactions into products that are then converted into microbial biomass.

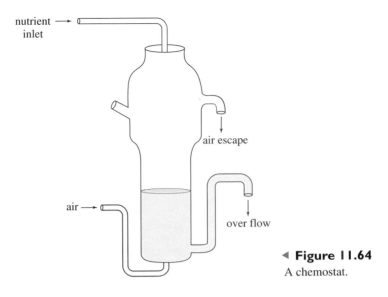

nutrient → inlet

air escape

air →

over flow

◀ **Figure 11.64**
A chemostat.

In the following, we will investigate a mathematical model for microbial growth in a chemostat. The growth of microbes will be limited by the availability of the substrate.

A chemostat is a growth chamber in which sterile medium with concentration s_0 of the substrate enters the chamber at a constant rate D. Air is pumped into the growth chamber to mix and aerate the culture. To keep the volume in the growth chamber constant, the content of the growth chamber is removed at the same rate D as new medium enters. A sketch of a chemostat can be seen in Figure 11.64.

We denote the microbial biomass at time t by $x(t)$, and the substrate concentration at time t by $s(t)$. Jacques Lucien Monod was very influential in the development of quantitative microbiology; in (1950) he derived the following system of differential equations to describe the growth of microbes in a chemostat:

$$\frac{ds}{dt} = D(s_0 - s) - q(s)x$$
$$\frac{dx}{dt} = Yq(s)x - Dx$$
(11.77)

where $s_0 > 0$ is the substrate concentration of the entering medium. $D > 0$ is the rate at which medium enters or leaves the chemostat, and $Y > 0$ is the yield constant. The function $q(s)$ is the rate at which microbes consume the substrate; this rate depends on the substrate concentration. The yield factor Y can thus be interpreted as a conversion factor of substrate into biomass.

Monod (1942) showed empirically that the uptake rate $q(s)$ fits the hyperbolic relationship

$$q(s) = \frac{v_m s}{K_m + s}$$
(11.78)

where v_m is the saturation level and K_m is the half-saturation constant (that is, $q(K_m) = v_m/2$). A graph of $q(s)$ is shown in Figure 11.65. It later occurred to Monod that (11.78) is identical to the Michaelis-Menten law (11.76), which might suggest that microbial growth is governed by enzymatic reactions.

In the following, we will determine possible equilibria of (11.77) and analyze their stability. There is always, of course, the trivial equilibrium, which is obtained when substrate enters a growth chamber void of microbes (that is, when $x(0) = 0$).

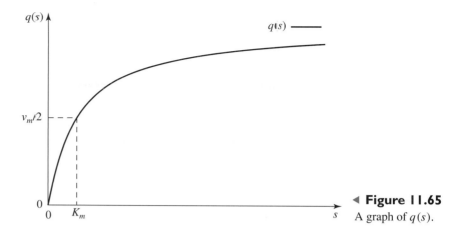

◀ **Figure 11.65**
A graph of $q(s)$.

In that case, there will be no microbes at later times and, hence, $\frac{dx}{dt} = 0$ for all times $t \geq 0$. The substrate equilibrium is then found by setting $\frac{ds}{dt} = 0$ with $x = 0$:

$$0 = D(s_0 - s)$$

which has solution $s = s_0$. Hence, one equilibrium is

$$(\hat{s}_1, \hat{x}_1) = (s_0, 0) \tag{11.79}$$

To obtain a nontrivial equilibrium $(\hat{s}_2, \hat{x}_2)$, we will look for an equilibrium with $\hat{x}_2 > 0$. To find this, we solve the simultaneous equations

$$\frac{ds}{dt} = 0 \qquad \text{and} \qquad \frac{dx}{dt} = 0$$

It follows from $dx/dt = 0$ that

$$q(\hat{s}_2) = \frac{D}{Y} \tag{11.80}$$

We see immediately from Figure 11.65 that (11.80) has a solution $\hat{s}_2 > 0$ if $D/Y < v_m$. Using (11.78) in (11.80), we can compute $\hat{s}_2$, namely

$$\frac{v_m \hat{s}_2}{K_m + \hat{s}_2} = \frac{D}{Y}$$

which yields

$$\hat{s}_2 = \frac{D K_m}{Y v_m - D} \tag{11.81}$$

which is indeed positive, provided that $D/Y < v_m$. Using (11.80) in $ds/dt = 0$, we find

$$D(s_0 - \hat{s}_2) = \frac{D}{Y}\hat{x}_2$$

or

$$\hat{x}_2 = Y(s_0 - \hat{s}_2) \tag{11.82}$$

from which we can see that $\hat{x}_2 > 0$, provided that $\hat{s}_2 < s_0$. Under the assumption that $D/Y < v_m$ and $\hat{s}_2 < s_0$, we therefore have a nontrivial equilibrium

$$(\hat{s}_2, \hat{x}_2) = \left(\frac{D K_m}{Y v_m - D}, \ Y(s_0 - \hat{s}_2) \right) \tag{11.83}$$

There are no other equilibria.

To analyze the stability of the two equilibria (11.79) and (11.83), we find the Jacobi matrix $D\mathbf{f}$ associated with the system (11.77),

$$D\mathbf{f}(s, x) = \begin{bmatrix} -D - q'(s)x & -q(s) \\ Yq'(s)x & Yq(s) - D \end{bmatrix}$$

We analyze the stability of the trivial equilibrium (11.79) first:

$$D\mathbf{f}(s_0, 0) = \begin{bmatrix} -D & -q(s_0) \\ 0 & Yq(s_0) - D \end{bmatrix}$$

Since the Jacobi matrix is in upper triangular form, the eigenvalues are the diagonal elements, and we find

$$\lambda_1 = -D < 0$$

$$\lambda_2 = Yq(s_0) - D < 0 \quad \text{provided that } \frac{D}{Y} > q(s_0)$$

Therefore, the equilibrium

$$(s_0, 0) \quad \text{is} \quad \begin{cases} \text{locally stable} & \text{if } \frac{D}{Y} > q(s_0) \\ \text{unstable} & \text{if } \frac{D}{Y} < q(s_0) \end{cases}$$

For the nontrivial equilibrium (11.83), we find

$$A = D\mathbf{f}(\hat{s}_2, \hat{x}_2) = \begin{bmatrix} -D - q'(\hat{s}_2)\hat{x}_2 & -q(\hat{s}_2) \\ Yq'(\hat{s}_2)\hat{x}_2 & Yq(\hat{s}_2) - D \end{bmatrix}$$

Using (11.80), this simplifies to

$$A = \begin{bmatrix} -D - q'(\hat{s}_2)\hat{x}_2 & -\frac{D}{Y} \\ Yq'(\hat{s}_2)\hat{x}_2 & 0 \end{bmatrix}$$

Now,

$$\text{tr} A = -D - q'(\hat{s}_2)\hat{x}_2 < 0$$

and

$$\det A = Dq'(\hat{s}_2)\hat{x}_2 > 0$$

for $\hat{x}_2 > 0$, since $q(s)$ is an increasing function. Therefore, if the nontrivial equilibrium exists, that is, if both $\hat{s}_2 > 0$ and $\hat{x}_2 > 0$, then it is locally stable.

We saw that the nontrivial equilibrium exists, provided that $D/Y < v_m$ and $s_0 > \hat{s}_2$. These two conditions can be summarized as

$$\frac{D}{Y} < v_m \quad \text{and} \quad \frac{DK_m}{Yv_m - D} < s_0$$

If the first inequality holds, then the denominator in the second inequality is positive. Solving the second inequality for D, we then find

$$D < \frac{Ys_0v_m}{s_0 + K_m} = Yq(s_0) \tag{11.84}$$

We can now summarize our results. The chemostat has two equilibria, a trivial one in which microbes are absent, and a nontrivial one that allows stable microbial growth. If $D > Yq(s_0)$, then the trivial equilibrium is the only biologically reasonable equilibrium and it is locally stable. If $D < Yq(s_0)$, both equilibria are biologically reasonable; the trivial one is now unstable and the nontrivial is the locally stable one. Stable microbial growth is therefore possible, provided that the rate at which medium enters and leaves the growth chamber is between 0 and $Yq(s_0)$.

◆ 11.4 PROBLEMS

[11.4.1]

1. Suppose that the densities of two species evolve according to the Lotka-Volterra model of interspecific competition. Assume that species 1 has intrinsic rate of growth $r_1 = 2$ and carrying capacity $K_1 = 20$, and that species 2 has intrinsic rate of growth $r_2 = 3$ and carrying capacity $K_2 = 15$. Furthermore, assume that 20 individuals of species 2 have the same effect on species 1 as 4 individuals of species 1 have on themselves, and 30 individuals of species 1 have the same effect on species 2 as 6 individuals of species 2 have on themselves. Find a system of differential equations that describes this situation.

2. Suppose the densities of two species evolve according to the Lotka-Volterra model of interspecific competition. Assume that species 1 has intrinsic rate of growth $r_1 = 4$ and carrying capacity $K_1 = 17$, and that species 2 has intrinsic rate of growth $r_2 = 1.5$ and carrying capacity $K_2 = 32$. Furthermore, assume that 15 individuals of species 2 have the same effect on species 1 as 7 individuals of species 1 have on themselves, and 5 individuals of species 1 have the same effect on species 2 as 7 individuals of species 2 have on themselves. Find a system of differential equations that describes this situation.

In Problems 3–6, use the graphical approach to classify the following Lotka-Volterra models of interspecific competition according to "coexistence," "founder control," "species 1 excludes species 2," or "species 2 excludes species 1."

3. $\dfrac{dN_1}{dt} = 2N_1\left(1 - \dfrac{N_1}{10} - 0.7\dfrac{N_2}{10}\right)$

$\dfrac{dN_2}{dt} = 5N_2\left(1 - \dfrac{N_2}{15} - 0.3\dfrac{N_1}{15}\right)$

4. $\dfrac{dN_1}{dt} = 3N_1\left(1 - \dfrac{N_1}{50} - 0.3\dfrac{N_2}{50}\right)$

$\dfrac{dN_2}{dt} = 4N_2\left(1 - \dfrac{N_2}{30} - 0.8\dfrac{N_1}{30}\right)$

5. $\dfrac{dN_1}{dt} = N_1\left(1 - \dfrac{N_1}{20} - \dfrac{N_2}{5}\right)$

$\dfrac{dN_2}{dt} = 2N_2\left(1 - \dfrac{N_2}{15} - \dfrac{N_1}{3}\right)$

6. $\dfrac{dN_1}{dt} = 3N_1\left(1 - \dfrac{N_1}{25} - 1.2\dfrac{N_2}{25}\right)$

$\dfrac{dN_2}{dt} = N_2\left(1 - \dfrac{N_2}{30} - 0.8\dfrac{N_1}{30}\right)$

In Problems 7–10, use the eigenvalue approach to analyze all equilibria of the given Lotka-Volterra models of interspecific competition.

7. $\dfrac{dN_1}{dt} = 3N_1\left(1 - \dfrac{N_1}{18} - 1.3\dfrac{N_2}{18}\right)$

$\dfrac{dN_2}{dt} = 2N_2\left(1 - \dfrac{N_2}{20} - 0.6\dfrac{N_1}{20}\right)$

8. $\dfrac{dN_1}{dt} = 4N_1\left(1 - \dfrac{N_1}{12} - 0.3\dfrac{N_2}{12}\right)$

$\dfrac{dN_2}{dt} = 5N_2\left(1 - \dfrac{N_2}{15} - 0.2\dfrac{N_1}{15}\right)$

9. $\dfrac{dN_1}{dt} = N_1\left(1 - \dfrac{N_1}{35} - 3\dfrac{N_2}{35}\right)$

$\dfrac{dN_2}{dt} = 3N_2\left(1 - \dfrac{N_2}{40} - 4\dfrac{N_1}{40}\right)$

10. $\dfrac{dN_1}{dt} = N_1\left(1 - \dfrac{N_1}{25} - 0.1\dfrac{N_2}{25}\right)$

$\dfrac{dN_2}{dt} = N_2\left(1 - \dfrac{N_2}{28} - 1.2\dfrac{N_1}{28}\right)$

11. Suppose that two species of beetles are reared together and separately. When species 1 is reared alone, it reaches an equilibrium of about 200. When species 2 is reared alone, it reaches an equilibrium of about 150. When both of them are reared together, they seem to be able to coexist; species 1 reaches an equilibrium of about 180 and species 2 reaches an equilibrium of about 80. If their densities follow the Lotka-Volterra equation of interspecific competition, find α_{12} and α_{21}.

12. Suppose that two species of beetles are reared together. Species 1 wins if there are initially 100 individuals of species 1 and 20 individuals of species 2. But species 2 wins if there are initially 20 individuals of species 1 and 100 individuals of species 2. When the beetles are reared separately, both species seem to reach an equilibrium of about 120. Based on this information and assuming that the densities follow the Lotka-Volterra model of interspecific competition, can you give lower bounds on α_{12} and α_{21}?

[11.4.2]

In Problems 13–14, use a graphing calculator to sketch solution curves of the given Lotka-Volterra predator-prey model in the N-P plane. Also graph $N(t)$ and $P(t)$ as functions of t.

13.

$$\frac{dN}{dt} = 2N - PN$$

$$\frac{dP}{dt} = \frac{1}{2}PN - P$$

with initial conditions

(a) $(N(0), P(0)) = (2, 2)$ **(b)** $(N(0), P(0)) = (3, 3)$
(c) $(N(0), P(0)) = (4, 4)$

14.

$$\frac{dN}{dt} = 3N - 2PN$$

$$\frac{dP}{dt} = PN - P$$

with initial conditions

(a) $N(0), P(0)) = (1, 3/2)$ **(b)** $(N(0), P(0)) = (2, 2)$
(c) $(N(0), P(0)) = (3, 1)$

In Problems 15–16, we investigate Lotka-Volterra predator-prey model.

15. Assume that

$$\frac{dN}{dt} = N - 4PN$$

$$\frac{dP}{dt} = 2PN - 3P$$

(a) Show that this system has two equilibria, the trivial equilibrium $(0, 0)$ and a nontrivial one, in which both species have positive densities.

(b) Use the eigenvalue approach to show that the trivial equilibrium is unstable.

(c) Determine the eigenvalues corresponding to the nontrivial equilibrium. Does your analysis allow you to infer anything about its stability?

(d) Use a graphing calculator to sketch curves in the N-P plane. Also sketch solution curves of the prey and the predator densities as functions of time.

16. Assume that

$$\frac{dN}{dt} = 5N - PN$$

$$\frac{dP}{dt} = PN - P$$

(a) Show that this system has two equilibria, the trivial equilibrium $(0, 0)$ and a nontrivial one, in which both species have positive densities.

(b) Use the eigenvalue approach to show that the trivial equilibrium is unstable.

(c) Determine the eigenvalues corresponding to the nontrivial equilibrium. Does your analysis allow you to infer anything about its stability?

(d) Use a graphing calculator to sketch curves in the N-P

plane. Also sketch solution curves of the prey and the predator densities as functions of time.

17. Assume that $N(t)$ denotes the density of an insect species at time t and $P(t)$ denotes the density of its predator at time t. The insect species is an agricultural pest and its predator is used as a biological control agent. Their dynamics are given by the system of differential equations

$$\frac{dN}{dt} = 5N - 3PN$$

$$\frac{dP}{dt} = 2PN - P$$

(a) Explain why

$$\frac{dN}{dt} = 5N \tag{11.85}$$

describes the dynamics of the insect in the absence of the predator. Solve (11.85). Describe what happens to the insect population in the absence of the predator.

(b) Explain why introducing the insect predator into the system can help to control the density of the insect.

(c) Assume that at the beginning of the growing season the insect density is 0.5 and the predator density is 2. You decide to control the insects by using an insecticide in addition to the predator. You are careful, and choose an insecticide that does not harm the predator. After spraying, the insect density drops to 0.01 and the predator density remains at 2. Use a graphing calculator to investigate the long-term implications of your decision to spray the field. In particular, investigate what would have happened to the insect densities if you had decided not to spray the field, and compare your results to the insect density over time that results from your application of the insecticide.

18. Assume that $N(t)$ denotes prey density at time t and $P(t)$ denotes predator density at time t. Their dynamics are given by the system of equations

$$\frac{dN}{dt} = 4N - 2PN$$

$$\frac{dP}{dt} = PN - 3P$$

Assume that initially $N(0) = 3$ and $P(0) = 2$.

(a) If you followed this predator-prey community over time, what would you observe?

(b) Assume that bad weather kills 90% of the prey population and 67% of the predator population. If you continued to observe this predator-prey community, what would you expect to see?

19. An unrealistic feature of the Lotka-Volterra model is that the prey exhibits unlimited growth in the absence of the predator. The model described by the following system remedies this. (Namely, we assume that the prey evolves according to logistic growth in the absence of the predator. The other model features are retained.)

$$\frac{dN}{dt} = 3N\left(1 - \frac{N}{10}\right) - 2PN$$

$$\frac{dP}{dt} = PN - 4P \tag{11.86}$$

(a) Explain why the prey evolves according to

$$\frac{dN}{dt} = 3N\left(1 - \frac{N}{10}\right) \tag{11.87}$$

in the absence of the predator. Investigate the long term behavior of solutions to (11.87).

(b) Find all equilibria of (11.86), and determine their stability using the eigenvalue approach.

(c) Use a graphing calculator to sketch the solution curve of (11.86) in the N-P plane when $N(0) = 2$ and $P(0) = 2$. Also, sketch $N(t)$ and $P(t)$ as functions of time, starting with $N(0) = 2$ and $P(0) = 2$.

20. An unrealistic feature of the Lotka-Volterra model is that the prey exhibits unlimited growth in the absence of the predator. The model described by the following system remedies this. (Namely, we assume that the prey evolves according to logistic growth in the absence of the predator. The other model features are retained.)

$$\frac{dN}{dt} = N\left(1 - \frac{N}{K}\right) - 4PN$$
$$\frac{dP}{dt} = PN - 5P \tag{11.88}$$

where $K > 0$ denotes the carrying capacity of the prey in the absence of the predator. In the following, we will investigate how the carrying capacity affects the outcome of this predator-prey interaction.

(a) Draw the zero isoclines of (11.88) for (i) $K = 10$, and (ii) $K = 3$.

(b) When $K = 10$, the zero isoclines intersect, indicating the existence of a nontrivial equilibrium. Analyze the stability of this nontrivial equilibrium.

(c) Is there a minimum carrying capacity required in order to have a nontrivial equilibrium? If yes, find it and explain what happens when the carrying capacity is below this minimum, and what happens when the carrying capacity is above this minimum.

21. An unrealistic feature of the Lotka-Volterra model is that the prey exhibits unlimited growth in the absence of the predator. The model described by the following system remedies this. (Namely, we assume that the prey evolves according to logistic growth in the absence of the predator. The other model features are retained.)

$$\frac{dN}{dt} = N\left(1 - \frac{N}{20}\right) - 5PN$$
$$\frac{dP}{dt} = 2PN - 8P \tag{11.89}$$

(a) Draw the zero isoclines of (11.89).

(b) Use the graphical approach of Subsection 11.3.2 to determine whether the nontrivial equilibrium is locally stable.

In Problems 22–26, we will analyze how a change in parameters in the modified Lotka-Volterra predator-prey model

$$\frac{dN}{dt} = aN\left(1 - \frac{N}{K}\right) - bPN$$
$$\frac{dP}{dt} = cPN - dP \tag{11.90}$$

affects predator-prey interactions.

22. (a) Find the zero-isoclines of (11.90), and determine conditions under which a nontrivial equilibrium exists (that is, an equilibrium in which both prey and predator have positive densities).

(b) Use the graphical approach of Subsection 11.3.2 to show that if a nontrivial equilibrium exists, it is locally stable.

In Problems 23–26, we use the results of Problem 22. Assume that the parameters are chosen so that a nontrivial equilibrium exists.

23. Use the results of Problem 22 to show that an increase in a (the intrinsic rate of growth of the prey) results in an increase of the predator density but leaves the prey density unchanged.

24. Use the results of Problem 22 to show that an increase in b (the searching efficiency) reduces the predator density but has no effect on the equilibrium abundance of the prey.

25. Use the results of Problem 22 to show that an increase in c (the predator growth efficiency) reduces the prey equilibrium abundance and increases the predator equilibrium abundance.

26. Use the results of Problem 22 to show that an increase in K (the prey carrying capacity in the absence of the predator) increases the predator equilibrium abundance but has no effect on the prey equilibrium abundance.

[11.4.3]

In Problems 27–34, classify each community matrix at equilibrium according to the five cases considered in Subsection 11.4.3, and determine whether the equilibrium is stable. (Assume in each case that the equilibrium exists.)

27. $\begin{bmatrix} -1 & -1.3 \\ 0.3 & -2 \end{bmatrix}$ **28.** $\begin{bmatrix} -3 & -1.2 \\ -1 & -2 \end{bmatrix}$

29. $\begin{bmatrix} -1.5 & 1.6 \\ 2.3 & -5.1 \end{bmatrix}$ **30.** $\begin{bmatrix} -0.3 & 0 \\ 0.4 & -0.7 \end{bmatrix}$

31. $\begin{bmatrix} -1 & 1.3 \\ 2 & -1.5 \end{bmatrix}$ **32.** $\begin{bmatrix} -2.7 & 0 \\ -1.3 & -0.6 \end{bmatrix}$

33. $\begin{bmatrix} -5 & -1.7 \\ -2.3 & -0.2 \end{bmatrix}$ **34.** $\begin{bmatrix} -2.3 & -4.7 \\ 1.2 & -3.2 \end{bmatrix}$

In Problems 35–40, we consider communities composed of two species. The abundance of species 1 at time t is given by $N_1(t)$, the abundance of species 2 at time t is given by $N_2(t)$. Their dynamics are described by

$$\frac{dN_1}{dt} = f_1(N_1, N_2)$$

$$\frac{dN_2}{dt} = f_2(N_1, N_2)$$

Assume that when both species are at low abundances, their abundances increase. In each problem, determine the sign structure of the community matrix at the nontrivial equilibrium (indicated by a dot), based on the graph of the zero isoclines. Determine the stability of the equilibria if possible.

35.

▲ **Figure 11.66**

36.

▲ **Figure 11.67**

37.

▲ **Figure 11.68**

38.

▲ **Figure 11.69**

39.

▲ **Figure 11.70**

40.

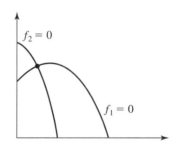

▲ **Figure 11.71**

41. Suppose that the diagonal elements a_{ii} of the community matrix of a species assemblage in equilibrium are negative. Explain why this implies that species i exhibits self-regulation.

42. Consider a community composed of two species. Assume that both species inhibit themselves. Explain why mutualistic and competitive interactions lead to qualitatively similar predictions about the stability of the corresponding equilibria. That is, show that if $A = [a_{ij}]$ is the community matrix at equilibrium for the case of mutualism, and if $B = [b_{ij}]$ is the community matrix at equilibrium for the case of competition, then the following holds: If $|a_{ij}| = |b_{ij}|$ for $1 \leq i, j \leq 2$, then either both equilibria are locally stable or both are unstable.

43. The classical Lotka-Volterra model of predation is given by

$$\frac{dN}{dt} = aN - bNP$$

$$\frac{dP}{dt} = cNP - dP$$

where $N = N(t)$ is the prey density at time t and $P = P(t)$ is the predator density at time t. The constants a, b, c, and d are all positive.

(a) Find the nontrivial equilibrium $(\hat{N}, \hat{P})$ with $\hat{N} > 0$ and $\hat{P} > 0$.

(b) Find the community matrix corresponding to the nontrivial equilibrium.

(c) Explain each entry of the community matrix found in (b) in terms of how individuals in this community affect each other.

44. The modified Lotka-Volterra model of predation is given by

$$\frac{dN}{dt} = aN\left(1 - \frac{N}{K}\right) - bNP$$

$$\frac{dP}{dt} = cNP - dP$$

where $N = N(t)$ is the prey density at time t and $P = P(t)$ is the predator density at time t. The constants a, b, c, d, and K are positive. Furthermore, assume that $d/c < K$.

(a) Find the nontrivial equilibrium $(\hat{N}, \hat{P})$ with $\hat{N} > 0$ and $\hat{P} > 0$.

(b) Find the community matrix corresponding to the nontrivial equilibrium.

(c) Explain each entry of the community matrix found in (b) in terms of how individuals in this community affect each other.

[11.4.4]

45. Use a graphing calculator to study the following example of the Fitzhugh-Nagumo model.

$$\frac{dV}{dt} = -V(V - 0.3)(V - 1) - w$$

$$\frac{dw}{dt} = 0.01(V - 0.4w)$$

Sketch the graph of the solution curve in the V-w plane when (i) $(V(0), w(0)) = (0.4, 0)$, and (ii) $(V(0), w(0)) = (0.2, 0)$.

46. Use a graphing calculator to study the following example of the Fitzhugh-Nagumo model.

$$\frac{dV}{dt} = -V(V - 0.6)(V - 1) - w$$

$$\frac{dw}{dt} = 0.03(V - 0.6w)$$

Sketch the graph of the solution curve in the V-w plane when (i) $(V(0), w(0)) = (0.8, 0)$, and (ii) $(V(0), w(0)) = (0.4, 0)$.

47. Assume the following example of the Fitzhugh-Nagumo model.

$$\frac{dV}{dt} = -V(V - 0.3)(V - 1) - w$$

$$\frac{dw}{dt} = 0.01(V - 0.4w)$$

Assume that $w(0) = 0$. For which initial values of $V(0)$ can you observe an action potential?

48. Assume the following example of the Fitzhugh-Nagumo model.

$$\frac{dV}{dt} = -V(V - 0.6)(V - 1) - w$$

$$\frac{dw}{dt} = 0.03(V - 0.6w)$$

Assume that $w(0) = 0$. For which initial values of $V(0)$ can you observe an action potential?

[11.4.5]

In Problems 49–52, use the mass action law to translate each chemical reaction into a system of differential equations.

49.
$$A + B \xrightarrow{k} C$$

50.
$$A + B \underset{k_-}{\overset{k_+}{\rightleftarrows}} C$$

51.
$$E + S \xrightarrow{k_1} ES \xrightarrow{k_2} E + P$$

52.
$$A + B \xrightarrow{k} A + C$$

53. Show that the following system of differential equations has a conserved quantity, and find it.

$$\frac{dx}{dt} = 2x - 3y$$

$$\frac{dy}{dt} = 3y - 2x$$

54. Show that the following system of differential equations has a conserved quantity, and find it.

$$\frac{dx}{dt} = -4x + 2y$$

$$\frac{dy}{dt} = -y + 2x$$

55. Show that the following system of differential equations has a conserved quantity, and find it.

$$\frac{dx}{dt} = -x + 2xy + z$$

$$\frac{dy}{dt} = -2xy$$

$$\frac{dz}{dt} = x - z$$

56. Suppose that $x(t) + y(t)$ is a conserved quantity. If

$$\frac{dx}{dt} = -3x + 2xy$$

find the differential equation for $y(t)$.

57. The Michaelis-Menten law (equation (11.76)) states that

$$\frac{dp}{dt} = \frac{v_m s}{K_m + s}$$

where $p = p(t)$ is the concentration of the product of the enzymatic reaction at time t, $s = s(t)$ is the concentration of the substrate at time t, and v_m and K_m are positive constants. Set

$$f(s) = \frac{v_m s}{K_m + s}$$

where v_m and K_m are positive constants.

(a) Show that

$$\lim_{s \to \infty} f(s) = v_m$$

(b) Show that

$$f(K_m) = \frac{v_m}{2}$$

(c) Show that for $s \geq 0$, $f(s)$ is (i) nonnegative, (ii) increasing, and (iii) concave down. Sketch a graph of $f(s)$. Label v_m and K_m on your graph.

(d) Explain why we said that the reaction rate dp/dt is limited by the availability of the substrate.

58. The growth of microbes in a chemostat was described by equation (11.77). Using the notation of (11.77) with

$$q(s) = \frac{v_m s}{K_m + s}$$

where v_m and K_m are positive constants, we will investigate how the substrate concentration $\hat{s}$ in equilibrium depends on the uptake rate Y.

(a) Assume that the microbes have a positive equilibrium density. Find the equilibrium concentration $\hat{s}$ algebraically, and investigate how the uptake rate Y affects $\hat{s}$.

(b) Assume that the microbes have a positive equilibrium density. Sketch a graph of $q(s)$, and explain how you would determine $\hat{s}$ graphically. Use your graph to explain how the uptake rate Y affects $\hat{s}$.

59. The growth of microbes in a chemostat was described by equation (11.77). Using the notation of (11.77) with

$$q(s) = \frac{v_m s}{K_m + s}$$

we will investigate how the substrate concentration $\hat{s}$ in equilibrium depends on D, the rate at which the medium enters the chemostat.

(a) Assume that the microbes have a positive equilibrium density. Find the equilibrium concentration $\hat{s}$ algebraically. Investigate how the rate D affects $\hat{s}$.

(b) Assume that the microbes have a positive equilibrium density. Sketch a graph of $q(s)$, and explain how you would determine $\hat{s}$ graphically. Use your graph to explain how the rate D affects $\hat{s}$.

In Problems 60–61, we investigate specific examples of microbial growth described by equation (11.77). We use the notation of Subsection 11.4.5. In each case, determine all equilibria and their stability.

60.

$$\frac{ds}{dt} = 2(4 - s) - \frac{3s}{2 + s} x$$

$$\frac{dx}{dt} = \frac{sx}{2 + s} - 2x$$

61.

$$\frac{ds}{dt} = 2(4 - s) - \frac{3s}{1 + s} x$$

$$\frac{dx}{dt} = \frac{3sx}{1 + s} - 2x$$

◆ 11.5 KEY TERMS

Chapter 11 Review: Topics   Discuss the following definitions and concepts:

1. Linear first order equation
2. Homogeneous
3. Direction field, slope field, direction vector
4. Solution of a system of linear differential equations
5. Eigenvalue, eigenvector, solution corresponding to an eigenvalue
6. Superposition principle
7. General solution
8. Stability
9. Sink or stable node
10. Saddle
11. Source or unstable node
12. Spiral
13. Euler's formula

14. Compartment model
15. Conserved quantity
16. Harmonic oscillator
17. Nonlinear autonomous system of differential equations
18. Critical point
19. Zero isoclines or null clines
20. Graphical approach to stability
21. Lotka-Volterra model of interspecific competition
22. Intraspecific competition, interspecific competition
23. Competitive exclusion, founder control, coexistence
24. Lotka-Volterra predator-prey model
25. Community matrix
26. Fitzhugh-Nagumo model
27. Action potential
28. Michaelis-Menten law

◆ 11.6 REVIEW PROBLEMS

1. Let $N_1(t)$ and $N_2(t)$ denote the respective sizes of two populations at time t, and assume that their dynamics are given by

$$\frac{dN_1}{dt} = r_1 N_1$$

$$\frac{dN_2}{dt} = r_2 N_2$$

where r_1 and r_2 are positive constants, denoting the intrinsic rate of growth of the two populations. Set $Z(t) = N_1(t)/N_2(t)$, and show that $Z(t)$ satisfies

$$\frac{d}{dt} \ln Z(t) = r_1 - r_2 \tag{11.91}$$

Solve (11.91), and show that $\lim_{t \to \infty} Z(t) = \infty$ if $r_1 > r_2$. Conclude from this that population 1 becomes numerically dominant when $r_1 > r_2$.

2. Let $N_1(t)$ and $N_2(t)$ denote the respective sizes of two populations at time t, and assume that their dynamics are given by

$$\frac{dN_1}{dt} = r_1 N_1$$

$$\frac{dN_2}{dt} = r_2 N_2$$

where r_1 and r_2 are positive constants, denoting the intrinsic rate of growth of the two populations. Denote the combined population size at time t by $N(t)$; that is, $N(t) = N_1(t) + N_2(t)$. Define the relative proportions

$$p_1 = \frac{N_1}{N} \quad \text{and} \quad p_2 = \frac{N_2}{N}$$

Use the fact that $p_1/p_2 = N_1/N_2$ to show that

$$\frac{dp_1}{dt} = p_1(1 - p_1)(r_1 - r_2)$$

Show that if $r_1 > r_2$ and $0 < p_1(0) < 1$, $p_1(t)$ will increase for $t > 0$, and population 1 will become numerically dominant.

3. An unrealistic feature of the Lotka-Volterra model is that the prey exhibits unlimited growth in the absence of the predator. The model described by the following system remedies this. (Namely, we assume that the prey evolves according to logistic growth in the absence of the predator. The other model features are retained.)

$$\frac{dN}{dt} = 2N\left(1 - \frac{N}{10}\right) - 3PN$$

$$\frac{dP}{dt} = PN - 3P \tag{11.92}$$

(a) Draw the zero isoclines of (11.92).

(b) Use the graphical approach of Subsection 11.3.2 to determine whether the nontrivial equilibrium is locally stable.

4. Tilman (1982) developed a theoretical framework for studying resource competition in plants. In its simplest form, one species competes for a single resource, for instance, nitrogen.

If $B(t)$ denotes the total biomass at time t and $R(t)$ is the amount of resource available at time t, then the dynamics are described by the following system of differential equations.

$$\frac{dB}{dt} = B[f(R) - m]$$

$$\frac{dR}{dt} = a(S - R) - cBf(R)$$

The first equation describes the rate of biomass change, the function $f(R)$ describes how the species growth rate depends on the resource, and m is the specific loss rate. The second equation describes the resource dynamics; the constant S is the maximal amount of resource in a given habitat. The rate of resource supply (dR/dt) is assumed to be proportional to the difference between the current resource level and the maximal amount of resource; the constant a is the constant of proportionality. The term $cBf(R)$ describes the resource uptake by the plants; the constant c can be considered as a conversion factor.

We assume in the following that $f(R)$ follows the Monod growth function

$$f(R) = \frac{dR}{k + R}$$

where d and k are positive constants.

(a) Find all equilibria. Show that if $d > m$ and $S > mk/(d - m)$, then there exists a nontrivial equilibrium.

(b) Sketch the zero isoclines for the case in which the system admits a nontrivial equilibrium. Use the graphical approach to analyze the stability of the nontrivial equilibrium.

5. The following describes a simple competition model in which two species of plants compete for vacant space. Assume that the entire habitat is divided into a large number of patches. Each patch can be occupied by at most one species. We denote by $p_i(t)$ the fraction of patches occupied by species i. Note that $0 \leq p_1(t) + p_2(t) \leq 1$. The dynamics are described by

$$\frac{dp_1}{dt} = c_1 p_1(1 - p_1 - p_2) - m_1 p_1$$

$$\frac{dp_1}{dt} = c_2 p_2(1 - p_1 - p_2) - m_2 p_2$$

where c_1, c_2, m_1, and m_2 are positive constants. The first term on the right-hand side of each equation describes the colonization of vacant patches; the second term on the right-hand side of each equation describes how occupied patches become vacant.

(a) Show that the dynamics of species 1 in the absence of species 2 are given by

$$\frac{dp_1}{dt} = c_1 p_1(1 - p_1) - m_1 p_1 \tag{11.93}$$

and find conditions on c_1 and m_1 so that (11.93) admits a nontrivial equilibrium (an equilibrium in which $0 < p_1 \leq 1$).

(b) Assume now that $c_1 > m_1$ and $c_2 > m_2$. Show that if

$$\frac{c_1}{m_1} > \frac{c_2}{m_2}$$

then species 1 will exclude species 2 if species 1 initially occupies a positive fraction of the patches.

6. (*Paradox of enrichment*) Rosenzweig (1971) analyzed a number of predator-prey models, and concluded that enriching the system by increasing the nutrient supply destabilizes the nontrivial equilibrium. We will think of the predator-prey model as a plant-herbivore system, in which plants represent prey and herbivores represent predators. The models analyzed were of the form

$$\frac{dN}{dt} = f(N, P) \tag{11.94}$$

$$\frac{dP}{dt} = g(N, P) \tag{11.95}$$

where $N = N(t)$ is the plant abundance at time t and $P = P(t)$ is the herbivore abundance at time t. The models all shared the property that the zero isocline for the herbivore was a vertical line and the zero isocline for the plants was a hump-shaped curve. We will look at one of the models, namely

$$\frac{dN}{dt} = aN\left(1 - \frac{N}{K}\right) - bP(1 - e^{-rN})$$

$$\frac{dP}{dt} = cP(1 - e^{-rN}) - dP \tag{11.96}$$

(a) Find the zero isoclines for (11.96), and show that (i) the zero isocline of the herbivore ($dP/dt = 0$) is a vertical line in the N-P plane, and (ii) the zero isocline for the plants ($dN/dt = 0$) intersects the N-axis at $N = K$.

(b) Plot the zero isoclines in the N-P plane for $a = b = c = r = 1$ and $d = 0.9$, and three levels of the carrying capacity: (i) $K = 1$, (ii) $K = 4$, and (iii) $K = 10$.

(c) For each of the three carrying capacities, determine whether a nontrivial equilibrium exists.

(d) Use the graphical approach of Subsection 11.3.2 to determine the stability of the existing nontrivial equilibria in (c).

(e) Enriching the community could mean increasing the carrying capacity of the plants. For instance, adding nitrogen or phosphorus to plant communities frequently results in an increase in biomass, which can be interpreted as an increase in the carrying capacity of the plants (the K value). Based on your answers in (d), explain why enriching the community (increasing the carrying capacity of the plants) can result in a destabilization of the nontrivial equilibrium. What are the consequences?

7. The growth of microbes in a chemostat was described by equation (11.77). We will investigate how the microbial abundance in equilibrium depends on the characteristics of the system.

(a) Assume that $q(s)$ is a nonnegative function. Show that the equilibrium abundance of the microbes, $\hat{x}$, is given by

$$\hat{x} = Y(s_0 - \hat{s})$$

where $\hat{s}$ is the substrate equilibrium abundance. When is $\hat{x} > 0$?

(b) Assume now that

$$q(s) = \frac{v_m s}{K_m + s}$$

Investigate how the uptake rate Y, and the rate D at which new medium enters the chemostat, affect the equilibrium abundance of the microbes.

8. (*Successional niche*) Pacala and Rees (1998) discuss a simple mathematical model of competition, to explain successional diversity by means of a successional niche mechanism. In this model, two species, an early successional and a late successional, occupy discrete patches. Each patch experiences disturbances (such as fire) at rate D. After a patch is disturbed, both species are present. Over time, however, the late successional species outcompetes the early successional species, which causes the early successional species to become extinct. This change, from a patch that is occupied by both species to a patch that is occupied by the late successional species only, happens at rate a. We keep track of the number of patches occupied by both species at time t, denoted by $x(t)$, and the number of patches occupied by just the late successional species at time t, denoted by $y(t)$. The dynamics are given by the following system of linear differential equations.

$$\frac{dx}{dt} = -ax + Dy$$

$$\frac{dy}{dt} = ax - Dy \tag{11.97}$$

where a and D are positive constants.

(a) Show that all equilibria are of the form $(x, ax/D)$.

(b) Find the eigenvalues and eigenvectors corresponding to each equilibrium.

(c) Show that

$$\begin{bmatrix} x(t) \\ y(t) \end{bmatrix} = C_1 \begin{bmatrix} u_1 \\ u_2 \end{bmatrix} + C_2 e^{\lambda_2 t} \begin{bmatrix} v_1 \\ v_2 \end{bmatrix}$$

is a solution of (11.97), where $\begin{bmatrix} u_1 \\ u_2 \end{bmatrix}$ is the eigenvector corresponding to the zero eigenvalue and $\begin{bmatrix} v_1 \\ v_2 \end{bmatrix}$ is the eigenvector corresponding to the nonzero eigenvalue λ_2.

(d) Show that $x(t) + y(t)$ does not depend on t. (*Hint*: Show that $\frac{d}{dt}(x(t) + y(t)) = 0$.) Show that the line $x + y = A$ (where A is a constant) is parallel to the line in the direction of the eigenvector corresponding to the nonzero eigenvalue.

(e) Show that the zero isoclines of (11.97) are given by

$$y = \frac{a}{D}x$$

and that this line is the line in the direction of the eigenvector corresponding to the zero eigenvalue.

(f) Suppose now that $x(t) + y(t) = c$, where c is a positive constant. Show that (11.97) can be reduced to just one equation, namely

$$\frac{dx}{dt} = -(a + D)x + Dc$$

Show that $\hat{x} = c\frac{D}{D+a}$ is the only equilibrium, and determine its stability.

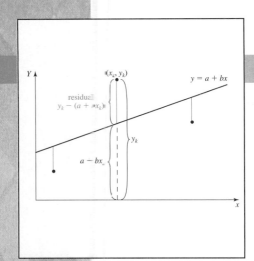

Probability and Statistics

We conclude this book with a chapter on probability and statistics. Though neither field is part of calculus, they both use it; more importantly, they provide indispensable tools for life scientists.

Most phenomena in nature are not deterministic. To name just a few examples, consider the number of eggs laid by a bird, the life expectancy of an organism, the inheritance of genes, or the number of people infected during an outbreak of a disease. To deal with the inherent randomness (or stochasticity) of natural phenomena, we need to develop special tools; these tools are provided by the areas of probability and statistics.

A short description of the role of probability and statistics in the life sciences might be as follows: Probability theory provides tools for modeling randomness and forms the foundation of statistics. Statistics provides tools for analyzing data from scientific experiments.

◆ 12.1 COUNTING

It is often necessary to count the ways in which a certain task can be performed. The mathematical field of **combinatorics** deals with such enumeration problems. Frequently, the total number of possible ways is very large, which makes it impractical to write down all possible choices. There are three basic counting principles that will help us to count in a more systematic way; the first is the multiplication principle; the other two follow from it.

12.1.1 The Multiplication Principle

To illustrate the first principle, imagine that we wish to experimentally manipulate growth conditions for plants, say the grass species big bluestem, *Andropogon gerardi*. We want to grow plants in pots in a greenhouse at two different levels of fertilizer (low and high) and four different temperatures (10°C, 15°C, 20°C, and 25°C). If we want three replicates of each possible combination of fertilizer and temperature treatment, how many pots will we need? We can answer this question using a tree

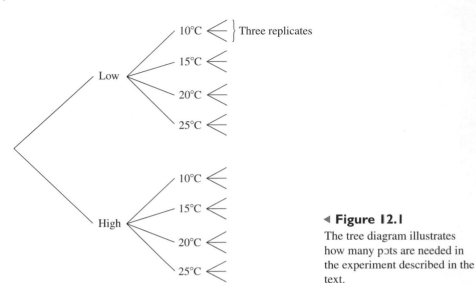

◀ **Figure 12.1**

The tree diagram illustrates how many pots are needed in the experiment described in the text.

diagram, as shown in Figure 12.1. We see from the tree that we will need

$$2 \cdot 4 \cdot 3 = 24$$

pots for our experiment.

The counting principle that we just used is called the multiplication principle, which we can summarize as follows.

> **Multiplication Principle**   Suppose that an experiment consists of m ordered tasks. Task 1 has n_1 possible outcomes, task 2 has n_2 possible outcomes, ..., and task m has n_m possible outcomes. The total number of possible outcomes of the experiment is
>
> $$n_1 \cdot n_2 \cdot n_3 \cdots n_m$$

Looking back at our example of growing big bluestem in pots, we see that the experiment consisted of three tasks: first, to select the fertilizer level; second, to select the temperature; and third, to replicate each combination of fertilizer and temperature three times. The successive tasks can be illustrated in the tree diagram, and the total number of pots required for our experiment can be read by counting the number of tips of the tree.

We present one more example that illustrates this counting principle.

▌**EXAMPLE 1**   Suppose that after a long day in the greenhouse you decide to order pizza. You call a local pizza parlor and learn that there are three choices of crust and five choices of toppings, and that you can order the pizza with or without cheese. If you only want one topping, how many different choices do you have for selecting a pizza?

Solution   Your "experiment," which consists of ordering a pizza, involves three tasks. The first task is to choose a crust, the second is to choose the topping, and the third is to decide whether or not you want cheese. Using the multiplication principle, we find that there are

$$3 \cdot 5 \cdot 2 = 30$$

different pizzas that you could order.   ▌

12.1.2 Permutations

Suppose that you grow plants in a greenhouse. To control for spatially varying environmental conditions, you rearrange the pots every other day. If you have six pots arranged in a row, in how many ways can you arrange the pots? To answer this question, imagine that you arrange the pots on the bench from left to right: you have six choices for the leftmost position on the bench, for the next position you can choose any of the remaining five pots, for the third position you can choose any of the remaining four pots, and so on, until there is one pot left that must go to the rightmost position. Using the multiplication principle, we find that there are

$$6 \cdot 5 \cdot 4 \cdot 3 \cdot 2 \cdot 1 = 720$$

ways to arrange the six pots on the bench.

As shorthand notation for such descending products, we define

$$n! = n(n-1)(n-2)\cdots 3 \cdot 2 \cdot 1$$

and read $n!$ as "n factorial." We can now write $6!$ instead of $6 \cdot 5 \cdot 4 \cdot 3 \cdot 2 \cdot 1$.
We present two important properties of $n!$:

$$(n+1)! = (n+1)n!$$
$$0! = 1$$

You should note that $n!$ grows very quickly. Suppose that instead of six pots you have seven; then there are $7! = 5040$ ways to arrange the seven pots. If you had twelve pots, then there would be $12! = 479,001,600$ possible ways to arrange them.

We will look at another example, and then state a general principle.

EXAMPLE 2   Suppose that a track team has ten sprinters, any four of whom can form a relay team. Assume that each person can run in any position on the team. How many teams can be formed, if teams that consist of the same four people in different running orders are considered different teams?

Solution   We select the members of the team in the order in which they run. There are ten available sprinters for the first position. After having chosen a person for the first position, there are nine left, and we can choose any of the nine for the second position. For the third position, we can choose among the eight remaining people and, finally, for the fourth position, we can select a person from the remaining seven. Using the multiplication principle, we find that there are

$$10 \cdot 9 \cdot 8 \cdot 7 = 5040$$

different relay teams.   ▮

In the preceding example, we selected $k = 4$ objects from a set of $n = 10$, where the order of selection was important. Such selections are called **permutations**. Using the multiplication principle, we can find the number of different ways such a selection can be made; we summarize this in the following.

> **Permutations**   A **permutation** of n different objects taken k at a time is an *ordered* subset of k out of the n objects. The number of ways that this can be done is denoted by $P(n, k)$, and is given by
>
> $$P(n, k) = n(n-1)(n-2)\cdots(n-k+1)$$

Note that the last term in the product defining $P(n, k)$ is of the form $(n - k + 1)$, because there are k descending factors and the first factor is n.

Returning to Example 2, where we wanted to select four out of ten objects in an ordered arrangement, we can now use our permutation rule to compute the number of ways that we can make the selection. Setting $n = 10$ and $k = 4$, we find

$$P(10, 4) = 10 \cdot 9 \cdot 8 \cdot 7$$

for the number of different relay teams. Since the product consists of four terms in descending order, starting with 10, the last term is $n - k + 1 = 10 - 4 + 1 = 7$.

Another way to compute $P(n, k)$ follows from the calculation

$$P(n, k) = n(n - 1)(n - 2) \cdot (n - k + 1) \frac{(n - k)(n - k - 1) \cdots 3 \cdot 2 \cdot 1}{(n - k)(n - k - 1) \cdots 3 \cdot 2 \cdot 1}$$

Simplifying this yields

$$P(n, k) = \frac{n!}{(n - k)!} \tag{12.1}$$

We present one more example before we introduce the third counting principle.

▮ **EXAMPLE 3**   How many five-letter words with no repeated letters can you form using the 26 letters of the alphabet? (Note that a "word" here need not be in the dictionary.)

Solution   This amounts to choosing 5 letters from 26, where the order is important. Hence, there are

$$P(26, 5) = \frac{26!}{21!} = 26 \cdot 25 \cdot 24 \cdot 23 \cdot 22 = 7,893,600$$

different words.   ▮

12.1.3 Combinations

When choosing a permutation, the order of the objects is important. Obviously, there must be examples where the order is not important. How would we then compute the number of arrangements?

We return to Example 2, where we chose a relay team. The order on the team is important when the members on the team actually run. But if we only wanted to know who was on the team, the order would no longer be important. We saw that there are $10 \cdot 9 \cdot 8 \cdot 7$ different relay teams. But since four people can be arranged in $4! = 4 \cdot 3 \cdot 2 \cdot 1 = 24$ different ways, each choice of four people appears in 24 different teams. If we divide $10 \cdot 9 \cdot 8 \cdot 7$ by $4 \cdot 3 \cdot 2 \cdot 1$, we obtain the number of ways that we can choose four people from a group of ten if the order does not matter. We find that this number is

$$\frac{10 \cdot 9 \cdot 8 \cdot 7}{4 \cdot 3 \cdot 2 \cdot 1} = 210$$

Such unordered selections are called **combinations**. Using the approach that we have employed, we can give a general formula for the number of combinations, summarized in the following.

Combinations   A **combination** of n different objects taken k at a time is an *unordered* subset of k out of n objects. The number of ways that this can be done is denoted by $C(n, k)$, and is given by

$$C(n, k) = \frac{n(n - 1)(n - 2) \cdots (n - k + 1)}{k!}$$

Instead of $C(n, k)$, we often write $\binom{n}{k}$, which we read "n choose k." Using (12.1), we find

$$C(n, k) = \binom{n}{k} = \frac{P(n, k)}{k!} = \frac{n!}{k!(n-k)!} \tag{12.2}$$

Looking at the rightmost expression, we see that this is equal to $\binom{n}{n-k}$. We therefore find the identity

$$\binom{n}{k} = \binom{n}{n-k} \tag{12.3}$$

The following counting argument also explains this identity. The expression $\binom{n}{k}$ denotes the number of ways that we can select an unordered subset of size k from a set of size n. Instead of choosing the elements that go into the set, we could also choose the elements, that do not go into the set. There are $n-k$ such elements, and we can select them in $\binom{n}{n-k}$ different ways.

Another identity follows from setting $k = 0$ in (12.3), and using (12.2) and $0! = 1$. Namely,

$$\binom{n}{0} = \binom{n}{n} = \frac{n!}{0!n!} = 1$$

This identity can also be understood from the following counting argument. The expression $\binom{n}{0}$ means that we select a subset of size 0 from a set of size n, where the order is not important. There is only one such set, namely, the empty set. Similarly, $\binom{n}{n}$ means that we select a subset of n objects from a set of size n, where the order is not important. There is only one way to do this, namely, we must take the entire set of n objects.

We can use similar reasoning to argue that $\binom{n}{1} = n$; this represents the number of ways that we can choose subsets of size 1, where the order is not important. There are n such subsets, namely, all the singletons. (Actually, the order does not play a role when considering sets with one element, reflected in the fact that $P(n, 1) = n$ as well.)

We look at one more example in which we use the rule for counting combinations.

EXAMPLE 4   Suppose that you wish to plant five grass species in a plot. You can choose among twelve different species. How many choices do you have?

Solution   Since the order is not important for this selection, there are

$$C(12, 5) = \frac{12!}{5!7!} = \frac{12 \cdot 11 \cdot 10 \cdot 9 \cdot 8}{5 \cdot 4 \cdot 3 \cdot 2 \cdot 1} = 792$$

different ways to make the selection. Alternatively, we could have written

$$C(12, 5) = \binom{12}{5} = \frac{P(12, 5)}{5!}$$

These expressions are equivalent, and are evaluated as above.   ▮

12.1.4 Combining the Counting Principles

Often, the difficult part of counting is deciding which rule to use. To gain experience with this, we present several examples in which we combine the three counting principles.

▌**EXAMPLE 5**   How many different eleven-letter words can be formed using the letters in the word MISSISSIPPI?

Solution   There are four S's, four I's, two P's and one M. There are 11! ways to arrange the letters, but some of the resulting words will be indistinguishable, since letters that repeat in a word can be swapped without creating a new word. Therefore, we need to divide by the order of the repeated letters. We then find that there are

$$\frac{11!}{4!4!2!1!} = 34,650$$

different words.   ▌

▌**EXAMPLE 6**   Returning to Example 1, suppose that you want two different toppings on your pizza. How does this change your answer?

Solution   As there are five toppings, and the order in which we choose them is not important, we have $\binom{5}{2}$ choices for the toppings. Everything else remains the same, and we find that there are

$$3 \cdot \binom{5}{2} \cdot 2 = 60$$

different pizzas to choose from.   ▌

▌**EXAMPLE 7**   Suppose that a license plate consists of three letters followed by three digits. How many license plates can there be, if repetition of letters but not of digits is allowed?

Solution   The order is important in this case. For each letter there are 26 choices, since repetition is allowed. There are ten choices for the first digit, nine choices for the second digit and eight choices for the third digit, since repetition of digits is not allowed. Hence, there are

$$26 \cdot 26 \cdot 26 \cdot 10 \cdot 9 \cdot 8 = 12,654,720$$

different license plates with the above restriction.   ▌

▌**EXAMPLE 8**   An urn contains six green and four blue balls. You take out three balls at random. How many different selections could contain exactly two green balls and one blue ball? (Assume that the balls are distinguishable.)

Solution   The order in this selection is not important. To obtain two green balls and one blue ball, we select two of the six green and one of the four blue balls and then combine our choices. That is, there are

$$\binom{6}{2}\binom{4}{1} = 60$$

different selections.   ▌

In the preceding example, we explicitly stated that all balls are distinguishable, and we will from now on always assume that the object we select are distinguishable, without explicitly stating this. Our counting principles apply only to distinguishable objects, as assumed in the definitions ("n different objects"). There are cases in physics, where objects are indistinguishable, for instance electrons.

▌**EXAMPLE 9**   A collection contains seeds for five different annual plants; two produce yellow flowers, and the others produce blue flowers. You plan a garden bed with three different annual plants from this selection, but you do not want both of the plants with yellow flowers in the bed. How many different selections can you make?

Solution   Possible selections contain either blue flowers only, or one of the yellow and two of the blue flowers but not both of the yellow. We must choose three different plants out of the available five. There are $\binom{3}{3}\binom{2}{0}$ choices for a flower bed with blue flowers only, and $\binom{3}{2}\binom{2}{1}$ choices for a flower bed with exactly one yellow flower. Hence, adding up the two cases, there are

$$\binom{3}{3}\binom{2}{0} + \binom{3}{2}\binom{2}{1} = (1)(1) + (3)(2) = 7$$

different selections.

Alternatively, we could have approached this problem in the following way. There are a total of $\binom{5}{3}$ ways to select three plants from the available five. Choices with two yellow flowers and one blue flower are not acceptable; there are $\binom{3}{1}\binom{2}{2}$ such choices. All other choices are acceptable. Hence, there are

$$\binom{5}{3} - \binom{3}{1}\binom{2}{2} = 10 - (3)(1) = 7$$

different selections.

◆ 12.1 PROBLEMS

[12.1.1]

1. Suppose that you want to investigate the influence of light and fertilizer levels on plant performance. You plan to use five fertilizer and two light levels. For each combination of fertilizer and light level, you want four replicates. What is the total number of replicates?

2. Suppose that you want to investigate the effects of leaf damage on the performance of drought stressed plants. You plan to use three levels of leaf damage and four different watering protocols. For each combination of leaf damage and watering protocol, you plan to have three replicates. What is the total number of replicates?

3. *Coleomegilla maculata*, a lady beetle, is an important predator of egg masses of *Ostrinia nubialis*, the European corn borer. *C. maculata* also feeds on aphids and maize pollen. To study its food preferences you choose two satiation levels for *C. maculata* and combinations of two of the three food sources (that is, either egg masses and aphids, or egg masses and pollen, or aphids and pollen). For each experimental protocol, you want twenty replicates. What is the total number of replicates?

4. To test the effects of a new drug, you plan the following clinical trial. Each patient receives the new drug, an established drug, or a placebo. You want to enroll fifty patients in each treatment group. How many patients will you need?

5. The Muesli-Mix is a popular breakfast hangout near a campus. A typical breakfast there consists of one beverage, one bowl of cereal, and a piece of fruit. If you can choose among three different beverages, seven different cereals, and four different types of fruit, how many choices for breakfast do you have?

6. To study sex differences in food preferences in rats, you offer one of three choices of food to each rat. You plan to have twelve rats for each food and sex combination. How many rats will you need?

[12.1.2]

7. You plan a trip to Europe, during which you wish to visit London, Paris, Amsterdam, Rome, and Heidelberg. Because you want to buy a railway ticket before you leave, you must decide on the order in which you will visit these five cities. How many different routes are there?

8. Five people line up for a photograph. How many different lineups are there?

9. You have just bought seven books. In how many ways can they be arranged on your bookshelf?

10. Four cars arrive simultaneously at an intersection. Only one car can go through at a time. In how many different ways can they leave the intersection?

11. How many four-letter words with no repeated letters can you form from the twenty-six letters of the alphabet?

12. A committee of three people, consisting of a president, a vice president and a treasurer, must be chosen from a group of ten. How many committees can be selected?

13. Three different awards are to be given to a class of fifteen students. Each student can receive at most one award. Count the number of ways these awards can be given out.

14. You have just enough time to play four songs out of ten from your favorite CD. In how many ways can you program your CD player to play the four songs?

[12.1.3]

15. A bag contains ten different candy bars. You are allowed to choose three. How many choices do you have?

16. During International Movie Week, sixty movies are shown. You have time to see five movies. How many different plans can you make?

17. A committee of three people must be formed from a group of ten. How many committees are there, if no specific tasks are assigned to the members of the committee?

18. A standard deck contains fifty-two different cards. In how many ways can you select five cards from the deck?

19. An urn contains fifteen different balls. In how many ways can you select four balls?

20. Twelve people wait in front of an elevator that only has room for five. Count the number of ways that the first group of people to take the elevator can be chosen.

21. Four A's and five B's are to be arranged into a nine-letter word. How many different words can you form?

22. Suppose that you want to plant a flowerbed with four different plants. You can choose from among eight plants. How may different choices do you have?

[12.1.4]

23. A box contains five red and four blue balls. You choose two balls.

(a) How many possible selections contain exactly two red balls; how many exactly two blue balls; and how many exactly one of each color?

(b) Show that the sum of the number of choices for the three cases in (a) is equal to the number of ways that you can select two balls out of the nine balls in the box.

24. Twelve children are divided up into three groups, of five, four, and three. In how many ways can this be done, if the order within each group is not important?

25. Five A's, three B's, and six C's are to be arranged into a fourteen-letter word. How many different words can you form?

26. A bag contains forty-five beans of three different varieties. Each variety is represented fifteen times in the bag. You grab nine beans out of the bag.

(a) Count the number of ways that each variety can be represented exactly three times in your sample.

(b) Count the number of ways that only one variety appears in your sample.

27. Let $S = \{a, b, c\}$. List all possible subsets, and argue that the total number of subsets is $2^3 = 8$.

28. Suppose that a set contains n elements. Argue that the total number of subsets of this set is 2^n.

29. In how many ways can Brian, Hilary, Peter, and Melissa sit on a bench, if Peter and Melissa want to be next to each other?

30. Paula, Cindy, Gloria, and Jenny have dinner at a round table. In how many ways can they sit around the table, if Cindy wants to sit to the left of Paula?

31. In how many ways can you form a committee of three people from a group of seven, if two of the people do not want to serve together?

32. In how many ways can you form two committees of three people each from a group of nine, if

(a) no person is allowed to serve on more than one committee?

(b) people can serve on both committees simultaneously?

33. A collection contains seeds for four different annual and three different perennial plants. You plan a garden bed with three different plants, and you want to include at least one perennial. How many different selections can you make?

34. In diploid organisms, chromosomes appear in pairs in the nuclei of all cells except gametes (sperm or ovum). Gametes are formed during meiosis, where the number of chromosomes in the nucleus is halved; that is, only one member of each pair of chromosomes ends up in a gamete. Humans have twenty-three pairs of chromosomes. How may kinds of gametes can a human produce?

◆ 12.2 WHAT IS PROBABILITY?

12.2.1 Basic Definitions

A **random experiment** is an experiment in which outcome is uncertain, and which can be repeated. Tossing a coin and rolling a die are examples of random experiments. The set of all possible outcomes of a random experiment is called the **sample space** and typically denoted by Ω. We look at some examples, where we describe random experiments and then give the associated sample space.

┃ EXAMPLE I   Suppose that we toss a coin labeled heads (H) on one side and tails (T) on the other. If we toss the coin once, the possible outcomes are H and T, and the sample space is therefore

$$\Omega = \{H, T\}$$

If we toss the coin twice, then each outcome is an ordered pair describing the outcome on the first toss followed by the outcome on the second toss, such as HT, which means heads followed by tails. The sample space is

$$\Omega = \{HH, HT, TH, TT\}$$

▮ EXAMPLE 2 Consider a population for which we keep track of the genotype at one locus. We assume that the genes at this locus occur in three different forms, called alleles, denoted by A_1, A_2, and A_3. Furthermore, we assume that the individuals in the population are diploid, that is, the chromosomes occur in pairs. This means that a genotype is described by a pair of genes, such as A_1A_1. If our random experiment consists of picking one individual out of the population and noting the genotype, then the sample space is given by

$$\Omega = \{A_1A_1, A_1A_2, A_1A_3, A_2A_2, A_2A_3, A_3A_3\}$$

▮ EXAMPLE 3 An urn contains five balls, numbered 1–5. We draw two balls from the urn without replacement, and note the numbers drawn.

There is some ambiguity in the formulation of this experiment. We can draw the two balls one after the other (without replacing the first in the urn after having noted its number), or we can draw the two balls simultaneously. In the first case, the sample space consists of ordered pairs (i, j) where the first entry is the number on the first ball and the second entry is the number on the second ball. Because the sampling is done without replacement, the two numbers are different. The sample space can be written

$$\Omega = \{(i, j) : 1 \le i \le 5, 1 \le j \le 5, i \ne j\}$$

In the second case, there is no first or second ball, because we draw the balls simultaneously. We can write the sample space as

$$\Omega\{(i, j) : 1 \le i < j \le 5\}$$

Now the first entry of (i, j) represents the smaller of the two numbers on the balls in our sample.

The specifics of the random experiment determine which description of the sample space we prefer.

When we perform random experiments, we often wish to know the likelihood of a particular outcome, or more generally, a particular subset of the sample space. We call subsets of the sample space **events**. Since an outcome is an element of the sample space, an outcome is an event as well, namely a subset that consists of just one element. We will use the basic set operations to deal with events.

Basic Set Operations Suppose that A and B are events of the sample space Ω. The **union** of A and B, denoted by $A \cup B$ (read: "A union B") is the set of all outcomes that belong to either A or B (or both). The **intersection** of A and B, denoted by $A \cap B$ (read: "A intersected with B") is the set of all outcomes that belong to both A and B. Figures 12.2 and 12.3 show these first two set operations. Figures 12.2 and 12.3, where sets are visualized as "bubbles," are called **Venn diagrams**.

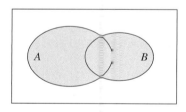

▲ **Figure 12.2**
The union of A and B, $A \cup B$, illustrated in a Venn diagram. The rectangle represents the set Ω.

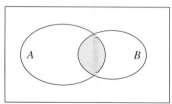

▲ **Figure 12.3**
The intersection of A and B, $A \cap B$, illustrated in a Venn diagram.

We can generalize the union and intersection of two events to a finite number of events. Let $A_1, A_2, \ldots, A_n$ be a finite number of events. Then

$$\bigcup_{i=1}^{n} A_i = A_1 \cup A_2 \cup \cdots \cup A_n = (A_1 \cup A_2 \cup \cdots \cup A_{n-1}) \cup A_n$$

$$= \left[\begin{array}{l} \text{the set of all outcomes that} \\ \text{belong to at least one set } A_i \end{array} \right]$$

$$\bigcap_{i=1}^{n} A_i = A_1 \cap A_2 \cap \cdots \cap A_n = (A_1 \cap A_2 \cap \cdots \cap A_{n-1}) \cap A_n$$

$$= \left[\begin{array}{l} \text{the set of all outcomes that} \\ \text{belong to all sets } A_i, \ i = 1, 2, \ldots, n \end{array} \right]$$

The **complement** of A, denoted by A^c, is the set of all outcomes contained in Ω that are not in A (see Figure 12.4). It follows that

$$\Omega^c = \emptyset \qquad \text{and} \qquad \emptyset^c = \Omega$$

where $\emptyset$ denotes the empty set. Furthermore,

$$(A^c)^c = A$$

When we take complements of unions or intersections, the following two identities are useful.

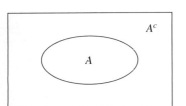

▲ **Figure 12.4**
The complement of A is A^c.

> **De Morgan's Laws**
>
> $$(A \cup B)^c = A^c \cap B^c$$
> $$(A \cap B)^c = A^c \cup B^c$$

▮**EXAMPLE 4**   Let us consider the coin tossing experiment of Example 1, in which we tossed a coin twice. The sample space is given by

$$\Omega = \{HH, HT, TH, TT\}$$

We denote by A the event that at least one head occurred:

$$A = \{HH, HT, TH\}$$

Let B denote the event that the first toss resulted in tails:

$$B = \{TH, TT\}$$

We see that

$$A \cup B = \{HH, HT, TH, TT\} \qquad \text{and} \qquad A \cap B = \{TH\}$$

Furthermore,

$$A^c = \{TT\} \qquad \text{and} \qquad B^c = \{HH, HT\}$$

To see how De Morgan's laws work, we compute

$$(A \cup B)^c = \emptyset \qquad \text{which is equivalent to} \qquad A^c \cap B^c = \emptyset$$

and

$$(A \cap B)^c = \{HH, HT, TT\} \quad \text{which is equivalent to} \quad A^c \cup B^c = \{HH, HT, TT\}$$

▮

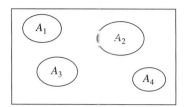

▲ **Figure 12.5**

The four sets A_1, A_2, A_3, and A_4 are disjoint.

We say that $A_1, A_2, \ldots, A_n$ are **disjoint** if

$$A_i \cap A_j = \emptyset \qquad \text{whenever} \qquad i \neq j$$

This is illustrated for four sets in the Venn diagram in Figure 12.5.

The Definition of Probability   We restrict our definition to finite sample spaces; that is, we assume that the set Ω has finitely many elements.

> **Definition**   A **probability** is a function that assigns values between 0 and 1 to events. The probability of an event A, denoted by $P(A)$, satisfies the following properties:
>
> **1.** For any event A, $0 \leq P(A) \leq 1$.
> **2.** $P(\emptyset) = 0$ and $P(\Omega) = 1$.
> **3.** For two disjoint events A and B,
>
> $$P(A \cup B) = P(A) + P(B)$$

Note that a probability is a number that is *always* between 0 and 1. If you compute a probability and get either a negative number or a number greater than 1, you immediately know that your answer must be wrong.

▮**EXAMPLE 5**   Assume that $\Omega = \{1, 2, 3, 4, 5\}$, and that

$$P(1) = P(2) = 0.2 \qquad P(3) = P(4) = 0.1 \qquad \text{and} \qquad P(5) = 0.4$$

where we wrote $P(i)$ for $P(\{i\})$. Set $A = \{1, 2\}$ and $B = \{4, 5\}$. Find $P(A \cup B)$, and show that $P(\Omega) = 1$.

Solution   Since A and B are disjoint ($A \cap B = \emptyset$), we find

$$P(A \cup B) = P(A) - P(B) = P(\{1, 2\}) + P(\{4, 5\})$$

Since $\{1, 2\} = \{1\} \cup \{2\}$ and $\{4, 5\} = \{4\} \cup \{5\}$, and both are unions of disjoint sets,

$$P(\{1, 2\}) = P(1) + P(2)$$
$$P(\{3, 4\}) = P(3) + P(4)$$

Hence,

$$P(A \cup B) = P(1) + P(2) + P(4) + P(5)$$
$$= 0.2 + 0.2 - 0.1 + 0.4 = 0.9$$

To show that $P(\Omega) = 1$, we observe that

$$\Omega = \{1, 2, 3, 4, 5\} = \{1\} \cup \{2\} \cup \{3\} \cup \{4\} \cup \{5\}$$

Since this is a union of disjoint sets,

$$P(\Omega) = P(1) + P(2) + P(3) + P(4) + P(5)$$
$$= 0.2 + 0.2 + 0.1 - 0.1 + 0.4 = 1$$

In the following, we will derive two additional basic properties of probabilities. The first is

$$P(A^c) = 1 - P(A)$$

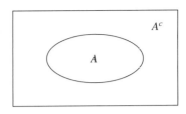

▲ **Figure 12.6**
The set Ω, represented by the rectangle, can be written as a disjoint union of the sets A and A^c.

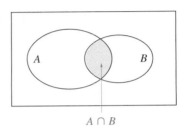

▲ **Figure 12.7**
To compute $P(A \cup B)$, we add $P(A)$ and $P(B)$, but since we count $A \cap B$ twice, we need to subtract $P(A \cap B)$.

To see why this is true, we look at Figure 12.6. We observe that $\Omega = A \cup A^c$, and that A and A^c are disjoint. Therefore,

$$1 = P(\Omega) = P(A \cup A^c) = P(A) + P(A^c)$$

from which the claim follows.

Note that in property 2 of the definition, we wrote $P(\emptyset) = 0$ *and* $P(\Omega) = 1$. It would have been sufficient to require just one of these two identities. For instance, since $\Omega^c = \emptyset$, we can write $\Omega = \Omega \cup \emptyset$. Now, Ω and $\emptyset$ are disjoint, and $P(\Omega) = 1$. It follows that

$$1 = P(\Omega) = P(\Omega \cup \emptyset) = P(\Omega) + P(\emptyset)$$

and therefore

$$P(\emptyset) = 1 - P(\Omega) = 0$$

The next property allows us to compute probabilities of unions of two sets (which are not necessarily disjoint, as in property 3 of the definition).

> For any sets A and B,
> $$P(A \cup B) = P(A) + P(B) - P(A \cap B)$$

This is illustrated in Figure 12.7, and follows from the fact that we count $A \cap B$ twice when we compute $P(A) + P(B)$. The proof of this property is relegated to Problem 15, but we give one example where we use both of these additional properties.

▌**EXAMPLE 6**   Assume that $\Omega = \{1, 2, 3, 4, 5\}$, and that

$$P(1) = P(2) = 0.2, \quad P(3) = P(4) = 0.1 \quad \text{and} \quad P(5) = 0.4$$

Set $A = \{1, 3, 4\}$ and $B = \{4, 5\}$. Find $P(A \cup B)$.

Solution   Observe that A and B are not disjoint. We find

$$A \cap B = \{4\}$$

Using the second of the additional properties, we obtain

$$\begin{aligned}
P(A \cup B) &= P(A) + P(B) - P(A \cap B) \\
&= P(\{1, 3, 4\}) + P(\{4, 5\}) - P(\{4\}) \\
&= (0.2 + 0.1 + 0.1) + (0.1 + 0.4) - (0.1) = 0.8
\end{aligned}$$

We could have obtained the same result by observing that

$$A \cup B = \{1, 3, 4, 5\} \quad \text{and therefore} \quad A \cup B = \{2\}^c$$

which yields

$$P(A \cup B) = P(\{2\}^c) = 1 - P(\{2\}) = 1 - 0.2 = 0.8 \qquad ▌$$

12.2.2 Equally Likely Outcomes

An important class of random experiments with finite sample spaces are those in which all outcomes are equally likely. That is, if $\Omega = \{1, 2, \ldots, n\}$, then $P(1) =$

$P(2) = \cdots = P(n)$, where we wrote $P(i)$ for $P(\{i\})$. Then

$$1 = P(\Omega) = \sum_{i=1}^{n} P(i) = nP(1)$$

which implies that

$$P(1) = P(2) = \cdots = P(n) = \frac{1}{n}$$

If we denote by $|A|$ the number of elements in A, and if $A \subset \Omega$ with $|A| = k$, then

$$P(A) = \frac{|A|}{|\Omega|} = \frac{k}{n}$$

In the following two examples, we will discuss random experiments in which all outcomes are equally likely. You should pay particular attention to the sample space, to make sure that you understand that all of its elements are indeed equally likely.

EXAMPLE 7 Toss a fair coin three times and find the probability of the event $A = \{\text{at least two heads}\}$.

Solution The sample space in this case is

$$\Omega = \{HHH, HHT, HTH, THH, HTT, THT, TTH, TTT\}$$

Note that all outcomes are equally likely since we assumed that the coin is fair. Since $|\Omega| = 8$, it follows that each possible outcome has probability $1/8$. Thus,

$$P(A) = P(HHH, HHT, HTH, THH)$$
$$= \frac{|A|}{|\Omega|} = \frac{4}{8} = \frac{1}{2}$$
∎

EXAMPLE 8 An urn contains five blue and six green balls. We draw two balls from the urn.

  a. Determine the sample space Ω and find $|\Omega|$.
  b. What is the probability that the two balls have different color?
  c. What is the probability that at least one of the two balls is green?

Solution

  a. As physical objects, the balls are distinguishable and we can imagine them being numbered from 1–11, where we assign the first five numbers to the five blue balls and the remaining six numbers to the six green balls. The sample space for this random experiment then consists of all subsets of size 2 that can be drawn from the set of eleven balls. (This is similar to the second case in Example 3.) Each subset of size 2 is then equally likely. The size of the sample space is

$$|\Omega| = \binom{11}{2}$$

  since the order in which the balls are removed from the urn is not important, and sampling is done without replacement.

  b. Let A denote the event that the two balls are of different color. To obtain an outcome where one ball is blue and the other is green, we must select one blue ball from the five blue balls, which can be done in $\binom{5}{1}$ different ways, and one green ball from the six green balls, which can be done in $\binom{6}{1}$ ways. Using the multiplication rule from Section 12.1, we find

$$|A| = \binom{5}{1}\binom{6}{1}$$

Hence,

$$P(A) = \frac{|A|}{|\Omega|} = \frac{\binom{5}{1}\binom{6}{1}}{\binom{11}{2}} = \frac{5 \cdot 6}{\frac{11 \cdot 10}{2}} = \frac{6}{11}$$

c. Let B denote the event that at least one ball is green. It can be written as a union of the two disjoint sets

$$B_1 = \{\text{exactly one ball is green}\}$$
$$B_2 = \{\text{both balls are green}\}$$

Since $B = B_1 \cup B_2$ with $B_1 \cap B_2 = \emptyset$, it follows that

$$P(B) = P(B_1) + P(B_2)$$

Using a similar argument as in (a), we find

$$P(B) = \frac{\binom{5}{1}\binom{6}{1}}{\binom{11}{2}} + \frac{\binom{5}{0}\binom{6}{2}}{\binom{11}{2}} = \frac{5 \cdot 6}{55} + \frac{15}{55} = \frac{9}{11} \qquad ▪$$

An Application from Genetics   Gregor Mendel, an Austrian monk, experimented with peas to study the laws of inheritance. He started these experiments in 1856, long before chromosomes were discovered. He was the first to describe the mechanisms of inheritance. His work was fundamental in the understanding of the laws of inheritance. It took over thirty-five more years until Mendel's original work was publicized and his conclusions were confirmed in additional experiments.

We will use the current knowledge of inheritance to determine the likelihood of outcomes of certain crossings. We describe an experiment that studies the inheritance of flower color in peas. Mendel had seeds that produced plants with either red or white flowers. Flower color in Mendel's peas is determined by a single locus on the chromosome. The genes at this locus occur in two forms, called alleles, which we denote by C and c. Since pea plants are diploid organisms, each plant has two genes that determine flower color, one from each parent plant. The following genotypes are thus possible: CC, Cc and cc. The genotypes CC and Cc result in red flowers, whereas the genotype cc results in white flowers.

EXAMPLE 9   Suppose that you cross two pea plants, both of type Cc. Determine the probabilities for each genotype in the next generation. What is the probability that a randomly chosen seed from this crossing results in a plant with red flowers?

Solution   We denote the offspring of this cross as a pair where the first entry is the maternal and the second entry is the paternal contribution. For instance, an offspring of type (c, C) inherited a c from the mother and a C from the father (see Figure 12.8). We list all possible outcomes of this cross in the sample space

$$\Omega = \{(C, C), (C, c), (c, C), (c, c)\}$$

The laws of inheritance imply that gametes form at random and that, therefore, all outcomes in Ω are equally likely. (This is Mendel's first law.) Since $|\Omega| = 4$, it follows that each outcome has probability $1/4$.

Though the sample space has four different outcomes, there are only three different genotypes, since (C, c) and (c, C) denote the same genotype, Cc. We

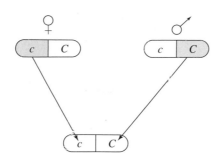

◀ **Figure 12.8**
The cross $Cc \times Cc$ resulted in an offspring of type Cc.

therefore find

$$P(CC) = \frac{1}{4} \qquad P(Cc) = \frac{1}{2} \quad \text{and} \quad P(cc) = \frac{1}{4}$$

Since both genotypes CC and Cc result in red flowers, it follows that

$$P(\text{red}) = P(\{(C, C), (C, c), (c, C)\}) = \frac{3}{4} \qquad \blacksquare$$

The Mark-Recapture Method   The mark-recapture method is commonly used to estimate population sizes. We illustrate this method using a fish population as an example. Suppose that N fish are in a lake, where N is unknown. To get an idea of how big N is, we capture K fish, mark them, and subsequently release them. We wait until the marked fish in the lake have had sufficient time to mix with the other fish. We then capture n fish. Suppose that k of the n fish are marked. Then, assuming that the ratio of the marked to unmarked fish in the sample of size n is approximately equal to the ratio of marked to unmarked fish in the lake, that is,

$$\frac{k}{n} \approx \frac{K}{N}$$

we conclude that there are about

$$N \approx K\frac{n}{k}$$

fish in the lake. Why does this approach makes sense?

EXAMPLE 10   Given the mark-recapture experiment described above, compute the probability of finding k marked fish in your sample.

Solution   There are N fish in the lake, K of which are marked. We choose a sample of size n. Each outcome is therefore a subset of size n, and all outcomes are equally likely. Using the counting techniques from Section 12.1, we find

$$|\Omega| = \binom{N}{n}$$

since the order in the sample is not significant.
　　We denote by A the event that the sample of size n contains exactly k marked fish. To determine how many outcomes contain exactly k marked fish, we argue as follows: We need to select k fish from the K marked ones and $n - k$ fish from the $N - K$ unmarked ones. Selecting the k marked fish can be done in $\binom{K}{k}$ ways; selecting the $n - k$ unmarked fish can be done in $\binom{N-K}{n-k}$ ways. Since each choice of k marked fish can be combined with any choice of the $n - k$ unmarked fish, we use the multiplication principle to find the total number of ways of obtaining a sample

of size n with exactly k marked fish, namely

$$|A| = \binom{K}{k}\binom{N-K}{n-k}$$

Therefore

$$P(A) = \frac{|A|}{|\Omega|} = \frac{\binom{K}{k}\binom{N-K}{n-k}}{\binom{N}{n}}$$

(This example is of the same basic type as the urn problem in Example 8.)   ▎

We will now give an argument that explains why the total number of fish in the lake can be estimated using the formula $N \approx Kn/k$.

▎EXAMPLE 11   Assume that there are K marked fish in the lake. We take a sample of size n and observe k marked fish. Show that the value of N that maximizes the probability of finding k marked fish in a sample of size n is the largest integer less than or equal to Kn/k. We use this value as our estimate for the population size N. Since N maximizes the probability of what we observe, it is called a **maximum likelihood estimate**.

Solution   We denote by A the event that the sample of size n contains exactly k marked fish. We found in Example 10 that

$$P(A) = \frac{|A|}{|\Omega|} = \frac{\binom{K}{k}\binom{N-K}{n-k}}{\binom{N}{n}}$$

We now consider $P(A)$ as a function of N and denote it by p_N. To find the value of N that maximizes p_N, we look at the ratio p_N/p_{N-1}. (The function p_N is not continuous, since it is only defined for integer values of N. Therefore, we cannot differentiate p_N to find its maximum.) The ratio is given by

$$\frac{p_N}{p_{N-1}} = \frac{\frac{\binom{K}{k}\binom{N-K}{n-k}}{\binom{N}{n}}}{\frac{\binom{K}{k}\binom{N-1-K}{n-k}}{\binom{N-1}{n}}} = \frac{\binom{N-K}{n-k}}{\binom{N-1-K}{n-k}}\frac{\binom{N-1}{n}}{\binom{N}{n}}$$

$$= \frac{(N-K)!(n-k)!(N-1-K-n+k)!}{(n-k)!(N-K-n+k)!(N-1-K)!}\frac{(N-1)!n!(N-n)!}{n!(N-1-n)!N!}$$

When we cancel terms, we find

$$\frac{p_N}{p_{N-1}} = \frac{N-K}{N-K-n+k}\frac{N-n}{N}$$

We will now investigate when this ratio is greater than or equal to 1, since this will allow us to find the values of N where p_N exceeds p_{N-1}. Values of N where p_N exceeds both p_{N-1} and p_{N+1} are local maxima. The ratio p_N/p_{N-1} is greater than or equal to 1 if

$$(N-K)(N-n) \geq N(N-K-n+k)$$

Multiplying this out, we find

$$N^2 - Nn - KN + Kn \geq N^2 - NK - Nn + Nk$$

Simplifying yields

$$Kn \geq kN$$

or

$$N \leq K\frac{n}{k}$$

Thus, $p_N \geq p_{N-1}$ as long as $N \leq Kn/k$. Therefore, the largest integer less than or equal to Kn/k maximizes the probability p_N. We use this value of N to estimate the total number of fish in the lake.   ▎

◆ 12.2 PROBLEMS

[12.2.1]

In Problems 1–4, determine the sample spaces for each random experiment.

1. The random experiment consisting of tossing a coin three times.

2. The random experiment consisting of rolling a six-sided die twice.

3. An urn contains five balls numbered 1–5. The random experiment consists of selecting two balls without replacement.

4. An urn contains six balls numbered 1–6. The random experiment consists of selecting five balls without replacement.

In Problems 5–8, assume that

$$\Omega = \{1, 2, 3, 4, 5, 6\}$$

$A = \{1, 3, 5\}$, *and* $B = \{1, 2, 3\}$.

5. Find $A \cup B$ and $A \cap B$.

6. Find A^c and show that $(A^c)^c = A$.

7. Find $(A \cup B)^c$.

8. Are A and B disjoint?

In Problems 9–12, assume that

$$\Omega = \{1, 2, 3, 4, 5\}$$

$P(1) = 0.1$, $P(2) = 0.2$, *and* $P(3) = P(4) = 0.05$. *Furthermore, assume that* $A = \{1, 3, 5\}$ *and* $B = \{2, 3, 4\}$.

9. Find $P(5)$.

10. Find $P(A)$ and $P(B)$.

11. Find $P(A^c)$.

12. Find $P(A \cup B)$.

13. Assume that $P(A \cap B) = 0.1$, $P(A) = 0.4$, and $P(A^c \cap B^c) = 0.2$. Find $P(B)$.

14. Assume that $P(A) = 0.4$, $P(B) = 0.4$, and $P(A \cup B) = 0.7$. Find $P(A \cap B)$ and $P(A^c \cap B^c)$.

15. We will show the second of the additional properties, namely

$$P(A \cup B) = P(A) - P(B) - P(A \cap B) \qquad (12.4)$$

(a) Use a diagram to show that B can be written as a disjoint union of the sets $A \cap B$ and $B \cap A^c$.

(b) Use a diagram to show that $A \cup B$ can be written as a disjoint union of the sets A and $B \cap A^c$.

(c) Use your results in (a) and (b) to show that

$$P(A \cup B) = P(A) + P(B \cap A^c)$$

and

$$P(B \cap A^c) = P(B) - P(A \cap B)$$

Conclude from this that (12.4) holds.

16. If $A \subset B$, we can define the difference between the two sets A and B, denoted by $B - A$ (read: "B minus A"). Namely,

$$B - A = B \cap A^c$$

as illustrated in Figure 12.9.

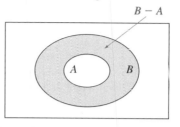

▲ **Figure 12.9**
The set A is contained in B. The shaded area is the difference of A and B, $B - A$.

The following difference rule holds:

$$P(B - A) = P(B) - P(A) \qquad (12.5)$$

To show (12.5), we go through the following steps.

(a) Use the diagram in Figure 12.9 to show that B can be written as a disjoint union of A and $B - A$.

(b) Use your result in (a) to conclude that

$$P(B) = P(A) + P(B - A)$$

and show that (12.5) follows from this.

(c) An immediate consequence of (12.5) is the following result. If $A \subset B$, then

$$P(A) \le P(B)$$

Use (12.5) to show this inequality.

[12.2.2]

17. Toss two fair coins and find the probability of at least one head.

18. Toss three fair coins and find the probability of no heads.

19. Roll a fair die twice and find the probability of at least one 4.

20. Roll two fair dice and find the probability that the sum of the two numbers is even.

21. Roll two fair dice, one after the other, and find the probability that the first number is larger than the second number.

22. Roll two fair dice and find the probability that the minimum of the two numbers will be greater than 4.

23. In Example 9, we considered a cross between two pea plants, each of genotype Cc. Find the probability that a randomly chosen seed from this cross has white flowers.

24. In Example 9, we considered a cross between two pea plants, each of genotype Cc. Now we cross a pea plant of genotype cc with a pea plant of genotype Cc.

(a) What are the possible outcomes of this cross?

(b) Find the probability that a randomly chosen seed from this cross results in red flowers.

25. Suppose that two parents are of genotype Aa. What is the probability that their offspring is of genotype Aa? (Assume Mendel's first law.)

26. Suppose that one parent is of genotype AA and the other is of genotype Aa. What is the probability that their offspring is of genotype AA? (Assume Mendel's first law.)

27. A family has three children. Assuming a 1:1 sex ratio, what is the probability that all are girls?

28. Color blindness is an X-linked inherited disease. A woman who carries the color blindness gene on one of her X chromosomes but not on the other has normal vision. A man who carries the gene on his only X chromosome is color blind. If a woman with normal vision who carries the color blindness gene on one of her X chromosomes has a son with a man who has normal vision, what is the probability that their son will be color blind?

29. A lake contains an unknown number of fish, denoted by N. You capture 100 fish, mark them and subsequently release them again. Later, you return and catch ten fish, three of which are marked.

(a) Find the probability that exactly three out of ten fish will be marked. This probability will be a function of N, the unknown number of fish in the lake.

(b) Find the value of N that maximizes the probability you computed in (a), and show that this value agrees with the value we computed in Example 11.

30. An urn contains five blue and three green balls. You remove three balls from the urn. What is the probability that at least two out of the three balls are green?

31. You select two cards from a standard deck of 52 cards. What is the probability that both cards are spades?

32. You select five cards from a standard deck of 52 cards. What is the probability that you get four aces?

33. An urn contains four green, six blue, and two red balls. You take three balls out of the urn. What is the probability that all three balls are of different colors?

34. An urn contains four green, six blue, and five red balls. You take three balls out of the urn. What is the probability that all three balls are of the same color?

◆ 12.3 CONDITIONAL PROBABILITY AND INDEPENDENCE

Before we define "conditional probability" and "independence," we will illustrate these concepts using the Mendelian crossing of peas that we considered in the previous section to study flower color inheritance.

Assume that two parents are of genotype Cc. Suppose you know that the offspring of the cross $Cc \times Cc$ has red flowers; what is the probability that it is of genotype CC? We can find this probability by noting that one of the three equally likely possibilities that produce red flowers (namely, (C, C), (C, c), and (c, C), if we list the types according to maternal and paternal contributions as in Example 9 of the previous section) is of type CC. Hence the probability that the offspring is of genotype CC is 1/3. Such a probability, where we condition on some prior knowledge (as flower color of offspring), is called a **conditional probability**.

Suppose now that the paternally transmitted gene in the offspring of the cross $Cc \times Cc$ is of type C. What is the probability that the maternally transmitted gene in the offspring is of type c? To answer this question, we note that the paternal gene has no impact on the choice of the maternal gene in this case. The probability that the maternal gene is of type c is therefore 1/2. We say that the maternal gene is **independent** of the paternal gene. Knowing which of the paternal genes was chosen, does not change the probability of the maternal gene.

12.3.1 Conditional Probability

As illustrated in the introduction of this section, conditional probabilities have to do with prior knowledge. Suppose we know that the event B has occurred, and that $P(B) > 0$. Then the conditional probability of A given B, denoted by $P(A \mid B)$, is the probability that A will occur given the fact that B has occurred. It is defined as

$$P(A \mid B) = \frac{P(A \cap B)}{P(B)} \tag{12.6}$$

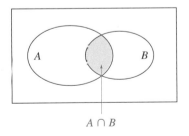

$A \cap B$

▲ **Figure 12.10**

The conditional probability of A given B is the proportion of A in the set $B(A \cap B)$ relative to the set B.

To explain this definition, we look at Figure 12.10. The probability of A given B is the proportion of A in the set B relative to B.

In the following example, we will use the definition (12.6) to repeat the introductory example, and find the probability that an offspring is of genotype CC given that its flower color is red.

▮**EXAMPLE 1**   Find the probability that the offspring of a $Cc \times Cc$ cross of pea plants is of type CC, given that its flowers are red.

Solution   Let A denote the event that the offspring is of genotype CC, and B be the event that the flower color of the offspring is red. We want to find $P(A \mid B)$. Using (12.6), we have

$$P(A \mid B) = \frac{P(A \cap B)}{P(B)}$$

The *unconditional* probabilities $P(B)$ and $P(A \cap B)$ are computed using the sample space $\Omega = \{(C, C), (C, c), (c, C), (c, c)\}$, whose outcomes all have the same probability. The probability $P(B)$ is the probability that the genotype of the offspring is in the set $\{(C, C), (C, c), (c, C)\}$. Since the sample space has equally likely outcomes, $P(B) = 3/4$. To compute $P(A \cap B)$, we note that $A \cap B$ is the event that the offspring is of genotype CC. Using the sample space Ω, we find $P(A \cap B) = 1/4$. Hence,

$$P(A \mid B) = \frac{P(A \cap B)}{P(B)} = \frac{\frac{1}{4}}{\frac{3}{4}} = \frac{1}{3}$$

which is the same answer that we obtained before.   ▮

Equation (12.6) can be used to compute conditional probabilities, as seen in Example 1. By rearranging terms, however, it can also be used to compute probabilities of the intersection of events. Namely, if we multiply both sides of (12.6) by $P(B)$, we obtain

$$P(A \cap B) = P(A \mid B)P(B) \qquad (12.7)$$

In equation (12.6), we conditioned on the event B. If we condition on A instead, we have the following identity:

$$P(B \mid A) = \frac{P(A \cap B)}{P(A)}$$

Rearranging terms as in (12.7), we find

$$P(A \cap B) = P(B \mid A)P(A) \qquad (12.8)$$

Formulas (12.7) and (12.8) are particularly useful for computing probabilities in two-stage experiments. We illustrate the use of these identities in the following example. We will also see that there is a natural choice for which of two events to condition on.

▮**EXAMPLE 2**   Suppose that we draw two cards from a standard deck of 52 cards. Compute the probability that both cards are spades.

Solution   This is a two-stage experiment. We first draw one card and then another. We define the two events

$$A = \{\text{the first card is spade}\}$$
$$B = \{\text{the second card is spade}\}$$

Then

$$A \cap B = \{\text{both cards are spades}\}$$

Should we use (12.7) or (12.8)? Since the first card is drawn first, it will be easier to condition on the outcome of the first draw than on the second draw; that is, we will compute $P(B \mid A)$ rather than $P(A \mid B)$, and then use (12.8). Now,

$$P(A) = \frac{13}{52}$$

since 13 out of the 52 cards are spades and each card has the same probability of being drawn. To compute $P(B \mid A)$, we note that if the first card is spade, then there are 12 spades left in the deck of the remaining 51 cards. Therefore,

$$P(B \mid A) = \frac{12}{51}$$

Using (12.8), we find

$$P(A \cap B) = P(B \mid A)P(A) = \frac{12}{51} \cdot \frac{13}{52} = \frac{1}{17}$$ ▮

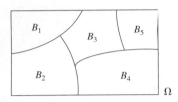

▲ **Figure 12.11**
The sets $B_1, B_2, \ldots, B_5$ form a partition of Ω.

12.3.2 The Law of Total Probability

We begin this subsection by defining a partition of the sample space. Suppose the sample space Ω is written as a union of n disjoint sets $B_1, B_2, \ldots, B_n$, that is,

(i) $B_i \cap B_j = \emptyset$ whenever $i \neq j$

(ii) $\Omega = \bigcup_{i=1}^{n} B_i$

we then say that the sets $B_1, B_2, \ldots, B_n$ form a **partition** of the sample space Ω (see Figure 12.11).

Let A be an event. We can write A as a union of disjoint sets using this partition of Ω:

$$A = (A \cap B_1) \cup (A \cap B_2) \cup \cdots \cup (A \cap B_n)$$

as illustrated in Figure 12.12.

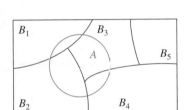

▲ **Figure 12.12**
The set A is a disjoint union of sets of the form $A \cap B_i$.

Since the sets $A \cap B_i$, $i = 1, 2, \ldots, n$, are disjoint, we find

$$P(A) = \sum_{i=1}^{n} P(A \cap B_i)$$

To evaluate $P(A \cap B_i)$, we might find it useful to condition on B_i, that is, $P(A \cap B_i) = P(A \mid B_i)P(B_i)$. Then

$$P(A) = \sum_{i=1}^{n} P(A \mid B_i)P(B_i) \tag{12.9}$$

Equation (12.9) is known as the **law of total probability**.

▮**EXAMPLE 3**   A test for the HIV virus shows a positive result in 99% of all cases when the virus is actually present, and in 5% of all cases when the virus is not present (a *false positive* result). If such a test is administered to a randomly chosen individual, what is the probability that the test result is positive? We assume that the prevalence of the virus in the population is $1/200$.

Solution   We set

$$A = \{\text{test result is positive}\}$$

Individuals in this population fall into two groups, those who are infected with the HIV virus and those who are not. These two sets form a partition of the population. If we pick an individual at random from the population, then the person belongs to one of these two groups. We define

$$B_1 = \{\text{person is infected}\}$$
$$B_2 = \{\text{person is not infected}\}$$

Using (12.9), we can write

$$P(A) = P(A \mid B_1)P(B_1) + P(A \mid B_2)P(B_2)$$

Now, $P(B_1) = 1/200$ and $P(B_2) = 199/200$. Furthermore, $P(A \mid B_1) = 0.99$ and $P(A \mid B_2) = 0.05$. Hence

$$P(A) = (0.99)\frac{1}{200} + (0.05)\frac{199}{200} = 0.0547 \qquad ▮$$

We can illustrated the last example using a tree diagram, as shown in Figure 12.13.

The numbers on the branches represent the respective probabilities. To find the probability of a positive test result, we multiply the probabilities along the paths that lead to tips labeled "test positive." (The two paths are marked in the tree.) Adding the results along the different paths then yields the desired probability. That is,

$$P(\text{test positive}) = \frac{1}{200}(0.99) + \frac{199}{200}(0.05) = 0.0547$$

as in Example 3. Tree diagrams are quite useful when using the law of total probability.

In the next example, we will return to our pea plants. Recall that red-flowering pea plants are of genotype CC or Cc and that white-flowering pea plants are of genotype cc.

EXAMPLE 4   Suppose that you have a batch of red-flowering pea plants, 20% of which are of genotype Cc and 80% of genotype CC. You pick one of the red-flowering plants at random and cross it with a white-flowering plant. Find the probability that the offspring will produce red flowers.

Solution   We use a tree diagram to illustrate the computation of the probability of a red-flowering offspring, as shown in Figure 12.14.

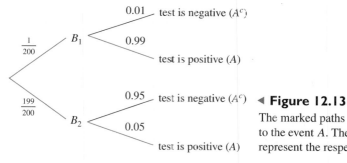

◀ **Figure 12.13**
The marked paths in this tree diagram lead to the event A. The numbers on the branches represent the respective probabilities.

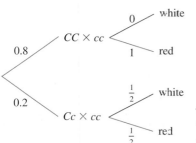

◀ **Figure 12.14**
The tree diagram for Example 4: The red paths lead to red flowering offspring.

The paths that lead to a red-flowering offspring are marked. Using the tree diagram (that is, using the law of total probability), we find

$$P(\text{red flowering offspring}) = (0.8)(1) + (0.2)(0.5) = 0.9$$ ▮

12.3.3 Independence

Suppose that you toss a fair coin twice. Let A be the event that the first toss results in heads and B the event that the second toss results in heads. Suppose that A occurrs. Does this change the probability that B will occur? The answer is obviously no. The outcome of the first toss does not influence the outcome of the second toss. We can express this mathematically using conditional probabilities. Namely,

$$P(B \mid A) = P(B) \tag{12.10}$$

and we say that A and B are independent. We will not use (12.10) as the definition of independence; rather, we will use the definition of conditional probabilities to rewrite (12.10). We find

$$\frac{P(A \cap B)}{P(A)} = P(B)$$

Multiplying both sides by $P(A)$, we find $P(A \cap B) = P(A)P(B)$. We use this as our definition.

> Two events A and B are independent if
> $$P(A \cap B) = P(A)P(B)$$

EXAMPLE 5   You draw one card from a standard deck of 52 cards. Let

$$A = \{\text{card is spade}\}$$
$$B = \{\text{card is king}\}$$

Show that A and B are independent.

Solution   To show that A and B are independent, we compute

$$P(A) = \frac{13}{52} \qquad P(B) = \frac{4}{52}$$

and

$$P(A \cap B) = P(\text{card is king of spade}) = \frac{1}{52}$$

Since

$$P(A)P(B) = \frac{13}{52}\frac{4}{52} = \frac{1}{52} = P(A \cap B)$$

it follows that A and B are independent. ▮

We will return to our pea plant example to illustrate how we can compute probabilities of intersections of events when we know that the events are independent.

EXAMPLE 6   What is the probability that the offspring of a $Cc \times Cc$ cross is of genotype cc?

Solution   We computed the answer previously using a sample space with equally likely outcomes. But we can also compute this probability using independence.

In order for the offspring to be of genotype cc, both parents must contribute a c gene. Let

$$A = \{\text{paternal gene is } c\}$$
$$B = \{\text{maternal gene is } c\}$$

Now, it follows from the laws of inheritance that A and B are independent and that $P(A) = P(B) = 1/2$. Hence

$$P(cc) = P(A \cap B) = P(A)P(B) = \frac{1}{2} \cdot \frac{1}{2} = \frac{1}{4}$$

This is the same result that we obtained when we looked at the sample space $\Omega = \{(C, C), (C, c), (c, C), (c, c)\}$ and observed that all four possible types were equally likely.   ▮

We can extend independence to more than two events. Events $A_1, A_2, \ldots, A_n$ are independent if for any $1 \le i_1 < i_2 < \cdots < i_k \le n$

$$P(A_{i_1} \cap A_{i_2} \cap \cdots \cap A_{i_k}) = P(A_{i_1})P(A_{i_2}) \cdots P(A_{i_k}) \qquad (12.11)$$

To see what this means when we have three events A, B, and C, we write the conditions explicitly. Three events A, B, and C are independent if

$$P(A \cap B) = P(A)P(B) \qquad P(A \cap C) = P(A)P(C) \qquad P(B \cap C) = P(B)P(C)$$
$$(12.12)$$
and
$$P(A \cap B \cap C) = P(A)P(B)P(C) \qquad (12.13)$$

That is, both (12.12) *and* (12.13) must hold.

The number of conditions one must verify increases quickly with the number of events. When there are just two sets, only one condition must be checked, namely $P(A \cap B) = P(A)P(B)$. When there are three sets, we just saw that there are four conditions, namely there are $\binom{3}{2}$ conditions involving pairs of sets and $\binom{3}{3}$ conditions involving all three sets. With four sets, there are a total of $\binom{4}{2} + \binom{4}{3} + \binom{4}{4} = 11$ conditions; with five sets, 26 conditions; and with ten sets, 1013 conditions.

We stress that it is *not* enough to check independence between pairs of events to determine whether a collection of sets is independent. However, independence between pairs of sets is an important property itself, and we wish to define it. Namely, events $A_1, A_2, \ldots, A_n$ are **pairwise independent** if

$$P(A_i \cap A_j) = P(A_i)P(A_j) \qquad \text{whenever } i \ne j$$

We provide an example where events are pairwise independent but not independent.

EXAMPLE 7   Roll two dice. Let

$$A = \{\text{the first die is even}\}$$
$$B = \{\text{the second die is even}\}$$
$$C = \{\text{the sum is odd}\}$$

Show that A, B, and C are pairwise independent but not independent.

Solution   To show pairwise independence, we compute

$$P(A) = \frac{3}{6} = \frac{1}{2} \qquad P(B) = \frac{3}{6} = \frac{1}{2} \qquad P(C) = \frac{18}{36} = \frac{1}{2}$$

$$P(A \cap B) = \frac{9}{36} = \frac{1}{4} = P(A)P(B)$$

$$P(A \cap C) = \frac{9}{36} = \frac{1}{4} = P(A)P(C)$$

$$P(B \cap C) = \frac{9}{36} = \frac{1}{4} = P(B)P(C)$$

However, the event $A \cap B \cap C = \emptyset$ since, if both dice are even, then their sum is even. Therefore

$$P(A \cap B \cap C) = P(\emptyset) = 0 \neq P(A)P(B)P(C)$$

which shows that A, B, and C are not independent.   ▮

When events are independent, we can use (12.11) to compute the probability of the intersections of events, as illustrated in the following example.

▮ **EXAMPLE 8**   Assume a 1:1 sex ratio. A family has five children. Find the probability that at least one of the children is a girl.

Solution   Instead of computing the probability that at least one of the children is a girl, we will look at the complement of this event. This is a particularly useful trick when we look at events that ask for the probability of "at least one." We denote by

$$A_i = \{\text{the } i\text{th child is a boy}\}$$

Then the events $A_1, A_2, \ldots, A_5$ are independent. Let

$$B = \{\text{at least one of the children is a girl}\}$$

Instead of computing the probability of B directly, we look at the complement of B. Now, B^c is the event that all children are boys, which is expressed as

$$B^c = A_1 \cap A_2 \cap \ldots \cap A_5$$

It follows that

$$
\begin{aligned}
P(B) &= 1 - P(B^c) = 1 - P(A_1 \cap A_2 \cap \ldots \cap A_5) \\
&= 1 - P(A_1)P(A_2) \cdots P(A_5) \\
&= 1 - \left(\frac{1}{2}\right)^5 = \frac{31}{32}
\end{aligned}
$$

If we had tried to compute $P(B)$ directly, we would have needed to compute the probability of exactly one girl, exactly two girls and so on, and then add the probabilities. In this situation, it is quicker and easier to compute the probability of the complement.   ▮

12.3.4 Bayes Formula

In Example 3 of this section, we computed the probability that the result of an HIV test of a randomly chosen individual is positive. For the individual, however, it is much more important to know whether a positive test result actually means that he or she is infected. Recall that we defined

$$A = \{\text{test result is positive}\}$$
$$B_1 = \{\text{person is infected}\}$$
$$B_2 = \{\text{person is not infected}\}$$

We are interested in $P(B_1 \mid A)$, that is, the probability that a person is infected given that the result is positive. We saw in Example 3 that $P(A \mid B_1)$ and $P(A \mid B_2)$ followed immediately from the characteristics of the test. However, we wish now to compute a conditional probability where the roles of A and B_1 are reversed.

Before we compute the probability for this specific example, we look at the general case. We assume that the sets $B_1, B_2, \ldots, B_n$ form a partition of the sample space Ω, A is an event, and the probabilities $P(A \mid B_i)$, $i = 1, 2, \ldots, n$ are known. We are interested in computing $P(B_i \mid A)$. This can be accomplished as follows.

Using the definition of conditional probabilities, we find

$$P(B_i \mid A) = \frac{P(A \cap B_i)}{P(A)} \tag{12.14}$$

To compute $P(A \cap B_i)$, we now condition on B_i, that is,

$$P(A \cap B_i) = P(A \mid B_i)P(B_i) \tag{12.15}$$

To evaluate the denominator $P(A)$, we use the law of total probability, that is,

$$P(A) = \sum_{j=1}^{n} P(A \mid B_j)P(B_j) \tag{12.16}$$

Combining (12.14), (12.15), and (12.16), we arrive at the following formula, known as the Bayes formula.

> **The Bayes Formula.** Let $B_1, B_2, \ldots, B_n$ form a partition of Ω and A be an event. Then
>
> $$P(B_i \mid A) = \frac{P(A \mid B_i)P(B_i)}{\sum_{j=1}^{n} P(A \mid B_j)P(B_j)}$$

We will now return to our example. We wish to find $P(B_1 \mid A)$, that is, the probability that a person is infected given a positive result. We partition the population into the two sets B_1 and B_2. Then, using the Bayes formula, we find

$$P(B_1 \mid A) = \frac{P(A \mid B_1)P(B_1)}{P(A \mid B_1)P(B_1) + P(A \mid B_2)P(B_2)}$$

$$= \frac{(0.99)\frac{1}{200}}{(0.99)\frac{1}{200} + (0.05)\frac{199}{200}} = 0.090$$

The denominator is equal to $P(A)$, which we already computed in Example 3.

This example is worth discussing in more detail. The probability is quite low that a person is infected given a positive result. But you should compare it to the unconditional probability that a person is infected, namely $P(B_1)$. If we look at the ratio of the conditional to the unconditional probability, we find

$$\frac{P(B_1 \mid A)}{P(B_1)} = 18.1$$

That is, if a test result is positive, the chance of actually being infected increases by a factor of 18 when compared to a randomly chosen individual in the population who has not been tested. (In practice, if a test result is positive, other tests are performed to see whether the person is indeed infected or the first result was a false positive.)

If a test result is negative, then the probability that the individual is not infected can also be computed using the Bayes formula. We find

$$P(B_2 \mid A^c) = \frac{P(B_2 \cap A^c)}{P(A^c)} = \frac{P(A^c \mid B_2)P(B_2)}{P(A^c)}$$

$$= \frac{(0.95)\frac{199}{200}}{1 - 0.0547} = 0.999947$$

where we used $P(A) = 0.547$, which we computed in Example 3. This result is rather reassuring. The reason that the probability of being infected is so low, given a positive result, comes from the fact that the prevalence of the disease is relatively low (1 in 200). To illustrate this, we treat the prevalence of the disease as a variable

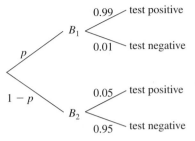

◀ **Figure 12.15**

The tree diagram for computing the probability that a person is infected given the test came back positive when the prevalence of the disease is p.

and compute $P(B_1 \mid A)$ as a function of the prevalence of the disease. That is, we set

$$p = P(\text{a randomly chosen individual is infected})$$

Using the same test characteristics as before, we find (see Figure 12.15),

$$f(p) = P(B_1 \mid A) = \frac{p(0.99)}{p(0.99) + (1 - p)(0.05)} = \frac{0.99p}{0.05 + 0.94p}$$

A graph of $f(p)$ is shown in Figure 12.16. We see that for small p, $f(p)$ (the probability of being infected given a positive result) is quite small. The ratio $f(p)/p$ is shown in Figure 12.17, from which we conclude that though $f(p)$ is small when p is small, the ratio $f(p)/p$ is quite large for small p.

The Bayes formula is also important in genetic counseling. Hemophilia is a blood disorder that is caused by a recessive gene on the X chromosome. That is,

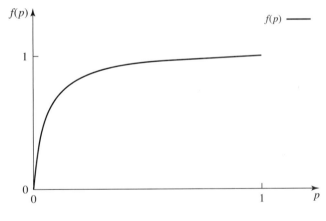

◀ **Figure 12.16**

The probability of being infected given the test came back positive as a function of the prevalence of the disease.

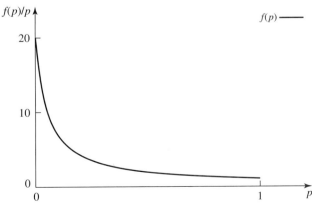

◀ **Figure 12.17**

The ratio $P(p)/p$ as a function of p illustrates by what factor the probability of being infected increases when the test result is positive compared to the prevalence of the disease in the population.

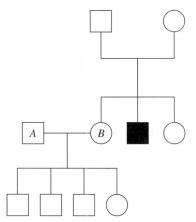

◀ **Figure 12.18**

The pedigree of a family where one member suffers from hemophilia. Squares indicate males, circles females, and the black square shows an affected individual.

a female carrying the gene for hemophilia on one of her X chromosomes and the healthy gene on the other X chromosome will not develop the disease; whereas a male carrying the hemophilia gene on his only X chromosome will have the disease. We will look at the pedigree in Figure 12.18, and determine the risk that a son of A and B will be affected. Males are denoted by squares and females by circles; affected males are indicated by solid black squares.

We see from the pedigree that B has a hemophilic brother. Since B's father does not have the disease, we conclude that B's mother must be a carrier.

Let E denote the event that B is a carrier. Then $P(E) = 1/2$. Now, assume that we are told that B has three sons. If F denotes the event that all three sons are healthy, then

$$P(F \mid E) = \frac{1}{2} \cdot \frac{1}{2} \cdot \frac{1}{2} = \frac{1}{8}$$

since if B is a carrier, each son has probability of $1/2$ of inheriting the disease gene and thus being hemophilic.

Individual B, who comes from a family where some male relatives are afflicted with hemophilia, might wish to know what her risk is of being a carrier of the disease. Taking into account her pedigree, in particular that B has a hemophilic brother, we can use the Bayes formula to compute the probability that B is a carrier given that none of her sons has the disease.

$$P(E \mid F) = \frac{P(E \cap F)}{P(F)} = \frac{P(F \mid E)P(E)}{P(F)}$$

To compute the denominator, we must use the law of total probability, as illustrated in the tree diagram in Figure 12.19.

We find

$$P(F) = \frac{1}{2} \cdot \frac{1}{8} + \frac{1}{2} \cdot 1 = \frac{9}{16}$$

Therefore, using the Bayes formula,

$$P(E \mid F) = \frac{\frac{1}{8} \cdot \frac{1}{2}}{\frac{9}{16}} = \frac{1}{9}$$

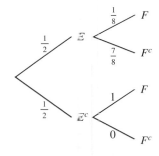

▲ **Figure 12.19**

The sample space is partitioned into two set, E and E^c, where E is the event that the individual B is a carrier of the hemophilia gene. Based on whether or not B is a carrier, the probability of the event that all three sons are healthy (F) can be computed as shown.

◆ 12.3 PROBLEMS

[12.3.1]

1. You draw two cards from a standard deck of 52 cards. Find the probability that the second card is a spade given that the first card is a club.

2. You draw three cards from a standard deck of 52 cards. Find the probability that the third card is a club given that the first two cards were clubs.

3. An urn contains five blue and six green balls. You take two balls out of the urn. Find the probability that the second ball is green given that the first ball is blue.

4. An urn contains five green, six blue, and four red balls. You take three balls out of the urn. Find the probability that the third ball is green given that the first two balls were red.

5. A family has two children. The younger one is a girl. Find the probability that the other child is a girl as well.

6. A family has two children. One of their children is a girl. Find the probability that both children are girls.

7. You roll two fair dice. Find the probability that the first die is a 4 given that the sum is 7.

8. You roll two fair dice. Find the probability that the first die is a 5 given that the minimum of the two numbers is a 3.

9. You toss a fair coin three times. Find the probability that the first coin is heads given that at least one head occurred.

10. You toss a fair coin three times. Find the probability that at least two heads occurred given that the second toss resulted in heads.

[12.3.2]

11. A screening test for a disease shows a positive test result in 90% of all cases when the disease is actually present and in 15% of all cases when it is not. Assume that the prevalence of the disease is 1 in 100. If the test is administered to a randomly chosen individual, what is the probability that the result is negative?

12. A screening test for a disease shows a positive test result in 95% of all cases when the disease is actually present and in 20% of all cases when it is not. When the test was administered to a large number of people, 21.5% of the results were positive. What is the prevalence of the disease?

13. A drawer contains three bags numbered 1–3. Bag 1 contains two blue balls, bag 2 contains two green balls, and bag 3 contains one blue and one green ball. You choose one bag at random and take out one ball. Find the probability that the ball is blue.

14. A drawer contains six bags numbered 1–6. Bag i contains i blue balls and 2 green balls. You roll a fair die and then pick a ball out of the bag with the number shown on the die. What is the probability that the ball is blue?

15. You pick two cards from a standard deck of 52 cards. Find the probability that the second card is an ace. Compare this to the probability that the first card is an ace.

16. You pick three cards from a standard deck of 52 cards. Find the probability that the third card is an ace. Compare this to the probability that the first card is an ace.

17. Suppose that you have a batch of red-flowering pea plants of which 40% are of genotype CC and 60% of genotype Cc. You pick one plant at random and cross it with a white-flowering pea plant. Find the probability that the offspring of this cross will have white flowers.

18. Suppose that you have a batch of red- and white-flowering pea plants where all three genotypes CC, Cc, and cc are equally represented. You pick one plant at random and cross it with a white-flowering pea plant. What is the probability that the offspring will have red flowers?

19. A bag contains two coins: one is a fair coin, the other one has two heads. You pick one coin at random and flip it. Find the probability that the outcome is heads.

20. A drug company claims that a new headache drug will bring instant relief in 90% of all cases. If a person is treated with a placebo, there is a 10% chance the person will feel instant relief. In a clinical trial, half the subjects are treated with the new drug, the other half receive the placebo. If an individual from this trial is chosen at random, what is the probability that the person will have experienced instant relief?

[12.3.3]

21. You are dealt one card from a standard deck of 52 cards. If A denotes the event that the card is a spade and if B denotes the event that the card is an ace, determine whether A and B are independent.

22. You are dealt two cards from a standard deck of 52 cards. If A denotes the event that the first card is an ace and B denotes the event that the second card is an ace, determine whether A and B are independent.

23. An urn contains five green and six blue balls. You take two balls out of the urn. If A denotes the event that the first ball is green and B denotes the event that the second ball is green, determine whether A and B are independent.

24. An urn contains five green and six blue balls. You take one ball out of the urn, note its color and replace it. You then take a second ball out of the urn, note its color and replace it. If A denotes the event that the first ball is green and B denotes the event that the second ball is green, determine whether A and B are independent.

25. Assume a 1:1 sex ratio. A family has three children. Find the probability of the event A, where

(a) $A = \{\text{all children are girls}\}$

(b) $B = \{\text{at least one boy}\}$

(c) $C = \{\text{at least two girls}\}$

(d) $D = \{\text{at most two boys}\}$

26. Assume that 20% of a very common insect species in your study area are parasitized. If you collect ten specimens of this species, what is the probability that at least one specimen in your sample is parasitized?

27. A multiple choice question has four choices, and a test has a total of ten multiple choice questions. A student passes the test if they answer at least nine questions correctly. If the student guesses the answers to all questions randomly, find the probability that they will pass.

28. Assume that A and B are disjoint and that both events have positive probability. Are they independent?

29. Assume that the probability that an insect species lives more than five days is 0.1. Find the probability that in a sample of size 10 of this species at least one insect will still be alive after five days.

30. **(a)** Use a Venn diagram to show that

$$(A \cup B)^c = A^c \cap B^c$$

(b) Use your result in (a) to show that if A and B are independent, then A^c and B^c are independent.

(c) Use your result in (b) to show that if A and B are independent, then

$$P(A \cup B) = 1 - P(A^c)P(B^c)$$

[12.3.4]

31. A screening test for a disease shows a positive result in 95% of all cases when the disease is actually present and in 10% of all cases when it is not. If the prevalence of the disease is 1 in 50, and an individual tests positive, what is the probability that the individual actually has the disease?

32. A screening test for a disease shows a positive result in 95% of all cases when the disease is actually present and in 10% of all cases when it is not. If a result is positive, the test is repeated. If the prevalence of the disease is 1 in 50, and an individual tests positive twice, what is the probability that they actually have the disease?

33. A bag contains two coins; one is a fair coin, the other one has two heads. You pick one coin at random and flip it. What is the probability that you picked the fair coin given that the outcome of the toss was heads?

34. You pick two cards from a standard deck of 52 cards. Find the probability that the first card was a spade given that the second card was a spade.

35. Suppose a woman has a hemophilic brother one healthy son. Find the probability that she is a carrier of the hemophilia gene.

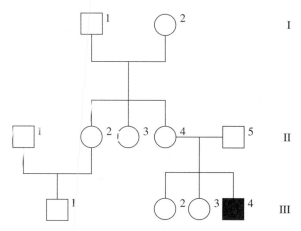

▲ **Figure 12.20**
The pedigree for Problems 36 and 37. The solid black square (individual III-4) represents an afflicted male.

The pedigree in Figure 12.20 shows a family in which one member (II-4) is hemophilic. In problems 36–37, refer to this pedigree.

36. **(a)** Given the pedigree, find the probability that the individual I-2 is a carrier of the hemophilia gene.

(b) Given the pedigree, find the probability that II-3 is a carrier of the hemophilia gene.

37. **(a)** Given the pedigree, find the probability that II-3 is a carrier of the hemophilia gene.

(b) Given the pedigree, find the probability that III-2 is a carrier of the hemophilia gene.

(c) Given the pedigree, find the probability that II-2 is a carrier of the hemophilia gene.

◆ 12.4 DISCRETE RANDOM VARIABLES AND DISCRETE DISTRIBUTIONS

Outcomes of random experiments frequently are real numbers, such as the number of heads in a coin tossing experiment, the number of seeds produced in a cross between two plants, or the lifespan of an insect. Such numerical outcomes can be described by **random variables**. A random variable is a function from the sample space Ω into the set of real numbers. Random variables are typically denoted by X, Y, or Z, or other capital letters chosen from the end of the alphabet. For instance,

$$X : \Omega \rightarrow \mathbf{R}$$

describes a random variable X, which is a map from the sample space Ω to the set of real numbers.

Random variables are classified according to their range. If the number of values that X can take is either finite or countably infinite,[1] X is called a **discrete random variable**. If X takes on a continuous range of values, for instance, every value in an interval, X is called a **continuous random variable**. (We discuss continuous random variables in the next section.)

12.4.1 Discrete Distributions

In the first example, we look at a random variable that takes on a finite number of values.

EXAMPLE 1 Toss a fair coin three times. Let X be a random variable that counts the number of heads in each outcome. The sample space is

$$\Omega = \{HHH, HHT, HTH, THH, HTT, THT, TTH, TTT\}$$

and the random variable

$$X : \Omega \to \mathbf{R}$$

takes on values 0,1,2, or 3. For instance,

$$X(HHH) = 3 \qquad X(TTH) = 1 \qquad \text{or} \qquad X(TTT) = 0 \qquad \blacksquare$$

In the next example, we look at a random variable that takes on a countably infinite number of values.

EXAMPLE 2 Toss a fair coin repeatedly until the first heads. Let Y be a random variable that counts the number of trials until the first heads shows up. The sample space is

$$\Omega = \{H, TH, TTH, TTTH, \ldots\}$$

and the random variable

$$Y : \Omega \to \mathbf{R}$$

takes on values $1, 2, 3, \ldots$. For instance,

$$Y(H) = 1 \qquad Y(TH) = 2 \qquad Y(TTH) = 3 \qquad \ldots \qquad \blacksquare$$

Now that we are more familiar with the concept of random variables, we will turn to the problem of how to assign probabilities to the different values of a random variable X. We illustrate this in relation to Example 1.

Since the coin is fair, each outcome in Ω has the same probability, namely $1/8$. We can translate this into probabilities for X. For instance,

$$P(X = 1) = P(\{HTT, THT, TTH\})$$

$$= P(HTT) + P(THT) + P(TTH)$$

$$= \frac{1}{8} + \frac{1}{8} + \frac{1}{8} = \frac{3}{8}$$

---

[1]A countably infinite set has infinitely many elements that can be enumerated, like the set of integers. (The set of rational numbers is also countably infinite; the set of real numbers is uncountably infinite.)

We can perform similar computations for all other values of X. The following table summarizes the results.

x	$P(X = x)$
0	1/8
1	3/8
2	3/8
3	1/8

The function $p(x) = P(X = x)$ is called a **probability mass function**. Note that $p(x) \geq 0$ and $\sum_x p(x) = 1$.

> **Definition**   A random variable is called a discrete random variable if it takes on only a finite or countably infinite number of values. The probability distribution of X can be described by the probability mass function $p(x)$, where the following hold.
>
> **1.** $p(x) = P(X = x) \geq 0$
>
> **2.** $\sum_x p(x) = 1$, where the sum is over all values of X with $P(X = x) > 0$

The probability mass function of X is one way to describe the probability distribution of X. Another important function that describes the probability distribution of X is the distribution function $F(x) = P(X \leq x)$.

> **Definition**   The **distribution function** $F(x)$ for a random variable X is defined as
>
> $$F(x) = P(X \leq x)$$

It is possible to alternate between the probability mass function and the distribution function, as illustrated in the following example.

▌**EXAMPLE 3**   Suppose that the probability mass function of a discrete random variable X is given by the following

x	$P(X = x)$
-1	0.1
0	0.2
1.5	0.05
3	0.15
5	0.5

Find and graph the corresponding distribution function $F(x)$.

Solution   The function $F(x)$ is defined for all values of $x \in \mathbf{R}$. For instance, $F(-2.3) = P(X \leq -2.3) = P(\emptyset) = 0$. Since $F(x) = P(X \leq x)$, we must be particularly careful when x is in the range of X. To illustrate this, we compute $F(1.4)$ and $F(1.5)$. We find

$$F(1.4) = P(X \leq 1.4) = P(X = -1 \text{ or } 0) = 0.1 + 0.2 = 0.3$$
$$F(1.5) = P(X \leq 1.5) = P(X = -1, 0 \text{ or } 1.5) = 0.1 + 0.2 + 0.05 = 0.35$$

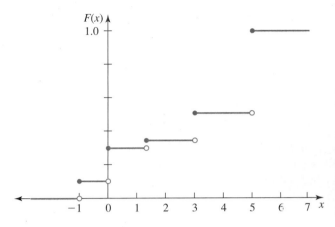

◀ **Figure 12.21**
The distribution function
$F(x)$ of Example 3. The
solid circles on the left ends
of the line segments indicate
that the distribution function
takes on this value at the
points where the function
jumps.

The distribution function $F(x)$ will be a piecewise defined function. We find

$$F(x) = \begin{cases} 0 & \text{for } x < -1 \\ 0.1 & \text{for } -1 \leq x < 0 \\ 0.3 & \text{for } 0 \leq x < 1.5 \\ 0.35 & \text{for } 1.5 \leq x < 3 \\ 0.5 & \text{for } 3 \leq x < 5 \\ 1 & \text{for } x \geq 5 \end{cases}$$

The graph of $F(x)$ is shown in Figure 12.21. ▮

Looking at the figure, we see that the graph of $F(x)$ is a nondecreasing and piecewise constant function that takes jumps at certain values. The function $F(x)$ is right-continuous. Furthermore,

$$\lim_{x \to -\infty} F(x) = 0 \qquad \text{and} \qquad \lim_{x \to \infty} F(x) = 1$$

This is typical for the distribution function of a discrete random variable.

It is easy to obtain the probability mass function from the distribution function. For instance, if we take the distribution function from Example 3, the function jumps at $x = 3$ and the jump height is 0.15. Since $F(x) = P(X \leq x)$, it follows that

$$p(3) = P(X = 3) = P(X \leq 3) - P(X < 3)$$
$$= F(3) - \lim_{x \to 3-} F(x) = 0.5 - 0.35 = 0.15$$

That is, the distribution function jumps at the values of X for which $P(X = x) > 0$. The jump height is then equal to the probability of X at this value.

12.4.2 Mean and Variance

Knowing the distribution of a random variable tells us everything about the random variable. In practice, however, it is often impossible or unnecessary to know the full probability distribution of a random variable that describes a particular random experiment. Instead, it might suffice to determine a few characteristic quantities, such as the average value and a measure that describes the spread around the average value. We discussed average values in Section 8.3, but now we will put this in the context of probability distributions.

▮ **EXAMPLE 4**   Clutch size can be thought of as a random variable. Let X denote the number of eggs per clutch laid by a bird, and let's assume that the distribution of X is described by

26. Assume that 20% of a very common insect species in your study area are parasitized. If you collect ten specimens of this species, what is the probability that at least one specimen in your sample is parasitized?

27. A multiple choice question has four choices, and a test has a total of ten multiple choice questions. A student passes the test if they answer at least nine questions correctly. If the student guesses the answers to all questions randomly, find the probability that they will pass.

28. Assume that A and B are disjoint and that both events have positive probability. Are they independent?

29. Assume that the probability that an insect species lives more than five days is 0.1. Find the probability that in a sample of size 10 of this species at least one insect will still be alive after five days.

30. **(a)** Use a Venn diagram to show that

$$(A \cup B)^c = A^c \cap B^c$$

(b) Use your result in (a) to show that if A and B are independent, then A^c and B^c are independent.

(c) Use your result in (b) to show that if A and B are independent, then

$$P(A \cup B) = 1 - P(A^c)P(B^c)$$

[12.3.4]

31. A screening test for a disease shows a positive result in 95% of all cases when the disease is actually present and in 10% of all cases when it is not. If the prevalence of the disease is 1 in 50, and an individual tests positive, what is the probability that the individual actually has the disease?

32. A screening test for a disease shows a positive result in 95% of all cases when the disease is actually present and in 10% of all cases when it is not. If a result is positive, the test is repeated. If the prevalence of the disease is 1 in 50, and an individual tests positive twice, what is the probability that they actually have the disease?

33. A bag contains two coins; one is a fair coin, the other one has two heads. You pick one coin at random and flip it. What is the probability that you picked the fair coin given that the outcome of the toss was heads?

34. You pick two cards from a standard deck of 52 cards. Find the probability that the first card was a spade given that the second card was a spade.

35. Suppose a woman has a hemophilic brother one healthy son. Find the probability that she is a carrier of the hemophilia gene.

▲ **Figure 12.20**
The pedigree for Problems 36 and 37. The solid black square (individual III-4) represents an afflicted male.

The pedigree in Figure 12.20 shows a family in which one member (II-4) is hemophilic. In problems 36–37, refer to this pedigree.

36. **(a)** Given the pedigree, find the probability that the individual I-2 is a carrier of the hemophilia gene.

(b) Given the pedigree, find the probability that II-3 is a carrier of the hemophilia gene.

37. **(a)** Given the pedigree, find the probability that II-3 is a carrier of the hemophilia gene.

(b) Given the pedigree, find the probability that III-2 is a carrier of the hemophilia gene.

(c) Given the pedigree, find the probability that II-2 is a carrier of the hemophilia gene.

◆ 12.4 DISCRETE RANDOM VARIABLES AND DISCRETE DISTRIBUTIONS

Outcomes of random experiments frequently are real numbers, such as the number of heads in a coin tossing experiment, the number of seeds produced in a cross between two plants, or the lifespan of an insect. Such numerical outcomes can be described by **random variables**. A random variable is a function from the sample space Ω into the set of real numbers. Random variables are typically denoted by X, Y, or Z, or other capital letters chosen from the end of the alphabet. For instance,

$$X : \Omega \to \mathbf{R}$$

describes a random variable X, which is a map from the sample space Ω to the set of real numbers.

Random variables are classified according to their range. If the number of values that X can take is either finite or countably infinite,[1] X is called a **discrete random variable**. If X takes on a continuous range of values, for instance, every value in an interval, X is called a **continuous random variable**. (We discuss continuous random variables in the next section.)

12.4.1 Discrete Distributions

In the first example, we look at a random variable that takes on a finite number of values.

EXAMPLE 1 Toss a fair coin three times. Let X be a random variable that counts the number of heads in each outcome. The sample space is

$$\Omega = \{HHH, HHT, HTH, THH, HTT, THT, TTH, TTT\}$$

and the random variable

$$X : \Omega \to \mathbf{R}$$

takes on values 0, 1, 2, or 3. For instance,

$$X(HHH) = 3 \qquad X(TTH) = 1 \qquad \text{or} \qquad X(TTT) = 0$$

In the next example, we look at a random variable that takes on a countably infinite number of values.

EXAMPLE 2 Toss a fair coin repeatedly until the first heads. Let Y be a random variable that counts the number of trials until the first heads shows up. The sample space is

$$\Omega = \{H, TH, TTH, TTTH, \ldots\}$$

and the random variable

$$Y : \Omega \to \mathbf{R}$$

takes on values 1, 2, 3, ... For instance,

$$Y(H) = 1 \qquad Y(TH) = 2 \qquad Y(TTH) = 3 \qquad \cdots$$

Now that we are more familiar with the concept of random variables, we will turn to the problem of how to assign probabilities to the different values of a random variable X. We illustrate this in relation to Example 1.

Since the coin is fair, each outcome in Ω has the same probability, namely, 1/8. We can translate this into probabilities for X. For instance,

$$P(X = 1) = P(\{HTT, THT, TTH\})$$

$$= P(HTT) + P(THT) + P(TTH)$$

$$= \frac{1}{8} + \frac{1}{8} + \frac{1}{8} = \frac{3}{8}$$

[1] A countably infinite set has infinitely many elements that can be enumerated, like the set of integers. (The set of rational numbers is also countably infinite; the set of real numbers is uncountably infinite.)

the following probability mass function.

x	$P(X = x)$
1	0.05
2	0.1
3	0.2
4	0.3
5	0.25
6	0.1

The average number of eggs per clutch can be computed as in Section 8.3, as the weighted sum

$$\text{average value} = \sum_{x} x P(X = x)$$
$$= (1)(0.05) + (2)(0.1) + (3)(0.2)$$
$$+ (4)(0.3) + (5)(0.25) + (6)(0.1) = 3.9$$

That is, the average clutch size is 3.9. ∎

The average value of X is called the **expected value** or **mean** of X, and denoted by EX. Since it is a very important quantity, we provide its definition in the following.

If X is a discrete random variable, then the expected value or mean of X is

$$EX = \sum_{x} x P(X = x)$$

where the sum is over all values of x with $P(X = x) > 0$.

When the range of X is finite, the sum in the definition is always defined. When the range of X is countably infinite, we must sum an infinite number of terms. Such sums can be infinite, depending on the distribution of X. The expected value of X is only defined if both $\sum_{x<0} x P(X = x)$ and $\sum_{x \geq 0} x P(X = x)$ are finite. We will not encounter any cases of discrete random variables where this sum is undefined.

We can easily extend the definition of the expected value of X to the expected value of a function of X. We will need this subsequently. Let $g(x)$ be a function of x. Then

$$Eg(X) = \sum_{x} g(x) P(X = x) \tag{12.17}$$

Another important quantity that characterizes the distribution of random variables is called the **variance**. It describes how spread out the range of the random variable is. To motivate the definition, let's look at the two random variables X and Y, with probability mass functions as follows.

k	$P(X = k)$	$P(Y = k)$
-10	0	0.2
-1	0.2	0
0	0.6	0.6
1	0.2	0
10	0	0.2

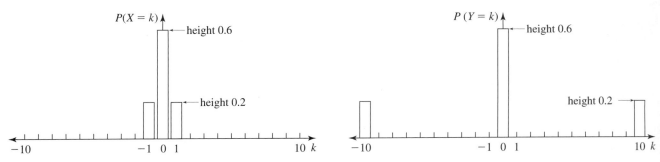

▲ **Figure 12.22**
The probability mass functions of X and Y. The distribution of Y is more spread out than the distribution of X.

We illustrate these two distributions in Figure 12.22. Both random variables have mean 0, but the range of Y is much more spread out than the range of X. To capture this in a single quantity, we will compute a weighted average of the squared distances to the mean, which is called the variance.

If X is a discrete random variable with mean μ, then the variance of X is

$$\text{var}(X) = E(X - \mu)^2 = \sum_x (x - \mu)^2 P(X = x)$$

Since the variance is an average value of a squared quantity, it is always nonnegative.

Returning to our random variables X and Y: since their means are both equal to 0, their variances are

$$\text{var}(X) = (-1)^2(0.2) + (0)^2(0.6) + (1)^2(0.2) = 0.4$$
$$\text{var}(Y) = (-10)^2(0.2) + (0)^2(0.6) + (10)^2(0.2) = 40$$

We see that the variance of Y is larger than the variance of X, reflecting the fact that the range of Y is more spread out than the range of X.

The variance of X is often denoted by σ^2. A quantity that is closely related to the variance is the **standard deviation**, denoted by s.d. or σ. It is defined as the square root of the variance.

$$\text{s.d.} = \sigma = \sqrt{\text{var}(X)}$$

Before we look at other examples, we will collect some important rules regarding expected values and variances. The first rule tells us how to compute the expected value and the variance of a linear transformation of X.

Let a and b be constants, then

$$E(aX + b) = a(EX) + b$$
$$\text{var}(aX + b) = a^2\text{var}(X)$$

The first property says that the expected value of a linear function of X is the linear function evaluated at the expected value of X. The second property tells us what happens to the variance when we multiply a random variable by a constant factor; it is important to note that the constant factor is squared when we pull it out of the variance. Furthermore, we see that the variance is unchanged when we shift a random variable by a constant term. We will prove these two properties in Problems 15 and 16.

It is often necessary to look at sums of random variables. We collect some rules without proof. Let X and Y be two discrete random variables, then $X + Y$ is also a discrete random variable. We have

$$E(X + Y) = EX + EY$$

We can use our rules to find an alternate formula for the variance. We start with

$$(X - \mu)^2 = X^2 - 2X\mu + \mu^2$$

Taking expectations on both sides, we find

$$E(X - \mu)^2 = E(X^2 - 2\mu X + \mu^2)$$

Since the expectation of a sum is the sum of the expectations, the right hand side simplifies to

$$EX^2 - E(2\mu X) + E\mu^2$$

Now, $E\mu^2 = \mu^2 = (EX)^2$ and $E(2\mu X) = 2\mu EX = 2(EX)^2$. Since $E(X - \mu)^2 = \text{var}(X)$, we have

$$\text{var}(X) = EX^2 - (EX)^2$$

This formula is often more convenient to use, since it leads to algebraically simpler expressions. Note that $EX^2 \neq (EX)^2$ unless $\text{var}(X) = 0$. We apply this formula to the random variable X in Example 4. There, we found $EX = 3.9$. To compute EX^2, we must evaluate

$$EX^2 = \sum_x x^2 P(X = x)$$
$$= (1)^2(0.05) + (2)^2(0.1) + (3^2)(0.2)$$
$$+ (4)^2(0.3) + (5)^2(0.25) + (6)^2(0.1) = 16.9$$

Hence,

$$\text{var}(X) = 16.9 - (3.9)^2 = 1.69$$

The rule for the expected value of a product of random variables is more complicated. Namely, we need the condition that the two random variables are independent. The definition of the independence of random variables is similar to that of the independence of events. Namely, when X and Y are discrete random variables, we say that X and Y are independent if

$$P(X = x, Y = y) = P(X = x)P(Y = y)$$

for all x and y. We can now state how to deal with the expected value of a product of independent random variables.

If X and Y are two independent random variables, then

$$E(XY) = (EX)(EY)$$

This rule allows us to compute the variance of the sum of two independent random variables. Namely, suppose that X and Y are independent; then

$$\begin{aligned}
\text{var}(X + Y) &= E(X + Y)^2 - [E(X + Y)]^2 \\
&= E(X^2 + 2XY + Y^2) - (EX + EY)^2 \\
&= EX^2 + 2E(XY) + EY^2 - (EX)^2 - 2(EX)(EY) - (EY)^2
\end{aligned}$$

Since X and Y are independent, $E(XY) = (EX)(EY)$, and the sum simplifies to

$$EX^2 - (EX)^2 + EY^2 - (EY)^2$$

But this we recognize as $\text{var}(X) + \text{var}(Y)$. Hence,

> If X and Y are independent, then
> $$\text{var}(X + Y) = \text{var}(X) + \text{var}(Y)$$

12.4.3 The Binomial Distribution

In this subsection, we will discuss a discrete random variable that models the number of successes among a fixed number of trials. Suppose that you perform a random experiment of repeated trials where each trial has two possible outcomes, success or failure. The trials are independent and the probability of success in each trial is p. We define the random variables X_k, $k = 1, 2, \ldots, n$, as

$$X_k = \begin{cases} 1 & \text{if the } k\text{th trial is successful} \\ 0 & \text{otherwise} \end{cases}$$

Then $P(X_k = 1) = p = 1 - P(X_k = 0)$ for $k = 1, 2, \ldots, n$.

If we repeat these trials n times, we might want to know the total number of successes. We set

$$S_n = \text{number of successes in } n \text{ trials}$$

We can define S_n in terms of the random variables X_k, namely

$$S_n = \sum_{k=1}^{n} X_k \tag{12.18}$$

Since the trials are independent, this representation shows that S_n can be written as a sum of independent random variables, all having the same distribution. We will use this representation subsequently.

The random variable S_n is discrete, and takes on values $0, 1, 2, \ldots, n$. To find its probability mass function $p(k) = P(S_n = k)$, we argue as follows. The event $\{S_n = k\}$ can be represented as a string of 0s and 1s of length n, where 0 represents failure and 1 represents success. For instance, if $n = 5$ and $k = 3$, then 01101 could be interpreted as the outcome of five trials, the first resulting in failure, followed by two successes, then a failure, and finally a success. The probability of this particular outcome is easy to compute, since the trials are independent. We find

$$P(01101) = (1 - p)pp(1 - p)p = p^3(1 - p)^2$$

The outcome 01101 is not the only one with three successes in five trials. Any other string of length 5 with exactly three 1s has the same property. To determine the number of different strings with this property, note that there are $\binom{5}{3}$ different ways of placing the three 1s in the five possible positions, and there is exactly one way to place the 0s in the remaining two positions. Hence there are $\binom{5}{3} \cdot 1 = \binom{5}{3}$ different strings of length 5 with exactly three 1s. There is another way to find this, which we will use in the following. There would be 5! ways of arranging the three 1s and the two 0s if the 0s and the 1s were distinguishable. Since the 0s and the 1s can be rearranged among themselves without changing the outcome, we must divide by the order. We then find that there are

$$\frac{5!}{3!2!} = \binom{5}{3}$$

different outcomes. As all outcomes are equally likely, we find

$$P(S_5 = 3) = \binom{5}{3} p^3 (1 - p)^2$$

We can use similar reasoning to derive the general formula. We summarize this in the following.

> **Binomial Distribution**   Let S_n be a random variable that counts the number of successes in n independent trials, each having probability p of success. Then the **binomial distribution** can be stated as
>
> $$P(S_n = k) = \binom{n}{k} p^k (1 - p)^{n-k}, \quad k = 0, 1, 2, \ldots, n$$

If we use the representation $S_n = \sum_{k=1}^{n} X_k$ in (12.18) for the binomial random variable, it will be straightforward to compute its mean and its variance. We find

$$EX_1 = (1)p + (0)(1 - p) = p$$

and

$$\text{var}(X_1) = EX_1^2 - (EX_1)^2 = p - p^2 = p(1 - p)$$

Since all $X_k, k = 1, 2, \ldots, n$, have the same distribution, we find

$$ES_n = E \sum_{k=1}^{n} X_k = \sum_{k=1}^{n} EX_k = np \tag{12.19}$$

In addition, since the X_k are independent,

$$\text{var}(S_n) = \text{var}\left(\sum_{k=1}^{n} X_k \right) = \sum_{k=1}^{n} \text{var}(X_k) = np(1 - p) \tag{12.20}$$

We give two more applications of the binomial distribution, to become familiar with other situations where this distribution plays an important role.

EXAMPLE 5   We consider the flowering pea plants again. Suppose that 20 independent offspring result from $Cc \times Cc$ crosses. Find the probability that at most two offspring have white flowers. Compute the expected value and the variance of the number of offspring with white flowers.

Solution   In a $Cc \times Cc$ cross, the probability of a white-flowering offspring (genotype cc) is 1/4, and the probability of a red-flowering offspring (genotype CC or Cc) is 3/4. The flower colors of different offspring are independent. We can therefore think of this as an experiment with twenty trials, each having success probability 1/4. We want to know the probability of at most two successes. With $n = 20, k \leq 2$

and $p = 1/4$, we find

$$P(S_{20} \leq 2) = P(S_{20} = 0) + P(S_{20} = 1) + P(S_{20} = 2)$$

$$= \binom{20}{0}\left(\frac{1}{4}\right)^0\left(\frac{3}{4}\right)^{20} + \binom{20}{1}\left(\frac{1}{4}\right)^1\left(\frac{3}{4}\right)^{19} + \binom{20}{2}\left(\frac{1}{4}\right)^2\left(\frac{3}{4}\right)^{18}$$

$$= 0.0913$$

The expected value of the number of white-flowering offspring is

$$E S_{20} = (20)\left(\frac{1}{4}\right) = 5$$

and the variance is

$$\text{var}(S_{20}) = (20)\left(\frac{1}{4}\right)\left(\frac{3}{4}\right) = \frac{15}{4} = 3.75$$ ▮

▮ **EXAMPLE 6**   Suppose that a woman who is a carrier for hemophilia has four daughters with a man who is not hemophilic. Find the probability that at least one daughter carries the hemophilia gene.

Solution   Each daughter has probability 1/2 of carrying the disease gene, independently of each other. We can think of this as an experiment with four independent trials and success probability 1/2 ("success" in this case is being a carrier). Therefore

$$P(S_4 \geq 1) = 1 - P(S_4 = 0)$$

$$= 1 - \binom{4}{0}\left(\frac{1}{2}\right)^0\left(\frac{1}{2}\right)^4$$

$$= 1 - \frac{1}{16} = \frac{15}{16} = 0.9375$$ ▮

12.4.4 The Multinomial Distribution

In the previous subsection, we considered experiments where each trial resulted in exactly one of two possible outcomes. We will now extend this to more than two possible outcomes. The distribution is then called the **multinomial distribution**.

▮ **EXAMPLE 7**   To study food preferences in the lady beetle *Coleomegilla maculata*, we present each beetle with three different food choices: maize pollen, egg masses of the European corn borer and aphids. We suspect that 20% of the time the beetle prefers the aphids, 35% of the time egg masses, and 45% of the time pollen. We carry out this experiment with 30 beetles, and find that 8 beetles prefer aphids, 10 egg masses, and 12 pollen. Compute the probability of this event, assuming that the trials are independent.

Solution   We define the random variables

$$N_1 = \text{number of beetles that prefer aphids}$$
$$N_2 = \text{number of beetles that prefer egg masses}$$
$$N_3 = \text{number of beetles that prefer pollen}$$

and the probabilities

$$p_1 = P(\text{beetle prefers aphids}) = 0.2$$
$$p_2 = P(\text{beetle prefers egg masses}) = 0.35$$
$$p_3 = P(\text{beetle prefers pollen}) = 0.45$$

We claim that

$$P(N_1 = 8, N_2 = 10, N_3 = 12) = \frac{30!}{8!10!12!}(0.2)^8(0.35)^{10}(0.45)^{12}$$

The term $\frac{30!}{8!10!12!}$ counts the number of ways that we can arrange 30 objects, 8 of which are of one type, 10 of another, and 12 of a third, which represent the beetles preferring aphids, egg masses, and pollen, respectively. The term $(0.2)^8(0.35)^{10}(0.45)^{12}$ comes from the probability of a particular arrangement, just as in the binomial case. ▮

A more involved example for the multinomial distribution is another of Mendel's experiments, in which he crossed pea plants that had round yellow seeds with plants that had green wrinkled seeds. Roundness and yellow color are dominant traits, and greenness and wrinkled texture are recessive traits. We denote by R the allele for round seeds, by r the allele for wrinkled seeds, by Y the allele for yellow seeds and by y the allele for green seeds. Then a cross between plants that are homozygous for round yellow seeds (genotype RR/YY) and plants that are homozygous for wrinkled green seeds (genotype rr/yy), is written as

$$RR/YY \times rr/yy$$

This results in offspring of type Rr/Yy. That is, all offspring are heterozygous with round yellow seeds. Crossing plants from this offspring generation then results in all possible combinations, as illustrated in the following table.

	RY	Ry	rY	ry
RY	RR/YY round yellow	RR/Yy round yellow	Rr/YY round yellow	Rr/Yy round yellow
Ry	RR/Yy round yellow	RR/yy round green	Rr/yY round yellow	Rr/yy round green
rY	rR/YY round yellow	rR/Yy round yellow	rr/YY wrinkled yellow	rr/Yy wrinkled yellow
ry	rR/yY round yellow	rR/yy round green	rr/yY wrinkled yellow	rr/yy wrinkled green

The laws of inheritance tell us that each outcome of this cross is equally likely. There are sixteen different genotypes, some of which give rise to the same morphological type of seed (for instance, round yellow). The morphological type of the seed is called the **phenotype**. We summarize the phenotypes and their probability distribution in the following table.

trait 1	trait 2	
	yellow	green
round	9/16	3/16
wrinkled	3/16	1/16

▮ **EXAMPLE 8**   Suppose that you obtain 50 independent offspring from a cross

$$Rr/Yy \times Rr/Yy$$

where 25 seeds are round yellow, 9 are round green, 12 are wrinkled yellow, and 4 are wrinkled green. Find the probability of this outcome.

Solution   This is another application of the multinomial distribution. Arguing as in the previous example, we find that the probability of this outcome is

$$\frac{50!}{25!9!12!4!}\left(\frac{9}{16}\right)^{25}\left(\frac{3}{16}\right)^{9}\left(\frac{3}{16}\right)^{12}\left(\frac{1}{16}\right)^{4}$$ ▮

◆ 12.4 PROBLEMS

[12.4.1]

1. Toss a fair coin twice. Let X be the random variable that counts the number of tails in each outcome. Find the probability mass function describing the distribution of X.

2. Roll a fair die twice. Let X be the random variable that gives the maximum of the two numbers. Find the probability mass function describing the distribution of X.

3. An urn contains three green and two blue balls. You remove two balls at random. Let X denote the number of green balls in your sample. Find the probability mass function describing the distribution of X.

4. You draw five cards from a standard deck of 52 cards. Let X denote the number of aces in your hand. Find the probability mass function describing the distribution of X.

5. Suppose that the probability mass function of a discrete random variable X is given by the following table.

x	$P(X = x)$
-3	0.2
-1	0.3
1.5	0.4
2	0.1

Find and graph the corresponding distribution function $F(x)$.

6. Suppose the probability mass function of a discrete random variable X is given by the following table.

x	$P(X = x)$
-1	0.1
-0.5	0.2
0.1	0.1
0.5	0.25
1	0.35

Find and graph the corresponding distribution function $F(x)$.

7. Let X be a random variable with distribution function

$$F(x) = \begin{cases} 0 & x < -1 \\ 0.2 & -2 \le x < 0 \\ 0.3 & 0 \le x < 1 \\ 0.7 & 1 \le x < 2 \\ 1 & x \ge 2 \end{cases}$$

Determine the probability mass function of X.

8. Let X be a random variable with distribution function

$$F(x) = \begin{cases} 0 & x < 0 \\ 0.1 & 0 \le x < 1.3 \\ 0.3 & 1.3 \le x < 1.7 \\ 0.8 & 1.7 \le x < 1.9 \\ 0.9 & 1.9 \le x < 2 \\ 1 & x \ge 2 \end{cases}$$

Determine the probability mass function of X.

9. Let $S = \{1, 2, 3, \ldots, 10\}$, and assume that

$$p(k) = \frac{k}{N}, \; k \in S$$

where N is a constant.

(a) Determine N so that $p(k)$, $k \in S$, is a probability mass function.

(b) Let X be a discrete random variable with $P(X = k) = p(k)$. Find the probability that X is less than 8.

10. (*Geometric distribution*) In Example 2, we tossed a coin repeatedly until the first heads showed up. Assume that the probability of heads is p with $p \in (0, 1)$. Let Y be a random variable that counts the number of trials until the first heads shows up.

(a) Show that $P(Y = 1) = p$, $P(Y = 2) = (1 - p)p$, and $P(Y = 3) = (1 - p)^2 p$.

(b) Explain why

$$P(Y = j) = (1 - p)^{j-1}p$$

for $j = 1, 2, \ldots$. This is called the **geometric distribution**.

(c) Prove that

$$\sum_{j \ge 1} P(Y = j) = 1$$

by showing the following steps.
(i) For $0 \le q < 1$, define

$$S_n = 1 + q + q^2 + \cdots + q^n$$

Show that

$$S_n - qS_n = 1 - q^{n+1}$$

and conclude from this that

$$S_n = \frac{1 - q^{n+1}}{1 - q}$$

(ii) Show that

$$P(Y \le k) = \sum_{j=1}^{k} P(Y = j)$$

$$= p \sum_{j=1}^{k} (1 - p)^{j-1}$$

Use your results in (i) to show that this simplifies to

$$1 - (1 - p)^k$$

and conclude from this that

$$\lim_{k \to \infty} P(Y \le k) = 1$$

which is equivalent to

$$\sum_{j \ge 1} P(Y = j) = 1$$

[12.4.2]

11. Suppose that the probability mass function of a discrete random variable X is given by the following table.

x	$P(X = x)$
-3	0.2
-1	0.3
1.5	0.4
2	0.1

Find the mean, the variance, and the standard deviation of X.

12. Suppose that the probability mass function of a discrete random variable X is given by the following table.

x	$P(X = x)$
-1	0.1
-0.5	0.2
0.1	0.1
0.5	0.25
1	0.35

Find the mean, the variance, and the standard deviation of X.

13. Let X be **uniformly distributed** on the set

$$S = \{1, 2, 3, \ldots, 10\}$$

that is,

$$P(X = k) = \frac{1}{10}, \quad k \in S$$

(a) Find EX.

(b) Find var(X).

14. Let X be uniformly distributed on the set

$$S = \{1, 2, 3, \ldots, n\}$$

where n is a positive integer; that is,

$$P(X = k) = \frac{1}{n}, \quad k \in S$$

(a) Find EX.

(b) Find var(X).

Hint: Recall that

$$\sum_{k=1}^{n} k = \frac{n(n + 1)}{2}$$

and

$$\sum_{k=1}^{n} k^2 = \frac{n(n + 1)(2n + 1)}{6}$$

15. Assume that X is a discrete random variable with finite range, and set

$$p(x) = P(X = x)$$

(a) Show that

$$E(aX + b) = \sum_{x} (ax + b) p(x)$$

(b) Use your result in (a) and the rules for finite sums to conclude that

$$E(aX + b) = aEX + b$$

16. Assume that X is a discrete random variable with finite range, and set

$$p(x) = P(X = x)$$

(a) Show that

$$\text{var}(aX + b) = a^2 \sum_{x} (x - EX)^2 p(x)$$

(b) Use your result in (a) and the rules for finite sums to conclude that

$$\text{var}(aX + b) = a^2 \text{var}(X)$$

17. Let X and Y be two independent random variables with probability mass function described by the following table.

k	$P(X = k)$	$P(Y = k)$
-2	0.1	0.2
-1	0	0.2
0	0.3	0.1
1	0.4	0.3
2	0.05	0
3	0.15	0.2

(a) Find EX and EY.

(b) Find $E(X + Y)$.

(c) Find var(X) and var(Y).

(d) Find var$(X + Y)$.

18. Let X and Y be two independent random variables with probability mass function described by the following table.

k	$P(X = k)$	$P(Y = k)$
−3	0.1	0.1
−1	0.1	0.2
0	0.2	0.1
0.5	0.3	0.3
2	0.15	0.1
2.5	0.15	0.2

(a) Find EX and EY.

(b) Find $E(X + Y)$.

(c) Find var(X) and var(Y).

(d) Find var$(X + Y)$.

19. We have two formulas for computing the variance of X; namely,

$$\text{var}(X) = E(X - EX)^2$$

and

$$\text{var}(X) = EX^2 - (EX)^2$$

(a) Explain why var$(X) \geq 0$.

(b) Use your results in (a) to explain why

$$EX^2 \geq (EX)^2$$

20. Assume that X is a discrete random variable with finite range. Show that if var$(X) = 0$, then $P(X = EX) = 1$.

[12.4.3]

21. Toss a fair coin ten times. Let X be the number of heads.

(a) Find $P(X = 5)$.

(b) Find $P(X \geq 8)$.

(c) Find $P(X \leq 9)$.

22. Toss a coin with probability 0.3 of heads five time. Let X be the number of tails.

(a) Find $P(X = 2)$.

(b) Find $P(X \geq 1)$.

23. Roll a fair die six times. Let X be the number of times you roll a 6. Find the probability mass function.

24. An urn contains four green and six blue balls. You draw a ball at random, note its color, and replace it. You repeat this four times. Let X denote the total number of green balls you obtain. Find the probability mass function of X.

25. Assume that 20% of all plants in a field are infested with aphids. Suppose that you pick twenty plants at random. What is the probability that none of them carried aphids?

26. To test for a disease that has a prevalence of 1 in 100 in a population, blood samples of 10 individuals are pooled and the pooled blood is then tested. What is the probability that the test result is negative (the disease is not present in the pooled blood sample)?

27. Suppose that a box contains ten apples. The probability that any one apple is spoiled is 0.1. (Assume that the apples are independent.)

(a) Find the expected number of spoiled apples per box.

(b) A shipment contains ten boxes. Find the expected number of boxes that contain no spoiled apples.

28. Toss a fair coin ten times. Let X denote the number of heads. What is the probability that X is within one standard deviation of its mean?

29. A multiple choice exam contains fifty questions. Each question has four choices. Find the expected number of correct answers if a student guesses the answers at random.

30. (*Sampling with and without replacement*) An urn contains K green and $N - K$ blue balls.

(a) You take n balls out of the urn. Find the probability that k of the n balls are green.

(b) You take a ball out of the urn, note its color, and replace it. You repeat this n times. Find the probability that k of the n balls are green.

[12.4.4]

31. Repeat Example 7 when $N_1 = 10$, $N_2 = 14$, and $N_3 = 6$.

32. Repeat Example 7 when $N_1 = 5$, $N_2 = 15$, and $N_3 = 10$.

33. Repeat Example 8 when twenty seeds are round yellow, ten seeds are round green, eight seeds are wrinkled yellow, and two seeds are wrinkled green.

34. Repeat Example 8 when seventeen seeds are round yellow, twenty-two seeds are round green, thirteen seeds are wrinkled yellow, and eight seeds are wrinkled green.

35. An urn contains six green, eight blue, and ten red balls. You take one ball out of the urn, note its color, and replace it. You repeat this six times. What is the probability that you sampled two of each color?

36. An urn contains eight green, four blue, and six red balls. You take one ball out of the urn, note its color, and replace it. You repeat this four times. What is the probability that you sampled two green, one blue, and one red ball?

37. In a $Cc \times Cc$ cross of peas, five offspring are of genotype CC, twelve are of genotype Cc, and six are of genotype cc. What is the probability of this event?

38. Assume a 1:1 sex ratio. A woman who is a carrier of hemophilia has four children with a man who is not hemophilic. What is the probability that she has one daughter who is not a carrier, one daughter who is a carrier, one son who is, and one son who is not hemophilic?

◆ 12.5 CONTINUOUS DISTRIBUTIONS

12.5.1 Density Functions

In the previous section, we discussed random variables that took on a finite (or countably infinite) number of values. In this section, we will discuss random variables that take on a continuum of possible values. Such examples arise, for instance, when we consider the length distribution of an organism. Within an appropriate (species-specific) interval, an individual's length can take on any value. Such random variables are then called continuous random variables.

To describe the probability distribution of a continuous random variable X, we will use the distribution function $F(x)$ that we introduced in the previous section. The distribution function for a continuous random variable is defined as

$$F(x) = P(X \leq x)$$

and has the following properties (see Figure 12.23).

1. $0 \leq F(x) \leq 1$.
2. $\lim_{x \to -\infty} F(x) = 0$ and $\lim_{x \to \infty} F(x) = 1$.
3. $F(x)$ is nondecreasing and continuous.

Note that the distribution function of a continuous random variable is continuous. You should contrast this with the distribution function of a discrete random variable, which is piecewise constant and takes jumps at those values of x where $P(X = x) > 0$.

EXAMPLE 1   Show that

$$F(x) = \begin{cases} 1 - e^{-2x} & \text{for } x > 0 \\ 0 & \text{for } x \leq 0 \end{cases}$$

is a distribution function of a continuous random variable.

Solution   (A graph of $F(x)$ is shown in Figure 12.24.) We must check the three properties of distribution functions of continuous random variables. 1. Since $0 \leq 1 - e^{-2x} \leq 1$ for $x > 0$, it follows that $0 \leq F(x) \leq 1$ for all $x \in \mathbf{R}$. 2. Since $F(x) = 0$ for $x \leq 0$ and $\lim_{x \to \infty} e^{-2x} = 0$, it follows that

$$\lim_{x \to -\infty} F(x) = 0 \qquad \text{and} \qquad \lim_{x \to \infty} F(x) = 1$$

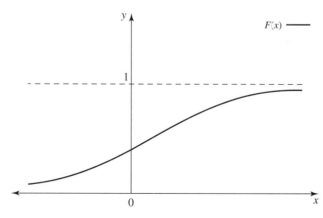

◀ **Figure 12.23**
The distribution function of a continuous random variable.

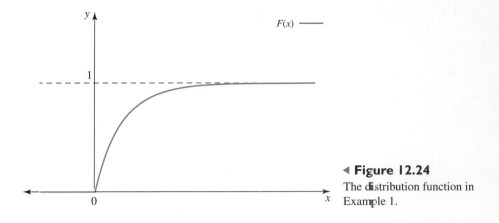

◀ **Figure 12.24**

The distribution function in Example 1.

3. To show that $F(x)$ is continuous for all $x \in \mathbf{R}$ note that $F(x)$ is continuous for $x > 0$ and for $x < 0$. To check continuity at $x = 0$, we compute

$$\lim_{x \to 0^+} F(x) = \lim_{x \to 0^+} (1 - e^{-2x}) = 0 = \lim_{x \to 0^-} F(x)$$

which is equal to $F(0) = 0$. Hence, $F(x)$ is continuous at $x = 0$.

To show that $F(x)$ is nondecreasing, we compute $F'(x)$.

$$F'(x) = \begin{cases} 2e^{-2x} > 0 & \text{for } x > 0 \\ 0 & \text{for } x < 0 \end{cases}$$

which implies that $F(x)$ is nondecreasing for $x \in \mathbf{R}$.   ▮

If $F(x)$ is differentiable with $F'(x) = f(x)$, then the fundamental theorem of calculus, part I, tells us that

$$F(x) = \int_{-\infty}^{x} f(u)\, du$$

The function $f(x)$ is called a (probability) density function (see Figure 12.25). (We discussed such functions in Section 8.2.)

Since $F(x)$ is a distribution function, it follows that

$$f(x) \geq 0 \qquad \text{and} \qquad \int_{-\infty}^{\infty} f(x)\, dx = 1$$

Any function that satisfies these two properties, can be interpreted as a density function. The function $f(x)$ need not be continuous. (In all of our applications,

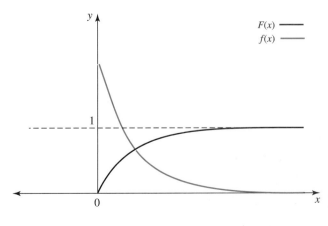

◀ **Figure 12.25**

The distribution function $F(x)$ and corresponding density function $f(x)$ of a continuous random variable.

$f(x)$ will be continuous except for possibly a finite number of points.) Furthermore, since

$$P(a < X \leq b) = P(X \leq b) - P(X \leq a) = F(b) - F(a)$$

we find

$$P(a < X \leq b) = \int_a^b f(x)\,dx$$

That is, the area under the curve $y = f(x)$ between a and b represents the probability that the random variable takes on values between a and b, as illustrated in Figure 12.26.

In contrast to discrete random variables, for which $P(X \leq b)$ and $P(X < b)$ can differ, there is no difference for continuous random variables, since

$$P(X = b) = \int_b^b f(x)\,dx = 0$$

Therefore

$$\int_a^b f(x)\,dx = P(a < X \leq b) = P(a \leq X \leq b)$$
$$= P(a < X < b) = P(a \leq X < b)$$

▮ EXAMPLE 2   The distribution function of a random variable X is given by

$$F(x) = \begin{cases} 0 & \text{for } x \leq 0 \\ x^2 & \text{for } 0 < x < 1 \\ 1 & \text{for } x \geq 1 \end{cases}$$

a. Find the corresponding density function.

b. Compute $P(-1 \leq X \leq 1/2)$.

Solution

a. The density function $f(x)$ is the derivative of the distribution function $F(x)$:

$$f(x) = F'(x) = \begin{cases} 2x & \text{for } 0 < x < 1 \\ 0 & \text{otherwise} \end{cases}$$

The distribution function together with its density function is shown in Figure 12.27.

b. Using the distribution function, we immediately find

$$P\left(-1 \leq X \leq \frac{1}{2}\right) = F\left(\frac{1}{2}\right) - F(-1) = \frac{1}{4} - 0 = \frac{1}{4}$$

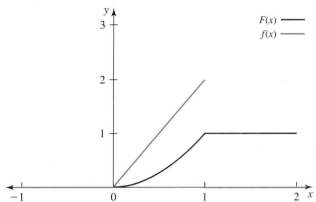

◀ **Figure 12.27**
The distribution function $F(x)$ and corresponding density function $f(x)$ in Example 2.

If, instead, we use the density function, we must evaluate

$$P\left(-1 \le X \le \frac{1}{2}\right) = \int_{-1}^{1/2} f(x)\,dx = \int_0^{1/2} 2x\,dx$$

$$= x^2\Big]_0^{1/2} = \frac{1}{4}$$

The formulas for the mean and the variance of a continuous random variable are analogous to the discrete case. The expectation EX of a continuous random variable with density function $f(x)$ was defined in Section 8.2:

$$EX = \int_{-\infty}^{\infty} x f(x)\,dx$$

The variance of a continuous random variable X with mean μ is

$$\text{var}(X) = E(X - \mu)^2 = \int_{-\infty}^{\infty} (x - \mu)^2 f(x)\,dx$$

The alternate formula that we gave in the previous section holds as well, namely

$$\text{var}(X) = EX^2 - (EX)^2 = \int_{-\infty}^{\infty} x^2 f(x)\,dx - \left(\int_{-\infty}^{\infty} x f(x)\,dx\right)^2$$

EXAMPLE 3   The density function of a random variable X is given by

$$f(x) = \begin{cases} 3x^2 & \text{for } 0 < x < 1 \\ 0 & \text{otherwise} \end{cases}$$

Compute the mean and the variance of X.

Solution   To compute the mean, we must evaluate

$$EX = \int_{-\infty}^{\infty} x f(x)\,dx = \int_0^1 3x^3\,dx = \frac{3}{4}x^4\Big]_0^1 = \frac{3}{4}$$

To compute the variance, we first evaluate

$$EX^2 = \int_{-\infty}^{\infty} x^2 f(x)\,dx = \int_0^1 3x^4\,dx = \frac{3}{5}x^5\Big]_0^1 = \frac{3}{5}$$

The variance of X is then given by

$$\text{var}(X) = EX^2 - (EX)^2 = \frac{3}{5} - \left(\frac{3}{4}\right)^2 = \frac{3}{80}$$

In the following subsections, we will introduce some continuous distributions and explain where they might be used.

12.5.2 The Normal Distribution

The normal distribution was first introduced by Abraham De Moivre (1667–1754) in the context of computing probabilities in binomial experiments when the number of trials is large. Gauss showed later that this distribution was important in the error analysis of measurements. It is the most important continuous distribution, and we discussed it in Section 8.4. We will restate some of its properties.

Recall that we say that a random variable X is normally distributed with parameters $\mu \in \mathbf{R}$ and $\sigma > 0$ if its density is given by

$$f(x) = \frac{1}{\sigma\sqrt{2\pi}}e^{-(x-\mu)^2/2\sigma^2}, \quad -\infty < x < \infty$$

When $\mu = 0$ and $\sigma = 1$, then the distribution is called the **standard normal distribution** (see Figure 12.28).

We cannot show that $\int_{-\infty}^{\infty} f(x)\,dx = 1$, but it is nevertheless useful to know that

$$\int_{-\infty}^{\infty} e^{-x^2/2}\,dx = \sqrt{2\pi}$$

We were able to identify the parameter μ as the mean in Section 8.4, that is, if X is normally distributed with parameters μ and σ, then

$$\mu = EX = \int_{-\infty}^{\infty} xf(x)\,dx$$

In Section 8.4, we said that the parameter σ describes how spread out the distribution is. We can now characterize σ; namely, σ^2 is the variance of X. That is,

$$\sigma^2 = \text{var}(X) = \int_{-\infty}^{\infty} (x-u)^2 f(x)\,dx$$

These calculations will be carried out in Problems 13 and 14.

We also mentioned in Section 8.4 that the distribution function $F(x) = P(X \le x)$ can only be evaluated numerically. The values of the distribution function of the standard normal distribution,

$$F(z) = \int_{-\infty}^{z} \frac{1}{\sqrt{2\pi}}e^{-u^2/2}\,du$$

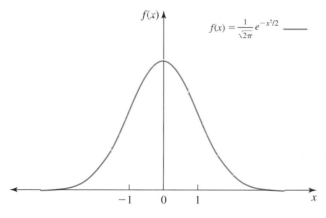

$$f(x) = \frac{1}{\sqrt{2\pi}}e^{-x^2/2}$$

◀ **Figure 12.28**

The density function of a standard normally distributed random variable.

are listed in standard tables. One such table, for $z \geq 0$, is provided in Appendix B. We can use this table to compute probabilities of events when X is normally distributed, with mean μ and variance σ^2; namely, if Z is standard normally distributed, then

$$P(X \leq x) = P\left(Z \leq \frac{x - \mu}{\sigma}\right) \tag{12.21}$$

▌**EXAMPLE 4**   Let X be normally distributed with mean 3 and variance 4. Find $P(1 \leq X \leq 4)$.

Solution   We use the identity (12.21) to compute this probability.

$$P(1 \leq X \leq 4) = P\left(\frac{1 - 3}{2} \leq \frac{X - \mu}{\sigma} \leq \frac{4 - 3}{2}\right)$$

$$= P\left(-1 \leq Z \leq \frac{1}{2}\right) = F\left(\frac{1}{2}\right) - F(-1)$$

$$= F\left(\frac{1}{2}\right) - (1 - F(1)) = 0.5199 - 1 + 0.8413 = 0.3612 \quad ▌$$

The importance of the normal distribution cannot be overstated. A large part of statistics is based on the assumption that observed quantities are normally distributed. You will probably ask why we can assume the normal distribution in the first place. The reason for this is quite deep, and unfortunately is beyond the scope of this book. But we wish to at least give you the gist of it.

Many quantities can be thought of as a sum of a large number of small contributions. For instance, when we measure a quantity, the total error of measurement can be thought of as a sum of a large number of independent contributions from different sources. One can show that the distribution of any sum of independent random variables, which all have the same distribution with a finite mean and a finite variance, converges to a normally distributed random variable when the number of terms in the sum increases. This result is known as the **central limit theorem**. This is the reason why it is often assumed that errors of measurements are normally distributed.

The central limit theorem is also evoked in quantitative genetics, namely, many quantitative traits (such as height or birth weight of an organism) are thought of as resulting from numerous genetic and environmental factors, which all act in an additive or multiplicative way. If these factors act in an additive way and are independent, the central limit theorem can be applied directly. If the factors act in a multiplicative way, then a logarithmic transformation reduces this to the additive case.

12.5.3 The Uniform Distribution

The uniform distribution is in some ways the simplest continuous distribution. We say that a random variable U is **uniformly distributed** over the interval (a, b) if its density function is given by

$$f(x) = \begin{cases} \frac{1}{b-a} & \text{for } x \in (a, b) \\ 0 & \text{otherwise} \end{cases}$$

as illustrated in Figure 12.29.

The reason for the term "uniform" can be seen when we compute the probability that the random variable U falls into the interval $(x_1, x_2) \subset (a, b)$. To compute probabilities of events for a uniformly distributed random variable, we compute the area of a rectangle. We therefore use the simple geometric formula "area is equal to the product of height and length," instead of formally integrating the density function

◄ Figure 12.29
The density function of a
uniformly distributed random
variable over the interval
(a, b).

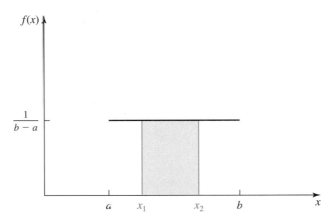

◄ Figure 12.30
The probability
$P(U \in (x_1, x_2))$ is equal to
the area of the shaded region.

(see Figure 12.30). We find

$$P(U \in (x_1, x_2)) = \begin{bmatrix} \text{area under } f(x) = \frac{1}{b-a} \\ \text{between } x_1 \text{ and } x_2 \end{bmatrix}$$

$$= \frac{x_2 - x_1}{b - a}$$

We see that this probability depends only on the length of the interval (x_1, x_2) relative to the length of the interval (a, b), and *not* on the location of (x_1, x_2) provided that (x_1, x_2) is a subset of (a, b). Therefore, intervals of equal lengths that are contained in (a, b) have equal chances of containing U.

To find the mean of a uniformly distributed random variable, we evaluate

$$EU = \int_{-\infty}^{\infty} x f(x)\, dx = \int_{a}^{b} x \frac{1}{b - c}\, dx = \frac{1}{b - a}\left[\frac{1}{2}x^2\right]_{a}^{b}$$

$$= \frac{1}{b - a}\frac{b^2 - a^2}{2} = \frac{a + b}{2}$$

which is the midpoint of the interval (a, b). To find the variance, we first compute

$$EU^2 = \int_{-\infty}^{\infty} x^2 f(x)\, dx = \int_{a}^{b} x^2 \frac{1}{b - a}\, dx = \frac{1}{b - a}\left[\frac{1}{3}x^3\right]_{a}^{b}$$

$$= \frac{1}{b - a}\frac{b^3 - a^3}{3} = \frac{b^2 + ab + a^2}{3}$$

then,

$$\text{var}(U) = EU^2 - (EU)^2 = \frac{b^2 + ab + a^2}{3} - \frac{(a+b)^2}{4}$$

$$= \frac{4b^2 + 4ab + 4a^2 - 3a^2 - 6ab - 3b^2}{12} = \frac{b^2 - 2ab + a^2}{12}$$

$$= \frac{(b-a)^2}{12}$$

Uniform distributions are frequently used in computer simulations of random experiments, as in the following example.

EXAMPLE 5   Suppose that you wish to simulate on the computer a random experiment that consists of tossing a coin five times with probability 0.3 of heads. The software that you want to use can generate uniformly distributed random variables in the interval $(0, 1)$. How do you proceed?

Solution   Each trial consists of flipping the coin and then recording the outcome. To simulate a coin flip with probability 0.3 of heads, we draw a uniformly distributed random variable U from the interval $(0, 1)$. The computer returns a number u in the interval $(0, 1)$. Since $P(U \le 0.3) = 0.3$, if

$$u = \begin{cases} \le 0.3 & \text{we record heads} \\ > 0.3 & \text{we record tails} \end{cases}$$

We repeat this five times.

To be more concrete, assume that successive values of the uniform random variable are 0.2859, 0.9233, 0.5187, 0.8124, and 0.0913. These numbers are then translated into $HTTTH$, where H stands for heads and T for tails.   ▮

In the next example, we will see what the distribution function of a uniformly distributed random variable looks like.

EXAMPLE 6

a.   Find the density and distribution functions of a uniformly distributed random variable on the interval $(1, 5)$, and graph both in the same coordinate system.

b.   Suppose that we draw a uniformly distributed random variable from the interval $(1, 5)$. Compute the probability that the first digit after the decimal point is a 2.

Solution

a.   Since the length of the interval $(1, 5)$ is 4, the density function is given by

$$f(x) = \begin{cases} \frac{1}{4} & \text{for } 1 < x < 5 \\ 0 & \text{otherwise} \end{cases}$$

The distribution function $F(x) = P(X \le x)$ is given by

$$F(x) = \int_{-\infty}^{x} f(u) \, du = \begin{cases} 0 & \text{for } x \le 1 \\ \int_1^x \frac{1}{4} \, du = \frac{1}{4}x - \frac{1}{4} & \text{for } 1 < x < 5 \\ 1 & \text{for } x \ge 5 \end{cases}$$

Graphs of $f(x)$ and $F(x)$ are shown in Figure 12.31.

b.   The event that the first digit after the decimal point is a 2 is the event that the random variable U falls into the set

$$A = [1.2, 1.3) \cup [2.2, 2.3) \cup [3.2, 3.3) \cup [4.2, 4.3)$$

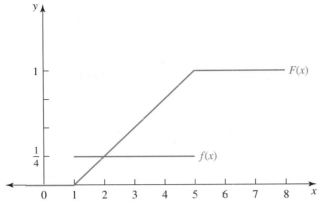

◀ **Figure 12.31**
The density function $f(x)$
and the distribution function
$F(x)$ of a uniformly
distributed random variable
over the interval $(1, 5)$ (see
Example 6).

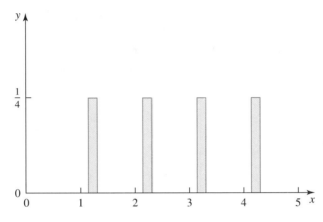

◀ **Figure 12.32**
The probability $P(U \in A)$ in
Example 6 is equal to the sum
of the areas of the shaded
regions.

Therefore,

$$P(U \in A) = (4)(0.1)\left(\frac{1}{4}\right) = 0.1$$

as illustrated in Figure 12.32.

12.5.4 The Exponential Distribution

We give the density function of the exponential distribution first, and then explain
where this distribution plays a role. We say that a random variable X is **exponentially
distributed** with parameter λ if its density function is given by

$$f(x) = \begin{cases} \lambda e^{-\lambda x} & \text{for } x > 0 \\ 0 & \text{for } x \le 0 \end{cases}$$

Since

$$\int_0^x \lambda e^{-\lambda u}\, du = -e^{-\lambda u}\Big|_0^x = 1 - e^{-\lambda x}$$

its distribution function $F(x) = P(X \le x)$ is given by

$$F(x) = \begin{cases} 1 - e^{-\lambda x} & \text{for } x > 0 \\ 0 & \text{for } x \le 0 \end{cases}$$

The expected value of X is

$$EX = \int_0^\infty x\lambda e^{-\lambda x}\, dx = \frac{1}{\lambda}$$

and the variance of X is

$$\text{var}(X) = \int_0^\infty \left(x - \frac{1}{\lambda}\right)^2 \lambda e^{-\lambda x}\, dx = \frac{1}{\lambda^2}$$

(The mean and the variance will be calculated in Problems 25 and 26.)

The exponential distribution is frequently used to model waiting times or lifetimes.

▮ **EXAMPLE 7** Suppose that the time between arrivals of insect pollinators to a flowering plant is exponentially distributed with parameter $\lambda = 0.3/\text{hr}$.

a. Find the mean and the standard deviation of the waiting time between successive pollinator arrivals.

b. A pollinator just left the plant, what is the probability that you will have to wait for more than three hours before the next pollinator arrives?

Solution

a. Since $\lambda = 0.3/\text{hr}$, we find that the mean is equal to $1/\lambda = 10/3$ hours, and the standard deviation is equal to $1/\lambda = 10/3$ hours.

b. If we denote the waiting time by T, then

$$P(T > 3) = 1 - F(3) = e^{-(0.3)(3)} = 0.4066$$ ▮

▮ **EXAMPLE 8** Suppose that the life time of an organism is exponentially distributed with parameter $\lambda = (1/200)\ \text{yrs}^{-1}$.

a. Find the probability that the organism will live for more than 50 years.

b. Given that the organism is 100 years old, find the probability that it will live for at least another 50 years.

Solution We denote the lifetime of the organism by T. Then T is exponentially distributed with parameter $\lambda = (1/200)\ \text{yrs}^{-1}$.

a. We want to find $P(T > 50)$. We compute

$$P(T > 50) = 1 - P(T \le 50) = 1 - (1 - e^{-50/200})$$

$$= e^{-50/200} = e^{-1/4} = 0.7788$$

Note that the units in the exponent cancelled out.

b. We want to find $P(T > 150 \mid T > 100)$. This is a conditional probability. We evaluate it in the following way:

$$P(T > 150 \mid T > 100) = \frac{P(T > 150 \text{ and } T > 100)}{P(T > 100)}$$

Since $\{T > 150\} \subset \{T > 100\}$, it follows that

$$\{T > 150\} \cap \{T > 100\} = \{T > 150\}$$

Therefore,

$$P(T > 150 \text{ and } T > 100) = P(T > 150)$$

We can now continue to evaluate

$$\frac{P(T > 150 \text{ and } T > 100)}{P(T > 100)} = \frac{P(T > 150)}{P(T > 100)} = \frac{e^{-150/200}}{e^{-100/200}}$$

$$= e^{-3/4+1/2} = e^{-1/4} = 0.7788$$

This is the same answer as in (a). That is, the fact that the organism has lived for 100 years does not change its probability of it living for another 50 years. Such an organism does not age; this non-aging property is a characteristic feature of the exponential distribution. Of course, most organisms age; the exponential distribution is still frequently used to model lifetimes, and you should consider this as an approximation of the real situation. ▌

We discuss this non-aging property in more detail. If T is an exponentially distributed lifetime, then T satisfies the property

$$P(T > t + h \mid T > t) = P(T > h)$$

In words, if the organism is still alive after t units of time, then the probability of living for at least another h units of time is the same as the probability that the organism survived the first h units of time. This implies that death does not become more (or less) likely with age.

This property follows immediately from the following calculation, which is essentially the same as the one that we carried out in Example 7.

$$P(T > t + h \mid T > t) = \frac{P(T > t + h \text{ and } T > t)}{P(T > t)}$$

$$= \frac{P(T > t + h)}{P(T > t)} = \frac{e^{-\lambda(t+h)}}{e^{-\lambda t}}$$

$$= e^{-\lambda h} = P(T > h)$$

For most organisms or objects, the exponential distribution would be a poor lifetime model. However, this distribution is particularly well-suited to describe radioactive decay.

▌ **EXAMPLE 9**   Assume that the lifetime of a radioactive atom is exponentially distributed with parameter $\lambda = 3/$days.

    a. Find the average lifetime of this atom.

    b. Find the time T_h such that the probability that the atom will not have decayed at time t is equal to $1/2$. (The time T_h is called the half-life.)

Solution

    a. The average lifetime is $1/\lambda = 1/3$ days.

    b. If the random variable T denotes the lifetime of the atom, then T_h satisfies

$$P(T > T_h) = \frac{1}{2}$$

that is,

$$e^{-\lambda T_h} = \frac{1}{2} \quad \text{or} \quad T_h = \frac{\ln 2}{\lambda}$$

With $\lambda = 3/$days, we find

$$T_h = \frac{\ln 2}{3} \text{ days} = 0.2310 \text{ days} \qquad \blacksquare$$

◆ 12.5 PROBLEMS

[12.5.1]

1. Show that

$$f(x) = \begin{cases} 3e^{-3x} & \text{for } x \geq 0 \\ 0 & \text{for } x < 0 \end{cases}$$

is a density function. Find the corresponding distribution function.

2. Show that

$$f(x) = \begin{cases} \frac{1}{2} & 0 \leq x \leq 2 \\ 0 & \text{otherwise} \end{cases}$$

is a density function. Find the corresponding distribution function.

3. Determine c so that

$$f(x) = \frac{c}{1 + x^2}, \quad x \in \mathbf{R}$$

is a density function.

4. Determine c so that

$$f(x) = \frac{c}{x^2}, \quad x \geq 1$$

is a density function.

5. Let X be a continuous random variable with density function

$$f(x) = \begin{cases} 2e^{-2x} & x \geq 0 \\ 0 & x < 0 \end{cases}$$

Find EX and var(X).

6. Let X be a continuous random variable with density function

$$f(x) = \frac{1}{2}e^{-|x|}$$

for $x \in \mathbf{R}$. Find EX and var(X).

7. Let X be a continuous random variable with distribution function

$$F(x) = \begin{cases} 1 - \frac{1}{x^3} & \text{for } x \geq 1 \\ 0 & \text{for } x < 1 \end{cases}$$

Find EX and var(X).

8. Let X be a continuous random variable with

$$P(X > x) = e^{-ax}, \quad x \geq 0$$

where a is a positive constant. Find EX and var(X).

9. Let X be a continuous random variable with density function

$$f(x) = \begin{cases} (a - 1)x^{-a} & \text{for } x > 1 \\ 0 & \text{for } x < 1 \end{cases}$$

(a) Show that $EX = \infty$ when $a \leq 2$.

(b) Compute EX when $a > 2$.

10. Suppose that X is a continuous random variable which takes on only nonnegative values. Set

$$G(x) = P(X > x)$$

(a) Show that

$$G'(x) = -f(x)$$

where $f(x)$ is the corresponding density function.

(b) Assume that

$$\lim_{x \to \infty} xG(x) = 0$$

and use integration by parts and (a) to show

$$EX = \int_0^\infty G(x) \, dx \tag{12.22}$$

(c) Let X be a continuous random variable with

$$P(X > x) = e^{-ax}, \quad x > 0$$

where a is a positive constant. Use (12.22) to find EX. (If you did Problem 8, compare your answers.)

[12.5.2]

11. Suppose that X is normally distributed with mean 2 and standard deviation 1. Find $P(0 \leq X \leq 3)$.

12. Suppose that X is normally distributed with mean -1 and standard deviation 2. Find $P(-3.5 \leq X \leq 0.5)$.

13. Suppose that X is normally distributed with mean μ and standard deviation σ. Show that $EX = \mu$. (You may use the fact that if Z is standard normally distributed, then $EZ = 0$ and var$(X) = 1$.)

14. Suppose that X is normally distributed with mean μ and standard deviation σ. Show that var$(X) = \sigma^2$. (You may use the fact that if Z is standard normally distributed, then $EZ = 0$ and var$(X) = 1$.)

15. Suppose that X is standard normally distributed. Find $E|X|$.

16. Suppose that the number of seeds a plant produces is normally distributed, with mean 142 and standard deviation 31. Find the probability that in a sample of 5 plants, at least one plant produces more than 200 seeds.

[12.5.3]

17. Suppose that you pick a number at random from the interval $(0, 4)$. What is the probability that the first digit after the decimal point is a 3?

18. Suppose that you pick a number X at random from the interval $(0, a)$. If $P(X \geq 1) = 0.2$, find a.

19. Suppose that you pick a number X at random from the interval (a, b). If $EX = 4$ and var$(X) = 3$, find a and b.

20. Suppose that you pick five numbers at random from the interval $(0, 1)$. Assume that the numbers are independent. What is the probability that all numbers are greater than 0.7?

21. Suppose that X_1, X_2, and X_3 are independent and uniformly distributed over $(0, 1)$. Define

$$Y = \max(X_1, X_2, X_3)$$

Find EY. (*Hint*: Compute $P(Y \leq y)$ and deduce from this the density of Y.)

22. Suppose that X_1, X_2, and X_3 are independent and uniformly distributed over $(0, 1)$. Define

$$Y = \min(X_1, X_2, X_3)$$

Find EY. (*Hint*: Compute $P(Y > y)$ and deduce from this the density of Y.)

23. Suppose that you wish to simulate a random experiment that consists of tossing a coin with probability 0.6 of heads ten times. The computer generates the following ten random variables: 0.1905, 0.4285, 0.9963, 0.1666, 0.2223, 0.6885, 0.0489, 0.3567, 0.0719, 0.8661. Find the corresponding sequence of heads and tails.

24. Suppose that you wish to simulate a random experiment that consists of rolling a fair die. The computer generates the following ten random variables: 0.7198, 0.2759, 0.4108, 0.7780, 0.2149, 0.0348, 0.5673, 0.0014, 0.3249, 0.6630. Find the corresponding sequence of numbers on the die.

[12.5.4]

25. Let X be exponentially distributed with parameter λ. Find EX.

26. Let X be exponentially distributed with parameter λ. Find $\mathrm{var}(X)$.

27. Suppose that the lifetime of a battery is exponentially distributed, with an average lifespan of three months. What is the probability that the battery will last for more than four months?

28. Suppose that the lifetime of a battery is exponentially distributed, with an average lifespan of two months. You buy six batteries. What is the probability that none of them will last more than two months. (Assume that the batteries are independent.)

29. Suppose that the lifetime of a radioactive atom is exponentially distributed, with an average lifetime of 27 days.

(a) Find the probability that the atom will not decay during the first twenty days after you start to observe it.

(b) Suppose that the atom does not decay during the first twenty days that you observe it. What is the probability that it will not decay during the next twenty days?

30. If X has distribution function $F(x)$, we can show that $F(X)$ is uniformly distributed over the interval $(0, 1)$. Use this fact to generate exponentially distributed random variables with mean 1 on a graphing calculator or a computer. (Assume that the computer generates the following four uniformly distributed random variables on the interval $(0, 1)$: 0.0371, 0.5123, 0.1370, 0.9865.)

◆ 12.6 STATISTICAL TOOLS

In the preceding sections of this chapter, we learned how to model various random experiments. By assuming the underlying probability distribution of the model, we were able to compute the probabilities of events, such as obtaining white-flowering pea plants.

To understand phenomena in nature, however, we often take the reverse approach, and infer the underlying probability distribution from the observation of events. More precisely, based on a collection of observations, called **data**, we will estimate characteristics of the underlying probability distribution. For instance, in the case of the normal distribution, our goal might be to estimate the mean and the variance, which are parameters that describe a normal distribution.

This section will provide some statistical tools to achieve this goal. We assume that all of our observations are presented as numerical data.

12.6.1 Collecting and Describing Data

To learn something about the distribution of a quantity in a population (clutch size, plant height, lifespan, and so forth), we cannot measure the quantity for every individual in the population. Instead, we take a subset of the population, called a **sample**, measure the quantity in the sample, and then try to infer the distribution of the quantity in the population from its distribution in the sample. In order to do so, the sample must be representative of the population. This is a difficult problem, which we cannot address satisfactorily here. We only mention that it is easy to include a bias in a sample if the observations are chosen from a section of the population with similar characteristics. For instance, if we want to assess fecundity in a plant that grows in a wide variety of soils but we sample plants on sandy soils only, then our sample might not reflect the fecundity of the entire population.

In the following, we will always assume that our sample is representative of the population; we call such a sample a **random sample**. All observations in a random sample are independent and come from the same distribution (namely, the distribution of the quantity in the entire population). A typical scheme to obtain a random sample of size n is to pick an individual at random from the population, record the quantity of interest, replace the individual, and then select the next individual. This procedure is repeated until a sample of size n is obtained. Replacing the sampled individual after recording the quantity of interest ensures that the population always has the same composition, and, hence, all observations have the same distribution. We denote the sample by the vector $(X_1, X_2, \ldots, X_n)$, where X_k is the kth observation. The X_k are independent random variables that all have the same distribution. We say that the X_k are **independent** and **identically distributed**.

The following data set describes the scores of eleven students in two exams. The maximum score on each exam was 10. We will use this data set to explain the definitions in this subsection.

Student	A	B	C	D	E	F	G	H	I	J	K
exam 1	6	4	9	7	8	7	8	5	7	8	7
exam 2	8	6	8	7	7	–	9	6	7	8	8

One way to summarize data is to give the number of times a certain category (in this case, exam score) occurs. These numbers are called frequencies. If we divide the frequencies by the total number of observations, we obtain relative frequencies. The list of (relative) frequencies is called a (relative) frequency distribution. (We encountered frequency distributions in Section 8.2.)

To obtain the frequency distribution of (say) exam 1, we simply count the number of times each score appears. For instance, score 6 appears once, so its frequency is 1; score 7 appears four times, so its frequency is 4; and so on. If we wanted relative frequencies, we would then divide each frequency by 11, because eleven students took the first exam.

The frequencies of the exam scores appear in the table below. (Scores that are not listed in the table have frequency 0.)

Score	4	5	6	7	8	9
exam 1	1	1	1	4	3	1
exam 2	0	0	2	3	4	1

A quantity that is computed from the observations in the sample is called a **statistic**.

The first statistic that we define is the **sample median**. It is the middle of the observations when we order the data according to size. When the number of observations is odd, there is one data point in the middle of the ordered data. If the number of observations is even, we take the average of the two observations in the middle. The list of the ordered exam data follows:

$$\text{exam 1:} \quad 4, 5, 6, 7, 7, \mathbf{7}, 7, 8, 8, 8, 9$$
$$\text{exam 2:} \quad 6, 6, 7, 7, \mathbf{7, 8}, 8, 8, 8, 9$$

Eleven students took the first exam; after ordering the scores, the sixth score is the median, which is 7. Ten students took the second exam; after ordering the scores, the average of the fifth and the sixth score is the median, which is $\frac{7+8}{2} = 7.5$.

The two most important statistics that are used to summarize data are the **sample mean** and the **sample variance**. To define these quantities, we recall that we denoted a sample of size n by the vector $(X_1, X_2, \ldots, X_n)$, where X_k is the kth observation. The sample mean and the sample variance are defined as follows.

$$\text{sample mean:}\quad \bar{X} = \frac{1}{n} \sum_{k=1}^{n} X_k$$

$$\text{sample variance:}\quad S^2 = \frac{1}{n-1} \sum_{k=1}^{n} (X_k - \bar{X})^2$$

The sample mean is thus the arithmetic average of the observations. The sample variance is the sum of the squared deviations from the sample mean, divided by $n-1$. (We will explain in the next subsection why we divide by $n - 1$ rather than n.)

This definition of the sample variance is not very convenient for computation. There is an alternate form that is typically easier to use. It follows from an algebraic manipulation of the definition of the sample variance:

$$S^2 = \frac{1}{n-1} \sum_{k=1}^{n} (X_k - \bar{X})^2 = \frac{1}{n-1} \sum_{k=1}^{n} \left(X_k^2 - 2X_k\bar{X} + \bar{X}^2 \right)$$

$$= \frac{1}{n-1} \left[\sum_{k=1}^{n} X_k^2 - 2\bar{X} \sum_{k=1}^{n} X_k + n\bar{X}^2 \right]$$

Using the fact that $\sum_{k=1}^{n} X_k = n\bar{X}$, this simplifies to

$$S^2 = \frac{1}{n-1} \left[\sum_{k=1}^{n} X_k^2 - n\bar{X}^2 \right] = \frac{1}{n-1} \left[\sum_{k=1}^{n} X_k^2 - \frac{1}{n} \left(\sum_{k=1}^{n} X_k \right)^2 \right]$$

If the sample is summarized in a frequency distribution, the sample mean and the sample variance have the following definitions: Assume that a sample of size n has l distinct values $x_1, x_2, \ldots, x_l$, where x_k occurs f_k times in the sample. Then the sample mean is given by the formula

$$\bar{X} = \frac{1}{n} \sum_{k=1}^{l} x_k f_k$$

and the sample variance has the form

$$S^2 = \frac{1}{n-1} \left[\sum_{k=1}^{l} x_k^2 f_k - \frac{1}{n} \left(\sum_{k=1}^{l} x_k f_k \right)^2 \right]$$

We compute the sample mean and the sample variance for the data of the first exam. We find

$$\sum_{k=1}^{11} X_k = 76 \qquad \text{and} \qquad \sum_{k=1}^{11} (X_k)^2 = 546$$

hence,

$$\bar{X} = \frac{76}{11} = 6.9$$

and

$$S^2 = \frac{1}{n-1}\left[\sum_{k=1}^{n}X_k^2 - \frac{1}{n}\left(\sum_{k=1}^{n}X_k\right)^2\right]$$

$$= \frac{1}{10}\left(546 - \frac{1}{11}(76)^2\right) = 2.09$$

If we use the frequency distribution to calculate the sample mean and the sample variance, we find the same answers,

$$\bar{X} = \frac{1}{n}\sum_{k=1}^{l}x_k f_k$$

$$= \frac{1}{11}[(4)(1) + (5)(1) + (6)(1) + (7)(4) + (8)(3) + (9)(1)] = \frac{76}{11}$$

and

$$S^2 = \frac{1}{n-1}\left[\sum_{k=1}^{l}x_k^2 f_k - \frac{1}{n}\left(\sum_{k=1}^{l}x_k f_k\right)^2\right]$$

$$= \frac{1}{10}\left[(16)(1) + (25)(1) + (36)(1) + (49)(4) + (64)(3) + (81)(1)\right.$$

$$\left. - \frac{1}{11}(76)^2\right]$$

$$= \frac{1}{10}\left[546 - \frac{1}{11}(76^2)\right] = 2.09$$

12.6.2 Estimating Means and Proportions

We saw in Section 12.4 that the mean and the variance are useful parameters for describing the probability distribution of a population. If we wish to learn something about the distribution of some quantity from a random sample of the population, we might want to know what the mean and variance of the population distribution are. We will use the sample mean and the sample variance to estimate the mean and the variance of the population.

Any statistic computed from a sample will vary from sample to sample, since the samples are random subsets of the population. Statistics are therefore random variables with their own probability distributions.

We assume that the population distribution has finite mean μ and finite variance σ^2. These two parameters are unknown to us, and we wish to estimate them by taking a random sample $(X_1, X_2, \ldots, X_n)$ of size n from the population. The X_k are independent and identically distributed according to the distribution of the population with

$$EX_k = \mu \quad \text{and} \quad \text{var}(X_k) = \sigma^2 \quad \text{for } k = 1, 2, \ldots, n \qquad (12.23)$$

To estimate the mean and the variance of the population distribution, we will use the sample mean and the sample variance defined in the previous subsection. The sample mean is the arithmetic average

$$\bar{X} = \frac{1}{n}\sum_{k=1}^{n}X_k$$

We stated that statistics are random variables. As such, we can compute their mean and variance. Using (12.23), the mean of the sample mean is

$$E\bar{X} = E\left(\frac{1}{n}\sum_{k=1}^{n} X_k\right) = \frac{1}{n}\sum_{k=1}^{n} EX_k = \frac{1}{n}(n\mu) = \mu$$

Using independence of the observations in addition to (12.23), the variance of the sample mean is

$$\mathrm{var}(\bar{X}) = \mathrm{var}\left(\frac{1}{n}\sum_{k=1}^{n} X_k\right) = \frac{1}{n^2}\sum_{k=1}^{n} \mathrm{var}(X_k) = \frac{1}{n^2}(n\sigma^2) = \frac{\sigma^2}{n}$$

We see from this that the expected value of the sample mean is equal to the population mean. The spread of the distribution of $\bar{X}$ is described by the variance of $\bar{X}$. Since the variance of $\bar{X}$ becomes smaller as the sample size increases ($\sigma^2/n \to 0$ as $n \to \infty$), we conclude that the sample mean of large samples shows less variation about its mean than the sample mean of small samples. This implies that the larger the sample size, the more accurately the mean of the population can be estimated.

The sample variance is

$$S^2 = \frac{1}{n-1}\sum_{k=1}^{n}(X_k - \bar{X})^2$$

To compute its mean, we need the following identity. For any $c \in \mathbf{R}$,

$$\sum_{k=1}^{n}(X_k - c)^2 = \sum_{k=1}^{n}(X_k - \bar{X})^2 + n(\bar{X} - c)^2 \tag{12.24}$$

To see why this is true, we expand the left-hand side of (12.24), after adding $0 = \bar{X} - \bar{X}$ inside the parantheses:

$$\sum_{k=1}^{n}(X_k - c)^2 = \sum_{k=1}^{n}(X_k - \bar{X} + \bar{X} - c)^2$$

$$= \sum_{k=1}^{n}[(X_k - \bar{X})^2 + 2(X_k - \bar{X})(\bar{X} - c) + (\bar{X} - c)^2]$$

$$= \sum_{k=1}^{n}(X_k - \bar{X})^2 + 2(\bar{X} - c)\sum_{k=1}^{n}(X_k - \bar{X}) + n(\bar{X} - c)^2$$

Since $\sum_{k=1}^{n}(X_k - \bar{X}) = 0$, the middle term is equal to 0, and (12.24) follows.

If we set $c = \mu$ in (12.24), and rearrange the equation, then

$$\sum_{k=1}^{n}(X_k - \bar{X})^2 = \sum_{k=1}^{n}(X_k - \mu)^2 - n(\bar{X} - \mu)^2$$

Taking expectations on both sides, and using on the right-hand side the fact that the expectation of a sum is the sum of the expectations, we find

$$E\left(\sum_{k=1}^{n}(X_k - \bar{X})^2\right) = \sum_{k=1}^{n} E(X_k - \mu)^2 - nE(\bar{X} - \mu)^2$$

Now, $\sum_{k=1}^{n}(X_k - \bar{X})^2 = (n-1)S^2$, $E(X_k - \mu)^2 = \sigma^2$, and $E(\bar{X} - \mu)^2 = \mathrm{var}(\bar{X}) = \frac{1}{n}\sigma^2$. Hence,

$$(n-1)ES^2 = n\sigma^2 - \sigma^2 = (n-1)\sigma^2$$

and, therefore,

$$ES^2 = \sigma^2$$

That is, the expected value of the sample variance is equal to the variance of the population. This is the reason that we divided by $n-1$ instead of n when computing the sample variance. (We will not compute the variance of the sample variance; it is given by a complicated formula. One can show, however, that the variance of the sample variance goes to 0 as the sample size becomes infinite.)

Let's look at an example to illustrate how we would estimate the mean and the variance of a characteristic of a population.

▌**EXAMPLE 1**   Suppose that a computer generates the following sample of independent observations from a population.

$$0.0201, 0.8918, 0.9619, 0.1713, 0.0357,$$
$$0.6325, 0.4276, 0.2517, 0.2330, 0.6754$$

Estimate the mean and the variance.

Solution   To estimate the mean, we compute the sample mean $\bar{X}$. We sum the ten numbers in the sample and divide the result by 10, which yields

$$\bar{X} = 0.4301$$

Thus, our estimate for the mean is 0.4301.

To estimate the variance, we compute the sample variance S^2. We square the difference between each sample point and the sample mean and add the results. We then divide this number by 9. We find

$$S^2 = 0.1176$$

Thus, our estimate for the variance is 0.1176.   ▌

In the preceding example, we estimated the population mean and the population variance from a sample of size 10. Since we don't know the population parameters, we have no idea how good our estimates are. How can we assess the quality of our estimates?

Confidence Intervals   In scientific publications, one frequently finds the sample mean reported under a heading of the form "Mean ± S.E." The expression "Mean" stands for the sample mean $\bar{X}$. The expression "S.E." stands for the **standard error**, which serves as an estimate for the standard deviation of the sample mean. It is denoted by $S_{\bar{X}}$ and defined as

$$S_{\bar{X}} = \frac{S}{\sqrt{n}} \tag{12.25}$$

where S is the square root of the sample variance, called the **sample standard deviation**. This definition is motivated by the following consideration. (Again, we assume that the mean and the variance of the population distribution are finite.) To estimate the population mean μ, we use the sample mean $\bar{X}$. The variance of $\bar{X}$ gives us an idea how much the distribution of $\bar{X}$ varies. Now,

$$\text{var}(\bar{X}) = \text{var}\left(\frac{1}{n}\sum_{k=1}^{n} X_k\right) = \frac{1}{n^2}\text{var}\left(\sum_{k=1}^{n} X_k\right)$$

Since the X_k are independent, the variance of the sum is the sum of the variances; in addition, all X_k have the same distribution. Hence

$$\text{var}(\bar{X}) = \frac{1}{n^2} n\sigma^2 = \frac{\sigma^2}{n} \tag{12.26}$$

The variance of $\bar{X}$ thus depends on another population parameter, namely, the variance σ^2. In problems where we wish to estimate the mean, we typically don't know the variance either. If the sample size is large, however, the sample variance will be close to the population variance. We can therefore approximate the variance of $\bar{X}$ by replacing σ^2 in (12.26) by S^2, which yields S^2/n. The standard error is then the square root of this expression.

Here is an example that illustrates how to determine a sample mean and its standard error.

▮ EXAMPLE 2   In a sample of six leaves from a morning glory plant that is infested with aphids, the following numbers of aphids per leaf are found: 12, 27, 17, 35, 14, and 18. Find the sample mean, the sample variance, and the standard error.

Solution   The sample size is $n = 6$. We construct the following table:

$\sum_{k=1}^{n} X_k$	$\sum_{k=1}^{n} X_k^2$	$\bar{X}$	S^2	$\frac{S}{\sqrt{n}}$
123	2907	20.5	77.1	3.58

where we used the formulas

$$\bar{X} = \frac{1}{n} \sum_{k=1}^{n} X_k \quad \text{and} \quad S^2 = \frac{1}{n-1} \left[\sum_{k=1}^{n} X_k^2 - \frac{1}{n} \left(\sum_{k=1}^{n} X_k \right)^2 \right]$$

with $n = 6$. If we wanted to report this result as "Mean ± S.E.", we would then write 20.5 ± 3.58.   ▮

What does "Mean ± S.E." mean? When we write "Mean ± S.E.", we specify an interval, namely, [Mean − S.E., Mean + S.E.]. Since we use "Mean" $(= \bar{X})$ as an estimate for the population mean μ, we would like this interval to contain μ. Surely, since $\bar{X}$ is a random variable, if we took repeated samples and computed such intervals for each sample, not *all* the intervals would contain μ. But maybe we can at least find out what fraction of these intervals (or similar intervals) contain the population mean μ. In other words, we might wish to know before taking the sample what the probability is that the interval [Mean − S.E., Mean + S.E.] or more generally, [Mean − aS.E., Mean + aS.E.], where a is a positive constant, will contain μ. To be concrete, we will try to determine a so that this probability is equal to 0.95.

If the sample size n is very large, one can show that

$$\frac{\bar{X} - \mu}{\sigma/\sqrt{n}}$$

is approximately standard normally distributed. This follows from the central limit theorem, whose importance we mentioned. If Z is standard normally distributed, then

$$P(-1.96 \le Z \le 1.96) = 0.95$$

Hence, the event

$$-1.96 \le \frac{\bar{X} - \mu}{\sigma/\sqrt{n}} \le 1.96 \qquad (12.27)$$

has probability approximately 0.95 for n sufficiently large.

Rearranging the terms in (12.27), we find

$$-1.96 \frac{\sigma}{\sqrt{n}} \le \bar{X} - \mu \le 1.96 \frac{\sigma}{\sqrt{n}}$$

or

$$\bar{X} - 1.96 \frac{\sigma}{\sqrt{n}} \le \mu \le \bar{X} + 1.96 \frac{\sigma}{\sqrt{n}}$$

We can thus write

$$P\left(\bar{X} - 1.96 \frac{\sigma}{\sqrt{n}} \le \mu \le \bar{X} + 1.96 \frac{\sigma}{\sqrt{n}}\right) \approx 0.95 \quad \text{for large } n \qquad (12.28)$$

This means that if we repeatedly draw random samples of size n from a population with mean μ and standard deviation σ, then in about 95% of the samples, the interval $[\bar{X} - 1.96 \frac{\sigma}{\sqrt{n}}, \bar{X} + 1.96 \frac{\sigma}{\sqrt{n}}]$ would contain the true mean μ. Such an interval is referred to as a **95% confidence interval**.

Notice that this interval contains the parameter σ. If we do not know μ, we probably do not know σ either. We might then wish to replace σ by the square root of the sample variance, denoted by S. Fortunately, when n is large, S will be very close to σ and (12.28) holds approximately when σ is replaced by S.

We thus find that for large n,

$$P\left(\bar{X} - 1.96 \frac{S}{\sqrt{n}} \le \mu \le \bar{X} + 1.96 \frac{S}{\sqrt{n}}\right) \approx 0.95$$

or, rewriting the event in interval notation,

$$P\left(\mu \in \left[\bar{X} - 1.96 \frac{S}{\sqrt{n}}, \bar{X} + 1.96 \frac{S}{\sqrt{n}}\right]\right) \approx 0.95$$

The interval in this expression is of the form

$$[\text{Mean} - 1.96 \text{S.E.}, \text{Mean} + 1.96 \text{S.E.}] \qquad (12.29)$$

We thus succeeded in determining what fraction of intervals of the form (12.29) contain the mean. Namely, if n is large, and we take repeated samples each of size n, then 95% of the intervals $[\text{Mean} - 1.96 \text{S.E.}, \text{Mean} + 1.96 \text{S.E.}]$ would contain the true mean. (Or, equivalently, before we take the sample, with probability 0.95 the interval in (12.29) will contain the actual value of μ.) If we wanted 99% of such intervals to contain the true mean, we would need to replace the factor 1.96 by 2.58, since if Z is standard normally distributed, then $P(-2.58 \le Z \le 2.58) = 0.99$ (95% and 99% are the most frequently used percentages for confidence intervals). Since $P(-1 \le Z \le 1) = 0.68$, we conclude that if the sample size is large, approximately 68% of intervals of the form $[\text{Mean} - \text{S.E.}, \text{Mean} + \text{S.E.}]$ contain the population mean μ.

We wish to emphasize that the preceding discussion requires that the sample size n is large. Whether or not a particular n can be considered large depends on the population distribution. If the X_k themselves are normally distributed with mean μ and standard deviation σ, then one can show that $\sqrt{n}(\bar{X} - \mu)/\sigma$ is standard normally distributed for *all* n. When we replace σ by S in (12.27) in this case, then n is considered large for $n \ge 40$. That is, for $n \ge 40$, $\sqrt{n}(\bar{X} - \mu)/S$ is then approximately standard normally distributed. The more the distribution of the X_k

differs from the normal distribution, the larger n needs to be to use the normal distribution as an approximation.

Estimating Proportions   Of particular interest is the case of estimating proportions (for instance, the proportion of white-flowering pea plants in a $Cc \times Cc$ cross). If we take a random sample of size n from a population, where a proportion p has a certain characteristic, then the number of observations in the sample with this characteristic is binomially distributed with parameters p and n. If we denote this quantity by S_n, then

$$P(S_n = k) = \binom{n}{k} p^k (1 - p)^{n-k} \quad \text{for } k = 0, 1, 2, \dots, n$$

We consider each single observation in the sample as a success or a failure, depending on whether the observation has the characteristic under investigation. We set

$$X_k = \begin{cases} 1 & \text{if the } k\text{th observation is a success} \\ 0 & \text{if the } k\text{th observation is a failure} \end{cases}$$

Then $S_n = \sum_{k=1}^{n} X_k$ is the total number of successes in the sample, and p is the success probability. The sample mean $\bar{X} = \frac{1}{n} \sum_{k=1}^{n} X_k$ is then the fraction of successes in the sample. We find

$$E\bar{X} = \frac{1}{n} \sum_{k=1}^{n} EX_k = \frac{1}{n}(np) = p$$

and

$$\text{var}(\bar{X}) = \text{var}\left(\frac{1}{n} \sum_{k=1}^{n} X_k \right) = \frac{1}{n^2} n(\text{var}(X_1))$$

$$= \frac{np(1 - p)}{n^2} = \frac{p(1 - p)}{n}$$

The sample mean $\bar{X}$ will serve as an estimate for the success probability p. We denote the estimate for p by $\hat{p}$ (read: "p hat"). That is, if we observe k successes in a sample of size n, then

$$\hat{p} = \frac{k}{n}$$

Now,

$$E\hat{p} = p \quad \text{and} \quad \text{var}(\hat{p}) = \frac{p(1 - p)}{n}$$

To find the standard error in the case of k successes in a sample of size n, note that

$$S^2 = \frac{1}{n - 1} \left[\sum_{j=1}^{n} X_j^2 - \frac{1}{n} \left(\sum_{j=1}^{n} X_j \right)^2 \right]$$

$$= \frac{1}{n - 1} \left(k - \frac{k^2}{n} \right)$$

$$= \frac{n}{n - 1} \frac{k}{n} \left(1 - \frac{k}{n} \right) = \frac{n\hat{p}(1 - \hat{p})}{n - 1}$$

The standard error for a binomial parameter p is therefore

$$S_{\hat{p}} = \sqrt{\frac{\hat{p}(1-\hat{p})}{n-1}} \tag{12.30}$$

In the literature, you will typically find the standard error for proportions as

$$S_{\hat{p}} = \sqrt{\frac{\hat{p}(1-\hat{p})}{n}} \tag{12.31}$$

For n large, the two numbers are very close, so that it does not matter which you use.

As an example, we consider Mendel's experiment of crossing pea plants.

EXAMPLE 3   To estimate the probability of white-flowering pea plants in a $Cc \times Cc$ cross, Mendel randomly crossed red-flowering pea plants of genotype Cc. He obtained 705 plants with red flowers and 224 plants with white flowers. Estimate the probability of a white-flowering pea plant in a $Cc \times Cc$ cross, and give a 95% confidence interval.

Solution   The sample mean is

$$\bar{X} = \frac{224}{224 + 705} = 0.24$$

The estimate for the probability of a white-flowering pea plant in a $Cc \times Cc$ cross is therefore $\hat{p} = 0.24$. The standard error is

$$\text{S.E.} = \sqrt{\frac{\hat{p}(1-\hat{p})}{n}} = \sqrt{\frac{(0.24)(0.76)}{705 + 224}} = 0.014$$

We can thus report the result as 0.24 ± 0.014. Since n is large, we find a 95% confidence interval as

$$[0.24 - (1.96)(0.014), 0.24 + (1.96)(0.014)] = [0.213, 0.267]$$

(We know from the laws of inheritance that the expected value of $\bar{X}$ is $p = 0.25$, which is contained in the confidence interval.)

12.6.3 Linear Regression

In textbooks, or in scientific literature, you will frequently see plots where a straight line has been fitted to data (as shown in Figure 12.33). The quantities on the horizontal and the vertical axes are linearly related, and a linear model is used to describe the

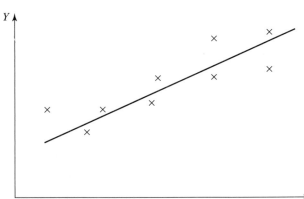

◄ **Figure 12.33**

relationship. We denote the quantity on the horizontal axis by x and the quantity on the vertical axis by Y. We think of x as a particular treatment, which is under control of the experimenter, and of Y as the response. When taking measurements of Y, errors are typically present, so that the data points will not lie exactly on the straight line (even if the linear model is correct) but will be scattered around it, that is, Y is not completely determined by x. The degree of scatter is an indication of how much random variation there is. A linear model can help in separating the random variation from the actual relationship between the two quantities. We will discuss one particular model in the following.

We assume that x is under the control of the experimenter to the extent that it can be measured without error. The response Y, however, shows random variation. We use the following linear model

$$Y = a + bx + \epsilon$$

where ϵ is a normal random variable, representing the **error**, with mean 0 and standard deviation σ. The standard deviation of the error does not depend on x, and is thus the same for all values of x.

Our goal is to estimate a and b from data. The data consist of points (x_i, y_i), $i = 1, 2, \ldots, n$. The approach will be to choose a and b so that the sum of the squared deviations

$$h(a, b) = \sum_{k=1}^{n} [y_k - (a + bx_k)]^2$$

is minimized. The deviations $y_k - (a + bx_k)$ are called **residuals**. The procedure of finding a and b is called the **method of least squares** and is illustrated in Figure 12.34. The resulting straight line is called the **least square line** (or **linear regression line**).

▌EXAMPLE 4   Given the three points $(0, 2)$, $(1, 0)$, and $(2, 1)$, use the method of least squares to find the least square line.

Solution   We wish to find a straight line of the form $y = a + bx$. For given values of a and b, the residuals are

$$2 - (a + 0b) \qquad 0 - (a + b) \qquad 1 - (a + 2b)$$

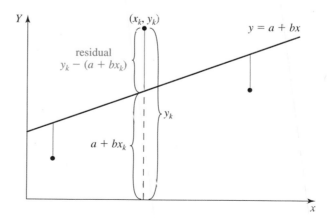

▲ **Figure 12.34**
The line $Y = a + bx$ is chosen so that the sum of the squared residuals is minimized.

and the sum of their squares is

$$(2 - a)^2 + (a + b)^2 + (1 - a - 2b)^2$$
$$= (4 - 4a + a^2) + (a^2 + 2ab + b^2) + (1 + a^2 + 4b^2 - 2a - 4b + 4ab)$$
$$= 5 - 6a + 3a^2 + 6ab + 5b^2 - 4b$$
$$= (2b^2 + 2b) + (3 + 3b^2 + 3a^2 + 6ab - 6a - 6b) + 2$$
$$= 2(b^2 + b) + 3(1 + b^2 + a^2 + 2ab - 2a - 2b) + 2$$

Grouping the terms in this way allows us to complete the squares; we find that this is then equal to

$$2\left(b + \frac{1}{2}\right)^2 + 3(1 - a - b)^2 + \frac{3}{2} \tag{12.32}$$

(If this looks like magic, don't worry; we will derive a general formula for a and b shortly.) Since (12.32) consists of two squares (plus a constant term), we see that the expression is minimized when the two squares are both equal to 0. We solve

$$b + \frac{1}{2} = 0$$
$$1 - a - b = 0$$

which yields $b = -1/2$ and $a = 3/2$. Therefore, the least square line is of the form

$$y = \frac{3}{2} - \frac{1}{2}x$$

This line together with the three points is shown in Figure 12.35. ▮

We will now derive the general formula for finding a and b. The basic steps will be similar to those in Example 4. We will first rewrite the residuals. We set

$$\bar{x} = \frac{1}{n}\sum_{k=1}^{n} x_k \qquad \text{and} \qquad \bar{y} = \frac{1}{n}\sum_{k=1}^{n} y_k$$

and

$$y_k - (a + bx_k) = (y_k - \bar{y}) + (\bar{y} - a - b\bar{x}) - b(x_k - \bar{x})$$

In the following, we will simply write $\sum$ instead of $\sum_{k=1}^{n}$. If we square the expression

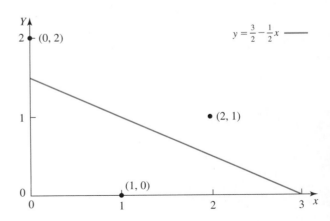

▲ **Figure 12.35**
The least square line together with the three data point, of Example 4.

and sum over k, we find

$$\sum[y_k - (a + bx_k)]^2 = \sum(y_k - \bar{y})^2 + n(\bar{y} - a - b\bar{x})^2$$
$$+ b^2 \sum(x_k - \bar{x})^2 - 2b\sum(x_k - \bar{x})(y_k - \bar{y})$$
$$+ 2(\bar{y} - a - b\bar{x})\sum(y_k - \bar{y})$$
$$- 2b(\bar{y} - a - b\bar{x})\sum(x_k - \bar{x}) \qquad (12.33)$$

The last two terms are equal to 0. We introduce notation to simplify our derivation:

$$SS_{xx} = \sum(x_k - \bar{x})^2 = \sum x_k^2 - \frac{\left(\sum x_k\right)^2}{n}$$

$$SS_{yy} = \sum(y_k - \bar{y})^2 = \sum y_k^2 - \frac{\left(\sum y_k\right)^2}{n}$$

$$SS_{xy} = \sum(x_k - \bar{x})(y_k - \bar{y}) = \sum x_k y_k - \frac{\left(\sum x_k\right)\left(\sum y_k\right)}{n}$$

Using this notation, the right-hand side of (12.33) can be written as

$$SS_{yy} + n(\bar{y} - a - b\bar{x})^2 + b^2 SS_{xx} - 2b SS_{xy}$$

The last two terms suggest that we should complete the square:

$$SS_{yy} + n(\bar{y} - a - b\bar{x})^2 + SS_{xx}\left(b^2 - 2b\frac{SS_{xy}}{SS_{xx}} + \left(\frac{SS_{xy}}{SS_{xx}}\right)^2\right) - \frac{(SS_{xy})^2}{SS_{xx}}$$

$$= n(\bar{y} - a - b\bar{x})^2 + SS_{xx}\left(b - \frac{SS_{xy}}{SS_{xx}}\right)^2 + SS_{yy} - \frac{(SS_{xy})^2}{SS_{xx}}$$

As in Example 4, we succeeded in writing the sum of the squared deviations as a sum of two squares plus an additional term. We can minimize the expression by setting each squared expression equal to 0:

$$\bar{y} - a - b\bar{x} = 0$$
$$b - \frac{SS_{xy}}{SS_{xx}} = 0$$

Solving for a and b yields

$$b = \frac{SS_{xy}}{SS_{xx}}$$
$$a = \bar{y} - b\bar{x}$$

These expressions serve as estimates for a and b, denoted by $\hat{a}$ and $\hat{b}$. Summarizing our results, we have the following.

The least square line (or linear regression line) is given by

$$y = \hat{a} + \hat{b}x$$

with

$$\hat{b} = \frac{\sum_{k=1}^{n}(x_k - \bar{x})(y_k - \bar{y})}{\sum_{k=1}^{n}(x_k - \bar{x})^2} \qquad (12.34)$$

$$\hat{a} = \bar{y} - \hat{b}\bar{x} \qquad (12.35)$$

We illustrate finding $\hat{a}$ and $\hat{b}$ in the following example.

▮EXAMPLE 5   Fit a linear regression line through the points

$$(1, 1.62), (2, 3.31), (3, 4.57), (4, 5.42), (5, 6.71)$$

Solution   To facilitate the computation, we construct the following table.

x_k	y_k	$x_k - \bar{x}$	$y_k - \bar{y}$	$(x_k - \bar{x})(y_k - \bar{y})$	$(x_k - \bar{x})^2$
1	1.62	−2	−2.706	5.412	4
2	3.31	−1	−1.016	1.016	1
3	4.57	0	0.244	0	0
4	5.42	1	1.094	1.094	1
5	6.71	2	2.384	4.768	4
$\bar{x} = 3$	$\bar{y} = 4.326$		$\sum(y_k - \bar{y})^2$ $=15.29$	$\sum(x_k - \bar{x})(y_k - \bar{y})$ $=12.29$	$\sum(x_k - \bar{x})^2$ $=10$

Now,

$$\hat{b} = \frac{12.29}{10} = 1.229$$
$$\hat{a} = 4.326 - (1.229)(3) = 0.639$$

hence, the linear regression line is given by

$$y = 1.23x + 0.64$$

This line and the data points are shown in Figure 12.36.   ▮

Now that we know how to fit a straight line to a set of points, we might want to know how good the fit is. To this purpose, we will define a quantity known as the **coefficient of determination**. We motivate its definition as follows. We start with a set of observations (x_k, y_k), $k = 1, 2, \ldots, n$, and assume the linear model $Y = a + bx + \epsilon$. We set

$$\hat{y}_k = \hat{a} + \hat{b}x_k \qquad \text{and} \qquad \bar{y} = \frac{1}{n}\sum_{k=1}^{n} y_k$$

We think of $\hat{y}_k$ as the expected response under the linear model if $x = x_k$. Now, $y_k - \bar{y}$ is the deviation from the observation to the sample mean, $y_k - \hat{y}_k$ is the deviation from the observation to the expected response under the linear model, and $\hat{y}_k - \bar{y}$

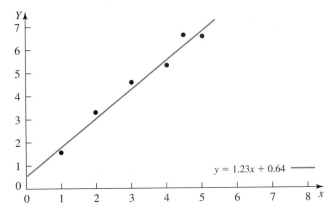

$y = 1.23x + 0.64$ ———

◀ Figure 12.36
The linear regression line and the data points of Example 5.

is the deviation from the expected response under the linear model to the sample mean. The deviation $\hat{y}_k - \bar{y}$ can be thought of as being explained by the model, and the deviation $y_k - \hat{y}_k$ as the unexplained part due to random variation (the stochastic error). We can write

$$y_k - \bar{y} = (\hat{y}_k - \bar{y}) + (y_k - \hat{y}_k) \tag{12.36}$$

If we look at the total sum of the squared deviations, $\sum(y_k - \bar{y})^2$, and use (12.36), we find

$$\begin{aligned}
\sum(y_k - \bar{y})^2 &= \sum[(\hat{y}_k - \bar{y}) + (y_k - \hat{y}_k)]^2 \\
&= \sum(\hat{y}_k - \bar{y})^2 + 2\sum(\hat{y}_k - \bar{y})(y_k - \hat{y}) \\
&\quad + \sum(y_k - \hat{y}_k)^2
\end{aligned} \tag{12.37}$$

We want to show that $\sum(\hat{y}_k - \bar{y})(y_k - \hat{y}) = 0$. To do so, we observe that

$$\hat{y}_k - \bar{y} = (\hat{a} + \hat{b}x_k) - (\hat{a} + \hat{b}\bar{x}) = \hat{b}(x_k - \bar{x})$$

$$y_k - \hat{y}_k = (y_k - \bar{y}) - (\hat{y}_k - \bar{y}) = (y_k - \bar{y}) - \hat{b}(x_k - \bar{x})$$

Therefore,

$$\begin{aligned}
\sum(\hat{y}_k - \bar{y})(y_k - \hat{y}_k) &= \sum \hat{b}(x_k - \bar{x})[(y_k - \bar{y}) - \hat{b}(x_k - \bar{x})] \\
&= \hat{b}\sum(x_k - \bar{x})(y_k - \bar{y}) - \hat{b}^2\sum(x_k - \bar{x})^2
\end{aligned}$$

Using (12.34) for one of the $\hat{b}$ in the term $\hat{b}^2$, this becomes

$$\begin{aligned}
&\sum(\hat{y}_k - \bar{y})(y_k - \hat{y}_k) \\
&= \hat{b}\sum(x_k - \bar{x})(y_k - \bar{y}) - \hat{b}\frac{\sum(x_k - \bar{x})(y_k - \bar{y})}{\sum(x_k - \bar{x})^2}\sum(x_k - \bar{x})^2 \\
&= 0
\end{aligned} \tag{12.38}$$

This allows us to partition the total sum of squares into the explained and the unexplained sums of squares. Namely, continuing with (12.37) and using (12.38), we find

$$\underbrace{\sum(y_k - \bar{y})^2}_{\text{total}} = \underbrace{\sum(\hat{y}_k - \bar{y})^2}_{\text{explained}} + \underbrace{\sum(y_k - \hat{y}_k)^2}_{\text{unexplained}}$$

The ratio

$$\frac{\text{explained}}{\text{total}} = \frac{\sum(\hat{y}_k - \bar{y})^2}{\sum(y_k - \bar{y})^2}$$

is therefore the proportion that is explained by the model. It is denoted by r^2 and called the coefficient of determination. With $\hat{y}_k - \bar{y} = \hat{b}(x_k - \bar{x})$ and $\hat{b}$ given in (12.34), this can be written as

$$\begin{aligned}
r^2 &= (\hat{b})^2 \frac{\sum(x_k - \bar{x})^2}{\sum(y_k - \bar{y})^2} = \left(\frac{\sum(x_k - \bar{x})(y_k - \bar{y})}{\sum(x_k - \bar{x})^2}\right)^2 \frac{\sum(x_k - \bar{x})^2}{\sum(y_k - \bar{y})^2} \\
&= \frac{[\sum(x_k - \bar{x})(y_k - \bar{y})]^2}{\sum(x_k - \bar{x})^2 \sum(y_k - \bar{y})^2}
\end{aligned}$$

We summarize this in the following.

> The coefficient of determination r^2 is given by
> $$r^2 = \frac{[\sum(x_k - \bar{x})(y_k - \bar{y})]^2}{\sum(x_k - \bar{x})^2 \sum(y_k - \bar{y})^2}$$
> and represents the proportion that is explained by the model.

Returning to Example 5, we find

$$r^2 = \frac{(12.29)^2}{(10)(15.29)} = 0.988$$

That is, 98.8% of the variation is explained by the model.

Since r^2 is the ratio of explained to total variation, it follows that $r^2 \leq 1$. Furthermore, since r^2 is the square of an expression, it is always nonnegative. That is, we have

$$0 \leq r^2 \leq 1$$

The closer r^2 is to 1, the more closely the data points follow the straight line resulting from the linear model. In the extreme case, when $r^2 = 1$, all points lie on the line; there is no random variation.

◆ 12.6 PROBLEMS

[12.6.1]

1. The following data represent the number of aphids per plant found in a sample of ten plants:

$$17, 13, 21, 47, 3, 6, 12, 25, 0, 18$$

Find the median, the sample mean, and the sample variance.

2. The following data represent the number of seeds per flower head in a sample of nine flowering plants:

$$27, 39, 42, 18, 21, 33, 45, 37, 21$$

Find the median, the sample mean, and the sample variance.

3. The following data represent the frequency distribution of seed numbers per flower head in a flowering plant.

Seed number	Frequency
9	37
10	48
11	53
12	49
13	61
14	42
15	31

Calculate the sample mean and the sample variance.

4. The following data represent the frequency distribution of the numbers of days that it took a certain ointment to clear up a skin rash.

Number of days	Frequency
1	2
2	7
3	9
4	27
5	11
6	5

Calculate the sample mean and the sample variance.

5. The following data represent the relative frequency distribution of clutch size in a sample of 300 laboratory guinea pigs.

Clutch size	Relative frequency
2	0.05
3	0.09
4	0.12
5	0.19
6	0.23
7	0.12
8	0.13
9	0.07

Calculate the sample mean and the sample variance.

6. The following data represent the relative frequency distribution of clutch size in a sample of 42 mallards.

Clutch size	Relative frequency
5	0.10
7	0.24
8	0.29
9	0.21
10	0.16

Calculate the sample mean and the sample variance.

7. Let $(X_1, X_2, \ldots, X_n)$ denote a sample of size n. Show that

$$\sum_{k=1}^{n}(X_k - \bar{X}) = 0$$

where $\bar{X}$ is the sample mean.

8. Let $(X_1, X_2, \ldots, X_n)$ denote a sample of size n. Show that

$$n\bar{X}^2 = \frac{1}{n}\left(\sum_{k=1}^{n} X_k\right)^2$$

where $\bar{X}$ is the sample mean.

9. Assume that a sample of size n has l distinct values $x_1, x_2, \ldots, x_l$, where x_k occurs f_k times in the sample. Explain why the sample mean is given by the formula

$$\bar{x} = \frac{1}{n}\sum_{k=1}^{l} x_k f_k$$

10. Assume that a sample of size n has l distinct values $x_1, x_2, \ldots, x_l$, where x_k occurs f_k times in the sample. Explain why the sample variance is given by the formula

$$S^2 = \frac{1}{n-1}\left[\sum_{k=1}^{l} x_k^2 f_k - \frac{1}{n}\left(\sum_{k=1}^{l} x_k f_k\right)^2\right]$$

[12.6.2]

11. Use a graphing calculator to generate three samples of size 10 each from a uniform distribution over the interval (0, 1).

(a) Compute the sample mean and the sample variance of each sample.

(b) Combine all three samples and compute the mean and the sample variance of the combined sample.

(c) Compare your answers in (a) and (b) with the true values of the mean and the variance.

12. Suppose that X is exponentially distributed with mean 1. A computer generates the following sample of independent observations from this population.

0.3159, 0.5531, 2.376,

1.150, 0.6174, 0.1563

2.936, 1.778, 0.7357

0.1024

Find the sample mean and the sample variance, and compare with the corresponding population parameters.

13. Compute the sample mean and the standard error for the sample in Problem 1.

14. Compute the sample mean and the standard error for the sample in Problem 2.

15. The following data represent a sample from a normal distribution with mean 0 and variance 1.

$$-0.68, 1.22, 1.33, -0.84$$
$$-0.06, 0.50, 0.03, -0.13$$
$$-0.29, -0.47$$

Construct a 95% confidence interval.

16. The following data represent a sample from a normal distribution with mean 0 and variance 1.

$$-1.18, 0.52, 0.36, -0.16$$
$$0.92, 0.68, -0.61, -0.54$$
$$0.15, 1.04$$

Construct a 95% confidence interval.

17. Use a graphing calculator to construct a 95% confidence interval for a sample of size 30 from a uniform distribution over the interval (0, 1). Take a class poll to determine the percentage of confidence intervals that contain the true mean. Discuss the result in class.

18. (a) If X has distribution function $F(x)$, we can show that $F(X)$ is uniformly distributed over the interval (0, 1). Use this fact, a graphing calculator, and the table for the standard normal distribution to generate fifteen standard normally distributed random variables.

(b) Use your data from (a) to construct a 95% percent confidence interval. Take a class poll to determine the percentage of confidence intervals that contain the true mean. Discuss the result in class.

19. To determine the germination success of seeds of a certain plant, you plant 162 seeds. You find that 117 of the seeds germinate. Estimate the probability of germination and give a 95% confidence interval.

20. To test a new drug for lowering cholesterol, 72 people with elevated cholesterol receive this drug; 51 of them show reduced cholesterol levels. Estimate the probability that this drug lowers cholesterol and give a 95% confidence interval.

[12.6.3]

In Problems 21–22, fit a linear regression line through the given points and compute the coefficient of determination.

21. $(-3, -6.3)$, $(-2, -5.6)$, $(-1, -3.3)$, $(0, 0.1)$, $(1, 1.7)$, $(2, 2.1)$

22. $(0, 0.1)$, $(1, -1.3)$, $(2, -3.5)$, $(3, -5.7)$, $(4, -5.8)$

23. Show that the sum of the residuals about the linear regression line is equal to 0.

24. Show that the last two terms in (12.33),

$$2(\bar{y} - a - b\bar{x})\sum(y_k - \bar{y})$$

and

$$2(\bar{y} - a - b\bar{x}) \sum (x_k - \bar{x})$$

are equal to 0.

25. To determine whether the frequency of chirping crickets depends on temperature, the following data were obtained (Pierce 1949).

Temperature (F)	Chirps/sec
69	15
70	15
72	16
75	16
81	17
82	17
83	16
84	18
89	20
93	20

Fit a linear regression line and compute the coefficient of determination.

26. The initial velocity v of an enzymatic reaction that follows Michaelis-Menten kinetics is given by

$$v = \frac{v_{max}s}{K_m + s} \tag{12.39}$$

where s is the substrate concentration, and v_{max} and K_m are two parameters that characterize the reaction. The following computer-generated table contains values of the initial velocity v when the substrate concentration s was varied.

s	v
1	4.1
2.5	6.1
5	9.3
10	12.9
20	17.1

(a) Invert (12.39) and show that

$$\frac{1}{v} = \frac{K_m}{v_{max}} \frac{1}{s} + \frac{1}{v_{max}} \tag{12.40}$$

This is the Lineweaver-Burk equation. If we plot $1/v$ as a function of $1/s$, a straight line with slope K_m/v_{max} and intercept $1/v_{max}$ results. Transform the data using (12.40), and fit a linear regression line to the transformed data. Find the slope and the intercept of the linear regression line, and determine K_m and v_{max}.

(b) Dowd and Riggs (1965) proposed to use the transformation

$$v = v_{max} - K_m \frac{v}{s} \tag{12.41}$$

and then plot v against v/s. The resulting straight line has slope $-K_m$ and intercept v_{max}. Use (12.41) to transform the data, and fit a linear regression line to the transformed data. Find the slope and the intercept of the linear regression line, and determine K_m and v_{max}.

◆ **12.7 KEY TERMS** _____

Chapter 12 Review: Topics   Discuss the following definitions and concepts:

1. Multiplication Principle
2. Permutation
3. Combination
4. Random experiment
5. Sample space
6. Basic set operations, Venn diagram, De Morgan's laws
7. Definition of probability
8. Equally likely outcomes
9. Mendel's pea experiments
10. Mark-recapture method
11. Maximum likelihood estimate
12. Conditional probability
13. Partition of a sample space
14. Law of total probability
15. Independence
16. Bayes formula
17. Random variable
18. Discrete distribution
19. Probability mass function
20. Distribution function of a discrete random variable
21. Mean and variance
22. Binomial distribution
23. Multinomial distribution
24. Continuous random variable
25. Density function
26. Distribution function of a continuous random variable
27. Uniform distribution
28. Normal distribution
29. Exponential distribution
30. Sample
31. Statistic
32. Sample median, sample mean, sample variance, standard error
33. Confidence interval
34. Estimating proportions
35. Linear regression line
36. Coefficient of determination

◆ 12.8 REVIEW PROBLEMS

1. (a) There are twenty-five students in a calculus recitation. What is the probability that no two students have the same birthday?

(b) Let p_n denote the probability that in a group of n people, no two people have the same birthday. Show that

$$p_1 = 1 \quad \text{and} \quad p_{n+1} = p_n \frac{365 - n}{365}$$

Use this formula to generate a table of p_n for $1 \le n \le 25$.

2. Thirty patients are to be randomly assigned to two different treatment groups. How many ways can this be done?

3. Fifteen different plants are to be equally divided among five plots. How many ways can this be done?

4. Assume that a certain disease is caused by a genetic mutation or appears spontaneously. The disease will appear in 67% of all people with the mutation, and in 23% of all people without the mutation. Assume that 3% of the population carries the disease gene.

(a) What is the probability that a randomly chosen individual will develop the disease?

(b) Given an individual who suffers from the disease, what is the probability that they have the genetic mutation?

5. Suppose that 42% of the seeds of a certain plant germinate.

(a) What is the expected number of germinating seeds in a sample of ten seeds?

(b) You plant ten seeds in one pot. What is the probability that none of the seeds will germinate?

(c) You plant five pots with ten seeds each. What is the expected number of pots with no germinating seeds?

(d) You plant five pots with ten seeds each. What is the probability that at least one pot has no germinating seeds?

6. Suppose that the amount of yearly rainfall in a certain area is normally distributed with mean 27 and standard variation 5.7 (measured in inches).

(a) What is the probability that in a given year the rainfall will exceed 35 inches?

(b) What is the probability that in five consecutive years the rainfall will exceed 35 inches in each year?

(c) What is the probability that in at least one out of ten years the rainfall will exceed 35 inches per year?

7. Suppose that each time a student takes a particular test they have a 20% chance of passing. (Assume that consecutive trials are independent.)

(a) What are the chances of passing the test on the second trial?

(b) Given that a student failed the test the first time, what are the chances of passing the test on the second trial?

8. Explain why

$$2^r = \sum_{k=1}^{n} \binom{n}{k}$$

9. A bag contains on average 170 chocolate covered raisins. Production standards require that in 95% of all bags the number

of raisins does not deviate from 170 by more than 10. Assume that the number of raisins is normally distributed with mean μ and variance σ^2.

(a) Determine μ and σ.

(b) A shipment contains 100 bags. What is the probability that no bag contains less than 160 raisins?

10. Suppose that two parents are carriers of a recessive gene causing a metabolic disorder. Neither parent has the disease. If they have three children, what is the probability that none of them will be afflicted by the disease. (Note that a recessive gene only causes a disorder if an individual has two copies of the gene.)

11. Suppose that you choose a plant from a large batch of red-flowering pea plants and cross it with a white-flowering pea plant. What percentage of the red-flowering parent plants are of genotype Cc if 90% of the offspring have red flowers?

12. Suppose that a random variable is normally distributed with mean μ and variance σ^2. How would you estimate μ and σ?

13. Let $(X_1, X_2, \ldots, X_n)$ be a sample of size n from a population with mean μ and variance σ^2. Define

$$V = \frac{1}{n} \sum_{k=1}^{n} (X_k - \bar{X})^2$$

where $\bar{X}$ is the sample mean.

(a) If S^2 is the sample variance, show that

$$V = \frac{n - 1}{n} S^2$$

(b) Compute EV.

14. Assume that the weight of a certain species is normally distributed with mean μ and variance σ^2. The following data represent the weight (measured in grams) of ten randomly selected individuals from this species.

$$171, 168, 151, 192, 175$$
$$163, 182, 157, 177, 169$$

(a) Find the median, the sample mean, and the sample variance.

(b) Construct a 95% confidence interval for the population mean.

15. (a) Generate five observations (x, y) from a random experiment, where

$$y = 2x + 1 + \epsilon$$

$x = 1, 2, 3, 4, 5$, and ϵ is normally distributed with mean 0 and variance 1.

(b) Use your data from (a) to find the least square line and compare your results to the linear model that describes this experiment.

(c) How much of your data is explained by the model?

Answers to Odd-numbered Solutions

Section 1.1

1. (a) $\{-4, 2\}$

(b) $\{-4, 2\}$

3. (a) $\{-\frac{2}{7}, 2\}$

(b) $\{\frac{2}{5}, 8\}$

(c) $\{-1, 6\}$

5. $x + 3y - 14 = 0$

7. $7x - 3y + 5 = 0$

9. $2y - 3 = 0$

11. $3x - y + 2 = 0$

13. $2x + 5y - 22 = 0$

15. (a) $y = kx$, $[x] = $ ft, $[y] = $ cm, $k = \frac{30.5\text{cm}}{1\text{ft}}$ implies $y = (30.5\frac{\text{cm}}{\text{ft}})x$

(b) (i) 183 cm, (ii) $\frac{1159}{12}$ cm, (iii) $\frac{1159}{24}$ cm

(c) (i) $\frac{346}{61}$ ft, (ii) $\frac{150}{61}$ ft, (iii) $\frac{96}{61}$ ft

17. $\frac{1}{(0.305)^2}$ ft^2

19. (a) $[y] = $ liter, $[x] = $ ounces, $y = \left(\frac{1}{33.81}\frac{\text{liter}}{\text{ounces}}\right)x$

(b) $\frac{12}{33.81}$ liters

21. $(x + 1)^2 + (y - 4)^2 = 9$

23. center: $(2, 0)$; radius: 4

25. (a) $\frac{5}{12}\pi$,

(b) $255°$

27. (a) $\alpha = -\frac{\pi}{3}$ or $\alpha = \frac{4\pi}{3}$

(b) $\alpha = \frac{\pi}{3}$ or $\alpha = \frac{4\pi}{3}$

29. Divide both sides by $\sin^2\theta$.

31. $\{0, \frac{\pi}{3}\}$

33. (a) $\ln 5$

(b) $\ln(x - y) + \ln(x + y) - \frac{1}{2}\ln x$

(c) $2x + 1$

35. (a) $x = 3 + e^5$

(b) $x = \sqrt{4 + e}$

(c) $x = 18$

37. (a) $C = K - 273.15$,

(b) $77.4\text{K} = -195.75°\text{C} = -320.35°\text{F}$, $90.2\text{K} = -182.95°\text{C} = -297.31°\text{F}$; nitrogen gets distilled first since it has the lower boiling point.

Section 1.2

1. range: $y \geq 0$

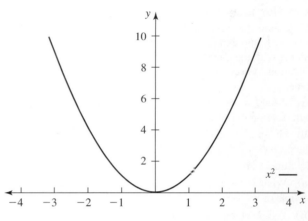

▲ **FIGURE 1.2.1**

3. range: $y \in [0, 1)$

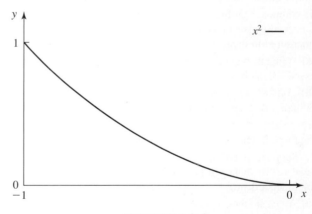

▲ **FIGURE 1.2.3**

5. (b) No, their domains are different.

7. $f(x)$ is odd. $f(-x) = -2x - f(x)$

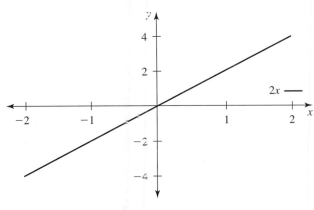

▲ **FIGURE 1.2.7**

9. $f(x)$ is even. $f(-x) = |3(-x)| = |3x| = f(x)$

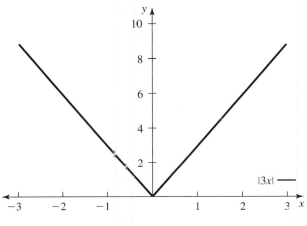

▲ **FIGURE 1.2.9**

11. $f(x)$ is even. $f(-x) = -|-x| = -|x| = f(x)$

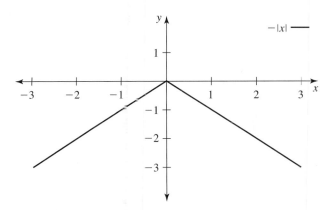

▲ **FIGURE 1.2.11**

15. **(a)** $(f \circ g)(x) = \frac{1}{2x^2+1}$, $x \in \mathbf{R}$

(b) $(g \circ f)(x) = 2\left(\frac{1}{+}\right)^2$, $x \neq -1$

17. $x^2 > x^4$ for $0 < x < 1$; $x^2 < x^4$ for $x > 1$

▲ **FIGURE 1.2.17**

19. They intersect at $x = 0$ or 1.

▲ **FIGURE 1.2.19**

21. **(a)**

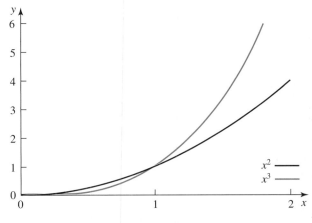

▲ **FIGURE 1.2.21**

23. **(a)** $f(-x) = f(x)$ **(b)** $f(-x) = -f(x)$

25. **(a)** $k = \frac{3}{2}$ **(b)** domain: $0 \le x \le 3$

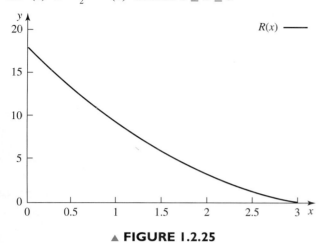

▲ **FIGURE 1.2.25**

27. $s(t) = t$, polynomial of degree 1

29. domain: $x \ne 1$; range: $y \ne 0$

31. domain: $x \ne -3, 3$; range: **R**

33. $\frac{1}{x} < \frac{1}{x^2}$ for $0 < x < 1$; $\frac{1}{x} > \frac{1}{x^2}$ for $x > 1$; they intersect at $x = 1$.

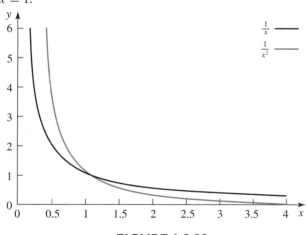

▲ **FIGURE 1.2.33**

35. **(a)**

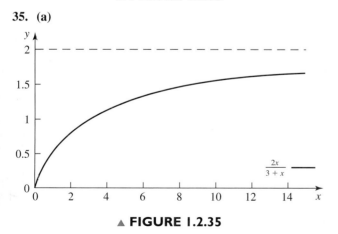

▲ **FIGURE 1.2.35**

(c) $x = 3$

37. 83.3%, 4.76%

39.

▲ **FIGURE 1.2.39**

41.

▲ **FIGURE 1.2.41**

43. **(a)**

▲ **FIGURE 1.2.43**

45. increases   **47.** increases

49. **(a)** 1,2,4,8,16

(b)

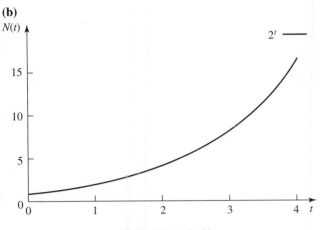

▲ FIGURE 1.2.49

51. $20 \exp\left[-\frac{\ln 2}{5730} 2000\right]$ **53.** $\lambda = \frac{\ln 2}{7\,\text{days}}$

55. (a) $W(t) = (300\,\text{gr})\exp\left[-\frac{\ln 2}{140\,\text{days}}t\right]$

(b) $t = \frac{\ln 5}{\ln 2} 140\,\text{days} \approx 325\,\text{days}$

(c)

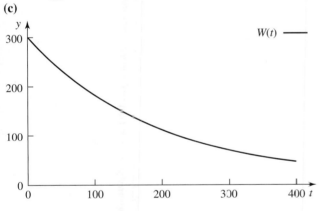

▲ FIGURE 1.2.55

57. $\frac{W(t)}{W(0)} = \exp\left[-\frac{\ln 2}{5730} 15{,}000\right] \approx 16.3\%$

59. (a) yes   **(b)** no   **(c)** yes   **(d)** yes   **(e)** no

61. (a) $f^{-1}(x) = \sqrt{x-1},\ x \geq 1$

(b)

$x^2 + 1$ —
$\sqrt{x-1}$ —
x – –

▲ FIGURE 1.2.61

63. (a) $\exp\left[(x^2 - 1)\ln 2\right]$

(b) $\exp[5x \ln 3]$ **(c)** $\frac{\ln(x-2)}{\ln 10}$ **(d)** $\frac{\ln(2x^2-1)}{\ln 2}$

65. (b) $f^{-1}(x) = \log_2 x,\ x > 0$

(c)

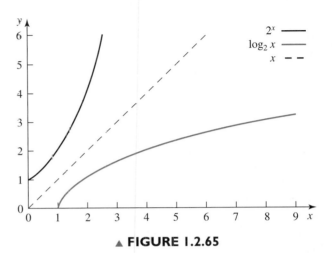

2^x —
$\log_2 x$ —
x – –

▲ FIGURE 1.2.65

67. Same period; $2\sin x$ has twice the amplitude of $\sin x$.

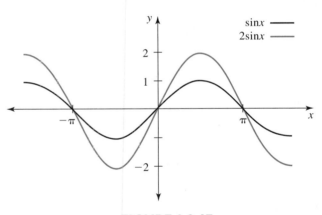

$\sin x$ —
$2\sin x$ —

▲ FIGURE 1.2.67

69. Same period; $2\cos x$ has twice the amplitude of $\cos x$.

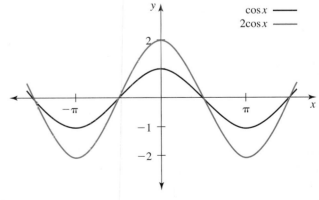

$\cos x$ —
$2\cos x$ —

▲ FIGURE 1.2.69

71. Same period; $y = 2 \tan x$ is stretched by a factor of 2.

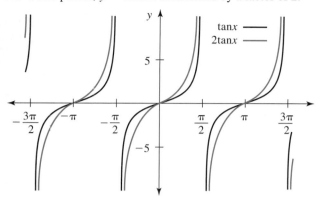

▲ **FIGURE 1.2.71**

73. (a)

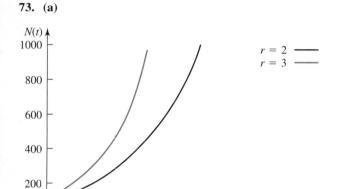

▲ **FIGURE 1.2.73**

The population with $r = 3$ grows faster.
(b) $r = \ln \frac{250}{200} \approx 0.223$

Section 1.3
1.

▲ **FIGURE 1.3.1**

3.

▲ **FIGURE 1.3.3**

5.

▲ **FIGURE 1.3.5**

7.

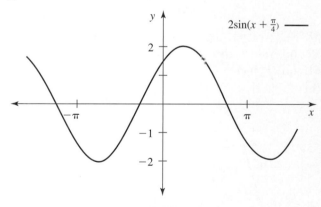

▲ **FIGURE 1.3.7**

9. (a) Shift two units down. **(b)** Shift $y = x^2$ one unit to the right and then one unit up. **(c)** Shift $y = x^2$ two units to the left, stretch by a factor of 2, and reflect about the x-axis.
13. $y = 10^{0.89} x^{0.37}$ **15.** $y = 10^{-2.3} x^{0.69}$
17. $y = 10^{2.3} 10^{-1.73x}$ **19.** $y = 10^{-.05} 10^{0.18x}$

21. $\log y = \log 3 + 1.7 \log x$, log-log transformation

23. $\log N(t) = \log 130 + (0.2t) \log 2$, log-linear transformation

25. $\log R(t) = \log 3.6 + 1.2 \log t$, log-log transformation

27. (a) $\log N = \log 2 + 3t \log e$ **(b)** slope: $3 \log e \approx 1.303$

29. $\log S = \log C + z \log A$, $z =$ slope of straight line

31. $v_{max} =$ horizontal line intercept, $\frac{v_{max}}{K_m} =$ vertical line intercept

33. (a) $\log y = \log 1.162 + 0.933 \log x$

Section 1.5

1. (a) $10^4, 1.1 \times 10^4, 1.22 \times 10^4, 1.35 \times 10^4, 1.49 \times 10^4$
(b) $t = 10 \ln 10 \approx 23.0$

3. (b) $R(x) = -4kx^3 - 4k(a+b)x^2 - kb(4a-b)x + kab^2$, polynomial of degree 3

(c) $R(x) = (0.3)(5-x)(5-2x)^2, 0 \le x \le 3$

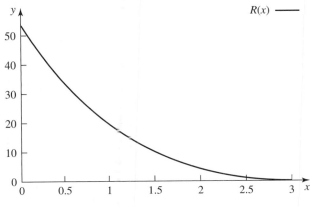

▲ **FIGURE 1.5.3**

5. (a) $L(t) = 0.69, 2.40, 4.62, 6.91, E(t) = 1.72, 2.20 \times 10^4, 2.69 \times 10^{43}, 10^{434}$ **(b)** 20.93 years, 3.13 ft

(c) $10^{536,000,000}$ years, 21.98 ft **(d)** $L = 19.92$ ft, $E = 10^{195 \times 10^6}$ ft

7. $T = \frac{\ln 2}{\ln(1 + \frac{q}{100})}$, T goes to infinity as q gets closer to 0.

9. (a)

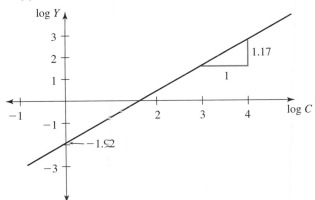

▲ **FIGURE 1.5.9**

(b) $Y = C^{1.17} 10^{-1.52}$ **(c)** $Y_p = 2.25 Y_c$ **(d)** 8.5%

11. (a) 400 days per year

(b) $y = 4.32 \times 10^9 - 1.8 \times 10^8 x$

(c) 376×10^6 to 563×10^6 years ago

13. (a) males: $S(t) = \exp[-(0.019t)^{3.41}]$; females: $S(t) = \exp[-(0.022t)^{3.24}]$

(b) males: 47.27 days; females: 40.59 days

(c) males should live longer

15. (a) $x = k, v = \frac{a}{2}$

(b) $x_{0.9} = 81x_{0.1}$

Section 2.1

1. -1 **3.** $-\frac{1}{2}$

5. $\frac{3}{2}\sqrt{2}$ **7.** $\frac{4}{3}\sqrt{3}$

9. $e^{-1/2}$ **11.** 0

13. 7 **15.** 0

17. 1 **19.** 0

21. $-\infty$ **23.** ∞

25. ∞ **27.** ∞

29. $\lim_{x \to \infty} f(x) = 0$; $\lim_{x \to -\infty} f(x) = 0$; $y = 0$ is a horizontal asymptote; $x = 0$ is a vertical tangent.

31. divergence by oscillations

33. $0.495 < x < 0.505$

35. (b) $1.995 < x < 2.005$

37. (a)

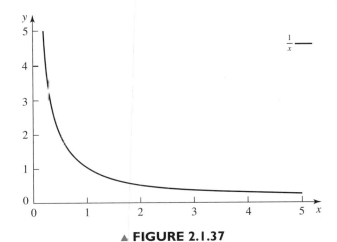

▲ **FIGURE 2.1.37**

(b) $0 < x < \frac{1}{4}$

39. $\hat{\delta} = \frac{\epsilon}{2}$

41. $0 < |x| < M^{-1/4}$ implies $\frac{1}{x^4} > M$

43. -9 **45.** $\frac{47}{2}$

47. $\frac{28}{3}$ **49.** 2

51. 4 **53.** -5

55. 1 **57.** 2 **59.** 2

61. $\frac{1}{6}$ **63.** $\frac{1}{2}$

718

Section 2.2

1. (a)

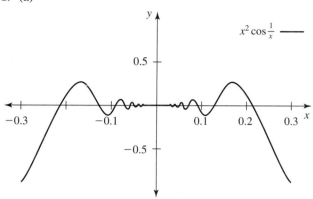

▲ **FIGURE 2.2.1**

3. (a)

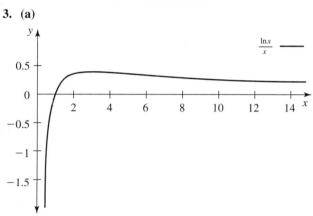

▲ **FIGURE 2.2.3**

(b) $x \geq e$ **5.** 1   **7.** 5   **9.** π
11. 0   **13.** 1   **15.** 0   **17.** 0

Section 2.3

1. 0   **3.** ∞ **5.** 3   **7.** ∞ **9.** 4
11. 2   **13.** 0   **15.** 0   **17.** a **19. (a)**

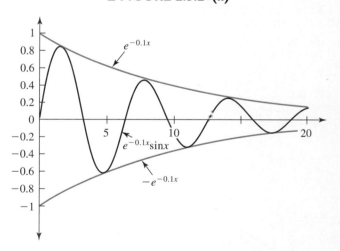

▲ **FIGURE 2.3.19**

(b) 500   **(c)** 250

21. (a)

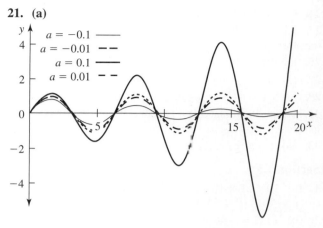

▲ **FIGURE 2.3.21(i)**

(b) The sine function. **(c)** $\lim_{x \to \infty} e^{ax} \sin x = 0$ for $a < 0$;
$\lim_{x \to \infty} e^{ax} \sin x$ does not exist for $a \geq 0$

(d)

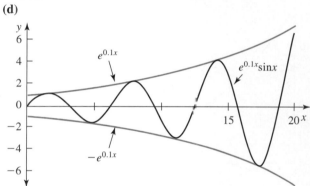

▲ **FIGURE 2.3.21(ii)**

▲ **FIGURE 2.3.21(iii)**

Section 2.4

5. $f(2) = 3$
7. $a = 6$
9. $\lim_{x \to 3} \frac{1}{x-3}$ does not exist

11. $f(1) \neq \lim_{x \to 1} f(x)$

15. (b)

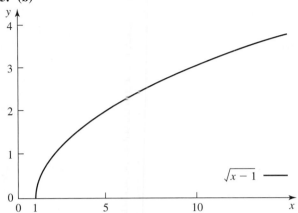

▲ **FIGURE 2.4.15**

(c) No   **17.** $x \in \mathbf{R}$ **19.** $x \neq 1$

21. $x \in \mathbf{R}$ **23.** $\{x : x < -1 \text{ or } x > 0\}$

25. $\{x \in \mathbf{R} : x \neq \frac{1}{4} + \frac{k}{2}, k \in \mathbf{Z}\}$

27. (a) $f(x)$ is not continuous at $x = 0$

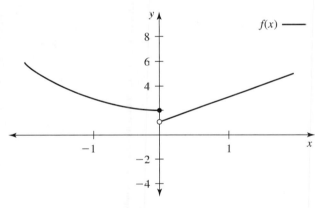

▲ **FIGURE 2.4.27**

(b) $c = 2$

Section 2.5

1. absolute minimum: $(0, -1)$; absolute maximum: $(1, 1)$

3. absolute minimum: $(-\frac{3\pi}{2}, -1)$; absolute maximum: $(\frac{\pi}{2}, 1)$

5. absolute minimum: $(0, 0)$; absolute maximum: $(-1, 1)$ and $(1, 1)$

7. absolute minimum: $(2, e^{-2})$; absolute maximum: $(0, 1)$

9. (b) $f(0) = -1$, $f(2) = 3$

11. (b) $f(1) = \sqrt{3} < 2 < f(2) = \sqrt{6}$

13. $f(0) = 1 > 0$, $f(1) = e^{-1} - 1 < 0$ **15.** $x \approx 0.57$

17. (a) $x \approx -0.67$ **(c)** No

19. (a) $N(10) = 23$ **(b)** $N(10) = 23.5$ (in million)

Section 2.7

1. $x \in \mathbf{R}$ **3.** $x \in \mathbf{R}$

5.

▲ **FIGURE 2.7.5**

7.

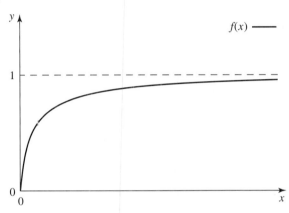

▲ **FIGURE 2.7.7**

9. $f(-2) = -2, \lim_{x \to -2^+} f(x) = -2, \lim_{x \to -2^-} f(x) = -3$

11. $a = 1.24 \times 10^6, k = 5$

13. (a) $g(t) = \begin{cases} 1 & \text{for } \frac{1}{6} + 2k \leq x \leq \frac{5}{6} + 2k, k = 0, 1, 2, \ldots \\ 0 & \text{otherwise} \end{cases}$

(b) $s(t)$ is continuous, $g(t)$ is not continuous

15. (c) $T_h = 0$: $g(N) = aTN = f(N)$

(d) $\lim_{N \to \infty} (aTN) = \infty$; $\lim_{N \to \infty} \frac{aTN}{1 + aT_hN} = \frac{T}{T_h}$

17. (a)

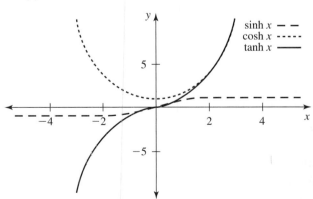

▲ **FIGURE 2.7.17**

(b) $\lim_{x\to\infty} \sinh x = \infty$, $\lim_{x\to-\infty} \sinh x = -\infty$, $\lim_{x\to\infty} \cosh x = \infty$, $\lim_{x\to-\infty} \cosh x = \infty$, $\lim_{x\to\infty} \tanh x = 1$, $\lim_{x\to-\infty} \tanh x = -1$

Section 3.1

1. 0

3. 2

5. 0

7. 0

9. $-2h$

11. $\sqrt{4+h} - 2$

13. **(a)** $f'(-1) = -10$ **(b)** $y = -10x - 5$
(c)

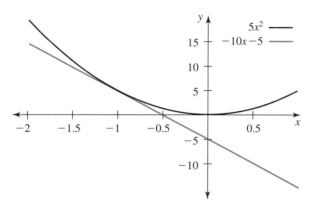

▲ **FIGURE 3.1.13**

15. **(a)** $f'(2) = -12$ **(b)** $y = \frac{1}{12}x - \frac{43}{6}$
(c)

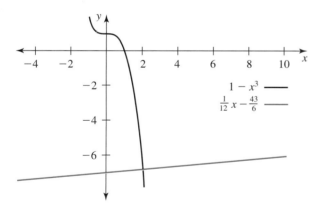

▲ **FIGURE 3.1.15**

17. $f'(x) = \frac{1}{2\sqrt{x}}$

19. $y = x - 3$

21. $f(x) = \frac{1}{x^2+1}$, $a = 2$

23. **(a)** $s\left(\frac{3}{4}\right) = 30$, $s(1) = \frac{160}{3}$

(b) $\frac{280}{3}$ **(c)** $v\left(\frac{3}{4}\right) = 80$, $\left|v\left(\frac{3}{4}\right)\right| = 80$

25. $R^\star = 1.25$

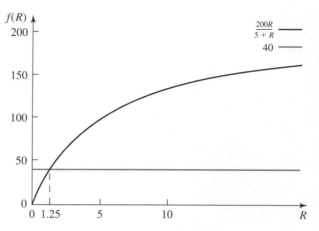

▲ **FIGURE 3.1.25**

27. $f(N) = 0$ for $N = 0$ and $N = 20$

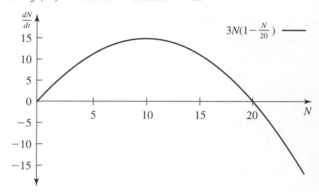

▲ **FIGURE 3.1.27**

29. A differentiable function is continuous. A continuous functions need not be differentiable.

31. -1 **33.** 0

35. $\frac{dN}{dt} = 0$ for $N = 0$ or $N = K$

Section 3.2

1. $6x^2 - 3$ **3.** $t - 3$ **5.** $3t^2 e^{-2} + 1$

7. $3ax^2$ **9.** $4(b-1)N^3 - \frac{2N}{b}$ **11.** $V_0\gamma$

13. $1 - \frac{2N}{K}$ **15.** $y - 8x + 5 = 0$

17. $8x - \sqrt{2}y - 18 = 0$ **19.** $x + 61y - 1588 = 0$

21. $2ax - y - a = 0$ **23.** $\left(\frac{3}{2}, \frac{9}{4}\right)$

25. $\left(\frac{1}{3}\sqrt{3}, \frac{7}{9}\sqrt{3} + 2\right)$, $\left(-\frac{1}{3}\sqrt{3}, -\frac{7}{9}\sqrt{3} + 2\right)$

27. $P'(x)$ is a polynomial of degree $k - 1$

Section 3.3

1. $f'(x) = 3x^2(3x - 14x^2) + (x^3 + 17)(3 - 28x)$

3. $h'(t) = \frac{t^2+2t-4}{(t+1)^2}$ **5.** $f'(x) = \frac{1}{2}\sqrt{x}(9x^3 - 25x)$

7. $f'(x) = 3x^2 + \frac{3}{x^4}$ **9.** $y = -8x - \frac{41}{3}$

11. $y = \frac{7}{16}x - 1$ **13.** $f'(x) = \frac{3a}{(3+x)^2}$

15. $g'(t) = 2a(at + 1)$ **17.** **(a)** $-\frac{1}{16}$ **(b)** 11

19. $y' = 6xf(x) + 3x^2 f'(x)$

21. $y' = 2f(x)f'(x) - \frac{f(x) - xf'(x)}{[f(x)]^2}$

23. $y' = f'(x)g(x) + f(x)g'(x) + 4g(x)g'(x)$

25. $y' = \frac{2x[f(x) - g(x)] - x^2[f'(x) - g'(x)]}{[f(x) - g(x)]^2}$

27. $g'(x) = -\frac{f'(x)}{[f(x)]^2}$ **29.** $w'(t) = \frac{f'(t)(C+t) - f(t)}{(C+t)^2}$

Section 3.4

1. $f'(x) = 6(2x - 1)^2$ **3.** $f'(x) = -\frac{12x^2}{(x^3 - 2)^5}$

5. $g'(t) = \frac{-9t^2}{(t-3)^4}$

7. $h'(s) = \frac{-4(3-s)(s^2 + (7s-1)^2 - 2(3-s)^2(2s + 14(7s-1))}{(s^2 + (7s-1)^2)^2}$

9. $g'(s) = \pi(3s^7 - 7s)^{\pi - 1}(21s^6 - 7)$

11. $f'(x) = 3(ax + 1)^2 c$ **13.** $g'(N) = \frac{bk - bN}{(k+N)^3}$

15. (a) -6 (b) $\frac{5}{4}$ **17.** $\frac{f'(x) + g'(x)}{2\sqrt{f(x) + g(x)}}$

19. $f'[\frac{1}{g(x)}]\frac{-g'(x)}{[g(x)]^2}$ **21.** $\frac{dy}{dx} = -\frac{x}{y}$

23. $\frac{dy}{dx} = -\left(\frac{y}{x}\right)^{1/4}$

25. $\frac{dy}{dx} = 4\sqrt{xy} - \frac{y}{x}$ **27.** $\frac{dy}{dx} = \frac{y}{x}$

29. (a) $y = \frac{4}{3}x - \frac{25}{3}$ (b) $y = -\frac{3}{4}x$

31. (a) $y = \frac{3}{4}x - \frac{9}{4}$ (b) $y = -\frac{4}{3}x + \frac{136}{9}$

33. (a) $(27)^{1/6}$

(b)

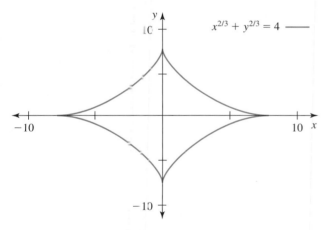

$x^{2/3} + y^{2/3} = 4$ ———

▲ **FIGURE 3.4.33**

35. $-\frac{2}{3}\sqrt{3}$ **37.** $-\frac{3}{4}$ **39.** $\frac{dV}{dt} = 3x^2\frac{dx}{dt}$

41. $\frac{dS}{dt} = 8\pi r\frac{dr}{dt}$ **43.** $\frac{dh}{dt} = \frac{1}{100\pi}\frac{m}{min}$

45. $\frac{183}{\sqrt{61}}\frac{mi}{hr}$ for both $t = 20$ min and $t = 40$ min

47. $f'(x) = 3x^2 - 6x$ $f''(x) = 6x - 6$

49. $g'(x) = 2(x + 1)^{-2}, g''(x) = -\frac{4}{(x+1)^3}$

51. $g'(t) = \frac{9t^2 + 2}{2\sqrt{3t^3 + 2t}}$ $g''(t) = \frac{27t^4 + 36t^2 - 4}{4(3t^3 + 2t)^{3/2}}$

53. $f'(s) = \frac{3}{4}\left(s^{1/2} - \frac{1}{s}\right)^{-1/2}$ $f''(s) = -\frac{3}{8}\frac{\frac{1}{2\sqrt{s}} + \frac{1}{s^2}}{(s^{1/2} - \frac{1}{s})^{3/2}}$

55. $g'(t) = -\frac{5}{2}t^{-7/2} - \frac{1}{2}t^{-1/2}$ $g''(t) = \frac{35}{4}t^{-9/2} + \frac{1}{4}t^{-3/2}$

57. $f(x) = x^5$, $f'(x) = 5x^4$, $f''(x) = 20x^3$, $f'''(x) = 60x^2$, $f^{(4)}(x) = 120x$, $f^{(5)}(x) = 120$, $f^{(6)} = \cdots = f^{(10)}(x) = 0$

59. $p(x) = 3x^2 + 2x + 3$

Section 3.5

1. $f'(x) = 3\cos x - 5\sin x - 2\sec x \tan x$

3. $f'(x) = 6\sin x \cos x$

5. $g'(t) = \frac{3\cos(3t)\cos(5t) + 5\sin(3t)\sin(5t)}{\cos^2(5t)}$

7. $f'(x) = -4x\cos(x^2 - 1)\sin(x^2 - 1)$

9. $f'(x) = -\frac{1}{1 + \sin x}$ **11.** $h'(x) = -\frac{6\sec^2(2x) - 3}{(\tan(2x) - x)^2}$

13. $h'(s) = 0$ **15.** $x = \frac{3}{2} + 3k, k \in \mathbf{Z}$

21. (a) $\frac{dc}{dt} = \frac{\pi}{2}\cos(\frac{\pi}{2}t)$

(b)

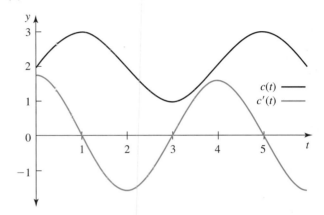

▲ **FIGURE 3.5.21**

(c) (i) $\frac{dc}{dt} = 0$, (ii) increasing, (iii) $c(t)$ has a horizontal tangent and either a maximum or a minimum.

Section 3.6

1. $f'(x) = 2e^{2x}$ **3.** $f'(x) = (-4x + 3)e^{-2x^2 + 3x - 1}$

5. $f'(x) = (\cos x)e^{\sin x}$

7. $g'(s) = \exp[\sec s^2](\sec s^2)(\tan s^2)(2s)$

9. $h'(t) = (2t)(\ln 2)2^{t^2 - 1}$

11. $g'(r) = -3(\ln 2)(\cos r)2^{-3\sin r}$

13. 2 **15.** 0 **17.** $\frac{1}{\ln 2}$

19. $\frac{dN}{dt} = (\ln 2)N(t)$ which implies that $\frac{dN}{dt}$ is proportional to $N(t)$

21. (a)

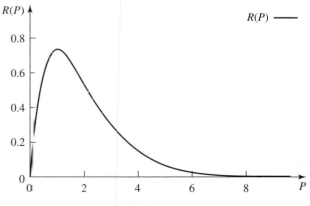

▲ **FIGURE 3.6.21**

(b) $\frac{dR}{dP} = \alpha e^{-\ell P}(1 - \beta P)$ **(c)** $P = \frac{1}{\beta}$

23. (a) $\frac{dW}{dt} = -\lambda W(t)$

(b) $W(5) = (10)\left(\frac{8}{10}\right)^5 \approx 3.28$

(c) $T_h = \frac{\ln 2}{-\ln(0.8)} \approx 3.11$

Section 3.7

1. $f^{-1}(x) = x^2 + 1, x \geq 0, \frac{d}{dx}f^{-1}(x) = 2x$

3. $f^{-1}(x) = \left(\frac{3-x}{2}\right)^{1/3}, x \leq 3, \frac{d}{dx}f^{-1}(x) = -\frac{1}{6}\left(\frac{2}{3-x}\right)^{2/3}$

5. $\frac{1}{4}$ **7.** $\frac{1}{2}$ **9.** $\frac{1}{4}$ **11.** 1

13. $f'(x) = \frac{3}{x}$ **15.** $f'(x) = \frac{x}{x^2+1}$

17. $g'(s) = (2s + \frac{1}{s})\exp(s^2 + \ln s)$

19. $h'(t) = \frac{\frac{1}{t}(1+t^2)-2t\ln t}{(1+t^2)^2}$ **21.** $h'(s) = \frac{1}{s\ln s}$

23. $f'(x) = \frac{3x^2-3}{(\ln 10)(x^3-3x)}$ **25.** $f'(u) = \frac{4u^3}{(\ln 3)(3+u^4)}$

29. $\frac{dy}{dx} = x^x(\ln x + 1)$

31. $\frac{dy}{dx} = x^{x\ln x}2(\ln x)\frac{1}{x}$

33. $\frac{dy}{dx} = \left[x^x(\ln x + 1)\ln x + x^{x-1}\right]x^{x^x}$

35. $\frac{dy}{dx} = x^{\cos x}\left[\frac{\cos x}{x} - (\sin x)(\ln x)\right]$

37. $\frac{1}{y}\frac{dy}{dx} = 2 + \frac{27}{5x-2} - \frac{x}{2(x^2+1)} - \frac{9x^2}{4(3x^3-7)}$

Section 3.8

1. $\sqrt{65} \approx 8.0625$, error $= 2.42 \times 10^{-4}$

3. $\sqrt[3]{124} \approx \frac{374}{75}$, error $\approx 3.57 \times 10^{-5}$

5. $(0.99)^{25} \approx 0.75$, error ≈ 0.0278

7. $\sin\left(\frac{\pi}{2} + 0.02\right) \approx 1$, error $\approx 2.00 \times 10^{-4}$

9. $\ln(1.01) \approx 0.01$, error $\approx 4.97 \times 10^{-5}$

11. $L(x) = 1 - x$ **13.** $L(x) = 1 + x$

15. $L(x) = 1 - nx$ **17.** $L(x) = x$

19. 100.3 **21.** $[1.8, 2.2]$

23. $[10.8, 13.2]$ **25.** $[7.24, 7.54]$

27. $\pm 6\%$ **29.** $\pm 0.668\%$

31. $\pm 9\%$ **33.** $\pm 2.4\%$

35. $\pm \frac{(a+b-2x)x}{(a-x)(b-x)}\left(100\frac{\Delta x}{x}\right)$

Section 3.10

1. $f'(x) = -12x^3 - \frac{1}{x^{3/2}}$

3. $h'(t) = \frac{1}{3}\left(\frac{1+t}{1-t}\right)^{2/3}\frac{-2}{(1+t)^2}$

5. $f'(x) = 2e^{2x}\sin\left(\frac{\pi}{2}x\right) + e^{2x}\frac{\pi}{2}\cos\left(\frac{\pi}{2}x\right)$

7. $f'(x) = \frac{\frac{1}{x+1}\ln x - \frac{1}{x}\ln(x+1)}{[\ln x]^2}$

9. $f'(x) = -xe^{-x^2/2}, f''(x) = e^{-x^2/2}(x^2 - 1)$

11. $h'(x) = \frac{1}{(x+1)^2}, h''(x) = -\frac{2}{(x+1)^3}$

13. $\frac{dy}{dx} = \frac{\cos x + y^2 - 2xy}{x^2 - 2xy}$ **15.** $\frac{dy}{dx} = 1 - 2(x - y)$

17. $\frac{dy}{dx} = -\frac{x}{y}, \frac{d^2y}{dx^2} = -\frac{16}{y^3}$

19. $\frac{dy}{dx} = \frac{1}{x\ln x}, \frac{d^2y}{dx^2} = -\frac{\ln x+1}{(x\ln x)^2}$

21. $5.70\frac{\text{ft}}{\text{sec}}$

23. (a) $\frac{dy}{dx} = f'(x)e^{f(x)}$ **(b)** $\frac{dy}{dx} = \frac{f'(x)}{f(x)}$

(c) $\frac{dy}{dx} = 2f(x)f'(x)$

25. (a)

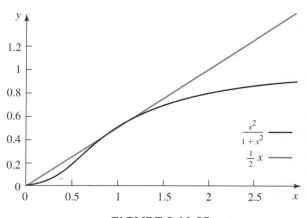

▲ **FIGURE 3.10.25**

(b) $c = 1, y = \frac{1}{2}x$

27. $y - \frac{1}{2}e^{-(\pi/3)^2} = -e^{-(\pi/3)^2}\left(\frac{1}{2}\sqrt{3} + \frac{\pi}{3}\right)\left(x - \frac{\pi}{3}\right)$

29. $y = x$ **31.** $p(x) = 2x^2 + 4x + 8$

33. (a) $s(5.5) \approx 54.84$ miles **(b)** $v(t) = \frac{ds}{dt} = 3\pi + 3\pi\sin(\pi t), a(t) = \frac{dv}{dt} = 3\pi^2\cos(\pi t)$ **(c)** $v(t) \geq 0$ **(d)** Three valleys and three hills

35. (b) $N(t) = 100 + 50\cos\left(\frac{\pi}{2}t\right)$

▲ **FIGURE 3.10.35**

(c) The population size shows oscillations.

37. 9.33%

Section 4.1

1. local maximum = global maximum = $(-1, 3)$, no local and global minima

3. local minimum = global minimum = $(0, -2)$, local maximum = global maximum = $(-1, -1)$ and $(1, -1)$

5. local minimum = $(-2, -3)$ and $(1, 0)$, global minimum = $(-2, -3)$, local maximum = global maximum = $(0, 1)$

7. $f(0) = 0$ and $f(x) = x^2 > 0$ for $x \neq 0$; $f'(0) = 0$

9. $f(0) = 0$ and $f(x) > 0$ for $x \neq 0$; $\lim_{x\to 1-} f'(x) = -1$, $\lim_{x\to 1+} f'(x) = 1$.

11. (a)

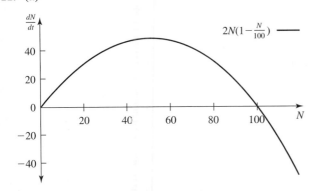

▲ **FIGURE 4.1.11**

$\dfrac{dN}{dt}$ is maximal for $N = 50$

(b) $f'(N) = r - \dfrac{2r}{K} N$ **13.** $c = \dfrac{1}{2}$

15.

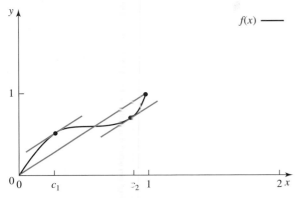

▲ **FIGURE 4.1.15**

17. The slope of the secant line through $(a, f(a))$ and $(b, f(b))$ is positive. Use the MVT.

19. (b) Since $1 \le f(1) \le 2$, $f(1) \ne 3$.

21. (b) $f'(x) = \begin{cases} -e^{-x} & \text{for } x \in (0, 2) \\ e^x & \text{for } x \in (-2, 0) \end{cases}$

(d) Rolle's theorem assumes that $f(x)$ is differentiable. But here, $f(x)$ is not differentiable at $x = 0$.

(e)

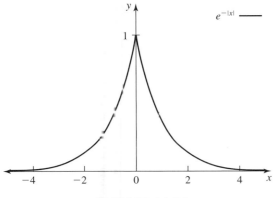

▲ **FIGURE 4.1.21**

Section 4.2

1. $y' = 3 - 2x$, $y'' = -2$; $y' > 0$ and y is increasing on $(-\infty, 3/2)$; $y' < 0$ and y is decreasing on $(3/2, \infty)$; $y'' < 0$ and y is concave down.

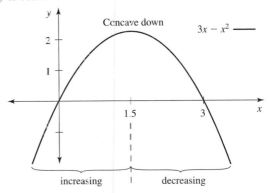

▲ **FIGURE 4.2.1**

3. $y' = 2x + 1$, $y'' = 2$; $y' > 0$ and y is increasing on $(-1/2, \infty)$; $y' < 0$ and y is decreasing on $(-\infty, -1/2)$; $y'' > 0$ and y is concave up.

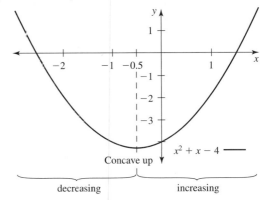

▲ **FIGURE 4.2.3**

5. $y' = -2x^2 + 7x - 3$, $y'' = -4x + 7$; $y' > 0$ and y is increasing on $(1/2, 3)$; $y' < 0$ and y is decreasing on $(-\infty, 1/2) \cup (3, \infty)$; $y'' > 0$ and y is concave up on $(-\infty, 7/4)$; $y'' < 0$ and y is concave down on $(7/4, \infty)$.

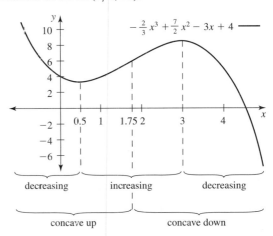

▲ **FIGURE 4.2.5**

7. $y' = \frac{1}{2\sqrt{x+1}}$, $x > -1$; $y'' = -\frac{1}{4(x-1)^{3/2}}$, $x > -1$; $y' > 0$ and y is increasing; $y'' < 0$ and y is concave down.

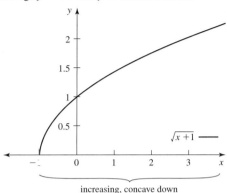

▲ **FIGURE 4.2.7**

9. $y' = -\frac{1}{x^2}$, $x \neq 0$; $y'' = \frac{2}{x^3}$, $x \neq 0$; $y' < 0$ and y is decreasing for $x \neq 0$; $y'' < 0$ and y is concave down for $x < 0$; $y'' > 0$ and y is concave up for $x > 0$.

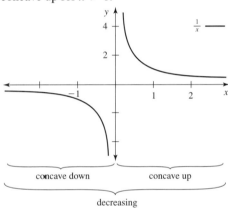

▲ **FIGURE 4.2.9**

11. $y' = \frac{1}{3}(x^2 - 1)^{-2/3} 2x$, $y'' = \frac{6-2x^2}{9(x^2+1)^{5/3}}$; $y' > 0$ and y is increasing on $(0, \infty)$; $y' < 0$ and y is decreasing on $(-\infty, 0)$; $y'' > 0$ and y is concave up on $(-\sqrt{3}, \sqrt{3})$; $y'' < 0$ and y is concave down on $(-\infty, -\sqrt{3}) \cup (\sqrt{3}, \infty)$.

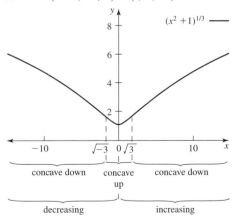

▲ **FIGURE 4.2.11**

13. $y' = -\frac{2}{(1+x)^3}$, $y'' = \frac{6}{(1+x)^4}$; $y' > 0$ and y is increasing on $(-\infty, -1)$; $y' < 0$ and y is decreasing on $(-1, \infty)$; $y'' > 0$ and y is concave up for $x \neq -1$.

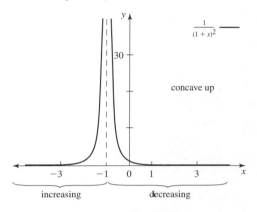

▲ **FIGURE 4.2.13**

15. $y' = \cos x$, $y'' = -\sin x$; $y' > 0$ and y is increasing on $(0, \pi/2) \cup (3\pi/2, 2\pi)$; $y' < 0$ and y is decreasing on $(\pi/2, 3\pi/2)$; $y'' > 0$ and y is concave up on $(\pi, 2\pi)$; $y'' < 0$ and y is concave down on $(0, \pi)$.

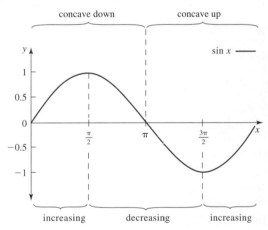

▲ **FIGURE 4.2.15**

17. $y' = e^x$, $y'' = e^x$; $y' > 0$ and y is increasing for $x \in \mathbf{R}$; $y'' > 0$ and y is concave up for $x \in \mathbf{R}$.

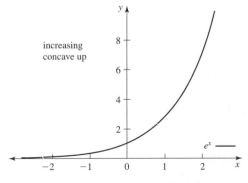

▲ **FIGURE 4.2.17**

19. $y' = -xe^{-x^2/2}$, $y'' = e^{-x^2/2}(x^2 - 1)$; $y' > 0$ and y is increasing for $x < 0$; $y' < 0$ and y is decreasing for $x > 0$; $y'' > 0$ and y is concave up for $x < -1$ and $x > 1$; $y'' < 0$ and y is concave down for $-1 < x < 1$.

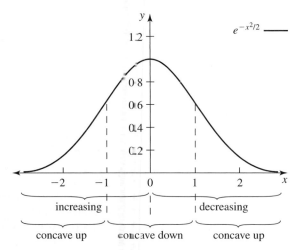

▲ **FIGURE 4.2.19**

21. (a)

▲ **FIGURE 4.2.21a**

(b)

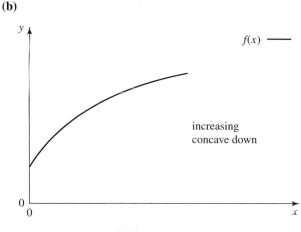

▲ **FIGURE 4.2.21b**

(c) In **(a)**, $y' > 0$ and $y'' > 0$; in **(b)**, $y' > 0$ and $y'' < 0$.

23. (b) $f(-1) > 0$ and $f(1) < 0$

25. $f'(x)$ is decreasing; use the definition of concave down.

27. (a) (i)

▲ **FIGURE 4.2.27ai**

(ii)

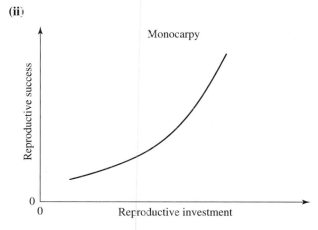

▲ **FIGURE 4.2.27aii**

(b) Polycarpy: $y' > 0$, $y'' < 0$; monocarpy: $y' > 0$, $y'' > 0$.

29. $\frac{dE}{dF} < 0$: The average number of pollen grain exported per flower decreases with the number of flowers.

31. (a) $\frac{1}{W}\frac{dN}{dx} = kW^{a-1}$; relative growth rate is increasing for $a > 1$ and decreasing for $a < 1$. **(b)** Exclude values of a for which $a \geq 1$. Juvenile fish have a larger percentage weight increase than adults relative to their respective body sizes.

Section 4.3

1. local max: $(0, 4)$; absolute max: $(0, 4)$; local min: $(-2, 0)$ and $(3, -5)$; absolute min: $(3, -5)$; y is increasing on $(-2, 0)$ and decreasing on $(0, 3)$.

3. local max: $(\pi/2, 1)$; absolute max: $(\pi/2, 1)$; local min: $(3\pi/2, -1)$; absolute min: $(3\pi/2, -1)$; y is increasing on $(0, \pi/2)$ and $(3\pi/2, 2\pi)$ and decreasing on $(\pi/2, 3\pi/2)$.

5. local max: $(0, 1)$; absolute max: $(0, 1)$; y is decreasing on $[0, \infty)$

7. no extrema; y is increasing for all $x \in \mathbf{R}$.

9. local max: $(0, 1)$; absolute max: $(0, 1)$; local min: $(\pi, -1)$;

absolute min: $(\pi, -1)$; y is decreasing on $(0, \pi)$.

11. local max: $(0, 1)$; absolute max: $(0, 1)$; y is increasing on $(-\infty, 0)$, and decreasing on $(0, \infty)$.

13. local max: $(-3, 15.5)$; local min: $(2, -16/3)$; y is increasing on $(-\infty, -3) \cup (2, \infty)$; y is decreasing on $(-3, 2)$.

15. local min: $(-1, 0), (0, 0)$, and $(1, 0)$; absolute min: $(-1, 0)$, $(0, 0)$, and $(1, 0)$; local max: $(-\sqrt{2}/2, 1)$ and $(\sqrt{2}/2, 1)$; absolute max: $(-\sqrt{2}/2, 1)$ and $(\sqrt{2}/2, 1)$; y is increasing on $(-1, -\sqrt{2}/2) \cup (0, \sqrt{2}/2)$; y is decreasing on $(-\sqrt{2}/2, 0) \cup (\sqrt{2}/2, 1)$.

19. $(0, -2)$ **21.** $(\sqrt{2}/2, e^{-1/2})$

23. $(0, 0)$

27. local min: $(-2, 2/3)$, $(3, -16)$; local max: $(-1, 16/3)$, $(5, 16/3)$; absolute min: $(3, -16)$; absolute max: $(-1, 16/3)$, $(5, 16/3)$; inflection point: $(1, -16/3)$ y is increasing on $(-2, -1) \cup (3, 5)$; y is decreasing on $(-1, 3)$; y is concave up on $(1, 5)$; y is concave down on $(-2, 1)$;

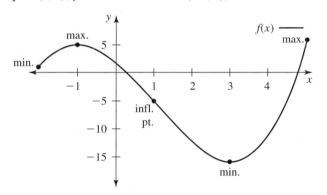

▲ **FIGURE 4.3.27**

29. local min: $(-3, 0)$, $(3, 0)$; local max: $(-4, 5)$, $(0, 9)$, $(5, 16)$; absolute min: $(-3, 0)$, $(3, 0)$; absolute max: $(5, 16)$; inflection points: $(-3, 0)$, $(3, 0)$; y is increasing on $(-3, 0) \cup (3, 5)$; y is decreasing on $(-4, -3) \cup (0, 3)$; y is concave up on $(-4, -3) \cup (3, 5)$; y is concave down on $(-3, 3)$;

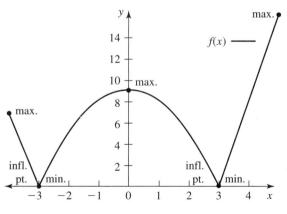

▲ **FIGURE 4.3.29**

31. no extrema; inflection points: $x = \pi/2 + k\pi$, $k \in \mathbf{Z}$; y is increasing on $\mathbf{R}$ y is concave up on $(\pi/2 + 2k\pi, 3\pi/2 + 2k\pi)$, $k \in \mathbf{Z}$; y is concave down on $(-\pi/2 + 2k\pi, \pi/2 + 2k\pi)$,

$k \in \mathbf{Z}$;

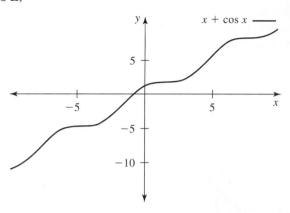

▲ **FIGURE 4.3.31**

33. local min: $(0, -1)$; absolute min: $(0, -1)$; inflection points at $x = -1/\sqrt{3}$ and $x = 1/\sqrt{3}$; y is increasing on $(0, \infty)$; y is decreasing on $(-\infty, 0)$; y is concave up on $(-1/\sqrt{3}, 1/\sqrt{3})$; y is concave down on $(-\infty, -1/\sqrt{3}) \cup (1/\sqrt{3}, \infty)$;

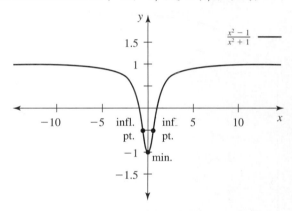

▲ **FIGURE 4.3.33**

35. **(c)** decreasing for all $x \neq -1$; no extrema; **(d)** $f(x)$ is concave down on $(-\infty, 1)$ and concave up on $(1, \infty)$; no inflection points;

(e)

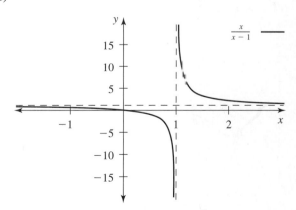

▲ **FIGURE 4.3.35**

37. (a) $\lim_{x\to -2^+} f(x) = \infty$, $\lim_{x\to -2^-} f(x) = -\infty$;
(b) $f(x)$ is increasing on $(-\infty, -2 - \sqrt{6}/2) \cup (-2 + \sqrt{6}/2, \infty)$; $f(x)$ is decreasing on $(-2-\sqrt{6}/2, -2)\cup(-2, -2+\sqrt{6}/2)$; local max at $x = -2 - \sqrt{6}/2$; local min at $-2+\sqrt{6}/2$;
(c) $f(x)$ is concave up on $(-2, \infty)$; $f(x)$ is concave down on $(-\infty, -2)$; **(d)** $y = 2x - 4$;
(e)

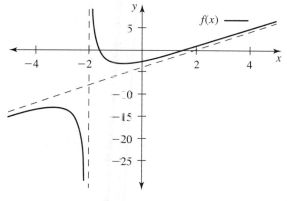

▲ **FIGURE 4.3.37**

Section 4.4

1. 20 in. **3.** 4
5. The field is 80 ft by 160 ft, where the length along the river is 160 ft.
7. 9 **9.** 4
11. (b) $(6/5, \sqrt{1.6})$ **(c)** local minimum at $x = 6/5$ as in (b).
13. $g'(x) = 2 f(x) f'(x)$ has the same sign change as $f'(x)$.
15. The height is equal to the diameter, namely $2 \left(\frac{500}{\pi}\right)^{1/3}$.
17. (a) $r = \sqrt{2}, \theta = 2$; **(b)** $r = \sqrt{10}, \theta = 2$
19. $r = \left(\frac{355}{4\pi}\right)^{1/3}$
21. (a) Local maximum at t^* satisfies $f'(t^*) = \frac{f(t^*)}{C+t^*}$, which is the slope of the straight line through $(-C, 0)$ and the point with $t = t^*$. **(b)** $y = \frac{1}{(1+\sqrt{2})^2}t + \frac{2}{(1+\sqrt{2})^2}$

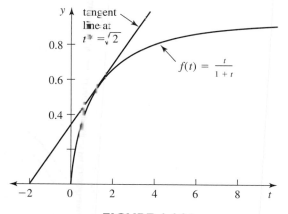

▲ **FIGURE 4.4.21**

23. (a) $-x\frac{dr}{dx} - r(x) - L + \frac{3ke^{-kx}}{1-e^{-kx}} - \frac{r'(x)e^{-(r(x)+L)}}{1-e^{-(r(x)+L)}} = 0$

Section 4.5

1. 8 **3.** -7 **5.** $\frac{1}{2}$ **7.** 1
9. $\frac{1}{2}$ **11.** ∞ **13.** $\frac{\ln 2}{\ln 3}$ **15.** $\frac{1}{2}$
17. 0 **19.** ∞ **21.** 0 **23.** 0
25. 1 **27.** 0 **29.** 0 **31.** 0
33. 1 **35.** 1 **37.** e^3 **39.** e^{-1} **41.** 0
43. ∞ **45.** 3 **47.** 0 **49.** 1 **51.** $\frac{\ln a}{\ln b}$

53. $\lim_{x\to\infty} \left(1 + \frac{c}{x^p}\right) = \begin{cases} \infty & \text{if } 0 < p < 1 \\ e^c & \text{if } p = 1 \\ 1 & \text{if } p > 1 \end{cases}$

57. Let $y = 121e^{-17/x}$. **(a)** $\lim_{x\to 0+} \frac{dy}{dx} = 0$, $\lim_{x\to\infty} \frac{dy}{dx} = 0$
(b) $\frac{17}{2}$ **(c)** Height is increasing at an accelerating rate for $0 < x < \frac{17}{2}$ and at a decelerating rate for $x > \frac{17}{2}$.
(d)

▲ **FIGURE 4.5.57i**

▲ **FIGURE 4.5.57ii**

Section 4.6

1. (a) $11.59 \approx 12$ **(b)** $23.45 \approx 23$
3. (a) $b = 1.02$ **(b)** $24.38 \approx 24$ **(c)** $\frac{\ln 2}{\ln 1.02} \approx 35$
5. (a) $b = 1 + \frac{x}{100}$ **(b)** doubling time: $\frac{\ln 2}{\ln(1+x/100)}$

x	0.1	0.5	1	2	5	10
k	639	139	69.7	35.0	14.2	7.27

7. (a)

k	1	2	3	4	5
N_k	135.7	177.7	223.2	267.8	307.3

k	6	7	8	9	10
N_k	339.0	362.3	378.3	388.7	395.3

(b)

k	1	2	3	4	5
N_k	539.2	471.7	441.5	425.9	417.3

k	6	7	8	9	10
N_k	412.4	409.5	407.9	406.9	406.3

(c) $N^* = \frac{\ln 1.5}{0.001} \approx 405.5$

(d)

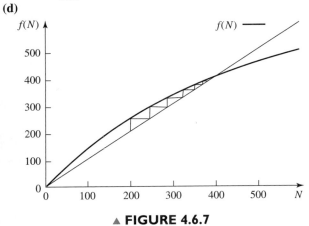

▲ **FIGURE 4.6.7**

(e) $f(N) = 1.5Ne^{-0.001N}$, $\lambda = f'\left(\frac{\ln 1.5}{0.001}\right) = 1 - \ln 1.5$; the eigenvalue λ satisfies $|\lambda| < 1$

9. (a) $N^* = 0$ and $N^* = 100(1 - 1/r)$ **(b)** $r = \frac{1}{2}$: population goes extinct; $r = 2$: population reaches size 50; $r = 3.1$: population oscillates between two values.

11. 0.567143

13. $\sqrt{5} \approx 2.236068$, $-\sqrt{5} \approx -2.236068$

15. (b) The method does not converge.

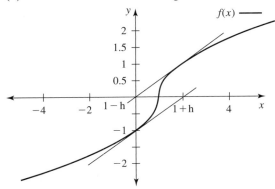

▲ **FIGURE 4.6.15**

17. (a) $x_1 = \infty$

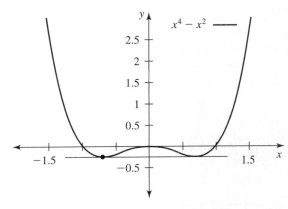

▲ **FIGURE 4.6.17(a)**

(c) When $x_0 = -0.71$, x_n converges to -1; when $x_0 = -0.70$, then x_n converges to 1; when $x_0 = -0.6$, then x_n converges to 0.

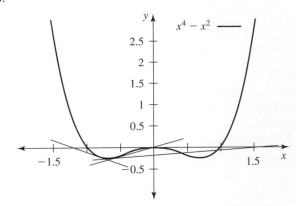

▲ **FIGURE 4.6.17(b)**

19. If $f(x_n) = 0$ and $f'(x_n) \neq 0$, then $x_{n+1} = x_n$.

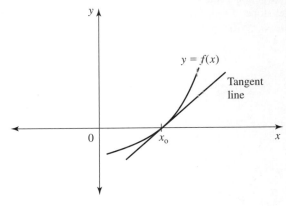

▲ **FIGURE 4.6.19**

Section 4.7

1. $F(x) = \frac{1}{5}x^5 - x^3 + x + C$

3. $F(x) = x^4 - x^2 + 3x + C$

5. $F(x) = \frac{1}{3}x^3 - 2\ln|x| + C$

7. $F(x) = -\frac{1}{3}e^{-3x} + C$ **9.** $F(x) = e^{2x} + C$

11. $F(x) = \frac{1}{2}\tan(2x) + C$

13. $y = 2\ln x - \frac{1}{2}x^2 + C$

15. $y = -\frac{1}{\pi}\cos(\pi s) + C$

17. $y = \frac{4}{3}x^{3/2} + \frac{2}{3}$

19. $W(t) = 1 - \frac{1}{3}e^{-3t}$

21. $y = \frac{e^x - e^{-x}}{2}$

23. $L(x) = 25 - 10e^{-0.1x}$, $L(0) = 15$

25. $t = 2.5$ sec, $v(2.5) = 80\frac{m}{sec}$

27. (a) The first term on the right-hand side describes evaporation; the second one describes watering.

(b) $a \approx 0.0417$

Section 4.9

1. (b) Absolute maximum at $(1, e^{-1})$; **(c)** inflection point at $(2, 2e^{-2})$;

(d)

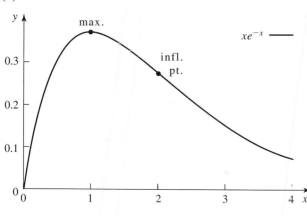

▲ **FIGURE 4.9.1**

3. (a) $f'(x) = \frac{4}{(e^x + e^{-x})^2} > 0$, hence, $f(x)$ is strictly increasing.

5. (e)

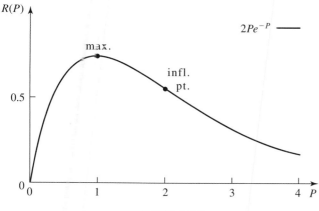

▲ **FIGURE 4.9.5**

7. (a) $\lim_{x \to \infty} f(x) = c$ **(b)** $f'(x) = \frac{ck}{(k+x)^2} > 0$, $f''(x) = -\frac{2ck}{(k+x)^3} < 0$ **(c)** $f(k) = \frac{c}{2}$

(d)

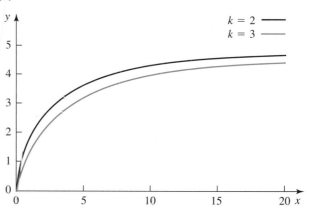

▲ **FIGURE 4.9.7**

9. (a) $L(\theta) = (\theta^2)^8(2\theta(1-\theta))^6((1-\theta)^2)^3$ **(b)** $\frac{d}{d\theta}\ln L(\theta) = \frac{L'(\theta)}{L(\theta)}$ and $L(\theta) > 0$. **(c)** $\hat{\theta} = \frac{11}{17}$

11. (a) $c(t) = \frac{1}{3}e^{-0.3t}$, $t \geq 0$ **(b)** $t = \frac{\ln 2}{0.3}$

Section 5.1

1. (a) $\frac{7}{32}$ **(b)** $\frac{21}{64}$ **(c)** $\frac{15}{22}$

3. $3^2 + 4^2 + 5^2 + 6^2$ **5.** $\frac{1}{2} + \frac{2}{3} + \frac{3}{4}$

7. $0^x + 1^x + 2^x + 3^x + 4^x$

9. $f(c_1)\Delta x_1 + f(c_2)\Delta x_2 + \cdots + f(c_n)\Delta x_n$

11. $\cos\left(\frac{\pi}{n}\right)\frac{\pi}{n} + \cos\left(2\frac{\pi}{n}\right)\frac{\pi}{n} + \cos\left(3\frac{\pi}{n}\right)\frac{\pi}{n} + \cdots + \cos\left(n\frac{\pi}{n}\right)\frac{\pi}{n}$

13. $\sum_{k=1}^{4}\frac{1}{\sqrt{k}}$ **15.** $\sum_{k=3}^{7}\frac{k}{k+2}$ **17.** $\sum_{k=0}^{n}\frac{1}{2^k}$

19. $\sum_{k=0}^{n}(-1)^k a^k$ **21.** -365

23. $\frac{3n(n+1)}{2}$

25. $\frac{n(n+1)(2n+1)}{6} - 4n$ **27.** -1

29. (a) -2 **(b)** -1 **(c)** -2

31. $\int_a^b x\, dx = \frac{1}{2}b^2 - \frac{1}{2}a^2$

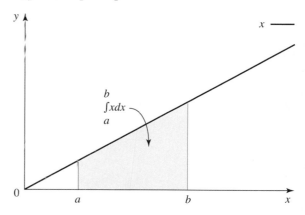

▲ **FIGURE 5.1.31**

33. $\int_0^1 x^2\, dx$ **35.** $\int_{-3}^{2}(2x-1)\, dx$

37. $\int_{-3}^{3}(x-1)(x+2)\, dx$ **39.** $\int_{-5}^{2} e^x\, dx$

41. $\lim_{\|P\| \to 0}\sum_{k=1}^{n}\sqrt{c_k + 1}\Delta x_k$ where $\|P\|$ is a partition of $[2, 6]$, $c_k \in [x_{k-1}, x_k]$ and $\Delta x_k = x_k - x_{k-1}$

730

43. $\lim_{\|P\|\to 0} \sum_{k=1}^{n} \ln c_k \Delta x_k$ where $\|P\|$ is a partition of $[1, e]$, $c_k \in [x_{k-1}, x_k]$ and $\Delta x_k = x_k - x_{k-1}$

45. $\lim_{\|P\|\to 0} \sum_{k=1}^{n} g(c_k) \Delta x_k$ where $\|P\|$ is a partition of $[0, 5]$, $c_k \in [x_{k-1}, x_k]$ and $\Delta x_k = x_k - x_{k-1}$

47.

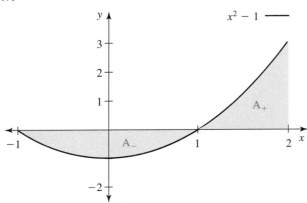

▲ **FIGURE 5.1.47**

$\int_{-1}^{2} (x^2 - 1) \, dx = -A_- + A_+$

49.

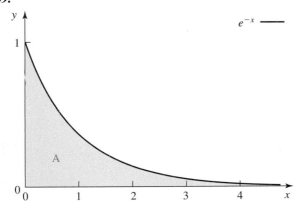

▲ **FIGURE 5.1.49**

$\int_{0}^{5} e^{-x} \, dx = A$

51.

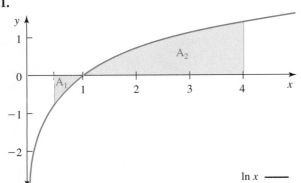

▲ **FIGURE 5.1.51**

$\int_{1/2}^{4} \ln x \, dx = A_2 - A_1$

53.

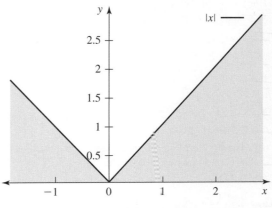

▲ **FIGURE 5.1.53**

$\int_{-2}^{3} |x| \, dx = \frac{13}{2}$

55.

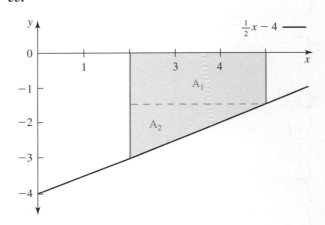

▲ **FIGURE 5.1.55**

$\int_{2}^{5} (\frac{1}{2} x - 4) \, dx = -\frac{27}{4}$

57.

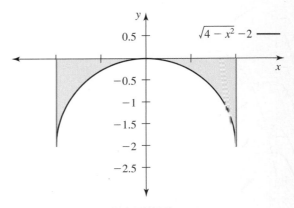

▲ **FIGURE 5.1.57**

$\int_{-2}^{2} \left(\sqrt{4 - x^2} - 2 \right) \, dx = 2\pi - 8$

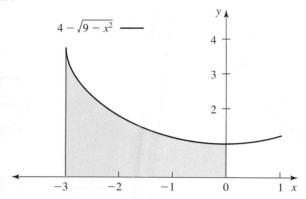

▲ FIGURE 5.1.59

$\int_{-3}^{0} \left(4 - \sqrt{9-x^2}\right) dx = 12 - \frac{9}{4}\pi$

61. 0 **63.** 4 **65.** C **67.** $x \geq x^2$ for $0 \leq x \leq 1$
69. $\sqrt{x} \geq 0$ for $x \geq 0$ and $\sqrt{x} \leq 2$ for $0 \leq x \leq 4$
71.

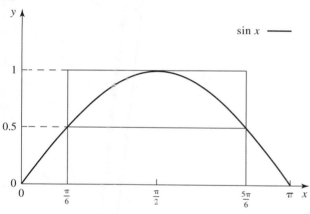

▲ FIGURE 5.1.71

The rectangle with height $1/2$ from $\pi/6$ to $5\pi/6$ is contained in the area under $y = \sin x$ from $\pi/6$ to $5\pi/6$, which is contained in the rectangle with height 1 from $\pi/6$ to $5\pi/6$.

73. $a = \pi/2$ **75.** $a = 3$

Section 5.2

1. $\sqrt{2+x}$ **3.** $\frac{1}{x+3}$ **5.** $2x^2$ **7.** 0
9. $-\frac{1}{x^2}$ **11.** $x^5 - x^2 + C$ **13.** $x^{3/2}\left(\frac{4}{5}x - \frac{2}{3}\right) + C$
15. $\frac{2}{9}x^{9/2} + \frac{7}{9}x^{9/7} + C$ **17.** $\frac{1}{3}x^3 - x + C$ **19.** $\frac{1}{2}e^{2x} + C$
21. $-e^{-x^2/2} + C$ **23.** $-\frac{1}{2}\cos(2x) + C$ **25.** $\sec x + C$
27. $\tan x + \ln|\sec x| + C$ **29.** $4\tan^{-1} x + C$
31. $x - \ln|x| + C$ **33.** $5(x - \tan^{-1} x) + C$
35. $\frac{1}{3}x^3 + \frac{2^x}{\ln 2} + C$ **37.** -6 **39.** $-\frac{1}{3}$ **41.** 3
43. $\frac{28}{3}$ **45.** $\frac{1}{2}$ **47.** $\frac{1}{2}$ **49.** $\frac{\pi}{4}$ **51.** $\frac{\pi}{6}$
53. $\frac{1}{2}\ln 2 - \frac{1}{2}\ln 3$ **55.** $\frac{1}{3}(1 - e^{-3})$ **57.** 1 **59.** 1
61. $\ln\frac{3}{2}$ **63.** $\frac{1}{2}$ **65.** $f(x) = 4x$

Section 5.3

1. $e^2 + 1$ **3.** $\frac{9}{2}$ **5.** $\frac{4}{3}$ **7.** $\frac{14}{3} - \ln 4$

9. $4 - \ln 3$ **11.** $\frac{3}{2}$ **13.** $\frac{2}{3}$ **15.** $\frac{16}{3}$
17. **(a)** $N(t) = 101 - e^{-t}$ **(b)** $1 - e^{-5}$ **(c)** $N(5) - N(0) = \int_0^5 e^{-t}\, dt$; shaded area.

▲ FIGURE 5.3.17

19. **(a)**

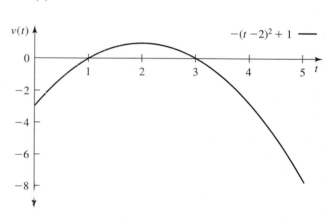

▲ FIGURE 5.3.19(a)

(b) Particle moves to the left for $0 \leq t \leq 1$ and $3 \leq t \leq 5$, it moves to the right for $1 \leq t \leq 3$.
(c) $s(t) = 2t^2 - \frac{1}{3}t^3 - 3t$; signed area between $v(u)$ and the horizontal axis from 0 to t
(d)

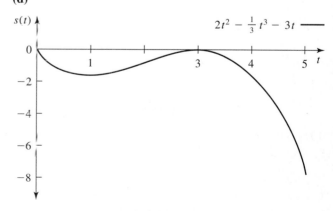

▲ FIGURE 5.3.19(d)

731

732

The rightmost position: $s(0) = s(3) = 0$; the leftmost position: $s(5) = -\frac{20}{3}$.

21. Cumulative growth between $t = 2$ and $t = 7$.

23. Cumulative change in biomass between $t = 1$ and $t = 6$.

25. $-\frac{2}{3}$

27. (a)

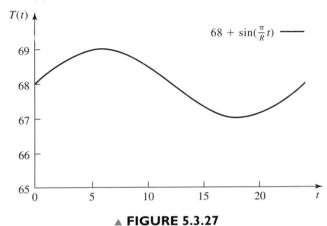

▲ **FIGURE 5.3.27**

(b) 68

29. $f(x)$ is symmetric about the origin.

31. average value of $f(x) = 2$; $f(1) = 2$

33. $\frac{1}{3}\pi hr^2$ **35.** $\frac{1115}{3}\pi$ **37.** 2π

39. $2\pi\sqrt{3}$ **41.** $\frac{2}{15}\pi$ **43.** $\frac{\pi}{2}\left(e^4 + e^{-4} - 2\right)$

45. $\pi\left(\frac{\pi}{2} - 1\right)$ **47.** $\frac{32}{5}\pi$ **49.** $\pi \ln 3$ **51.** $\frac{3}{10}\pi$

Section 5.5

1. $3.38\frac{\text{m}^3}{\text{sec}}$

3. (a)

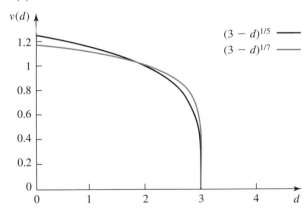

▲ **FIGURE 5.5.3**

(b) $v(D) = 0$, $v'(d) < 0$ for $d \in [0, D)$

5. (a) $\bar{v} = v(d_1)$

Section 6.1

1. $\frac{2}{3}(x^2 + 3)^{3/2} + C$ **3.** $-\frac{7}{12}\cos(4x^3) + C$

5. $-e^{-x^2/2} + C$ **7.** $3\ln|x + 4| + C$

9. $\frac{2}{3}(x + 3)^{3/2} + C$

11. $\frac{2}{3}(2x^2 - 3x + 2)^{3/2} + C$

13. $-\frac{2}{3\pi}\cos\left(\frac{3\pi}{2}x + \frac{\pi}{4}\right) + C$

15. $\frac{1}{2}\tan^2 x + C$

17. $\frac{1}{3}(\ln x)^3 + C$

19. $\frac{2}{5}(5 + x)^{5/2} - \frac{10}{3}(5 + x)^{3/2} + C$

21. $\ln|ax^2 + bx + c| + C$

23. $\frac{1}{n+1}[g(x)]^{n+1} + C$

25. $-e^{-g(x)} + C$

27. $\frac{1}{3}(10^{3/2} - 1)$

29. $\frac{7}{2025}$ **31.** $-e^{-9/2} + 1$

33. $\frac{3}{8}$ **35.** $\frac{1}{2}$

37. $4 + 3\ln 3$ **39.** $\frac{1}{2}$

41. $2(e^{-1} - e^{-3})$

43. $\ln|\sin x| + C$

Section 6.2

1. $x\sin x + \cos x + C$

3. $-2xe^{-x} - 2e^{-x} + C$

5. $\frac{1}{2}x^2\ln x - \frac{1}{4}x^2 + C$

7. $x\tan x + \ln|\cos x| + C$

9. $\frac{1}{2}\left(\sqrt{3} - \frac{\pi}{3}\right)$ **11.** $2\ln 4 - \frac{3}{2}$ **13.** $2 - 17e^{-3}$

15. $\frac{2e^{-3x}}{36+\pi^2}\left[\pi\sin\left(\frac{\pi}{2}x\right) - 6\cos\left(\frac{\pi}{2}x\right)\right] + C$

17. $\frac{1}{2}x[\sin(\ln x) - \cos(\ln x)] + C$

19. $\int\cos^2 x\,dx = \frac{1}{2}\sin x\cos x + \frac{1}{2}x + C$

21. (b) $\int\arcsin x\,dx = x\arcsin x + \sqrt{1 - x^2} + C$

23. (b) $\int x^3 e^x\,dx = x^3 e^x - 3x^2 e^x + 6xe^x - 6e^x + C$

25. (b) $\int(\ln x)^3\,dx = x(\ln x)^3 - 3x(\ln x)^2 + 6x\ln x - 6x + C$

27. $-2\sqrt{x}\cos\sqrt{x} + 2\sin\sqrt{x} + C$

29. $\frac{1}{2}x^4 e^{x^2} - x^2 e^{x^2} + e^{x^2} + C$

31. $(\sqrt{x} + 1)^2\ln(\sqrt{x} + 1) - \frac{1}{2}(\sqrt{x} + 1)^2 - 2(\sqrt{x} + 1)\ln(\sqrt{x} + 1) + 2(\sqrt{x} + 1) + C$

Section 6.3

1. $-\frac{1}{2}xe^{-2x} - \frac{1}{4}e^{-2x} + C$ **3.** $\ln|\sin x| + C$

5. $-\cos(x^2) + C$ **7.** $\frac{1}{4}\tan^{-1}\left(\frac{x}{4}\right) + C$

9. $x - 3\ln|x + 3| + C$ **11.** $\frac{1}{2}\ln|x^2 + 3| + C$

13. $\int\ln x\,dx = x\ln x - x + C$

15. $\frac{4}{9}(x - 2)^{9/4} + \frac{8}{5}(x - 2)^{5/4} + C$

17. $2e^2$ **19.** $\frac{\pi}{2}$ **21.** $\frac{1}{2}$

23. (a) $\frac{1}{x(x-2)} = \frac{1}{2}\left[\frac{1}{x-2} - \frac{1}{x}\right]$ **(b)** $\frac{1}{2}\ln\left|\frac{x-2}{x}\right| + C$

25. $\frac{1}{3}\tan^{-1}\left(\frac{x-2}{3}\right) + C$ **27.** $\frac{1}{5}\ln\left|\frac{x-3}{x+2}\right| + C$

29. $\frac{1}{6}\ln\left|\frac{x-3}{x+3}\right| + C$ **31.** $\frac{1}{3}\ln\left|\frac{x-2}{x+1}\right| + C$

33. $x - 5\ln|x + 2| + 2\ln|x + 1| + C$

35. $x + 2\ln\left|\frac{x-2}{x+2}\right| + C$ **37.** $2 - \ln 5 + \ln 3$

39. $\frac{1}{2}\ln 2$ **41.** $-\ln 2$ **43.** $\frac{\pi}{4} - \frac{1}{2}\ln 2$

45. $\frac{1}{1+x} + \ln\left|\frac{x}{x+1}\right|$ **47.** $-\frac{2}{x+1} + \ln\left|\frac{x+1}{x-1}\right|$

49. $\frac{x}{18(9+x^2)} + \frac{1}{54}\tan^{-1}\left(\frac{x}{3}\right)$ **51.** $-\frac{1}{x} - \tan^{-1} x$

(b)

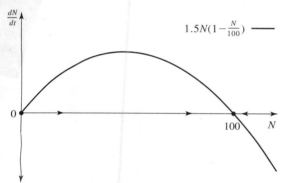

▲ **FIGURE 7.2.3**

$y = 0$ and $y = 2$ are unstable, $y = 1$ is locally stable.

(c) Eigenvalue associated with $y = 0$ is $2 > 0$, hence $y = 0$ is unstable; eigenvalue associated with $y = 1$ is $-1 < 0$, hence $y = 1$ is locally stable; eigenvalue associated with $y = 2$ is $2 > 0$, hence $y = 2$ is unstable.

5. (a) $\frac{dN}{dt} = 1.5N\left(1 - \frac{N}{100}\right)$

(b)

▲ **FIGURE 7.2.5**

$N = 0$ is unstable; $N = 100$ is stable

(c) Eigenvalue associated with $N = 0$ is $1.5 > 0$, hence $N = 0$ is unstable; eigenvalue associated with $N = 100$ is $-1.5 < 0$, hence $N = 100$ is locally stable. Same results as in (b).

7. (a) $K = 2000$ **(b)** $t = \frac{1}{2}\ln 199 \approx 2.65$ **(c)** 2000

9. (a) $N \approx 52.79$ is unstable; $N \approx 947.21$ is locally stable **(b)** The maximal harvesting rate is $rK/4$.

11.

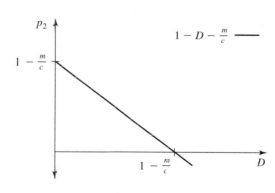

▲ **FIGURE 7.2.11**

The equilibrium C_I is locally stable.

13. (a) $\frac{dC}{dt} = \frac{0.2}{400}(3 - C)$

(b) $C(t) = 3 - 3e^{-t/2000}$, $t \geq 0$; $\lim_{t \to \infty} C(t) = 3$

(c) $C = 3$ is locally stable.

15. (a) Equilibrium concentration: $C_I = 254$

(b) $T_R = \frac{1}{0.37} \approx 2.703$

(c) $T_R = \frac{1}{0.37} \approx 2.703$

(c) They are the same.

19. $T_R = \frac{12.3 \times 10^9}{220}$ seconds $\approx$ 647.1 days; $C(T_R) \approx 0.806\frac{\text{mg}}{\text{l}}$

21. (a)

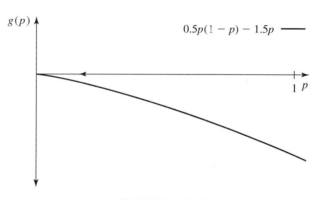

▲ **FIGURE 7.2.21**

(b) $p = 0$ is locally stable

(c) $g'(0) = -1 < 0$, which implies that 0 is locally stable.

23. (a) $\frac{dp}{dt}$ describes the rate of change of $p(t)$; $cp(1 - p - D)$ describes colonization of vacant undestroyed patches; $-mp$ describes extinction.

(b)

▲ **FIGURE 7.2.23**

(c) $D < 1 - \frac{m}{c}$; $p_1 = 0$ is unstable; $p_2 = 1 - D - \frac{m}{c}$ is locally stable

25. (a) $N = 0$, $N = 17$, and $N = 200$ **(b)** $N = 0$ is locally stable; $N = 17$ is unstable; $N = 200$ is locally stable

736

(c)

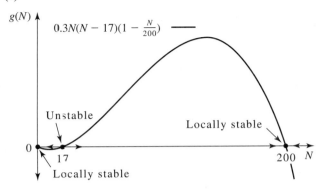

▲ **FIGURE 7.2.25**

Section 7.3

1. (a) $R_0 = 1.5 > 1$, the disease will spread **(b)** $R_0 = \frac{1}{2} < 1$, the disease will not spread

3. $R_0 = 0.9999 < 1$, the disease will not spread

5. (a)

$$\frac{dN}{dt} = N_I - 5N - 0.02NX + X$$
$$\frac{dX}{dt} = 0.02NX - 2X$$

(b) equilibrium: $(\hat{N}, \hat{X}) = (100, N_I - 500)$, this is a nontrivial equilibrium provided $N_I > 500$

7. (a)

$$\frac{dN}{dt} = 200 - N - 0.01NX + 2X$$
$$\frac{dX}{dt} = 0.01NX - 3X$$

(b)

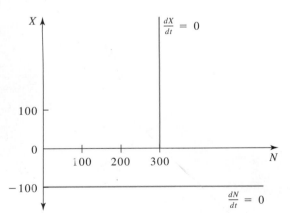

▲ **FIGURE 7.3.7**

(c) No nontrivial equilibria

9. (a) Equilibria: $(0, 0)$, $(0, 2/3)$, $(1/2, 0)$

(b) Since $\frac{dp_2}{dt} < 0$ when $p_1 = 1/2$ and p_2 is small, species 2 cannot invade.

11. (a)

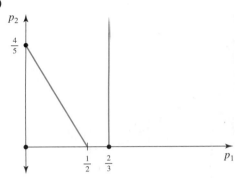

▲ **FIGURE 7.3.11**

(b) equilibria: $(0, 0)$, $(2/3, 0)$, $(0, 4/5)$

Section 7.5

1. (a) $\frac{dT}{dt}$ is proportional to the difference between the temperature of the object and the temperature of the surrounding medium.

(b) $t = \frac{1}{0.013} \ln \frac{9}{4} \approx 62.38$ minutes

3. (a) $N(t) = N(0)e^{r_e t}$ **(b)** $N(t) = \frac{K}{1 + \left(\frac{K}{N_0} - 1\right)e^{-r_l t}}$

(c) $r_e \approx 0.691$; $K = 1001$: $r_l \approx 1.382$; $K = 10,000$: $r_l \approx 0.701$

5. (a)

▲ **FIGURE 7.5.5(a)**

(c)

▲ **FIGURE 7.5.5(c)**

$N = 0$ is unstable; $N = 1000$ is locally stable
K is the carrying capacity

7. (b)

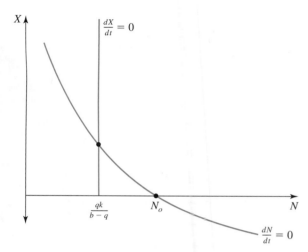

▲ **FIGURE 7.5.7**

(c) $\hat{N} = \frac{qk}{b-q} > 0$ if $b > q$, $\hat{X} = \left(k + \frac{qz}{b-q}\right)\frac{q}{b}\left(\frac{N_0(b-q)}{qk} - 1\right)$

Section 8.1

1. (a) $2\sqrt{5}$ **(b)** $2\sqrt{5}$

3. $\frac{8}{27}\left[(10)^{3/2} - \left(\frac{13}{4}\right)^{3/2}\right]$ **5.** $\frac{14}{3}$

7. $\int_{-1}^{1}\sqrt{1 + 4x^2}\,dx$ **9.** $\int_{0}^{1}\sqrt{1 + e^{-2x}}\,dx$

11. (a) $\frac{\pi}{2}$ **(b)** $\frac{\pi}{2}$ **13.** $\frac{3}{2}$

Section 8.2

1. Show that $f(x) \geq 0$ and $\int_{-\infty}^{\infty} f(x)\,dx = 1$

3. Show that $f(x) \geq 0$ and $\int_{-\infty}^{\infty} f(x)\,dx = 1$

5. $c = a/2$

7. (a) $F(r) = 1 - e^{-ar}$; $g(r) = ar$, straight line with slope a
(b) Find $-\ln(1 - F(r))$ for the given values of r, graph $-\ln(1 - F(r))$ as a function of r, the slope of the resulting straight line is $a = 3/20$

▲ **FIGURE 8.2.7**

9. (a)

▲ **FIGURE 8.2.9(a)**

(b)

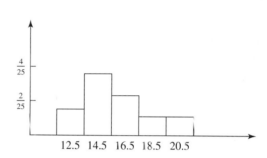

▲ **FIGURE 8.2.9(b)**

Section 8.3

1. (a)

k	12	13	14	15	16
p_k	$\frac{1}{25}$	$\frac{4}{25}$	$\frac{5}{25}$	$\frac{3}{25}$	$\frac{1}{25}$

k	17	18	19	20	21
p_k	$\frac{5}{25}$	$\frac{2}{25}$	$\frac{1}{25}$	$\frac{2}{25}$	$\frac{1}{25}$

(b) average value: $\frac{396}{25} = 15.84$

3. (a)

k	2	3	4	5	6	7	8	9	10
p_k	$\frac{1}{25}$	$\frac{1}{25}$	$\frac{2}{25}$	$\frac{3}{25}$	$\frac{4}{25}$	$\frac{2}{25}$	$\frac{6}{25}$	$\frac{3}{25}$	$\frac{3}{25}$

(b) average value: $\frac{171}{25} = 6.84$

5. mean of X: $\frac{1}{2}$ **7.** $\frac{a-1}{a-2}$

9. (a) $c = \frac{1}{\pi}$ **(b)** $\frac{1}{\pi}\int_{-\infty}^{\infty}\frac{x\,dx}{1+x^2}$ does not exist, hence the mean does not exist.

11. 0

738

13.

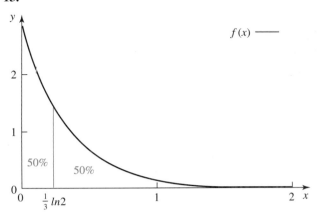

▲ **FIGURE 8.3.13**

15. The mean income is inflated by the income of the very rich people.

Section 8.4

1. (d)

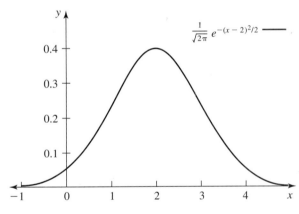

▲ **FIGURE 8.4.1**

3. 95%: (7.4, 18.2), 99%: (4.7, 20.9)

5. 50%   **7.** 99.5%   **9.** 2.5%

11. (a) 0.6915   **(b)** 0.383   **(c)** 0.1587
(d) 0.0668

13. (a) $x = 3.56$ **(b)** $x = 1.5$ **(c)** $x = 0.5$
(d) $x = 1.34$

15. (a) 0.0228   **(b)** $x = 628$

17. 0.76%

Section 8.5

1. $\frac{1}{3}e^{3t} - t - \frac{1}{3}$

3. $\frac{125}{6}$

5. (a) $N(x)\,dx$ is the number of individuals in the infinitesimal interval $[x, x + dx)$
(b) $\int_0^D N(x)\,dx$

7. (a) Euler's equation: $1 = \frac{m}{r}(1 - e^{-2r})$ **(b)** $m > \frac{1}{2}$

9. $r \approx 0.0792$

Section 8.7

1. (a)

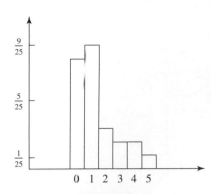

▲ **FIGURE 8.7.1**

(b)

k	0	1	2	3	4	5
$f(k)$	$\frac{8}{25}$	$\frac{9}{25}$	$\frac{3}{25}$	$\frac{2}{25}$	$\frac{2}{25}$	$\frac{1}{25}$

(c) $\frac{34}{25} = 1.36$

3. 0.5711

5. (a) $B = \int_T^\infty f(x)\,dx$

(c)

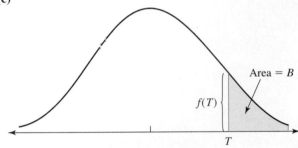

▲ **FIGURE 8.7.5**

Section 9.1

1. $\{(4, 3)\}$

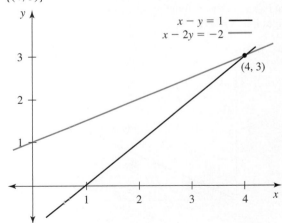

▲ **FIGURE 9.1.1**

3. No solution; the lines are parallel.

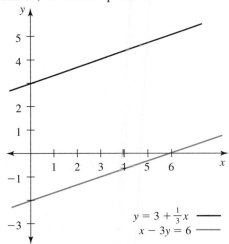

▲ **FIGURE 9.1.3**

5. (a) $c = 10$ (b) $c \neq 10$ (c) No

7. Zach bought five fish and six plants.

11. $x = 1, y = -1, z = 0$ **13.** $x = 2, y = 0, z = -3$

15. $x = 1, y = 1, z = -2$

17. $\{(x, y, z) : x = 2 - t, y = 1 + t, z = t, t \in \mathbf{R}\}$

19. underdetermined; $\{(x, y, z) : x = 7 + t, y = t + 2, z = t, t \in \mathbf{R}\}$

21. overdetermined; no solution

23. 750 gr of SL 24-4-8; 1000 gr of SL 21-7-12; $\frac{11000}{17}$ gr of SL 17-0-0.

Section 9.2

1. $\begin{bmatrix} 1 & -3 \\ 0 & -9 \end{bmatrix}$

3. $D = \begin{bmatrix} 1 & 0 \\ 4 & 11 \end{bmatrix}$

7. $\begin{bmatrix} 19 & 3 & 6 \\ 9 & 11 & 4 \\ 3 & -13 & -7 \end{bmatrix}$

9. $D = \begin{bmatrix} -4 & -2 & -7 \\ -5 & -1 & -3 \\ -1 & 5 & -1 \end{bmatrix}$

13. $A' = \begin{bmatrix} -1 & 2 \\ 0 & 1 \\ 3 & -4 \end{bmatrix}$

15. (a) $\begin{bmatrix} -2 & -3 \\ 0 & 5 \end{bmatrix}$

(b) $\begin{bmatrix} 1 & 6 \\ 2 & 2 \end{bmatrix}$

17. $AC = \begin{bmatrix} -1 & -2 \\ 1 & 0 \end{bmatrix}$

$CA = \begin{bmatrix} 1 & 4 \\ -1 & -2 \end{bmatrix}$

$AC \neq CA$

19. $(A + B)C = AC + BC = \begin{bmatrix} 1 & -1 \\ 0 & -3 \end{bmatrix}$

21. 3×2

23. (a) 1×4 (b) 3×3 (c) 4×3

25. (a) $\begin{bmatrix} 7 & 5 & 9 & -1 \\ -4 & -2 & -6 & 0 \end{bmatrix}$

(b) $\begin{bmatrix} 1 & -1 \\ 2 & 4 \\ 0 & -6 \\ -1 & -3 \end{bmatrix}$

27. $A^2 = \begin{bmatrix} 3 & -1 \\ 1 & 8 \end{bmatrix}$

$A^3 = \begin{bmatrix} 7 & 6 \\ -6 & -23 \end{bmatrix}$

$A^4 = \begin{bmatrix} 8 & -11 \\ 11 & 63 \end{bmatrix}$

31. $\begin{bmatrix} 2 & 3 & -1 \\ 0 & 2 & 1 \\ 1 & 0 & -2 \end{bmatrix} \begin{bmatrix} x_1 \\ x_2 \\ x_3 \end{bmatrix} = \begin{bmatrix} 0 \\ 1 \\ 2 \end{bmatrix}$

33. $\begin{bmatrix} 2 & -3 \\ -1 & 1 \\ 3 & 0 \end{bmatrix} \begin{bmatrix} x_1 \\ x_2 \end{bmatrix} = \begin{bmatrix} 4 \\ 3 \\ 4 \end{bmatrix}$

35. $AB = \begin{bmatrix} 1 & 0 \\ 0 & 1 \end{bmatrix}, BA = \begin{bmatrix} 1 & 0 \\ 0 & 1 \end{bmatrix}$

37. $A^{-1} = \begin{bmatrix} -3/5 & 1/5 \\ 2/5 & 1/5 \end{bmatrix}$

39. $A^{-1} = \begin{bmatrix} -3/5 & 1/5 \\ 2/5 & 1/5 \end{bmatrix}$ and show that $(A^{-1})^{-1} = A$

41. C does not have an inverse.

43. (a) $x_1 = 2, x_2 = 3$ (b) $A^{-1} = \begin{bmatrix} -1 & 0 \\ -2/3 & -1/3 \end{bmatrix}$, $x_1 = 2, x_2 = 3$

45. $\det A = 7$, A is invertible

47. $\det A = 0$, A is not invertible

49. (a) $\det A = 0$, A is not invertible

(b)

$$2x + 4y = b_1$$
$$3x + 6y = b_2$$

(c) solution: $\{(x, y) : x = \frac{3}{2} - 2t, y = t, t \in \mathbf{R}\}$

(d) The system has no solutions when $\frac{b_1}{2} \neq \frac{b_2}{3}$

51. $\det A = 2$, A is invertible

$A^{-1} = \begin{bmatrix} 1 & 1/2 \\ 0 & 1/2 \end{bmatrix}$

$X = \begin{bmatrix} 0 \\ 0 \end{bmatrix}$

53. C^{-1} does not exist. $\{(x, y) : x = -3t, y = t, t \in \mathbf{R}\}$

55. $\begin{bmatrix} 1/4 & 1/4 & 0 \\ -1/8 & 3/8 & 1/2 \\ -3/8 & 1/8 & -1/2 \end{bmatrix}$

57. $\begin{bmatrix} -2/3 & -1/6 & -1/3 \\ 0 & -1/2 & 0 \\ -1/3 & 1/6 & 1/3 \end{bmatrix}$

59. $L = \begin{bmatrix} 0 & 3.2 & 1.7 \\ 0.2 & 0 & 0 \\ 0 & 0.7 & 0 \end{bmatrix}$

$N(2) = \begin{bmatrix} 2232 \\ 580 \\ 280 \end{bmatrix}$

61. $L = \begin{bmatrix} 0 & 0 & 4.6 & 3.7 \\ 0.7 & 0 & 0 & 0 \\ 0 & 0.5 & 0 & 0 \\ 0 & 0 & 0.1 & 0 \end{bmatrix}$

$N(2) = \begin{bmatrix} 1242 \\ 934 \\ 525 \\ 25 \end{bmatrix}$

63. four age classes; 60% of one-year old survive until the end of the following breeding season; 2 is the average number of female offspring of a two-year-old.

65. four age classes; 20% of two-year old survive until the end of the following breeding season; 2.5 is the average number of female offspring of a one-year-old.

67. $q_0(t)$ and $q_1(t)$ seem to converge to 2.3; it appears that 74% of females will be of age 0 in the stable age distribution.

Section 9.3

3.

▲ **FIGURE 9.3.3**

length: $2\sqrt{2}$, angle $\frac{\pi}{4}$

5.

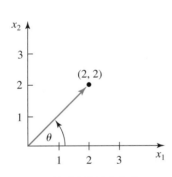

▲ **FIGURE 9.3.5**

length: 3, angle $\frac{\pi}{2}$

7.

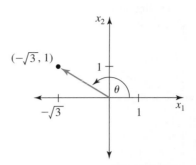

▲ **FIGURE 9.3.7**

length: 2, angle $\frac{5\pi}{6}$

9.

▲ **FIGURE 9.3.9**

$\begin{bmatrix} 4 \\ 4 \end{bmatrix}$

11.

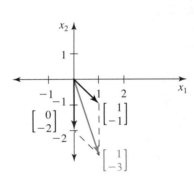

▲ **FIGURE 9.3.11**

$\begin{bmatrix} 1 \\ -3 \end{bmatrix}$

13.

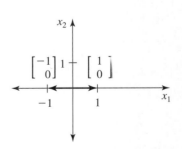

▲ **FIGURE 9.3.13**

$\begin{bmatrix} 0 \\ 0 \end{bmatrix}$

15.

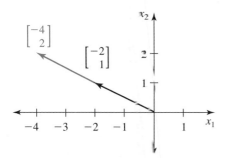

▲ **FIGURE 9.3.15**

$\begin{bmatrix} -4 \\ 2 \end{bmatrix}$

17.

▲ **FIGURE 9.3.17**

19.

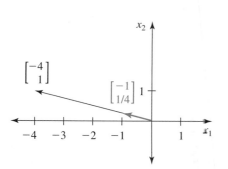

▲ **FIGURE 9.3.19**

$\begin{bmatrix} -1 \\ 1/4 \end{bmatrix}$

21.

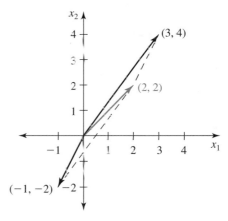

▲ **FIGURE 9.3.21**

$\begin{bmatrix} 2 \\ 2 \end{bmatrix}$

23.

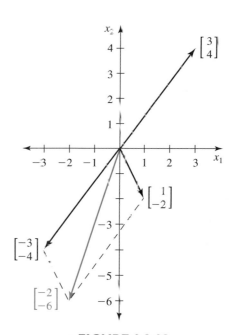

▲ **FIGURE 9.3.23**

$\begin{bmatrix} -2 \\ -6 \end{bmatrix}$

742

25.

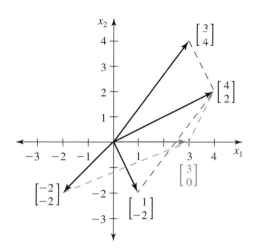

▲ **FIGURE 9.3.25**

$\begin{bmatrix} 3 \\ 0 \end{bmatrix}$

27. leaves **x** unchanged

29. counterclockwise rotation by $\theta = \frac{\pi}{2}$

31. counterclockwise rotation by $\theta = \frac{\pi}{6}$

33. $\lambda_1 = 2$, $\mathbf{v}_1 = \begin{bmatrix} 1 \\ 0 \end{bmatrix}$, $\lambda_2 = -1$, $\mathbf{v}_2 = \begin{bmatrix} -1 \\ 1 \end{bmatrix}$

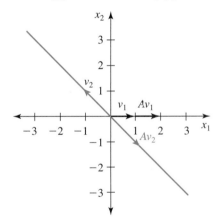

▲ **FIGURE 9.3.33**

35. $\lambda_1 = 1$, $\mathbf{v}_1 = \begin{bmatrix} 1 \\ 0 \end{bmatrix}$, $\lambda_2 = -1$, $\mathbf{v}_2 = \begin{bmatrix} 0 \\ 1 \end{bmatrix}$

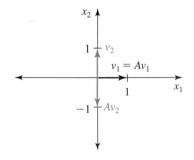

▲ **FIGURE 9.3.35**

37. $\lambda_1 = -2$, $\mathbf{v}_1 = \begin{bmatrix} 1 \\ 1 \end{bmatrix}$, $\lambda_2 = -1$, $\mathbf{v}_2 = \begin{bmatrix} 2 \\ 3 \end{bmatrix}$

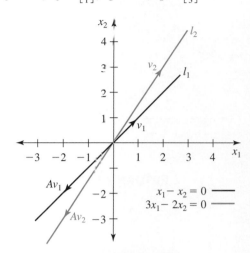

▲ **FIGURE 9.3.37**

39. $\lambda_1 = 1$, $\mathbf{v}_1 = \begin{bmatrix} 1 \\ -4 \end{bmatrix}$, $\lambda_2 = 4$, $\mathbf{v}_2 = \begin{bmatrix} 1 \\ 2 \end{bmatrix}$

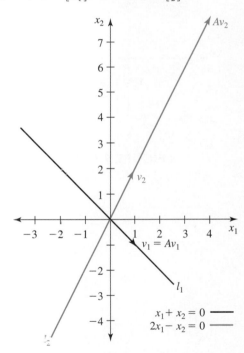

▲ **FIGURE 9.3.39**

41. The real parts of both eigenvalues are negative.

43. The real parts of both eigenvalues are not negative.

45. The real parts of both eigenvalues are negative.

47. (a) $l_1 \colon x_2 = 0$; $l_2 \colon -3x_1 + x_2 = 0$; since l_1 and l_2 are not identical, $\mathbf{u}_1$ and $\mathbf{u}_2$ are linearly independent.

(b) $\mathbf{x} = 2\mathbf{u}_1 - \mathbf{u}_2$

(c) $A^{20}\mathbf{x} = \begin{bmatrix} -1048574 \\ -3145728 \end{bmatrix}$

49. $\begin{bmatrix} -2 \\ 6 \end{bmatrix}$

51. (a) $\lambda_1 = 1 + \sqrt{2.2}, \lambda_2 = 1 - \sqrt{2.2}$

(b) The larger eigenvalue corresponds to the growth rate.

(c) 89.2% are in age class 0, 10.8% are in age class 1 in the stable age distribution.

53. (a) $\lambda_1 = \frac{1}{2}(7 + \sqrt{50.2}), \lambda_2 = \frac{1}{2}(7 - \sqrt{50.2})$

(b) The larger eigenvalue corresponds to the growth rate.

(c) 98.6% are in age class 0, 1.4% are in age class 1 in the stable age distribution.

55. (a) $\lambda_1 = \sqrt{0.45}, \lambda_2 = -\sqrt{0.45}$

(b) The larger eigenvalue corresponds to the growth rate.

(c) 88.17% are in age class 0, 11.83% are in age class 1 in the stable age distribution.

Section 9.4

1. (a) $\begin{bmatrix} -1 \\ 5 \\ -1 \end{bmatrix}$ **(b)** $\begin{bmatrix} 2 \\ 8 \\ -2 \end{bmatrix}$ **(c)** $\begin{bmatrix} 6 \\ -3 \\ 0 \end{bmatrix}$

3. $\begin{bmatrix} 2 \\ -2 \end{bmatrix}$ **5.** $\begin{bmatrix} -1 \\ -2 \\ 5 \end{bmatrix}$

7. $\sqrt{10}$ **9.** $\sqrt{26}$

11. $\begin{bmatrix} 1/\sqrt{11} \\ 3/\sqrt{11} \\ -1/\sqrt{11} \end{bmatrix}$ **13.** $\begin{bmatrix} 1 \\ 0 \\ 0 \end{bmatrix}$

15. 1 **17.** 2 **19.** $\sqrt{5}$ **21.** $\sqrt{30}$

23. $\cos\theta = 1/\sqrt{50}, \theta \approx 1.429$

25. $\cos\theta = 2/\sqrt{110}, \theta \approx 1.379$

27. $\begin{bmatrix} 1 \\ 1 \end{bmatrix}$ **29.** $\begin{bmatrix} -2 \\ 1 \\ 1 \end{bmatrix}$

31. $x + 2y = 4$ **33.** $4x + y = 2$

35. $-y + z = 1$ **37.** $x = 0$

Section 9.6

1. (a) $Ax = \begin{bmatrix} 0 \\ 2 \end{bmatrix}$

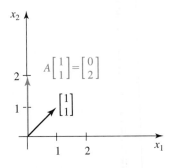

▲ **FIGURE 9.6.1(a)**

(b) $\lambda_1 = -1, \mathbf{u}_1 = \begin{bmatrix} 1 \\ 0 \end{bmatrix}, \lambda_2 = 2, \mathbf{u}_1 = \begin{bmatrix} 1 \\ 3 \end{bmatrix}$

(c)

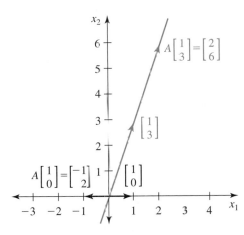

▲ **FIGURE 9.6.1(c)**

(d) $a_1 = 2/3, a_2 = 1/3$

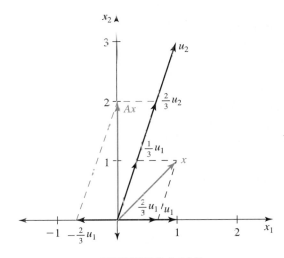

▲ **FIGURE 9.6.1(d)**

3. Growth rate: $\lambda_1 = 1.75$; a stable age distribution: $\begin{bmatrix} 35 \\ 10 \end{bmatrix}$

5. $\begin{bmatrix} -2 & 5 \\ -2 & 9 \end{bmatrix}$

7. 1. Gaussian elimination; 2. Write in matrix form $AX = B$ and find the inverse of A, then compute $X = A^{-1}B$

9. $a = -3$

11. For $\frac{5}{13} < a \le 1$, the population will grow.

744

1.

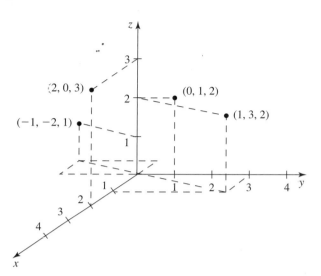

▲ **FIGURE 10.1.1**

3. $\frac{4}{13}$

5. $e^{-1/10}$

7. domain: $\{(x, y) : x \in \mathbf{R}, y \in \mathbf{R}\}$; range: $\{z : z \geq 0\}$; level curves: $x^2 + y^2 = c$, circle with radius $\sqrt{c}$ centered at $(0, 0)$

9. domain: $\{(x, y) : y > x^2, x \in \mathbf{R}\}$; range: $\{z : z \in \mathbf{R}\}$; level curves: $y = e^c + x^2$, parabolas shifted in $y > 0$ direction by e^c

11. Figure 10.22

13. Figure 10.24

15. level curve: $x^2 + y^2 = c$, circle centered at origin with radius $\sqrt{c}$; intersection with x-z plane: $z = x^2$, intersection with y-z plane: $z = y^2$

(b)

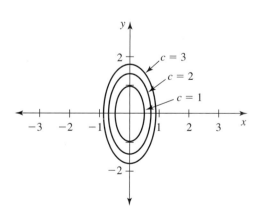

▲ **FIGURE 10.1.15(b)**

intersection with x-z plane: $z = 4x^2$, intersection with y-z plane: $z = y^2$

(c)

▲ **FIGURE 10.1.15(c)**

intersection with x-z plane: $z = \frac{1}{4}x^2$, intersection with y-z plane: $z = y^2$

(d) The intersection with the y-z plane is always $z = y^2$. When $a = 1$, the resulting surface could be obtained by rotating the curve $z = y^2$ about the z-axis. When $0 < a < 1$, then the resulting surface is still a paraboloid but longer along the x-axis than along the y-axis. When $a > 1$, the paraboloid is longer along the y-axis than along the x-axis.

17. Day 180: 22 m; day 200: 18 m; day 220: 16 m

Section 10.2

1.

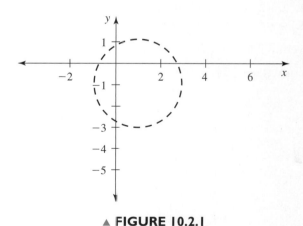

▲ **FIGURE 10.2.1**

$\{(x, y) : (x - 1)^2 + (y + 1)^2 < 4\}$

3. The boundary is a circle with radius 3, centered at $(0, 2)$. The boundary is not included.

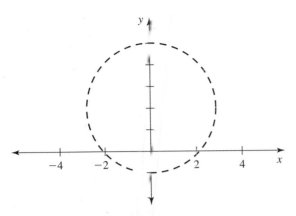

▲ **FIGURE 10.2.3**

5. Choose $2\delta^2 = \epsilon$. **7.** 1 **9.** $-\frac{1}{2}$

11. $-\frac{3}{2}$ **13.** $\frac{2}{3}$

15. Along positive x-axis: 1; along positive y-axis: -2

17. Along x-axis: 0; along y-axis: 0; along $y = x$: 2

19. Along $y = mx$, $m \neq 0$: 2; along $y = x^2$: 1; the limit does not exist.

21. 1. $f(x, y)$ is defined at $(0, 0)$. 2. $\lim_{(x,y)\to(0,0)}(x^2 + y^2)$ exists. 3. $f(0, 0) = 0 = \lim_{(x,y)\to(0,0)}(x^2 + y^2)$

23. In Problem 17, we showed that $\lim_{(x,y)\to(0,0)} \frac{4xy}{x^2+y^2}$ does not exist. Hence $f(x, y)$ is discontinuous at $(0, 0)$

25. The limit along $y = x^2$ as $(x, y) \to (0, 0)$ is different from the limit along $y = mx$ as $(x, y) \to (0, 0)$. Hence the function is discontinuous at $(0, 0)$.

27. (a) $h(x, y) = g[f(x, y)]$ with $f(x, y) = x^2 + y^2$ and $g(z) = \sin z$; (b) the function is continuous for all $(x, y) \in \mathbf{R}^2$

29. (a) $h(x, y) = g[f(x, y)]$ with $f(x, y) = xy$ and $g(z) = e^z$; (b) the function is continuous for all $(x, y) \in \mathbf{R}^2$

Section 10.3

1. $\frac{\partial f}{\partial x} = 2xy + y^2$, $\frac{\partial f}{\partial y} = x^2 + 2xy$

3. $\frac{\partial f}{\partial x} = \frac{3}{2}y\sqrt{xy} - \frac{2y}{3(xy)^{1/3}}$, $\frac{\partial f}{\partial y} = \frac{3}{2}x\sqrt{xy} - \frac{2x}{3(xy)^{1/3}}$

5. $\frac{\partial f}{\partial x} = \cos(x + y)$, $\frac{\partial f}{\partial y} = \cos(x + y)$

7. $\frac{\partial f}{\partial x} = -4x\cos(x^2 - 2y)\sin(x^2 - 2y)$, $\frac{\partial f}{\partial y} = 4\cos(x^2 - 2y)\sin(x^2 - 2y)$

9. $\frac{\partial f}{\partial x} = e^{\sqrt{x+y}}\frac{1}{2\sqrt{x+y}}$, $\frac{\partial f}{\partial y} = e^{\sqrt{x+y}}\frac{1}{2\sqrt{x+y}}$

11. $\frac{\partial f}{\partial x} = e^x\sin(xy) + e^x\cos(xy)y$, $\frac{\partial f}{\partial y} = e^x\cos(xy)x$

13. $\frac{\partial f}{\partial x} = \frac{2}{2x+y}$, $\frac{\partial f}{\partial y} = \frac{1}{2x+y}$

15. $\frac{\partial f}{\partial x} = \frac{-2x}{(\ln 3)(y^2 - x^2)}$, $\frac{\partial f}{\partial y} = \frac{2y}{(\ln 3)(y^2 - x^2)}$

17. 6 **19.** $3e^5$ **21.** 1 **23.** $\frac{2}{9}$

25. $f_x(1, 1) = -2$, $f_y(1, 1) = -2$

27. (a) $\frac{\partial P_e}{\partial a} > 0$: the number of prey items eaten increases with increasing attack rate.

(b) $\frac{\partial P_e}{\partial T} > 0$: the number of prey items eaten increases with increasing T.

29. $\frac{\partial f}{\partial x} = 2xz - y$, $\frac{\partial f}{\partial y} = z^2 - x$, $\frac{\partial f}{\partial z} = 2yz + x^2$

31. $\frac{\partial f}{\partial x} = 3x^2y^2z + \frac{1}{yz}$, $\frac{\partial f}{\partial y} = 2x^3yz - \frac{x}{y^2z}$, $\frac{\partial f}{\partial z} = x^3y^2 - \frac{x}{yz^2}$

33. $\frac{\partial f}{\partial x} = e^{x+y+z}$, $\frac{\partial f}{\partial y} = e^{x+y+z}$, $\frac{\partial f}{\partial z} = e^{x+y+z}$

35. $\frac{\partial f}{\partial x} = \frac{1}{x+y+z}$, $\frac{\partial f}{\partial y} = \frac{1}{x+y+z}$, $\frac{\partial f}{\partial z} = \frac{1}{x+y+z}$

37. $2y$ **39.** e^y **41.** $2\sec^2(u+w)\tan(u+w)$

43. $-6x\sin y$ **45.** $\frac{2}{(x+y)^3}$

47. (a) $\frac{\partial f}{\partial N} > 0$: the number of prey encounters per predator increases as the prey density increases.

(b) $\frac{\partial f}{\partial T} > 0$: the function increases as the time for search increases.

(c) $\frac{\partial f}{\partial T_h} < 0$: the function decreases as the handling time T_h increases.

(d)

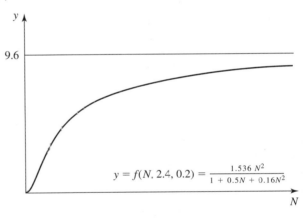

$$y = f(N, 2.4, 0.2) = \frac{1.536\, N^2}{1 + 0.5N + 0.16N^2}$$

▲ **FIGURE 10.3.47**

49.

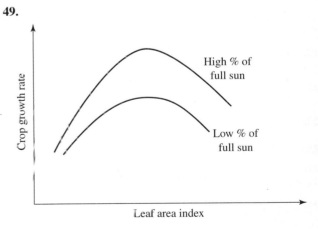

▲ **FIGURE 10.3.49**

Section 10.4

1. $7 = 6x + 4y - z$ **3.** $z - 2ex = -e$ **5.** $x - z = 1$

7. $f(x, y)$ is defined in an open disk centered at $(1, 1)$ and is continuous at $(1, 1)$.

9. $f(x, y)$ is defined in an open disk centered at $(0, 0)$ and is continuous at $(0, 0)$.

11. $f(x, y)$ is defined in an open disk centered at $(-1, 2)$ and is continuous at $(-1, 2)$.

13. $L(x, y) = \frac{1}{2}x + 2y + \frac{1}{2}$ **15.** $L(x, y) = x + y$

17. $L(x, y) = x + \frac{1}{2}y - \frac{3}{2} + \ln 2$ **19.** $L(x) = 1 + x$

21. $Df(x, y) = \begin{bmatrix} 1 & 1 \\ 2x & -2y \end{bmatrix}$

23. $Df(x, y) = \begin{bmatrix} e^{x-y} & -e^{x-y} \\ e^{x+y} & e^{x+y} \end{bmatrix}$

25. $Df(x, y) = \begin{bmatrix} -\sin(x - y) & \sin(x - y) \\ -\sin(x + y) & -\sin(x + y) \end{bmatrix}$

27. $Df(x, y) = \begin{bmatrix} 4xy + 1 & 2x^2 - 3 \\ e^x \sin y & e^x \cos y \end{bmatrix}$

29. $L(x, y) = \begin{bmatrix} 4x + 2y - 4 \\ -x - y + 3 \end{bmatrix}$

31. $L(x, y) = \begin{bmatrix} e(2x - y) \\ 2x - y - 1 \end{bmatrix}$

33. $L(x, y) = \begin{bmatrix} x - y + 1 \\ y - x + 1 \end{bmatrix}$

Section 10.5

1. $18 \ln 2 + 8$ **3.** $\dfrac{\frac{\pi}{3} + \frac{\sqrt{3}}{4}}{\sqrt{\frac{\pi^2}{9} + \frac{3}{4}}}$ **5.** 0

7. $\frac{dz}{dt} = \frac{\partial f}{\partial x}u'(t) + \frac{\partial f}{\partial y}v'(t)$

9. $-\frac{2x}{2y + x^2 + y^2}$

11. $-\frac{2x - 3y(x^2 + y^2)}{2y - 3x(x^2 + y^2)}$

13. $\frac{dy}{dx} = -\frac{1}{\sqrt{1 - x^2}}$ for $0 \le y \le \pi$ and $-1 \le x \le 1$

15. The growth rate decreases over time.

17. $\operatorname{grad} f = \begin{bmatrix} 3x^2 y^2 \\ 2x^3 y \end{bmatrix}$

19. $\operatorname{grad} f = \frac{1}{2\sqrt{x^3 - 3xy}} \begin{bmatrix} 3x^2 - 3y \\ -3x \end{bmatrix}$

21. $\operatorname{grad} f = \frac{\exp[\sqrt{x^2 + y^2}]}{\sqrt{x^2 + y^2}} \begin{bmatrix} x \\ y \end{bmatrix}$

23. $\operatorname{grad} f = \frac{x^2 - y^2}{x^2 + y^2} \begin{bmatrix} \frac{1}{z} \\ -\frac{1}{y} \end{bmatrix}$

25. $\frac{2}{3}\sqrt{3}$ **27.** $-\sqrt{2}$ **29.** $\frac{3}{2}\sqrt{10}$

31. $f(x, y)$ increases most rapidly in the direction of $\begin{bmatrix} 5 \\ -3 \end{bmatrix}$ at the point $(-1, 1)$.

33. $f(x, y)$ increases most rapidly in the direction of $\begin{bmatrix} 5/4 \\ -3/4 \end{bmatrix}$ at the point $(5, 3)$.

35. $\begin{bmatrix} 3/5 \\ 4/5 \end{bmatrix}$ **37.** $\frac{1}{\sqrt{733}} \begin{bmatrix} 2 \\ -27 \end{bmatrix}$

39. The amoeba move in the direction of $\begin{bmatrix} -1 \\ -3 \end{bmatrix}$

43. (b)

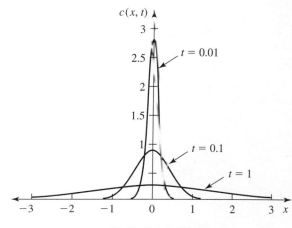

▲ **FIGURE 10.5.43**

Section 10.7

1.

▲ **FIGURE 10.7.1(i)**

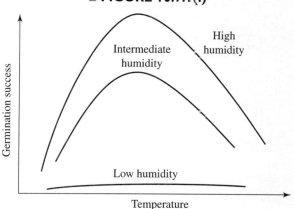

▲ **FIGURE 10.7.1(ii)**

3. (a) $\frac{\partial A_i}{\partial F} > 0$, $\frac{\partial A_i}{\partial D} < 0$

(b) $\frac{\partial A_e}{\partial F} > 0$: area covered by introduced species increased with the amount of fertilizer added.

$\frac{\partial A_e}{\partial D} > 0$: area covered by introduced species increased with disturbance intensity (note that this is the opposite to (a)).

(c) Fertilization had a positive effect in both cases. That is, fertilization increased the total area covered of both introduced and indigenous species. Disturbance intensity had a negative effect on indigenous species and a positive effect on introduced species.

5. $D\mathbf{f}(x, y) = \begin{bmatrix} 2x & -1 \\ 3x^2 & -2y \end{bmatrix}$

7. (a)

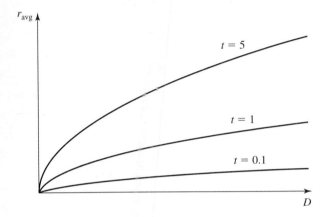

▲ **FIGURE 10.7.7**

(b) Use (a): $r_{avg}^2 = \pi D t$ and solve for D

(c) r_{avg} = arithmetic average = $\frac{1}{N} \sum_{i=1}^{N} d_i$ and use formula for D in (b)

Section 11.1

1. $\frac{d\mathbf{x}}{dt} = \begin{bmatrix} 2 & 3 \\ -4 & 1 \end{bmatrix} \mathbf{x}(t)$ **3.** $\frac{d\mathbf{x}}{dt} = \begin{bmatrix} 0 & 0 & 1 \\ -1 & 0 & 0 \\ 1 & 1 & 0 \end{bmatrix} \mathbf{x}(t)$

5. $(1, 0)$: $\begin{bmatrix} -1 \\ 1 \end{bmatrix}$; $(0, 1)$: $\begin{bmatrix} 2 \\ 0 \end{bmatrix}$; $(-1, 0)$: $\begin{bmatrix} 1 \\ -1 \end{bmatrix}$; $(0, -1)$: $\begin{bmatrix} -2 \\ 0 \end{bmatrix}$; $(1, 1)$: $\begin{bmatrix} 1 \\ 1 \end{bmatrix}$; $(0, 0)$: $\begin{bmatrix} 0 \\ 0 \end{bmatrix}$; $(-2, 1)$: $\begin{bmatrix} 4 \\ -2 \end{bmatrix}$;

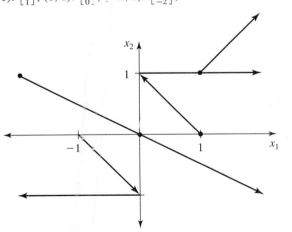

▲ **FIGURE 11.1.5**

7. $(1, 0)$: $\begin{bmatrix} 1 \\ -1 \end{bmatrix}$; $(0, 1)$: $\begin{bmatrix} 3 \\ 2 \end{bmatrix}$; $(-1, 1)$: $\begin{bmatrix} 2 \\ 3 \end{bmatrix}$; $(0, -1)$: $\begin{bmatrix} -3 \\ -2 \end{bmatrix}$; $(-3, 1)$: $\begin{bmatrix} 0 \\ 5 \end{bmatrix}$; $(0, 0)$: $\begin{bmatrix} 0 \\ 0 \end{bmatrix}$; $(-2, 1)$: $\begin{bmatrix} 1 \\ 4 \end{bmatrix}$;

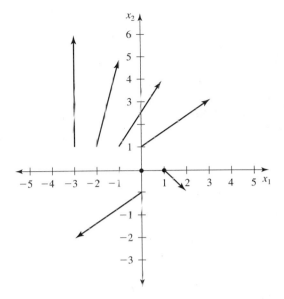

▲ **FIGURE 11.1.7**

9. Figure 11.18: (d); Figure 11.19: (a); Figure 11.20: (c); Figure 11.21: (b)

11.

▲ **FIGURE 11.1.11**

13. $\mathbf{x}(t) = c_1 e^{6t} \begin{bmatrix} 3 \\ 5 \end{bmatrix} + c_2 e^{-2t} \begin{bmatrix} 1 \\ -1 \end{bmatrix}$

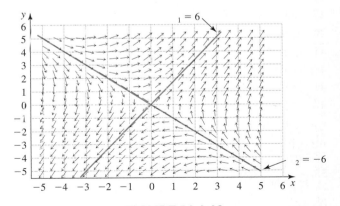

▲ **FIGURE 11.1.13**

748

15. $\mathbf{x}(t) = c_1 e^{-5t} \begin{bmatrix} -3 \\ 2 \end{bmatrix} + c_2 e^{6t} \begin{bmatrix} 1 \\ 3 \end{bmatrix}$

(c)

▲ **FIGURE 11.1.15**

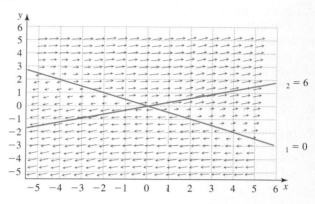

▲ **FIGURE 11.1.57**

All direction vectors are parallel. The solutions move parallel to the eigendirection of $\lambda_2 = 6$.

17. $\mathbf{x}(t) = c_1 e^{-2t} \begin{bmatrix} 1 \\ 1 \end{bmatrix} + c_2 e^{t} \begin{bmatrix} 0 \\ 1 \end{bmatrix}$

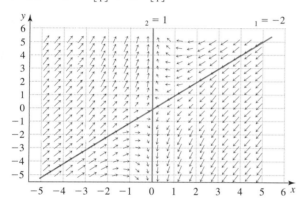

▲ **FIGURE 11.1.17**

19. $\mathbf{x}(t) = e^{-3t} \begin{bmatrix} -5 \\ 4 \end{bmatrix} + e^{2t} \begin{bmatrix} 0 \\ 1 \end{bmatrix}$

21. $\mathbf{x}(t) = e^{t} \begin{bmatrix} 1 \\ 1 \end{bmatrix}$

23. $\mathbf{x}(t) = \frac{13}{8} e^{-3t} \begin{bmatrix} 1 \\ -1 \end{bmatrix} - \frac{3}{8} e^{5t} \begin{bmatrix} 7 \\ 1 \end{bmatrix}$

27. saddle

29. unstable node

31. stable node

33. saddle

35. saddle

37. stable spiral

39. unstable spiral

41. neutral spiral

43. neutral spiral

45. stable spiral

47. saddle

49. stable spiral

51. saddle

53. stable node

55. unstable spiral

57. (a) $\lambda_1 = 0, \mathbf{v}_1 = \begin{bmatrix} 2 \\ -1 \end{bmatrix}; \lambda_2 = 6, \mathbf{v}_1 = \begin{bmatrix} 4 \\ 1 \end{bmatrix}$

(b) $\mathbf{x}(t) = c_1 \begin{bmatrix} 2 \\ -1 \end{bmatrix} + c_2 e^{6t} \begin{bmatrix} 4 \\ 1 \end{bmatrix}$

Section 11.2

1.
$$\frac{dx_1}{dt} = -0.55x_1 + 0.1x_2$$
$$\frac{dx_2}{dt} = 0.5x_1 - 0.12x_2$$

$(0, 0)$ is a stable node.

3.
$$\frac{dx_1}{dt} = -2.5x_1 + 0.7x_2$$
$$\frac{dx_2}{dt} = 2.5x_1 - 0.8x_2$$

$(0, 0)$ is a stable node.

5.
$$\frac{dx_1}{dt} = -0.1x_1$$
$$\frac{dx_2}{dt} = -0.3x_2$$

$(0, 0)$ is a stable node.

7. $a = 0.1, b = 0.3, c = 0.3, d = 0.2$

9. $a = 0.3, b = 0, c = 0.9, d = 0.2$

11. $a = 0, b = 0, c = 0.2, d = 0.3$

13. $x_1(t) = 4e^{-0.3t}, x_2(t) = 4(1 - e^{-0.3t})$

15. (a) $a = 0.2, b = 0.1, c = 0, d = 0$ **(b)** The constant is the total area.

(d) $x_2(t) = 20 - x_1(t) = \frac{40}{3} + \frac{14}{3} e^{-0.3t}$, $\lim_{t \to \infty} x_1(t) \frac{20}{3}$, $\lim_{t \to \infty} x_2(t) \frac{40}{3}$

17. $x(t) = 3\sin(2t)$

19. $\frac{dx}{dt} = v, \frac{dv}{dt} = 3x$

21. $\frac{dx}{dt} = v, \frac{dv}{dt} = x - v$

Section 11.3

1. saddle **3.** saddle

5. $(0, 0)$: unstable node; $(0, 4/5)$: saddle; $(1/2, 0)$: saddle; $(0.5, 0.3)$: stable node

7. $(0, 0)$: unstable node; $(0, 1)$: saddle; $(1, 0)$: unstable node; $(1/2, 1)$: saddle

9. $(0, 0)$: saddle; $(1, 1)$: unstable spiral

11. (a)

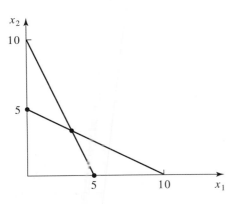

▲ **FIGURE 11.3.11**

(b) $(10/3, 10/3)$ is a stable node.

13. tr < 0, det $=$?

15. tr $=$?, det $=$?

17. tr < 0, det > 0; equilibrium is locally stable

19. (a)

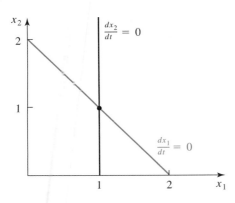

▲ **FIGURE 11.3.19**

(b) tr < 0, det > 0; $(1, 1)$ is locally stable

Section 11.4

1.

$$\frac{dN_1}{dt} = 2\left(1 - \frac{N_1}{20} - \frac{N_2}{100}\right)$$

$$\frac{dN_2}{dt} = 3\left(1 - \frac{N_2}{15} - \frac{N_1}{75}\right)$$

3. Species 2 outcompetes species 1.

5. Founder control

7. $(0, 0)$: unstable (source); $(18, 0)$: unstable (saddle); $(0, 20)$: stable (sink)

9. $(0, 0)$: unstable (source); $(35, 0)$: stable (sink); $(0, 40)$: stable (sink); $(85/11, 100/11)$: unstable (center)

11. $(\alpha_{12}, \alpha_{21}) = (1/4, 7/18)$

13.

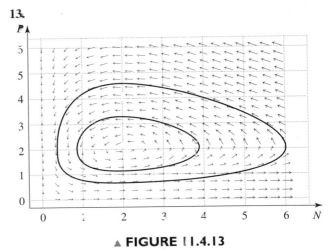

▲ **FIGURE 11.4.13**

15. (a) trivial equilibrium: $(0, 0)$; nontrivial equilibrium: $(3/2, 1/4)$

(b) $\lambda_1 = 1$, $\lambda_2 = -3$: $(0, 0)$ is an unstable saddle

(c) $\lambda_1 = i\sqrt{3}$, $\lambda_2 = -i\sqrt{3}$: purely imaginary eigenvalues, linear stability analysis cannot be used to infer stability of equilibrium.

(d)

▲ **FIGURE 11.4.15(i)**

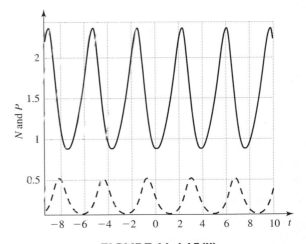

▲ **FIGURE 11.4.15(ii)**

17. (a) $\frac{dN}{dt} = 5N$, $N(t) = N(0)e^{5t}$; in the absence of the predator, the insect species grows exponentially fast.

(b) If $P(t) > 0$, then $N(t)$ stays bounded.

(c)

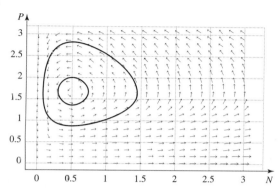

▲ **FIGURE 11.4.17**

By spraying the field, the solution moves to a different cycle; this results in a much larger insect outbreak later in the year compared to before the spraying.

19. (a) When $P = 0$, then $\frac{dN}{dt} = 3N\left(1 - \frac{N}{10}\right)$; equilibria: $\hat{N} = 0$ (unstable) and $\hat{N} = 10$ (locally stable). If $N(0) > 0$, then $\lim_{t\to\infty} N(t) = 10$

(b) $(0, 0)$: unstable (saddle); $(10, 0)$: unstable (saddle); $(4, 0.9)$: stable spiral

(c)

▲ **FIGURE 11.4.19(i)**

▲ **FIGURE 11.4.19(ii)**

21. (a)

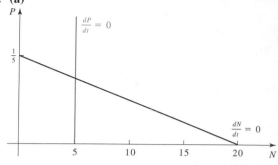

▲ **FIGURE 11.4.21**

(b) $\mathrm{tr}D\mathbf{f}(\hat{N}, \hat{P}) < 0$ and $\det D\mathbf{f}(\hat{N}, \hat{P}) > 0$: the nontrivial equilibrium is locally stable.

23. $\hat{N} = \frac{d}{c}$ does not depend on a; hence, it remains unchanged if a changes. $\hat{P} = \frac{a}{b}(1 - \frac{d/c}{K})$ is an increasing function of a; hence, the predator equilibrium increases when a increases.

25. $\hat{N} = \frac{d}{c}$ is a decreasing function of c; hence, the prey abundance decreases as c increases. $\hat{P} = \frac{a}{b}(1 - \frac{d/c}{K})$ is an increasing function of c; hence, the predator abundance increases as c increases.

27. predation; locally stable

29. mutualism; locally stable

31. mutualism; unstable

33. competition; unstable

35. $\begin{bmatrix} - & - \\ - & - \end{bmatrix}$; trace negative; determinant undetermined

37. $\begin{bmatrix} - & - \\ - & - \end{bmatrix}$; trace negative; determinant undetermined

39. $\begin{bmatrix} - & - \\ - & - \end{bmatrix}$; trace negative; determinant undetermined

41. If $a_{ii} < 0$, then the growth rate of species i is negatively affected by an increase in the density of species i. This is referred to as self regulation.

43. (a) $(\hat{N}, \hat{P}) = (\frac{d}{c}, \frac{a}{b})$ **(b)** $\begin{bmatrix} 0 & -b\frac{d}{c} \\ c\frac{a}{b} & 0 \end{bmatrix}$

(c) $a_{11} = a_{22} = 0$: neither species has an effect on itself; $a_{12} = -b\frac{d}{c} < 0$: prey is affected negatively by predators; $a_{21} = c\frac{a}{b} > 0$: predators are affected positively by prey.

45.

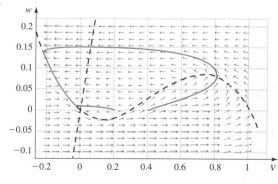

▲ **FIGURE 11.4.45**

47. $V(0) > 0.3$

49. $\frac{dc}{dt} = kab$

51. $e = [E], s = [S], c = [ES], p = [P], \frac{de}{dt} = -k_1es + k_2c,$ $\frac{ds}{dt} = -k_1es, \frac{dc}{dt} = k_1es - k_2c, \frac{dp}{dt} = k_2c$

53. $\frac{dx}{dt} + \frac{dy}{dt} = 0, x(t) + y(t)$ is constant.

55. $\frac{dx}{dt} + \frac{dy}{dt} + \frac{dz}{dt} = 0, x(t) + y(t) + z(t)$ is constant.

57. **(c)**

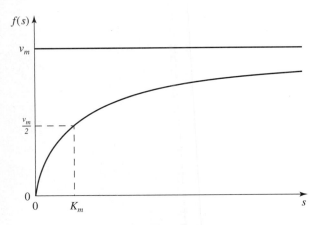

▲ FIGURE 11.4.57

(d) Since $\frac{dp}{dt} = f(s)$, the reaction rate is a function of s and hence, the availability of s determines the reaction rate.

59. **(a)** $\hat{s} = \frac{DK_m}{Yv_m - D}$; $\hat{s}$ is an increasing function of D.

(b)

▲ FIGURE 11.4.59

The s coordinate of the point of intersection of the graph of $q(s)$ and the horizontal line $f(s) = D/Y$ is the equilibrium.

61. $(4, 0)$: unstable; $(2, 2)$: stable

Section 11.6

1. $Z(t) = Z(0)e^{(r_1 - r_2)t}$

3. **(a)**

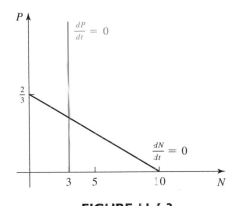

▲ FIGURE 11.6.3

(b) The nontrivial equilibrium is locally stable.

5. **(a)** If $c_1 > m_1$, then there exists a nontrivial equilibrium in which species 1 has positive density and species 2 is absent.

7. **(a)** $\hat{x} = Y(s_0 - \hat{s}) > 0$ when $\hat{s} < s_0$

(b) $\frac{\partial \hat{s}}{\partial D} > 0; \frac{\partial \hat{s}}{\partial Y} < 0$

Section 12.1

1. 40

3. 120

5. 84

7. 120

9. 5040

11. 358, 800

13. 2730

15. 120

17. 120

19. 1365

21. 126

23. **(a)** exactly two red balls: 10; exactly two blue balls: 6; one of each: 20

(b) total: 36

25. 168, 168

27. $\emptyset, \{a\}, \{b\}, \{c\}, \{a, b\}, \{a, c\}, \{b, c\}, \{a, b, c\}$; to form a subset, we need to decide for each element whether it should be in the subset. There are $2^3 = 8$ choices.

29. 12

31. 30

33. 31

Section 12.2

1. $\Omega = \{HHH, HHT, HTH, THH, HTT, THT, TTH, TTT\}$

3. $\Omega = \{(i, j) : 1 \le i < j \le 5\}$

5. $A \cup B = \{1, 2, 3, 5\}, A \cap B = \{1, 3\}$

7. $\{4, 6\}$

9. 0.6

11. 0.25

13. 0.5

15. (a)

▲ **FIGURE** 12.2.15(a)

(b)

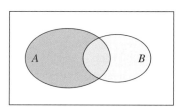

▲ **FIGURE** 12.2.15(b)

17. $\frac{3}{4}$

19. $\frac{5}{36}$

21. $\frac{5}{12}$

23. $\frac{1}{4}$

25. $\frac{1}{2}$

27. $\frac{1}{8}$

29. (a) $\frac{\binom{N-100}{7}\binom{100}{3}}{\binom{N}{10}}$

(b) 333

31. $\frac{1}{17}$

33. $\frac{12}{55}$

Section 12.3

1. $\frac{12}{51}$

3. $\frac{2}{5}$

5. $\frac{2}{5}$

7. $\frac{1}{5}$

9. $\frac{4}{7}$

11. 0.8425

13. $\frac{1}{2}$

15. $P(\text{first card is an ace}) = P(\text{second card is an ace}) = \frac{1}{13}$

17. 0.3

19. $\frac{3}{4}$

21. A and B are independent.

23. A and B are not independent.

25. (a) $\frac{1}{8}$ (b) $\frac{7}{8}$ (c) $\frac{1}{2}$ (d) $\frac{7}{8}$

27. $\left(\frac{1}{4}\right)^{10} + (10)\left(\frac{3}{4}\right)\left(\frac{1}{4}\right)^{9}$

29. $1 - (0.9)^{10}$

31. 0.1624

33. $\frac{1}{3}$

35. $\frac{1}{3}$

37. (a) $\frac{1}{2}$ (b) $\frac{1}{2}$ (c) $\frac{1}{3}$

Section 12.4

1. $P(X = 0) = \frac{1}{4}$, $P(X = 1) = \frac{1}{2}$, $P(X = 2) = \frac{1}{4}$

3. $P(X = 0) = \frac{\binom{3}{0}\binom{2}{2}}{\binom{5}{2}}$, $P(X = 1) = \frac{\binom{3}{1}\binom{2}{1}}{\binom{5}{2}}$, $P(X = 2) = \frac{\binom{3}{2}\binom{2}{0}}{\binom{5}{2}}$

5. $F(x) = \begin{cases} 0, & x < -3 \\ 0.2, & -3 \le x < -1 \\ 0.5, & -1 \le x < 1.5 \\ 0.9, & 1.5 \le x < 2 \\ 1, & x \ge 2 \end{cases}$

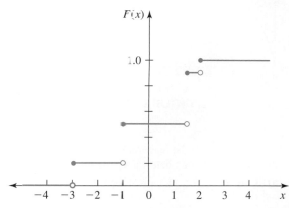

▲ **FIGURE** 12.4.5

7. $P(X = -1) = 0.2$, $P(X = 0) = 0.1$, $P(X = 1) = 0.4$, $P(X = 2) = 0.3$

9. (a) $N = 55$ (b) $\frac{28}{55}$

11. $EX = -0.1$, $\text{var}(X) = 3.39$, s.d. = 1.8412

13. (a) $EX = \frac{55}{10}$ (b) $\text{var}(X) = 8.25$

17. (a) $EX = 0.75$, $EY = 0.3$ (b) $E(X + Y) = 1.05$ (c) $\text{var}(X) = 1.7875$, $\text{var}(Y) = 3.01$ (d) $\text{var}(X + Y) = \text{var}(X) + \text{var}(Y) = 4.7975$

19. (a) Since $(X - EX)^2 \ge 0$, it follows that $E(X - EX)^2 \ge 0$; therefore, $\text{var}(X) \ge 0$ (b) Since $\text{var}(X) = EX^2 - (EX)^2 \ge 0$, it follows that $EX^2 \ge (EX)^2$

21. (a) $\binom{10}{5}(0.5)^{10}$ (b) $(0.5)^{10}\left[\binom{10}{8} + \binom{10}{9} + \binom{10}{10}\right]$ (c) $1 - (0.5)^{10}$

23. $P(X = k) = \binom{6}{k}\left(\frac{1}{6}\right)^k\left(\frac{5}{6}\right)^{6-k}$, $k = 0, 1, 2, \ldots, 6$

25. $(0.8)^{20}$

27. (a) 1 (b) $(10)(0.9)^{10}$

29. 12.5

31. $\frac{30!}{10!14!6!}(0.2)^{10}(0.35)^{14}(0.45)^{6}$

33. $\frac{40!}{20!10!8!2!}\left(\frac{9}{16}\right)^{20}\left(\frac{3}{16}\right)^{10}\left(\frac{3}{16}\right)^{8}\left(\frac{1}{16}\right)^{2}$

35. $\frac{6!}{2!2!2!} \left(\frac{6}{24}\right)^2 \left(\frac{8}{24}\right)^2 \left(\frac{10}{24}\right)^2$

37. $\frac{23!}{5!12!6!} \left(\frac{1}{4}\right)^5 \left(\frac{1}{2}\right)^{12} \left(\frac{1}{4}\right)^6$

Section 12.5

1. distribution function: $F(x) = 1 - e^{-3x}$ for $x \geq 0$, $F(x) = 0$ for $x \leq 0$

3. $c = \frac{1}{\pi}$

5. $EX = \frac{1}{2}$, $\text{var}(X) = \frac{1}{4}$

7. $EX = \frac{3}{2}$, $\text{var}(X) = \frac{3}{4}$

9. (b) $EX = \frac{a-1}{a-2}$

11. 0.8185

15. $E|X| = 2$

17. 0.1

19. $a = 1$, $b = 7$

21. $EY = \frac{3}{4}$

23. $HHTHHTHHHT$

25. $EX = \frac{1}{\lambda}$

27. $e^{-4/3}$

29. (a) $e^{-20/27}$ **(b)** $e^{-20/27}$

Section 12.6

1. median: 15; sample mean: 16.2; sample variance: 180.2

3. $\bar{X} = 11.93$; $S^2 = 3.389$

5. $\bar{X} = 5.69$; $S^2 = 3.465$

11. (c) true values: $\mu = 0.5$, $\sigma^2 = \frac{1}{12}$

13. $\bar{X} = 16.2$; S.E. $= 4.245$

15. $[-0.3993, 0.5213]$

19. $\hat{p} = 0.72$; $[0.651, 0.789]$

21. $y = 1.92x - 0.92$; $r^2 = 0.9521$

25. $y = 0.201x + 0.481$; $r^2 = 0.841$

Section 12.8

1. (a) 0.431

3. 168, 168, 000

5. (a) $EX = 4.2$ **(b)** 0.0043 **(c)** 0.0215 **(d)** $1 - (0.0043)^5$

7. (a) 0.16 **(b)** 0.2

9. (a) $\mu = 170$, $\sigma = 6.098$ **(b)** 0.4013

11. 20%

13. (b) $EV = \frac{n-1}{n}\sigma^2$

Appendices

◆ A FREQUENTLY USED SYMBOLS

A.1 Greek Letters

Lowercase Letters

α	alpha	η	eta	ν	nu	τ	tau
β	beta	θ	theta	ξ	xi	υ	upsilon
γ	gamma	ι	iota	o	omicron	ϕ	phi
δ	delta	κ	kappa	π	pi	χ	chi
ϵ	epsilon	λ	lambda	ρ	rho	ψ	psi
ζ	zeta	μ	mu	σ	sigma	ω	omega

Uppercase Letters

Γ	Gamma	Λ	Lambda	Σ	Sigma
Δ	Delta	Π	Pi	Ω	Omega

A.2 Mathematical Symbols

$<$	less than	$\subset$	subset	$=$	equal to
$\leq$	less than or equal to	$\in$	element of	$\neq$	not equal to
$>$	greater than	$\perp$	perpendicular	$\approx$	approximately
$\geq$	greater than or equal to	$\parallel$	parallel	$\propto$	proportional to
$\cup$	union	$\cap$	intersection	Σ	sum

◆ B TABLE OF THE STANDARD NORMAL DISTRIBUTION

Areas under the Standard Normal Curve from $-\infty$ to z (see Figure B.1).

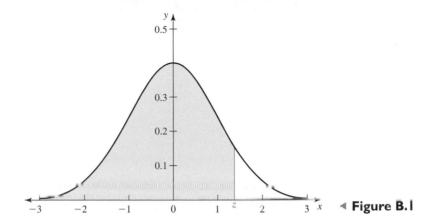

◄ **Figure B.I**

z	0	1	2	3	4	5	6	7	8	9
0.0	.5000	.5040	.5080	.5120	.5160	.5199	.5239	.5279	.5319	.5359
0.1	.5398	.5438	.5478	.5517	.5557	.5596	.5636	.5675	.5714	.5754
0.2	.5793	.5832	.5871	.5910	.5948	.5987	.6026	.6064	.6103	.6141
0.3	.6179	.6217	.6255	.6293	.6331	.6368	.6406	.6443	.6480	.6517
0.4	.6554	.6591	.6628	.6664	.6700	.6736	.6772	.6808	.6844	.6879
0.5	.6915	.6950	.6985	.7019	.7054	.7088	.7123	.7157	.7190	.7224
0.6	.7258	.7291	.7324	.7357	.7389	.7422	.7454	.7486	.7518	.7549
0.7	.7580	.7612	.7642	.7673	.7704	.7734	.7764	.7794	.7823	.7852
0.8	.7881	.7910	.7939	.7967	.7996	.8023	.8051	.8078	.8106	.8133
0.9	.8159	.8186	.8212	.8238	.8264	.8289	.8315	.8340	.8365	.8389
1.0	.8413	.8438	.8461	.8485	.8508	.8531	.8554	.8577	.8599	.8621
1.1	.8643	.8665	.8686	.8708	.8729	.8749	.8770	.8790	.8810	.8830
1.2	.8849	.8869	.8888	.8907	.8925	.8944	.8962	.8980	.8997	.9015
1.3	.9032	.9049	.9066	.9082	.9099	.9115	.9131	.9147	.9162	.9177
1.4	.9192	.9207	.9222	.9236	.9251	.9265	.9279	.9292	.9306	.9319
1.5	.9332	.9345	.9357	.9370	.9382	.9394	.9406	.9418	.9429	.9441
1.6	.9452	.9463	.9474	.9484	.9495	.9505	.9515	.9525	.9535	.9545
1.7	.9554	.9564	.9573	.9582	.9591	.9599	.9608	.9616	.9625	.9633
1.8	.9641	.9649	.9656	.9664	.9671	.9678	.9686	.9693	.9699	.9706
1.9	.9713	.9719	.9726	.9732	.9738	.9744	.9750	.9756	.9761	.9767
2.0	.9772	.9778	.9783	.9788	.9793	.9798	.9803	.9808	.9812	.9817
2.1	.9821	.9826	.9830	.9834	.9838	.9842	.9846	.9850	.9854	.9857
2.2	.9861	.9864	.9868	.9871	.9875	.9878	.9881	.9884	.9887	.9890
2.3	.9893	.9896	.9898	.9901	.9904	.9906	.9909	.9911	.9913	.9916
2.4	.9918	.9920	.9922	.9925	.9927	.9929	.9931	.9932	.9934	.9936
2.5	.9938	.9940	.9941	.9943	.9945	.9946	.9948	.9949	.9951	.9952
2.6	.9953	.9955	.9956	.9957	.9959	.9960	.9961	.9962	.9963	.9964
2.7	.9965	.9966	.9967	.9968	.9969	.9970	.9971	.9972	.9973	.9974
2.8	.9974	.9975	.9976	.9977	.9977	.9978	.9979	.9979	.9980	.9981
2.9	.9981	.9982	.9982	.9983	.9984	.9984	.9985	.9985	.9986	.9986

References

Allee, W.C. *Animal Aggregations: A Study in General Sociology*. Chicago: University of Chicago Press, 1931.

Andrewartha, H.G. and Birch, L.C. *The Distribution and Abundance of Animals*. Chicago: University of Chicago Press, 1954.

Bellows, T.S. Jr. "The descriptive properties of some models for density dependence." *Journal of Animal Ecology*, 1981, 50: 139–156.

Benton, M.J. "Models for the Diversification of Life." *TREE*, 1997, vol. 12: 490–495.

Benton, M. and Harper, D. *Basic Paleontology*. Addison Wesley and Longman, 1997.

Beyer, W.H., ed. *CRC Standard Probability and Statistics Tables and Formulae*. CRC Press, 1991.

Briggs, G.E. and Haldane, J.B.S. "A note on the kinematics of enzyme actions." *Biochemical Journal*, 1925, 19: 338–339.

Burke, M.J.W. and Grime, J.P. "An experimental study of plant community invasibility." *Ecology*, 1996, 77: 776–790.

Darwin, C. *The Origin of Species*. Reprinted in Penguin Classics 1985. England : Penguin Books, Ltd., 1959.

Cain, M.L., Pacala, S.W., Silander, J.A. Jr., and Fortin, M.-J. "Neighborhood models of clonal growth in the white clover *Trifolium repens*." *The American Naturalist*, 1995, 145: 888–917.

Campbell, G.S. *An Introduction to Environmental Biophysics*. New York: Springer, 1986.

Carlos, M. Duarte and Agustí, Susana. "The CO_2 balance of unproductive aquatic ecosystems." *Science*, 1998, 281: 234–236.

Crawley, M.J. "Plant-Herbivore Dynamics." In *Plant Ecology*, ed. M.J. Crawley, pp. 401–474. Blackwell Science, 1997.

DeAngelis, D.L. *Dynamics of Nutrient Cycling and Food Webs*. Chapman and Hall, 1992.

de Roos, A.M. "A gentle introduction to physiologically structured population models." In *Structured-Population Models in Marine, Terrestrial, and Freshwater Systems*, eds. S. Tuljapurkar and H. Caswell. Chapman & Hall, 1996.

Dott, R.H. Jr. and Batten, R.L. *Evolution of the Earth*. Second edition. McGraw-Hill Book Company, 1976.

Dowd, J.E. and Riggs, D.S. "A comparison of estimates of Michaelis-Menten kinetic constants from various linear transformations." *The Journal of Biological Chemistry*, 1965, 240: 863–869.

Fitzhugh, R. "Impulses and physiological states in theoretical models of nerve membrane." *Biophysics Journal*, 1961, 1: 445-466.

Falconer, D.S. and Mackay, T.F.C. *Introduction to Quantitative Genetics*. Fourth Edition. Longman, 1996.

Futuyama, D.J. *Science on Trial*. Second edition. Sinauer, 1995.

Gaastra, P. Mededelinger van de Landbouwhogeschool Te Wageningen, The Netherlands, 1959. In F.B. Salisbury and C.W. Ross *Plant Physiology*. Wadsworth, 1978.

Gause, G.F. *The Struggle for Existence*. Baltimore: Williams & Wilkins, 1934. (Reprinted 1964 by Hafner, New York.)

Haldane, J.B.S. "The cost of natural selection." *J. Genet.*, 1957, 55: 511–524.

Hall, D.J. "An experimental approach to the dynamics of a natural population of *Daphnia galeata mendota*." *Ecology*, 1964, 45: 94–111.

Hartl, D.L. and Clark, A.G. *Principles of Population Genetics*. Second Edition. Sinauer, 1989.

Herschy, R.W. *Streamflow Measurements*. Second edition. E&FN Spon, 1995.

Holling, C.S. "Some characteristics of simple types of predation and parasitism." *Can. Ent.*, 1959, 91: 385–398.

Horn, H.S. *The adaptive geometry of trees*. Monographs in Population Biology 3. Princeton, NJ: Princeton University Press, 1971.

Iwasa, Y., deJong, T.J., and Klinkhamer, P.G.L. "Why Pollinators visit only a fraction of the open flowers on a plant: The plant's point of view." J. evol. Biol., 1995, 8: 439-453.

Kareiva, P.M. "Local movement in herbivorous insects: applying a passive diffusion model to mark-recapture field experiments." *Oecologia*, 1983, 57: 322–327.

Kermack, W.O. and McKendrick, A.G. "Contributions to the mathematical theory of epidemics." *Proc. Roy. Soc.* A 115: 700–721 (1927); 138: 55–83 (1932); 141: 94–122 (1933).

Krauskopf, K.B. and Bird, D.K. *Introduction to Geochemistry*. New York: McGraw-Hill, Inc., 1995.

Landsberg, H.E. *Weather and Health, an Introduction to Biometerology*. Garden City, New Jersey: Doubleday, 1969.

Leopold, A.C. and Kriedemann, P.E. *Plant growth and development*. New York: McGraw Hill, 1975.

Leslie, P. "On the use of matrices in certain population mathematics." *Biometrika*, 1945, 33: 183–212.

Levins, R. "Some demographic and genetic consequences of environmental heterogeneity for biological control." *Bull. Entomol. Soc. Am.*, 1969, 15: 237–240.

Levins, R. "Community equilibria and stability, and an extension of the competitive exclusion principle." *Am. Nat.*, 1970, 104: 413–423.

Lloyd, D.C. "Selection of offspring size at independence and other size- versus number-strategies." *American Naturalist*, 1987, 129: 800–817.

Lotka, A.J. "The growth of mixed populations: two species competing for a common food supply." *Journal of the Washington Academy of Sciences*, 1932, 22: 461-469.

MacArthur, R.H. and Wilson, E.O. "An equilibrium theory of insular zoogeography." *Evolution*, 1963, 17: 373–387.

Mackay, T.F.C. "Jumping genes meet abdominal bristles: Hybrid dysgenesis-induced quantitative variation in *Drosophila melanogaster*." *Genet. Res.*, 1984, 44: 231–237.

May, R. *Stability and Complexity in Model Ecosystems*. Princeton, NJ: Princeton University Press, 1975.

Michaelis, L. and Menten, M.I. "Die Kinetik der Invertinwirkung." *Biochem. Z.*, 1913, 49: 333-369.

Monod, J.L. *Recherches sur la Croissance des Cultures Bacteriennes*. Paris: Hermann, 1942.

Monod, J.L. "La technique de culture continué: theorie et applications." *Ann. Instit. Pasteur*, 1950, 79: 390–410.

Moss, B. *Ecology of Freshwaters*. Blackwell Scientific Publication, 1980.

Nagumo, J.S, Arimoto, S. and Yoshizawa, S. "An active pulse transmission line simulating nerve axon." *Proc. IRE.*, 1962, 50: 2061–2071.

Nee, S. and May, R.M. "Dynamics of metapopulation: habitat destruction and competitive coexistence." *Journal of Animal Ecology*, 1992, 61: 37–40.

Nicholson, A.J. "The balance of animal populations." *Journal of Animal Ecology*, 1933, 2: 131–178.

Nicholson, A.J. and Bailey, V.A. "The balance of animal populations." *Proceedings of the Zoological Society of London*, 1935, 3: 551–598.

Niklas, K.J. *Plant Allometry. The Scaling of Form and Process*. Chicago: The University of Chicago Press, 1994.

Oglesby, R.T. "Relationships of fish yield to lake phytoplankton, standing crop, production, and morphoedaphic factors." *J. Fish. Res. Bd. Can.*, 1977, 34: 2271–2279.

Pacala, S.W. and Rees, M. "Models suggesting field experiments to test two hypotheses explaining successional diversity." *American Naturalist*, 1998, 152: 729-737.

Pearl, R. and Reed, L.J. "On the rate of growth of the population of the United States since 1790 and its mathematical representation." *Proceedings of the National Academy of Sciences*, 1920, 6: 275–288.

Pierce, G.W. *The Songs of Insects*. Cambridge, Mass.: Harvard University Press, 1949.

Pisek, A., Larcher, W., Moser, W. and Pack, I. "Kardinale Temperaturbereiche der Photosynthese und Grenztemperaturen des Lebens der Blätter verschiedener Spermatophyten. III. Temperaturabhängigkeit und optimaler Temperaturbereich der Netto-Photosynthese." *Flora, Jena*, 1969, 158: 608–630.

Preston, F.W. "The canonical distribution of commonness and rarity." *Ecology*, 1962, 43: 185–215, 410–431.

Reiss, M.J. *The Allometry of Growth and Reproduction*. Cambridge University Press, 1989.

Roff, D.A. *The Evolution of Life Histories.* Chapman & Hall, 1992.

Rosenzweig, M.L. "Paradox of Enrichment: Destabilization of Exploitation Ecosystems in Ecological Time." *Science*, 1971, 171: 385-387.

Roughgarden, J. *Theory of Population Genetics and Evolutionary Ecology.* Second edition. Prentice Hall, 1996.

Stanley, S.M. *Macroevolution: Pattern and Process.* San Francisco: W.H. Freeman & Co, 332, 1979.

Statistical Abstract of the United States 1994, U.S. Bureau of the Census, 14th edition. Washington, DC, 1994.

Taylor, C.R., Maloiy, G.M.O., Weibel, E.R., Langman, U.A., Kaman, J.M.Z., Seeherman, H.J., and Heglund, B.N.C. "Design of the mammalian respiratory system. III. Scaling maximum aerobic capacity to body mass: wild and domestic mammals." *Respir. Physiol*, 1980, 44: 25–37.

Tilman, D. *Resource Competition and Community Structure* Princeton, NJ: Princeton University Press, 1982.

Tilman, D. "Competition and biodiversity in spatially structured habitats." *Ecology*, 1994, 75: 2–16.

Valentine, J.W. "Biotic diversity and clade diversity." In *Phanerozoic Diversity Patterns: Profiles in Macroevolution*, Valentine, J.W., ed., pp. 419–424. Princeton University, 1985.

Vitousek, P.M. and Farrington, H. "Nutrient limitation and soil development: Experimental test of a biogeochemical theory." *Biogeochemistry*, 1997, 37: 63–75.

Volterra, V. "Fluctuations in the abundance of species considered mathematically." *Nature*, 1926, 118: 558–560.

Walker, T.D. "Diversification functions and the rate of taxonomic evolution." In *Phanerozoic Diversity Patterns: Profiles in Macroevolution*, Valentine, J.W., ed., pp. 311–334. Princeton, NJ: Princeton University Press, 1985.

Walker, T.D. and Valentine, J.W. " Equilibrium models of evolutionary species diversity and the number of empty niches." *American Naturalist*, 1984, 124: 887–899.

Ward, P.D. *On Methuselah's Trail-Living Fossils and the Great Extinctions.* Freeman, 1992.

Yeargers, E.K., Shonkwiler, R.W. and Herod, J.V. *An Introduction to the Mathematics of Biology: with Computer Algebra Models.* Boston: Birkhäuser, 1996.

Index

Photo Credits

Algebra

Quadratic Formula

The solutions of the quadratic equation $ax^2 + bx + c = 0$ are given by

$$x = \frac{-b \pm \sqrt{b^2 - 4ac}}{2a}.$$

Factorial notation

For each positive integer n,

$$n! = n(n-1)(n-2)\cdots 3 \cdot 2 \cdot 1 \; ;$$

by definition, $0! = 1$.

Radicals

$$\sqrt[n]{x^m} = \left(\sqrt[n]{x}\right)^m = x^{m/n}$$

Exponents

$$(ab)^r = a^r b^r \qquad a^r a^s = a^{r+s} \qquad x^{-n} = \frac{1}{x^n}$$

$$(a^r)^s = a^{rs} \qquad \frac{a^r}{a^s} = a^{r-s}$$

Binomial Formula

$$(x + y)^2 = x^2 + 2xy + y^2$$
$$(x + y)^3 = x^3 + 3x^2 y + 3xy^2 + y^3$$
$$(x + y)^4 = x^4 + 4x^3 y + 6x^2 y^2 + 4xy^3 + y^4$$

In general, $(x + y)^n = x^n + \binom{n}{1}x^{n-1}y + \binom{n}{2}x^{n-2}y^2$

$$+ \cdots + \binom{n}{k}x^{n-k}y^k$$

$$+ \cdots + \binom{n}{n-1}xy^{n-1} + y^n,$$

where the binomial coefficient $\binom{n}{m}$ is the integer $\frac{n!}{m!(n-m)!}$

Special Factors

$$x^2 - a^2 = (x - a)(x + a)$$
$$x^3 + a^3 = (x + a)(x^2 - ax + a^2)$$
$$x^3 - a^3 = (x - a)(x^2 + ax + a^2)$$
$$x^4 - a^4 = (x^2 - a^2)(x^2 + a^2)$$

Geometry

Distance Formulas

Distance on the real number line:

$$d = |a - b|$$

Distance in the coordinate plane:

$$d = \sqrt{(x_1 - x_2)^2 + (y_1 - y_2)^2}$$

Equations of Lines and Circles

Slope-intercept equation:
$$y = mx + b$$

Point-slope equation:
$$y - y_1 = m(x - x_1)$$

Circle with center (h, k) and radius r:

$$(x - h)^2 + (y - k)^2 = r^2$$

Triangle area:
$$A = \frac{1}{2}bh$$

Trapezoid area:
$$A = \frac{b_1 + b_2}{2}h$$

Sphere volume:
$$V = \frac{4}{3}\pi r^3$$
Surface area:
$$A = 4\pi r^2$$

Cone volume:
$$V = \frac{1}{3}\pi r^2 h$$
Curved surface area:
$$A = \pi r \sqrt{r^2 + h^2}$$

Rectangle area:
$$A = bh$$

Circle area:
$$A = \pi r^2$$
Circumference:
$$C = 2\pi r$$

Sector of Circle
(θ in radians)
$$\text{Area} = \frac{\theta r^2}{2}$$
$$s = r\theta$$

Cylinder volume:
$$V = \pi r^2 h$$
Curved surface area:
$$A = 2\pi r h$$